AA001071

2023 IEEE Asian Solid-State Circuits Conference (A-SSCC 2023)

Haikou, China
5-8 November 2023

IEEE Catalog Number: CFP23SSC-POD
ISBN: 979-8-3503-3004-5

**Copyright © 2023 by the Institute of Electrical and Electronics Engineers, Inc.
All Rights Reserved**

Copyright and Reprint Permissions: Abstracting is permitted with credit to the source. Libraries are permitted to photocopy beyond the limit of U.S. copyright law for private use of patrons those articles in this volume that carry a code at the bottom of the first page, provided the per-copy fee indicated in the code is paid through Copyright Clearance Center, 222 Rosewood Drive, Danvers, MA 01923.

For other copying, reprint or republication permission, write to IEEE Copyrights Manager, IEEE Service Center, 445 Hoes Lane, Piscataway, NJ 08854. All rights reserved.

*** This is a print representation of what appears in the IEEE Digital Library. Some format issues inherent in the e-media version may also appear in this print version.

IEEE Catalog Number: CFP23SSC-POD
ISBN (Print-On-Demand): 979-8-3503-3004-5
ISBN (Online):   979-8-3503-3003-8

Additional Copies of This Publication Are Available From:

Curran Associates, Inc
57 Morehouse Lane
Red Hook, NY 12571 USA
Phone: (845) 758-0400
Fax: (845) 758-2633
E-mail: curran@proceedings.com
Web: www.proceedings.com

TABLE OF CONTENTS

MSN: Battery-Less Multiple Subcarrier Multiple Access Sensor Node with Full-Duplex Backscatter Concurrent Data Streaming 1
Yuxiao Zhao, Hanyang Wang, Yongming Xu, Zhongyuan Ying, Yu Lu, Jinlin Liu, Tuo Hu, Jin Mitsugi, Hao Min

A 25kHz-BW 97.4dB-SNDR 100.2dB-DR 3rd-Order SAR-Assisted CT DSM with 1–0 MASH and DNC 4
Kent Edrian Lozada, Dong-Hun Lee, Ye-Dam Kim, Ho-Jin Kim, Youngjae Cho, Michael Choi, Seung-Tak Ryu

A Ka-Band Mutual Coupling Resilient Balanced PA with Magnetic Coupling Self-Cancelling Inductor Achieving 21.2dBm OP$_{1dB}$ and 27.6% PAE$_{1dB}$ 7
Jian Zhang, Wei Zhu, Dawei Wang, Xiangjie Yi, Ruitao Wang, Yan Wang

A 80Gb/s/pin Single-Ended PAM-4 Transmitter with an Edge Boosting Auxiliary Driver and a 4-Tap FFE in 28-Nm CMOS 10
Dae-Won Rho, Jae-Koo Park, Seung-Jae Yang, Woo-Young Choi

A 28nm 49.7TOPS/W Sparse Transformer Processor with Random-Projection-Based Speculation, Multi-Stationary Dataflow, and Redundant Partial Product Elimination 13
Yubin Qin, Yang Wang, Dazheng Deng, Xiaolong Yang, Zhiren Zhao, Yang Zhou, Yuanqi Fan, Jingchuan Wei, Tianbao Chen, Leibo Liu, Shaojun Wei, Yang Hu, Shouyi Yin

A 4ns Settling Time FVF-Based Fast LDO Using Bandwidth Extension Techniques for HBM3 16
Jinook Jung, Jun-Han Choi, Kyoung-Jun Roh, Jaewoo Park, Won-Mook Lim, Tae-Sung Kim, Han-Ki Jeong, Myoungbo Kwak, Jae-Youn Youn, Jeong-Don Ihm, Changsik Yoo, Youngdon Choi, Jung-Hwan Choi, Hyungjong Ko

An On-Chip DC-DC Converter and Power Management System Achieving Zero Standby-To-Active Transition Time in MCU 19
Toyohiro Shimogawa, Makoto Nonaka, Toma Ogata, Toshiki Kiryu, Yasuto Igarashi, Kosuke Yayama, Masahiro Kitamura, Hiromichi Ishikura, Mitsuru Hiraki, Masao Ito, Takashi Kono

A 6.4Gbps/pin NAND Flash Memory Multi-Chip Package Employing a Frequency Multiplying Bridge Chip for Scalable Performance and Capacity Storage Systems 22
Shinichi Ikeda, Akira Iwata, Goichi Otomo, Tomoaki Suzuki, Hiroaki Iijima, Mikio Shiraishi, Shinya Kawakami, Masatomo Eimitsu, Yoshiki Matsuoka, Kiyohito Sato, Shigehiro Tsuchiya, Yoshinori Shigeta, Takuma Aoyama

A Fully Integrated Bit-To-Bit 24/48Gb/s QPSK/16-QAM D-Band Transceiver with Mixed-Signal Baseband in 28nm CMOS Technology 25
Pingda Guan, Haikun Jia, Wei Deng, Ruichang Ma, Mingxing Deng, Jiamin Xue, Angxiao Yan, Shiyan Sun, Zhihua Wang, Baoyong Chi

A 0.12-V 200-Hz-BW 10-Bit ADC Using Quad-Channel VCO and Interpolation Linearization 28
Shea Smith, Taylor Barton, Yen-Cheng Kuan, Armin Tajalli, Mau-Chung Frank Chang, Shiuh-Hua Wood Chiang

A 5GS/s 38.04dB SNDR Single-Channel TDC-Assisted Hybrid ADC with λ/4 Transmission Line Based Time Quantizer Achieving a PVT Robustness 416.6fs Time Step 31
Hongzhi Liang, Yi Shen, Jun Chang, Shubin Liu, Ruixue Ding, Zhangming Zhu

CIMFormer: A 38.9TOPS/W-8b Systolic CIM-Array Based Transformer Processor with Token-Slimmed Attention Reformulating and Principal Possibility Gathering 34

    Ruiqi Guo, Yang Wang, Xiaofeng Chen, Lei Wang, Hao Sun, Jingchuan Wei, Leibo Liu, Shaojun Wei, Yang Hu, Shouyi Yin

A 5.2GHz Trifilar Transformer-Based Class-F_{23} Noise Circulating VCO with FoM of 192.6 dBc/Hz 37

    Hanzhang Cao, Tongde Huang, Xiaolong Liu, Hao Wang, Jin Jin, Wen Wu

A Wide Frequency Range, Small Area and Low Supply Memory Interface PLL Using a Process and Temperature Variation Aware Current Reference in 3 nm Gate-All Around CMOS 40

    Kyungmin Lee, Jaehong Jung, Gyusik Kim, Joomyoung Kim, Seungjin Kim, Seunghyun Oh, Sung Min Park, Jongwoo Lee

A 26-30GHz Digitally-Controlled Variable Gain Power Amplifier with Phase Compensation and Third Order Nonlinearity Cancellation Technique 43

    Haoqi Qin, Junjie Gu, Hao Xu, Weitian Liu, Kefeng Han, Rui Yin, Zongming Duan, Hao Gao, Na Yan

LOG-CIM: A 116.4 TOPS/W Digital Computing-In-Memory Processor Supporting a Wide Range of Logarithmic Quantization with Zero-Aware 6T Dual-WL Cell 46

    Soyeon Um, Sangjin Kim, Seongyon Hong, Sangyeob Kim, Hoi-Jun Yoo

A 2 X 24Gb/s Single-Ended Transceiver with Channel-Independent Encoder-Based Crosstalk Cancellation in 28nm CMOS 49

    Hongzhi Wu, Weitao Wu, Liping Zhong, Xuxu Cheng, Xiongshi Luo, Zhenghao Li, Dongfan Xu, Quan Pan

A 10TFLOPS Datacenter-Oriented GPU with 4-Corner Stacked 64GB Memory by the Means of 2.5D Packaging Technology 52

    Shuang Wang, Weiliang Chen, Xueqing Li, Leibo Liu, Huazhong Yang

An 890 μW Multichannel Injection-Locked OOK Transmitter with 23% Global Efficiency and 22 pJ/bit Energy Efficiency 55

    Sheng-Kai Chang, Shao-Ting Chang, Zhi-Wei Lin, Kuang-Wei Cheng

A 110-160 GHz ASK Receiver with Isomorphic Low Noise Power Amplifier and Multi-Mode Interface in 28-Nm CMOS 58

    Dawei Tang, Zekun Li, Leyang Huang, Rui Zhang, Liqun Lu, Chun Yang, Zheng Yan, Peigen Zhou, Zhe Chen, Jixin Chen, Wei Hong

A 0.5-To-1.5GHz BW-Extended Gain-Boosted N-Path Filter Using a Switched g_m-C Network Achieving 50MHz BW and 18.2dBm OB-IIP$_3$ 61

    Gengzhen Qi, Pui-In Mak

A D-Band ASK Transmitter with 50GHz RF Bandwidth and Multi-Mode Interface Implemented in 28nm CMOS.. 64

    Rui Zhang, Dawei Tang, Zhe Chen, Liqun Lu, Peigen Zhou, Jixin Chen, Wei Hong

A PAM4 Level Mismatch Adjustment Scheme for 48-Gb/s PAM4 Memory Tester Bridge............................ 67

    Daeho Yun, Minsu Park, Kahyun Kim, Kyungmin Baek, Eonhui Lee, Woo-Seok Choi, Deog-Kyoon Jeong

An 8.73GΩ Input Impedance VCO-Based Neural Front-End with Autonomous Time-Domain Impedance Boosting Technique and Maximum 0.215s Re-Calibration Time 70

    Huaiyu Liu, Yang Lin, Guoxing Wang, Yan Liu

GCFP-ACIM: A 40nm 4.74TFLOPS/W General Complex Float-Point Analog Compute-In-Memory with Adaptive Power-Saving for HDR Signal Processing Applications 73

Zizhao Ma, Xianwu Hu, Gan Wen, Yihao Wang, Zeming Wang, Yukai Lin, Yu Wang, Yuhao Guo, Yanlei Li, Xingdong Liang, Xiaoyang Zeng, Yufeng Xie

A 0.000261 mm^2 Single-Channel 1 GS/s 8-Bit 3-Stage Capacitor Array-Assisted Charge-Injection DAC-Based SAR ADC in 28 nm CMOS 76

Chan-Ho Kye, Kyojin Choo

A 1.01V 8.5Gb/s/pin 16Gb LPDDR5x SDRAM with Self-Pre-Emphasized Stacked-Tx, Supply Voltage Insensitive Rx, and Optimized Clock Using 4th-Generation 10nm DRAM Process for High-Speed and Low-Power Applications 79

Hyun-A Ahn, Yoo-Chang Sung, Yong-Hun Kim, Janghoo Kim, Kihan Kim, Donghun Lee, Young-Gil Go, Jae-Woo Lee, Jae-Woo Jung, Yong-Hyun Kim, Ga-Ram Choi, Jun-Seo Park, Bo-Hyeon Lee, Jinhyeok Baek, Daesik Moon, Daihyun Lim, Seung-Jun Bae, Young-Soo Sohn, Changsik Yoo, Tae-Young Oh

A Low Noise 8Mpixel CMOS Image Sensor with 5.36GHz Global Counter and Dual Latch Skew Canceler for Surveillance AI Camera System 82

Yoichi Iizuka, Akihide Maezono, Wataru Saito, Atsushi Yamane, Kazuhiko Takami, Fukashi Morishita

A 28-nW Noise-Robust Voice Activity Detector with Background Aware Feature Extraction 85

Jingsen Yang, Liangjian Lyu, Zirui Dong, Heyu Ren, C.-J. Richard Shi

An 83.6dB-SNDR 101.6dB-SFDR 4th-Order Noise-Shaping SAR with 2nd-Order Nonlinearity Error Shaping... 88

Yanbo Zhang, Xianghui Zhang, Li Tian, Shubin Liu, Zhangming Zhu

A Compact 1,257-Gbps/W Byte-Serial AES Accelerator for IoT Applications in 22 nm 91

Shutao Zhang, Malte Wabnitz, Tobias Gemmeke

A 200MHz Bandwidth 84% Peak Efficiency AC-Coupled Envelope Tracking Supply Modulator with 10MHz Constant Switching Frequency... 94

Peng Xu, Xueli Zhang, Tao Wang, Peng Cao, Jiawei Xu, Zhiliang Hong

An Energy-Efficient Heterogeneous Fourier Transform-Based Transformer Accelerator with Frequency-Wise Dynamic Bit-Precision 97

Jingu Lee, Sangjin Kim, Wooyoung Jo, Hoi-Jun Yoo

An 8.5ps Resolution, 2000μm^2 Phase-Domain Delta-Sigma TDC for Lidar Applications 100

Yoondeok Na, Seokho Yun, Myung-Jae Lee, Youngcheol Chae

3D-ISC: A 65nm 3D Compatible In-Sensor Computing Accelerator with Reconfigurable Tile Architecture for Real-Time DVS Data Compression 103

Gokul Krishnan, Gopikrishnan Raveendran Nair, Jonghyun Oh, Anupreetham Anupreetham, Pragnya Sudershan Nalla, Ahmed Hassan, Injune Yeo, Kishore Kasichainula, Jae-Sun Seo, Mingoo Seok, Yu Cao

A 23.9 μW 13.6-Bit Period Modulation-Based Capacitance-To-Digital Converter with Dynamic Current Mirror Front-End Achieving Capacitor Range of 1 to 68 pF 106

Hyeyeon Lee, Donguk Seo, Young-Jin Woo, Yoonmyung Lee, Inhee Lee, Youngcheol Chae

MorphBungee: A 65nm 7.2mm^2 27μJ/Image Digital Edge Neuromorphic Chip with On-Chip 802 Frame/s Multi-Layer Spiking Neural Network Learning 109

Tengxiao Wang, Min Tian, Zhengqing Zhong, Haibing Wang, Junxian He, Fang Tang, Xichuan Zhou, Shuangming Yu, Nanjian Wu, Liyuan Liu, Cong Shi

SESOMP: A Scalable and Energy-Efficient Self-Organizing Map Processor with Computing-In-Memory and Dead Neuron Pruning..112

Yuncheng Lu, Xin Zhang, Zehao Li, Bo Wang, Tony Tae-Hyoung Kim

A Time-Based PAM-4 Transceiver Using Single Path Decoder and Fast-Stochastic Calibration Techniques..115

Dong-Hyun Yoon, He Junsen, Kwang-Hyun Baek, Youngdon Choi, Jung-Hwan Choi, Tony Tae-Hyoung Kim

A 160x120 Indirect Time-Of-Flight Sensor with Pixel-Level Adaptive $\Delta\Sigma$-Operations for Background Light Cancellation..118

Yongjae Park, Jubin Kang, Dahwan Park, Insang Son, Jung-Hye Hwang, Seong-Jin Kim

An 11bit 360MS/s Pipelined SAR ADC with Dynamic Negative-C Assisted Residue Amplifier 121

Yigi Kwon, Sangwoo Lee, Changuk Lee, Hyunchul Yoon, Byounghan Min, Youngcheol Chae

A Jitter-Programmable Bang-Bang Phase-Locked Loop Using PVT Invariant Stochastic Jitter Monitor... 124

Yong-Jo Kim, Taekwang Jang, Seonghwan Cho

TEPD: A Compound Timing Detection of Both Data-Transition and Path-Activation for Reliable In-Situ Timing Error Detection and Correction in 28nm CMOS 127

Zhengguo Shen, Junyi Qian, Keran Li, Ziyu Li, Lishuo Deng, Weiwei Shan

Flexible and Efficient Implementation of CRYSTALS-KYBER SIMD RISC-V Coprocessor Based on Customized Vector Instruction-Set Extension 130

Jiaming Zhang, Jiahao Lu, Dongsheng Liu, Aobo Li, Xiang Li, Shuo Yang, Ang Hu, Xuecheng Zou

A 1.2 V 2.3 µW 94.7 dB DR Delta-Sigma Modulator with Dynamic-Range Enhancement and Tri-Level CDAC... 133

Cong Wei, Rongshan Wei, Lijie Huang, Gongxing Huang, Jinze Lai, Zhichao Tan

Integrated Circuit to Compensate Parasitic Leakage Component for WL Leakage Current in NAND Flash Memory... 136

Bu-Il Nam, Jayang Yoon, Kyunghea Lee, Sol Kim, Junhong Park, Chi-Weon Yoon, Eunkyoung Kim

A 12.75-To-16-GHz Spur-Jitter-Joint-Optimization SS-PLL Achieving -94.55-DBc Reference Spur, 31.9-Fs Integrated Jitter and -260.1-DB FoM 139

Yixi Li, Zhao Zhang, Yong Chen, Xinyu Shen, Zhaoyu Zhang, Nan Qi, Jian Liu, Nanjian Wu, Liyuan Liu

A Wideband Low-Noise Linear LiDAR Analog Front-End Achieving 1.6 GHz Bandwidth, 2.7 pA/Hz$^{0.6}$ Input Referred Noise, and 103 dBΩ Transimpedance Gain... 142

Yidan Zhang, Zhao Zhang, Yiqing Xu, Xinyu Shen, Nan Qi, Nanjian Wu, Jian Liu, Liyuan Liu

An Adaptive-Sampling Digital LDO with Statistical Comparator Selection Achieving 99.99% Maximum Current Efficiency and 0.25ps FoM in 65nm.. 145

Shun Yamaguchi, Takashi Hisakado, Osami Wada, Mahfuzul Islam

A Resource-Efficient Super-Resolution FPGA Processor with Heterogeneous CNN and SNN Core Architecture.. 148

Jiwon Choi, Sangyeob Kim, Wonhoon Park, Wooyoung Jo, Hoi-Jun Yoo

A 142.8-µW 98.1dB-SNDR Power/Bandwidth Configurable Fully Dynamic Discrete-Time Zoom ADC with Interstage Leakage Shaping 151

    Yuke Shen, Shubin Liu, Kui Wen, Yanbo Zhang, Yi Shen, Ruixue Ding, Zhangming Zhu

A 71-To-86GHz, 20.4dBm P_{out}, 6.0dB NF Transceiver with Quadrature Direct-Modulated Transmitter and Reflection-Less Heterodyne Receiver in 40nm CMOS 154

    Jie Zhou, Changxuan Han, Wen Chen, Bingzheng Yang, Xun Luo

K/Ka-Band 4-Element 4-Beam Hybrid Phased-Array Transmitter and Receiver Front-Ends with Compact Layout Floor-Plans and Fault-Tolerant Digital Circuits 157

    Mengru Yang, Chenyu Xu, Peng Gu, Yongran Yi, Mohan Guo, Liangliang Liu, Xiangxi Yan, Xiaohu You, Dixian Zhao

A 54-To-69-GHz Wideband 2T2R FMCW Radar Transceiver Employing Cascaded-PLL Topology and PTAT-Enhanced Temperature Compensation in 40-Nm CMOS 160

    Chenyu Xu, Xiaofei Liao, Feifan Hong, Mengru Yang, Peijuan Ju, Wendi Chen, Pengfei Diao, Hao Gong, Xiang Liu, Xiaohu You, Dixian Zhao

A Low-Voltage Bias-Current-Free Pseudo-Differential Hybrid PLL Using a Time-Interleaving Flip-Flop Phase Detector 163

    Liqun Feng, Qianxian Liao, Woogeun Rhee, Zhihua Wang

A Cryo-CMOS 4.5~7GHz Dual-Qubit Homodyne Reflectometer Array with High Q Degenerate Parametric Amplifier Through Dynamic Mode Coupling 166

    Yujie Geng, Haichuan Lin, Bo Wang, Cheng Wang

A 2-W, 90%-Efficiency Single-Stage Dual-Output Wireless Power Receiver with 0.1 to 700-MA Output Current Range Through Dynamic Delay Compensation and Bootstrap Adaptive Body Biasing Circuit 168

    Yutang Chen, Yuxuan Luo, Jianping Guo, Xian Tang, Dihu Chen

An 18-NW, 170°C Temperature Range, Voltage and Current Reference Circuit with Low Line Sensitivity 171

    I-Fan Lin, Yu-Chu Tsai, Heng-Li Lin, Yu-Te Liao

A 400MHz 249.1TOPS/W 64Kb Fully-Reconfigurable SRAM-Based Digital Compute-In-Memory Macro for Accelerating CNNs 174

    Xin Zhang, Vishal Sharma, Yuncheng Lu, Yongjun Jo, Tony Tae-Hyoung Kim

A 73.8K Inference/mJ SVM Learning Accelerator for Brain Pattern Recognition 177

    Tzu-Wei Tong, Yi-Yen Hsieh, Tai-Jung Chen, Chia-Hsiang Yang

An Instant-Setup On-Delay Compensation Scheme Based on Phase Detection for Series-Resonant Wireless Power Receiver with Up-To-4.41% Efficiency Enhancement 180

    Yuxuan Luo, Yutang Chen, Jianping Guo, Xian Tang, Dihu Chen

A 4 X 112-Gb/s PAM-4 Silicon-Photonic Transceiver Front-End for Linear-Drive Co-Packaged Optics 183

    Han Liu, Nan Qi, Donglai Lu, Zizheng Dong, Zhihan Zhang, Jian He, Guike Li, Leliang Li, Ye Liu, Ziyue Dang, Daigao Chen, Zhao Zhang, Jian Liu, Nanjian Wu, Xi Xiao, Liyuan Liu

A 33.6 FPS Embedding Based Real-Time Neural Rendering Accelerator with Switchable Computation Skipping Architecture on Edge Device 186

    Jongjun Park, Donghyeon Han, Junha Ryu, Dongseok Im, Gwangtae Park, Hoi-Jun Yoo

A 26.9-GHz 4-Element Code-Domain Hybrid Beamforming Phased-Array Receiver 189
Ziyi Lin, Haikun Jia, Chuanming Zhu, Wei Deng, Huabing Liao, Bao Shi, Lujie Hao, Xiangrong Huang, Baoyong Chi

A Triple-Band Radio for WLAN 11b/g/n/ax in 45nm CMOS 192
D Sahu, V Srinivas, R Chatterjee, M Agrawal, P Agrawal, R Juluri, M. Mukherjee, Vimal E, A. Yerramsetty, G Bakalzuk, O Rahmanony, K Rajmohan, A Sancheti, R Anand

An Area and Power Efficient Fully Nonlinear 10-Bit Column Driver with Time-Shared Multi-
Gamma-Slope DAC and Time-Interleaved Sampling Buffer for Mobile AMOLEDs 195
Seung Hun Choi, Jeongmin Kim, Jaewoong Ahn, Junyeol An, Jiwoong Kim, Ohjo Kwon, Ki-Duk Kim, Hyung-Min Lee

A 5.37-TSOPS/W Reconfigurable Neuron Array with Dual-Mode Neurons and Asynchronous
Synapses for Energy-Efficient Inference and Biological Neural Network Simulation 198
Xiangao Qi, Yuqing Lou, Yongfu Li, Guoxing Wang, Kea-Tiong Tang, Jian Zhao

A 28nm 386.5GOPS/W Coarse-Grained DSP Using Configurable Processing Elements for Always-
On Computation with FPGA Implementation 201
Hedi Wang, Zengwei Wang, Yaolei Li, Chen Tang, Jinxu Gao, Huazhong Yang, Yongpan Liu

A 92.7%-Efficiency 6.78-MHz Dual-Output Energy-Resuscitating Resonant Regulating Rectifier
for Wirelessly Powered Systems 204
Hyun-Su Lee, Hyung-Min Lee

2-20-V Wide Input Range Active Rectifier with Single-Multiplexing Adaptive Delay Compensation
and Low-Voltage Enhancement for Wireless Power Transmission 207
Chao Xie, Guangshu Zhao, Junliang Wei, Man-Kay Law, Milin Zhang

A 1.08ms Ultrafast Scanning Capacitive Touch-Screen Sensor Interface with Charge-Interpolated
Common-Mode Compensation and Host-Based Adaptive Median Filtering 210
Jonghang Choi, Subin Kim, Yongjun Lee, Sanghyun Heo, Keum-Dong Jung, Young-Ha Hwang, Jun-Eun Park

A Primary Driver with Real-Time Resonance Tracking for Wireless-Powered Implantable Medical
Devices 213
Xiaodong Meng, Xing Li, Yuan Yao, Chi-Ying Tsui, Wing-Hung Ki

Fully Integrated Reconfigurable Solar Energy Harvester for 100μA Burst Output Current Delivery
with 78.6% Peak Energy Extraction Efficiency and Minimum Startup Incident Light Power of
0.27mW/cm^2 216
Jiangchao Wu, Guangshu Zhao, Litao Zhang, Yu Jia, Yang Jiang, Pui In Mak, Rui P. Martins, Man Kay Law

An 80-DB Dynamic Range Hybrid Pulse-Phase Analog Front-End Circuit Cooperating with a Low-
Resolution TDC for LiDAR 219
Zeli Li, Tianrui Lyu, Kaiyou Li, Haoxin Zheng, Zhuohang Ye, Shengzhao Su, Jianping Guo, Yang Liu

A 4-Element 4-Beam Ka-Band Phased-Array Receiver Using Mesh Topology in 65 nm CMOS 222
Xiangrong Huang, Haikun Jia, Wei Deng, Chuanming Zhu, Zhihua Wang, Xuzhi Liu, Zhiming Chen, Baoyong Chi

A Compact E-Band Load-Modulation Balanced Power Amplifier Using Coupled Transmission-
Line Output Network Achieving 22.1-DBm Psat and 34.9%/12.2% Efficiency at Psat/6-DB PBO 225
Xiangrong Huang, Haikun Jia, Wei Deng, Zhihua Wang, Baoyong Chi

A Crystal-Less Clock Generator for Low-Power and Low-Cost Sensor Transceivers with 12.9MHz-To-3.3GHz Range, 16.67ppm/°C Inaccuracy from −25°C to 85°C, and 0.25us Settle-Time 228

    Sangdon Jung, Jehyeok Yu, Minsu Park, Jung-Hoon Chun

A 6.78-MHz Wireless Power Transfer System with Voltage-/Current-Mode 0X/1X Regulating Rectifier and Global-Loop Power Control 231

    Bing-Jen Wu, Dao-Han Yao, Po-Hung Chen

A 40nm 2.76μJ/Op Energy-Efficient Secure Post-Quantum Crypto-Processor for CRYSTALS-KYBER on Module-LWE 234

    Aobo Li, Jiahao Lu, Dongsheng Liu, Xiang Li, Shuo Yang, Tianze Huang, Jiaming Zhang, Siqi Xiong, Chenjun Yang

A Resistor/Trimming-Less Self-Biased Current Reference Class with Area Down to 3500 μm², 42.8 pW Power and 10.4% Accuracy Across Corner Wafers in 180 nm 237

    Luigi Fassio, Hoang Hong Hanh, Massimo Alioto

A 11.6-to-13.6GHz, 47-fsrms Jitter Integer-N PLL with Simultaneous Differential 2ND-Harmonic Output using a 128dBc/Hz Phase Noise at 1MHz offset Quad-Core VCO in 65nm CMOS 240

    Ruichang Ma, Haikun Jia, Hongzhuo Liu, Wei Deng, Zhihua Wang, Baoyong Chi

A 27-To-31.6 GHz 8-Element Phased-Array Transmitter Front-End with Inter-Element-Interference Cancellation Scheme in 65 nm CMOS 243

    Dongze Li, Wei Deng, Xintao Li, Ruiheng Qiu, Haikun Jia, Xiangrong Huang, Ziyuan Guo, Baoyong Chi

A 69μW Dual-Mode Electrochemical miR-21 Detection SoC with 3rd-Harmonic Direct-Solution Technique 246

    Xiangdong Feng, Xin Liu, Yangfan Xuan, Fengheng Li, Yuxuan Luo, Ping Wang, Bo Zhao

A Transient Enhancement Digital LDO with Adaptive Ripple Cancelation Based on Optimal Compensation Period Approximation 249

    Angxiao Yan, Wei Deng, Haikun Jia, Shiwei Zhang, Baoyong Chi

Task-Specific Low-Power Beamforming MIMO Receiver Using 2-Bit Analog-To-Digital Converters 252

    Timur Zirtiloglu, Peter Crary, Eyyup Tasci, Arslan Riaz, Yonina C. Eldar, Nir Shlezinger, Rabia Tugce Yazicigil

A 2D Beam-Steerable 252-285-GHz 25.8-Gbit/s CMOS Receiver Module 255

    Takeshi Yoshida, Shinsuke Hara, Tatsuo Hagino, Mohamed H. Mubarak, Akifumi Kasamatsu, Kyoya Takano, Yoshiki Sugimoto, Kunio Sakakibara, Shuhei Amakawa, Minoru Fujishima

A 2pA/√Hz Current-Conveyor-Assisted Ultrasound Receiver with 25pF CMUT Parasitic Capacitance 258

    Gichan Yun, Kyeongwon Jeong, Haidam Choi, Seunghyeon Nam, Chaerin Oh, Hyunjoo Jenny Lee, Sohmyung Ha, Minkyu Je

A 226 GHz Coupled Harmonic VCO with 9.34% Tuning Range Utilizing Three-Coil Transformer with Switched Inductor in 65nm CMOS 261

    Yue Liang, Qin Chen, Xu Wu, Xiangning Fan, Lianming Li

A 74.0 dB-SNDR 175.4 dB-FoM Pipelined-SAR ADC Using a Cyclically Charged Floating Inverter Amplifier 264

    Changjoo Park, Jeongmyeong Kim, Kyounghun Kang, Minkyu Yang, Byeongmin Moon, Siheon Lee, Wanyeong Jung

A Hybrid Integrated W-Band 4-Element Phased-Array Transceiver Front-End Achieving 21.6% Full TX Peak PAE at 14.8dBm Output Power and <1°/dB Phase/Gain Resolution in 65-Nm CMOS Technology 267

Wei Zhu, Jian Zhang, Jiazhi Ying, Xiangjie Yi, Yan Wang, Houjun Sun

248 GHz Compact Mixer-Last Direct-Conversion Transmitter with I/Q Imbalance and LO Feedthrough Calibration Capability 270

Seunghoon Lee, Junhyeong Kim, Ho-Jin Song

A Cross-Coupled Hybrid SC Converter with Extended VCR Range and Intrinsic Loss Balance Achieving 90% Average Efficiency with 1.5% Variation Over Full Li-Ion Battery Input Range and 0.95A/mm^2 Peak Current Density 273

Qiaobo Ma, Huihua Li, Xiongjie Zhang, Anyang Zhao, Yang Jiang, Man-Kay Law, Rui P. Martins, Pui-In Mak

A DRAM Bandwidth-Scalable Sparse Matrix-Vector Multiplication Accelerator with 89% Bandwidth Utilization Efficiency for Large Sparse Matrix 276

Hyunji Kim, Eunkyung Ham, Sunyoung Park, Hana Kim, Ji-Hoon Kim

ESD Protection is About Circuit Design: Practices and Perspectives 279

Weiquan Hao, Xunyu Li, Zijin Pan, Runyu Miao, Albert Wang

5G Lessons Learned and an Outlook on 6G 282

Bongtae Kim

Semiconductor Chip Design in a Legoland 284

Chih-Ming Hung

Architecture Challenges for Heterogeneous Processors in Embedded SoCs 288

Masayuki Ito

Author Index

2023 IEEE Asian Solid-State Circuits Conference(A-SSCC)

ADVANCE PROGRAM

November 5th - 8th, 2023
Starbay Haikou Pullman Hotel
Haikou, Hainan, China

Sponsored by
IEEE Solid-State Circuits Society (IEEE SSCS)

Welcome Message

On behalf of the organizing committee, it is my great pleasure to extend to you all a very warm welcome to the IEEE Asian Solid-State Circuits Conference (A-SSCC) on November 5 to 8, 2023 in Haikou, China. A-SSCC is an international electronics forum that takes place in Asia with the support of the IEEE Solid-State Circuits Society. Since its inception in 2005, A-SSCC 2023 will be the 19th edition of the conference where the most updated and advanced chips and circuit designs in solid-state and semiconductor fields will be presented. As the global pandemic comes to the end, we have eventually decided to hold the fully in-person conference again.

The theme of A-SSCC 2023 is "Silicon System with Open Platform for Heterogeneous Integration". To embrace the conference theme of this year, a full and rich four-day program consisting of 4 outstanding plenary talks, 4 key technology tutorials, 1 panel discussion, 1 industry session and various regular paper sessions covering more than 90 regular papers in the areas of analog circuits, data converters, digital circuits and systems, emerging technology and applications, memory, radio-frequency circuits, system-on-chip and signal processing, wireline and mixed signal circuits have been organized. The Student Design Contest and FPGA exhibition will also include demonstrations from the best student papers as well as the presentation of advanced technologies with real-time discussion. In addition, the warm welcome reception, the social activities, the banquet with performances will enable the interactive communication among experts and students from all over the world.

We are pleased to have the presence of a number of world-renowned experts joining this conference and sharing their precious experience and knowledge with us. A-SSCC 2023 will provide a valuable opportunity for academics, students, research scientists, industry specialists and decision-makers to engage and interact with each other, and hopefully to come up with new innovations to collaborate.

Many people have made their indelible contributions to the success of A-SSCC 2023. I would like to take this opportunity to express my sincere gratitude and appreciation to the members of the organizing committee chaired by Prof. Chong Shen, the steering committee chaired by Prof. Hoi-Jun Yoo, and the technical program committee chaired by Prof. Ken Takeuchi and assisted by Prof. Pei-Yun Tsai (Co-Chair), Tetsuya Iizuka (Vice-Chair), as well as all speakers, authors, and sponsors.

Last, but not least, the Conference is hosted in the beautiful city of Haikou on Hainan island, which is known as the "Hawaii of China". It is a tropical island paradise where you can enjoy sandy beaches, five-star resorts and duty-free shopping. The beauty of Haikou will add to the pleasure of the Conference and provide lasting memories beyond science.

Once again, thank you very much for joining us in A-SSCC 2023. I am sure that you will have fruitful and rewarding exchanges in this conference. Please enjoy it and share your positive experience with colleagues and friends. Your continuous support is of paramount importance for the continuing success of A-SSCC.

Leibo Liu

Conference Chair, A-SSCC 2023
Tsinghua University, China

Foreword

Welcome to the IEEE Asian Solid-State Circuits Conference (A-SSCC) 2023. This year, A-SSCC is held on-site in Hainan, China from Nov. 5th to Nov. 8th. Being one of the five conferences fully sponsored by the IEEE Solid-State Circuits Society, A-SSCC is one of the leading conferences in the field of integrated circuits and systems design.

The conference theme for this year is "Silicon System with Open Platform of Heterogeneous Integration". As the IC design is becoming more complicated as well as the applications are diversified like AI and IoT, the domain specific computing with heterogeneously integrated ICs with open platforms are becoming important. A-SSCC 2023 will highlight these heterogeneously integrated ICs that provide unique and innovative solutions for such diversified emerging applications.

This year, we received 235 submissions from around the world. After rigorous review process, the Technical Program Committee (TPC) selected 93 high quality papers with an acceptance rate of 39.6%. In addition, we invited 10 student papers from ESSCIRC 2023 and CICC 2023, having 103 technical paper presentations.

ASSCC will continue its 4-day program similar to last year. Four plenary speeches will be given by distinguished leaders of the industry. First on Nov. 6th, Dr. Chih-Ming Hung, AVP of Technology MediaTek will give a talk on "Semiconductor Chip Design in a Legoland", and Dr. Bongtae Kim, ETRI Fellow will share his view on "Envisioning 6G Mobile New World". On Nov. 7th, Dr. Masayuki Ito, Director, NSITEXE Inc. will give a talk on "Architecture Challenges for Heterogeneous Processors in Embedded SoCs", followed by Professor Albert Wang, University of California, Riverside, who will talk about "Listen: ESD Protection is About Circuit Design".

Before the regular technical sessions, we have 4 tutorials on Sunday, Nov. 5th, offered by Professor Yan Lu of University of Macau on Hybrid DC-DC Converters, Professor Jae-Yoon Sim of POSTECH on Circuit Design for Scalable Quantum Computing, Professor Masum Hossain of Carleton University on Digital Equalization for Multilevel Signaling in High-speed SerDes and Professor Shouyi Yin of Tsinghua University on Reconfigurable ML Processor: Fundamental Concepts, Application, and Future Trends. A panel discussion is held on Tuesday, Nov. 7th with the topic Heterogeneous Integration: Does it call the end of the foundry-fabless business model?". The panel invites 6 experts to discuss issues and challenges in heterogeneous integration. The industry session, also held on Tuesday, highlights advances in Design Techniques for Industrial Applications, where four outstanding industry papers are presented. The regular conference papers are grouped in 22 sessions in three parallel tracks. Student Design Contest provides live demos from the top 10 student-authored papers including several FPGA papers. Three winners will be selected and recognized at the conference.

Continuing our diverse programs from last year, we have seven invited special sessions. To expand the knowledge of IC designers beyond circuits, two special sessions, "IT Vision in Asia" and "Convergence Workshop" are organized. To offer advanced and practical technologies in the real world, two special sessions, "Industry Forum" and "Start-up!" is prepared. Besides these technical talks, two mentoring sessions, "Learn to Establish and Achieve Professionalism (LEAP)" and "RiSE (Rising Star Express) Forum" will give inspiring talks to young scholars and students. We also established "ASSCC-CICC-ESSCIRC (ACE) Joint Session" where 5 student papers from CICC and 5 from ESSCIRC are presented. The ACE program is supported by IEEE Solid-State Circuits Society.

The A-SSCC 2022 TPC consists of 118 members divided into 10 technical subcommittees. The members come from both industry and academia around the world. This year, TPC members gathered in virtual platform in late July to select excellent papers. Their contributions to maintain a high-quality conference are highly appreciated. Furthermore, I would like to acknowledge the leadership of the technical subcommittee chairs: Prof. Po-Chiun Huang (Analog Circuits and Systems), Dr. Kazuko Nishimura (Data Converters), Prof. Jun Zhou (Digital Circuits and Systems), Prof. Joo-Young Kim (SoC and Signal Processing), Prof. Minoru Fujishima (Wireless), Prof. Byungsub Kim (Wireline), Prof. Jerald Yoo (Emerging Technology and Applications), Dr. Chiweon Yoon (Memory), Dr. Jun Deguchi (FPGA), and Dr. Stefan Rusu (Industry Program).

I would also like to acknowledge Prof. Jung-Hoon Chun, Prof. Sai-Weng Sin, and Prof. Nan Qi for organizing the Student Design Contest, the FPGA Demo, and the Industry Demo programs, Prof. Zhihua Wang for preparing the plenary program and panel, and Prof. Seung-Tak Ryu for the tutorial planning. I would like to give special thanks to the TPC Co-Chair, Prof. Pei-Yun Tsai and the TPC Vice-Chair, Prof. Tetsuya Iizuka for their full support.

I would like to extend my sincere appreciation to all authors and speakers, conference organizers, committee members, moderators, panelists, and, last but not least, all the participants. I hope you will enjoy the technical program of the A-SSCC 2023, take this opportunity to network with experts around the world, and bring back good memories with you!

竹内 健

Ken Takeuchi
Technical Program Committee Chair of A-SSCC 2023
The University of Tokyo, Japan

Committees

Steering Committee

Chair
Hoi-Jun Yoo, KAIST, Korea

Members
Tzi-Dar Chiueh, National Taiwan University, Taiwan
Lawrence Loh, MediaTek, USA
Hidetoshi Onodera, Keio University, Japan
Seonghuan Cho, KAIST, Korea
Nicky Lu, Etron Technology Inc., Taiwan
Stefan Rusu, TSMC, USA
Makoto Ikeda, The University of Tokyo, Japan
Yoshio Masubuchi, Toshiba Corp., Japan
Seng-Pan U, University of Macau, China
Deog-Kyoon Jeong, Seoul National University, Korea
Eun-Seung Jeong, Samsung Electronics, Korea
Zhihua Wang, Tsinghua University, China
Woogeun Rhee, Tsinghua University, China

Advisors
Tadahiro Kuroda, The University of Tokyo, Japan
Takayasu Sakurai, The University of Tokyo, Japan
Chorng-kuang Wang, National Taiwan University, Taiwan

Liaison Members
Anantha Chandrakasan, Massachusetts Institute of Technology, USA
Bram Nauta, Twente University, Holland

Organization Committee

Conference Chair
Leibo Liu, Tsinghua University, China

Organizing Committee Chair
Chong Shen, Hainan University, China

Organizing Committee Vice Chair
Yu Zhou, Hainan Normal University, China
Jianwei Zhang, Dalian University of Technology, China

Publicity
Chenchen Deng, Tsinghua University, China

Publication
Bohan Yang, Tsinghua University, China

Exhibit
Guanjun Wang, Hainan University, China

Local Arrangement
Mingxin Song, Hainan University, China

Treasurer
Rongzhao He, Hainan University, China

Technical Program Committee

Technical Program Chair: Ken Takeuchi, The University of Tokyo
Technical Program Co-Chair: Pei-Yun Tsai, National Central University
Technical Program Vice-Chair: Tetsuya Iizuka, The University of Tokyo
Industry Program Chair: Stefan Rusu, TSMC
Educational Program Chair: Seung-Tak Ryu, KAIST
Invited Program Chair: Zhihua Wang, Tsinghua University
Special Programs Chair: Makoto Takamiya, The University of Tokyo
Student Design Contest Co- Chair: Jung-Hoon Chun, Sungkyunkwan University
Student Design Contest Co- Chair: Sai-Weng Sin, University of Macao

Analog Circuits and Systems (ACS)

Chair: Po-Chiun Huang, National Tsing Hua University
Members: Po-Hung Chen, National Chiao Tung University
Hyun-Sik Kim, KAIST
Wanyuan Qu, Zhejiang University
Sai-Weng Sin, University of Macau
Makoto Takamiya, The University of Tokyo
Taekwang Jang, ETH, Swiss
Pieter Harpe, Eindhoven University of Technology
Christoph Sandner, Infineon
Min-Jae Seo, Gachon University
Hirshi Fuketa, Nat. Inst. of Adv. Ind. Sci and Tech (AIST)
Sining Pan, Tsinghua University
Garg Paras, STMicroelectronics

Data Converters (DC)

Chair: Kazuko Nishimura, Panasonic Corporation
Members: Chih-Cheng Hsieh, National Tsing Hua University
Qiang Li, University of Electronic Science and Technology of China
Yong Lim, Samsung Electronics
Seung-Tak Ryu, KAIST
Takashi Oshima, Hitachi
Shuang Zhu, NVDIA
Hsin-Shu Chen, National Taiwan University
Hayun Chung, Korea University
Hyungil Chae, Konkuk university
Lu Jie, Tsinghua University
Chi-Hang Chan, Ivor, University of Macau
Zhichao Tan, Zhejiang University

Digital Circuits & Systems (DCS)

Chair: Jun Zhou, University of Electronic Science and Technology of China
Members: Massimo Alioto, National University of Singapore
Mototsugu Hamada, University of Tokyo
Yoonmyung Lee, Sungkyunkwan University
Koyo Nitta, University of Aizu
Shouyi Yin, Tsinghua University
Xiaoyang Zeng, Fudan University
Robert Chen-Hao Chang, National Chung Hsing University
Shuou Nomura, Kioxia Corporation
Yongtae Kim, Kyungbook National University
Amit Agarwal, Intel
Yu-Guang Chen, National Central University
Fengwei An, Southern University of Science and Technology

SoC and Signal Processing Systems (SOC)

Chair: Joo-Young Kim, KAIST
Members: Kazutami Arimoto, Okayama Prefectural Univ.
Kyuho Jason Lee, UNIST
Tsung-Te Liu, National Taiwan University
Daisuke Mizoguchi, Renesas
JunYoung Park, UX Factory
Pei-Yun Tsai, National Central University
Sachin Taneja, Intel
Leibo Liu, Tsinghua University
Kun-Chin Chen, National Yang Ming Chiao Tung University
Weiwei Shan, Southeast University

Wireless (WLS)

Chair: Minoru Fujishima, Hiroshima University
Members: Baoyong Chi, Tsinghua University
Chien-Nan Kuo, National Chao Tung University
Minjae Lee, Gwangju Institute of Science and Technology
Giovanni Mangraviti, IMEC
Dixian Zhao, Southeast University
Jaehyouk Choi, Seoul National University
Kenichi Okada, Tokyo Institute of Technology

Mustafijur Rahman, Indian Institute of Technology Delhi
Kuang-Wei Cheng, National Cheng Kung University
Hao Gao, Eindhoven University of Technology
Yun Yin, Fudan University

Wireline Communications (WLN)

Chair: Byungsub Kim, POSTECH
Members: Jung-Hoon Chun, Sungkyunkwan University
Tetsuya Iizuka, University of Tokyo
Peng Liu, Zhejiang University
Ziqiang Wang, Tsinghua University
Pen-Jui Peng, National Tsing Hua University
Yong Chen, University of Macau
Masum Hossain, U. of Alberta
Jihwan Kim, Intel
Huaide Wang, M31 Technology
Ilmin Yi, GIST
Takashi Toi, Kioxia
Nan Qi, Institute of Semiconductors, Chinese Academy of Science
Wei Deng, Tsinghua University

Emerging Technologies and Applications (ETA)

Chair: Jerald Yoo, National University of Singapore
Members: Ping-Hsuan Hsieh, National Tsing Hua University
Inhee Lee, University of Pittsburgh
Chao Wang, Huazhong University of Science and Technology
Joonsung Bae, Kangwon National University
Takuji Miki, Kobe University
Bo Zhao, Zhejiang University
Chihiro Okada, Sony Semiconductor Solutions Corporation
Kea-Tiong (Samuel) Tang, National Tsing-Hua Univ
Wanyeong Jung, KAIST
Jongwoo Lee, Samsung Electronics
Haikun Jia, Tsinghua University

Memory (MEM)

Chair: Chiweon Yoon, Samsung Electronics
Members: Junghwan Choi, Samsung Electronics

November 5, 2023 (Sunday)

Time				
09:00-10:20(80')	Tutorial 1 Hybrid DC-DC Converters			On-site Registration (Hotel Lobby)
10:20-10:40(20')	Break			
10:40-12:00(80')	Tutorial 2 Circuit Design for Scalable Quantum Computing			
12:00-13:20(80')	Lunch			
13:20-14:40(80')	Tutorial 3 Digital Equalization for Multilevel Signaling in High-speed SerDes			
14:40-15:00(20')	Break			
15:00-16:20(80')	Tutorial 4 Reconfigurable ML Processor: Fundamental Concepts, Application, and Future Trends			
16:40-18:20(100')	Learn to Establish and Achieve Professionalism (LEAP)		Start-up!	IT Vision in Asia
18:20-19:20(60')	TPC Meeting			
19:30-21:00(90')	Welcome Reception (Riverside Lawn)			

November 6, 2023 (Monday)

Time				
08:40-9:00(20')	Opening			
09:00-09:45(45')	Plenary Speech 1 Semiconductor Chip Design in a Legoland Speaker:Chih-Ming Hung, MediaTek			
09:45-10:30(45')	Plenary Speech 2 Envisioning 6G Mobile New World Speaker: Bongtae Kim, ETRI			
10:30-10:50(20')	Break			
10:50-12:30(100')	Session 1 Advanced Designs for Industrial Applications	Rising Star Express (RiSE)	WiC & YP Panel	
12:30-14:00(90')	Lunch			
14:00-15:40(100')	Session 2 Analog Techniques	Session 3 Energy-Efficient SoCs: From Tiny Accelerator to Huge GPU	Session 4 Bio and Neuromorphic System	SDC Exhibition &FPGA Demo(Ballroom Foyer)
15:40-16:00(20')	Break			
16:00-17:40(100')	Session 5 Computing and Analog Circuit for Next Generation Memories	Session 6 THz Transceivers	Session 7 Emerging Sensors and Displays	
17:40-18:00(20')	Break			
18:00-19:30(90')	Social Event (Outdoor area outside the corridor of the conference center)			

November 7, 2023 (Tuesday)

Time			
09:00-09:45(45')	Plenary Speech 3 Architecture Challenges for Heterogeneous Processors in Embedded SoCs Speaker:Masayuki Ito, Director, NSITEXE Inc.		
09:45-10:30(45')	Plenary Speech 4 Listen: ESD Protection is About Circuit Design Speaker:Albert Wang, University of California, Riverside		
10:30-10:50(20')	Break		
10:50-12:30(100')	Session 8 Voltage Regulators	Session 9 Energy-Efficient Application-Specific Processors	Session 10 Advanced Wireline Transceiver Techniques
12:30-14:00(90')	Lunch		
14:00-15:40(100')	Session 11 Innovative Nyquist ADCs	Session 12 Phased Arrays	Session 13 AI on FPGA
15:40-16:00(20')	Break		Industry Demo (Entrance Foyer)
16:00-17:40(100')	Panel Discussion Heterogeneous Integration: Does it call the end of the foundry-fabless business model?		Industry Forum
17:40-18:00(20')	Break		
18:00-20:00(120')	Banquet		

November 8, 2023 (Wednesday)

Time			
09:00-10:40(100')	Session 14 High Precision Oversampling ADCs	Session 15 High-Performance Transceivers	Session 16 Intelligent Memory and Logic Circuit Techniques
10:40-11:00(20')	Break		
11:00-12:40(100')	Convergence Workshop	Highlights from ESSCIRC	Highlights from CICC
12:40-14:00(80')	Lunch		
14:00-15:40(100')	Session 17 Power Management Techniques	Session 18 High-Performance PA and VCO	Session 19 Emerging LiDAR Sensor and Security Chips
15:40-16:00(20')	Break		
16:00-17:40(100')	Session 20 Wireless Power Transfer	Session 21 Application-oriented ADCs	Session 22 High-performance Clock Generation

☐ Tianjin Hall Grand Ballroom A   ☐ Tianjin Hall Grand Ballroom B   ☐ Coral Hall   ☐ Tianjin Hall Grand Ballroom

MSN: Battery-Less Multiple Subcarrier Multiple Access Sensor Node with Full-Duplex Backscatter Concurrent Data Streaming

Yuxiao Zhao[1], Hanyang Wang[12], Yongming Xu[1], Zhongyuan Ying[1], Yu Lu[1], Jinlin Liu[2], Tuo Hu[12], Jin Mitsugi[3], Hao Min[1]

[1]Fudan University, Shanghai, China

[2]Quanray Electronics, Shanghai, China

[3]Keio University, Fujisawa, Japan

To scale to trillions of Internet-of-Things (IoT) nodes, devices must be battery-less by achieving energy autonomy, having a µW power floor, few off-chip components, and a small form factor. These features are beneficial for short-range high-density IoT applications such as structural health monitoring (SHM) for wings and gears, human motion capture (HMC), and air conditioning system monitoring. This work focuses on the synchronous transmission of sensor data streams, called concurrent data streaming, from multiple sensor nodes for multi-points sensing. Traditional wireless communication protocols need construct wireless networks for time synchronization, increasing system complexity [1]. However, the recent development of Multiple Subcarrier Multiple Access (MSMA), enables concurrent backscatter communication by assigning multiple uplink subcarrier (USC) channels to sensor nodes [2][3][4]. To expand the deployment in the mentioned IoT applications, MSMA should be compatible with existing communication protocols and achieve real-time control of starting and stopping of concurrent data streaming without wasting extra sensing system energy. To avoid channel interference among multiple nodes and catastrophic failure of concurrent data transmission, the USC frequency must always keep stable from the influence of PVT.

This paper proposes a system-on-chip (SoC) compatible with MSMA and EPC Generation-2 Protocol (Gen2), which enables a multi-physical quantity detection sensor platform or node called MSMA Sensor Node (MSN). This silicon implementation allows in-band full-duplex (FD) operation in concurrent backscatter and supports wireless frequency calibration for the on-chip oscillator and sub-µW temperature readout system while maintaining small package footprint. Even in the harsh environment, this SoC can allow robust concurrent data transmission without batteries and crystals, which significantly expands the application range of radio frequency identification technology (RFID), while keeping low cost.

Fig. 1 shows the SoC integrated into a less than 20cm^3 MSN with a stub antenna and an optional battery. The node integrates the SoC with an on-PCB matching network, several sensors such as temperature/relative humidity sensor, and an accelerometer. The SoC's primary goals are to 1) enable concurrent data streaming based on MSMA, and FD communication for seamless switching between Gen2 and MSMA to conduct real-time control of starting and stopping of concurrent data streaming, 2) support continuous wireless frequency calibration for the on-chip oscillator, provide sufficient frequency accuracy for backscatter concurrent data transmission even in the extreme conditions, and 3) allow on-chip sub-µW physical quantity sensing compatible with Gen2, such as temperature sensing. The rest of the paper describes 3 main components that accomplish these goals, namely: in-band full-duplex communication system with ASK & full-duplex demodulator (AFD), uplink subcarrier frequency calibration system (UFC), and wireless on-chip temperature sensing system (WTS).

To execute backscatter concurrent data streaming, called STRM mode, and seamless switching with Gen2, this paper uses STRM_START to start up STRM mode of multiple nodes and STRM_STOP to stop STRM mode for suppressing the nodes in wrong USC channels causing by PVT. Furthermore, the full-duplex operation must demodulate the downlink signal, e.g., STRM_STOP by AFD in STRM mode. Fig. 2 shows that the reader sends custom Gen2 commands, STRM_START, to multiple MSNs simultaneously. Then reader sends STRM_SYNC or downlink subcarrier (DSC) to MSNs for on-chip oscillator calibration (Fig. 3) and sends STRM_STOP making MSNs can respond to Gen2. To avoid RF energy loss caused by oversize modulation depth, this work achieved 10-30% downlink ASK modulation(Fig. 2), while avoiding

reducing sensitivity. The design of AFD enables self-interference cancellation (SIC) based on a typical ASK demodulator. Because of the modulation of backscatter switch M1, the uplink signal creates the aliasing of the downlink signal. The basic idea of SIC is non-uniform sampling in the time domain. The pulse generator gives USC(with uplink data)-related narrow pulse and enables every sampling point to be the high level of USC(with uplink data). Thus, the S/H can block non-ideal transmission of the uplink signal from S_2 to S_3 and accomplish the downlink signal recovery (Fig. 2). The signal shaper in AFD can reshape the low-amplitude signal V_{SH} to recover the digital domain data V_{DATA}. Fig. 5(c) shows the full-duplex test results of STRM_STOP, the waveform received by USRP2952R shows that STRM_STOP is valid, STRM mode stops, and MSNs return to Gen2 mode. The demodulated STRM_STOP waveform matches with the reference signal of reader, verifying valid full-duplex operation.

To allocate and keep concurrent data streaming channels in extreme environments, this paper proposes UFC and generates high-frequency accuracy USC for MSMA communication. The UFC includes a two-step USC calibration process, the first USC pre-calibration supports wireless frequency calibration for a single MSN by obtaining the USC response of the MSN, which allocates multiple USC channels sequentially and prepares for STRM Mode. The second USC tracking, enables real-time and continuous frequency calibration for on-chip oscillators in multiple MSNs to suppress USC frequency offset caused by the environment (Fig. 3). The basic idea of the proposed USC tracking is to use 10kHz DSC or 10Hz STRM_SYNC as a wireless reference source and construct on-chip digital frequency-locked loop to tune digitally controlled ring oscillator. Thus the USC divided by oscillator output can track DSC and keep PVT-robust frequency accuracy. The full-duplex demodulation of DSC and STRM_SYNC also depends on AFD mentioned in Fig. 2. Since the frequency-locked loop is a first-order loop, there is no loop stability problem. Fig. 5(b) shows USC tracking test results of 4 MSNs under extreme temperature conditions. Even rapidly heating up at +134 °C/min and cooling down at -33.5 °C/min, all 4 MSNs can continuously maintain 1kHz (0.25%) frequency accuracy from 27°C to 90 °C.

To achieve sub-µW or even sub-0.1µW on-chip sensing of physical quantities (i.e., temperature), the proposed SoC supports WTS embedded with Gen2 protocol (Fig. 4). The WTS consists of an embedded time-domain temperature sensor and backscatter transmitter. The reader sends Gen2 Write to the REG defined in USER bank, and MSN returns 4-10ms USC waveform. Then the reader can measure the frequency of USC and calculate the temperature of the MSN point. The designed temperature sensor uses a single-ended, three-stage, current-starved ring oscillator, enabling <500nW typical power (including buffers and dividers). Compared with commercial Gen2 tags, the WTS only adds a ring oscillator, buffer, and a few digital logic circuits with sub-µW additional power consumption. The WTS is the first wireless temperature sensing system perfectly compatible with Gen2 communication and can easily be embed into any Gen2 chip. Fig. 5(c) shows the temperature sensing test of a single MSN, which obtained average sensitivity of 5.826kHz/°C from 25 °C to 125 °C.

MSN SoC is manufactured in 130nm EEPROM CMOS technology (Fig. 7) and provides a platform for developing various battery-free multi-node wireless sensing systems. Fig. 6 compares the SoC with other ultra-low power, self-powered, and RFID-based sensing SoC. The table shows that 1) This work is the only SoC, which implements backscatter in-band full-duplex communication, 2) The proposed subcarrier calibration system enables robust multi-node concurrent data sensing without complex networking, and 3) The proposed WTS can easily be embed into any RFID tags to achieve temperature sensing with minimum power consumption.

Acknowledgment: This work is a result of backscatter communications research consortium in Keio University activity.

References:

[1] J. K. Brown et al., "27.1 A 65nm Energy-Harvesting ULP SoC With 256kB Cortex-M0 Enabling an 89.1µW Continuous Machine Health Monitoring Wireless Self-Powered System," ISSCC, 2020.

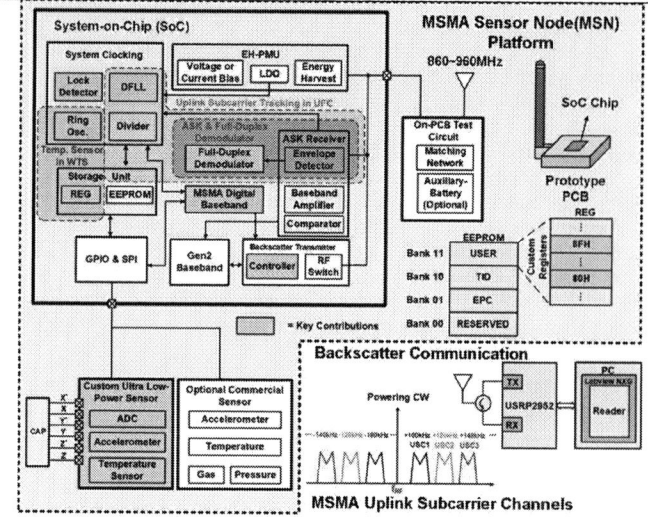

Fig. 1. Block diagram of the proposed MSN platform including standard Gen2 blocks, peripherals, and key modules for full-duplex communication, UFC and WTS.

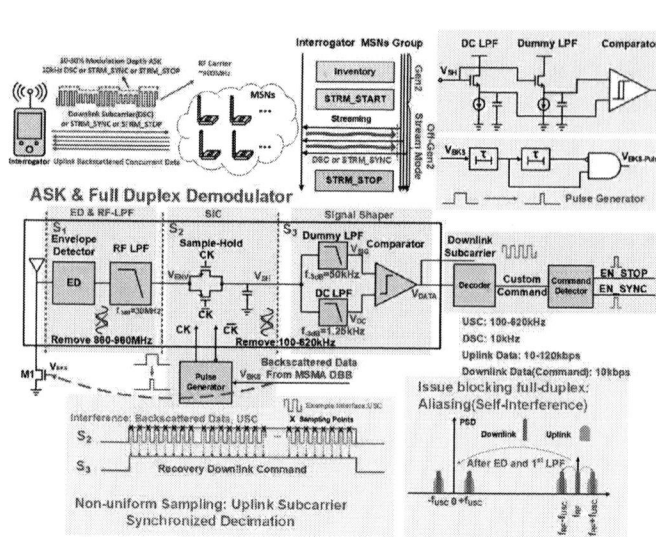

Fig. 2. Full-duplex communication system architecture under backscatter concurrent data transmission mode.

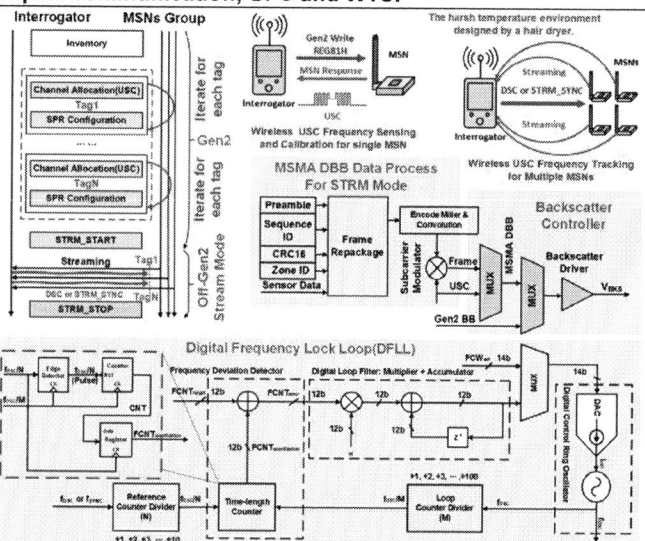

Fig. 3. UFC: single MSN frequency calibration in Gen2 mode and frequency tracking for multiple MSNs in STRM mode.

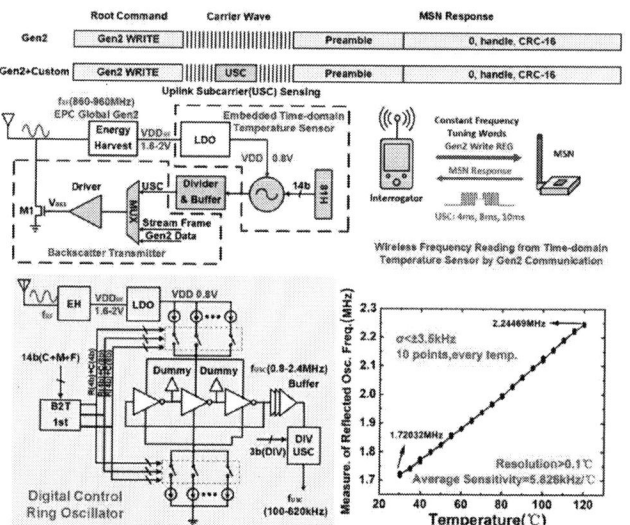

Fig. 4. Architecture of sub-µW time-domain WTS compatible with the Gen2 protocol.

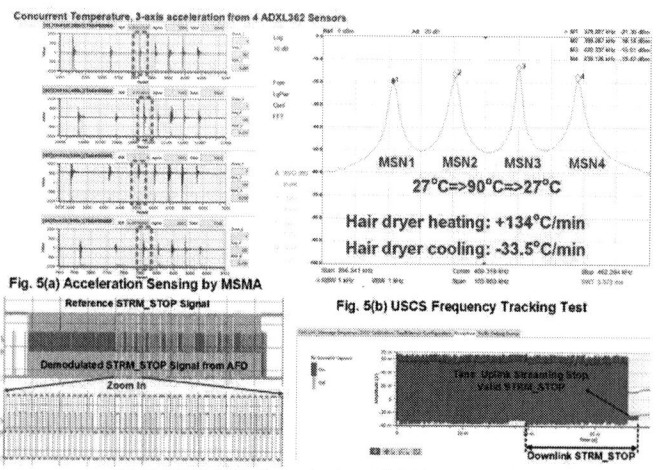

Fig. 5. Measured concurrent acceleration sensing, USC frequency accuracy under extreme temperature conditions, and full-duplex downlink command reception.

Fig. 6. Performance and comparison table to other ultra-low power systems, self-powered systems, and state-of-the-art RFID sensor nodes.

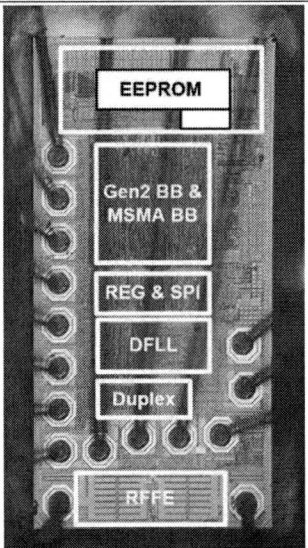

Fig. 7. Die photo of the 130nm, 0.51×1.02mm² SoC.

Additional References:

[2] Y. Zhao et al., "USCoCa: A Subcarrier Frequency Continuous Calibration Method for Backscattered Concurrent Data Mode in UHF RFID System," RFID, 2023.

[3] J. Mitsugi et al., "Perfectly Synchronized Streaming From Multiple Digitally Modulated Backscatter Sensor Tags," IEEE JRFID, Sept. 2019.

[4] K. Tang et al., "A Highly Integrated Passive Wireless Sensing System With Synchronized Data Streaming of Multiple Tags," IEEE JIOT, Sept. 2022

[5] L. Lin et al., "Battery-Less IoT Sensor Node with PLL-Less WiFi Backscattering Communications in a 2.5-μW Peak Power Envelope," VLSIC, 2021.

[6] Y. Yano et al., "An IoT Sensor Node SoC with Dynamic Power Scheduling for Sustainable Operation in Energy Harvesting Environment," A-SSCC, 2019.

[7] J. Yin et al., "A System-on-chip EPC Gen-2 Passive UHF RFID Tag With Embedded Temperature Sensor," ISSCC, 2010.

[8] D. Yeager et al., "A 9.2μA Gen2 Compatible UHF RFID Sensing Tag With −12dBm Sensitivity and 1.25μVrms Input-referred Noise Floor," ISSCC, 2010.

A 25kHz-BW 97.4dB-SNDR 100.2dB-DR 3rd-order SAR-Assisted CT DSM with 1-0 MASH and DNC

Kent Edrian Lozada[1*], Dong-Hun Lee[1*], Ye-Dam Kim[1],

Ho-Jin Kim[2], Youngjae Cho[2], Michael Choi[2], and Seung-Tak Ryu[1]

*Equally-Credited Authors (ECAs)

[1]KAIST, Daejeon, Korea

[2]Samsung Electronics, Hwaeseong, Korea

Continuous-time delta-sigma modulators (CT DSMs) have been a popular choice for high-resolution applications such as audio devices and sensors due to their inherent anti-aliasing, easy driving capability, and relaxed amplifier requirements [1-2]. In typical DSMs, the noise-shaping order is directly proportional to the number of integrators used. As an alternative, a single-loop DSM with a digital noise coupling (DNC) topology [3] can be used to provide aggressive noise-shaping without increasing the order of the loop filter (LF) by moving the analog burden to the digital side. In [3-5], the DNC structure utilizes a single successive-approximation-register (SAR) quantizer and a robust NC filter composed of simple digital cells (Fig. 1a). Inevitably, however, the DNC architecture displays a low maximum stable amplitude (MSA) owing to the high out-of-band gain (OBG) and the large quantization error of the main ADC. The latter issue can be alleviated using a MASH architecture in which the quantization error of the first stage is eliminated by a digital filter, resulting in a high MSA and more reliable stability.

In CT MASH DSMs, proper extraction of the quantization error is challenging compared to their discrete-time counterparts. As the input to the quantizer works in a continuous-time fashion and the output signal is in discrete time, a proper time delay must be inserted into the input before quantization error extraction to match the quantizer delay [6]. Alternatively, a SAR quantizer can be utilized to eliminate the delay issue as it inherently generates the residue on its capacitor array. However, a residue amplifier is necessary to drive the following stage [7].

To address the abovementioned problems, we propose a CT DSM architecture with 1-0 MASH and DNC, as shown in Fig. 1(b), where E_{Q1} is cancelled out and the mismatch issue between the analog and digital filter is relaxed with low-order LFs. In the proposed architecture, the quantization error of the first-stage DSM with a first-order LF (E_{Q1} from ADC_1) is fed into the zeroth-order second-stage ADC (ADC_2). While the second-stage itself does not have any noise-shaping capability, E_{Q2} goes through the DNC path for additional noise shaping. In this work, a second-order DNC filter (H_{DNC}) is utilized. The output of both stages goes through the digital filter for the final output generation with cancelled E_{Q1}, resulting in overall third-order noise shaping in E_{Q2}. In contrast to the conventional MASH, owing to the second-order DNC, the digital filter for the second-stage output only requires a first-order noise transfer function (NTF_1). Therefore, the matching requirements between the analog LF and digital filter can be relaxed. It should be noted that ADC_1, ADC_2, and NC-ADC are all implemented with a single SAR quantizer, similar to [3]. Thus, E_{Q1} from ADC_1 is readily available as input to the ADC_2. Therefore, the explicit extraction of the quantization error, including timing matching, is no longer needed, resulting in a simple and hardware-efficient implementation.

Figure 2 presents a comparison between earlier DNC work and the proposed 1-0 MASH with DNC. For a fair comparison, a 10b quantizer, first-order LF, and second-order DNC were used in a behavioral simulation. For the 10b quantizer, the main-ADC (Fig. 1a) and ADC_1 (Fig. 1b) both had 4-bit resolutions considering DWA. The remaining 6 bits were allocated to the NC-ADC on Fig. 1(a), while in Fig. 1(b), they were divided into 3 bits each for ADC_2 and NC-ADC. In Fig. 2(a), it can be observed that E_{Q1} from the prior work and E_{Q2} from the proposed work have the same order of noise shaping in the final output. However, the noise power of the proposed work is lower since its quantization error has a higher effective resolution (7 bits) than that of the prior work. Moreover, in Fig. 2(b), the prior work exhibits a high peak-to-peak swing at the input of the quantizer, reaching nearly full scale, due to the third-order noise shaping applied to the quantization error in the main loop. In contrast, the proposed work exhibits much lower swing as its main loop is subjected to first-order shaping only. As a result, the proposed work will have higher MSA and a relaxed linearity requirement for the amplifier. It is important to note that E_{Q1} in the proposed work is cancelled due to the MASH implementation. The comparison shown in Fig. 2(c) illustrates the effect of extending MSA, where the proposed architecture achieves a higher MSA by up to 2 dB.

Figure 3 presents a detailed implementation (single-ended version) of the proposed CT DSM with 1-0 MASH and DNC. A single 10-bit asynchronous SAR quantizer is used with 4-bit MSBs for ADC_1, 3-bit ISBs for the ADC_2, and 3-bit LSBs for the NC-ADC. The DNC filter is set to $H_{DNC} = 2z^{-1} - z^{-2}$. During the sampling phase, the output of the DNC filter is connected to the 4b LSBs of the sampling capacitors. This 4b LSB represents the NC-DAC. After the sampling phase is finished, the 4b LSBs are reset and the information from the DNC filter is combined from the sampled input to obtain the noise-coupled input (i.e. $V_{LF\_OUT} - H_{DNC} \cdot D_{EQ2}$). Afterwards, the first-stage A/D conversion is done by utilizing the 4b MSBs. After the first-stage conversion, E_{Q1} now becomes the direct input to the ADC_2 to obtain 3b ISBs. In the same manner, as soon as ADC_2 is complete, E_{Q2} remains on the top node of the capacitor DAC and is further quantized for the 3b LSBs by the NC-ADC. The 4b MSBs of the SAR quantizer are sent to the DWA and then to the main FB RDAC, while the 3b ISBs go to the digital filter. The output of the digital filter is combined with the 4b MSBs and this becomes the final output of the proposed CT MASH DSM with DNC. The first-order LF is implemented using an active-RC integrator with chopping. A two-stage opamp is used with class-AB output to reduce the static power. Considering the mismatch between the analog and digital filters, the opamp was designed with sufficient margin for PVT variations.

The proposed modulator with a 25kHz bandwidth (BW) is fabricated in the 28nm CMOS process, occupying an area of 0.24mm^2 (Fig. 7). Clocked at a sampling frequency of 12.8MHz, the modulator consumes 260μW with a 1.1V supply. Figure 4 shows the measured spectrum of the proposed modulator with input of 7kHz. When DWA and CHOP are both deactivated, the modulator achieves an SNDR of 68.1dB, which is limited by the non-linearity of the FB DAC, whereas the SNDR improves to 85.6dB when DWA is activated. When both the DWA and CHOP are turned on, the modulator achieves a peak SNDR and SNR of 97.4dB and 98.4dB, respectively. Figure 5 shows the measured SNDR with respect to the input amplitude. The measured dynamic range (DR) is 100.2dB. Figure 6 presents a performance summary and a comparison with other state-of-the-art DSMs with a similar NTF order, BW, and SNDR. This work achieves a competitive performance with a FoM$_{S,SNDR}$ value of 177.2dB. These results demonstrate that a high-order CT DSM can be implemented with a lower analog burden with low-order LFs through the combination of MASH and DNC.

Acknowledgment:

This work was supported by Samsung Electronics Co. Ltd (IO201209-07917-01), and the CAD tools were supported by IDEC of Korea.

References:

[1] Y. H. Leow et al., "A 1 V 103 dB 3rd-Order Audio Continuous-Time ΔΣ ADC with Enhanced Noise Shaping in 65 nm CMOS," JSSC, vol. 51, no. 11, pp. 2625–2638, Nov. 2016.

[2] S. Billa et al., "Analysis and Design of Continuous-Time Delta–Sigma Converters Incorporating Chopping," JSSC, vol. 52, no. 9, pp. 2350-2361, Sept. 2017.

[3] I. -H. Jang et al., "A 4.2mw 10MHz BW 74.4db SNDR Fourth-Order CT DSM with Second-Order Digital Noise Coupling Utilizing an 8b SAR ADC," VLSI, pp. C34-C35, June 2017.

[4] Y.-D. Kim et al., "A 4th-Order CT I-DSM with Digital Noise Coupling and Input Pre-conversion Method for Initialization," A-SSCC, pp. 1-3, Nov. 2021.

[5] K. E. Lozada et al., "A 4th-Order Continuous-Time Delta-Sigma Modulator with Hybrid Noise-Coupling," IEEE TCAS II, vol. 69, no. 9, pp. 3635-3639, Sept. 2022.

[6] D. -Y. Yoon et al., "An 85dB-DR 74.6dB-SNDR 50MHZ-BW CT MASH ΔΣ modulator in 28nm CMOS," ISSCC, pp. 1-3, Feb. 2015.

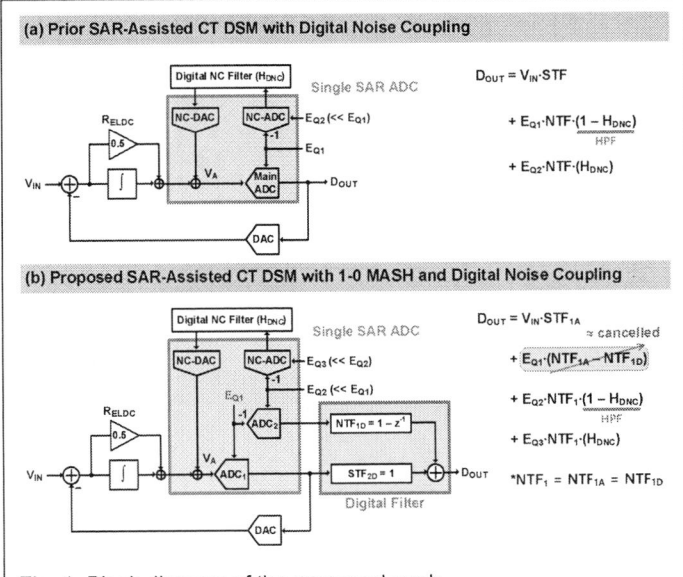

Fig. 1. Block diagram of the proposed work.

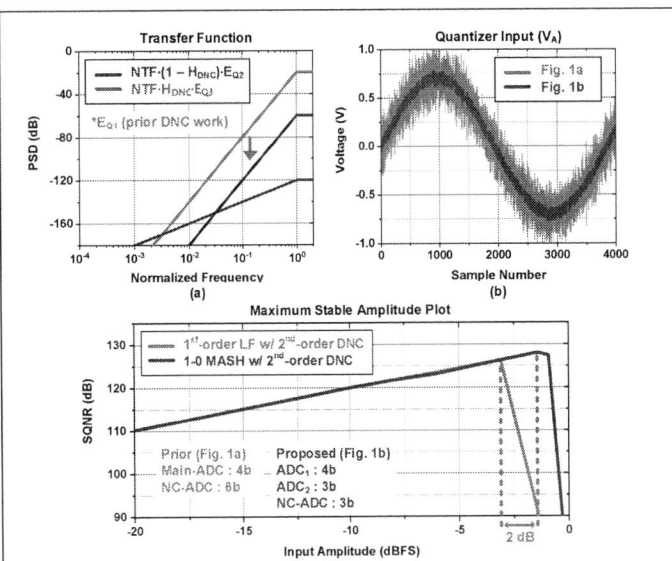

Fig. 2. Comparison between the prior work and proposed work.

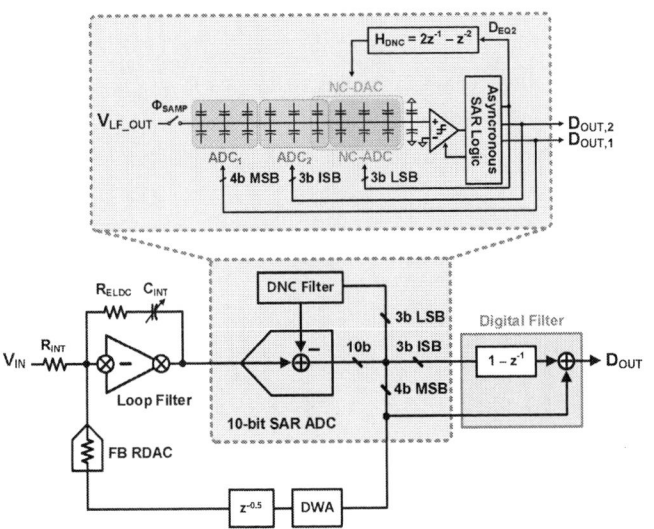

Fig. 3. Overall implementation of the proposed work.

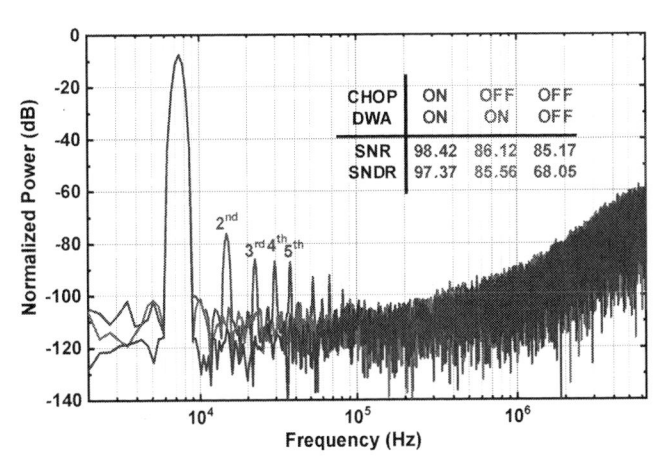

Fig. 4. Measured FFT spectrum.

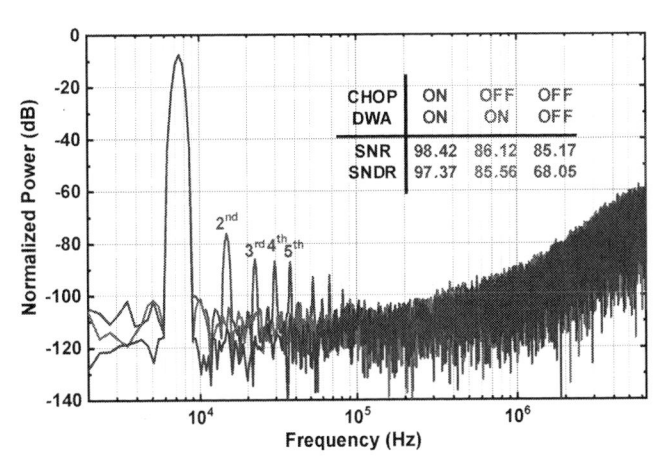

Fig. 5. Measured SNDR versus the input amplitude and power breakdown.

	This Work	JSSC 22 Kumar [10]	JSSC 21 Jang [9]	JSSC 19 Liao [8]	JSSC 17 Billa [2]	JSSC 16 Leow [1]
Architecture	CT MASH DSM with Digital NC	DT DSM with FIA	CT DSM with Negative-R and FIR DAC	DT MASH DSM	CT DSM with FIR DAC	CT DSM with Analog NC
Total NTF	3	3	3	3	3	3
Process [nm]	28	180	65	65	180	65
BW [kHz]	25	24	24	25	24	25
SNR$_{PK}$ [dB]	98.4	96.4	101.0	96.1	99.3	100.1
SNDR$_{PK}$ [dB]	97.4	96.2	99.4	94.6	98.5	95.2
DR [dB]	100.2	98.0	102.8	98.5	103.6	103
Power [μW]	260	340	134	175	280	800
FoM$_{SNDR}$ [dB]	177.2	174.7	181.9	176.2	177.8	170.1

FoM$_{SNDR}$ = SNDR (dB) + 10log(BW/P)

Fig. 6. Performance summary and comparison with state-of-the-art DSMs with similar NTF order, BW, and SNDR.

979-8-3503-3004-5/23 $31.00 © 2023 IEEE

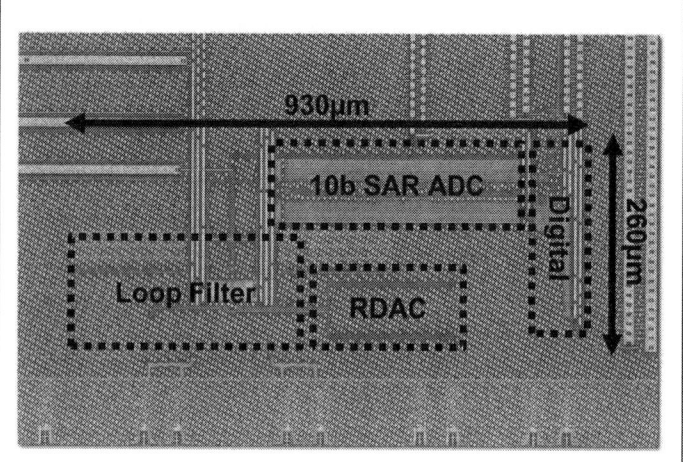

Fig. 7. Die Micrograph.

Additional References:

[7] H. Liu *et al.*, "An 85-MHz-BW ASAR-Assisted CT 4-0 MASH $\Delta\Sigma$ Modulator With Background Half-Range Dithering-Based DAC Calibration in 28-nm CMOS," IEEE TCAS-I, vol. 66, no. 7, pp. 2405-2414, July 2019.

[8] S. -H. Liao *et al.*, "A 1-V 175-μW 94.6-dB SNDR 25-kHz Bandwidth Delta-Sigma Modulator Using Segmented Integration Techniques," JSSC, vol. 54, no. 9, pp. 2523-2531, Sept. 2019.

[9] M. Jang *et al.*, "A 134-μW 99.4-dB SNDR Audio Continuous-Time Delta-Sigma Modulator With Chopped Negative-R and Tri-Level FIR-DAC," JSSC, vol. 56, no. 6, pp. 1761-1771, June 2021.

[10] R. S. A. Kumar *et al.*, "Analysis and Design of a Discrete-Time Delta-Sigma Modulator Using a Cascoded Floating-Inverter-Based Dynamic Amplifier," JSSC, vol. 57, no. 11, pp. 3384-3395, Nov. 2022.

A Ka-band Mutual Coupling Resilient Balanced PA with Magnetic Coupling Self-cancelling Inductor Achieving 21.2dBm OP$_{1dB}$ and 27.6% PAE$_{1dB}$

Jian Zhang[1], Wei Zhu[2], Dawei Wang[1], Xiangjie Yi[1], Ruitao Wang[1], and Yan Wang[1]

[1]School of Integrated Circuits, Tsinghua University, Beijing, China

[2]School of Integrated Circuits and Electronics, Beijing Institute of Technology, Beijing, China

The performance and robustness of millimeter-wave (mm-wave) power amplifiers (PAs) define the quality of a high-data-rate communication system to a large extent [1]. However, the performance of the PAs is strongly affected by mutual coupling among the close element in the phased-array antenna, yielding a beam-steering angle-dependent and time-varying loading or VSWR condition rather than ideal 50Ω as shown in Fig. 1 (top). Furthermore, mm-wave communication systems typically demand PAs with high output power and power-added efficiency (PAE), worsening the output reflection coefficient (Γ_{PA}) [2-7]. Therefore, the unwanted "element to element" coupled signals reflect back to the antenna, deteriorating the phased array beam pattern and the linearity of PAs. Recently, a method of tuning the output matching network (self-healing) based on load mismatch detection has been proposed to deal with this problem [8]. However, this method requires a reconfigurable and inevitably lossy output matching network, which will degrade the performance of PAs. Besides, the design of PA with high output power and efficiency meets many challenges as the operating frequency moves into the mm-wave range mainly due to breakdown issues when high voltages are handled, especially in modern downscaled CMOS technologies. In addition, the increased routing parasitic when increasing the device size results in higher loss and makes the design of matching networks more difficult due to a high impedance transformation ratio.

To address the challenges, we present a power-enhanced balance PA (PEBPA) as shown in Fig. 1 (bottom left). Two high Pout PAs with 3-stack transistor cells are combined through a quadrature hybrid coupler (QHC). To facilitate a balanced layout and ensure the consistency of the two signals, the QHC is implemented by quarter-wavelength microstrip lines with 50Ω characteristic impedance. Floating metal strips are placed to shield the fields from the substrates, maintaining a reasonable transmission efficiency. Fig. 1 (bottom right) illustrates the VSWR condition of the proposed PEBPA through two aspects. In the mutual-coupling scenario, the unwanted signals from other antennas will be split by the QHC and reflected by the two sub-PAs with a 180° phase difference at the antenna port. Therefore, reflected signals are canceled at the antenna port while combined at the isolation port and absorbed in the 50Ω termination. As for the PAs loading conditions, the reflected signal can also be spilt by the QHC, which will provide a significant large-signal VSWR robustness. Based on this theory, this proposed PEBPA can offer a perfect termination to any unwanted element-to-element coupled signal.

Fig. 2 (top) shows the schematic of the proposed PEBPA, which consists of QHCs, common source driver amplifiers (DAs), inter matching transformers with magnetic coupling cancelling Linter (TF with MCC Linter), and 3-stack transistor PAs that are shown in Fig. 2 (bottom left). To address the deterioration of gain and Pout caused by large parasitic resistance and inductance at the source network of the standard transistor cells, custom transistor cells with optimized layouts are utilized to improve the performance of the proposed PEBPA. The Pad-Open-Short method is used for de-embedding and measurement results show that the Gmax, Psat, and peak PAE of the custom cell are 0.4 dB, 0.8 dBm, and 3.3% higher than that of the standard cell respectively as shown in Fig. 2 (inset). Different from the cascode PA where the gate of the common-gate transistor is AC ground, the capacitors C2 and C3 need to be elaborately designed to avoid transistor breakdown. The gate capacitor and Cgs form a capacitive voltage divider that allows an RF swing on the gate in phase to enhance reliability. Fig. 2 (bottom center) shows the simulated drain-to-source voltages (Vds) of each stacked FET, which are distributed uniformly, and reliability issues are relieved.

Furthermore, a shunt inductor Linter is introduced to suppress the degradation of PAE and Pout caused by the parasitic capacitances seen from the drain of each stacked FET. Fig. 2 (bottom right) shows the simulated PAE$_{max}$ and Gain of the PA cell versus different values of the intermediate inductance Linter and 225 pH is chosen in this design based on this simulation.

However, the conventional implementation of Linter will consume a large chip area to avoid unwanted self-coupling and mutual coupling, which is not beneficial for achieving large-scale phased arrays as shown in Fig. 3 (top left). In order to address this problem, a magnetic coupling cancelling (MCC) inductor is used in this design as shown in Fig. 3 (top). The conventional square inductor is twisted into 2-turn 8-shapes that can self-cancel their magnetic coupling to other transformers and then the Linter can be stacked into the inter-matching network, sharing the same footprints, avoiding self-coupling and mutual-coupling. Fig. 3 (inset) shows the coupling current density of the inter-matching transformer in detail and the coupling current of the Linter is ultra-low thanks to the 8-shape topology. Although most of the magnetic coupling can be cancelled, it is worth mentioning that the transformer coupling coefficient (K) is not zero when the center position d0 is zero because the arms of the transformers contribute extra inductance, which is not self-cancelled by the 8-shape. However, the coupling coefficient can be easily manipulated by shifting the central axis of the 8-shape transformer as shown in Fig. 3 (inset center). This proposed MCC Linter has perfect coupling cancellation (K=0) at 30 GHz when d0 is about 4.2 um and maintains an ultra-low coupling coefficient (K<0.05) within 10 to 35 GHz through electromagnetic simulation. Fig. 3 (bottom) shows the parameters of the interstage transformer with and without MCC Linter regarding 30 GHz. The variation of most parameters is very small (Δ<1%) and the quality factor of Lp gets worse a lot (ΔQp=3.2%) due to strong capacitive parasitic between Lp and Linter. However, the performance of the amplifier will hardly deteriorate because the tank loss is dominated by the source impedance and the load impedance in this scenario.

Both the single-way high Pout PA and the PEBPA are fabricated in the 45nm CMOS SOI process. Fig. 4 shows the measured small and large signal performance of these two PAs. The single-way high Pout PA achieves 25.2 dB maximum small-signal gain, 19.7 dBm OP$_{1dB}$, 20.2 dBm P$_{sat}$, 29% PAE$_{1dB}$, and 30.4% peak PAE. The proposed PEBPA achieves 21.2 dBm OP$_{1dB}$ with 27.6% PAE$_{1dB}$ and 22.2 dBm OP$_{1dB}$ with 28.8% peak PAE. The maximum small-signal gain is 26.5 dB with <-19/-14 dB S11/S22. The measurement results also show that OP$_{1dB}$ and PAE$_{1dB}$ are beyond 21 dBm and 22% for a frequency range from 25GHz to 32GHz, respectively. Finally, the performance of the two PAs under VSWR=3 has been simulated. The P$_{sat}$ of the proposed PEBPA is relatively constant whose variation is less than 0.5 dBm over a range of 0-to-360° angle but the max variation of the single-way high Pout PA is 2.5 dBm. Fig. 5 shows the modulation measurements of the proposed PEBPA and single-carrier 64-QAM (PAPR=7.15 dB) and 256-QAM (PAPR=6.73 dB) signals are used. For a 3.6 Gb/s (600 MSym/s) 64-QAM signal, the proposed PEBPA achieves -29.2 dB RMS EVM with an average Pout/PAE of 16.3 dBm/13.8%. For a 4.8 Gb/s (600 MSym/s) 256-QAM signal, the proposed PEBPA achieves -34.8 dB RMS EVM with an average Pout/PAE of 12.8 dBm/8.3%.

The performance of the proposed PEBPA is summarized in Fig. 6 and compared to that of the state-of-the-art two/three-way combining PAs. The proposed PEBPA maintains excellent performance while handling 3.6 Gb/s 64QAM and 4.8 Gb/s 256-QAM signals without using any DPD. This work also provides a better output impedance termination than the previous works and generates more than 21 dBm OP$_{1dB}$ and 27.6% PAE$_{1dB}$, which achieves the highest ITRS FoM1.

References:

[1] Y. Yi et al., " A 24-29.5-GHz Highly Linear Phased-Array Transceiver Front-End in 65-nm CMOS Supporting 800-MHz 64-QAM and 400-MHz 256-QAM for 5G New Radio," IEEE JSSC, vol. 57, no. 9, pp. 2702-2718, Sept. 2022.

[2] P. Guan et al., "An Ultra-Compact 16-to-45 GHz Power Amplifier within A Single Inductor Footprint Using Folded Transformer Technique," CICC 2021, pp. 1-2.

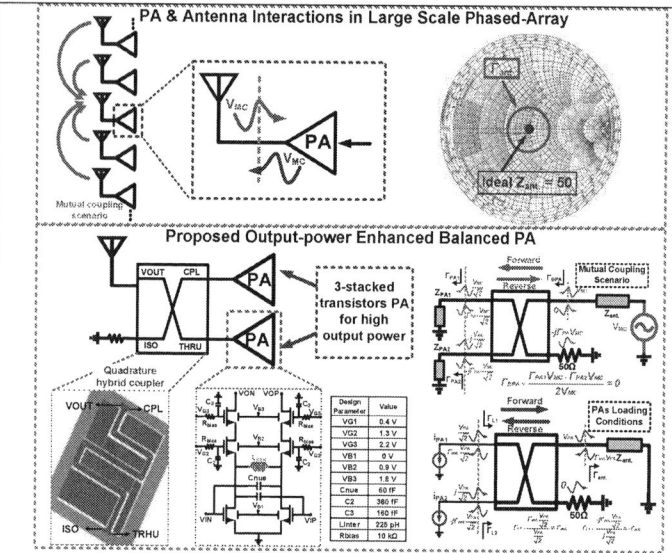

Fig. 1. The interactions of PA & antenna in large-scale phased arrays and the illustration of the proposed power-enhanced balanced PA.

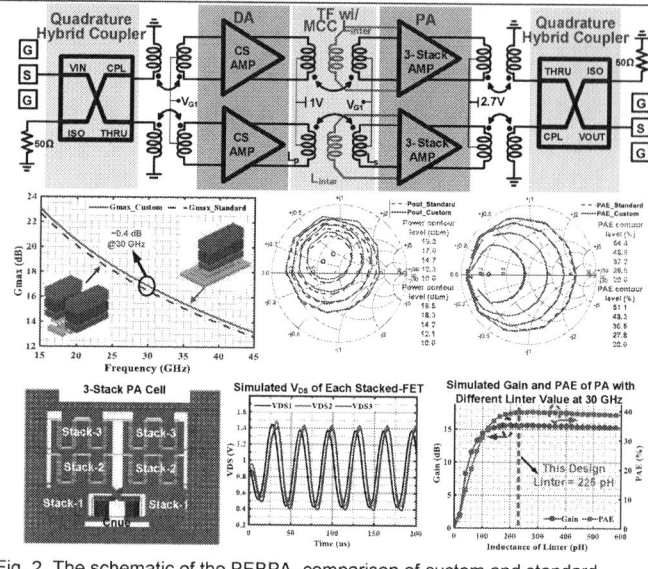

Fig. 2. The schematic of the PEBPA, comparison of custom and standard transistor cell, and the layout and design parameters of the 3-stack PA cell.

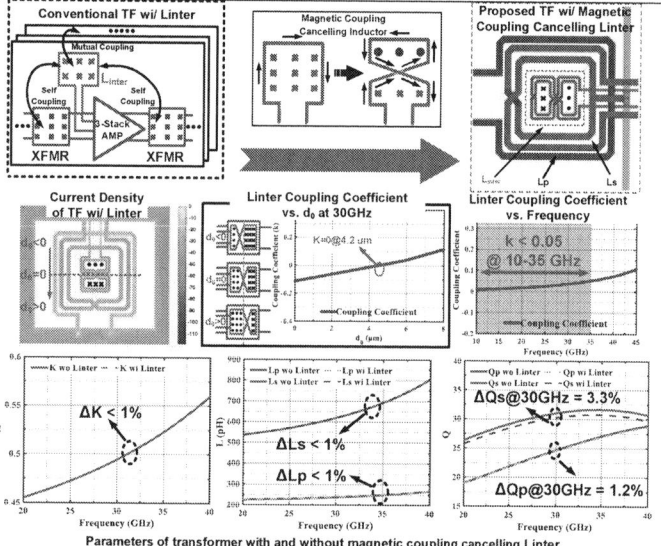

Fig. 3. The implementation of the magnetic coupling cancelling (MCC) inductor and parameters comparison of the transformer wi/ and wo/ MCC Linter.

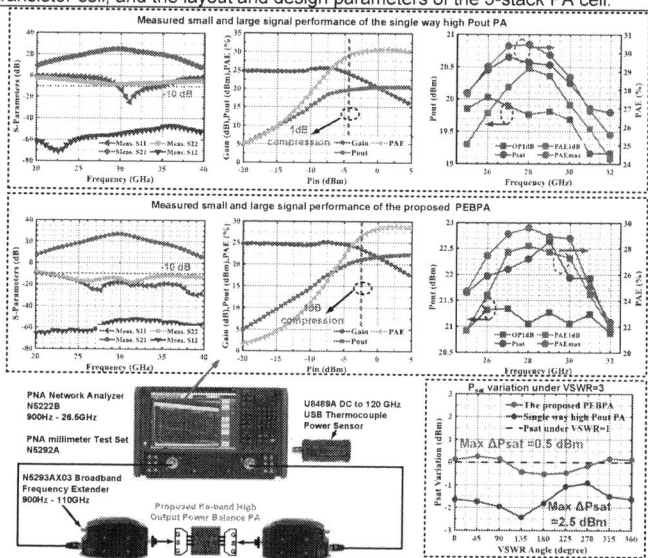

Fig. 4. Measured S-parameters, large-signal results, P_{sat} variation under VSWR=3 of the single-way PA and the PEBPA, and the measurement setup.

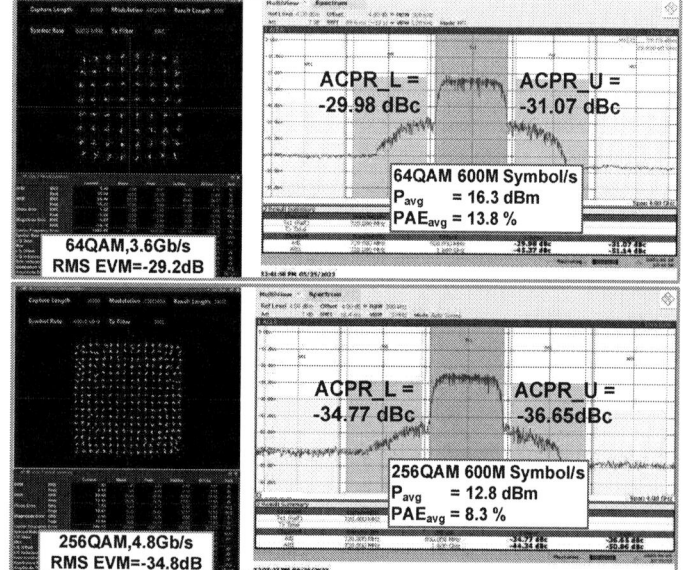

Fig. 5. Measured constellation, spectrum, and EVM for 3.6 Gb/s 64-QAM signal and 4.8 Gb/s 256-QAM signal of the proposed PEBPA.

Comparison With the State-of-the-art Two/Three-way Combining PAs.

	CICC'20 [4]	ASSCC'21 [5]	JSSC'23 [6]	JSSC'20 [7]	This Work	
Technology	22nm FD-SOI	40nm CMOS	45nm SOI	45nm SOI	45nm SOI	
Frequency (GHz)	28	28	38	30	30	
Architecture	2-way Current Combination	2-way Series Doherty PA	3-way Doherty PA	2-way Outphasing PA	Single-way Stack-FET PA	2-way Balanced Stack-FET PA
Gain (dB)	27	26.1	15	7	25.2	26.5
VSWR at Antenna	4.4*	5.8*	2.4*	3.6*	2.3	1.5
Psat (dBm)	21.7	22.5	18.9	17	20.2	22.2
OP1dB (dBm)	19.1	21.1	18.4	N/A	19.7	21.1
PAEpeak (%)	27.1	28.5	23.3	44.5*	30.4	28.8
PAE1dB (%)	23	26.6	23.0	N/A	29.2	27.6
ITRS FoM1**	96.1	95.1	86.7	74.9	98.1	106.8
QAM Modulation	64	256	64	64(OFDM)	64	N/A	64	256
Data rate (Gb/s)	3	0.8	2.4	0.8	0.6	3	N/A	3.6	4.8
EVM (dB)	-25	-30.6	-25	-30	-25	-28	N/A	-29.2	-34.8
Pout @EVM (dBm)	12.9	14.3	10.9	10.2	11.3	10.2	N/A	16.3	12.8
PAE @EVM (%)	9	11.8	9.2	9	14.7	31.2	N/A	13.8	8.3

*Graphically estimated.
**ITRS FoM1=Psat(dBm)+Gain(dB)+20log(fc[GHz])+10log(PAE$_{max}$(%))+20log[(VSWR+1)/(VSWR-1)]

Fig. 6. Performance comparison table.

979-8-3503-3004-5/23 $31.00 © 2023 IEEE

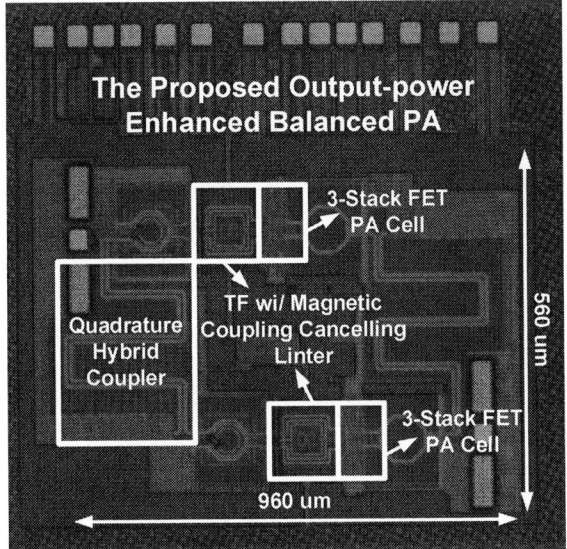

Fig. 7. Die micrograph of the proposed PEBPA.

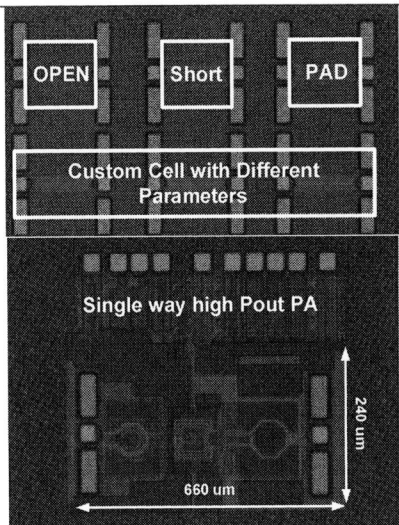

Fig. S1. Custom transistor cells with de-embedding structures and the die micrograph of the single way of the proposed PEBPA.

Additional References:

[3] M. Pashaeifar et al., "14.4 A 24-to-30GHz Double-Quadrature Direct-Upconversion Transmitter with Mutual-Coupling-Resilient Series-Doherty Balanced PA for 5G MIMO Arrays," ISSCC 2021, pp. 223-225.

[4] Z. Zong et al., "A 28GHz Two-Way Current Combining Stacked-FET Power Amplifier in 22nm FD-SOI," CICC 2020, pp. 1-4.

[5] M. Pashaeifar et al., "A 24-to-32GHz series-Doherty PA with two-step impedance inverting power combiner achieving 20.4dBm Psat and 38%/34% PAE at Psat/6dB PBO for 5G applications," A-SSCC 2021, pp. 1-3.

[6] X. Zhang et al., " A Millimeter-Wave Three-Way Doherty Power Amplifier for 5G NR OFDM," IEEE JSSC, vol. 58, no. 5, pp. 1256-1270, May 2023.

[7] K. Ning et al., "A 30-GHz CMOS SOI Outphasing Power Amplifier With Current Mode Combining for High Backoff Efficiency and Constant Envelope Operation," in IEEE JSSC, vol. 55, no. 5, pp. 1411-1421, May 2020.

[8] S. M. Bowers et al, "Integrated Self-Healing for mm-Wave Power Amplifiers," IEEE TMTT, vol. 61, no. 3, pp. 1301-1315, March 2013.

A 80Gb/s/pin Single-Ended PAM-4 Transmitter With an Edge Boosting Auxiliary Driver and a 4-Tap FFE in 28-nm CMOS

Dae-Won Rho[1*], Jae-Koo Park[12*], Seung-Jae Yang[1] and Woo-Young Choi[1]

[1]Yonsei University, Seoul, Korea

[2]Samsung Electronics, Hwaseong, Korea

*Equally-Credited Authors (ECAs)

The demands for the higher-bandwidth memory access are continuously increasing for many applications such as data centers, HPC, and AI processors. With this, the importance of high-speed memory interfaces is increasing as well and a number of technical approaches are being pursed in order to overcome the channel bandwidth limitation. Especially, the pulse amplitude modulation 4 (PAM-4) technique is actively investigated [1-4]. PAM-4 can reduce the symbol rate by half but at the cost of the reduced signal-to-noise ratio (SNR), resulting in the increased bit error rate (BER)[5]. To minimize the SNR decrease, the level-separation mismatch ratio (RLM) should be optimized. Furthermore, the performance of equalizers should be well optimized for inter-symbol interference (ISI) reduction. This article presents the technique of achieving a 80Gb/s transmitter (TX) with a PAM-4 single-ended voltage mode (VM) driver implemented in 28-nm CMOS technology, which includes a reconfigurable 4-tap feed-forward equalizer (FFE) and an edge-boosting auxiliary driver for channel equalization.

Fig. 1 illustrates the overall TX structure. It receives 10GHz clock from an external source, which goes through the poly phase filter (PPF) and is divided into four-phase clocks ($C4_{I/Q/IB/QB}$). Each of these clocks passes through a duty cycle corrector (DCC) and a quadrature error corrector (QEC), resulting in the quadrature clock with the precise phase difference and duty cycle. Subsequently, these clocks are divided into the octal-rate clocks ($C8_{I/Q/IB/QB}$) and applied to each data path. The data path begins with the generation of PRBS31 data in the pattern generator, which undergo thermometer encoding, resulting in three distinct data signals. Prior to being introduced into the serializers, re-timers are employed to ensure an optimal timing margin. The serialized data then pass through an 8:4 multiplexer (MUX) and a 4:1 MUX, with each output signal directed to five bundles of drivers. Within these bundles, one driver is designed to be tunable with the calibration code, enabling the precise adjustment for the RLM of PAM-4 data and ensuring 50-ohm impedance matching. In the case of the FFE, the data selection from the 8:4 MUX permits the generation of a total of four-tap data, each of which can be directed into a segmented driver. To boost the data output bandwidth, an auxiliary driver having the same structure as the single slice of the bundle driver is implemented. This auxiliary driver accepts data with transition information encoded within, thereby reducing rising and falling times during transitions without distorting the output level.

To generate PAM-4 data, the driver can be configured with MSB and LSB drivers. However, for the structural symmetry, it is common to utilize two instances of the MSB driver and one instance of the LSB driver [1-3]. Moreover, when applying binary-to-thermometer encoding to the MSB and the LSB data, the data toggle, which is the main contributor to power consumption in the VM driver, can be reduced. Fig. 2 (a) depicts a conventional low-voltage swing terminated logic (LVSTL) PAM-4 driver, which allows the efficient driver size and the impedance control. However, its critical path stage, operating at the high speed, consists of three stages, leading to poor power-supply-induced jitters (PSIJ). In contrast, Fig. 2 (b) illustrates the driver structure of a single slice used in our design. It employs a 2-stack structure for the impedance control, reducing the critical path to two stages and improving PSIJ. Within the driver's data path, quarter-rate data ($D4_{(PU/PD)\_(A/B/C)}$) are fed into the pull-up and the pull-down drivers in opposite signs. Similarly, the ZQ calibration code is split into pull-up and pull-down codes for each of A, B, C cases, with a resolution of 5 bits each ($ZQ_{(PU/PD)\_(A/B/C)}[4:0]$). By calibrating each thermometer-encoded ZQ code, it becomes possible to achieve both 50-ohm impedance matching and RLM control simultaneously [6].

Fig. 3 illustrates the components for constructing an edge-boosting auxiliary driver, including an encoder, timing examples, and the 4:1 MUX at the front end of the auxiliary driver. The encoder is composed of logic gates, shown in Fig. 3. By sequentially inputting the parallel data of 9 bits ($D8[7:0]$, $D8_{PRE}[7]$) into the logic gates, we can obtain data that carry information about data transitions ($D8_{R/F}[7:0]$). The obtained data are then serialized through the 4:1 MUX. During this process, the pulse width of the 4-phase quadrature clock ($C4_{I/Q/IB/QB}$) can be adjusted using delay cells, allowing the control over the pulse width time (t_{PW}) of $D8_{R/F}[7:0]$. This enables the generation of input data for the edge-boosting auxiliary driver and provides the ability to finely adjust the pulse width, ensuring the precise control over the timing characteristics.

Fig. 4 illustrates the specific architecture of the 8:4 MUX, its timing diagrams, and the table for the relevant output of the data selector. 8 data signals with 1/8 speed of the output signal are generated in parallel. By selecting the appropriate data in the data selector for each case of 4 taps and sampling them at the correct timing, quarter-rate data can be obtained. To enhance the bandwidth, 4 UI pulse generators are used instead of a 3-stacked MUX. Furthermore, by not using a clock selector for timing margin optimization, the loading of the octal rate clock is reduced, which results in power consumption reduction.

Fig. 5 (a) and (b) display the measured eye diagrams of PRBS31 40Gb/s NRZ and PRBS31 80Gb/s PAM-4 single-ended outputs, respectively. The PAM-4 output exhibits a voltage swing of 297mV, with an eye opening of approximately 45mV in the worst case. Fig. 5 (c) illustrates a closed 64Gb/s eye diagram without FFE, passing through a channel with approximately -6.16dB loss at around 16GHz, while Fig. 5 (d) shows the output with equalization achieved with the FFE for the same channel loss. Although the swing is reduced about 15%, the worst eye opening is 37mV, indicating an improved performance. The total power consumption is 246mW, resulting in the energy efficiency of 3.07pJ/bit at 80Gb/s.

In Fig. 6 (a), the measured S21 of the channel is shown. Fig. 6. (b) provides a detailed power break-down. 4:1 MUX and driver consume the largest portion of power (137.5mW), followed by 10GHz clock distribution (43.3mW), 8:4 MUX (33.5mW), and pattern generator with encoder (31.7mW). Fig. 6 (c) presents a comparison table of our measurement results with the published state-of-the-art TXs. Our work achieves the highest data rate among all single-ended TXs.

Acknowledgments

This work was supported by Samsung Electronics Co., Ltd(IO201218-08228-01). The EDA tool was supported by the IC Design Education Center(IDEC), Korea.

References:

[1] Z. Toprak-Deniz et al., "A 128-Gb/s 1.3-pJ/b PAM-4 Transmitter With Reconfigurable 3-Tap FFE in 14-nm CMOS," *IEEE JSSC,* vol. 55, no. 1, pp. 19-26, Jan. 2020.

[2] P. J. Peng, et al., "A 112-Gb/s PAM-4 Voltage-Mode Transmitter with Four-Tap Two-Step FFE and Automatic Phase Alignment Techniques in 40-nm CMOS," *IEEE JSSC,* vol. 56, no. 7, pp. 2123–2131, 2021.

[3] J. Kim et al., "A 224-Gb/s DAC-Based PAM-4 Quarter-Rate Transmitter with 8-Tap FFE in 10-nm FinFET," *IEEE JSSC,* vol. 57, no. 1, pp. 6–20, 2022.

[4] Z. Wang et al., "An Output Bandwidth Optimized 200-Gb/s PAM-4 100-Gb/s NRZ Transmitter with 5-Tap FFE in 28-nm CMOS," *IEEE JSSC,* vol. 57, no. 1, pp. 21–31, 2022.

[5] K. R. Lakshmikumar et al., "A Process and Temperature Insensitive CMOS Linear TIA for 100 Gb/s/λ PAM-4 Optical Links," *IEEE JSSC,* vol. 54, no. 11, pp. 3180–3190, 2019.

[6] Y. U. Jeong et al., "A 0.64-pJ/Bit 28-Gb/s/Pin High-Linearity Single-Ended PAM-4 Transmitter with an Impedance-Matched Driver and Three-Point ZQ Calibration for Memory Interface," *IEEE JSSC,* vol. 56, no. 4, pp. 1278–1287, 2021.

[7] J. -H. Park et al., "A 32Gb/s/pin 0.51 pJ/b Single-Ended Resistor-less Impedance-Matched Transmitter with a T-Coil-Based Edge-Boosting Equalizer in 40nm CMOS," *ISSCC,* pp. 410-412, 2023.

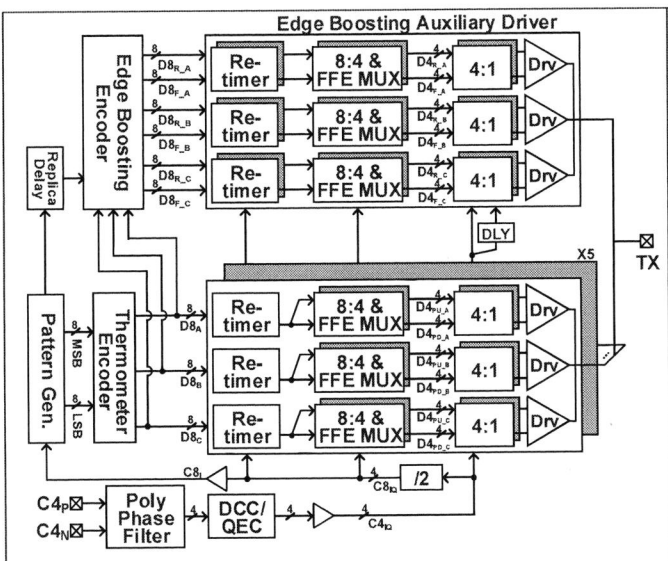

Fig. 1. Top block diagram of transmitter.

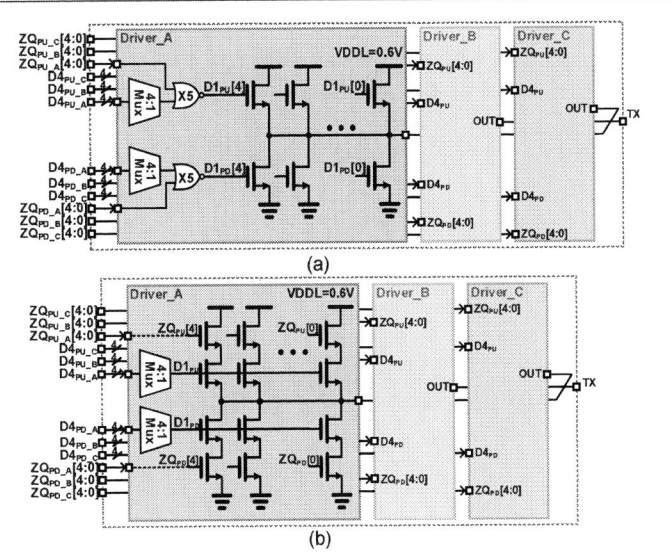

Fig. 2. (a) Structure of the conventional LVSTL driver, (b) proposed design.

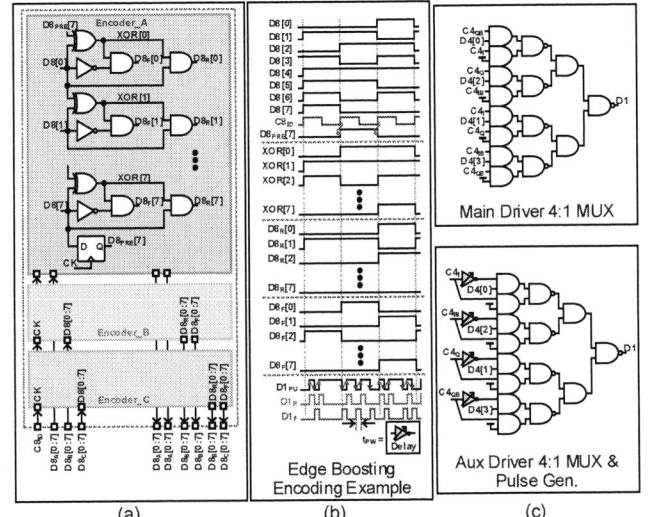

Fig. 3. (a) Implementations for the edge boost encoder, (b) Timing diagram, and (c) 4:1 MUX with pulse generator.

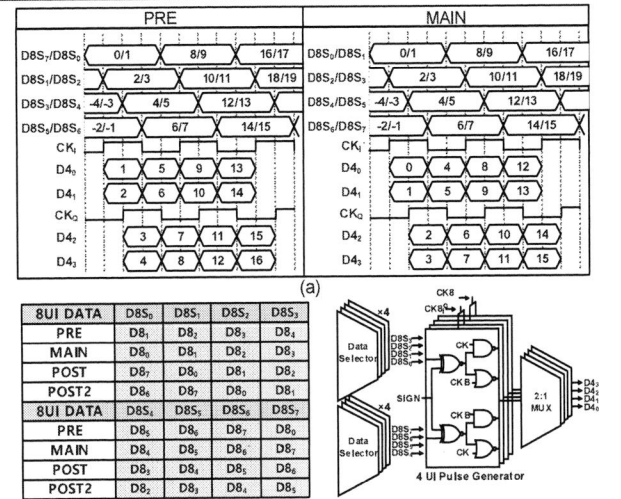

Fig. 4. (a) Comparison of timing diagrams for pre and main tap, (b) Data selection table based on tap types, and (c) Architecture of 8:4 MUX.

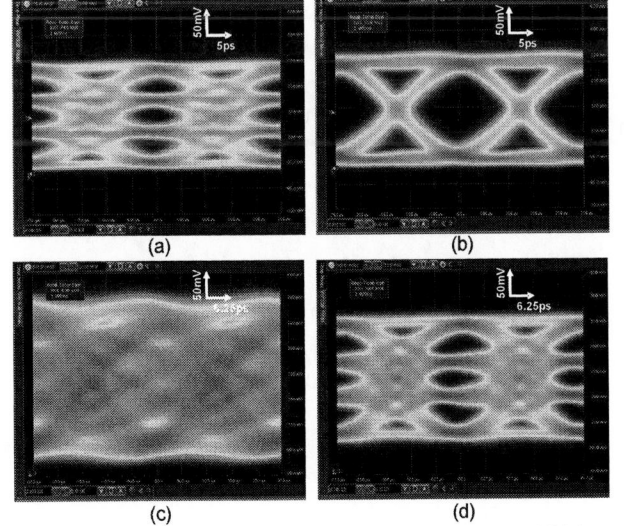

Fig. 5. (a) Measured 80Gb/s PAM-4 eye diagram, (b) 40Gb/s NRZ, (c) 64Gb/s eye without FFE, with -6.16dB loss channel, (d) 64Gb/s eye with FFE, with -6.16dB loss channel.

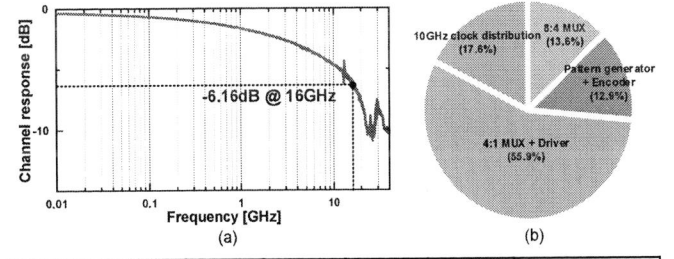

	[1] JSSC'20	[2] JSSC'21	[3] JSSC'22	[6] JSSC'21	[7] ISSCC'23	This work
Technology	14nm FinFET	40nm CMOS	10nm FinFET	65nm CMOS	40nm CMOS	28nm CMOS
Supply(V)	1.2	1.0/1.2	0.85/1.0/1.5	1.0/0.6	1.0/0.6	1.0/0.6
Data rate per pin (Gb/s/pin)	64	56	112	28	32	80
Signaling	Differential PAM-4	Differential PAM-4	Differential PAM-4	Single-ended PAM-4	Single-ended NRZ	Single-ended PAM-4
Driver type	CML (tailless)	VM (SST)	CML	VM (LVSTL)	VM (PN-over-NP)	VM (LVSTL)
Equalization	3-tap	4-tap	8-tap	2-tap assymetric	2-tap edge boosting	4-tap reconfigurable + edge boosting
Clock source	External	On-chip PLL	On-chip PLL	External	External	External
Output swing(V)	1.0	1.0	1.0	0.3	0.3	0.3
Energy efficiency (pJ/bit)	1.3	3.89	1.88	0.64	0.51*	3.1
TX Area(mm²)	0.048	0.56	0.088	0.03	0.005*	0.045

*with external input data

(c)

Fig. 6. (a) S21 of -6.16dB loss channel, (b) Precise power break

979-8-3503-3004-5/23 $31.00 © 2023 IEEE

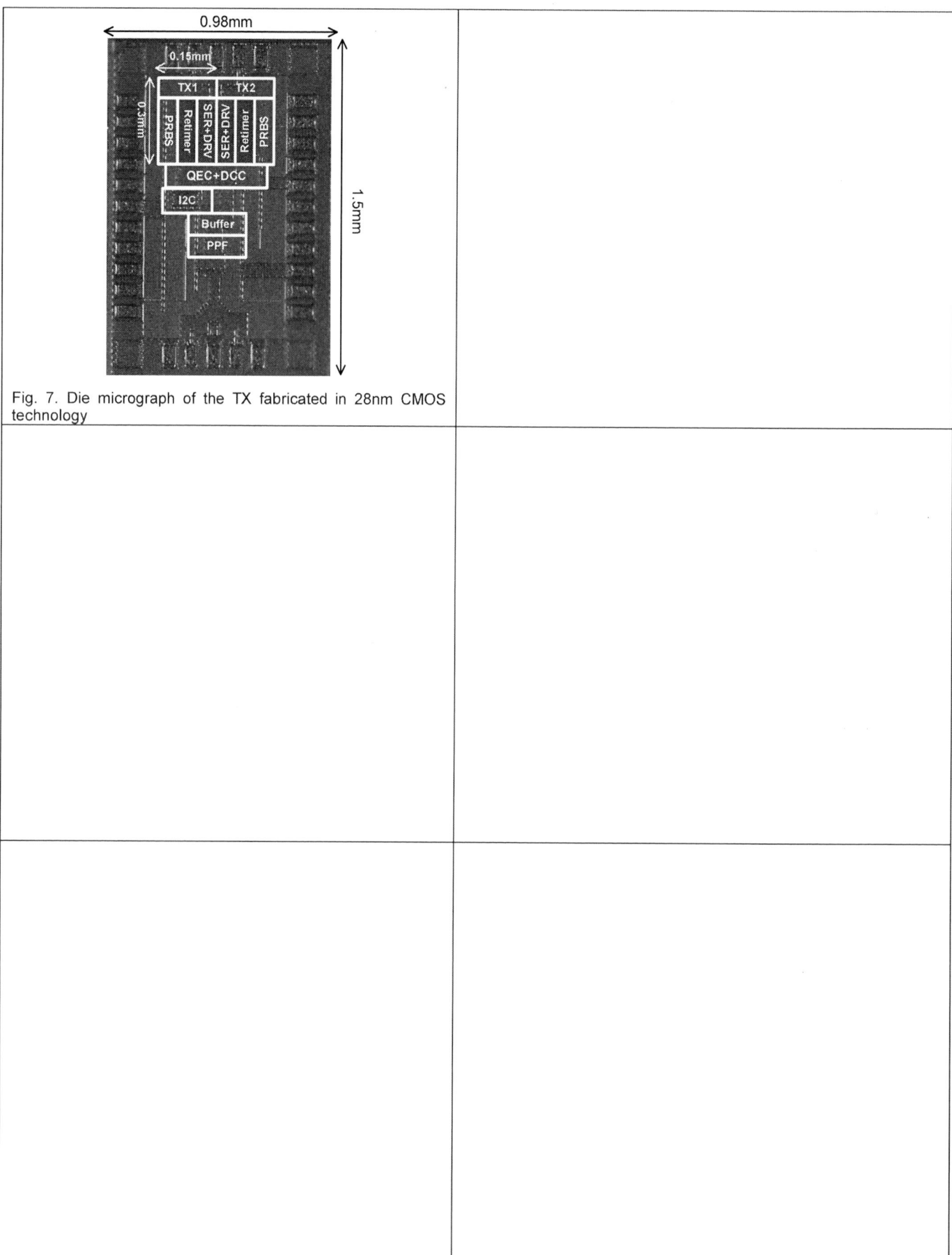

Fig. 7. Die micrograph of the TX fabricated in 28nm CMOS technology

A 28nm 49.7TOPS/W Sparse Transformer Processor with Random-Projection-based Speculation, Multi-Stationary Dataflow, and Redundant Partial Product Elimination

Yubin Qin[1*], Yang Wang[1*], Dazheng Deng[1], Xiaolong Yang[1], Zhiren Zhao[1], Yang Zhou[1], Yuanqi Fan[1], Jingchuan Wei[1], Tianbao Chen[2], Leibo Liu[1], Shaojun Wei[1], Yang Hu[1], Shouyi Yin[1]

[1]School of Integrated Circuit, BNRist, Tsinghua University, China.

[2]Tsing Micro, China.

[*]Equal contribution.

Transformer models have shown remarkable performance in various fields with significant accuracy improvement compared to traditional artificial intelligence models [1,2]. However, the high computational and memory complexity of Transformer models limits their deployment on power-constrained edge devices. Dynamic sparse attention (DSA) is a feasible method to improve throughput and energy efficiency by predicting significant attention computation during inference and then skipping insignificant ones [3,4]. Although a high sparsity ratio has the potential for significant energy savings, realizing high energy efficiency via DSA is challenging due to three factors, as shown in Fig.1. Firstly, traditional processors apply single-stage sparsity speculation method, which incurs low speculation benefits in DSA due to a dilemma: the sparsity speculation of attention relies on computing the entire QK matrix [3], which already accounts for 30-48.9% of attention block computation. Secondly, DSA introduces dynamic sparse data distribution and matrix sizes for the input/weight/output (IWO), resulting in redundant memory access when computing with traditional single X-stationary dataflow (X stands for I/W/O) [4, 5, 6]. Each of the IWO matrices can be the ideal stationary object to achieve minimum memory access. Finally, the attention scores computed after DSA speculation usually have similar significance, involving several identical partial products and sums (PP and PS), which have no contribution since adding the same value to each input of softmax does not affect its outputs.

This paper proposes a sparse Transformer processing unit, STPU, with three features to address the above challenges. 1) A random projection attention speculation (RPAS) unit speculates the sparsity QKV generation, $Q \times K^T$, and $P \times V$ using random projection before all computation to avoid any computation requirements. 2) A multi-stationary PE array (MSPA) improves data reuse by selecting the optimal stationary matrix according to the matrix size. 3) An identical partial product removed multiplier (IPRM) detects identical partial products and avoids redundant shift-and-add operations. Fig.2 shows the overall architecture of STPU. At runtime, RPAS first reduces the dimension of the token and weight, estimates the attention matrix, and prunes insignificant attention through a threshold. Then, 4 MSPAs perform sparse matrix multiplication with 16×16 PEs. Each PE contains an IPRM, enabling identical PP detection and skipping.

Fig.3 shows the principle of sparsity speculation with random projection (RP) in RPAS. Its goal is to predict the sparsity of the whole attention computation, including QKV generation, attention computation ($Q \times K^T$, softmax, and PxV) at the very beginning of the attention block. RP is an effective method to estimate output with ignorable overhead by reducing matrix dimension. Given a matrix multiplication A×B=C, RP calculates low-dimensional matrices A' and B', where C'=A'×B' is similar to C in terms of value. The A' and B' are obtained by multiplying A and B with a random ternary matrix (containing only ±r and 0, where r is a constant). RPAS uses RP to speculate Q,K and the attention matrix. This process is carried out through a random add operation since DSA only relies on relative values (such as top-k) instead of absolute ones, denoting ignorable magnitude r. Therefore, the ternary matrix (±r, 0) can be replaced with (±1, 0), allowing add-only computing. We design a 32-way XOR-Shift random number generator to generate the selection mask, indicating the random position of non-zero values. 8 sets of select-adder trees compute the low-dimension matrix. The result is sent to the PE array to judge the significant attention with a threshold obtained by 32 parallel comparators. Further, RPAS detects row-wise or column-wise sparsity in attention to skip unnecessary Q/K/V generation. A bitmap indicating sparsity data is generated then, to control the memory access and PE array during computation. RPAS reduces QKV and attention computation by 34.6% and 89.5% within a 1% accuracy loss.

Fig.4 presents the MSPA, which reconfigures the stationary dataflow to reduce the memory footprint of DSA computation. Theoretically, setting the sparsest or matrix with the fewest parameters as the stationary object achieves minimum memory access since it requires the fewest re-fetch for other operands. However, it is not applicable in DSA since the matrix sizes and sparsity distribution dynamically vary among the IWO matrices. STPU solves the problem by MSPA, with PEs and data routing supporting multi-stationary dataflows to adapt to dynamic sparsity. Each PE is connected to its neighboring PEs with a 2D PE grid through 4-way routing and has a 64b local register to store data. Before computing, MSPA gets the matrix size and sparsity feature, and reconfigures the PE array to the optimal X-stationary mode, where X indicates the matrix with the minimum size or minimum counts of sparse parameters. In W-St (I-st is analogous), the weight router first distributes the data in parallel to the PE register, then the input router broadcasts the input matrix to each row of PEs. The output in PE is passed horizontally to the output register. In O-St mode, the local register of the PE stores partial sums, and the input router and weight router synchronously distribute the broadcasted inputs and weights to the PEs. By combining different stationary modes, MSPA achieves an energy saving of 21.9% on average when processing wikitext-2 language modelling task with BERT-base, with only 12.2% area overhead introduced.

Fig.5 demonstrates the working principle of IRPM, which exploits invariance of softmax function to save bit-wise computation. In softmax, adding or subtracting all inputs with the same value does not affect its output result. Instead of computing full result of QxKT for softmax, our IRPM only computes the difference of QxKT and ignore the identical bit-wise PP and PS computation during the MAC operation. IRPM derives from a booth multiplier. After each multiplier (key) is booth encoded, IPRM calculates a 16×9 table recording shared Booth values, each number representing the weights and signs of the multiplicand (query). If a booth value is shared by all keys, it means the PPs are identical for the group of input, which can be safely skipped from accumulation. A reconfigurable Wallace Tree is designed to support this dynamic skipping. Unlike the traditional Wallace tree with sequential 3-2 compressors, the IPRM only activates necessary compressors by routing the PP to the input of independent compression stages. IPRM reduces the computation power by up to 1.91×.

Fig.6 shows the comparison of STPU with other state-of-the-art processors, and Fig.7 shows the die chart and detailed specifications. STPU is fabricated in 28nm 1P8M CMOS and occupies 10.76 mm². It has a total of 544KB SRAM on chip, and the operating voltage is 0.68-1.0V. STPU achieves a maximum frequency of 430MHz, which yields its peak performance of 6.53TOPS (INT8). STPU reaches its peak energy efficiency of 49.7TOPS/W (INT8) or 23.9TOPS/W (INT16) @0.69V, 100MHz when computing sparse attention matrix with 90% output sparsity. Compared to transformer processor [6], STPU has a 1.20× higher energy efficiency because it fully exploits the sparsity of the whole attention block, including attention and QKV generation. Additionally, since STPU is the only one processor that accelerates the whole attention computation, it costs only 16.2mJ per BERT inference with 512 input tokens.

Acknowledgement:

This work was supported in part by the National Key R&D project (No.2021ZD0114400), NSFC Grant U19B2041, 62125403, and 92164301, the BNRist, and the Beijing Advanced Innovation Center for Integrated Circuits. The corresponding author is Shouyi Yin.

References:

[1] A. Vaswani et al., "Attention is All you Need, " NeurIPS 2017.

[2] A. Dosovitskiy et al., "An Image is Worth 16x16 Words: Transformers for Image Recognition at Scale," ICLR 2021.

[3] N. Kitaev et al., "Reformer: The Efficient Transformer," ICLR 2020.

[4] L. Lu et al., " Sanger: A Co-Design Framework for Enabling Sparse Attention Using Reconfigurable Architecture," MICRO-54 2021, pp. 977-991.

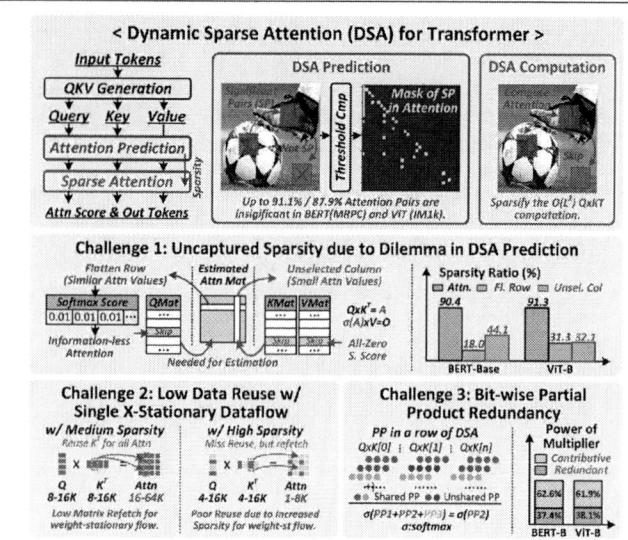

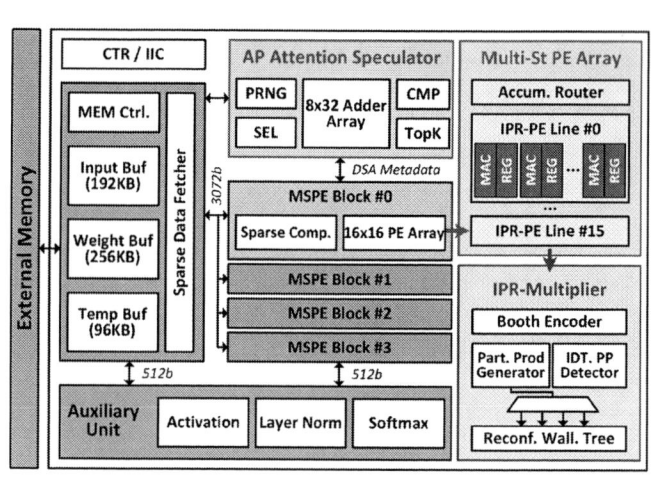

Fig. 1. Dynamic sparsity attention computation and challenges.

Fig. 2. Overall architecture.

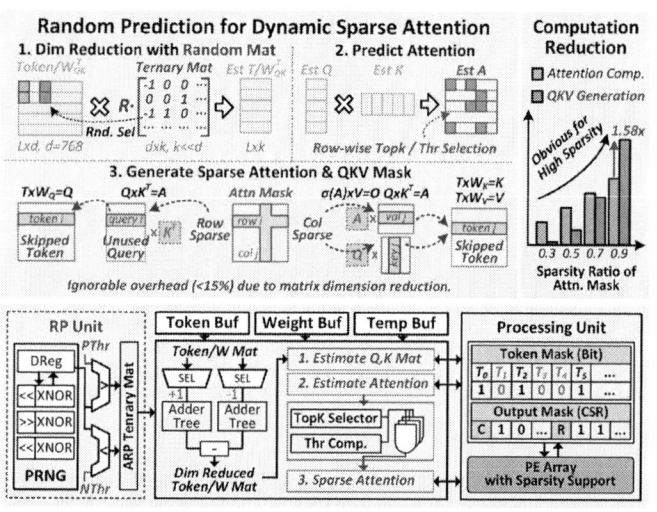

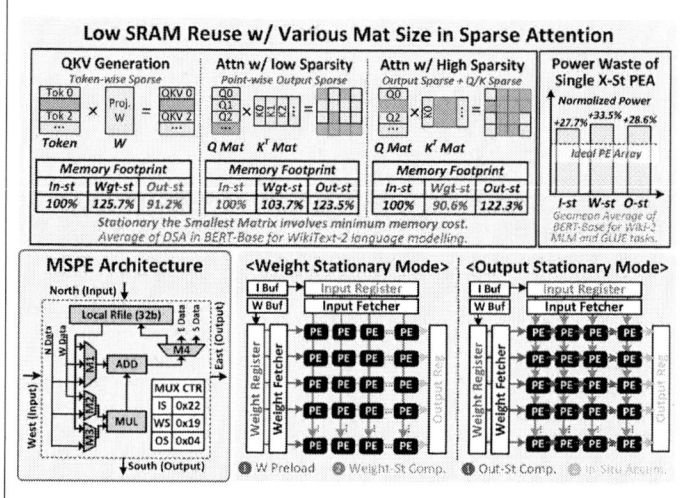

Fig. 3. Random projection speculation mechanism and design.

Fig. 4. Multi-stationary PE array design.

Fig. 5. Identical partial product removed multiplier design.

Fig. 6. Comparison of STPU with state-of-the-art processors.

	H100 GPU	MICRO' 21 [4]	ISSCC' 22 [5]	ISSCC' 22 [6]	This Work
Sparsity Utilization	Input Only	Attention Only[1]	Attention Only[1]	Attention Only[1]	QKV, Attn and PxV
Technology (nm)	4	55	28	28	28
Die Area (mm²)	814	16.9	6.83	6.82	10.76
Supply Voltage (V)	NA	NA	0.6-1.0	0.56-1.1	0.68-1.0
Frequency (MHz)	NA	500	80-240	50-510	85-430
Precision	FP64/32/16/8; INT32/8	INT8	INT16/8	INT12	INT16/8
Power (mW)	7x10⁵	2760	27.04-118.21	12.06-272.8	38.08-395.6
Peak Performance (TOPS or TFLOPS)	3958[2] @INT8	0.529	0.37@INT16 1.48@INT8	0.52-4.07[4]	0.17-6.53[5]@INT8 0.09-3.17[5]@INT16
Energy Efficiency (TOPS/W or TFLOPS/W)	5.71[2] @INT8	0.192[3]	5.1@INT16 20.5@INT8	1.91-27.56[4]	2.22-49.7[5]@INT8 1.11-23.9[5]@INT16

1) "Attention" refers to QxKᵀ's result (output sparsity). 2) With 50% input sparsity exploitation. 3) Sparse attention with 66.9-91.2% sparsity. 4) PxV with 90% tokens being weakly related. 5) Sparse attention with 90% attention sparsity and related QKV skipped (achieved peak energy efficiency in BERT).

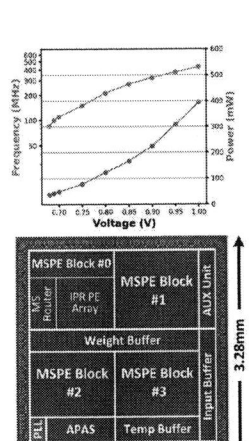

Chip Summary

	Specifications
Technology	28nm
Die Area	3.28mm x 3.28mm (10.76mm²)
SRAM	544KB
Supply Voltage	0.68-1.0V
Frequency	85-430MHz
Data Precision	INT16/INT8
Power	38.08-395.6mW
Throughput[1]	0.09-3.17[2] TOPS @INT16
	0.17-6.53[2] TOPS @INT8
Energy Efficiency	1.11-23.9[2][3] TOPS/W @INT16
	2.22-49.7[2][3] TOPS/W @INT8

1) One operation refers to one multiplication or addition.
2) Sparse Attention computation with output sparsity = 90% and unused QK matrices removed.
3) Peak efficiency achieved at 0.69V, 100MHz.

Fig. 7. Chip layout and summary.

Additional Reference:

[5] F. Tu et al., "A 28nm 15.59µJ/Token Full-Digital Bitline-Transpose CIM-Based Sparse Transformer Accelerator with Pipeline/Parallel Reconfigurable Modes," ISSCC 2022, pp. 466-468.

[6] Y. Wang et al., "A 28nm 27.5TOPS/W Approximate-Computing-Based Transformer Processor with Asymptotic Sparsity Speculating and Out-of-Order Computing," ISSCC 2022, pp. 1-3.

A 4ns Settling Time FVF-based Fast LDO Using Bandwidth Extension Techniques for HBM3

Jinook Jung, Jun-Han Choi, Kyoung-Jun Roh, Jaewoo Park, Won-Mook Lim, Tae-Sung Kim, Han-Ki Jeong, Myoungbo Kwak, Jae-youn youn, Jeong-Don Ihm, Changsik Yoo, Youngdon Choi, Jung-Hwan Choi, Hyungjong Ko

Samsung Memory, Hwaseong, Korea

As the demand for high-performance computers and systems including machine learning increases, high-bandwidth memory (HBM) is preferred due to its large capacity, low power consumption, and high bandwidth. The latest HBM3 standard supports a 6.4 Gbps transmission speed per PIN and a total of 1024 IO PINs are used to support a total data rate of up to 819 GB/s. The high data rate and a lot of IOs inevitably increase the power consumption in the IO circuit which increases the internal power supply noise. Since this power noise is converted into power supply induced jitter (PSIJ) in the clock buffer circuit and decrease the IO timing margin, a method to reduce PSIJ for high data rates effectively is required. Conventionally, a decoupling capacitor is used to reduce power noise. And as shown in Fig. 1 (top), HBM uses the cell-capacitors in the core-die as a decoupling capacitor across the through-silicon-via (TSV) because of lack of the area in the buffer-die (B-die). However, the stacked core-dies increase the equivalent series resistor (ESR), which is composed of RTSV and Rrouting. The increased ESR induces the power noise, as shown in Fig. 1 (bottom-left), and makes worsen PSIJ. To isolate the power supply noise, LDO can be applied in HBM3 for the DQS buffer. However, several operating modes with different current profile in HBM impose several critical constraints on the LDO. In this paper, we propose the compact ultra-high speed LDO that specialized in the DRAM structure with several operation modes with active inductor, self-adaptive bias and 2-step technique.

Fig. 2 (top) shows the operation of multiple modes of HBM and DQS buffer and shows the output voltage according to the bandwidth of LDO. During the initial power-up and STDBY, write-DQS (WDQS) does not toggle, and the DQS buffer waits until ACTCMD is activated, consuming minimal current 2uA. When the ACTCMD is activated, DQS buffer consumes 3mA and due to the command delay, WDQS starts to toggle after tSETTLE(below 5ns). In the HBM, these ACT and STDBY are switched continuously, resulting in undershoot and overshoot, and the duration of STDBY(tSTDBY) is uncertain, and the minimum duration is 1tCK of the input clock of the HBM. Therefore, if the transient response of the LDO is not sufficient to settle within tSETTLE and tSTDBY, two major problems occur when WDQS starts to toggle. First, since the ACTCMD rises and the WDQS is toggle in 5ns, the VOUT is not settled, resulting in a voltage droop(VDROOP1) and degrading the PSIJ. Second, when the mode changes back from STDBY to ACT before LDO settles after overshoot, the voltage droop (VDROOP2) occurs due to the different operating point, worsening PSIJ. In the proposed ultra-high-bandwidth (UHB) LDO, three techniques are employed to address these issues, and each technique will be described in detail in each paragraph. As shown in Fig. 2 (bottom), HBM3 has 16 channels and each channel has 2 pseudo channels. In a pseudo channel, a single LDO is located locally for every two DQS buffers.

In order to settle within few ns, the LDO should fulfil the bandwidth from a several hundred MHz to GHz. Due to the lower output impedance, the Flipped Voltage Follower (FVF) structure can increase the bandwidth without additional current. Fig. 3 (top-left) shows the schematic of the FVF-based LDO which consist of two feedback loops: slow-loop and fast-loop. During power-up, VOUT settles to VREF which is set with VTARGET, through the slow-loop. The fast-loop compensates for the instantaneous error of VOUT due to the instantaneous current change between ACT and STDBY. The loop-gain of the fast-loop comes from transconductance (gm) ratio of M0 and M2 transistors (Tr). However, as shown in Fig. 3 (bottom-left), the R1 and C1 form an active inductor that increases the impedance of diode-connected M2 at high frequencies, extending the bandwidth of the fast-loop (Enhanced fast-loop). Fig. 3 (right) shows the loop-gain of the individual loops and the effect of active inductor in the FVF structure. The active inductor increases the unity gain bandwidth of fast-loop to nearly 1 GHz without additional

current. As a result, VOUT can be settled before tWDQS,TOGGLE, which further opens the eye diagram.

As shown in Fig. 4, in the enhanced fast-loop, the current of M0 and M2 has same aspect. Therefore, the source follower current (ID1) cannot be constant, making Vout unstable. Fig. 4 (bottom) shows the FVF schematic with the self-adaptive bias to solve this problem, and shows the entire schematic of the proposed LDO. The self-adaptive bias circuit consisting of M4, M5, and M6 maintains ID6 as ID2, resulting in a constant current of ID1. Fig. 4 (top-right) shows the waveform of VOUT before and after applying the self-adaptive bias. Without the self-adaptive bias, the voltage droop occurs because of difference of operating point and is compensated slowly by slow-loop, causing PSIJ to deteriorate. However, the VOUT level becomes constant due to the self-adaptive bias. In addition, the operating range of the LDO increases considerable since, VOUT remains constant even when Iload increases.

The LDO which is equipped in HBM should manage the overshoot voltage to maintain VOUT when WDQS stops to toggle. Adding a bleeding current can be a method, but since there is a total of 32 LDOs in HBM, it is impossible to add 32 bleeding current in STDBY mode, which requires minimal current. Therefore, the proposed LDO uses 2-step technique. Fig.5 (top-right) and Fig.5 (bottom-right) shows the timing diagram for each mode in the case without 2-step technique and with 2-step technique, respectively. when the mode is switched from ACT to STDBY, overshoot occurs as Iload drops immediately, and VOUT settles very slowly as the ILDO drops from IACT to ISTDBY. In addition, if the mode is switched from STDBY to ACT before the voltage settles, voltage droop occurs and PSIJ deteriorates as the operating point changes. Fig.5 (top-left) shows the schematic of FVF which is implemented with 2-step technique and Fig.5 (bottom-left) shows the state diagram of the digital logic. ILDO is divided into IACT, ISTDBY, and I2STEP (i.e., during STDBY, ILDO=ISTDBY and during ACT, ILDO=ISTDBY+I2STEP+IACT). If ACTCMD is transitioned from High to Low, unlike the conventional method of switching from IACT to ISTDBY, the digital logic maintains I2STEP for t2STEP and then changes to ISTDBY, settling the overshoot and ensuring constant VOUT at tWDQS,TOGGLE. Consequently, the power consumption decreases as current is consumed during minimum time to settle overshoot without increasing the current consumption during the STDBY.

Fig. 6 (top) shows the measurement shmoo result of the WR clock of the HBM3, with which the proposed UHB-LDO is equipped. While the eye width in the LDO-OFF state is 303mUI, the eye width in the LDO-ON state is 424mUI, which is an improvement of more than 0.1UI. Fig.6 (bottom) shows the performance table of the proposed LDO and its comparison with the state-of-the-art works. Since the proposed LDO is equipped with HBM3, it focuses on performance rather than current, so the quiescent current (IQ) looks high, but as shown in FoM2 and FoM3, which normalize the minimum load current, the proposed LDO has the best performance because it should maintain VOUT even when the leakage current transitions to the maximum load current. Moreover, the proposed LDO has the best performance in the other performance characteristics such as Area and load capacitor, especially power supply rejection (PSR). The total area of the HBM3 is 115.5625 mm2, and LDO is 0.0061 mm2, and 32 LDOs are arranged in the HBM3.

References:

[1] G. Cai et al., "A Fully Integrated FVF LDO With Enhanced Full-Spectrum Power Supply Rejection," in IEEE Transactions on Power Electronics, vol. 36, no. 4, pp. 4326-4337, April 2021

[2] S. Kundu et al., "A fully integrated 40pF output capacitor beat-frequency-quantizer-based digital LDO with built-in adaptive sampling and active voltage positioning," ISSCC, pp.308-310, 2018

[3] Y. -H. Hwang et al., "A Residue-Current-Locked Hybrid Low-Dropout Regulator Supporting Ultralow Dropout of Sub-50 mV With Fast Settling Time Below 10 ns," in IEEE JSSC, vol. 57, no. 7, pp. 2236-2249, July 2022

[4] J. -E. Park et al., "32.4 A 0.4-to-1.2V 0.0057mm2 55fs-Transient-FoM Ring-Amplifier-Based Low-Dropout Regulator with Replica-Based PSR Enhancement," ISSCC, pp. 492-494, 2020

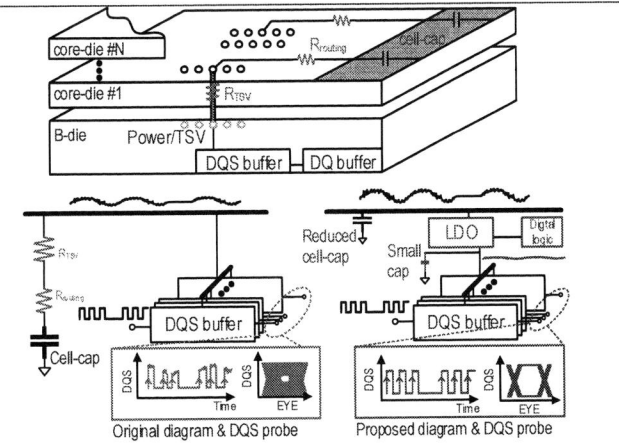

Fig. 1. In a HBM structure, ESR problem when using cell-capacitor as power coupling capacitor on core-die through TSV(top); PSIJ is worse even if a power coupling capacitor is added due to ESR(bottom-left); To reduce PSIJ, a LDO with small capacitor is added to the clock path instead of the power coupling capacitor(bottom-right)

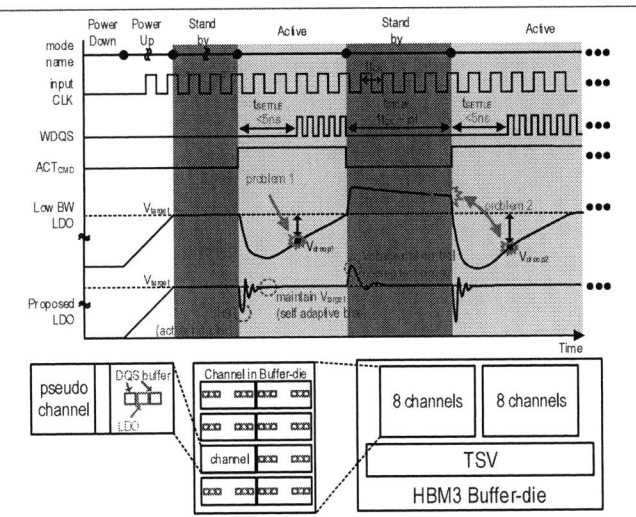

Fig. 2. Timing diagram of main signals, V_{OUT} with low-speed LDO and proposed LDO (top); Layout diagram of 32 pseudo channels and LDOs in overall HBM3

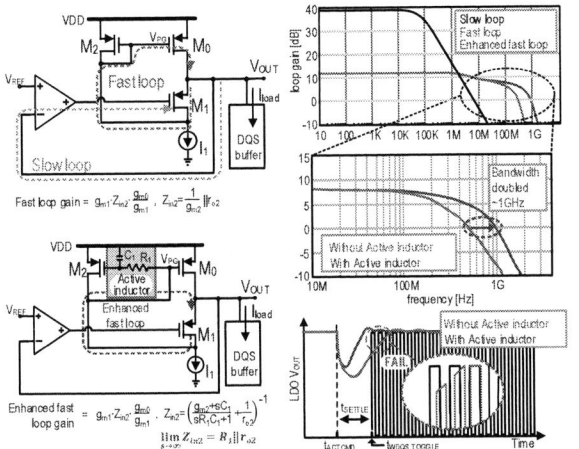

Fig. 3. Conventional FVF-based LDO and loop gain equation of fast-loop (top-left); FVF-based LDO with active inductor and loop gain equation of enhanced fast loop (bottom-left); AC gain plot of individual loops (top-right); LDO can perfectly settle, improving PSIJ due to active inductor (bottom-right)

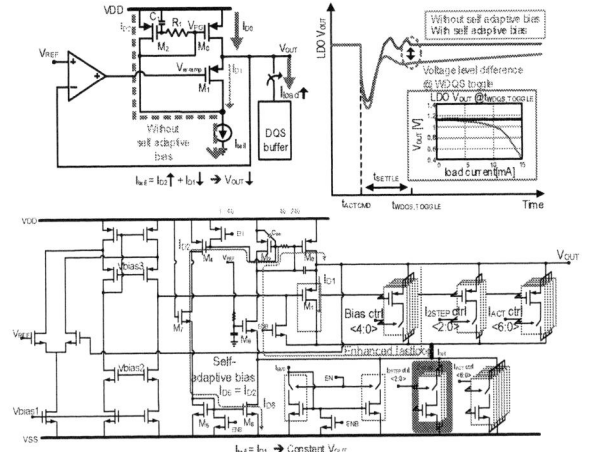

Fig. 4. FVF-based LDO without self-adaptive bias and VOUT error due to current change(top-left); transient waveform of VOUT with/without self-adaptive bias and VOUT according to the load current at tWDQS,TOGGLE (top-right); The entire schematic of the proposed LDO(bottom)

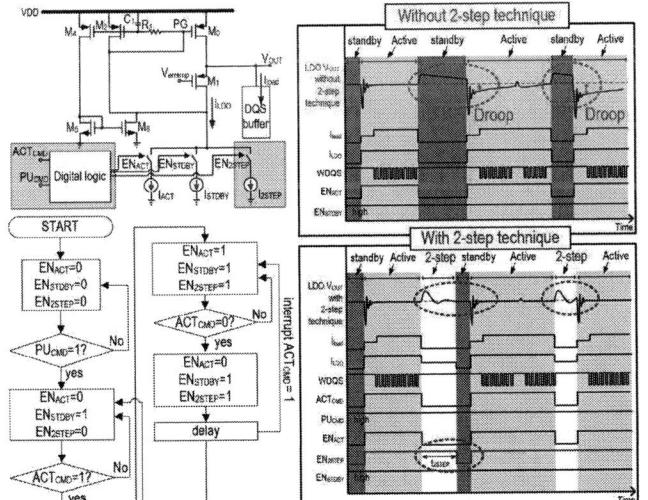

Fig. 5. Proposed FVF structure with 2-step technique and digital logic (top-left); Flow diagram of proposed digital logic (bottom-left); Timing diagram with and without 2-step technique applied (right)

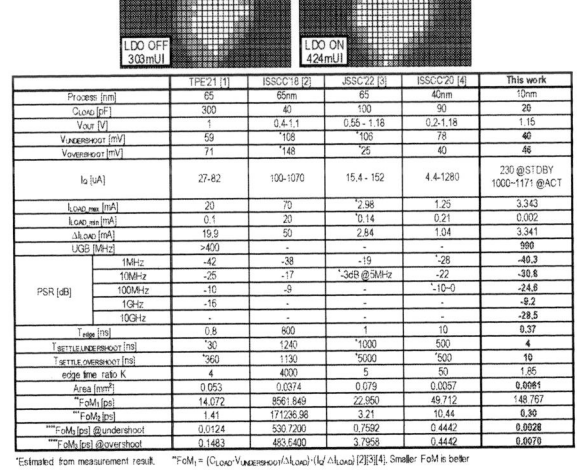

Fig. 6. Read/Write clock shmoo plot in Socket level test according to LDO ON and OFF (top); performance comparison table with state-of-the-art LDOs

979-8-3503-3004-5/23 $31.00 © 2023 IEEE

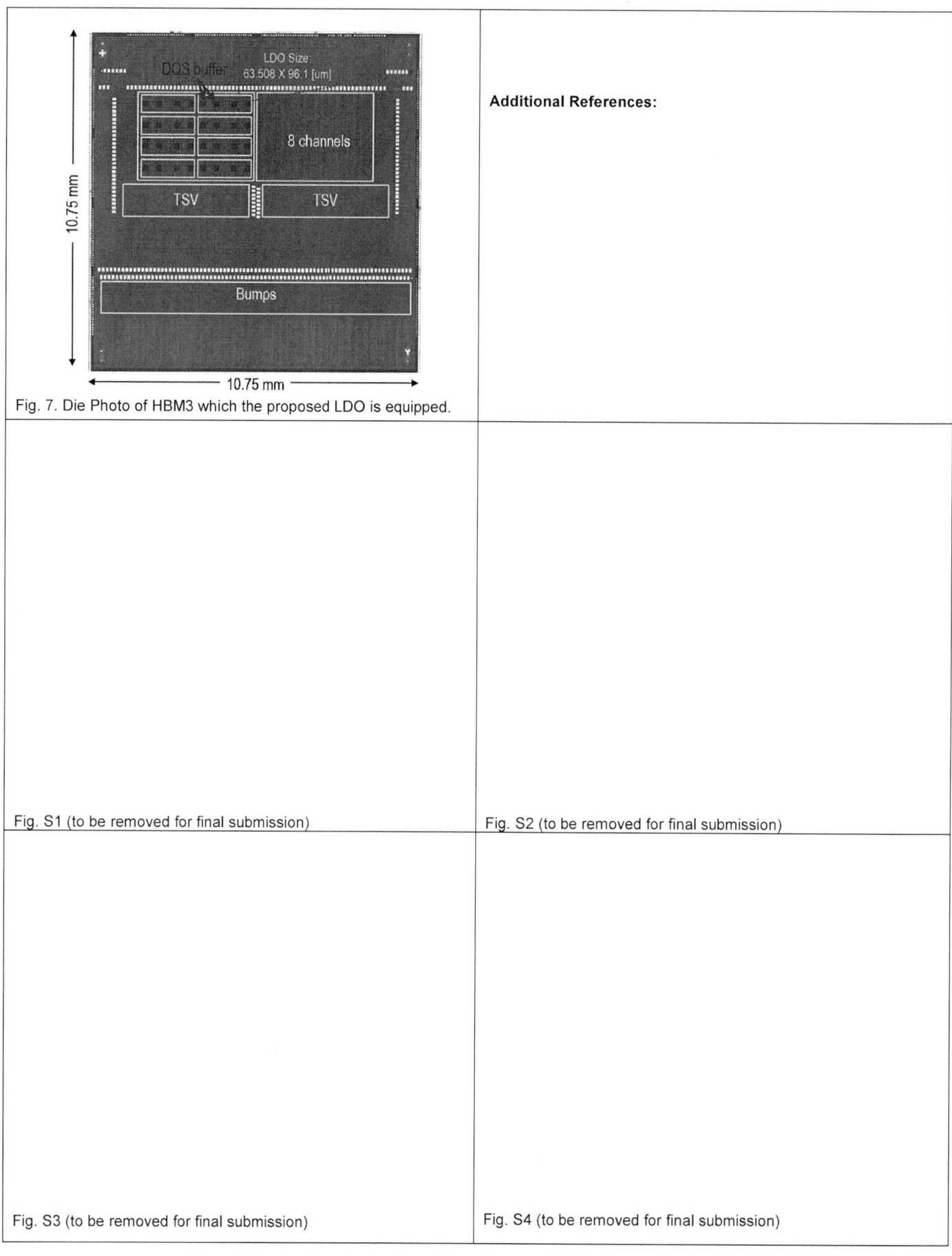

Fig. 7. Die Photo of HBM3 which the proposed LDO is equipped.

Additional References:

Fig. S1 (to be removed for final submission)

Fig. S2 (to be removed for final submission)

Fig. S3 (to be removed for final submission)

Fig. S4 (to be removed for final submission)

An On-Chip DC-DC Converter and Power Management System Achieving Zero Standby-to-Active Transition Time in MCU

Toyohiro Shimogawa, Makoto Nonaka, Toma Ogata, Toshiki Kiryu, Yasuto Igarashi, Kosuke Yayama, Masahiro Kitamura, Hiromichi Ishikura, Mitsuru Hiraki, Masao Ito, and Takashi Kono

Renesas Electronics Corporation, Tokyo, Japan

Nowadays, reducing micro controller unit (MCU) power consumption is increasingly important while high-performance processing is also required [1]. In battery-powered applications, MCUs are often used to go back and forth between active mode and standby mode: CPUs and clocks are working in active mode, stopped in standby mode.

In an MCU having an on-chip regulator to generate internal power supply, using a DC-DC converter is one of the best ways to reduce power consumption because of its high conversion efficiency in active mode. In standby mode, however, a low dropout regulator (LDO) is conventionally used instead of the DC-DC converter since the efficiency of the DC-DC converter deteriorates in light load range, as shown in Fig. 1. For the MCU to move from standby mode to active mode therefore, standby-to-active transition time (on the order of ten microseconds) is inevitably needed for switching the internal power supply circuits from the LDO to the DC-DC converter. Generally, this overhead time is not desirable for applications requiring real-time control because MCUs need to wake up quickly as soon as they receive the wake-up request. Thus, for the MCU that cannot meet the system needs of the quick wake-up operation, there is no other choice than to stay in active mode and consume useless power even though the actual workload is extremely light. To break through this problem, we propose a DC-DC converter achieving high efficiency even in light load range, which can therefore be used not only in active mode but also in standby mode. By using our proposed DC-DC converter, MCUs can move from standby mode to active mode without the transition time from which the conventional scheme was suffering.

Fig.2 shows the architecture of the proposed DC-DC converter. A comparator for upper limit (CMP2) of our DC-DC converter operates in a dynamic way, and therefore the comparator current flows only in a short period of time. This makes it possible for our DC-DC converter to minimize the quiescent current and to improve efficiency in light load. Thus, the LDO, which was conventionally essential for using in standby mode, is not needed. The circuit operation of our DC-DC converter is as follows. When the output voltage (V$_{OUT}$) falls below the lower limit, a comparator for lower limit (CMP1) detects it, and a charging phase starts by turning on a high-side switch (M$_P$). M$_P$ stays on for a time length determined by a variable delay generator. When the on-time is over, M$_P$ is turned off and a low-side switch (M$_N$) is turned on to move into a regeneration phase. In the regeneration phase, an external coil (L$_O$) current (I$_L$) lowers gradually. When I$_L$ reaches zero, a reverse current protection (RCP) detects it and turns M$_N$ off to move into an idle state. Triggered by RCP, CMP2 operates in a dynamic way, and samples the comparison result. Depending on the sampled comparison result, the on-time length set in the variable delay generator is updated. The updated time length is used in the next charging phase. In the idle state, since the output of switches (LX) is HiZ and an external capacitor (C$_O$) is discharged by a load current (I$_{LOAD}$), V$_{OUT}$ lowers gradually. When V$_{OUT}$ falls below the lower limit, CMP1 detects it, and the charging phase starts again. Fig. 3 shows the circuit implementation of CMP2 and the variable delay generator. Triggered by RCP, CMP2 is activated, compares the feedback signal (FBIN) with a reference voltage corresponding to the upper limit (VREFU), and latches the comparison result. At the timing that RCP detects zero-crossing of I$_L$, the V$_{OUT}$ level remains near its peak and is little affected by the equivalent series resistance (ESR) of C$_O$ due to approximately zero I$_L$. Thus, the upper side of V$_{OUT}$ can be detected accurately. The variable delay generator consists of an inverter chain and an up/down counter. The delay is decremented when CMP2 output is high (V$_{OUT}$ > upper limit). The delay is incremented when CMP2 output is low (V$_{OUT}$ < upper limit). As a result, the on-time of M$_P$ is dynamically controlled so that the V$_{OUT}$ ripple is within a certain range. In this operation, the current in CMP2 flows only transiently. No current flows before CMP2 is activated or after the comparison result is latched. In addition, since the RCP signal is used as a trigger of CMP2, the periods of CMP2 operation are limited to suitable moments for comparing V$_{OUT}$ with the upper limit. Owing to this feature, the quiescent current of our DC-DC converter is drastically reduced suppressing the V$_{OUT}$ ripple accurately, while, in a conventional DC-DC converter, its comparator for the upper limit consumes considerably large quiescent current to ensure quick circuit response for suppressing the output voltage ripple. In CMP2 of our DC-DC converter, there is no such crucial tradeoff between the comparator response time and the output voltage ripple as is in the comparator of a conventional DC-DC converter. We estimate that one order of magnitude smaller quiescent current can be achieved in our DC-DC converter compared with a conventional DC-DC converter.

Lowering internal power supply voltage (VDD) is one of the most effective ways to reduce power consumption. Meanwhile, special care must be taken without causing overhead time as the lowered target VDD gets closer to the lower limit voltage of internal circuits. Fig. 4 is provided to explain that. Here, we assume that the lowered target VDD and the lower limit voltage of the internal circuits are 0.8V and 0.72V, respectively. Right after external power supply is turned on, the generated VDD varies due to process variation. Since 0.8V is too close to the lower limit voltage of 0.72V to set this voltage as target from the beginning considering the process variation before voltage trimming execution, we raise the target VDD to 0.92V temporarily until the VDD trimming operation is finished. In the VDD trimming operation, on-chip Non-Volatile Memory (NVM) provides trimming code to the voltage reference buffer (VREFBUF) in response to the trimming request issued by the startup control circuit. After providing the trimming code, NVM notifies the startup control circuit of trimming end. Then, the startup control circuit switches the target VDD from 0.92V (temporal target) to 0.8V, the value set in the Dynamic Voltage and Frequency Scaling (DVFS) register. Our proposed scheme with the startup control circuit eliminates the overhead time (transition time) that would have been caused by overwriting the DVFS register by software without this scheme.

A test chip implementing the proposed DC-DC converter was fabricated in 22nm-CMOS process with a 3.3V I/O device option. The off-chip components employ L$_O$ 2.2uH and C$_O$ 47uF. The measured steady state waveforms, efficiency and quiescent current are shown in Fig.5. V$_{OUT}$ is well stabilized both in case of 10uA output light load and 40mA output heavy load, and its ripple is suppressed to 8mV. In addition, two different amplitude ripples are alternately repeated, showing that the proposed DC-DC converter scheme in which the V$_{OUT}$ ripple converges to two points close to the upper limit is working properly. The efficiency and quiescent current at 20uA output which is assumed as a standby current of the MCU employing the proposed DC-DC converter are 65% and <3uA, respectively. The achieved light load efficiency (65%) of the proposed DC-DC converter greatly exceeds an ideal LDO efficiency, which is 28% assuming that V$_{IN}$ is 3.3V and V$_{OUT}$ is 0.925V. Therefore, the proposed DC-DC converter can be used in standby mode as well as in active mode, making it possible to achieve zero standby-to-active transition time.

Fig.6 shows a comparison with prior works. The proposed DC-DC converter realizes high efficiency in a dozen uA output light load suppressing its V$_{OUT}$ ripple. Fig.7 shows a micrograph of the test chip.

References:

[1] " Renesas Samples Its First 22-nm Microcontroller," https://www.renesas.com/us/en/about/press-room/renesas-samples-its-first-22-nm-microcontroller, Apr. 2023 (accessed Jun. 2023)

[2] S.Sugahara, "Low Power Consumption and High Power Density Integrated DC-DC Converter for Portable Equipments," IEEE ASSCC,pp.150-152, Nov. 2008.

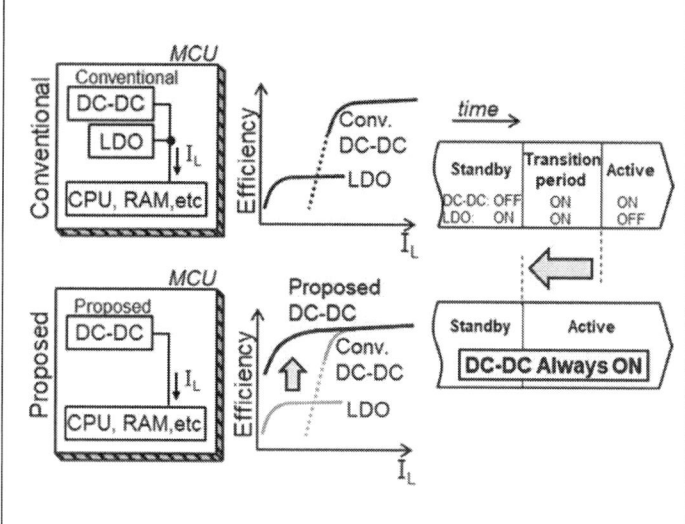

Fig. 1. Reduction of transition time between standby mode and active mode in MCU.

Fig. 2. Architecture of the proposed DC-DC converter.

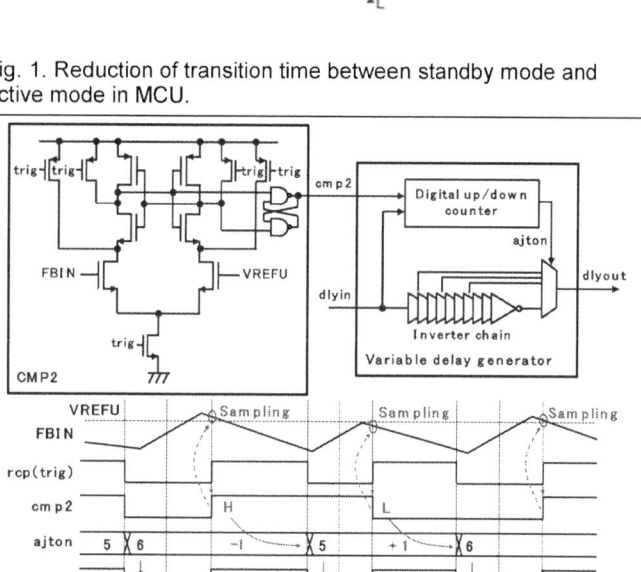

Fig. 3. Circuit implementation of CMP2 and Variable delay generator.

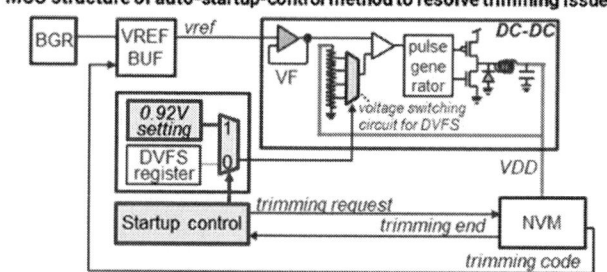

Fig. 4. Countermeasure against trimming issue with low voltage.

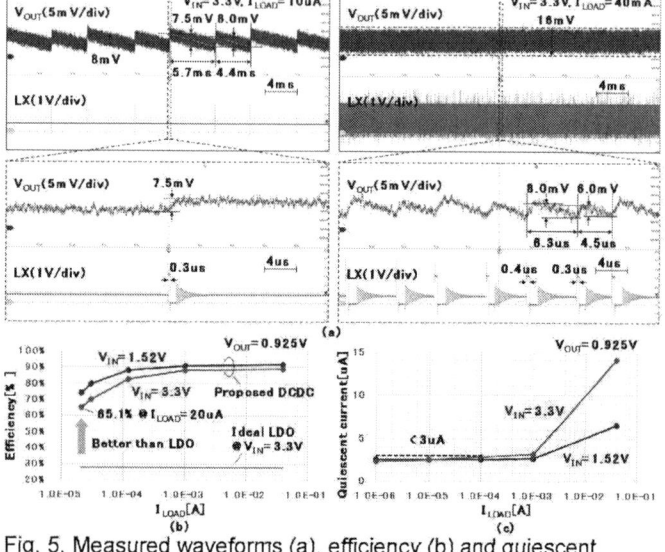

Fig. 5. Measured waveforms (a), efficiency (b) and quiescent current (c) of the proposed DC-DC converter.

Fig. 6. Comparison table with prior works.

	ASSCC 2008 [2]	IEEE TPEL 2022 [3]	IEEE JSSC 2018 [4]	This work
Process	0.35um	0.35um	0.13um	22nm
Operating mode	Buck,PFM	Buck,PFM	Buck,HYS	Buck,PFM
Input voltage	2.5V–5.5V	2.7V–4.2V	1.8V–3.3V	1.52V–3.63V
Output voltage	1.8V	1.2V–1.8V	1.2V	0.925V
Maximum Load current	500mA	50mA	2.65mA	40mA
Lo	0.7uF,1.4uH	15uH	18uH	2.2uH
Co	10uF	100uF	0.056uF	47uF
Voltage ripple	5mV	3mV	29mV	8mV
Efficiency measurement point	V_I=3.6V Vo=1.8V	V_I=3.6V Vo=1.8V	V_I=3.3V Vo=1.2V	V_I=3.3V Vo=0.925V
Efficiency peak	90.0%	93.4%	88.0%	88.7%
Efficiency @ Iload=1mA	82.5%	91.9%	86.0%	87.9%
Efficiency @ Iload=100uA	n/a	81.1%	75.0%	82.4%
Efficiency @ Iload=20uA	n/a	n/a	53.0%	65.1%
Quiescent current	14uA	n/a	n/a	2.5uA

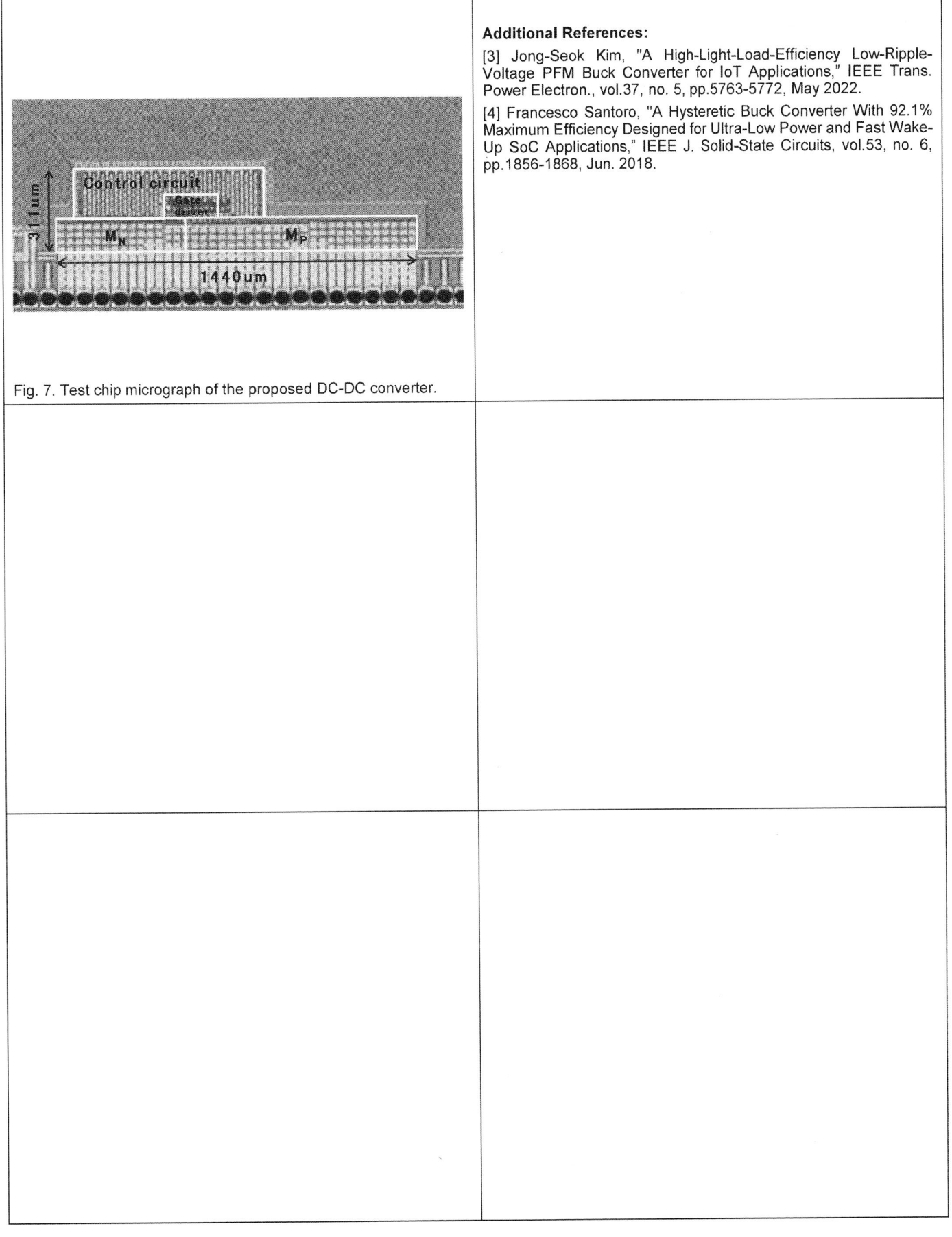

Fig. 7. Test chip micrograph of the proposed DC-DC converter.

Additional References:

[3] Jong-Seok Kim, "A High-Light-Load-Efficiency Low-Ripple-Voltage PFM Buck Converter for IoT Applications," IEEE Trans. Power Electron., vol.37, no. 5, pp.5763-5772, May 2022.

[4] Francesco Santoro, "A Hysteretic Buck Converter With 92.1% Maximum Efficiency Designed for Ultra-Low Power and Fast Wake-Up SoC Applications," IEEE J. Solid-State Circuits, vol.53, no. 6, pp.1856-1868, Jun. 2018.

A 6.4Gbps/pin NAND Flash Memory Multi-Chip Package Employing a Frequency Multiplying Bridge Chip for Scalable Performance and Capacity Storage Systems

Shinichi Ikeda, Akira Iwata, Goichi Otomo, Tomoaki Suzuki, Hiroaki Iijima, Mikio Shiraishi, Shinya Kawakami, Masatomo Eimitsu, Yoshiki Matsuoka, Kiyohito Sato, Shigehiro Tsuchiya, Yoshinori Shigeta, Takuma Aoyama

Kioxia Corporation, Yokohama, Japan

The growing demand for big data and cloud computing requires high-speed, high-capacity storage systems. While the speed of host interfaces (IFs), such as the PCIe employed in SSDs, is steadily increasing, it is becoming more difficult to increase the IF speed of NAND Flash Memory (NAND). This is partly because signal integrity is severely degraded when a high-capacity SSD requires multiple packages with multi-stacked NAND dies on each PCB channel, and partly because the increased number of word line stack layers and multi-levels (3b/cell or 4b/cell [1]) of NAND tend to increase the NAND read latency (tR). If the tR exceeds the data output period, the read throughput is limited by the tR. Thus, only increasing the speed of the NAND IF does not produce higher throughput. To reduce the tR, a single-level cell and higher multi-plane architecture, such as 16 instead of 2 or 4, are employed [2]. This architecture inflates the system cost. It has been reported that reducing the channel load by using an LSI IF between the SSD controller (Controller) and NAND [3] is one method to increase the SSD capacity. However, it does not provide speed beyond that of the NAND IF and still requires an increasing the number of NAND IF channels on the PCB in accordance with advancing PCIe development.

To realize scalable performance and capacity in a storage system, it is key to enhance the IF speed between the Controller and NAND Multi-Chip Package (MCP). This paper describes a NAND MCP incorporating a developed LSI IF Chip (Bridge Chip) in which the IF to and from the Controller has twice the speed as that of the NAND IF even with multiple packages on each PCB channel, retaining or even reducing the number of NAND IF channels on the PCB. The Bridge Chip employs a 2:1 frequency multiplying function, a Fast-Lock PLL (FL-PLL) with an extended pull-in range and 16-cycle lock time, and equalizers to compensate for Inter Symbol Interference (ISI) and reflected noise in up to a 4-drop configuration. The Bridge Chip implemented in a 12nm CMOS process is demonstrated at 6.4Gbps/pin with 2.85pJ/b I/O energy efficiency in a read operation. The NAND MCP incorporating the Bridge Chip and eight 1-Tb NAND dies achieves data transmission to and from FPGA at twice the speed of the NAND IF in a 2-drop configuration.

Fig. 1 (a) shows an example of a storage system configuration using the Bridge Chip. This example shows the 2-drop channel configuration, where two Bridge Chips are connected to the same NAND IF channel on the PCB. Fig. 1 (b) shows the Bridge Chip architecture. The IF to and from the Controller (BCIF) doubles the speed of the existing Toggle DDR standard. The IF to and from the NAND (BMIF) fully supports the Toggle DDR standard. The data processing part provides a frequency multiplying function to and from the NAND. The FL-PLL generates the read enable (RE)/data strobe (DQS) for the BMIF from the RE/DQS of the BCIF.

Fig. 2 shows a timing diagram of the data input and output operation to and from the NAND using two Bridge Chips per BCIF channel (2-drop) to access four BMIF channels simultaneously. During data input operation, the Bridge Chip in MCP0 hands data input commands and internally separated write data to both NAND channels (CH0_0, CH1_0). The Bridge Chip pushes out the remaining data in the data path after the DQS is stopped. During data output operation, the Bridge Chip hands data output commands to CH0_0 and CH1_0. After the tWHR2_0 period, the Bridge Chip starts to pre-fetch read data from the NAND channels. Subsequently, the pre-fetched data is immediately output from the Bridge Chip upon receiving the RE issued by the Controller. These operations enable a 2:1 speed ratio between the Controller and NAND to bridge the speed gap. The Bridge Chip in MCP0 exerts control over the NAND during the tADL_0 and tWHR2_0 periods. The Controller can issue commands to the Bridge Chip in MCP1 during these periods. All training is completed before the data input and output operation.

Hence, the efficiency of data transfer and capacity per channel can be enhanced. For the frequency multiplying function, pre-fetch operation is required even during the period while the Toggle DDR standard does not provide any reference clock. The FL-PLL provides its free-running clock for the read data pre-fetch and write date push-out operations as well as it provides a phase-locked clock to the DQS from the Controller for write operations and to the RE for read operations to enhance the performance with its jitter filtering effect and its short lock time.

Fig. 3 (a) shows the proposed FL-PLL. The FL-PLL employs an injection-locked VCO architecture to simultaneously manage a wide pull-in range and a short lock time. The developed injection current scheme at both the rising and falling edges of the input signal greatly extends its pull-in range compared to that of previous work [4], as shown in Fig. 3 (b). Due to this wide pull-in range, the free-running frequency of the VCO can be initially set lower than that from the RE/DQS frequency, which leads to lowering the -3dB bandwidth of the FL-PLL input period jitter transfer function, as shown in Fig. 3 (c). Thus, this FL-PLL architecture has an advantage over a typical DLL-based clock generation architecture regarding the input RE/DQS jitter rejection performance and improves the timing margin between the Bridge Chip and NAND. The FL-PLL also achieved the targeted lock time within 16 cycles of the RE/DQS from the Controller.

Fig. 4 (a) shows 2-drop and 4-drop channel configurations with MCPs. To compensate ISI and the reflected noises, it is indispensable to have proper target On Die Termination (ODT) and non-target ODT settings together with a continuous time linear equalizer (CTLE) and a multi-tap decision feedback equalizer (DFE) in the targeted RX of BCIF. The ODTs are set to 100Ω and 200Ω for 2-drop and 4-drop configuration, respectively. Fig. 4(b) and Fig. 4(c) show the simulated single-bit responses and eye diagrams at the output of DFE in the 2-drop and 4-drop configuration up to 9.6Gbps/pin.

Fig. 5 (a) shows the measured RX timing margins and its shmoo plots for the BCIF in the 2drop channel configuration, which are well consistent with simulation results in Fig.4 (c). Fig. 5 (b) shows the measured TX eye width and eye diagrams for the BCIF and BMIF. The TX of BCIF has a 3-tap feed-forward equalizer (FFE). In addition, the system operation with data input and output to and from the NAND was confirmed under the condition that the BCIF operates at 4.3Gbps/pin and the BMIF operates at 2.15Gbps/pin because of the limitation of the evaluation environment (FPGA). Fig. 6 shows the performance comparisons with other single-ended I/O for a multi-drop memory IF ([5] , [6]), and prior NAND MCP [3]. To reduce power consumption, the BCIF adopts Power Isolated Low-Tapped Termination (PI-LTT) interface whose supply voltage is 0.5V. The Bridge Chip achieves both 3.6x higher speed and 0.36x I/O energy efficiency compared to prior NAND MCP [3]. Thus, the proposed Bridge Chip demonstrates potential of scalability of the IF speed to and from the Controller and the capacity per PCB channel for storage systems.

References:

[1] J. Yuh, et al.,"A 1-Tb 4b/Cell 4-Plane 162-Layer 3D Flash Memory with a 2.4-Gb/s I/O Speed Interface," ISSCC, pp.130-132, Feb. 2022.

[2] T. Kouchi, et al., "A 128Gb 1b/Cell 96-Word-Line-Layer 3D Flash Memory to Improve Random Read Latency with tPROG=75μs and tR=4μs," ISSCC, pp. 226-228, Feb. 2020.

[3] D. NA, et al., "A 1.8-Gb/s/Pin 4-Tb NAND Flash Memory Multi-Chip Package With F-Chip for High-Performance and High-Capacity Storage," JSSC, vol. 56, no. 4, pp.1129-1140, Apr. 2021.

[4] D. Dunwell, et al., "Modeling Oscillator Injection Locking Using the Phase Domain Response," TCAS-I, vol. 60, no. 11, pp. 2823-2833, Nov. 2013.

[5] S. Lee et al., "A 7.8Gb/s/pin 1.96pJ/b compact single-ended TRX and CDR with phase-difference modulation for highly reflective memory interfaces," ISSCC, pp.272-274, Feb. 2018.

[6] Y. Jeong, et al., "A 10 Gb/s/pin Single-Ended Transmitter With Reflection-Aided Duobinary Modulation for Dual-Rank Mobile Memory Interfaces," TCAS-I, vol. 69, no. 3, pp.1125-1134, Mar. 2022.

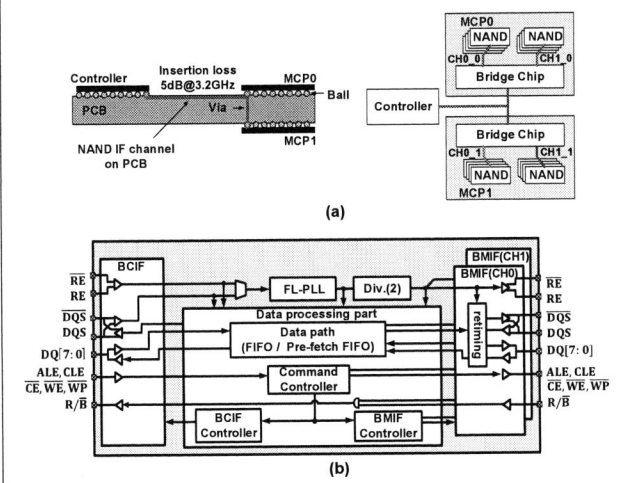

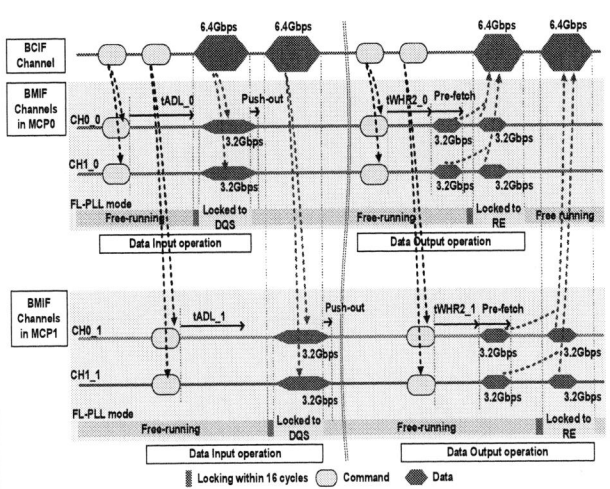

Fig. 1. (a) A storage system configuration (2-drop, 16-die/Ch.) and (b) Bridge Chip architecture.

Fig. 2. Data Input / Output operation with proposed architecture.

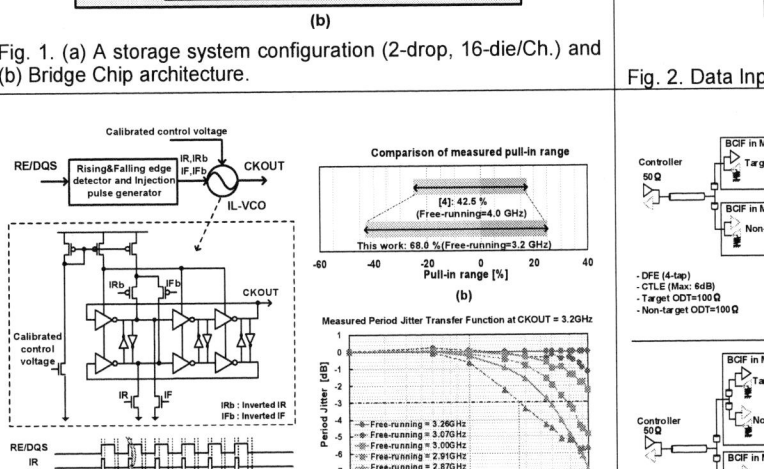

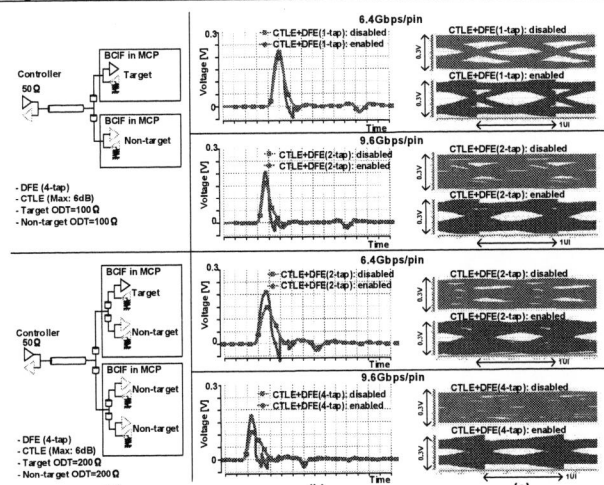

Fig. 3. (a) The proposed injection-locked VCO-based FL-PLL, (b) measured pull-in range, and (c)measured period jitter transfer function.

Fig. 4. (a) 2-drop and 4-drop channel configurations, (b) simulated Single-Bit Responses, and (c) simulated eye diagrams after DFE.

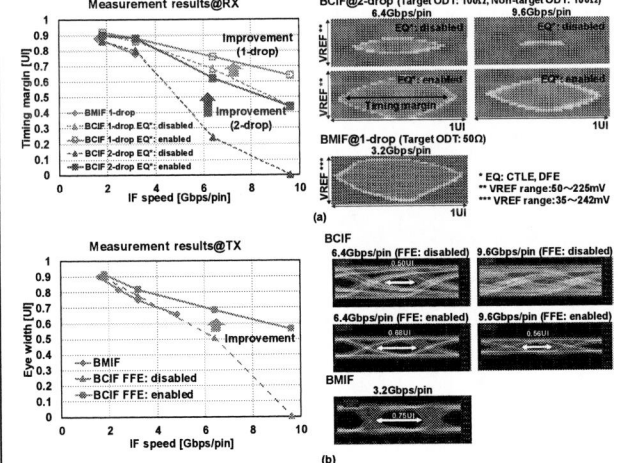

Fig. 5. (a) Measured RX timing margin and horizontal eye opening shmoo plots vs VREF, (b) measured TX eye width and eye diagrams.

		ISSCC 2018 [5]	TCAS-I 2022 [6]	JSSC2021 [3]	This work	
Application		DRAM	DRAM	NAND MCP	NAND MCP	
Technology		65nm CMOS	65nm CMOS	20nm CMOS	12nm CMOS	
Supply voltage [V]		1.0 / 0.9	1.0 / 0.6	1.2	0.8 / 0.5	
Data rate [Gbps/pin]		7.8	10	1.8	6.4	9.6
IF type		SST (Single-ended)	PI-LTT (Single-ended)	CTT (Single-ended)	PI-LTT (Single-ended)	
SI improvement technique		Phase-difference modulation, DFE(10-tap)	Reflection-aided duobinary	Non-target ODT	Non-target ODT, FFE(3-tap) CTLE, DFE(4-tap)	
Timing margin [UI]		RX: 0.34 @10^-12(PRBS7)	TX: 0.54 @10^-12(PRBS7)	0.80 (DQs valid window)	RX: 0.62 *** @10^-12(PRBS7)	RX: 0.44 *** @10^-12(PRBS7)
Single-ended I/O energy efficiency [pJ/b]	TX	1.393	1.24 *	N.A.	0.49	0.41
	RX	0.568	N.A.	N.A.	0.63 *** (EQ: 0.29)	0.49 *** (EQ: 0.21)
	total	1.961	N.A.	N.A.	1.12	0.90
Total I/O power [mW] (read operation)		N.A.	N.A.	75.63 **	145.98	N.A.
Total I/O energy efficiency [pJ/b] (read operation)		N.A.	N.A.	7.88 **	2.85	N.A.

* includes PRBS generator, CK Buffer   ** measured at 1.2Gbps/pin   *** measured with DFE(1-tap) in 2-drop Ch.

Fig. 6. Comparison with other multi-drop memory IF and prior NAND MCP.

979-8-3503-3004-5/23 $31.00 © 2023 IEEE

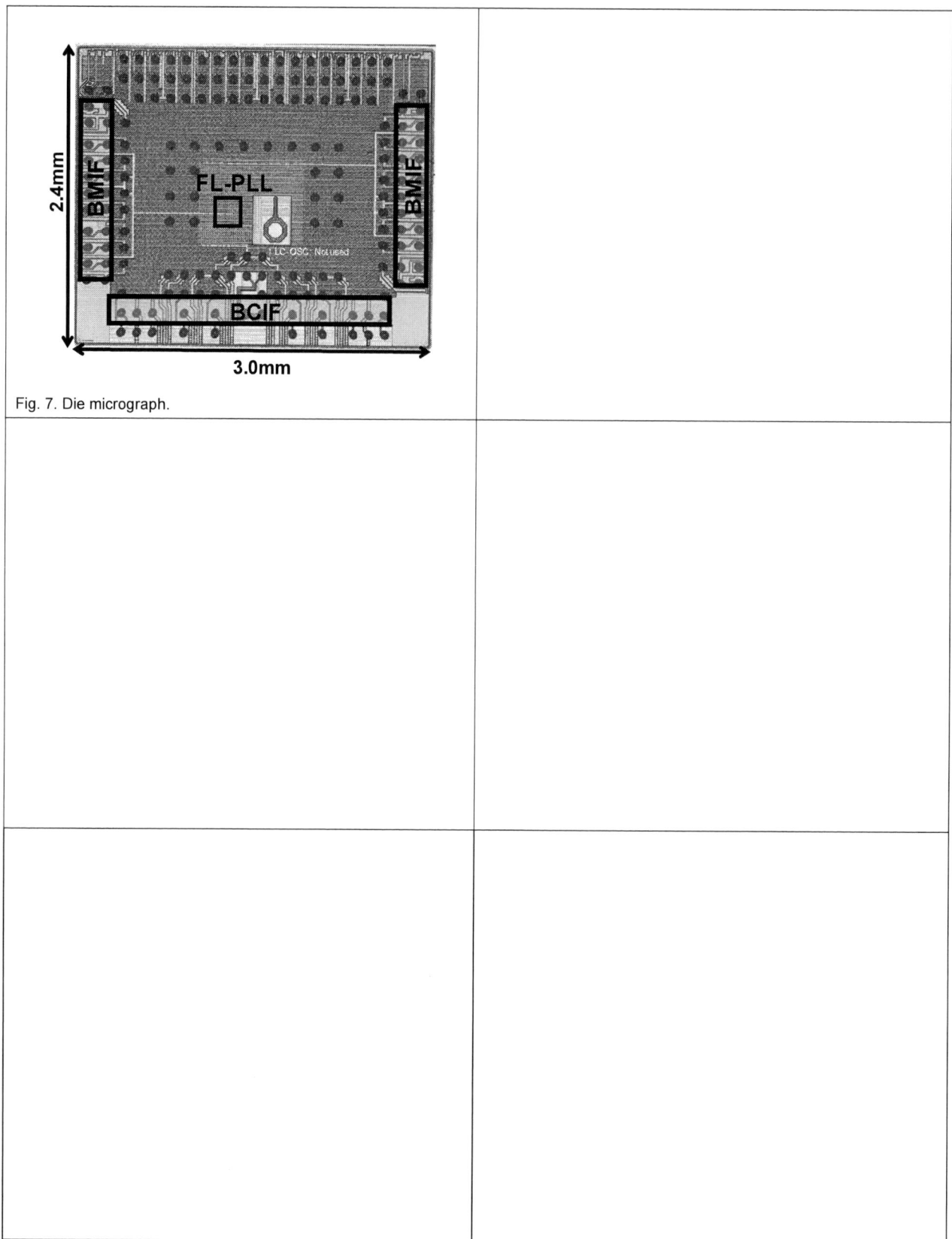

Fig. 7. Die micrograph.

A Fully Integrated Bit-to-Bit 24/48Gb/s QPSK/16-QAM D-Band Transceiver with Mixed-Signal Baseband in 28nm CMOS Technology

Pingda Guan, Haikun Jia, Wei Deng, Ruichang Ma, Mingxing Deng, Jiamin Xue, Angxiao Yan, Shiyan Sun, Zhihua Wang, and Baoyong Chi

School of Integrated Circuits, BNRist, Tsinghua University

Email: jiahaikun@tsinghua.edu.cn

D-band (110-170GHz) transceivers (TRXs) have great potential to satisfy the fast-growing demand for ultra-high data rates, thanks to the abundant spectrum in D-band. The continuous scaling process enables D-band TRXs in low-cost CMOS technologies. Recently reported D-band TRXs have successfully achieved wireless data rates beyond 40Gb/s targeting various applications, including short-range wireless [1-2], medium-range links over low-cost dielectric waveguides [3-4], etc. However, designing a high data rate D-band TRX in CMOS technologies is still challenging. The f_{MAX} of CMOS transistors limits the sub-THz RF front-end (FE) performances. More importantly, at such high data rates, the digital baseband (BB) and ADC/DAC are power-hungry and complex to design, especially considering the 2x or 4x oversampling for pulse shaping. To alleviate the high-speed digital BB and ADC/DAC issue, a fully integrated bit-to-bit D-band TRX with mixed-signal BB in 28nm CMOS technology is presented in this paper, achieving a data rate of 24/48Gb/s for QPSK/16-QAM and supporting on-chip modulation/demodulation. Compared with prior works, the proposed TRX achieves the performance using on-chip modulation/demodulation, fully integrating all critical blocks from bits to RF and from RF to bits, including RF FE, mixed-signal BB, a frequency synthesizer, etc.

Fig. 1 shows the top-level architecture of the proposed TRX. A two-stage conversion topology is adopted for better overall RF FE performances and relaxed local oscillation (LO) requirements. For the transmitter (TX), the BB signal is first converted to 47GHz using a direct digital modulator and then converted to 141GHz. The situation is similar for the receiver (RX), except that a passive IQ mixer replaces the digital modulator for better linearity. The LO frequencies of two-stage up/down-conversion are 47GHz and 94GHz. All critical components are integrated on-chip, including RF FE (TX, RX, and LO), mixed-signal BB for on-chip modulation/demodulation, all-digital PLL (ADPLL), etc.

Fig. 2 shows schematics of critical blocks of the RF FE. The same common-source (CS) differential pair unit with neutralization capacitors is used in both the power amplifier (PA) and the low-noise amplifier (LNA). The combination of wideband matching and stagger tuning techniques expands the bandwidth. The PA comprises a power divider, a driver stage, and a power combiner, with a 3-stage buffer amplifying output from the current choking mixer, generating high output power and linearity, thanks to the transformer-based power divider and combiner with compact layout, good symmetry, low insertion loss, and wide bandwidth. In the LNA, the input matching network optimizes the noise figure (NF), and the first and second stage provides wideband gain, achieving a low NF over a wide bandwidth. In the TX's two-stage up-conversion path, the digital modulator directly modulates NRZ/PAM4 data from the TX BB into QPSK/16-QAM signal. Then, the following mixer up-converts the modulated signal to D-band, with the current choking technique improving linearity and providing a consistently high conversion gain (CG) over a wide LO power range. Due to the high cost of traditional pulse shaping, the digital modulator embeds a 4-tap FIR to suppress side lobes based on current summing. The basic tap coefficient ratio of PRE, MAIN, POST1, and POST2 is 1/4/2/1, corresponding to the size ratio. Every tap has three digital bits (1x, 2x, and 4x) to realize modulation and adjust the tap coefficient.

Fig. 3 shows diagrams of the mixed-signal BB. In the RX BB, there are four mixed-signal loops, including the carrier recovery (CR) loop, clock data recovery (CDR) loop, decision feedback equalizer (DFE) adaptive loop, and reference voltage (V_{REF}) adaptive loop. The VGA and the continuous-time linear equalizer (CTLE) provide a dynamic range of 0–18.1dB and a peaking range of 0–13.7dB at the Nyquist frequency (6GHz). After the CTLE, the phase rotator calibrates the carrier offset between TX and RX, controlled by the CR loop. Then,

the 4-tap DFE compensates for the channel loss of all data paths. The half-rate bang-bang CDR controls the sampling instance and recovers the sampling clock. The V_{REF} generator provides V_{REF} for the slicers. All four loops keep running in the background and adaptively track the variation of carrier frequency, sampling phases, channel loss, and the level of the received signal. All digital modules work at a 1/16 sample clock rate, reducing power consumption. In the TX BB, two 24Gb/s PAM4 RXs receive the raw data, relaxing the clock frequency requirement. The data from two PAM4 RXs are retimed at a 1/64 sub-rate for sufficient timing margin. Then, the data are serialized into 4×12Gb/s and fed to the 4-tap FIR module with configurable polarity and digital weight, using a 12GHz quadrature-phase clock. Fig. 3 also shows diagrams of the type-II fractional-N ADPLL and the digital-controlled oscillator (DCO). The parallel 8-shaped inductor-based dual-core DCO improves the phase noise and rejects magnetic pulling. Tunable tail filtering LC resonators suppress the flicker noise up-conversion. The divided-by-8 output of the DCO is sent to a counter, and a digital-to-time converter (DTC) assisted time-to-digital converter (TDC) for precise phase detection.

Fig. 6 shows the die micrograph of the proposed TRX fabricated in a 28nm CMOS process. The whole chip area is 3.24×1.80mm^2, including all pads. For continuous-wave (CW), the proposed TRX's RF FE is measured using WR6.5 waveguide (WG) probes and a VDI frequency extension system. As shown in Fig. 4, the RX gain is higher than 24.0dB from 132.2GHz to 150.2GHz. The RX NF is lower than 12.3dB from 132.3GHz to 150.1GHz, reaching 7.1dB at 136.2GHz. The TX gain is higher than 24.0dB from 131.0GHz to 151.9GHz. The TX saturated output power (P_{sat}) is higher than 9.0dBm from 126.0GHz to 166.3GHz, reaching 13.0dBm at 151.5GHz. The image rejection ratio (IRR) is better than 40dB at frequencies over 131.1GHz. The LO leakage is lower than -50dB at frequencies over 126.9GHz.

Fig. 5's top right shows the over-the-air (OTA) measurement setup. The half-wavelength open bonding wires are adopted as antennas to achieve wireless RF signal transmission. On the TX side, raw bits generated by an on-chip PRBS are fed to the digital modulator and then transmitted. On the RX side, a BERT scope analyzes the demodulated BB bits. The BERs of 24Gb/s QPSK and 48Gb/s 16-QAM bit-to-bit links are 9.4×10^{-9} and 2.8×10^{-3}, respectively. Fig. 5's left shows the corresponding eye diagrams at the RX BB's output.

Fig. 6 shows the RF FE and communication performance summary and comparison with state-of-the-art D-band TRXs. Regarding RF FE performance, the proposed TRX achieves the highest TX P_{sat} of 13.0dBm and OP$_{1dB}$ of 7.5dBm compared with others and demonstrates a 0.4dB lower RX NF than the others. Regarding communication performance, the proposed TRX reports the highest integration level among those works, realizing 24/48Gb/s QPSK/16-QAM bit-to-bit links using on-chip modulation/demodulation.

Acknowledgement:

This work was supported by the National Key R&D Program of China under Grant No. 2019YFB2204700 and Beijing Innovation Center for Future Chip.

References:

[1] N. Dolatsha et al., "17.8 A compact 130GHz fully packaged point-to-point wireless system with 3D-printed 26dBi lens antenna achieving 12.5Gb/s at 1.55pJ/b/m," ISSCC, pp. 306-307, 2017.

[2] A. Townley et al., "A Fully Integrated, Dual Channel, Flip Chip Packaged 113GHz Transceiver in 28nm CMOS supporting an 80 Gb/s Wireless Link," IEEE CICC, 2020, pp. 1-4.

[3] A. Singh et al., "A D-Band Radio-on-Glass Module for Spectrally-Efficient and Low-Cost Wireless Backhaul," IEEE RFIC, 2020, pp. 99-102.

[4] T. W. Brown et al., "A 50Gbps 134GHz 16QAM 3m Dielectric Waveguide Transceiver System Implemented in 22nm CMOS," IEEE Symp. VLSI Circuits, 2021, pp. 1-2.

[5] S. Callender et al., "A Fully Integrated 160Gb/s D-Band Transmitter with 1.1pJ/b Efficiency in 22nm FinFET Technology," ISSCC, pp. 78-80, 2022.

[6] A. Visweswaran et al., "9.4 A 145GHz FMCW-Radar Transceiver in 28nm CMOS," ISSCC, pp. 168-170, 2019.

979-8-3503-3004-5/23 $31.00 © 2023 IEEE

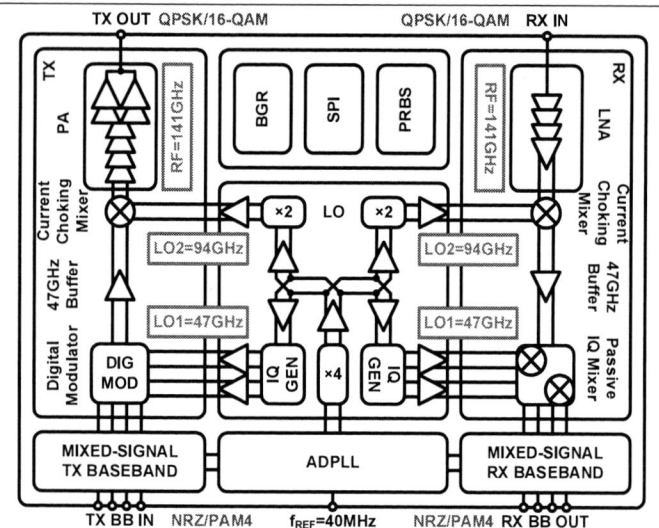

Fig. 1. Top-level architecture of the proposed TRX.

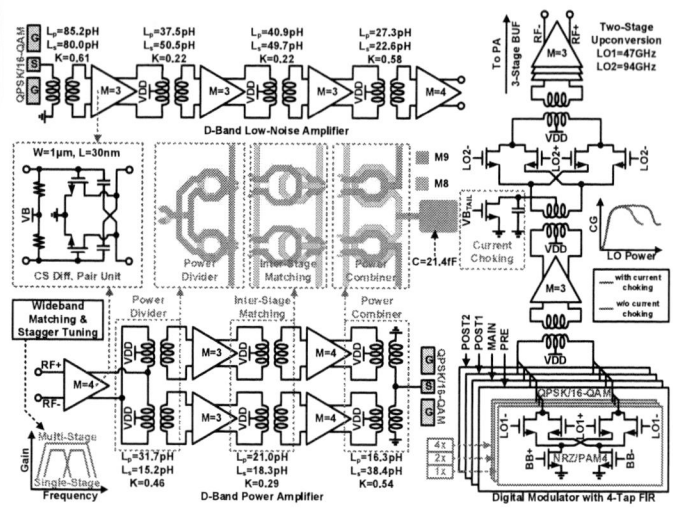

Fig. 2. Schematics of D-band power combining PA, LNA, TX's direct modulator, and up-conversion mixer.

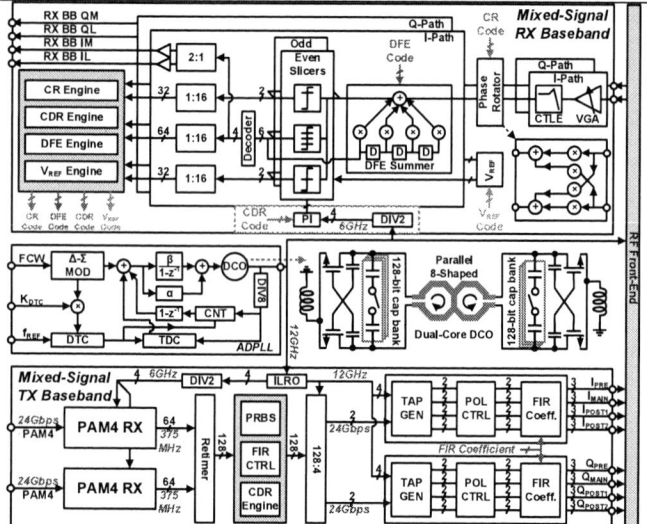

Fig. 3. Diagrams of the mixed-signal BB, ADPLL, and parallel 8-shaped dual-core DCO.

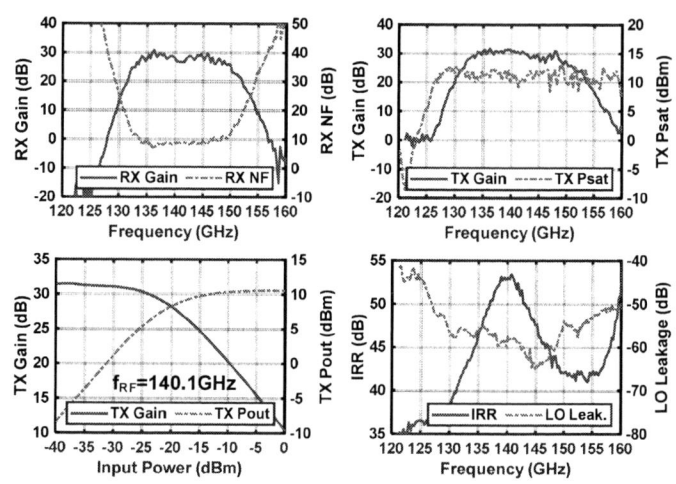

Fig. 4. CW measurement results of the RF FE.

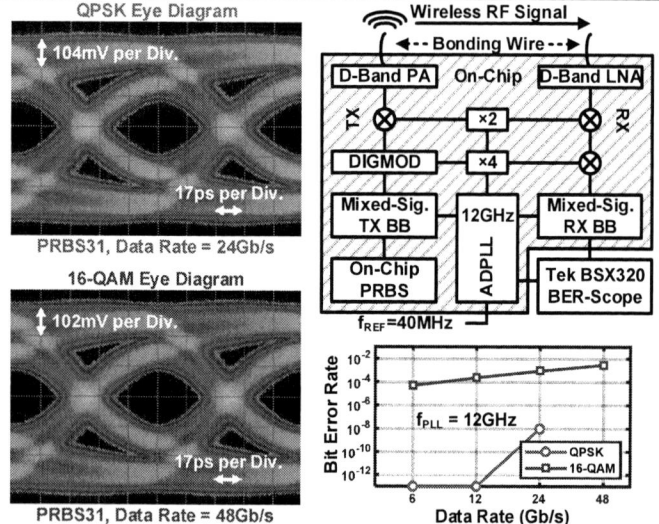

Fig. 5. OTA measurement setup, measured eye diagram at RX's output, and BER of QPSK/16-QAM OTA test.

D-Band TRX RF FE Performance Comparison

	Tech. Node	TRX Mode	Freq. (GHz)	BW (GHz)	TX OP (dBm)	RX NF (dB)
This Work	28nm CMOS	Comm. TRX	141	TX: 21 RX: 18	OP₁dB: 7.5 P_sat: 13.0	7.1
[1]	55nm BiCMOS	Comm. TRX	130	15	OP₁dB: 9.3 (nom.) EIRP: 16.3	N/A
[3]	130nm BiCMOS	Comm. TRX	145	40 (LB) 45 (HB)	P_sat: 13.0 (LB) P_sat: 13.0 (HB)	7.5-10.0 (LB) 8.0-11.0 (HB)
[5]	22nm FinFET	Comm. TX	140	N/A	OP₁dB: 2.6 (sim.) P_sat: 7.5 (sim.)	N/A
[6]	28nm CMOS	Radar TRX	145	13	EIRP: 11.5 P_sat: 7.0	8.0

D-Band TRX Communication Performance Comparison

	Tech. Node	Integration Level	Freq. (GHz)	DR (Gb/s)	Mod.	EVM (dB)	BER*	P_DC (mW)	Area (mm²)
This Work	28nm CMOS	Fully Integrated TRX (wireless)	141	24	QPSK	-15.0†	9.4e-9	1566	5.83
				48	16-QAM	-15.5†	2.8e-3	1677	
[1]	55nm BiCMOS	Fully packaged OOK TRX (wireless)	130	12.5	OOK	-12.5†	1.0e-8	97	3.21
[2]	28nm CMOS	TRX FE & analog BB (wireless)	113	50	QPSK	-12.6	N/A	970	4.72
				80	16-QAM	-15.0	N/A		
[3]	130nm BiCMOS	TRX FE & analog BB (waveguide link)	145	36	64-QAM	-23.0	N/A	2650	11.7
				8	256-QAM	-30.0	N/A		
[4]	22nm FinFET	TRX FE & analog BB (waveguide link)	134	28	QPSK	-16.5	N/A	494	16.0
				56	16-QAM	-18.4	N/A		
[5]	22nm FinFET	Fully Integrated TX (TX-only)	140	80	QPSK	-18.7	N/A	173	4.00
				160	16-QAM	-17.9	N/A		

*On-chip modulation/demodulation BER.
†Calculated from the BER, assuming an AWGN channel.

Fig. 6. RF FE/communication performance summary and comparison with state-of-the-art D-band TRXs.

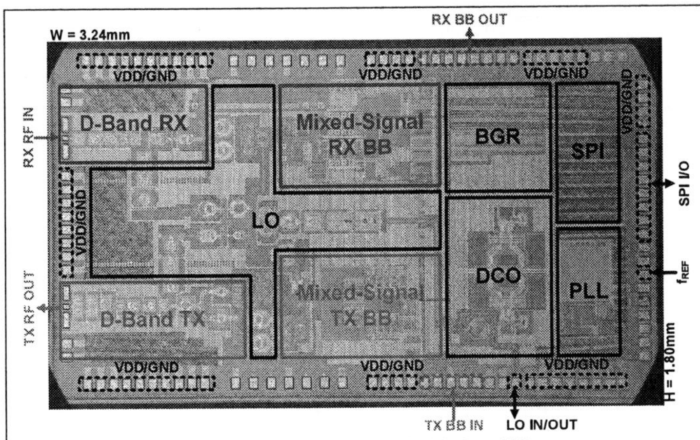

Power Breakdown Table (Unit: mW)

	TX	RX	LO	PLL	BBTX	BBRX	Total
24Gb/s QPSK	233	198	376	140	170	449	1566
48Gb/s 16-QAM	267	195	376	140	158	541	1677

Fig. 7. Die micrograph and power breakdown table of 24Gb/s QPSK and 48Gb/s 16-QAM.

Additional References:

[7]

[8]

[9]

[10]

Fig. S1. (to be removed for final submission)

Fig. S2. (to be removed for final submission)

Fig. S3. (to be removed for final submission)

Fig. S4. (to be removed for final submission)

A 0.12-V 200-Hz-BW 10-bit ADC Using Quad-Channel VCO and Interpolation Linearization

Shea Smith[1], Taylor Barton[1], Yen-Cheng Kuan[2], Armin Tajalli[3], Mau-Chung Frank Chang[4], and Shiuh-hua Wood Chiang[1]

[1]Brigham Young University, Utah, USA, [2]National Yang Ming Chiao Tung University, Hsinchu, Taiwan, [3]University of Utah, Utah, USA, [4]University of California, Los Angeles, California, USA

Wireless IoT devices operating remotely benefit from energy harvesting to ensure continuous operation. Energy harvesters typically generate output voltages in the range of a few 10's mV to 100's mV, thus requiring a voltage booster to power the CMOS circuits that operate at a nominal supply of ~1 V. However, this voltage conversion incurs loss and its efficiency decreases as the voltage differential increases. Therefore, from a systems perspective, there is a need for CMOS circuits that can operate close to the native energy harvester output voltage. Most IoT devices require low-power ADCs to digitize signals and the data rate may be low in many applications. Several previous works have demonstrated deep-subthreshold ADCs. VCO-based quantizers [1]-[3], and a low-supply comparator [4] are used in 0.2-V ADCs. The delta-sigma ADC [5] operates under a 0.15-V supply but can support only sub-Hz bandwidth. In this work, we propose a 0.12-V 10-bit 200-Hz-BW ADC based on a quad-channel VCO architecture. This work represents the lowest supply reported for an ADC to the best of our knowledge.

The proposed ADC (Fig. 1) utilizes a VCO-based architecture to leverage time-based processing to break the severe trade-offs in the subthreshold. Each of the 4 channels consists of a 4X bootstrapped sample-and-hold (S/H) with a bias control, a wide-input VCO driver, a VCO quantizer, fine-coarse VCO readout, and a digital interpolation linearizer. Since VCOs generally exhibit highly nonlinear V-F characteristics, the design exploits the natural features of the V-F curve to calibrate the nonlinearity. The ADC biases V_{CTRL-A} and V_{CTRL-B} to operate VCO_A and VCO_B at the compressive and expansive regions of the V-F characteristic, respectively. The nonlinearities of these two VCOs cancel each other out through the superposition of their readouts, thereby extending the input range for a given linearity target. For fully differential operation, the design incorporates a total of 4 VCOs. The bias control merges a 7-bit CDAC with the S/H. By switching the appropriate number of capacitors in the CDAC after sampling, the input bias is shifted to the desired operation region of the V-F characteristic for each VCO. A wide-input VCO driver buffers the sampled voltage and generates V_{ctrl}. Each VCO operates at 200-kHz and employs a counter for the coarse readout and a bank of comparators for the fine readout. The dual-counter tolerates metastability by using the fine code to select the coarse counter further from the critical edge. The fine and coarse codes are combined and interpolated and the output goes through a differencer for the first-order noise shaping.

Fig. 2 shows the details of the circuit implementation. A 4X bootstrapped switch enhances the S/H linearity. The circuit utilizes NMOS pull-up reset switches for the boosting capacitors instead of the conventional PMOS switches. This technique prevents the leakage current through the pull-up reset transistors when the top plate voltages are boosted above V_{DD}. A charge pump clocks the NMOS pull-up reset transistors. Because the charge pump itself suffers leakage through the pull-down transistor at the charge pump output, it realizes the pull-down switch using the leakage-lock switch. The same type of switch is employed for the pull-down reset switches for the boosting capacitors. To minimize leakage through the sampling transistor during the hold phase, we utilize a negative gate voltage generated using an auxiliary 2X bootstrapped switch to turn off the sampling switch. After sampling, the CDAC shifts the bias

point for the VCO interpolation linearization. The bias can range from $-V_{DD}$ to $2V_{DD}$ requiring the VCO driver to accept a wide input range. The wide-input VCO driver consists of a complementary input stage acting as a variable current source connected to a PMOS current mirror to generate V_{ctrl}. Two parallel transistors provide an additional current path to set the sensitivity. Clocked transistors deactivate the driver and pull V_{ctrl} high during the sampling phase to set the VCO frequency to a constant value. The VCO consists of 7 stages of current-starved inverters in a pseudo-differential configuration and the outputs are sampled differentially.

Fig. 3 shows the circuit details of the VCO readout. Each comparator in the fine readout bank uses a modified StrongARM latch. The leakage-lock tail switch reduces the leakage during reset. During the comparison phase, the PMOS reset transistors exhibit substantial leakage preventing the differential pair from properly pulling its drains (nodes P and Q) down. Simply increasing the differential pair size increases their leakage preventing proper reset. To solve this circular problem, we add the switched resistors implemented using NMOS devices to produce a canceling current during the comparison phase to pull the nodes P and Q low. The switched resistors are disabled during the reset phase, allowing the reset switches pull P and Q high. The coarse readout utilizes a Johnson counter that maximizes the pulse width of the counter output to tolerate the slow rise/fall times in subthreshold. Additionally, the flip-flops omit transmission gates which perform poorly in subthreshold. The digital outputs are level-shifted to 0.9 V by an ultra-low-voltage level shifter which incorporates switched resistors similar to those in the comparator. To determine the proper biases for the interpolation linearization, a ramp is applied to the input of the ADC to extract the input-output characteristics as shown in Fig. 4 and a gradient decent algorithm finds the optimal biases to linearize the output. The peak INLs before and after interpolation linearization are 15.1 LSB and 1.68 LSB, respectively, demonstrating the linearization effectiveness.

Fabricated in a 28-nm CMOS technology, the proposed ADC occupies an active area of 90 x 90 μm^2. It operates under a 0.12-V supply and samples at 30 kHz with an OSR of 75 for a BW of 200 Hz. Fig. 5 shows the measured SNDR and SFDR versus the input frequency as well as the input dynamic range of the ADC. The ADC achieves an SNDR, SFDR, and DR of 57.2 dB, 66.3 dB, and 57.2 dB, respectively. The measured spectra show a 26.1-dB improvement in SFDR resulting from the proposed linearization. The ADC consumes a total of 72 nW from the 0.12-V supply, including 22.3, 15.1, 15.0, 12.2, and 7.2 nW in the comparators, VCOs, VCO drivers, counters, and bootstrapped switches, respectively. Fig. 6 shows the performance comparison. Note that the 0.2-V ADCs are fundamentally in a different design space from ours due to the exponential dependency of the circuit performance on the supply in subthreshold, but they are included here for completeness. The proposed ADC operates at the lowest supply reported, has a BW that is more than 600X higher than [5], and achieves an FoM of 307 fJ/c-s which is more than a 4X improvement with respect to [5].

References:

[1] V. Nguyen *et al.*, "A 0.2-V 30-MS/s 11b-ENOB open-loop VCO-based ADC in 28-nm CMOS," *SSCL*, 2018, pp. 190-193.

[2] U. Wismar, *et al.*, "A 0.2V, 7.5 µW, 20 kHz ΣΔ modulator with 69 dB SNR in 90 nm CMOS," *ESSCIRC*, 2007, pp. 206-209.

[3] V. Nguyen *et al.*, "A deep-subthreshold variation-aware 0.2-V open-loop VCO-based ADC," *JSSC*, 2021, pp. 1684-1699.

[4] A. Petrie, *et al.*, "A 0.2-V 10-bit 5-kHz SAR ADC with dynamic bulk biasing and ultra-low-supply-voltage comparator," *CICC*, 2020, pp. 1-4.

[5] A. Catania *et al.*, "A 150 mV, sub-1 nW, 0.75%-full-scale INL delta-sigma ADC for power-autonomous sensor nodes," *ESSCIRC*, 2022, pp. 257-26

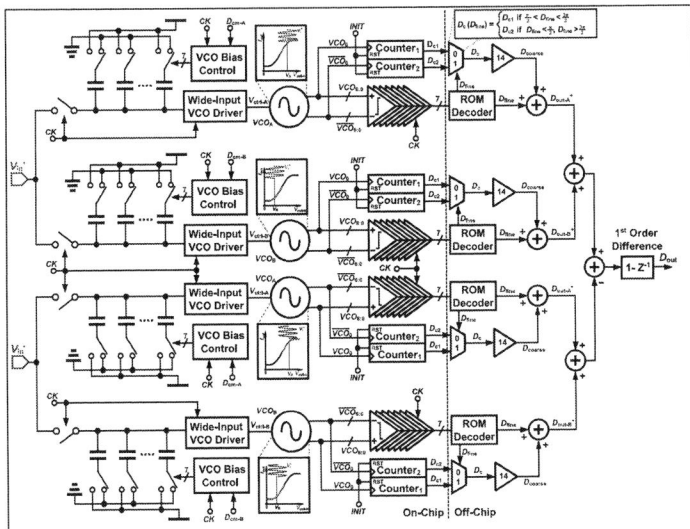

Fig. 1. Proposed architecture of quad-channel VCO-based ADC with interpolation linearization.

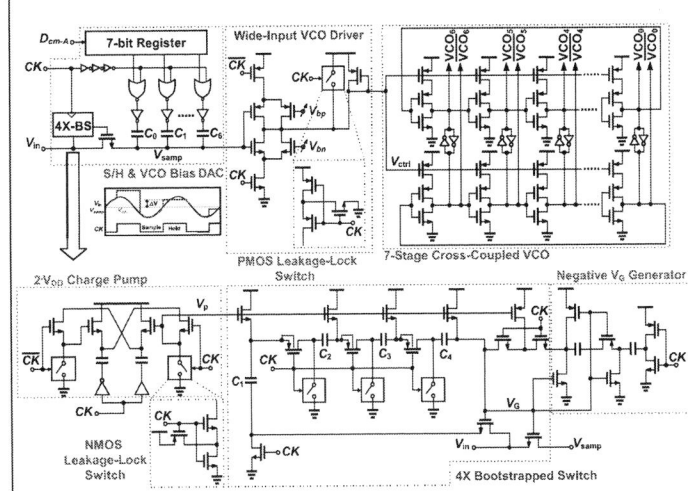

Fig. 2. Circuit implementation of S/H, VCO bias DAC, wide-input VCO driver, ring VCO, and 4X bootstrapped switch.

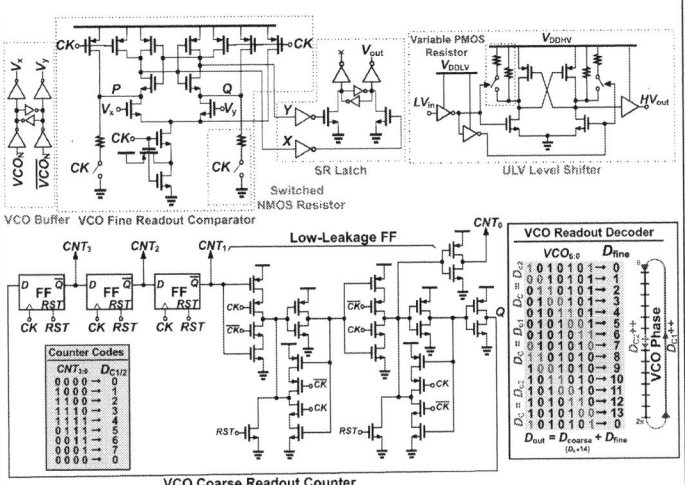

Fig. 3. Circuit implementation of VCO fine readout comparator, VCO coarse readout counter, and ultra-low-voltage level shifter.

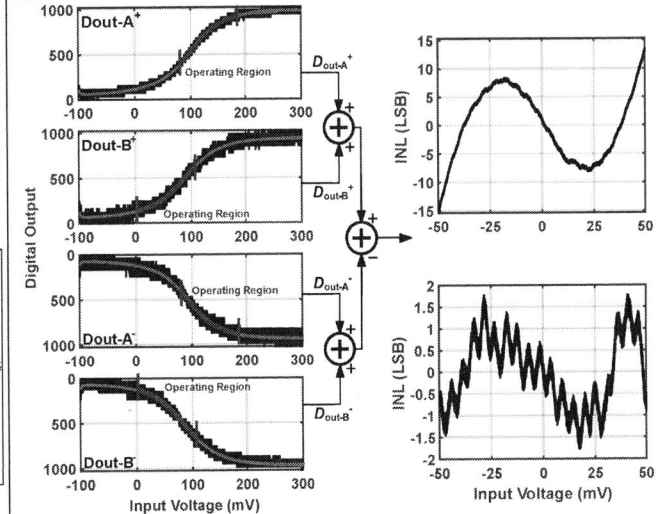

Fig. 4. Proposed interpolation linearization exploiting compressive/expansive regions of ADC characteristic. Measured ADC INL before and after interpolation linearization.

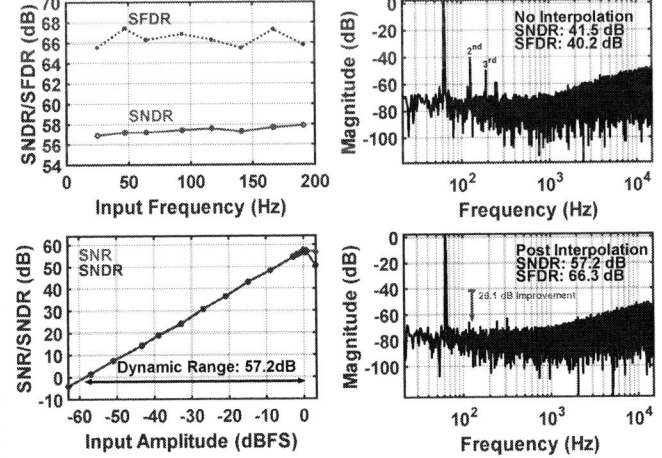

Fig. 5. Measured ADC performance: SNDR/SFDR versus input frequency; SNDR/SNDR versus input amplitude; spectrum before/after interpolation linearization.

	This Work		Catania [5] ESSCIRC22	Nguyen [1] SSCL18	Wismar [2] ESSCIRC07	Petrie [4] CICC20	Nguyen [3] JSSC21
Architecture	Open-loop VCO		DT-ΔΣ	Open-loop VCO	Open-loop VCO	SAR	Open-loop VCO
Process (nm)	28		180	28	90	180	28
Supply (V)	0.12		0.15	0.2	0.2	0.2	0.21
Area (mm²)	0.0081		0.03	0.12	0.016	0.13	0.12
Power	72.6 nW		0.6 nW	7 μW	7.5 μW	22 nW	15.9 μW
fs	30 kHz		85 Hz	30 MHz	12 MHz	5 kHz	34.6 MHz
BW	200 Hz	800 Hz	0.3 Hz	60 kHz	20 kHz	2.5 kHz	80 kHz
SNDR (dB)	57.2	52.1	59.3	67.4	60.3	54.7	64.4
SFDR (dB)	66.3	64.4	66.3	71.2	NA	72.6	71.2
ENOB (bits)	9.21	8.36	9.55	10.9	9.7	8.79	10.4
FoM (fJ/c-s)	307	138	1334	29.9	225	12.3	73.3

FoM = Power/(2^ENOBx2BW), ENOB = (SNDR-1.76)/6.02

Fig. 6. Performance summary and comparison with deep subthreshold ADCs.

979-8-3503-3004-5/23 $31.00 © 2023 IEEE

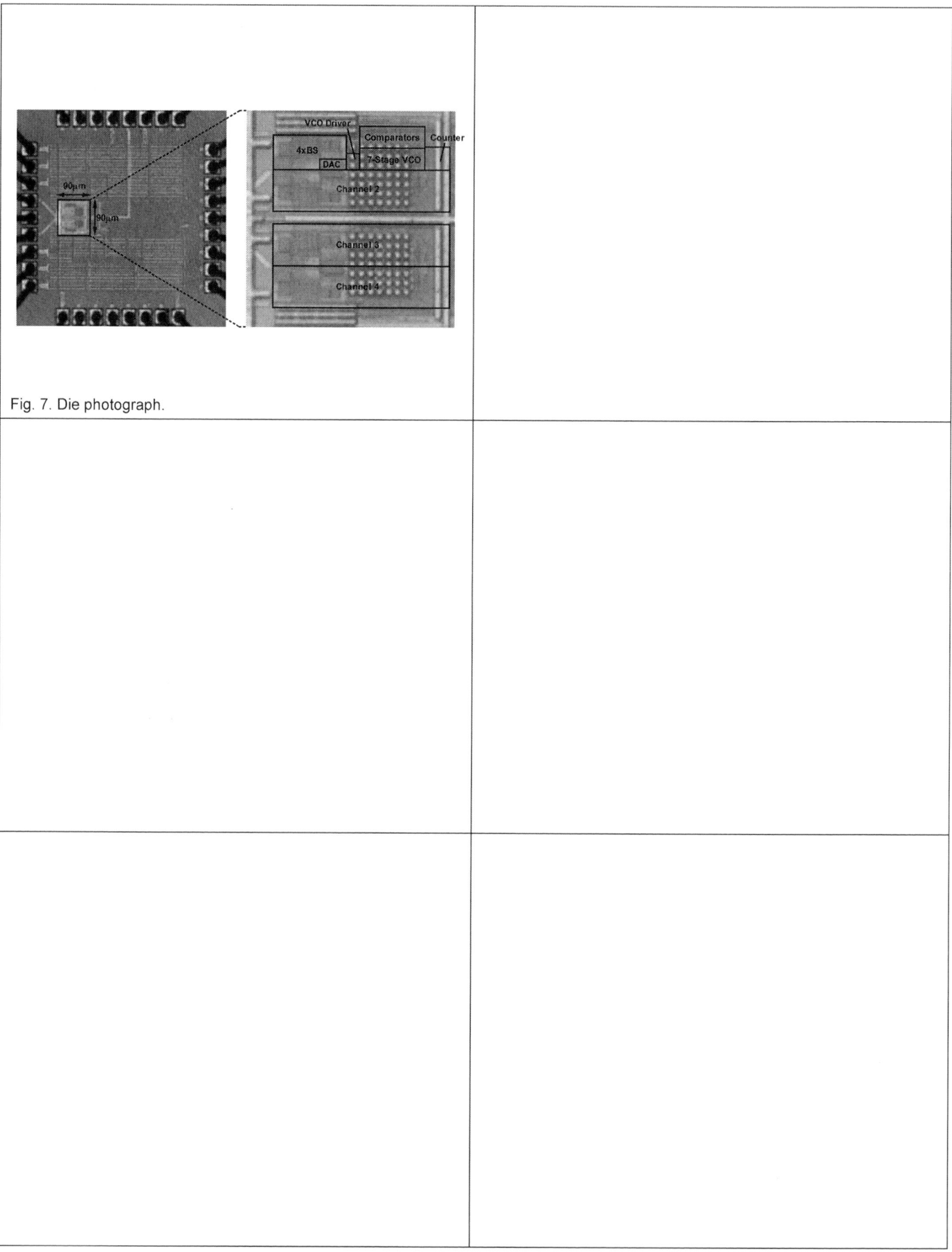

Fig. 7. Die photograph.

A 5GS/s 38.04dB SNDR Single-Channel TDC-Assisted Hybrid ADC with λ/4 Transmission Line Based Time Quantizer Achieving a PVT Robustness 416.6fs Time Step.

Hongzhi Liang, Yi Shen, Jun Chang, Shubin Liu, Ruixue Ding and Zhangming Zhu

Key Laboratory of Analog Integrated Circuits and Systems (Ministry of Education), School of Microelectronics, Xidian University, Xi'an, China

Time-based ADC is gaining popularity in medium-resolution high-speed (GS/s) analog-to-digital conversion, due to its capability to achieve faster conversion rates by scaling the time step T_{LSB} through advanced process technology. The conversion speed (f_s) of an N-bit resolution TDC is primarily dominated by the T_{LSB}, represented by $f_s < 1/(2^{N+1} \times T_{LSB})$. Several methods have been proposed to accelerate the speed by reducing T_{LSB}. The vernier delay line design in [1] produces the T_{LSB} by adjusting delay skew ($T_{d1}-T_{d2}$). Ref. [2] utilizes phase interpolation to generate multi-phase clock outputs with uniform time intervals (1.375 ps). Additionally, the selective delay tuning cell [4] achieves a 500fs T_{LSB}. These circuit implementations have enabled a significant reduction in T_{LSB}. However, further reducing T_{LSB} in traditional active inverter-based delay line of TDC poses several challenges: 1) The jitter budget of T_{LSB} incorporates the random jitter (RJ) introduced by noise in voltage-to-time converter (VTC) and active delay line, as well as determination jitter (DJ) generated by the power supply ripple. In particular, the jitter budget should satisfy the bit error ratio (BER) requirement in communication system. 2) From the perspective of PVT variation calibration for T_{LSB}, reducing T_{LSB} requires higher calibration accuracy, leading to an increased complexity of the calibration circuit. 3) As the T_{LSB} is reduced, a larger transconductance (gm) of input pair becomes critical for enhancing the precision of the dynamic time comparator. Consequently, this leads to an increase in power consumption, particularly during low slew-rate edge transitions.

This work offers a new perspective to address the challenges mentioned above. By utilizing the passive transmission line (TL) as a replacement of the traditional inverter-based delay line in TDC, as shown in Fig. 1. Benefiting from the inherent characteristics of TL, the proposed TDC features with the following advantages: 1) The unit delay in TL-based delay line is primarily determined by phase shift of traveling wave, not the active device delay. Thus, it allows for further reduction in T_{LSB} without introducing additional noise and jitter. 2) Due to the phase shift of traveling wave is only influenced by the properties of metal and dielectric materials [5], leading to T_{LSB} in proposed TDC is insensitivity to PVT variation. Meanwhile, a proposed phase opposition propagation TL-based time quantizer is adopted to break the bottlenecks between the T_{LSB} and power efficiency.

Figure 1 presents a TL-based generation scheme of a multi-phase reference clocks with the uniform delay units. By evenly spacing 2^N-1 taps along a λ/4-wavelength TL with matched termination impedance, the total phase of π/2 is equally distributed among each tap outputs. Based on the lumped model [5], the delay unit between the adjacent taps within the λ/4-wavelength TL can be expressed to $T_{LSB} = 1/(2^N-2)/(4 \cdot f_0)$, where f_0 is the traveling frequency. To increase chip integration and optimize area utilization, a high f_0 is employed to decrease the wavelength of TL.

Figure 2 depicts the proposed TDC-assisted hybrid ADC, which comprises a fronted 3-bit coarse flash ADC, a fully differential charge pump (CP) based VTC and a backend fine λ/4-wavelength TL-based TDC. The differential input signals VIP and VIN are simultaneously sampled on the 7 flash slices and CDAC. Note that the tunable delay chain is used to calibrate the bandwidth mismatch between 3-bit flash and CDAC. The output codes of the 3-bit flash ADC are utilized to switch the CDAC and generate the residue signal. The detail circuit implementation of the proposed fully differential CP-based VTC is also presented in Fig.2. In the traditional dynamic VTC, the drain voltage (V_B) of the current source varies with the operating state, worsening the linearity. Meanwhile, the parasitic C_{gs} of clock-controlled switch transistor will further exacerbates current instability during the initial phase of φ_{VTC}, especially in high discharge current

scenarios. In this work, the extra replica path is used to guarantee the high linearity of VTC. When the enable clock φ_{VTC} is low, unlike the conventional VTC, the auxiliary replica path maintains the appropriate V_B in the off period. The transient waveform is given in Fig.2. With the replica path, the slew rate variation is developed from 37.5% to 11.9%.

The λ/4-wavelength TL-based TDC is consists of four TLs, CMOS clock driver, the programmable termination matching and an open drain driver. After a pair of 20GHz differential clocks ($OSC_{CLKP/M}$) is driver and injected into a coupler TLs, a sequence of 31-phase differential clocks is generated, with a reference time delay (416.6fs) between adjacent differential taps. In order to mitigate swing and slew rate variations among the 31 taps, referring to low insertion loss, the back-end process of the TL incorporates a thick layer of aluminum (AP_{RDL}) along with 10 metal layers (M1-M10). The simplified transverse section and vertical section of TLs are shown in Fig.2. Based on electromagnetic (EM) analysis and SP simulation, the coupler TLs demonstrate an insertion loss (S12) of -1.43dB at 20GHz and a return loss (S11) of -18.88dB.

Figure 3 illustrates the operation principle of the 2-bit phase opposition propagation TL-based time quantizer. The output taps of the transmission line, labeled as $T_P<2:0>$, the waveform variations over time as depicted on the left. The corresponding output of phase comparator gradually transitions from 000 to 111 as time progresses. The pulse modulation signals $VTC_{P/N}$ are injected into TLs with a opposition propagation direction referred $OSC_{P/N}$. Resulting from the same wave velocity of TLs, the time detection out is varies with Δt generated by VTC. Note that the correlation between the digital output and the initial phase in the quantization encoding scheme, random states may occur when the phase difference is around T_{LSB}. The detailed circuit implementation of time detector and phase comparator is given in Fig.3. Since the input pair of phase comparator is differential clock with larger swing instead of thin delay edges, the precision demand is relaxed, and achieving a high power efficiency for the proposed TDC.

The ADC is fabricated in 28nm CMOS technology, occupying an active area of 0.012mm². Fig. 7 shows the chip micrograph and the core size of the ADC is 0.085mm² with the proposed λ/4-wavelenth TL spans 706µm in length. The measured spectrums at 5GS/s are shown in Fig. 4, and the ADC outputs are decimated by 64 for test. From the measured 16384-point FFT spectrum, the 5GS/s ADC shows 38.04dB SNDR and 46.24dB SFDR at Nyquist and achieves 36.41dB SNDR and 47.93dB SFDR with an over Nyquist 4.4GHz input. In Fig. 5, the ADC shows the measured SNDR < 0.34dB and 0.74B across -25°C to 125°C and ±5% supply variations, respectively. The total power consumption at 5GS/s is 19.81mW at Nyquist. The INL and DNL curves are shown in Fig. 7. For 5GS/s, the DNL range is [-0.689, 0.844] LSBs, and the INL range is [-1.495,1.405] LSBs. Figure 6 summarizes the performance of the proposed ADC with other state-of-the-art high-speed ADCs. Notably, the proposed quarter-wavelength TL-based TDC achieves the smallest time step of 416.6fs. With the proposed phase opposition propagation TL-based time quantizer, achieving Walden FoM is 60.7fJ/conv-step energy efficiency and the Schreier FoM is 149.1dB.

Acknowledgement:

This work was funded by the National Key R&D Program of China (2022YFB4401900), and the National Natural Science Foundation of China (U19A2053, 62021004, 62227816, 62261160649, 6216116030).

Corresponding Author: Shubin Liu and Zhangming Zhu.

References:

[1] B. Xu et al., " A 23-mW 24-GS/s 6-bit Voltage-Time Hybrid Time-Interleaved ADC in 28-nm CMOS," JSSC 2017:1091-2017.

[2] M. Zhang et al., " 16.2 A 4× Interleaved 10GS/s 8b Time-Domain ADC with 16× Interpolation-Based Inter-Stage Gain Achieving >37.5dB SNDR at 18GHz Input," ISSCC 2020:252-253.

[3] J. Liu et al., " 10.1 A 10GS/s 8b 25fJ/c-s 2850um² Two-Step Time-Domain ADC Using Delay-Tracking Pipelined-SAR TDC with 500fs Time Step in 14nm CMOS Technology," ISSCC 2022:160-161.

[4] M. Hassanpourghadi et al., " A 2-way 7.3-bit 10 GS/s Time-based Folding ADC with Passive Pulse-Shrinking Cells," CICC 2019:1-4.

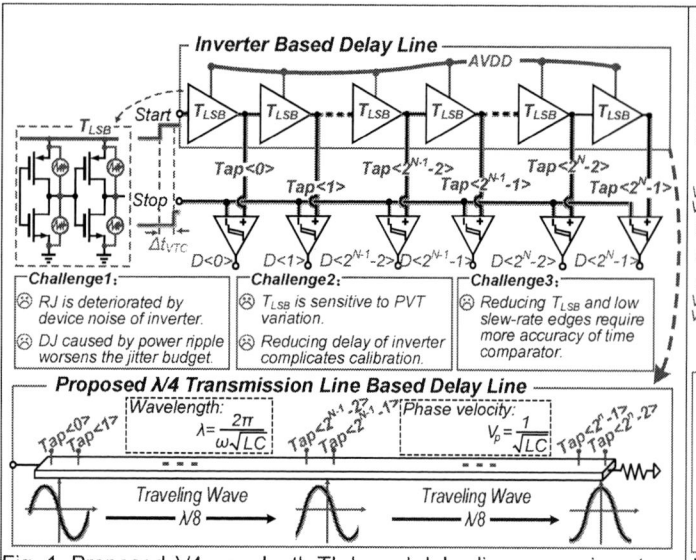

Fig. 1. Proposed λ/4-wavelenth TL-based delay line versus inverter-based delay line.

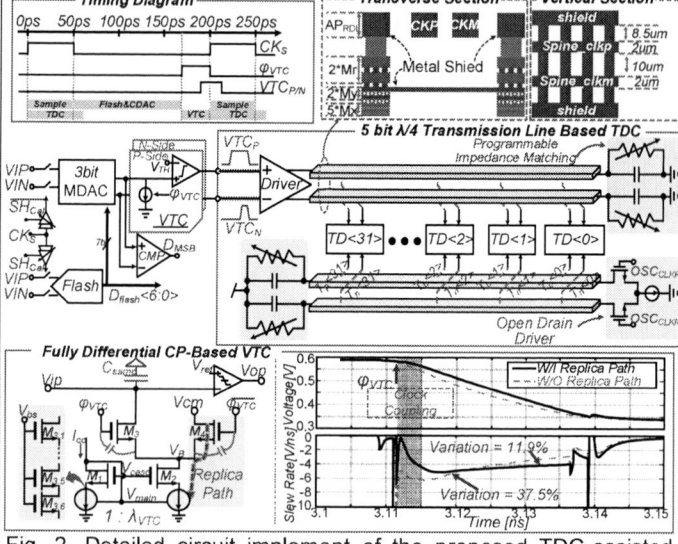

Fig. 2. Detailed circuit implement of the proposed TDC-assisted hybrid ADC (top) and fully differential CP-based VTC (bottom).

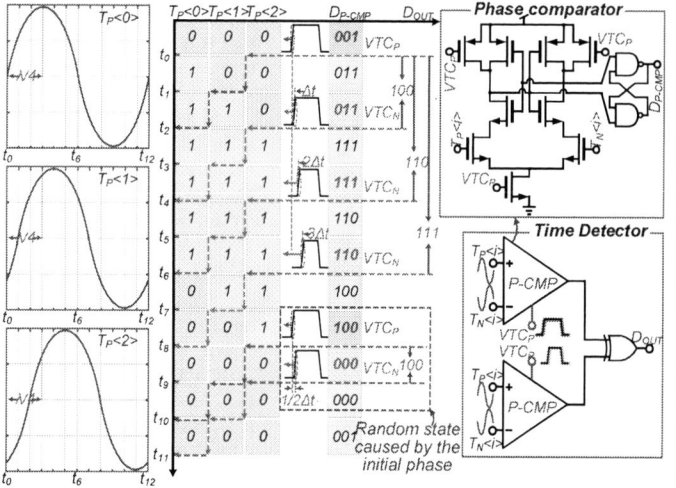

Fig. 3. Operation principle of the 2-bit phase opposition propagation TL-based time quantizer (left) and phase detector circuit details (right).

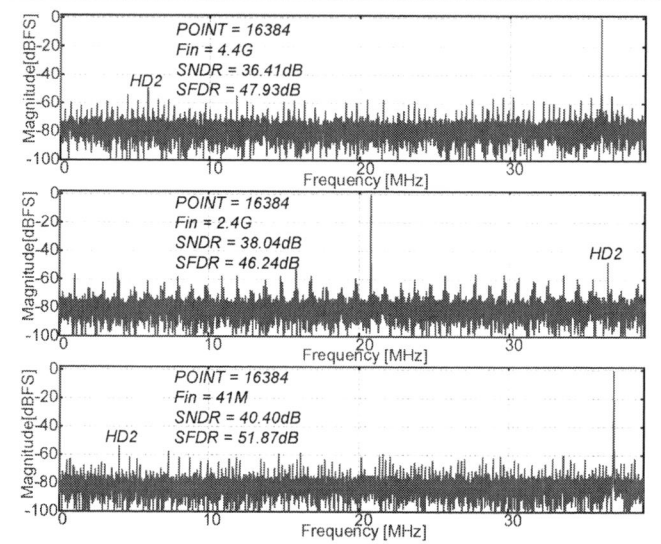

Fig. 4. Measured 16384-point output spectra at 5GS/s with different input frequencies (decimated by 64).

Fig. 5. Measured SNDR and SFDR versus sampling rates, input frequencies and SNDR variation versus supply and temperature variation at 5GS/s with low input frequency(375MHz)

Fig. 6. Measured performance summary and comparison with state-of-the art ADCs.

Reference	This work	JSSC'18 S. Zhu	ISSCC'22 J. Liu[3]	ISSCC'20 M. Zhang[2]	CICC'19 [4] M.H.Pourghadi	JSSC'22 A.Whitcombe
Architecture		Single Channel			Time Interleaved	
	SW-TL Based TDC	RNS TDC	SDT TDC	Phase Interpolation	Pulse-Shrinking TDC	TI-Hybrid TDC
Technology	28nm CMOS	65nm CMOS	14nm FinFET	65nm CMOS	65nm CMOS	22nm FinFET
Supply	0.9	1	0.8	0.9	1	1
# of TI channels	1	1	1	4	2	4
Resolution[bits]	7.5	8	8	8	7.3	7
T_{LSB}[ps]	0.416(Passive)	NA(Active)	0.5(Active)	1.5(Active)	1(Passive)	8(Active)
Fs [GS/s]	5	2	5	10 (2.5/CH)	10 (5/CH)	3.4
Power(w/o clk)[mW]	19.8	21	7.4	50.8	29.7	3.4
SFDR@Nyq[dB]	46.24	48.36	53.12	52.8	40.7	53.2
SNDR@Nyq[dB]	38.04	40.68	40.8	40.1	32.5	37.7
ΔSNDR@5% Supply Variation[dB]	0.74	<2.5	<2*	<0.6	NA	NA
ΔSNDR@ Temperature[dB]	0.34 (-25~125°C)	<3 (0~100°C)	<2 (20~80°C)	<0.5 (-55~125°C)	NA	NA
FoM$_{Walden}$@Nyq [fJ/conv.-step]	60.7	118.8	16.5	61.4	86.2	20.2
FoM$_{Sechreier}$ @Nyq[dB]	149.1	147.5	156.1	150	144.76	154.68
Active area[mm²]	0.012(W/O T-Line) 0.085((W/I T-Line)	0.08	0.0014	0.016	0.02	0.0045

979-8-3503-3004-5/23 $31.00 © 2023 IEEE

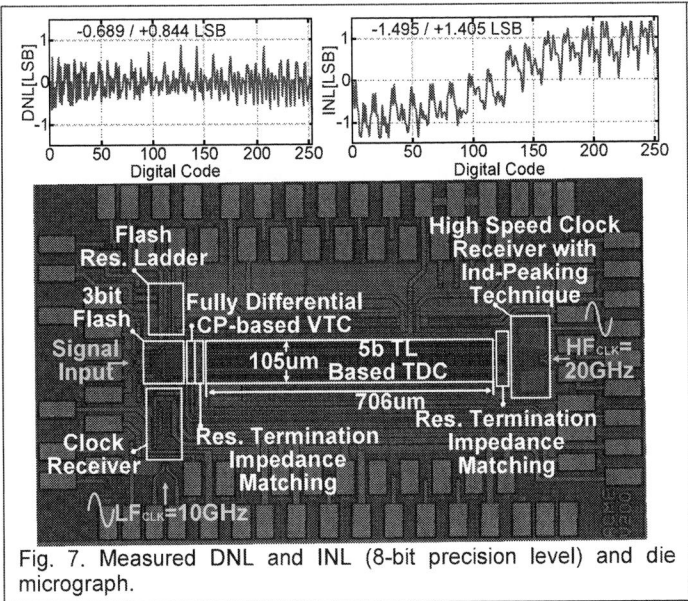

Fig. 7. Measured DNL and INL (8-bit precision level) and die micrograph.

Additional References:

[5] G. Li *et al.*, " 13.5 Standing Wave Based Clock Distribution Technique with Application to a 10 × 11 Gbps Transceiver in 28 nm CMOS," A-SSCC 2015:1-4.

[6] B. Ginsburg *et al.*, " 500-MS/s 5-bit ADC in 65-nm CMOS With Split Capacitor Array DAC," JSCC 2007:739-747.

[7] M. LaCroix *et al.*, " 8.4 A 116Gb/s DSP-Based Wireline Transceiver in 7nm CMOS Achieving 6pJ/b at 45dB Loss in PAM-4/Duo-PAM-4 and 52dB in PAM-2," ISSCC 2021:132-134.

CIMFormer: A 38.9TOPS/W-8b Systolic CIM-array based Transformer Processor with Token-slimmed Attention Reformulating and Principal Possibility Gathering

Ruiqi Guo, Yang Wang, Xiaofeng Chen, Lei Wang, Hao Sun, Jingchuan Wei, Leibo Liu, Shaojun Wei, Yang Hu, Shouyi Yin*

BNRist, Tsinghua University, Beijing, China; *Corresponding Author

Transformer models shows state-of-the-art results in natural language processing and computer vision, leveraging a multi-headed self-attention mechanism. In each head, the operation is defined as Attn=Softmax($Q \cdot K^T$)$\cdot$V, where $Q=X \cdot W_Q$, $K=X \cdot W_K$ and $V=X \cdot W_V$ are linear transformations for Query (Q), Key (K) and Value (V) with weight W_Q, W_K and W_V, respectively. $Q \cdot K^T$ is responsible to learn relevance scores between tokens (X). Previous CIM chips faces new challenges within and across attention heads (Fig. 1): (1) Computing-in-memory (CIM) shows great advantages only if fixing pre-trained weights. However, since Q and K are both generated at runtime, loading K in CIM macros consumes more energy in the computing of $Q \cdot K^T$. (2) A CIM macro performs multiply-accumulate (MAC) operations with bit-serial inputs, so the latency is determined by input precisions. The attention scores are normalized by Softmax to Probability (P). Most elements of P are close to or exactly zero. For the P$\cdot$V of Bert-T, only 10% elements with large effective bit-width (EBW) exacerbate the computing latency of the rest 90% of inputs. (3) On-chip SRAM-CIM cannot hold weights of all heads. As completing the computation of the stored weights, all macros need to reload new weights, causing a significant performance loss. This work designs a CIM processor for Transformer, called CIMFormer, with three features to solve above challenges: (1) A token-slimmed $Q \cdot K^T$ reformulation is proposed to reduce the loading of intermediate data and redundant computations in CIM macros. Besides, a column-partitioned X|W-CIM with flexible set-aggregate adder tree (Flex-AT) is designed to efficiently match the reformulated $Q \cdot K^T$ with high utilization. (2) A principal possibility gather-scatter scheduler (PPGSS) collects the elements with large EBWs, called principal possibility elements (PPEs), as simultaneous activations for a CIM macro, reducing the compute latency of the rest activations with small EBWs. (3) A systolic CIM array supports bidirectional matrix multiplications with macro-interleaved broadcast for activations. In the systolic CIM macro array, array-level weight reloading is divided into macro-level and hidden in macro-level systolic computing.

Fig. 2 shows the overall architecture, consisting of 2 attention cores with an X|W-CIM array, a CIM-output distribution network, 3 accumulator arrays, a PPGSS, a systolic activation broadcast unit (SABU), a BiW-CIM array, a quantizer and 64KB SRAM. The X|W-CIM array contains 16 4Kb SRAM-CIM macros with two operating modes: CIM mode and SRAM mode. Each macro contains 4 sub-CIM arrays and an accumulator. Each subarray contains 4x32 8b data, 32 activations and 4 Flex-ATs. The X|W-CIM array generates Q, K, V and performs the reformulated $Q \cdot K^T$. The generated scores are sent to the PPGSS, which gathers PPEs before sending them to the BiW-CIM macro array as activations of P$\cdot$V. Each BiW-CIM macro supports both horizontal and vertical writing, which loads simultaneously generated rows and columns of V as weights. The SABU broadcasts systolic activations to the X|W-CIM systolic array.

To reduce the K-related CIM loading, $Q \cdot K^T$ is reformulated as $Q \cdot W_K^T \cdot X^T$, which is performed in the cascaded W_K^T-/X^T-CIM after storing W_K and X in a unified CIM array. Conventionally, W_Q is stored in macros as weights, so X^T-CIM provides W_Q-CIM token-wise inputs to generate Q by row. However, if the complete W_Q, W_K and X cannot be stored, W_Q-CIM needs to be refreshed. To avoid it, W_Q-buffer will provide X^T-CIM data to generate Q by column. To fully utilize W_Q-CIM and avoid its refresh, in Q generation, this chip reformulates it as the combination of $X \cdot W_{Q,A}$ and $W_{Q,B}^T \cdot X^T$, where W_Q is divided into 2 parts and mapped in W_Q-CIM and SRAM buffer as weights and activations respectively. Due to quadratic time complexity in $Q \cdot K^T$, a token-slimming framework is introduced to prune over 60% input tokens that contribute little to the final prediction progressively in layers [1]. Due to progressively redundant tokens in layers, a unified CIM macro array with fixed-capacity needs to be dynamically divided into X^T/W_K^T/W_Q-CIM for the best reformulation. Conventionally, the global-configured accumulation network with $O(N^2)$ complexity [2] is required to construct respective sub-arrays for X, W_Q and W_K. To avoid it, this chip proposes an equivalent X|W-CIM macro partition with 3 column-sets for respective tiles of X, W_Q and W_K. In each period, activations are taken from 4 buffers appropriately, and sent to the corresponding column-sets. For correct results, the Flex-AT is implemented by placing links between different levels of adders, aggregating the products of a column-set with $O(log_2N)$ complexity.

The PPGSS normalizes attention scores by Softmax, and amplifies large scores to PPEs that have a large EBW. Dynamic PPEs are picked out by the normalization characteristic of Softmax function, which averages N elements of a row of P (P_K) to 1/N, represented by MV_P_K. The number of PPEs is small, but greatly affects MV_P_K. However, due to represented by binary, the EBW of PPEs have little impact on the mean of EBW of P_K, represented by MV[EBW_P_K]. As shown in Fig. 4, as PPEs are removed in P_K, the EBW of M_P_K, represented by EBW[MV_P_K], converges to MV[EBW_P_K]. The PPGSS consists of a Softmax unit, a principal possibility selection unit (PPSU), a sliced-based gather unit (SGU), an output aggregator, where the PPSU selects PPEs by monitoring the convergence of EBW[MV_P_K] and MV[EBW_P_K]. The SGU includes a PPE smoother, a vector slicer and a slice splice unit. The slicer cuts the front smoothed PPE vectors along the start-point of each PPE vector. The sliced PPEs from different P_K are gathered in the spice unit, and then sent to the BiW-CIM as activations in the out-of-order computing of P$\cdot$V, improving over 2x throughput and 1.8x energy efficiency. The BiW-CIM is also equipped with Flex-ATs for correct dot-products, which are scattered in the computing of non-PPEs for result recovery.

The systolic CIM array divides array-level weight reloads into macro-level and hides them in parallel multi-head computation. Taking $V=X \cdot W_V$ as example (Fig. 5), each CIM macro stores K tokens and J W_V-columns (Wcols). In one-dimensional X|W-CIM macro array, each macro acts as a temporary input buffer sequentially from left to right. The X part of each macro broadcasts token-wise X to the W parts to its right, generating a part of row of V. Simultaneously, the W part of each macro focuses on broadcasting column-wise W_V to the X parts that not yet read, generating a part of column in V. By combining the parts of rows and columns in a period, J columns and K rows of V are generated completely, which means that the involved K tokens and J Wcols are useless to compute V. Therefore, the involved X part of a macro can perform $W_Q^T \cdot X^T$ of the next head by reusing the stored tokens. Since each X part is released for token-reuse in sequence, the systolic columns of W_Q are input to them for maximizing throughput. Moreover, the W part of a macro is released early, which can load new weights immediately to improve over 3× throughput. Each BiW-CIM macro consists of 256×32 dual-split-wordline-controlled 8T-cells to align the writing of the simultaneously generated row and column of V, reducing SRAM access to 3x.

Under a voltage of 0.6-to-1.0V with a frequency of 80-to-240MHz, the power is 20.4-97.3mW (Fig. 6). The highest energy efficiency is 38.9TOPS/W at 0.65 V, 120MHz. Compared with traditional digital processing elements [3], the digital CIM unit achieves 1.83× higher energy efficiency. Compared with [4, 5], the X|W-CIM avoids to reload K as weights, increasing energy efficiency by 1.9×. Compared with [5, 6] that only optimize the computation pipeline for a head, the systolic CIM array increases simultaneous weight-update and token-reuse for adjacent heads, achieving 1.27x speedup. Furthermore, the PCGSS contributes 2.01x speedup in P$\cdot$V without accuracy loss, which can also be configured to handle the ReLU-based sparsity in FFNs. Figure 7 shows a summary and die photo.

Acknowledgements: This work was supported in part by NSFC Grant U19B2041, Grant 62125403, and Grant 92164301, the National Key Research and Development Program under Grant (No.2021ZD0114400), the BNRist, and the Beijing Advanced Innovation Center for Integrated Circuits.

979-8-3503-3004-5/23 $31.00 © 2023 IEEE

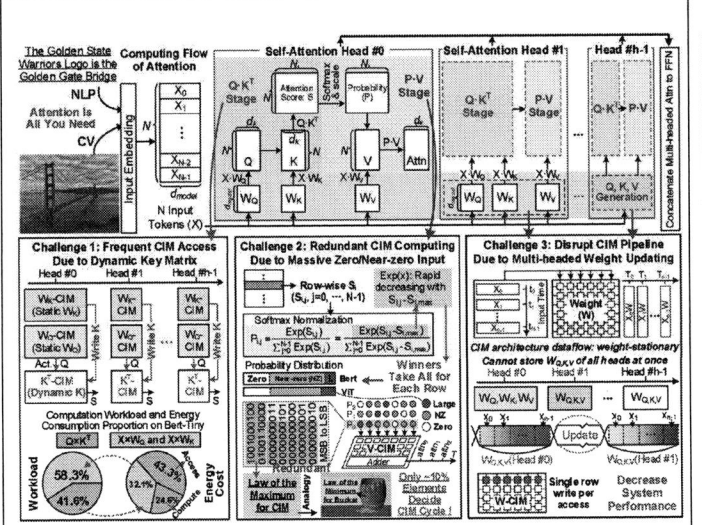

Fig. 1. Challenges of CIM processor for Transformer models.

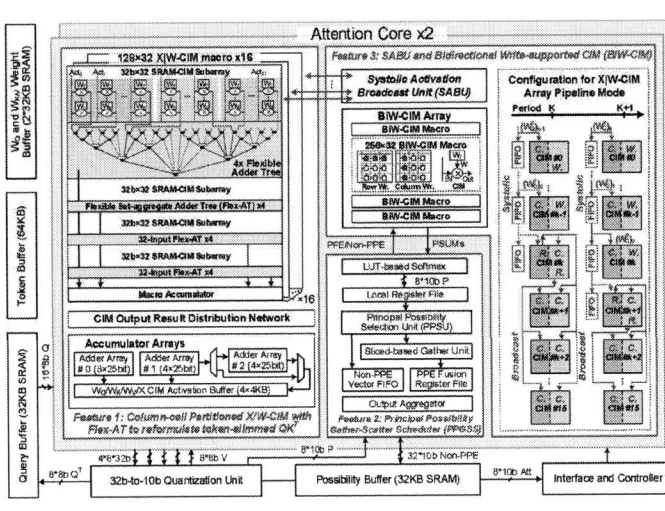

Fig. 2. Overall architecture of the chip with three design features.

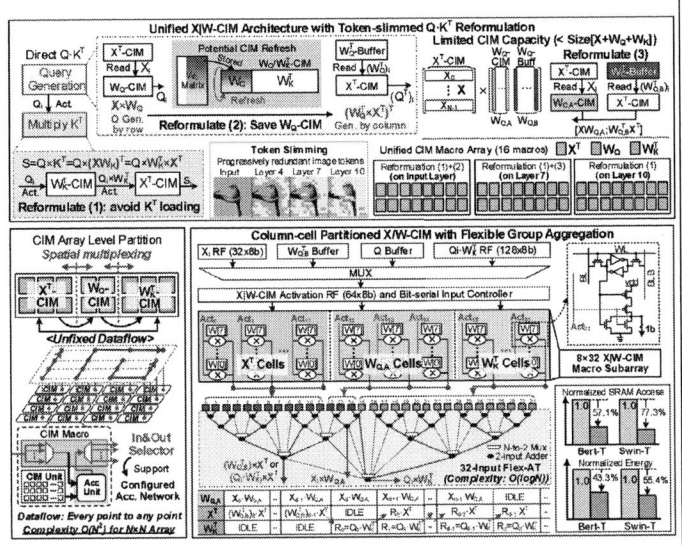

Fig. 3. Token-slimmed Q·K^T reformulation with X|W-CIM design.

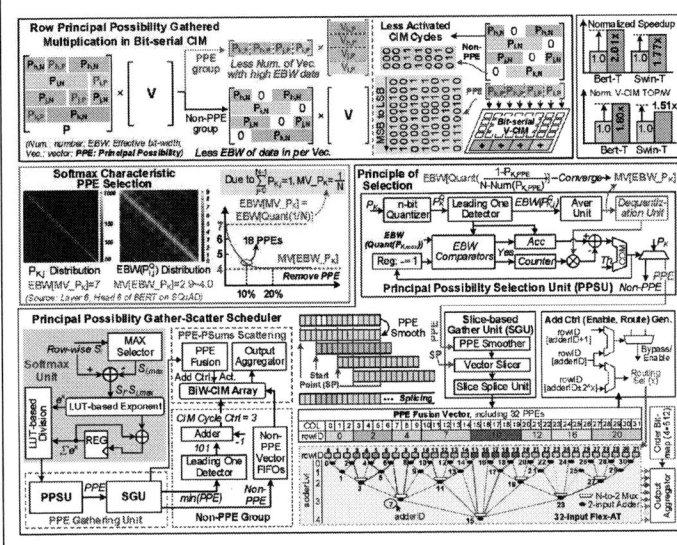

Fig. 4. PPGSS collects PPEs as simultaneous input for BiW-CIM.

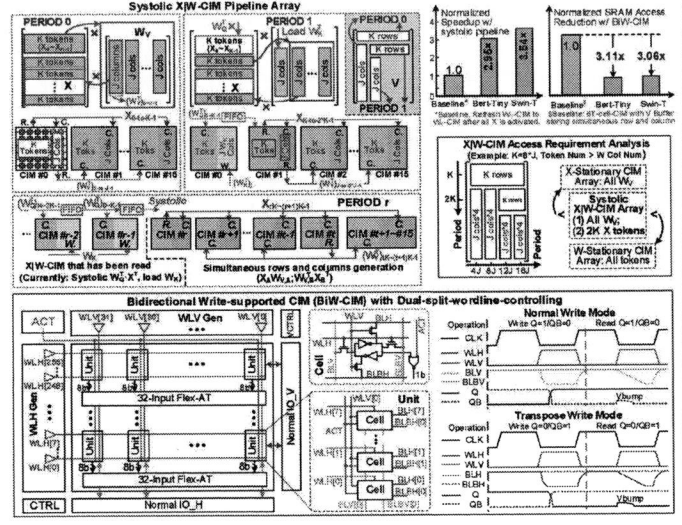

Fig. 5. Systolic X|W-CIM pipeline array and BiW-CIM design

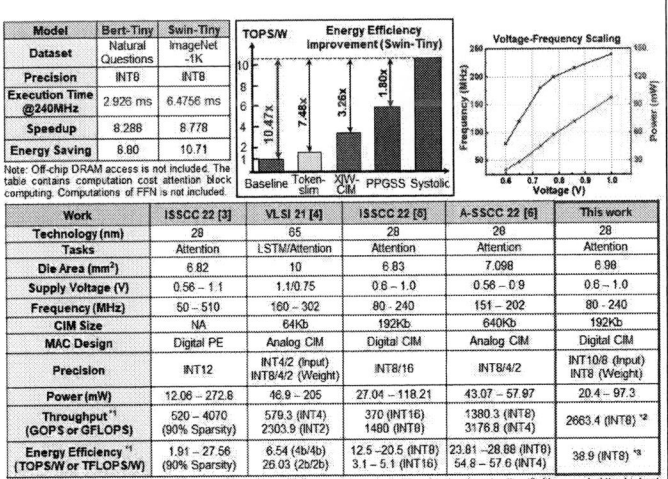

Fig. 6. Measurement results, and comparisons.

Metric		This work
Technology		28nm CMOS
Die Area		2.90x2.41 (mm²)
Supply Voltage		0.6V – 1.0V
Frequency		80MHz – 240MHz
CIM Size		192 Kb
SRAM Size		192 KB
Data Precision		INT8, INT10
Power		20.4 mW @0.6V, 80MHz 97.3 mW @1.0V, 240MHz
Peak Throughput (TOPS)	Q·K^T	2.12[1]) @1.0V, 240MHz
	P·V	2.63[1]) @1.0V, 240MHz with ~10% PPEs
Energy Efficiency (TOPS/W)	Q·K^T	29.95[1]) @0.65V, 120MHz
	P·V	38.85[1]) @0.65V, 120MHz

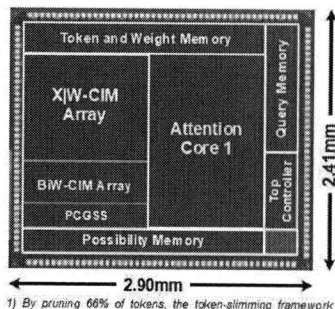

1) By pruning 66% of tokens, the token-slimming framework reduces 31%~37% FLOPs and improves the throughput by over 40% within 0.5% accuracy loss.

Fig. 7. Chip specification and die photo.

References:

[1] Y. Rao et al., "DynamicViT: Efficient Vision Transformers with Dynamic Token Sparsification," NeurIPS 2021.

[2] H. Jia et al., "A Programmable Neural-Network Inference Accelerator Based on Scalable In-Memory Computing," ISSCC 2021.

[3] Y. Wang et al., "A 28nm 27.5TOPS/W Approximate-computing-based Transformer processor with asymptotic sparsity speculating and out-of-order computing," ISSCC 2022.

[4] R. Guo et al., "A 6.54-to-26.03 TOPS/W Computing-In-Memory RNN Processor using Input Similarity Optimization and Attention-based Context-breaking with Output Speculation," VLSI 2021.

[5] F. Tu et al., "A 28nm 15.59uJ/Token Fully-Digital Bitline-Transpose CIM-based Sparse Transformer Accelerator with Pipeline/Parallel Reconfigurable Modes," ISSCC 2022.

[6] R. Guo et al., "A 28nm 57.6TOPS/W Attention-based NN Processor with Correlative Computing-in-Memory Ring and Dataflow-reshaped Digital-assisted Computing-in-Memory Array", A-SSCC, 2022.

A 5.2GHz Trifilar Transformer-Based Class-F_{23} Noise Circulating VCO with FoM of 192.6 dBc/Hz

Hanzhang Cao[1], Tongde Huang[1], Xiaolong Liu[2], Hao Wang[1], Jin Jin[1], Wen Wu[1]

[1]Ministerial Key Laboratory of JGMT, Nanjing University of Science and Technology, China

[2]School of Microelectronics, Southern University of Science and Technology, China

With the development of wireless communication, high-speed data transmissions require a pure local oscillation signal produced by a voltage-controlled oscillator (VCO). A variety of circuit technologies is developed for improving the LC-VCO's phase noise. The harmonic shaping technology represented by class-F_{23} VCO in Fig.1 (top left) utilizes the transformer-based multi-order load in differential mode (DM) to shape the oscillation waveform [1]. From the perspective of the impulse sensitivity function (ISF) [2], the harmonic shaping technology reshapes the ISF waveform to inhibit noise conversion to phase noise. As a result, the low root-mean square (RMS) value of the effective ISF decreases the flicker noise up-conversion to $1/f^3$ phase noise. However, the phase noise in the harmonic shaping oscillators is limited by the magnitude of the noise sources injected into the oscillator. Specifically, during the switching transition, much more noise flows into the tank, deteriorating the phase noise performance. Yet, the noise circulating (NC) technology shown in Fig.1 (top middle) can effectively lower the noise source injected into the oscillator by source degeneration [3]-[4]. During the zero-crossing of the tank voltage, the noise current partially flows into the resonant tank while most of them still circulates within the transistors, thus reducing the magnitude of the noise source effectively. However, the noise circulating oscillators fail to guarantee a resistively loaded CM path. The parasitic capacitance at source nodes introduces an extra capacitive noise flow path to the resonant tank, which causes flicker noise up-conversion. Besides, the technology usually adopts a cross-coupled pair as the active core, bringing in a unit gain between the gate and drain of the transistor and contributing little to reduce the phase noise. The drain and the gate share the same DC bias voltage, leading to a severe frequency pushing effect due to the voltage-dependent gate parasitic capacitance [5]. For the LC oscillator, the total effective noise produced by the ith devices can be expressed as

$$\sum_i N_{L,i} = \frac{1}{2N^2\pi} \sum_i \int_0^{2\pi} \Gamma_i^2(\theta)\overline{i_{n,i}^2(\theta)}d\theta \qquad (1)$$

where $\overline{i_{n,i}^2(\theta)}$ is the power spectral density (PSD) of the noise source, and $\Gamma_i(\theta)$ is the corresponding ISF. Equation (1) indicates that the phase noise of the oscillator is mainly determined by the magnitude of the noise sources and the noise sources' conversion to the phase noise.

In this study, by fully considering the side effect of the uncontrolled parasitical CM path in the conventional noise circulating technology and the noise current injecting to the resonant tank in class-F_{23} VCO design, we proposed a trifilar transformer-based class-F_{23} noise circulating VCO in Fig.1 (top right). The design optimizes the phase noise by reducing both the noise source and the conversion from the noise source to the phase noise simultaneously.

Figure 2 depicts the whole schematic of the proposed VCO. The active core consists of NMOS transistors pair N_1 and N_2 degenerated by PMOS transistors pair P_1 and P_2. The trifilar transformer couples the NMOS pair's drain nodes to both its own gate nodes and the PMOS pair's gate nodes. The equivalent half schematic is shown in Fig.3 (left). Because of the positive coupling between L_P and L_T, the NMOS and the PMOS transistors at the left or right half are switching on/off simultaneously during the same half period. The simulated transient voltage waveforms are also shown in Fig. 3 (right top). The transformer also serves as part of the resonant tank. To meet the requirements for class-F_{23} operation, the first DM resonant frequency of the resonant tank is designed at the fundamental frequency, and the second is tuned to $3f_o$. Regarding the CM return path, the adjacent current in L_S flows in the opposite direction, as a result, the CM coupling coefficient $k_{PS,cm} \approx 0$ since a subtractive magnetic flux is produced. The CM resonant frequency is mainly governed by the

$L_{P,cm}$ and the CM capacitance $C_{D,se}$ connected to the drain nodes of the NMOS pair. By adjusting the ratio of $C_{D,diff}$ and $C_{D,se}$ while keeping $C_{D,total}$ constant, the CM resonant frequency can be designed to $2f_0$ without affecting the DM resonant tank. The DM and CM magnitude of the tank impedances are plotted in Fig. 3 (right bottom).

Different from constructing the dynamic voltage feedback between NMOS pair and PMOS by a 1:1 transformer in [4] resulting in a unit gain feedback and high supply pushing, the proposed design establishes the voltage feedback by a 1:2:1 trifilar transformer-based resonant tank. This provides an extra passive voltage gain from the NMOS pair's drain nodes to its own gate nodes (A_{V1}), benefiting in suppressing the transistors noise. the gain offered by the trifilar transformer-based resonant tank reduces the effective transconductance to ($1+A_{V1}$) times, which helps to reduce transistor noise. Moreover, because of the noise circulating operation, most of the noise current still circulates within the transistors, the PSD of the active core $\overline{i_{n,ActiveCore}^2}$ is expressed as

$$\overline{i_{n,ActiveCore}^2} = 2KT\gamma G_{m,eff} = 2KT\gamma \frac{g_{mn}g_{mp}}{g_{mn}+g_{mp}} \frac{1}{(1+A_{V1})} \qquad (2)$$

which indicates the proposed oscillator effectively achieves noise source reduction. The simulated ISF, NMF, and the effective ISF (Γ_{eff}) waveform in one cycle are simulated in Fig. 3 (right middle). The proposed topology takes both DM/CM path and noise sources into account. Therefore, it succeeds in presenting a much smaller Γ_{eff} and its RMS value. As a result, the effective noise produced by the active core ($N_{L,ActiveCore}$) is greatly reduced by lowering both the magnitude of the noise sources and the noise sources' conversion to the phase noise. Figure 1 (lower) compares the simulated ISF, Γ_{eff}, and phase noise of the VCOs. The proposed design shows smaller Γ_{eff} and its RMS value, phase noise and the lowest $1/f^3$ phase noise corner compared with the other two VCOs only considering either noise source or DM/CM path.

Figure 7 shows the photography of the proposed VCO fabricated in 65-nm CMOS technology which has a core area of 0.12 mm^2 mainly composed of a trifilar transformer. An open drain buffer is designed for testing. The phase noise is measured by Rohde Schwarz FSV3044 spectrum analyzer. Figure 4 shows the measured phase noise at f_{min} and f_{max}, consuming 8.64 mW and 7.82mW at a 0.9 V power supply, respectively. At 5.22 GHz, the measured phase noise are -96.47/-124.84/-147.56 dBc/Hz, and the corresponding Figure-of-Merit (FoM) is 181.5/189.8/192.6 dBc/Hz at 100 k/1M/10-MHz offsets. The $1/f^3$ phase noise corner is about 380 kHz. The proposed VCO has a 14.9% tuning range (TR) covering 5.22-6.05 GHz. Figure 5 (top) depicts the measured phase noise and the corresponding FoM across the whole TR. At 10-MHz offset, the proposed VCO exhibits a FoM between 189 and 192.6 dBc/Hz. Figure 5 (bottom) shows the frequency pushing results, and a very low frequency pushing of 11.2 and -18.3 MHz/V is measured at f_{min} and f_{max}, respectively.

The proposed VCO's performance is outlined in Fig.6 along with a comparison to the state-of-the-art VCOs. By utilizing the trifilar transformer in noise circulating topology, the VCO exhibits competitive PN normalized to 5GHz, superior FoM in the thermal range and the best supply pushing compared with the other works.

Acknowledgment:

This work was partially funded by Jiangsu Province Science & Technology Department under Grant BE2021017.

References:

[1] M. Shahmohammadi et al., "A 1/f Noise Upconversion Reduction Technique for Voltage-Biased RF CMOS Oscillators," *IEEE JSSC*, vol. 51, no. 11, pp. 2610-2624, Nov. 2016.

[2] A. Hajimiri et al. "A general theory of phase noise in electrical oscillators," *IEEE JSSC*, vol. 33, no. 2, pp. 179-194, Feb. 1998.

[3] A. Mostajeran et al., "25.8 A 2.4GHz VCO with FOM of 190dBc/Hz at 10kHz-to-2MHz offset frequencies in 0.13μm CMOS using an ISF manipulation technique," *IEEE ISSCC*, 2015, pp. 1-3.

[4] F. Wang et al., "A Noise Circulating Oscillator," *IEEE JSSC*, vol. 54, no. 3, pp. 696-708, March 2019.

979-8-3503-3004-5/23 $31.00 © 2023 IEEE

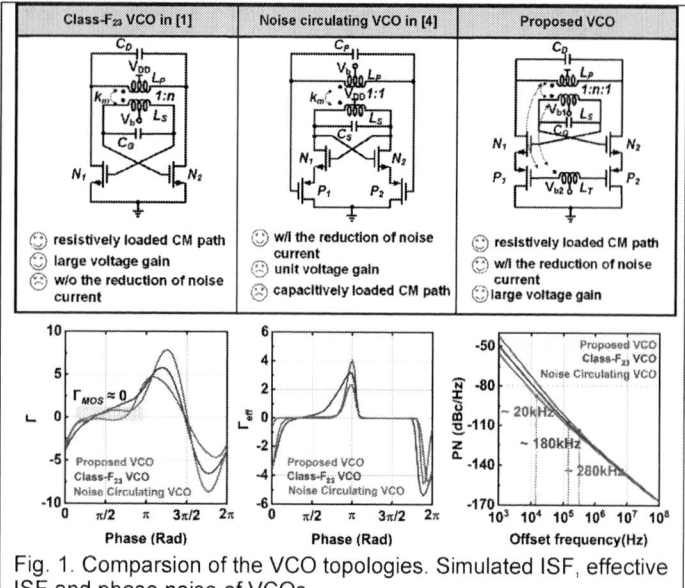

Fig. 1. Comparsion of the VCO topologies. Simulated ISF, effective ISF and phase noise of VCOs.

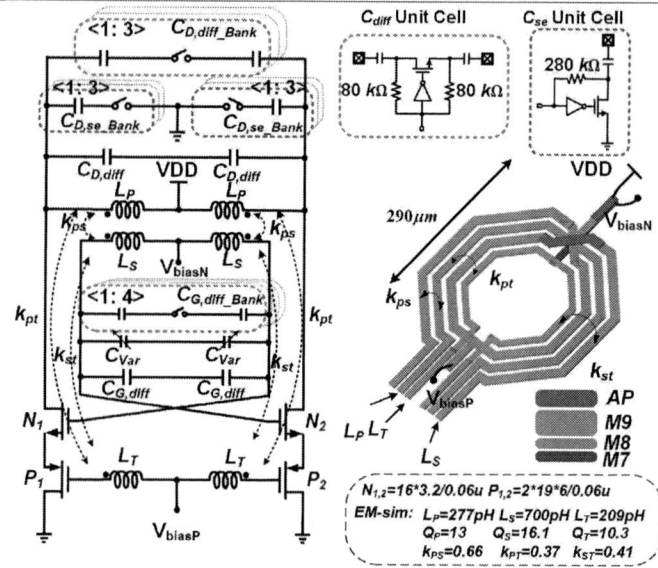

Fig. 2. The whole schematic of the proposed VCO.

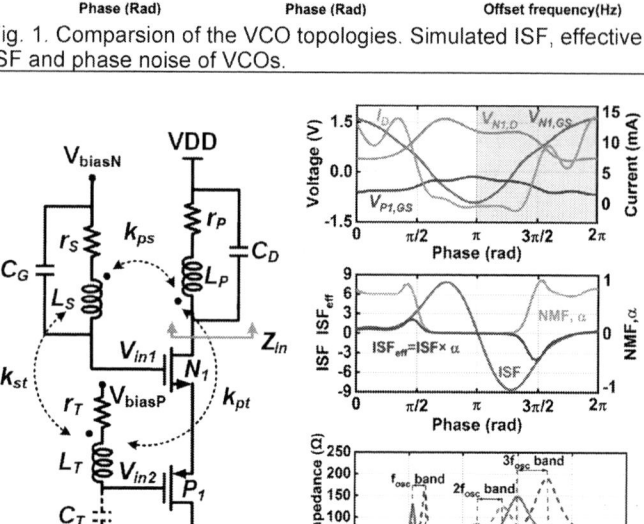

Fig. 3. Equivalent half schematic. Simulated tank impedance, transient voltage, current, ISF, NMF, ISF$_{eff}$ waveforms.

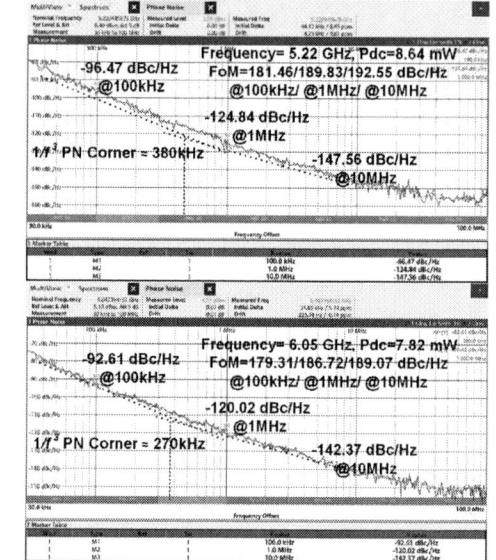

Fig. 4. Measured PN at f_{min} and f_{max}.

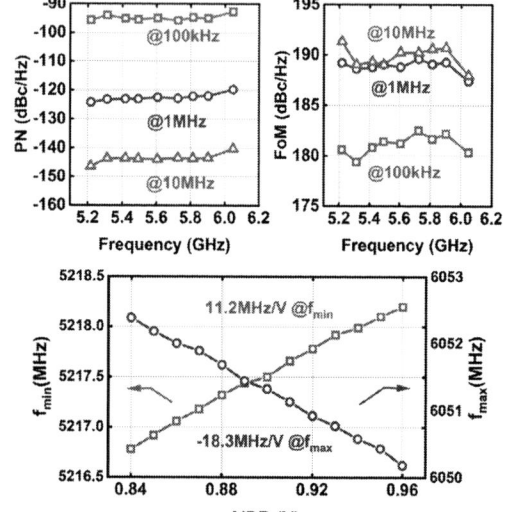

Fig. 5. Measured PNs, FoMs across the TR, and the measured frequency pushing at f_{min} and f_{max}.

Parameters		This Work	JSSC'16 [1]	ISSCC'15 [3]	JSSC'19 [4]	TMTT'22 [6]	RFIC'22 [7]
Key technology		Class-F23 with trifilar TF based NC	Class-F$_{23}$	Cap. coupled NC	TF based NC	Ind. feedback	NMOS biased switchable capacitor
VDD (V)		0.9	1	1.4	1.2	1.2	1.2
Tuning Range (TR)		5.2-6.1 (14.9%)	5.4-7 (25%)	2.4GHz (1.7%)	2.1-2.5 (18.6%)	6.3-8.5 (29.9%)	8.2-10.2 (24.3%)
Power (mW)		8.6	12	4.2	2.6	7.7	13
PN (dBc/Hz)	100K	-96.5	-105.3	-109	-111.1	-88.2	-88.8
	1M	-124.8	-126.7	-128.4	-131.6	-115.4	-115.1
	10M	-147.6	-146.7	N/A	-152.3	-136.0	-133.2
PN referred to 5GHz$^\&$	1M	-125.2	-127.3	-122.0	-123.9	-117.3	-119.4
	10M	-147.9	-147.3	N/A	-144.6	-138.0	-137.5
FoM* (dBc/Hz)	1M	189.8	190.5	189.8	195	183.2	183
	10M	192.6	190.5	N/A	195.6	183.9	181.15
Freq. pushing (MHz/V)		11.2	12	N/A	-13	N/A	N/A
Die Area (mm^2)		0.12	0.13	0.09	0.36	0.06	0.13
CMOS Technology		65-nm	40-nm	130-nm	130-nm	65-nm	28-nm

*FoM=|PN|+20 log$_{10}$(f$_0$/Δf) -10 log$_{10}$ (P$_{DC}$/1mW) $^\&$PN1=PN +20 log$_{10}$(5GHz /f$_{osc}$)

Fig. 6. Performance summary and comparison with recently reported related works.

979-8-3503-3004-5/23 $31.00 © 2023 IEEE

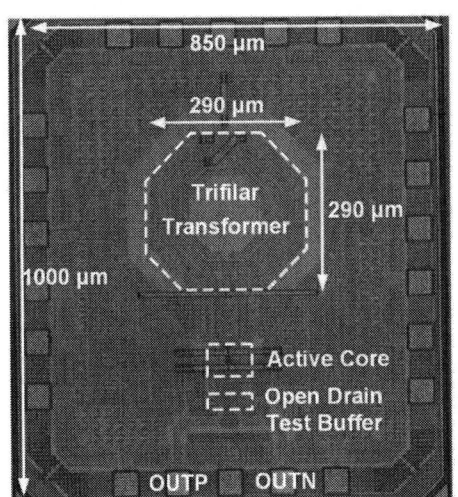

Fig. 7. Photography of the as-fabricated VCO.

Additional References:

[5] C. -C. Li *et al.*, "19.6 A 0.2V trifilar-coil DCO with DC-DC converter in 16nm FinFET CMOS with 188dB FOM, 1.3kHz resolution, and frequency pushing of 38MHz/V for energy harvesting applications," *IEEE ISSCC*, 2017, pp. 332-333.

[6] C. Wan *et al.*, "A Current-Reused VCO With Inductive-Transformer Feedback Technique," *IEEE TMTT*, vol. 70, no. 5, pp. 2680-2689, May 2022.

[7] L. Wang *et al.*, "An 8.2-10.2 GHz Digitally Controlled Oscillator in 28-nm CMOS Using Constantly-Conducting NMOS Biased Switchable Capacitor," *IEEE RFIC*, 2022, pp. 207-210.

A Wide Frequency Range, Small Area and Low Supply Memory Interface PLL Using a Process and Temperature Variation Aware Current Reference in 3 nm Gate-All Around CMOS

Kyungmin Lee[1,2], Jaehong Jung[1], Gyusik Kim[1], Joomyoung Kim[1], Seungjin Kim[1], Seunghyun Oh[1], Sung Min Park[2] and Jongwoo Lee[1]

[1]Samsung Electronics, Hwaseong, Korea

[2]Ewha Womans University, Seoul, Korea

Since the bandwidth and clock frequency of a memory interface have been increased for 5G communication, its power consumption should be tightly managed to extend battery time. To improve the energy efficiency, the dynamic voltage and frequency scaling (DVFS) technique [1] that optimally adjusts the operating voltage and frequency depending on the system scenarios has been used so that a wide frequency range of the system clock is required. For example, the advanced double data rate 5 (DDR5) mandates a clock frequency range from 2.1 to 9.6 GHz. Therefore, ring oscillator (RO) based PLLs are commonly utilized in memory interfaces due to their wider frequency tuning range when compared to LC-based PLLs. In an RO-based PLL, the output frequency can be tuned by controlling transconductance (g_m) of delay cells. Typically, the current of RO adjusts g_m through a current-DAC (I-DAC) with a foreground automatic frequency calibration (AFC) [2, 3]. However, as the output frequency of RO varies with respect to the process, voltage, and temperature (PVT) variations, the dynamic range of the I-DAC should cover not only the target frequency range, but also the frequency deviation over PVT variations, hence requiring large area of the I-DAC. This paper presents an RO-based PLL which uses a current reference that compensates for the frequency deviation of the RO over PVT variations; thus, the proposed PLL supports a wide frequency range from 0.019 GHz to 10.6 GHz, while occupying a small active area of 0.012mm^2. The proposed PLL achieves 0.32-ps$_{rms}$ period jitter and 0.53-mW/GHz energy efficiency at 10.6-GHz output frequency.

Figure 1 shows the schematic diagrams of a 3-stage current-controlled RO and its pseudo-differential delay cell with a cross-coupled latch. Although this current-controlled RO is independent of the supply voltage (V_{DD}), a frequency of the RO varies due to the threshold voltages of PMOS and NMOS (V_{THP} and V_{THN}) changes by the process and temperature variations as shown in Fig. 1(a). Moreover, the results of the skewed corners (FS and SF) indicate that the impacts of V_{THP} and V_{THN} are different. In consequence, I_{VCO} that can track the variations of V_{THP} and V_{THN} with a weight parameter (α) (0.75 in this work), as shown in Fig. 1(b), is necessary to achieve a constant frequency over the process and temperature variations by compensating the frequency deviation, as shown in Fig. 1(c).

Figure 2 depicts a conventional current reference (I-REF) for the RO [2] and its current simulation results over PVT variations (Fig. 2(a)). As the reference current (I_B) is solely determined by V_{DD} and a resistor (R_0), it cannot properly compensate for the frequency deviation over PVT variations. To generate I_B which tracks V_{THP} and V_{THN} simultaneously, a concept of a process and temperature variation aware current reference (PT-VAR) is proposed in Fig. 2. The PT-VAR combines the currents from PMOS and NMOS diode-connected I-REFs so that I_B is given by '$α\ V_{GSP}\ /\ R_1 + (1-α)\ V_{GSN}\ /\ R_2$' in Fig. 2(a). As a result, the frequency deviation of the RO over PVT variations at 9.6 GHz decreases from 45 % to 10 % with a fixed AFC code, as shown in Fig. 2(b). However, this PT-VAR needs a 2x large active area and power consumption by the requirement of both PMOS and NMOS diode-connected I-REFs.

To mitigate these issues, the optimized PT-VAR is proposed as shown in Fig. 3 (top). Since one amplifier is eliminated and the resistor is shared by using a 3-input amplifier, this optimized PT-VAR improves the area and power consumption by 60 % and 50 %, respectively, when compared to the PT-VAR in Fig. 2. The weights of V_{THP} and V_{THN} are controlled by g_m of the input transistors. Owing to the PT-VAR, the frequency deviation over PVT variations is reduced by 60 % when compared to the conventional current reference [2] for the same dynamic range of I-DAC as shown in Fig. 3(a). Namely, the minimum output frequency, f_{MIN} (i.e., AFC = 1) decreases from 1.5 GHz to 1.2 GHz whereas the maximum output frequency, f_{MAX} (i.e.,

AFC = 63) increases from 9.3 GHz to 10.6 GHz, such that the overall frequency range is improved by 20 %. The simulation results in Fig. 3 (b) show that the PT-VAR significantly reduces the difference of AFC over PVT variations from 28 to 7 (75 %) at the 9.6-GHz target frequency. Hence, the PT-VAR circuit generates I_B which compensates for the frequency deviation of the RO over PVT variations.

Figure 4 illustrates the overall block diagram of the proposed RO-based integer-N PLL with a pre-divider to control the resolution of the output frequency. To improve the noise from a resistor in a conventional single-path loop filter, a dual-path loop filter is used by splitting into the proportional and integral paths [4]. As shown in Fig. 4(a), the phase noise is improved by 3 dB at the 10-MHz offset frequency, when compared to using the conventional single-path loop filter. The dual-path loop filter can also reduce the size of an integrator in the loop filter by adjusting the currents of charge pumps (CP1 and CP2) independently. Besides, an NMOS common source amplifier with a source-degeneration resistor is used to improve the linearity of K_{VCO} and the noise, and to operate at a low V_{DD} for DVFS. Consequently, the simulation results of the ratio of K_{VCO}s for two paths (K_{VCO_I} and K_{VCO_P}) over PVT variations are highly constant for a wide dynamic range of the control voltages (V_{CI} and V_{CP}), as shown in Fig. 4(b). In the PLL operations, the AFC selects the frequency band first as a foreground calibration. After the AFC operation (AFC_END = 1), the frequency is adjusted finely by the control voltages which is switched to the loop filter paths. It should be noted that the noise of PT-VAR is ignored as the PT-VAR is only used for the AFC operation, alleviating the design burden of the PT-VAR.

The proposed RO-based PLL using the PT-VAR circuit was fabricated in a 3-nm GAA CMOS process. Figure 5 demonstrates the measured AFC code distributions for 50 samples (i.e., 10 samples per each 5-corner wafer) over temperature variation. When the target frequency is 2.1 GHz and 9.6 GHz, the maximum difference of AFC code is less than 4 and 10, respectively. Accordingly, since the proposed PT-VAR reduces the dynamic range of I-DAC, the active area of the PLL is 0.012mm^2 (Fig. 7). Figure 5 also shows that the measured period jitter (PJ) at f_{MIN} (1.2 GHz) is 1.94 ps$_{rms}$ and 18.3 ps$_{p2p}$ for 10^{-6} bit-error-rate (BER). The measured PJ at 5.3 GHz is 0.64 ps$_{rms}$ and 6.09 ps$_{p2p}$, which can be translated to 0.32 ps$_{rms}$ and 3.05 ps$_{p2p}$ at 10.6 GHz, when considering the phase noise slope of -30 dB/decade from 1 MHz to 10 MHz as shown in Fig. 5. Here, PJ at f_{MAX} (10.6 GHz) is measured at the divided-by-2 output (5.3 GHz) due to the insufficient output driving strength. Therefore, the proposed RO-based PLL satisfies with the wide frequency range as well as the period jitter specification (± 2 % of one clock period) of the advanced DDR5. The total power consumption of the proposed PLL at 10.6 GHz is 5.6 mW from a 0.8-V supply, which leads to the energy efficiency of 0.53 mW/GHz.

Figure 6 compares the performance of the proposed RO-based PLL with the prior arts, where the proposed PLL achieves the widest frequency range and the lowest period jitter with small active area.

References:

[1] C-K. Lee et al., "Dual-Loop Two-Step ZQ Calibration for Dynamic Voltage-Frequency Scaling in LPDDR4 SDRAM," JSSC 2018:2906-2916.

[2] S. Jung et al., "A 9.4MHz-to-2.4GHz Jitter-Power Reconfigurable Fractional-N Ring PLL for Multi-Standard Applications in 7nm FinFET CMOS Technology," A-SSCC 2019:87-90.

[3] M. Song et al., "A 0.009mm^2 2.06mW 32-to-2000MHz 2nd-Order ΔΣ Analogous Bang-Bang Digital PLL with Feed-Forward Delay-Locked and Phase-Locked Operations in 14nm FinFET Technology," ISSCC 2015:266-267.

[4] Y. Koo et al., "A Fully Integrated CMOS Frequency Synthesizer with Charge-Averaging Charge Pump and Dual-Path Loop Filter for PCS- and Cellular-CDMA Wireless Systems," JSSC 2002:536-542.

[5] J. Yun et al., "A 0.4-1.7GHz Wide Range Fractional-N PLL Using a Transition-Detection DAC for Jitter Reduction," A-SSCC 2020:S15-3.

[6] Z. Zhang et al., "A 0.25-0.4V, Sub-0.11mW/GHz, 0.15-1.6GHz PLL Using an Offset Dual-Path Loop Architecture with Dynamic Charge Pumps," Symp. VLSI 2019:C14-1.

979-8-3503-3004-5/23 $31.00 © 2023 IEEE

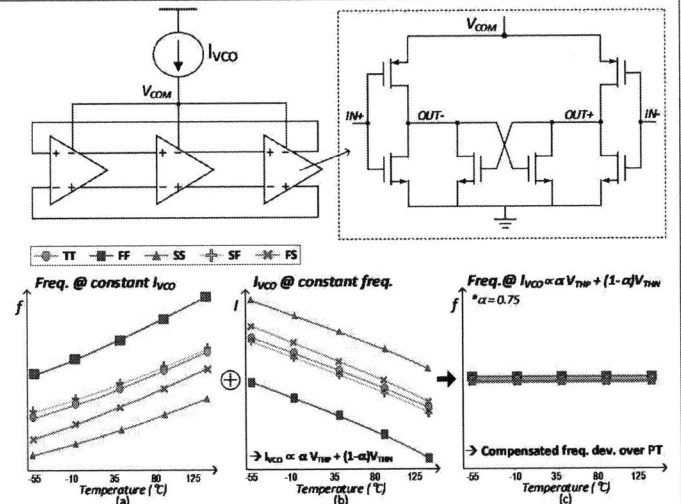

Fig. 1. Schematic diagram of the 3-stage RO, and its simulated PVT variations for (a) frequency at the constant I$_{VCO}$, (b) I$_{VCO}$ at the constant frequency, and (c) compensated frequency with (b)

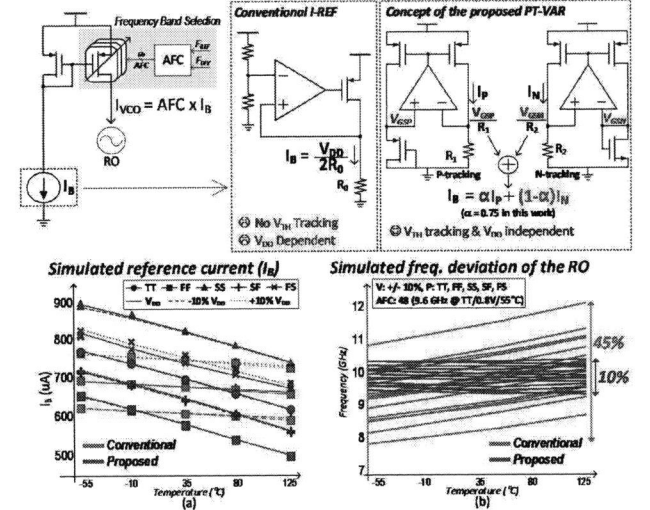

Fig. 2. Schematic diagrams of the conventional I-REF and the proposed PT-VAR; (a) the simulation results for each current, and (b) frequency deviation over PVT variations

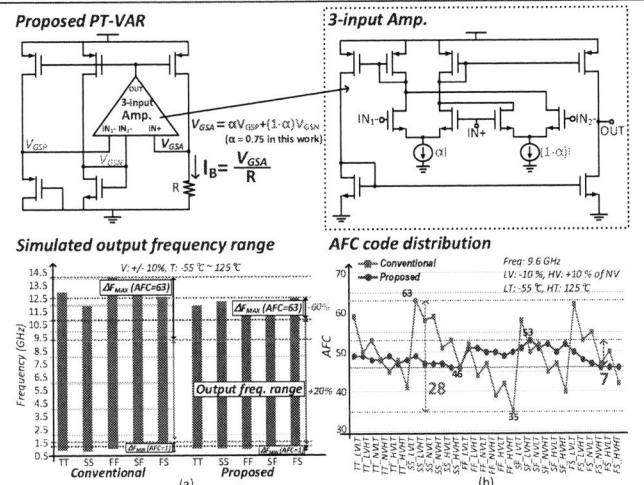

Fig. 3. Schematic diagrams of the proposed PT-VAR, and its simulation results compared with the conventional I-REF; (a) the output frequency range extension, and (b) the AFC code distributions over PVT variations

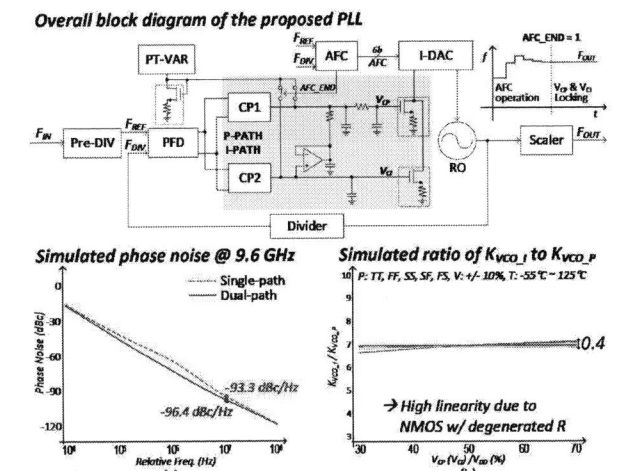

Fig. 4. Block diagram of the proposed PLL, and the simulated (a) phase noise of the RO with a dual-path loop filter compared to a single-path loop filter, and (b) the linearity of K$_{VCO}$ over PVT variations

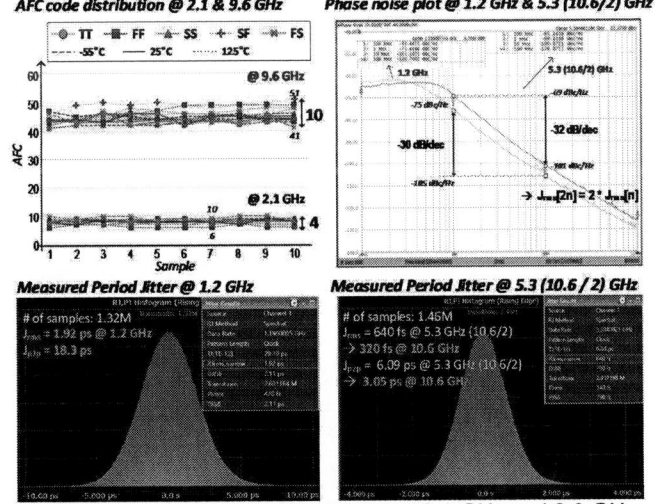

Fig. 5. Measured AFC code distribution at 2.1-GHz and 9.6-GHz frequency, and measured phase noise and period jitter at 1.2 GHz (f$_{MIN}$) and 5.3 GHz (f$_{MAX}$ / 2 = 10.6 GHz /2)

	This work	[6] J. Yun '20 ASSCC	[7] Z. Zhang '19 VLSI	[8] F. Ahmad '16 ISSCC	[9] K-Y Shen '18 T-CAS I
CMOS Tech.	3nm GAA	110nm	40nm	16nm	14nm FinFET
Supply (V)	0.8	3.3 / 1.5	0.25-0.4	N/A	0.6-0.95
PLL Type	Integer-N	Fractional-N	Integer-N	Fractional-N	Integer-N
Ref. Freq. (MHz)	26[1]	25	100	60	100
Output Freq. (GHz)	0.019[2]-10.6	0.4-1.7	0.2-1.6	0.5-9.5	0.8-4
Power Consumption (mW)	5.6 @ 10.6 GHz	21.1 @ 1.6125 GHz	0.17 @ 1.6 GHz	7.1 @ 3.25 GHz	2.6 @ 4 GHz
Energy Efficiency (mW/GHz)	0.53	13.09	0.11	2.18	0.65
Period Jitter (rms, ps)	0.32 @ 10.6 GHz	3.52 @ 1.6125 GHz	2.15 @ 1.6 GHz	0.45 @ 3.25G	0.91 @ 4 GHz
Active Area (mm^2)	0.012	0.284	0.0088	0.03	0.021
FoM$_{PJ}$ (dB)	-242.5	-215.9	-241.1	-238.5	-236.7
FoM$_{PJ,N}$ (dB)	-258.6	-234	-253.1	-255.8	-252.7

[1]Divided-by-2 from 52 MHz input clock for the frequency tuning resolution of the PLL
[2]Divided-by-64 from 1.2 GHz by the scaler
FoM$_{PJ}$ = 10 log$_{10}$ (Period Jitter^2 * Power)
FoM$_{PJ,N}$ = 10 log$_{10}$ (Period Jitter^2 * Power / (Output Freq. / Ref. Freq.))

Fig. 6. Performance summary

979-8-3503-3004-5/23 $31.00 © 2023 IEEE

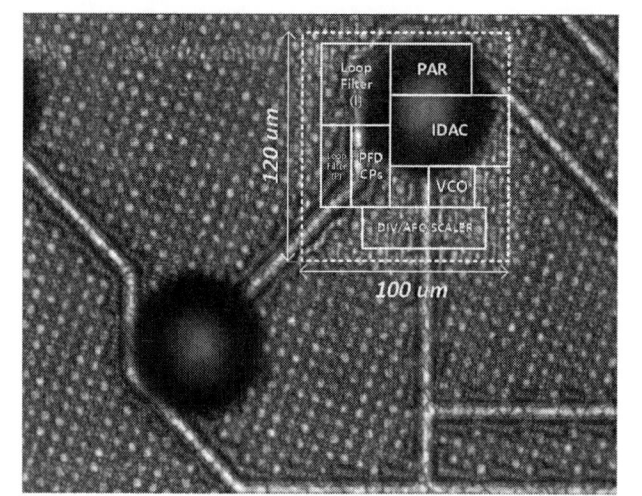

Fig. 7. Die micrograph

Additional References:

[7] F. Ahmad et al., "A 0.5-to-9.5GHz 1.2us-Lock-Time Fractional-N DPLL with ±1.25% UI Period Jitter in 16nm CMOS For Dynamic Frequency and Core-Count Scaling in SoC," ISSCC 2016:324-326

[8] K-Y Shen et al., "A Flexible, Low-Power Analog PLL for SoC and Processors in 14nm CMOS," Trans. Circuits and Systems-I 2018:2109-2117.

Fig. S1 (to be removed for final submission)

Fig. S2 (to be removed for final submission)

Fig. S3 (to be removed for final submission)

Fig. S4 (to be removed for final submission)

A 26-30GHz Digitally-Controlled Variable Gain Power Amplifier with Phase Compensation and Third Order Nonlinearity Cancellation Technique

Haoqi Qin[*], Junjie Gu[*], Hao xu[*], Weitian Liu[*], Kefeng Han[#], Rui Yin[*], Zongming Duan[^], Hao Gao[$], Na Yan[*]

[*]Fudan University, Shanghai 200433, China.

[#]Jiashan-Fudan Joint Research Institute, China

[^]East China Research Institute of Electronic Engineering, China

[$]Eindhoven University of Technology, the Netherlands

Phased array transceivers at Ka-band frequencies are of greater importance with the rapid development of satellite communications. This paper presents a two-stage variable gain power amplifier (VGPA) with high linearity and low phase variation in 40nm CMOS process. Accurate gain control is realized through digitally controlled current steering. The inter-stage compensation inductor suppresses phase variations among different gain states. Bias of each stage is chosen such that the third order distortions generated by each stage are of opposite polarities, which greatly improves the overall linearity of the cascading amplifier. Without additional trimming, the 26-30GHz VGPA achieves a total 3.5dB gain tuning range with 0.5dB resolution. The measured OP1dB is 14.4dBm and the RMS phase error is 0.29° at 28GHz.

Recent works have advanced Ka-band VGPA designs in various perspectives. Asymmetric neutralization capacitors and cross-coupled transistors are adopted in [1] that implements a VGA with 5-bit gain control and 0.2dB step. Accurate gain control is achieved with current steering scheme using replica transistors within a compact area [2]. However, both designs have limited linearity. Another two gain stages are placed afterwards to improve the linearity of a phase-inverting VGPA in [3], yet the more sophisticated structure increases area and power consumption.

Figure 1 illustrates the overall schematic and the corresponding passive EM devices of the proposed design. The variable gain stage utilizes common-gate configuration for the purpose of input matching with the phase compensation inductor at its output. The power stage is a common-source amplifier with neutralization capacitors. The two input traces in the output transformer of the PA are twisted, which intentionally utilize the coupling between inter-stage and output transformer. While the input and output transformers are tightly coupled to minimize insertion loss, the inter-stage matching transformer is loosely coupled to achieve a wider bandwidth.

Thanks to replica transistors, the current flowing through the variable gain stage remains constant across different gain states, leading to a linear relationship between the transconductance of current-steering transistors and the control code. Figure 2 compares the simulated gain step with and without the replica transistors. The replica transistors improve the gain accuracy by 0.2dB. However, the relationship between gain step and current steering ratio is only valid for small variations in channel width and will deviate from the original step size with larger transconductance adjustment. Demonstrated in Fig.2, the size of 0.5dB gain step at first code gives a larger RMS gain error due to the forementioned deviation. Thus, the sizes of current steering transistors are fine tuned to achieve an accurate linear-in-dB gain control.

Figure 2 also shows the equivalent circuits of the variable gain stage, and the transfer function of interest V_O/V_{in} is desired to exhibit a consistent phase response in different gain states. The common-gate input transistor is modeled as a controlled current source with parallel capacitance C_S, while the transistor output resistance is ignored for simplicity. Capacitor C_1 includes C_{gd} and C_{gs} of the subsequent stage. With a complementary operation between replica and current steering transistors, the numbers of on/off-state transistors remain unchanged, ensuring a constant C_1. Thus, the dominant source of phase variations in different gain states arises from the fluctuations in $1/g_{m,eff}$ that alters the transfer function V_m/V_{in}. Without L_C, the equivalent circuit would contain only a real pole that is dictated by $1/g_{m,eff}$, resulting in significant phase variations across different gain states. The introduction of L_C converts the one-pole network to a third-order network with conjugate poles that by proper design, can shield the phase response of the transfer function from the variation of $1/g_{m,eff}$ at certain desired frequencies. The simulation results in Fig.2 show that a properly chosen L_C minimizes the phase variation at 28GHz and the RMS phase error across frequencies of interest (26-30GHz) remains below 1°. This is over 4° improvement compared to the case without the compensation inductor.

Whereas the common-source amplifier often exhibits limited linearity, the linearity co-optimization adopted in this design allows the cascading amplifier to achieve higher linearity by purposely pre-distorting each stage. As illustrated by Fig.3, the effective third order transconductance can be neutralized by the different polarities of its counterpart in each stage. With the bias of the variable gain stage mainly determined by the requirement of gain control, the bias of the power stage is chosen for the desired nonlinearity cancellation. Simulation results in Fig.3 demonstrate the effectiveness of this approach that the overall gain becomes much less sensitive to the input power level. The output compression point is significantly improved as well. Moreover, with both bias voltages generated from on-chip current mirrors, this neutralization technique also has a sufficient PVT resilience without any extra calibration. Seen from Fig.3, the variation of OP1dB under typical corner and normal supply is less than 0.7dB among 0 to 100°C, while the deviation from typical corner is less than 1.5dB under different corners and supply voltages.

Besides circuit innovations, this design utilizes the mutual coupling between adjacent inductors that is often undesired to expand the bandwidth. Illustrated in Fig.4, there exists slight coupling between the secondary coil of inter-stage transformer and the primary coil of output transformer[4]. Since the signal is amplified after the power stage, k_{par} allows the signal at the output to influence the inter-stage transformer. This mutual coupling raises the effective quality factor of the inter-stage matching network and compensates the gain droop of frequency response around 26GHz. As shown in Fig.4, the transimpedance gain of the inter-stage matching network is boosted at lower frequencies when $k_{par}>0$, while a negative k_{par} leads to more loss at lower frequencies. In this design, the inputs of the output transformer are intentionally twisted to create the needed k_{par} for the desired pass-band frequency response with less ripple.

Fabricated in a 40nm CMOS process, the core area of the proposed VGPA is 0.64mm×0.22mm. Measurement results are summarized in Fig.5 with S-parameters, RMS gain and phase error, efficiency and the linearity. The measured and simulated S-parameters across 26-30GHz match closely. Normalized to the maximum gain state, 3.5dB gain tuning range with 0.5dB step is achieved. The RMS gain error is less than 0.35dB across the entire bandwidth with the minimum value of 0.05dB at 30GHz. The RMS phase error across the bandwidth remains less than 3.3° with the minimum value of 0.29° at 28GHz. The VGPA has an OP1dB of 14.4dBm and a peak PAE of 27.8%. The proposed design achieves state-of-art performance of high linearity and low phase error with moderate area and power consumption as shown in the comparison table in Fig.6.

References:

[1] Q. Zhang et al., "A Ka-Band CMOS Phase-Invariant and Ultralow Gain Error Variable Gain Amplifier With Active Cross-Coupling Neutralization and Asymmetric Capacitor Techniques," TMTT 2022.

[2] Y. Yi et al., "A Ka-band CMOS Digital-Controlled Phase-Invariant Variable Gain Amplifier with 4-bit Tuning Range and 0.5-dB Resolution," RFIC 2018.

[3] X. Zhang et al., "A 39-GHz Phase-Inverting Variable Gain Power Amplifier in 65-nm CMOS for 5G Communication," MWCL 2022.

[4] M. Vigilante and P. Reynaert, "On the Design of Wideband Transformer-Based Fourth Order Matching Networks for E-Band Receivers in 28-nm CMOS," JSSC 2017.

[5] S. Shakib et al., "2.7 A wideband 28GHz power amplifier supporting 8×100MHz carrier aggregation for 5G in 40nm CMOS," ISSCC 2017.

[6] J. -H. Tsai and C. -L. Lin, "A 40-GHz 4-Bit Digitally Controlled VGA With Low Phase Variation Using 65-nm CMOS Process," MWCL 2019.

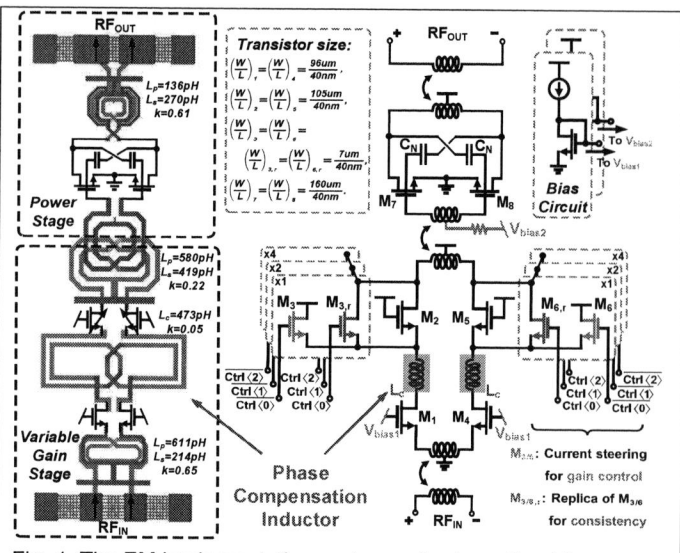

Fig. 1. The EM implementation and overall schematic of the proposed VGPA.

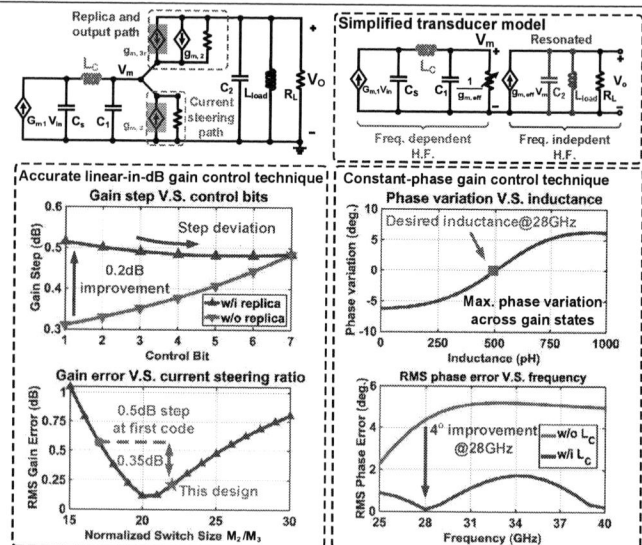

Fig. 2. Small-signal equivalent model of the variable gain stage and accurate linear-in-dB constant-phase gain control technique.

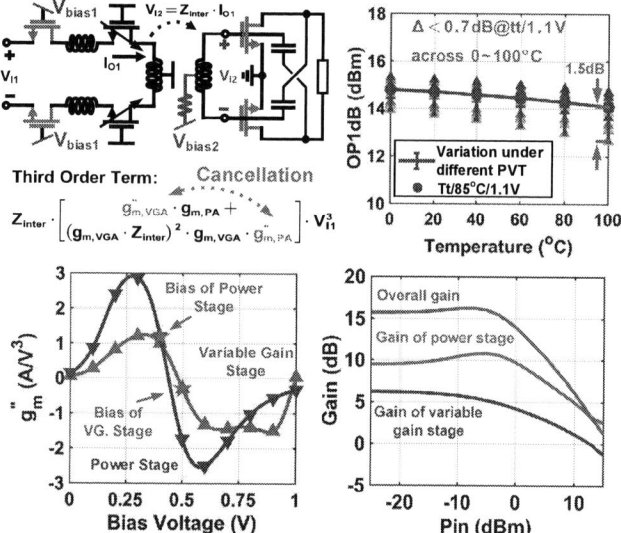

Fig. 3. Co-design of power stage with third order nonlinearity cancellation technique and simulation result under PVT conditions.

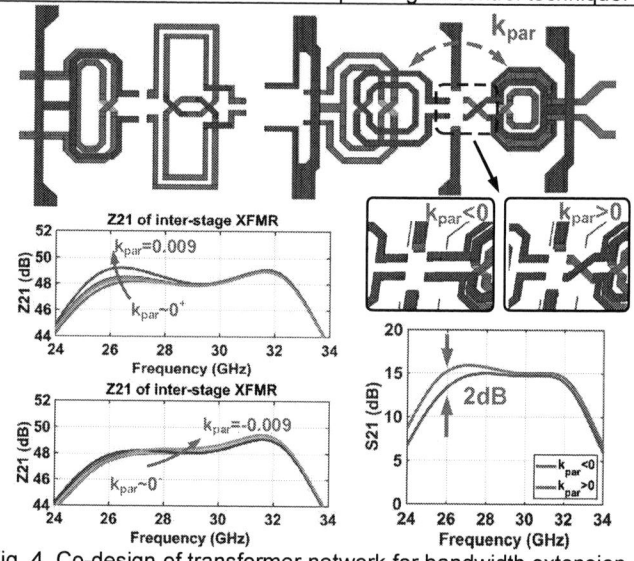

Fig. 4. Co-design of transformer network for bandwidth extension and simulation results under different k_{par}.

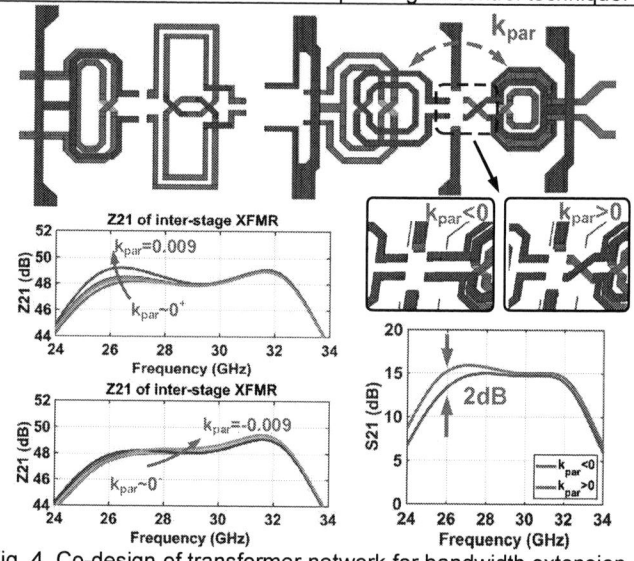

Fig. 5. Measured S-parameter, RMS error and linearity.

Ref.	TMTT '22[1]	RFIC '18[2]	MWCL '22[3]	ISSCC '17[5]	MWCL '19[6]	This work
Process	65 nm	65 nm	65 nm	40 nm	65 nm	**40 nm**
Frequency (GHz)	24-28	27-42	35-43.5	26-36	38-40	**26-30**
Gain/Range /Step (dB)	29.4/6.2 /0.2	9.6/7.8 /0.5	38.9/5.6 /0.8*	22.4/9 /1	22/16 /1	**15.9/3.5 /0.5**
RMS Gain error (dB)	<0.13	<0.3	N/A	<0.3*	N/A	**<0.35**
RMS Phase error (degree)	<0.92	<3.5	<3.6	<6*	<2.67	**<3.3**
OP$_{1dB}$ (dBm)	6.4	2.5	13.2	13.7	2.5	**14.4**
PAE$_{peak}$(%)	N/A	18*	34.5	33.7	N/A	**27.8**
P$_{dc}$ @ P$_{sat}$(mW)	103	15.6	150	N/A	38	**121**
P$_{dc}$ @ Static(mW)	N/A	N/A	N/A	30.3	N/A	**22.6**
Area (mm²)	0.14	0.083	0.294	0.23	0.37**	**0.18**

*Estimated from measurement result. **Including the pads.

Fig. 6. Comparison with the state-of-art Ka-band CMOS VGPAs.

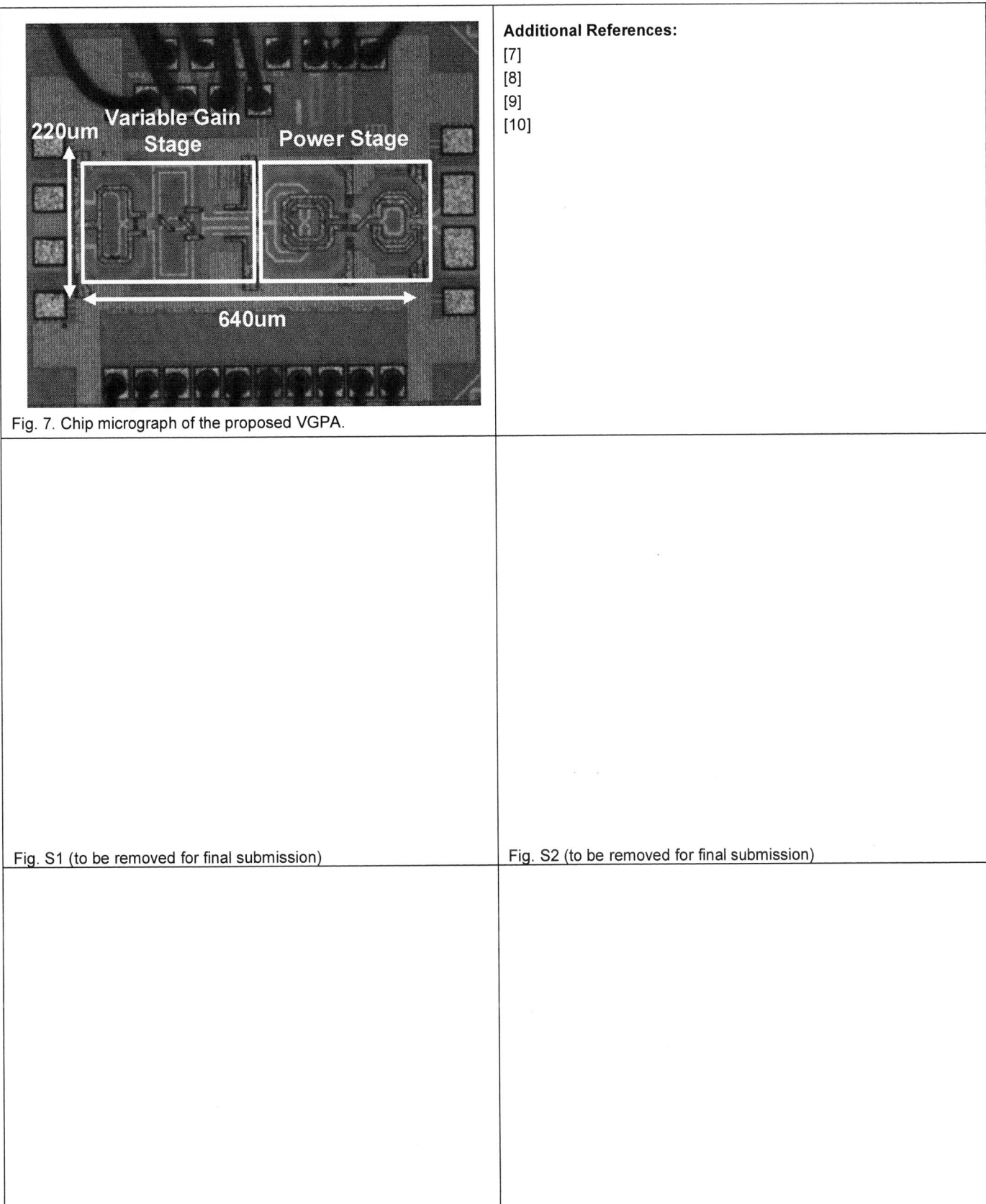

Fig. 7. Chip micrograph of the proposed VGPA.

Additional References:
[7]
[8]
[9]
[10]

Fig. S1 (to be removed for final submission)

Fig. S2 (to be removed for final submission)

Fig. S3 (to be removed for final submission)

Fig. S4 (to be removed for final submission)

LOG-CIM: A 116.4 TOPS/W Digital Computing-In-Memory Processor Supporting a Wide Range of Logarithmic Quantization with Zero-Aware 6T Dual-WL Cell

Soyeon Um, Sangjin Kim, Seongyon Hong, Sangyeob Kim, and Hoi-Jun Yoo

Korea Advanced Institute of Science and Technology (KAIST)

Recently, low-bit precision quantization for DNN has been proposed to reduce both data movement and computation power [1, 2]. This approach achieves high accuracy by employing logarithmic quantization (LOGQ) for input activation or weight rather than integer quantization (INT), enabling coverage of a wide range. Within Computing-In-Memory (CIM), LOGQ ensures that only 1 bit of the bit-serial CIM input is set to '1', resulting in lower power consumption. However, a physical shift alignment within the CIM is required each time to compute all CIM inputs '1' to allow the accumulation of bit-wise multiplications at once. Otherwise, as in previous CIM works [3, 4, 5], converting exponent CIM inputs into integer data leads to extended alignment cycle latency. In [5], it reduced the latency through the double-bit cells, but this approach wastes power as the bit-line (BL) pre-charge cannot be reused.

In this paper, we propose a LOG-CIM system that outperforms limited INT operations by additionally supporting LOGQ operations in CIM to achieve higher energy efficiency in inference and reduce total energy consumption in training while maintaining high accuracy. The LOG-CIM processor incorporates three key features: 1) Exponential Parallel Data Mapping (EPDM) supports shifted-alignment and accumulation for high energy efficiency, 2) a Zero-Aware DUal word-line (ZADU) cell uses only 6T cells to increase intra-local cell array (LCA) parallelism and reduces computation power by reusing pre-charge both BLs. 3) The Shift-Addition Peripheral (SAP) logic boosts energy efficiency by 1.6× when using the ZADU cell. As a result, the proposed LOG-CIM system achieves 116.4 TOPS/W using LOGQ4 weight and INT8 input activation while maintaining 75.44% inference accuracy on ImageNet @ResNet-50. This achieves 1.4× higher energy efficiency than when using INT4 weight, although the accuracy is similar to 75.87%. Furthermore, using INT4 and LOGQ4 during training reduced total energy by 85.6% while achieving the same accuracy as using only INT8 on ImageNet @ResNet-50.

The proposed LOG-CIM processor of Fig. 2 has an aggregation & activation core to accumulate partial-sums, a top controller, and 6 clusters where each cluster has 4 CIM macros. Each CIM macro is composed of 2 KB (512×32) ZADU cells, SAP, and adder trees. The LCA group contains 16 LCAs, and each LCA consists of 32 ZADU cells. Input activations are stored in ZADU cells and weights are fed horizontally through word-line (WLP, WLN). The input activation and weight are multiplied bit-wise at each LCA and the results are transferred to SAP. The resulting sign 1bit shift-addition of SAPs is accumulated in the 3bit 16-input adder tree. The sign controller sends the sign bit product of the input activation and weight to SAP for sign-magnitude addition. For LOGQ weight, the INT input activation is mapped bit-wise first and input channel-wise to accumulate shifted bits in an LCA group, and then spatial-wise to different LCA groups. For INT weight, INT input activation is mapped input channel-wise first in an LCA group as a conventional data mapping.

Fig.3 shows the read operation of the shifted-alignment input activation bit by weight within the LCA to accumulate the same output bit position in the adder tree in LOGQ. The corresponding cells are read from each LCA through the WL decoder in the order of small output bit positions, rather than read address sequentially. For example, when the smallest weight w_0, w_7, and w_8 is 2^2, 2^0, and 2^1, respectively, $a_{0[N]}$ should be accumulated with $a_{7[N+2]}$ and $a_{8[N+1]}$ in a CIM macro to obtain the result of $a_0 \times 2^2 + a_7 \times 2^0 + a_8 \times 2^1$. $a_{0[0]}$, $a_{7[2]}$, and $a_{8[1]}$ are read from each LCA in cycle 2 and sent to the adder tree as the 3rd LSB of partial-sum. Therefore, even if different weight exponent values are transferred to LCAs, it enables accumulation between different bit positions of each input activation by turning on different WL in each LCA. At this time, the input activation is mapped bit-wise first in intra-LCA, it is possible to accumulate different bit position data between LCAs. To perform shifted alignment only by adjusting the address, the WL decoder finds the smallest weight and determines which address in LCA should be turned on. It starts from the LSB of the output bit position and goes up to MSB sequentially. By matching weight shift values between LCAs similarly, EPDM enables parallel accumulation of shifted data, resulting in a minimal increase in operation cycles. As a result, EPDM reduces the execution cycle by 37% and increases energy efficiency by 51.5% compared to the previous data mapping with only 9.5% WL decoder area overhead in a macro.

The ZADU cell of Fig. 4 uses 6T cells to allow simultaneous computations of two cells within an intra-LCA, resulting in increased parallelism without requiring additional cell transistors such as [3]. WLP and WLN turn on two different cells simultaneously, and computation results are loaded onto the BLP and BLN, respectively. These results are converted into the AND of weight (w_0) and input activation (a_0) in cell 0 for BLP, and the NAND of weight (w_1) and input activation (a_1) in cell 1 for BLN. The two results are loaded into SAP and added so that the throughput of inter-LCA can be doubled. The ZADU cell uses both PMOS and NMOS in its pass-gates and pre-charges them with GND and VDD to reduce computation power, unlike traditional SRAM [5], all pass-gates are made of NMOS and pre-charged with only VDD. This enables the reuse of BL pre-charging for zero data cells in both BLP and BLN, as opposed to traditional SRAM enables only BLN. The ZADU cell reduces computation power by 88.9% with 89.3% of bit sparsity using sign-magnitude on ImageNet @ResNet-50 and reduces the operation cycle by 50% for INT and 27.1% for LOGQ.

Fig. 5 shows circuit schematics of SAP and its operation. SAP can enhance energy efficiency in LOGQ using ZADU cells. The SAP is designed with logic that receives the shift and sign bit from the WL decoder and sign bit controller, respectively, and adds the results of the BLP and BLN with or without a shift as needed. It enables reading input activation bit from LCA with a 1bit shift operation to reduce the operation cycle by 25.8%. For example, $a_{6[1]}$ could be shift added using SAP in cycle 1 because the bit difference between $a_{6[1]}$ and $a_{4[0]}$, $a_{5[0]}$ $a_{6[0]}$ is 1. Consequently, by pre-reading the unreadable cells caused by incorrect output bit positions, the overall operation cycle can be reduced using SAP. In the case of INT, a 1bit shift is not required, the shift signal is always zero. SAP increases energy efficiency from 2.55× to 3.17× with only 6.3% area and 8.5% power overhead. SAP was designed in between to fit the adder tree pitch.

Fig. 6 shows the ImageNet measurement result @ResNet-50 and the comparison of the LOG-CIM with previous digital CIM hardware. When using LOG4 weight for inference, the energy efficiency was 3.54× higher than using INT8 weight with similar accuracy. On the other hand, the accuracy increased by 1.8%p at a similar energy efficiency to using INT3 weight. During training, using INT4 and LOG4 to reach the same accuracy increases epoch compared to using only INT8, but consequentially saves 85.6% in relative cumulative energy (energy/epoch × total epoch). The LOG-CIM system using 6T cells for increased throughput achieved 5.9× higher memory density than [3] using 12T cells. Furthermore, the LOG-CIM system uniquely supports adaptive mapping INT and LOGQ within CIM.

Fig. 7 shows the chip photo of LOG-CIM, which is fabricated in a 28-nm CMOS technology. It occupies 5.832 mm² with a total of 48 KB on-chip CIMs. It can achieve 192 TOPS/W using conventional data mapping at INT4 and 116.4 TOPS/W using EPDM at LOGQ4 weight and INT8 input activation at 120 MHz clock and 0.85 V supply voltage. The LOG-CIM processors support a wide range of LOGQ in CIM to achieve high accuracy using low-bit precision results in 1.4× higher energy efficiency or 85.6% lower total energy in inference and training, respectively, compared to CIM with only supporting INT.

Acknowledgment: This work was supported by Samsung Electronics Co.,Ltd (IO201207-07799-01). The chip fabrication and EDA Tool were supported by the IC Design Education Center (IDEC).

References:

[1] M. Elhoushi *et al.*, "Deepshift: Towards multiplication-less neural networks," CVPR 2021:2359-2368.

[2] B. Chmiel *et al.*, " Accurate Neural Training with 4-bit Matrix Multiplications at Standard Formats," ICLR 2023.

[3] F. Tu *et al.*, "ReDCIM: Reconfigurable Digital Computing-In-Memory Processor With Unified FP/INT Pipeline for Cloud AI Acceleration," JSSC 2022: 243-255.

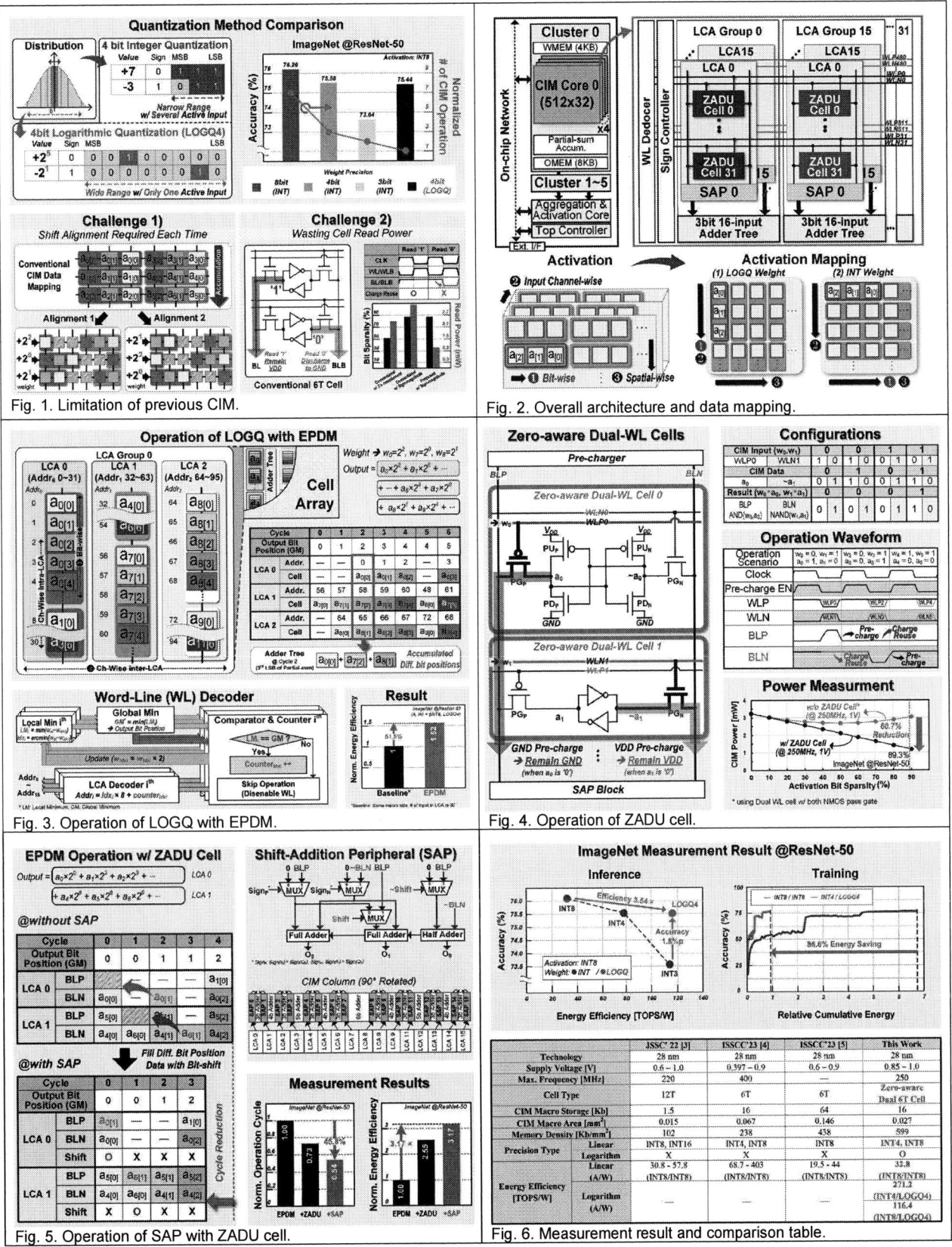

Fig. 1. Limitation of previous CIM.

Fig. 2. Overall architecture and data mapping.

Fig. 3. Operation of LOGQ with EPDM.

Fig. 4. Operation of ZADU cell.

Fig. 5. Operation of SAP with ZADU cell.

Fig. 6. Measurement result and comparison table.

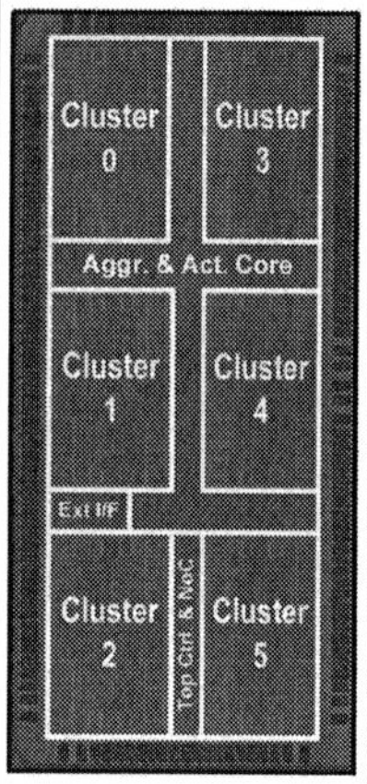

		Specifications
Technology		28nm Logic CMOS
Die Area	Core	103.4 μm × 258.4 μm
	Total	1.62 mm × 3.6 mm
Precision	INT	INT4, INT8
	Logarithm	LOGQ4
Supply Voltage		0.85 – 1.0 V
Frequency		~ 250 MHz
Power Consumption [mW]	Macro	0.4 – 1.5[*] (@ 120 MHz, 0.85V)
	System	23.4 – 49.8 (@ 120 MHz, 0.85V)
Energy Efficiency** [TOPS/W]	Integer	192 (INT4/INT4)
	Logarithm	92 – 306[**] (INT4/LOGQ4) 66 – 131[**] (INT8/LOGQ4)

[*] Activation Sparsity 0 ~ 90%, [**] Weight Difference 0 ~ 100%

Fig. 7. Chip photograph and performance summary.

Additional References:

[4] J. Yue *et al.*, "A 28nm 16.9-300TOPS/W computing-in-memory processor supporting floating-point NN inference/training with intensive-CIM sparse-digital architecture." ISSCC 2023: 1-3.
[5] A. Guo *et al.*, "A 28nm 64-kb 31.6-TFLOPS/W digital-domain floating-point-computing-unit and double-bit 6T-SRAM computing-in-memory macro for floating-point CNNs," ISSCC 2023: 128-130.

A 2 × 24Gb/s Single-Ended Transceiver with Channel-Independent Encoder-Based Crosstalk Cancellation in 28nm CMOS

Hongzhi Wu, Weitao Wu, Liping Zhong, Xuxu Cheng, Xiongshi Luo, Zhenghao Li, Dongfan Xu, and Quan Pan

Southern University of Science and Technology

The demand for ever-increasing throughput has impelled the swift growth of I/Os with finite areas and higher data rate per pin. To meet this demand, the design of passive links needs to be more complex, leading to a considerable rise in hardware expenses [1]. The single-ended multiple input multiple output (SE-MIMO) scheme discussed in [2], thus becomes an attractive solution, which aims to transmit two single-ended (SE) signals in a pair of tightly coupled differential channels. This scheme halves the bandwidth requirement and allows a 2× increase in routing density in a compact area. It means that higher data rates can be achieved with the existing hardware infrastructure, thus reducing costs. However, it would induce significant crosstalk. The crosstalk among different pairs of differential channels was typically eliminated through careful hardware layout [3]. Nevertheless, the crosstalk within a pair of closely coupled differential channels is still a thorny challenge for the existing advanced crosstalk cancellation (XTC) techniques. The crosstalk cancellation and signal reutilization technique in [5] is sensitive to the physical characteristics of different channels. In [4], [7], realizing pre-equalization with limited precision at TX to eliminate crosstalk will lead to excess noise tails. The Chord signaling, which transmits N-bit data over N+1 channels, is proposed to suppress crosstalk [6]. However, it lowers the pin efficiency by a factor of (N+1)/N. Therefore, an efficient and channel-independent XTC method is highly desirable to support SE-MIMO applications in various scenarios.

This paper proposes an encoder-based crosstalk cancellation scheme to support SE-MIMO applications. The encoder-based XTC (EB-XTC) scheme decomposes the signals into two orthogonal transverse electromagnetic (TEM) waves through the TX encoder and RX decoder. Therefore, crosstalk can be suppressed without channel dependence. Besides, the EB-XTC scheme relieves the pressure of the equalizer and is compatible with other equalization techniques, facilitating low hardware costs. Combining fractional-spaced FFE (FS-FFE) and CTLE, this EB-XTC scheme is validated in a 2×24Gb/s SE-MIMO transceiver with tightly coupled differential channels.

The proposed link in Figure 1(a) decomposes 2-bit signals into two orthogonal TEM waves that propagate on separate channels and the crosstalk can be eliminated without compromising the pin efficiency. To evaluate the scheme's effectiveness, this paper conducts a mathematical analysis using the transformation matrix H and the channel's S-parameters in Figure 1(b). As shown, the output signal matrix is a product of a coefficient diagonal matrix and the input signal matrix. It means that two signals are orthogonal during transmission and theoretically the crosstalk can be canceled, regardless of the characteristics of SE channels. Therefore, the proposed scheme demonstrates channel independence. Moreover, it can be easily extended to scenarios involving multiple channels.

The simulated S-parameters are provided to further assess the performance of the proposed scheme in Figure 1(c). Two channels are adopted for evaluation, where channel1 (CH1) is well designed and exhibits good symmetry, while channel2 (CH2) is intentionally designed to have asymmetry. This allows us to verify the proposed scheme's crosstalk cancellation capability while further assessing channel symmetry's impact on the XTC performance. With the proposed scheme, the crosstalk is decreased from −2dB to −65dB for CH1, and from −10dB to −26dB for CH2, respectively. Moreover, the SE IL is decreased from 15dB to 2dB for CH1, and from 17dB to 10dB for CH2, respectively. It proves that the EB-XTC scheme can not only eliminate the crosstalk but also alleviate the SE IL even under CH2 with poor symmetry so that achieving a differential-like transmission, and the equalization requirement for the system can be relaxed dramatically.

Figure 2 shows the TRX architecture. In the TX, an external SE clock is divided into quarter-rate clocks to feed the 4:1 MUXs. The 4:1

MUXs in A and B paths serialize the data from the on-chip PRBS generator and provide full-rate data to the pre-driver. The source-series-terminated (SST) driver consists of 32 parallel-connected slices, which helps coefficient controls of the 3-tap FFE. In the RX, the data path includes an S2D, a decoder, a two-stage CTLE, a VGA, and an output buffer. The RX employs the S2D as the first stage. Besides, the succeeding 5-stage differential circuits are adopted to suppress undesired noise. The CTLE only needs to provide a 6dB equalization tuning range at 12GHz since equalization pressure is relieved by the proposed scheme. The VGA provides a 6dB tuning range for the system gain.

Figure 3 shows the details of encoding and decoding circuits. In the encoder, the SST driver encodes data by setting the weight of signals A and B to 1:1 and selecting the polarity of signal B through a switch at the low-speed node without signal integrity degeneration. The decoder uses a current-mode logic adder to perform the decoding operation and suppress undesired noise. The polarity of the A-B path data can be selected by controlling the tail current.

The details of the TX data path and the 4:1 MUX are shown in Figure 4(a). The reconfigurable FS-FFE provides a flexible compensation range, thus making up for the limited accuracy of the SST driver-based FFE. Additionally, the reconfigurable FFE tap scheme further increases the adaptability of the TX equalizer by adopting phase-adjustable sample clocks. The timing diagram of the 4:1 MUX operation and an example of FFE post-tap data generation are shown in Figure 4(b). The high-resolution voltage-controlled delay line adjusts the phase spacing of the sample clock with a 60ps tuning range across different PVT corners. As shown in Figure 4(c), benefiting from the reconfigurable FS-FFE, the TX output minimum eye height and width are increased by 22mV and 7.1ps, respectively.

Both TX and RX chips are fabricated in 28nm CMOS. The TRX performance is measured via CH1 and CH2, with 15dB and 17dB SE IL at 12GHz, respectively. The cables and PCB traces of the measurement setup introduce an additional loss of 3dB. The IL to crosstalk ratio of the two channels at Nyquist frequency is −13dB and −7dB, respectively. To demonstrate the XTC capability, the Arbitrary Waveform Generator (Keysight M8196A) is utilized to transmit two ideal NRZ signals into the channels. Subsequently, the output eyes with crosstalk are measured. As shown in Figure 5(a), due to the crosstalk invasion from the adjacent channel, the eyes are completely closed. In contrast, the 24Gb/s TRX output eyes with the proposed EB-XTC scheme are shown in Figure 5(b). It realizes a clear eye-opening with 1e-12 BER even under CH2 with poor symmetry, and the eye heights and widths are 68mV and 0.34UI for CH1, and 32mV and 0.33UI for CH2, respectively. The nearly equal eye width indicates no additional jitter caused by crosstalk measured with CH2 (asymmetry channel). It means that the asymmetry of the channel does not degrade the performance of the proposed XTC.

The total active area of TRX is 0.03mm² per lane, and no passive inductor is used in TX. The entire TRX consumes 69.5mW per lane, and its power efficiency is 2.9pJ/b. Figure 6 summarizes the TRX performance and compares it with state-of-the-art works. The SE MIMO TRX with EB-XTC scheme demonstrates superior crosstalk cancellation capability with channel independence, the best FoM value, and provides a promising method for high-density I/O communications.

Acknowledgement:

This work was supported in part by National Key Research and Development Program under Grant 2022YFB2803303, and in part by the National Natural Science Foundation of China under Grant 62074074.

References:

[1] B. Dehlaghi and A. C. Carusone, "A 0.3 pJ/bit 20 Gb/s/wire parallel interface for die-to-die communication," JSSC, Nov. 2016.

[2] L. Zhong et al., "A 2×50 Gb/s Single-Ended MIMO PAM-4 Crosstalk Cancellation and Signal Reutilization Receiver in 28 nm CMOS," ESSCIRC, Sep. 2022.

[3] O. Taehyoun, H. Ramesh, "A 12-Gb/s Multichannel I/O Using MIMO Crosstalk Cancellation and Signal Reutilization in 65-nm CMOS," JSSC, Jun. 2013.

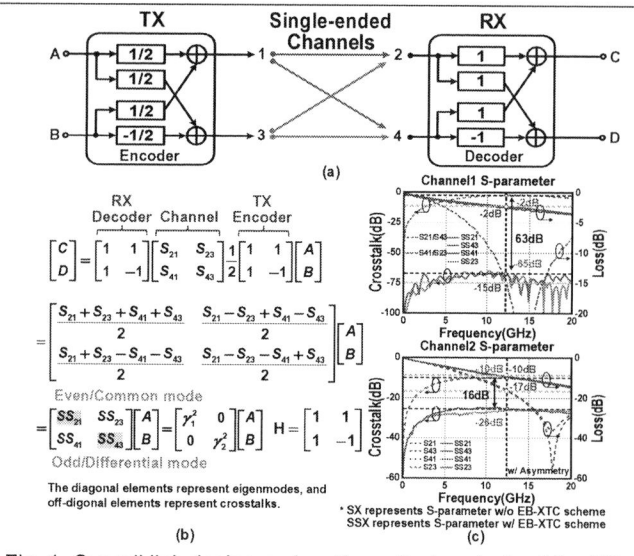

Fig. 1. Overall link design and mathematical analysis of the EB-XTC scheme.

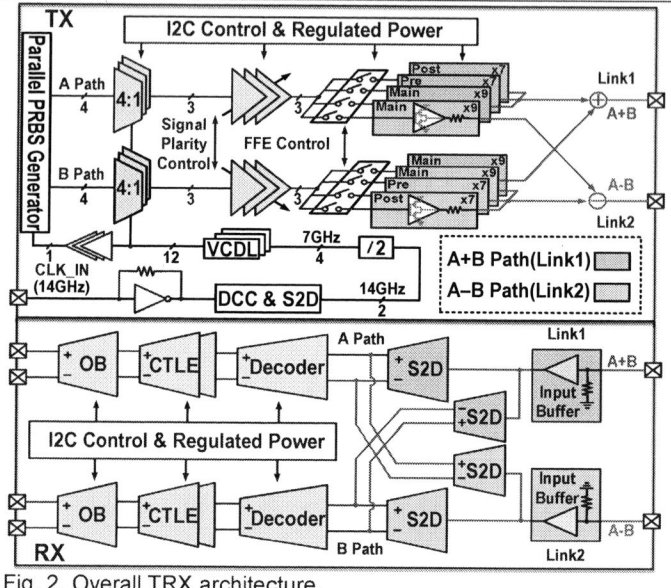

Fig. 2. Overall TRX architecture.

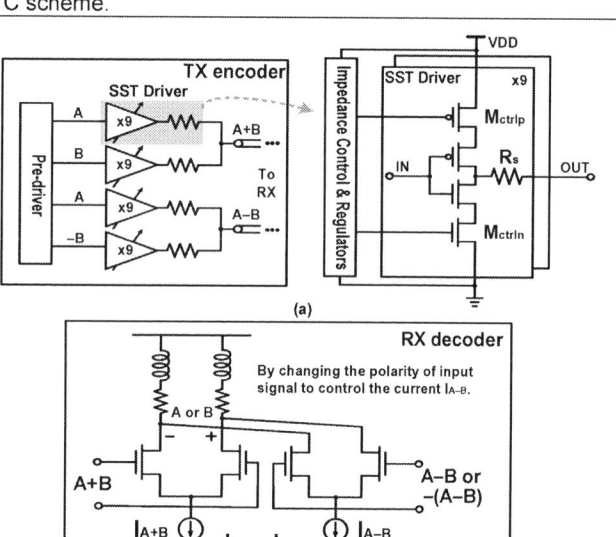

Fig. 3. Schematics of (a) the TX encoder and (b) the RX decoder.

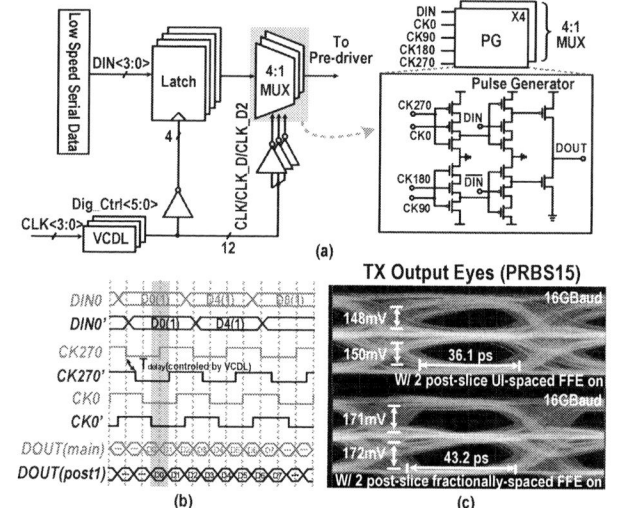

Fig. 4. (a) Schematics of the TX data path and the 4:1 MUX, (b) timing diagram of the 4:1 MUX operation and FFE post-tap data generation, and (c) measured TX output eyes w/ and w/o FS-FFE.

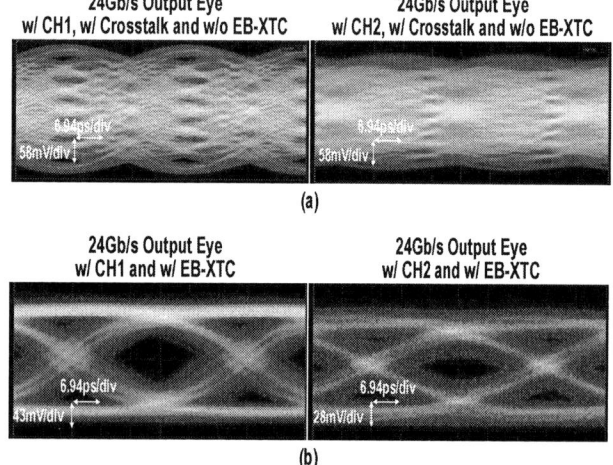

Fig. 5. (a) Measured output eyes with two channels, crosstalk, and without EB-XTC scheme, (b) measured output eyes with two channels and EB-XTC scheme.

	JSSC'13 [4]	JSSC'18 [5]		JSSC'20 [6]	ISSCC'20 [7]	This work
Process	65nm CMOS	32nm SOI CMOS		16nm FinFET	65nm CMOS	28nm CMOS
Supply (V)	1.2	1		1.1	1.2	1
Data rate per pin (GBaud/pin)	7.5	7		20.83	4	24
Signaling	NRZ	NRZ		CNRZ	NRZ	NRZ
Pin Efficiency	100%	100%		83.30%	100%	100%
SE IL at Nyquist Frequency (dB)	5.9	30	28	6	10.2	18* 20**
Crosstalk at Nyquist Frequency (dB)	−4.5	−30	−34	/	−16.4	−2* −10**
IL-to-Crosstalk Ratio (dB)	−1.4	0	6	/	6.2	−13* −7**
Equalization	2-tap FFE	CTLE + DFE		CTLE	2-tap sub-UI FFE	3-tap FS-FFE + CTLE
XTC Type	TX FIR	RX IIR +DFE		Encoded	TX FIR	Encoded
CIJ reduction ratio***	75%	63%		/	78%	100%
BER	<1e-12	<1e-12		<1e-15	<1e-12	<1e-12
Energy Efficiency (pJ/bit)	3.5	8		1	1.5	2.9
FoM**** (pJ/bit/dB)	0.60(TX)	0.27(RX)		0.17	0.15	0.14

*CH1 **CH2
***Estimated from the reduction ratio of the crosstalk noise amplitude
****Figure of Merit: Energy Efficiency (pJ/bit) / SE IL at Nyquist frequency (dB)

Fig. 6. Comparison table.

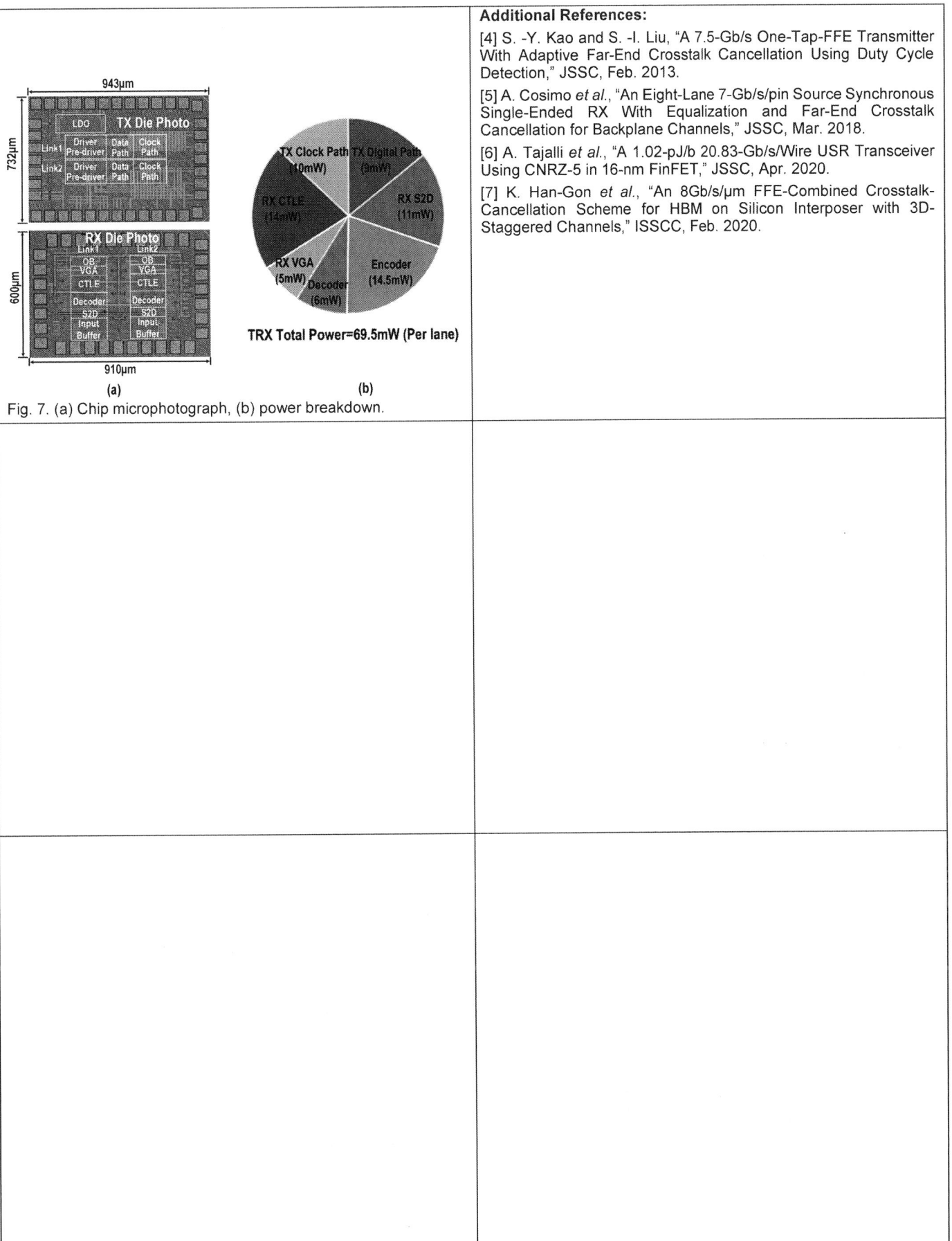

Fig. 7. (a) Chip microphotograph, (b) power breakdown.

Additional References:

[4] S. -Y. Kao and S. -I. Liu, "A 7.5-Gb/s One-Tap-FFE Transmitter With Adaptive Far-End Crosstalk Cancellation Using Duty Cycle Detection," JSSC, Feb. 2013.

[5] A. Cosimo et al., "An Eight-Lane 7-Gb/s/pin Source Synchronous Single-Ended RX With Equalization and Far-End Crosstalk Cancellation for Backplane Channels," JSSC, Mar. 2018.

[6] A. Tajalli et al., "A 1.02-pJ/b 20.83-Gb/s/Wire USR Transceiver Using CNRZ-5 in 16-nm FinFET," JSSC, Apr. 2020.

[7] K. Han-Gon et al., "An 8Gb/s/µm FFE-Combined Crosstalk-Cancellation Scheme for HBM on Silicon Interposer with 3D-Staggered Channels," ISSCC, Feb. 2020.

A 10TFLOPS Datacenter-Oriented GPU with 4-Corner Stacked 64GB Memory by The Means of 2.5D Packaging Technology

Shuang Wang[1,2], Weiliang Chen[1,2], Xueqing Li[1],Leibo Liu[1],Huazhong Yang(IEEE Fellow) [1]

[1]Tsinghua University, Beijing, China,

[2]MetaX Integrated Circuits (Shanghai) Co., Ltd.

GPU has a large number of parallel computing units and fast performance growth, which is suitable for datacenter[1]. In order to improve GPU performance, many schemes have been proposed such as adding computing core, increasing frequency and memory capacity[2][3]. But as Moore's Law slows down, GPU becomes larger and larger. The ultra large chip is very expensive to be manufactured because the yield is too low. This paper demonstrates that MCM(Multi-Chip Module) design and 2.5D packaging technology could be applied to address the issue of large chip area. Compared to the traditional monolithic GPU, 4-corner memory-stacked GPU by the means of 2.5D packaging technology(Die on Wafer on Substrate with Interposer) is adopted. New scheme also uses HBM2 based on TSV(Through Silicon Via) technology with area benefits. Secondly, 3+1 levels(DIE-Interposer-PKG-PCB) circuit design and 4 mark areas alignment methods are used to solve the SI(Signal Integrity) problem. And packaging IR-Drop is optimized by cell flipping and special LVT(Low Voltage Threshold) cell selection. A data center GPU is designed with performance exceeding 10TFLOPS (FP32) and 64GB memory. The result illustrates these technologies can reduce area and increase yield of datacenter GPU effectively.

Figure.1. shows the proposed 4-corner memory-stacked MCM scheme. Using the negative binomial yield model, the yield of an individual Die can be calculated with critical area A as[6]:

$$Y_{DIE} = \frac{1}{(1+\frac{AD}{\alpha})^\alpha} \qquad (1)$$

Where, Y_{DIE} represents Die yield, A represents the area of Die, D represents defect density. where α is process dependent clustering parameter, generally ≥ 10. It can be seen that the yield is inversely proportional to the area. To solve the problem of large area, it is proposed to adopt MCM design[5]. The monolithic GPU chip is divided into three Dies: Core Die, Memory Die and Interconnect Die. X represents the core die, M represents the memory Die, and Z represents the interconnection die between the core die and memory Die. Then, multiple Die integration is achieved through 2.5D advanced packaging technology.

The ideal scheme is tiling the memory and placing it around the Core Die shown in left figure. This solution is convenient for physical implement. But this scheme takes two versions of HBM2 Die simultaneously and makes it more expensive than using one version. In theory, stacking memory on top of the DIE has the smallest area. However, thermal management will be more difficult to handle. Thus, 4-corner memory-stacked scheme is adopted. We lay the HBM2 Die flat in the left and right corners. This scheme splits the monolithic DIE into 7DIEs.DIE0 is core DIE while DIE1 and DIE2 are interconnection DIEs. Place one HBM2 Die on each of the four corners. Each HBM2 Die has 8 stacked HBM DRAMs with higher bandwidth and smaller size. Due to splitting a single chip into multiple Dies, additional interconnection between Dies is required. DIE1 and DIE2 are the interconnection functions between the core Die and HBM2 Die.

Figure.2. 2.5D packaging technology based on high density interposer[4]. An interposer has been added between the DIE and substrate to achieve multi DIE integration. Due to its flexibility and scalability, it is widely used in GPU, CPU, AI(artificial intelligence) and other fields.

Figure. 3. shows top view and side view of 2.5D packaging with four corner stacking of memory. It can be seen that the packaging is divided into three layers including Substrate, interposer and DIE. There are many bumps on the Substrate, connected to the second layer of interposer. There are many connections inside the interposer, which are connected to the multiple DIEs on the third layer. The DIEs on the third layer are laid flat and HBM DIE placed in the four corners.

Each HBM DIE is stacked with 8 layers of DRAM. Each layer of DRAM is interconnected through TSV.

Figure.4.shows the 3+1 levels(DIE-Interposer-PKG-PCB) circuit design for SI. Beside DIE, PKG and PCB levels, It is necessary to place resistors and capacitors on the interposer. These circuits will control the transmission quality of the signal by adjusting the parameters of the resistors and capacitors.

Meanwhile, location alignment between the DIE and interposer must be ensured. we set up 4 alignment mark areas on DIE and interposer. it is necessary to place the GDS(Graphic Database System) samples in the four alignment point areas of the DIE and interposer. The coordinates of these alignment points include:1) At the horizontal X coordinate of DIE0,such as L2, L5, L7, and L10. 2) In the vertical Y coordinate of DIE0,such as L1, L4, L8, and L11.3) At the vertical Y coordinate of interposer, such as L3, L6, L9, and L12.After integrating DIE into the interposer, we must check whether the sample patterns in the alignment point area of DIE and interposer are consistent. If consistent, it indicates that the DIE and interposer are aligned.

Figure.5. shows test result of chip SI. The signal is sent out from the DIE ,passing through the interposer, substrate, PCB board, and arriving the HBM finally. By detecting the size of the eye diagram, the eye width was obtained, which is approximately 187ps. The signal transmission is good and meets SI requirements.

Figure.8. shows how to solve the problem of chip IR. We use a power supply(VDD) of 0.75V.Through cell libraries selection, deviation of total IR is $\pm$ 10% and power IR is $\pm$ 5%. It means range of VDD is [0.675V, 0.825V].In order to meet the high performance requirements, we have adopted a high frequency design, with frequency exceeding 1GHz. Deviation of DIE IR is controlled at $\pm$ 3%.Thus,range of IR_{PKG} is 10% -5% -3%=2%. Due to the use of 2.5D packaging technology, interposer IR was introduced:

$$IR_{PKG} = IR_{Interposer} + IR_{C4bump} + IR_{BGA} \qquad (2)$$

$IR_{Interposer}$ must be controlled at $\pm$ 1%, and IR_{C4bump}+IR_{BGA} at $\pm$ 1%.1)we distribute cells on the interposer to correct location avoiding high density of cell. 2)break up cells that are flipped at the same cycle and place cells that are not flipped at the same cycle together to avoid local IR issues. However, it doesn't increase cell density. 3) Choose the special cell type. Due to the limited number of non-functional cells in interpose, From the actual physical implementation, SVT cell is more conducive to reducing IR than LVT cell.

This paper demonstrates three key technologies for data center GPU deign: MCM design, 2.5D packaging technology, and integrate the high bandwidth storage chip HBM2 integration. A monolithic GPU chip is split into multi Dies by MCM architecture. At the same time, interposer is used between the wafer and substrate to interconnect multi Dies and HBM2, while HBM2 uses TSV technology for interconnection.

The area of monolithic GPU chip exceeds 800mm ^ 2 approximately. The new solution decreases the area of core Die to 700mm ^ 2. And the area of interconnect function DIE area is about 40mm ^ 2, while interposer area reaches 1500mm ^ 2. According to the yield model, the new scheme yield needs to consider the yield of interposer[7],thus we have

$$Y_{chip} = Y_{core\ DIE} * Y_{interconnet\ DIE} * Y_{Interposer} \qquad (3)$$

The theoretical yield has increased from 20% to nearly 30%. The results show that the 4-corner memory-stacked GPU architecture can effectively reduce chip area and increase yield.

References:

[1] John, Mike, *et al.*, "GPU Computing." Proceedings of the IEEE 2008.

[2] Jack, Edward, *et al.*, "The A100 Datacenter GPU and Ampere Architecture." ISSCC 2021.

[3]Sal Dasgupta,Teja Singh,Ashish Jain. "Radeon RX 5700 Series: The AMD 7nm Energy-Efficient High-Performance GPUs. " ISSCC 2020.

979-8-3503-3004-5/23 $31.00 © 2023 IEEE

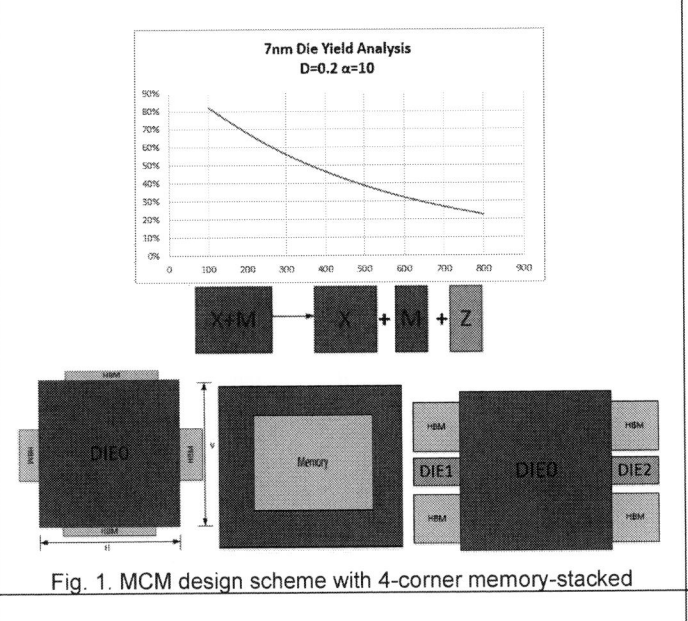

Fig. 1. MCM design scheme with 4-corner memory-stacked

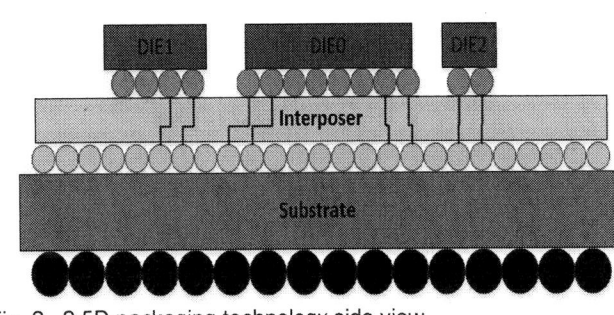

Fig. 2. 2.5D packaging technology side view

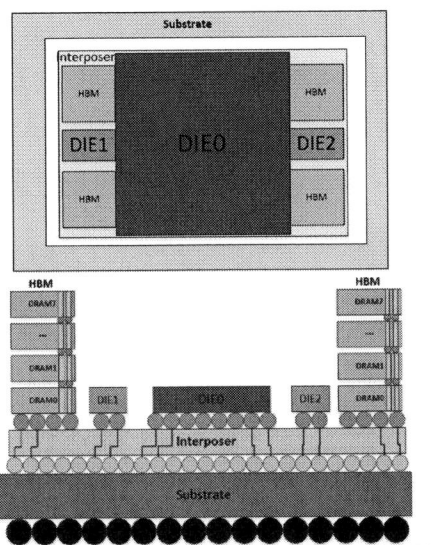

Fig. 3. 2.5D packaging with 4-corner memory-stacked GPU

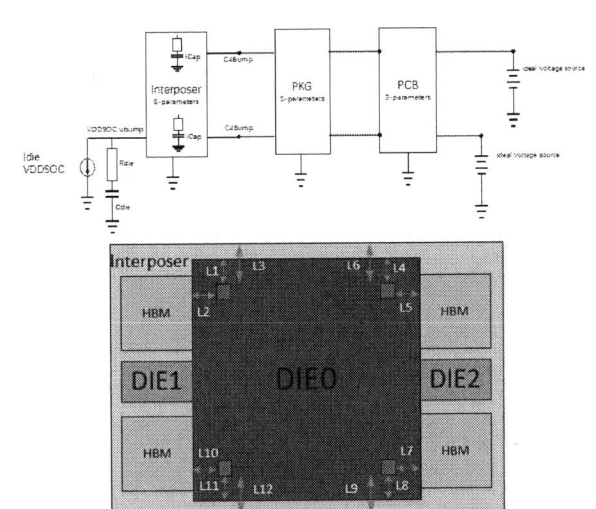

Fig. 4. 3+1 levels(DIE-Interposer-PKG-PCB) Circuit design and 4 mark areas alignment for SI design

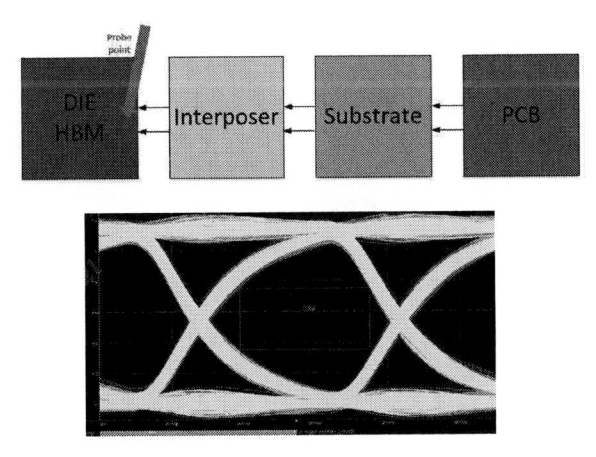

Fig. 5. Eye diagram of HBM in GPU system

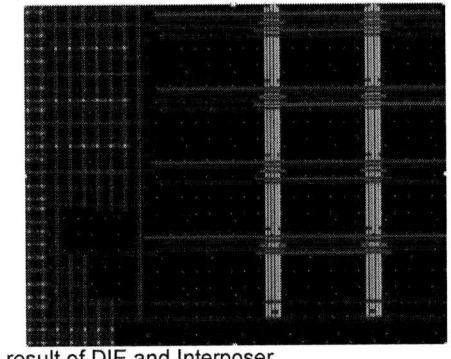

Fig. 6. IR result of DIE and Interposer

Frequency	1.1GHz
Die Size	700mm^2
Interposer Size	1500mm^2
FP32	>10TFLOPS
Memory Size	64GB HBM2

Fig. 7. DIE packaging photo and summary of performance

Additional References:

[4] P. K. Huang, *et al,* "Wafer Level System Integration of the Fifth Generation CoWoS-S with High Performance Si Interposer at 2500 mm2." IEEE 71st Electronic Components and Technology Conference 2021.

[5] Gabriel H. Loh, *et al.,* "Understanding Chiplets Today to Anticipate Future Integration Opportunities and Limits." DATE 2021.

[6] Dylan Stow; Yuan Xie; Taniya Siddiqua; Gabriel H. Loh. "Cost-Effective Design of Scalable High-Performance Systems Using Active and Passive Interposers." ICCAD 2017.

[7] Mudasir Ahmad; Javi DeLaCruz; Anu Ramamurthy. "Heterogeneous Integration of Chiplets: Cost and Yield Tradeoff Analysis." 23rd International Conference on Thermal, Mechanical and Multi-Physics Simulation and Experiments in Microelectronics and Microsystems (EuroSimE).

An 890 µW Multichannel Injection-Locked OOK Transmitter with 23% Global Efficiency and 22 pJ/bit Energy Efficiency

Sheng-Kai Chang, Shao-Ting Chang, Zhi-Wei Lin, and Kuang-Wei Cheng

National Cheng Kung University, Tainan, Taiwan

The Internet of Things (IoT) has witnessed significant advancements, facilitating wireless connectivity among numerous sensor nodes and edge devices. The demand for energy-efficient wireless communication has led to the development of ultra-low-power (ULP) wireless transmitters (TXs) to address the power consumption concerns associated with edge devices. Frequency-multiplying TXs [1-4] minimize power dissipation by employing edge-combining power amplifiers (ECPAs) for frequency translation, thereby reducing the number of blocks operating at the carrier frequency. Injection-locked ring oscillators (ILROs) are utilized to achieve equally-spaced edges required for the ECPA, ensuring area/energy efficiency and favorable phase noise characteristics [1-2]. However, conventional injection locking supports only a single channel due to the use of a fixed reference. Moreover, the power consumption of frequency synthesis in ULP TXs can be comparable to or even greater than the power consumed by the PA. This work presents a ULP OOK TX with a channelization capability and high global efficiency.

The edge-combined frequency variation caused by single-phase injection locking is shown in Fig. 1 for the illustrative case of a 3-stage RO. The pulse train (P_{ref}) generated from the clean reference signal is injected to realign the phase edge (E_1). However, the injection stage introduces an additional phase shift into the oscillation loop, causing phase deviation from the desired symmetry and resulting in significant frequency variations in the edge-combined output (V_o). The unequal spacing of phases leads to large reference spurs and frequency variations after frequency multiplication in the downstream ECPA. In [1], cascaded multiphase ILROs are used to address phase and amplitude mismatches. However, 3-phase symmetrical injection in a 9-stage RO still has limitations in suppressing spurs. To overcome this issue, the proposed approach suggests using symmetric injection signals produced by the multiphase N-path filter to substantially reduce phase imbalances in the ILRO outputs.

Fig. 2 presents a block diagram of the proposed 433 MHz OOK frequency-multiplying TX, featuring a fractional-N frequency synthesizer and dual multiphase injection-locked oscillators. The injection-locked phase rotator (ILPR)-based fractional-N frequency synthesizer enables multichannel support in the low-frequency stage. The synthesizer primarily comprises ILRO$_1$, serving as a 16-phase interpolator for a 45 MHz crystal oscillator. It also includes a PR and a delta-sigma modulator ($\Delta\Sigma$M). The PR combines the 16 input phases ($V_{RO1}<0:15>$) from ILRO$_1$ to achieve sub-integer frequency division with a step size of 1/16. When the phase control signal (K) is set to 1, an effectively divided ratio of 17/16 is achieved by sequentially rotating the phase state through the phase controller. When K is 0, the selected phase state remains unchanged, resulting in an output frequency equal to the input frequency. The second-order $\Delta\Sigma$M enables effective fractional frequency division, providing a finer frequency resolution of 2.6 kHz. Multiphase ILROs and an N-path filter are incorporated to mitigate quantization noise from the frequency synthesizer and provide balanced multiple phases for the ECPA. Dual injection stages enable frequency doubling. The TX employs a differential current mode class-D ECPA with a frequency multiplication factor of 5 to enhance the overall power efficiency. OOK modulation is embedded in the driver stage of the ECPA.

Typically, an ILRO requires sufficient bandwidth to preserve the superior phase noise of the reference signal. However, this wide bandwidth can also diminish the filtering effect on the quantization noise from the $\Delta\Sigma$M. To address this challenge, we introduce a multiphase N-path filter prior to ILRO$_2$ that performs additional band-pass filtering and effectively attenuates out-of-band phase noise. The center frequency of the N-path filter is determined by ILRO$_2$, while the bandwidth is adjusted based on the RC pole frequency and clock duty cycle. By generating multiphase outputs from the N-path filter for symmetric injection into ILRO$_2$, we successfully mitigate phase imbalance issues commonly associated with single-phase injection techniques. Fig. 3 presents the proposed N-path filter that carries out down-conversion and low-pass filtering using switches and a series of capacitors (C_{1-5}). The resulting voltages (V_{C1-5}) are then up-converted by a second set of switches to the output nodes ($V_F<1:5>$). Each output node is connected to a single capacitor at any given phase S_{1-5}. By synchronizing the clock frequency with the input frequency through injection locking of ILRO$_2$, we can produce multiphase outputs by sequentially rearranging the sampling phases in the second set of switches. As the illustrated 5-phase non-overlapping clocks for switches S_{1-5}, the corresponding filtered multiple output signals, $V_F<1:5>$, are subsequently injected into ILRO$_2$. The ILRO$_3$ is injection-locked to the second harmonic frequency of the ILRO$_2$ output for frequency doubling. This reduces the burden of a high multiplication ratio in the downstream ECPA and effectively mitigates phase asymmetry resulting from complex layout routing.

To generate an RF carrier frequency of 433 MHz, the ECPA utilizes the 10-phase outputs from ILRO$_3$ to perform frequency multiplication. Traditional ECPAs suffer from low power efficiency due to significant switching power loss. To address this issue, a high-efficiency current mode class-D ECPA with zero voltage switching (ZVS) characteristic is employed, reducing power loss and improving power efficiency in the output stage. Notably, only the final stage of the TX operates at the radio frequency, resulting in a significant overall improvement in power efficiency. Fig. 4 shows the schematic and timing diagram of the adopted current mode class-D ECPA. In the proposed configuration, each composite switch unit comprises two power transistors connected in series, performing a NAND operation with adjacent phases and generating short current pulses. By combining five sets of parallel-connected pulses, the input frequency is multiplied by a factor of 5. The power source is supplied to the ECPA through external inductors (L_{F1-2}), while a differential tapped-capacitor network (C_{F1-2} and C_S) transforms the load impedance (R_L) to match the 50-Ω antenna. The tank resonant frequency is chosen about 1.29 times of 433 MHz to meet the ZVS condition. The symmetric arrangement involves the two composite switches driven by 180° out-of-phase 433-MHz signals. As a result, the output waveforms at V_p and V_n exhibit half-sinusoidal waveforms with a 180° phase shift. OOK modulation is achieved by utilizing AND gates to control the output buffers of ILRO$_3$. An additional switch (SW) is introduced between the differential PA outputs to increase the data rate with faster turned-off responses.

The TX's output power at a carrier frequency of 433.9 MHz was measured to be −7 dBm. Fig. 5 displays the measured phase noise of the TX output, exhibiting a value of −87 dBc/Hz at a 1-MHz offset. The filtering bandwidth can be reduced to further suppress the far-out noise resulting from the shaped quantization noise of the $\Delta\Sigma$M. A 6-dB improvement is observed at high-frequency offsets by reducing the N-path filter's bandwidth from 9 to 5 MHz. The measured OOK-modulated output waveform at a data rate of 40 Mbps is also provided. With a supply voltage of 0.75 V, the TX consumes a total power of 890 µW. Fig. 6 compares the proposed TX and other state-of-the-art ULP sub-GHz TXs, demonstrating the prototype's good global efficiency of 23% and energy efficiency of 22 pJ/bit. The die micrograph as well as the power breakdown are shown in Fig. 7.

References:

[1] J. Pandey *et al.*, "A Sub-100µW MICS/ISM Band Transmitter based on Injection-Locking and Frequency Multiplication," JSSC 2011: 1049-1058.

[2] Y.-L. Tsai *et al.*, "A 330-µW 400-MHz BPSK Transmitter in 0.18-µm CMOS for Biomedical Applications," TCAS-II 2016: 448-452.

[3] H. Huang *et al.*, "A 2Mbps sub-100µW Crystal-less RF Transmitter with Energy Harvesting for Multi-Channel Neural Signal Acquisition," A-SSCC 2019: 157-160.

[4] C.-C. Lin *et al.*, "A 66.97pJ/bit, 0.0413mm^2 Self-Aligned PLL-Calibrated Harmonic-Injection-Locked TX with >62dBc Spur Suppression for IoT Applications," RFIC 2020: 323-326.

[5] A. Srivastava *et al.*, " Design and Implementation of 0.23 nJ/bit Reference-Spur-Free FSK/OOK Transmitter at 400 MHz for Wearable Health Monitoring," BioCAS 2021: 1-5.

979-8-3503-3004-5/23 $31.00 © 2023 IEEE

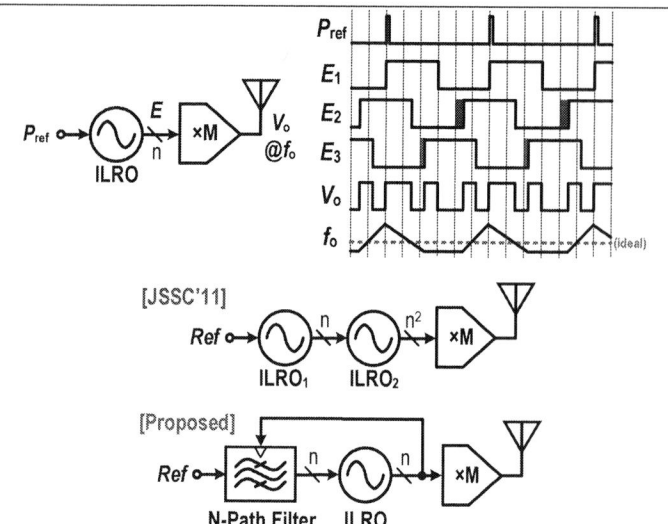

Fig. 1. Limitation of conventional frequency-multiplying TX and the proposed method to overcome it by using an N-path filter.

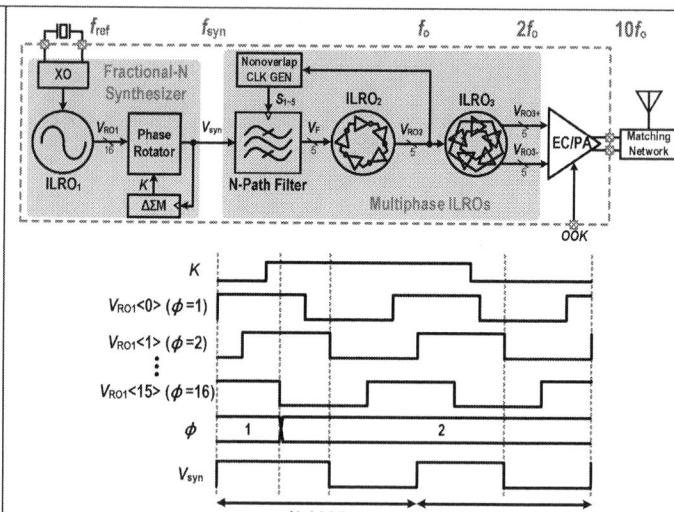

Fig. 2. Block diagram of the proposed TX and the timing diagram of fractional-N frequency synthesis.

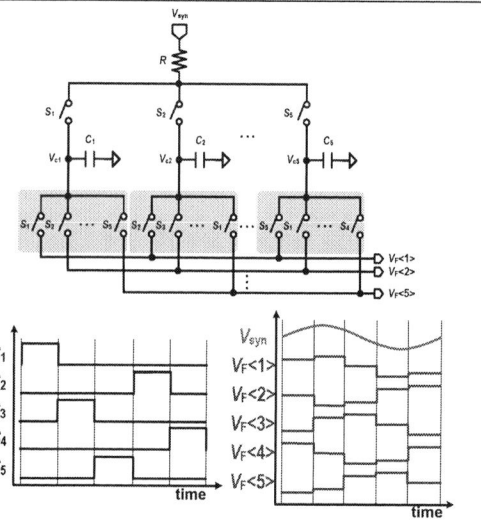

Fig. 3. Proposed N-path filter with multiphase outputs and associated timing diagram.

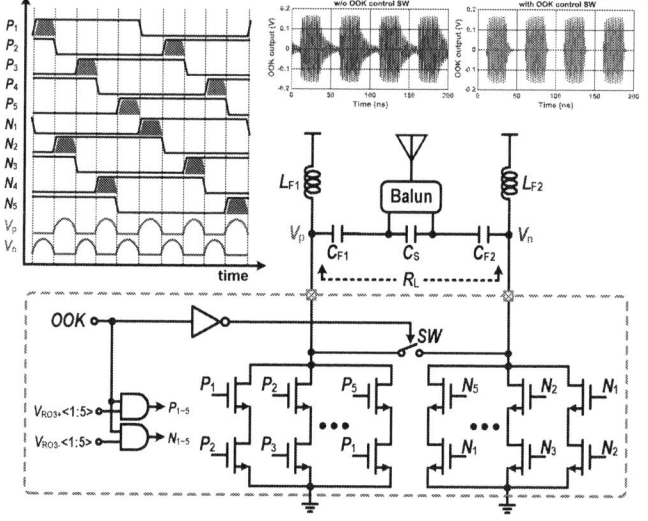

Fig. 4. Proposed current mode class-D ECPA and associated timing diagram.

Fig. 5. Measured TX output phase noise and OOK-modulated waveform at 40 Mbps.

Reference	Our Work	[3]	[4]	[5]	[2]
Modulation	OOK	OOK	OOK	OOK/FSK	BPSK
Frequency Synthesis	Yes	No	No	No	No
Carrier Frequency (MHz)	433	427	915	403.5	400
Output Power (dBm)	-7	-19	-14	-17	-15
Data Rate (Mbps)	40	2	3	2/0.25	20
Power Consumption (µW)	890	96	258.6	450	330
Energy Efficiency (pJ/b)	22	48	67	225/1800	17
Global Efficiency (%)	23	13	15	4	10
Process (nm)	90	180	180	180	180
FoM	99.6	84.2	94.4	79.5/70.4	92.8

FoM = P_{out} dBm - 10 log Energy Efficiency J/bit

Fig. 6. Performance summary and comparison with the prior arts.

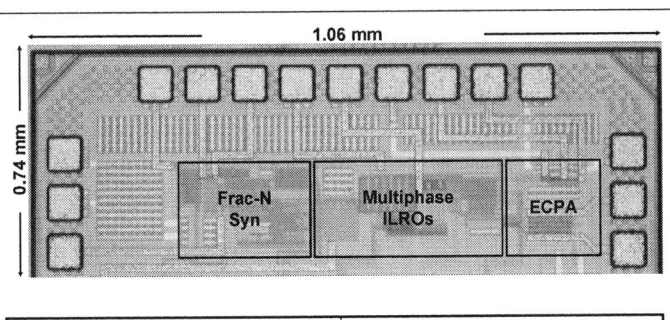

Block	Power Consumption (µW)
Frac-N Synthesizer	100
Multiphase ILROs	242
ECPA	548
Total	890

Fig. 7. Die micrograph of the proposed TX and power breakdown.

A 110-160 GHz ASK Receiver with Isomorphic Low Noise Power Amplifier and Multi-Mode Interface in 28-nm CMOS

Dawei Tang, Zekun Li, Leyang Huang, Rui Zhang, Liqun Lu, Chun Yang, Zheng Yan, Peigen Zhou, Zhe Chen, Jixin Chen, Wei Hong

Southeast University, Nanjing, China

In recent years, there has been a growing interest on the terahertz (THz) frequency to enhance communication rate, particularly in 100-200 GHz band. Within this research domain, silicon-based circuits have garnered significant attention due to their low cost and high integration. Additionally, Simple modulation scheme, such as amplitude shift keying (ASK), have also gain considerable traction owing to their compact architectures and minimal requirements on ADC and DAC. Several silicon-based D-band terahertz transceiver front-ends [1-3] and wideband amplifier [4-7] have been reported, showing remarkable achievements in terms of RF bandwidths exceeding 30 GHz and data rates over 38 Gbaud.

However, to meet the demands for future network with data rate over 50 Gbps, it is preferred to increase the RF bandwidth to higher than 50 GHz in PAM4 system. This poses a significant challenge for conventional silicon-based THz front-ends, particularly in the design of broadband low noise amplifiers (LNA) and power amplifiers (PA). On one hand, the accuracy of model in terahertz is limited, making it challenging to ensure the frequency response in transmitter and receiver after fabrication especially for wideband designs. This can lead to mismatches and even frequency band misalignments between transmitter and receiver, decrease the system bandwidth, and introduce out-of-band noise. On the other hand, for better system performance, differential interface like ground-signal-signal-ground is preferred. However, this configuration can be inconvenient for probing measurement, unless additional single-ended versions are designed, which significantly increases costs.

To address these issue, this paper design and implemented a 110-160 GHz multi-mode ASK receiver with a wideband isomorphic low noise power amplifier (LNPA). The system architecture is shown in Fig. 1, which consists of a 110-160 GHz LNPA with a multi-mode RF interface, a wideband envelop detector (ED), and a DC-35 GHz baseband amplifier (BBA). The LNPA serves as both LNA and PA, ensuring consistent frequency response between the transmitter and receiver to improve available bandwidth. Reconfigurable biasing state in the LNPA makes the gain and power adjustable, for maximum output power in PA state and saving power consumption in LNA state. A multi-mode balun and its corresponding GSGSG interface are introduced, allowing operation in both single-ended and differential modes while providing similar performance. It resolves the testing issues for differential ports and is compatible with two different antenna interface. This ASK receiver achieves the maximum bandwidth reported in the 100-200 GHz range in silicon-based process. A 16Gbps OOK demodulation is completed, which is limited by the rate of the BERT instrument.

Fig. 2 shows the schematic of the 110-160 GHz four-mode LNPA. It comprises six common-source stages. The first four stages utilize smaller transistors to minimize power consumption, while the last two stages employ larger transistors to enhance output power. Cross-coupled capacitors are employed to boost the gain of each stage. Broadband matching between stages is achieved through the use of coupled transmission line transformers, enabling wideband high-gain performance and improved common-mode rejection. Both the input and output stages employ the proposed dual-mode balun and their corresponding GSGSG interfaces. The dual-mode balun consists of two asymmetrically coupled lines. In single-ended input mode, the balun operates similarly to a classic Marchand balun but with an open-circuit RF pad at the open-circuit end. In differential input mode, the balun works as a differential coupled-line transformer. Analysis based on coupled-line theory shows that this structure provides the similar operating frequency range in both differential and single-ended configurations. Consequently, the amplifier can operate in four modes: Differential-In-Single-Out (DISO), Single-In-Single-Out (SISO), Diff-In-Diff-Out (DIDO), and Single-In-Diff-Out (SIDO). The figure showcases the simulated S-parameters and noise figure in these modes. In all four modes, the LNPA delivers approximately 25 dB broadband gain within the 100-160 GHz range.

Fig. 3 shows the schematic of the envelope detector and baseband amplifier. The BBA consists of four amplification stages, utilizing a inverter-based amplifier structure comprising differential NMOS and PMOS transistors to achieve high gain and provide DC bias. To ensure accurate DC operation, current mirrors and tail current control are used. The reduction of low-frequency gain and enhancement of high-frequency gain are achieved through negative feedback resistors and series peaking inductors [8], significantly widening the bandwidth. The baseband amplifier output is configured with a differential 100-ohm port. Simulation results demonstrate that the BBA exhibits approximately 30 dB flat gain from DC to 30 GHz, with a 3 dB bandwidth exceeding 35 GHz. The ED employs a classic self-mixing structure and is matched to the output impedance of the LNPA using coupled-line transformers. Additionally, an unloaded replica ED and a bias circuit are introduced to provide bias control and ensure the symmetry of the differential BBA's DC bias conditions.

The fabricated receiver chip is shown in Fig. 7 with a size of 1.19×0.65 mm^2. Fig. 4 shows the measurement setups for receiver and S-parameters measured results of the LNPA, which has been fabricated separately. The LNPA is measured by on-wafer probing, and the results show that it has approximately 22 dB gain from 110 to 160 GHz. To meet the testing requirements for receiver, a wideband modulator is also designed to generate wideband ASK modulation signals. These modulation signals are transmitted to the receiver through wire bonding. The local oscillator (LO) source for the modulator is provided by a 12x frequency multiplication from the Keysight E8267D signal source. An adjustable attenuator is used to control the LO input power. The input data is provided by the Keysight BERT M8041, and the baseband output of the receiver is either terminated to an oscilloscope for observing the eye diagram or connected to the BERT for measuring the bit error rate (BER). Due to the data rate limitation of the BERT, this test setup can only test up to 16 Gbps, but further testing will be conducted soon.

Fig. 5 presents the measured eye diagram and BER and sensitivity measurement when given 1Gbps modulated clock. For the eye diagram measurement, it show that at 135 GHz LO frequency and a 16 Gbps OOK data rate, the output swing reaches 700 mV and eye opens clearly with BER < 10^{-12}, indicating that it can operate at higher rates. The sensitivity measurement results show that when V_{Demod} is set to be 1.2 V, 0.6 V, and 0 V, the receiver can achieve a dynamic range exceeding 15 dB when the Vpp threshold is set as 400mV, with a minimum input power lower than -40 dBm. By dynamically adjusting V_{Demod}, the dynamic range will exceed 30 dB.

Fig. 6 shows the performance comparison with the state-of-the-art. Compared with the reported D-band silicon-based amplifiers, the proposed LNPA achieves better RF bandwidth, while providing substantial gain and noise figure, along with the capability for multi-mode operation. Furthermore, compared to the reported D-band ASK receivers, this receiver achieves the widest system bandwidth, exhibiting potential for ultra-high-speed communication.

References:

[1] Y. Kim et al., "A Millimeter-Wave CMOS Transceiver With Digitally Pre-Distorted PAM-4 Modulation for Contactless Communications," JSSC, vol. 54, no. 6, pp. 1600-1612, Jun 2019.

[2] M. De Wit et al., "Analysis and Design of a Foam-Cladded PMF Link With Phase Tuning in 28-nm CMOS," JSSC, vol. 54, no. 7, pp. 1960-1969, Jul 2019.

[3] K. Dens et al., "A PAM4 Dielectric Waveguide Link in 28 nm CMOS," ESSCIRC 2021, pp. 479-482.

[4] A. Haman et al., "A 125.5-157 GHz 8 dB NF and 16 dB of Gain D-band Low Noise Amplifier in CMOS SOI 45 nm," IMS 2020.

[5] S. W. Li et al., "A 134-149 GHz IF Beamforming Phased-Array Receiver Channel with 6.4-7.5 dB NF Using CMOS 45nm RFSOI," RFIC 2020.

[6] B. Yun et al., "A D-Band High-Gain and Low-Power LNA in 65-nm CMOS by Adopting Simultaneous Noise- and Input-Matched G(max)-Core," TMTT, vol. 69, no. 5, pp. 2519-2530, May 2021.

[7] S. Park et al., "A D-Band Low-Power and High-Efficiency Frequency Multiply-by-9 FMCW Radar Transmitter in 28-nm CMOS," JSSC, vol. 57, no. 7, pp. 2114-2129, Jul 2022.

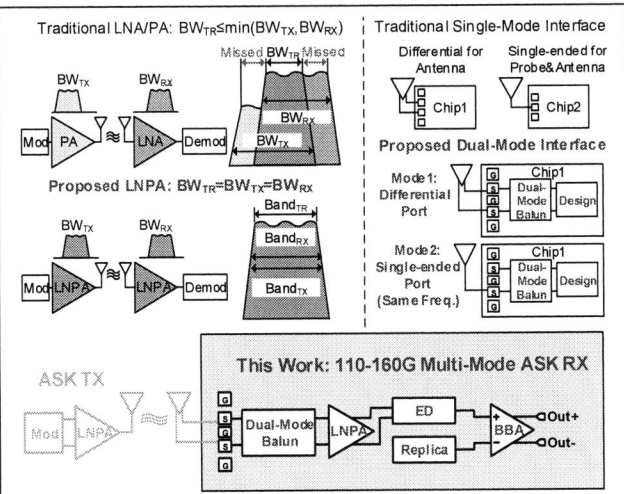

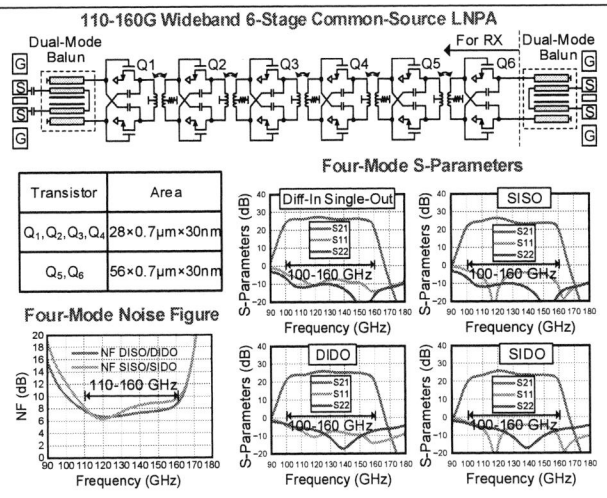

Fig. 1. Block diagram of the 110-160 GHz multi-mode ASK receiver and the comparison with other architectures.

Fig. 2. Schematic of the four-mode low noise power amplifier and the simulated results of the four-mode performance.

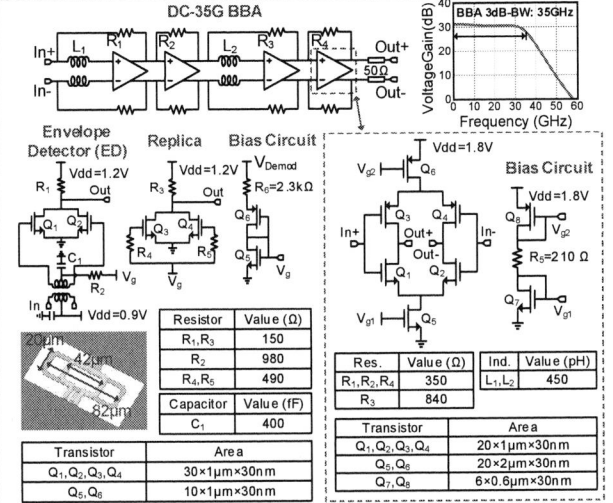

Fig. 3. Schematic of the envelope detector and baseband amplifier.

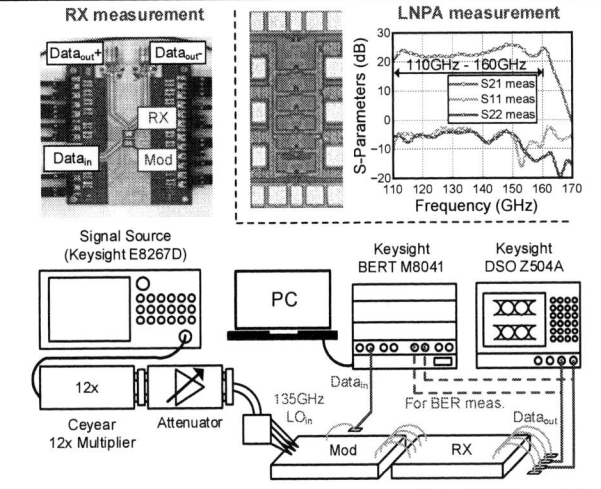

Fig. 4. The small signal measured results of the LNPA and the measurement setups for RX.

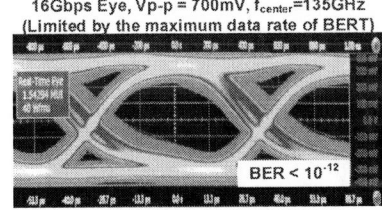

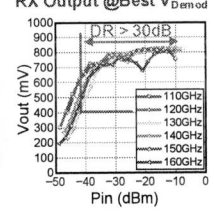

Fig. 5. Measured eye diagram and BER and the baseband output versus RX input (1Gbps modulated clock) power in different bias.

Reference	This LNPA	[4]JSSC22	[5]TMTT21	[6]IMS20	[7]RFIC20
Technology	28nm CMOS	28nm CMOS	65nm CMOS	45nm SOI	45nm SOI
Freq (GHz)	110-160	135-155	148-160	125.5-157	138-163
BW (GHz)	50	20	11	31.5	25
Gain (dB)	22	18.5	17.9	16	17.2
NF (dB)	6.2-8.5*	N.A.	4.7-6.2	8	6.4-7.5
DC power (mW)	81	N.A.	13.73	75	30
RF Interface	Four-Mode	Single	Differential	Single	Single

Reference	This RX	[1]JSSC19	[2]JSSC19	[3]ESSCIRC21
Technology	28nm CMOS	65nm CMOS	28nm CMOS	28nm CMOS
Topology	LNPA+ED+BBA	LNA+ED	LNA+PD+BBA	LNA+ED+BBA
Modulation	OOK / PAM4	PAM4	FSK	OOK / PAM4
RF Frequency (GHz)	110-160	127	140	30.6
LNA BW (GHz)	50	20#	20	30.6
BBA BW (GHz)	35	N.A.	N.A.	23
LNA Gain (dB)	22	10	12	35
BBA Gain (dB)	30	N.A.	N.A.	10.5
NF (dB)	6.2-8.5*	9	7-8.5	N.A.
Symbol Rate (Gbaud)	16 (BERT Limit)	10	12	38 (DSP)
BER	<10⁻¹²	<10⁻⁴	<10⁻¹²	<10⁻⁶
Output Voltage (mV)	400-800	30	51	N.A.
Dynamic Range (dB)	30	N.A.	N.A.	N.A.
Sensitivity (dBm)	<-40	<-14.5	N.A.	N.A.
RF Interface	Single & Diff.	Differential	Single	Single
Power (mW)	160	28.7	230	130

* Simulation Results   # Estimated by Figures

Fig. 6. Performance comparison with state-of-the-art.

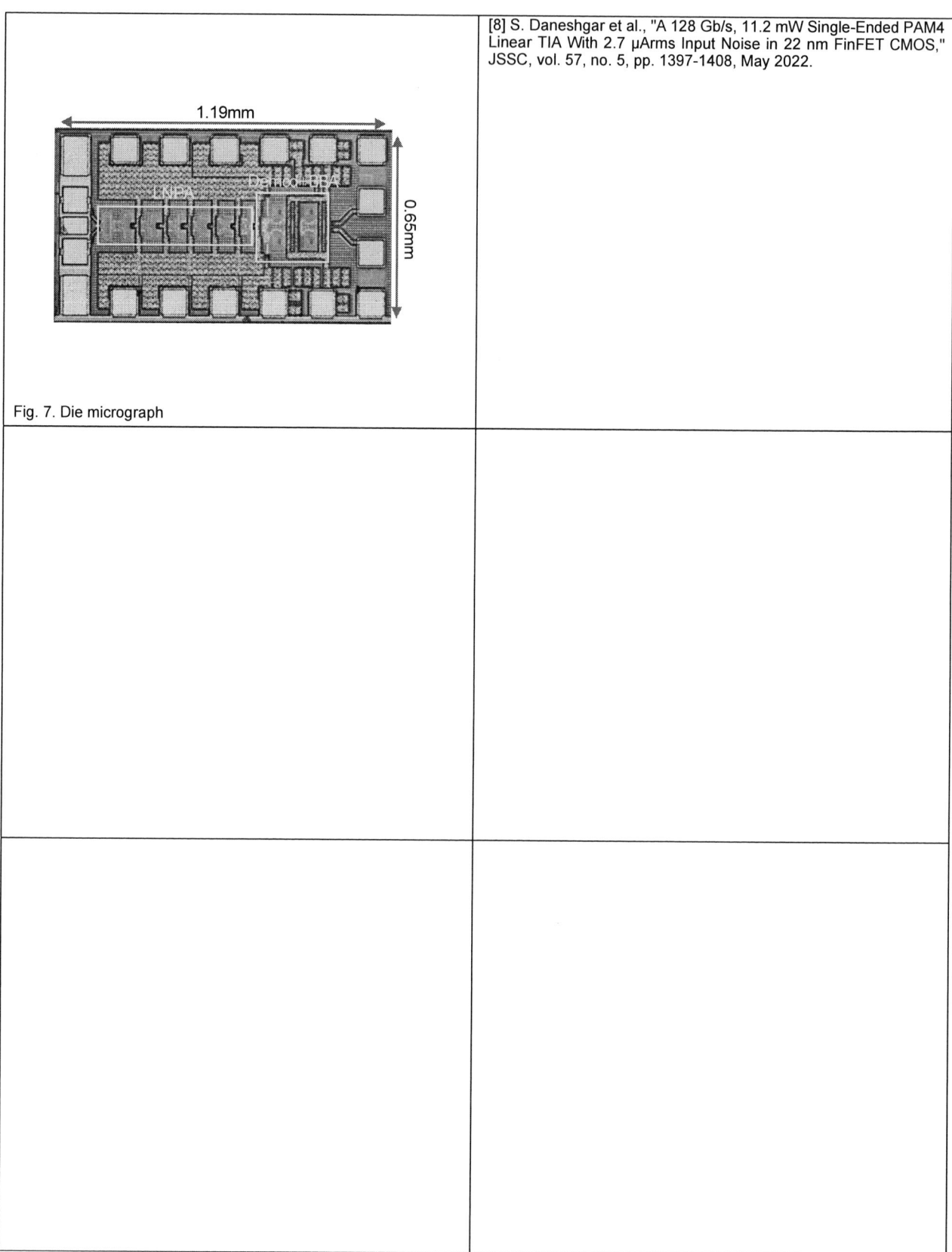

Fig. 7. Die micrograph

[8] S. Daneshgar et al., "A 128 Gb/s, 11.2 mW Single-Ended PAM4 Linear TIA With 2.7 µArms Input Noise in 22 nm FinFET CMOS," JSSC, vol. 57, no. 5, pp. 1397-1408, May 2022.

A 0.5-to-1.5GHz BW-Extended Gain-Boosted N-path Filter Using a Switched g_m-C Network Achieving 50MHz BW and 18.2dBm OB-IIP$_3$

Gengzhen Qi[1] and Pui-In Mak[2]

[1]Sun Yat-sen University, Zhuhai, Guangdong, China

[2]University of Macau, Macau, China

Classical N-path filters incorporate a passive switched-capacitor (SC) N-path network and a multi-phase local oscillator (LO) to generate a high-Q bandpass response at a flexible RF [1]. This feature and their high linearity render them as a promising RF technique for multiband radios to achieve SAW-less integration. In [2], the feasibility of the N-path filter leaps forward by operating the passive SC network under a negative feedback created by an RF gain stage, offering not only high-Q bandpass channel selection at flexible RF, but also in-band amplification and input-impedance matching. Yet, the trade-off of N-path filters between passband BW and out-of-band (OB) rejection severely hinders its application on wideband standards such as 5G New Radio (NR). To alleviate such trade-off of the N-path filter, [3] and [4] increase the order of the bandpass filter: the "subtraction" method [3] achieves a 4th-order response by shifting the baseband (BB) admittance with poly-phase g_m cells in the BB, but with a penalized NF of 9.5dB due to the loss of RF gain, and the BW is limited to 21MHz. [4] creates a 13th-order RF filter based on passively coupled N-path resonators, yet sacrificing the NF (8.6dB), power (97mW) and area (1.9mm^2) budgets.

Herein we report a BW-extended gain-boosted N-path filter with a switched g_m-C network. It features not only in-band gain and high-Q bandpass response, but also other valuable properties such as wide passband BW and high linearity. Prototyped in 65nm CMOS, our filter achieves a 10.7-to-12.7dB voltage gain and a 5.3-to-6.2dB NF over a 0.5-to-1.5GHz range, which covers ~50% of the sub-6GHz 5G-NR bands. The power consumption is 11.1 to 15.9mW and die area is 0.083mm^2 thanks to the Miller effect. With bias-tuning the class-A/B $G_{m,RF}$, the g_m linearity is optimized and our filter measured at 1GHz exhibits an OB-IIP$_3$ of 18.2dBm and a passband BW of 50MHz.

The subtraction of two filters with a slightly up/down-shifted response can allow both wide passband BW and steep roll-off [3]. Here, we apply a $G_{m,RF}$ stage around the tunable 2nd-order RLC resonator to realize such subtraction (Fig. 1). Two 2nd-order RLC resonators are set with an equal BW but with different center frequencies. At in-band, the resonators exhibit 45° offset from -180°, while at OB they are almost in-phase. After subtraction, the in-band signals add up while the OB signals cancel each other. Thus, we get a BW-extended 4th-order bandpass filter with Q enhancement and in-band amplification.

The schematic of our filter is shown in Fig. 2, in which the RLC resonators are replaced by 4-path SC networks (switch SW$_{NP}$ and capacitor C$_{NP}$). SW$_{NP}$ in each path is driven by a 25%-duty-cycle non-overlapping LO and the center frequency is defined by f_{LO}. To shift f_{LO} of the two N-path filters, a g_m-C network is used, with 4-phase g_m cells to shift the baseband admittance from $j\omega \cdot C_{NP}$ to $j(\omega\pm\Delta\omega)\cdot C_{NP}$ with $\Delta\omega=g_m/C_{NP}$, where $\Delta\omega$ is independent of $G_{m,RF}$. Similar to [3], two series capacitors C$_S$ (0.7pF) are used to split the input source, also isolating the two gain-boosted N-path filters. The class-A/B $G_{m,RF}$ stage of the filter is shown in Fig. 2 (right), which determines the linearity of the filter. A set of highly accurate (5mV step) tunable bias voltages (V$_{GN}$ and V$_{GP}$) is applied to the gates of $G_{m,RF}$ to optimize the small-signal linearity. A downscaled $G_{m,RF}$ replica sets the output DC voltage of $G_{m,RF}$ with a feedback loop. For a large signal, the bias voltage of switch (SW$_{NP}$) should be considered for both OFF and ON periods. Thus, a set of accurate (50mV step) tunable bias voltages are applied to the gate of SW$_{NP}$. With a 58.3mS transconductance, the class-A/B $G_{m,RF}$ consumes 2.8mA current. The differential g_m stages of the filter are shown in Fig. 2 (right), in which the effective transconductance is $g_{m1}/(1+g_{m1}\cdot R_{M1})$.

The steep filter shape can be tuned by changing C$_{NP}$. In simulation, when the transconductance of the BB g_m cell is 0.35mS and C$_{NP}$ is tuned from 1 to 5pF, the passband BW decreases from 85 to 31MHz while the OB rejection is increased by ~18dB at 80MHz offset (see Fig. 3, upper). The center-frequency shifting is due to the input series capacitor C$_S$. The transconductance of the g_m cell is also slightly tuned to optimized the passband shape. The NF can be improved by enlarging $G_{m,RF}$. In simulation, the NF improves from 8.3 to 5.1dB, when the transconductance of $G_{m,RF}$ is upscaled from 8.4 to 64.3mS. When removing the noise contribution from the g_m cells, the NF drops to 2.07dB. The simulated far-out rejection characteristics up to 5·f_{LO} is shown in Fig. 3 (lower), with R$_{SW}$=20Ω and C$_{NP}$=3pF, which features signal suppression at the even-order harmonics of f_{LO}, since there is no frequency shift and the signal is subtracted to zero.

Our filter was prototyped in 65nm CMOS (Fig. 7). The static power consumption due to $G_{m,RF}$ and $G_{m,BB}$ is 8.7mW. An on-chip divider-by-2 generates the 4-phase 25%-duty-cycle LO. The power efficiency is ~4.8mW/GHz for a -161.5dBc/Hz phase noise at 80MHz offset. A set of LC networks is added at the input to aid the input-impedance matching. Fig. 4 (upper) presents the LO-defined gain response and S$_{11}$, in which the passband gain is 10.7 to 12.7dB with a 50MHz BW, and an S$_{11}$<-10dB is obtained in the 0.5-to-1.5GHz RF range. The stopband rejection is 25dB at 100MHz offset (Δf=2·BW). The variation of the group delay of the filter is <13ns (Fig. 4, lower). The NF is 5.3 to 6.2dB for a 0.5-to-1.5GHz RF range (Fig. 5, upper). When injecting a 0dBm CW signal into the filter at Δf=4·BW, the blocker NF reaches 8.3dB at 1GHz, in which ~1.8dB is due to gain compression, and the rest is due to the reciprocal mixing of the phase noise in the LO generator (estimated in simulation). The OB-IIP$_3$ and OB-IIP$_2$ are 18.2 and 58dBm, respectively, at Δf=5·BW offset (Fig. 5, lower). The B$_{-1dB}$ is 1.5dBm at Δf=5·BW offset (Fig. 5, lower). Note the B$_{-1dB}$ is a large-signal linearity, which can be optimized by slightly tuning the bias voltage of the 4-phase LO. The measured LO leakage power to the input port is <-58dBm.

Fig. 6 benchmarks this work with the prior art [2]-[6]. With a comparable linearity as the N-path passive LNA in [6], our filter shows a 2.5x wider passband BW and >10x smaller area. Although the high-order filters in [3, 4] achieve better linearity, our filter has >12dB more gain and >3dB lower NF. Compared to the active gain-boosted N-path RXs in [2, 5], our filter enables a 6.5dB higher B$_{-1dB}$ and >10dB larger OB-IIP$_3$. Our filter achieves both wide passband BW and steep roll-off, also with comparable linearity and NF, while consuming less power and area.

Acknowledgements:

This work is funded in part by the National Natural Science Foundation of China (NSFC) under Grant 62104263; and in part by the GuangDong Basic and Applied Basic Research Foundation under Grant 2022A1515-012295. Corresponding author: Gengzhen Qi.

References:

[1] A. Ghaffari et al., "Tunable high-Q N-path band-pass filters: modeling and verification," IEEE JSSC, vol. 46, no. 5, pp. 998-1010, May 2011.

[2] Z. Lin et al., "A 0.028mm2 11mW single-mixing blocker-tolerant receiver with double-RF N-path filtering, S11 centering, +13dBm OB-IIP3 and 1.5-to-2.9dB NF," in IEEE ISSCC Dig. Tech. Papers, Feb. 2015, pp. 36-37.

[3] M. Darvishi et al., "Widely tunable 4th order switched G_m-C band-pass filter based on N-path filters," IEEE JSSC, vol. 47, no. 12, pp. 3105-3119, Dec. 2012.

[4] P. Song et al., "RF filter synthesis based on passively coupled N-path resonators," IEEE JSSC, vol. 54, no. 9, pp. 2475-2486, Sep. 2019.

[5] M. A. Montazerolghaem et al., "A 3dB-NF 160MHz-RF-BW blocker-tolerant receiver with third-order filtering for 5G NR applications," in IEEE ISSCC Dig. Tech. Papers, 2021, pp. 98–100.

[6] H. Shao et al., "A 1.7-3.6 GHz 20 MHz-bandwidth channel-selection N-path passive-LNA using a switched-capacitor-transformer network achieving 23.5 dBm OB-IIP3 and 3.4-4.8 dB NF," IEEE JSSC, vol. 57, no. 2, pp. 413-422, Feb. 2022.

979-8-3503-3004-5/23 $31.00 © 2023 IEEE

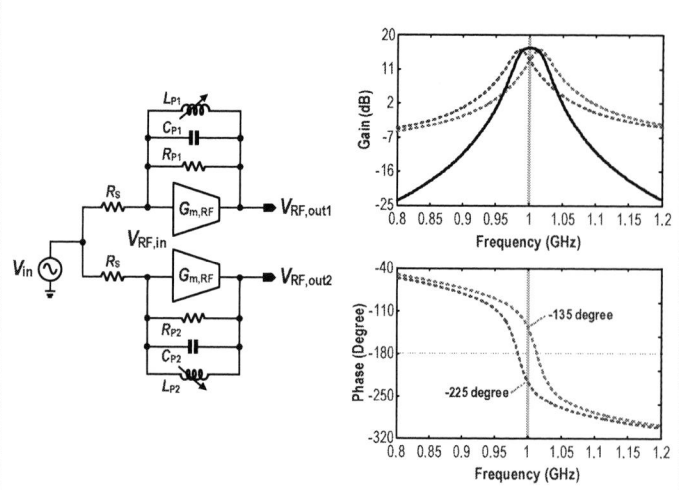

Fig. 1. Applying a $G_{m,RF}$ stage around the tunable 2nd-order RLC resonator to achieve the "subtraction" method with $V_{RF,out1} - V_{RF,out2}$ (*left*), and the in-band signals add up while the OB signals cancel each other (*right*).

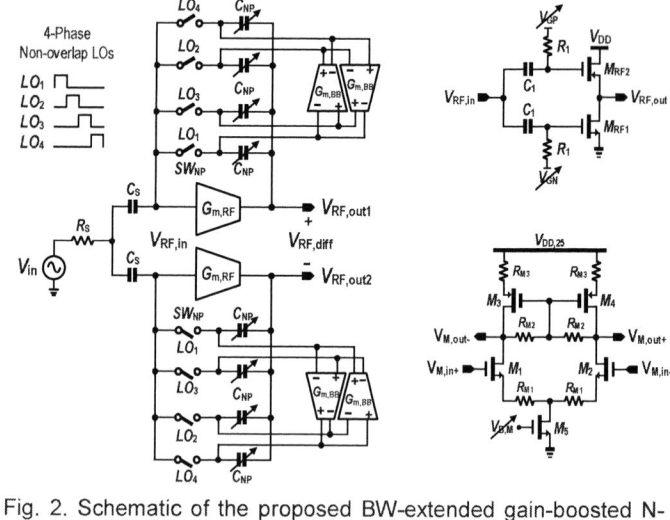

Fig. 2. Schematic of the proposed BW-extended gain-boosted N-path filter (*left*), the class-A/B gain stage and the baseband g_m-cell stage (*right*).

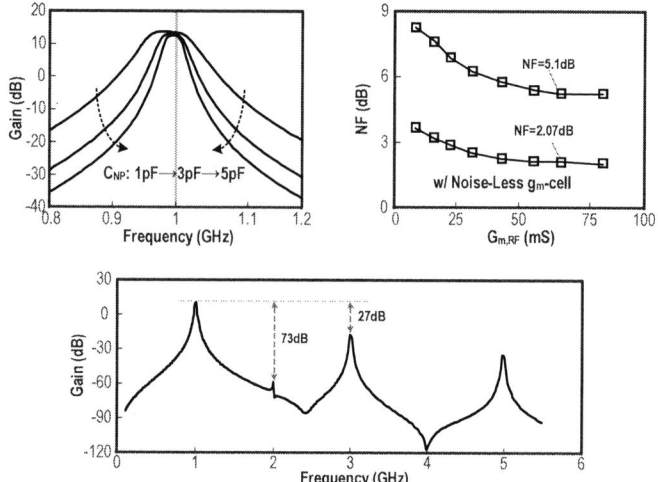

Fig. 3. Simulated gain response with different C_{NP} and NF optimization by enlarging $G_{m,RF}$ (*upper*), and far-out rejection characteristic (*lower*).

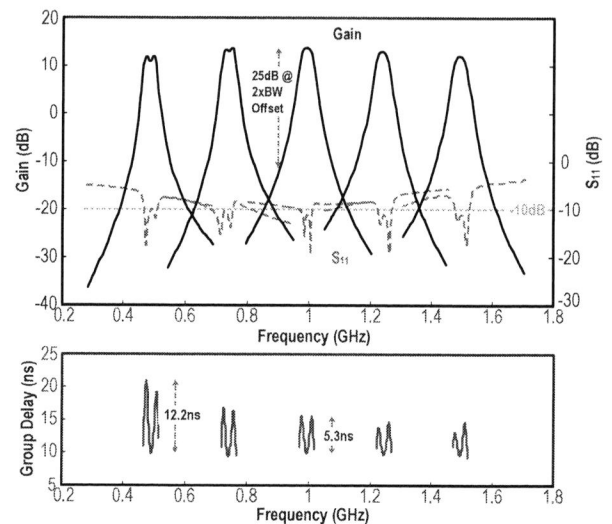

Fig. 4. Measured gain response and S_{11} at different RF (*upper*), and the group delay at different RF (*lower*).

Fig. 5. Measured NF at different RF and blocker-NF at $\Delta f=4\times BW$ (*upper*), IIP$_2$, IIP$_3$ and B$_{1dB}$ under different frequency offsets (*lower*).

	This Work	JSSC'22 [6]	ISSCC'21 [5]	JSSC'19 [4]	ISSCC'15 [2]	JSSC'12 [3]
Topology	**Gain-Boosted SC N-Path Filter w/ G_m-C Network**	N-Path pLNA w/ SC-XFMR Network	Gain-Boosted N-Path RX w/ Two-Zeros Network	High-Order Passive N-Path Filter	Gain-Boosted Mixer-First N-path RX	4th-Order N-Path Filter w/ G_m-C Network
Order of Filter	**4**	2	3	13	2	4
RF Range (GHz)	**0.5 to 1.5**	1.7 to 3.6	0.4 to 3.2	0.8 to 1.1	0.1 to 1.5	0.4 to 1.2
RF BW (MHz)	**50**	20	160	30 to 50	4	21
Gain (dB)	**10.7 to 12.7**	9.8	36	-4.6 to -3.8	38	-3.2 to -2
DSB NF (dB)	**5.3 to 6.2**	3.4 to 4.8	2.7 to 3.6	5 to 8.6	1.5 to 2.9	9.7 to 10.5
0dBm Blocker NF (dB) @ Offset (MHz)	**8.3 @$\Delta f=4\times BW$**	4.7 @$\Delta f=4\times BW$	8.4 @$\Delta f=6.25\times BW$	6.6 @$\Delta f=1\times BW$	13.5 @ $\Delta f=20\times BW$	No Data
B$_{-1dB}$ (dBm) @$\Delta f=5\times BW$ (MHz)	**1.5**	1.7	-5*	14*	-6	No Data
OB-IIP$_3$ (dBm) @$\Delta f=1\times BW$ (MHz)	**4**	20*	-7*	24	-11*	9
OB-IIP$_3$ (dBm) @$\Delta f=5\times BW$ (MHz)	**18.2**	23.5	13	25*	8*	29
OB-IIP$_2$ (dBm)	**58**	64	70	61	50	No Data
Power (mW)	**11.1 to 15.9**	17.8 to 38.2	65.6 to 114.8	80 to 97	11	21.4
Supply (V)	**1.2, 2.5**	1	1.2, 1.3	2, -1	0.7, 1.2	1.2, 2.5
Active Area (mm²)	**0.083**	0.84	0.6	1.9	0.028	0.127
Technology	**65nm CMOS**	28nm CMOS	40nm CMOS	65nm CMOS	65nm CMOS	65nm CMOS

Fig. 6. Performance comparison with the state-of-the-art.
*Estimated from the figure

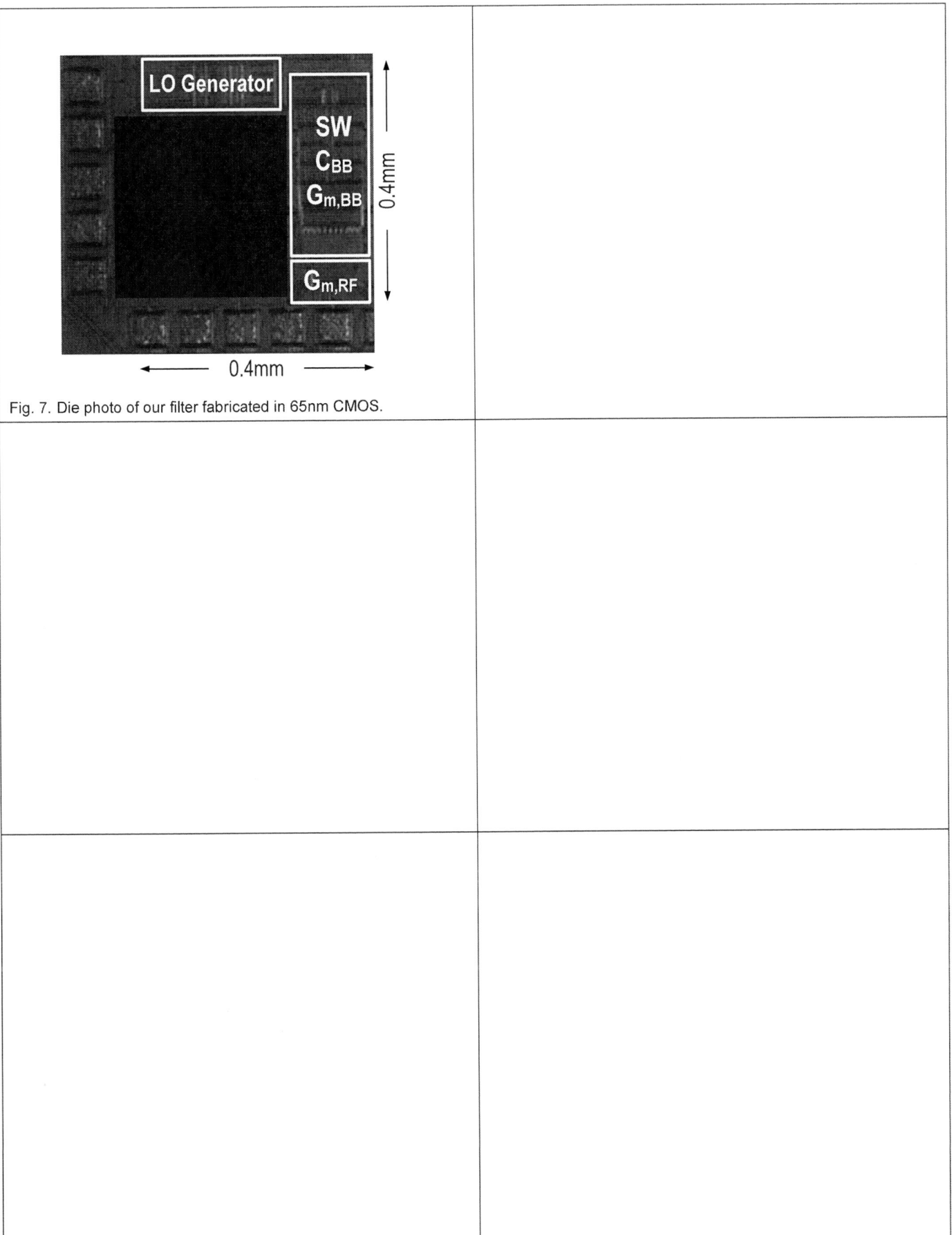

Fig. 7. Die photo of our filter fabricated in 65nm CMOS.

A D-band ASK Transmitter with 50GHz RF Bandwidth and Multi-Mode Interface Implemented in 28nm CMOS

Rui Zhang, Dawei Tang, Zhe Chen, Liqun Lu, Peigen Zhou, Jixin Chen, Wei Hong

Southeast University, Nanjing, China

As the demand for communication rates increases, the existing Sub-6GHz frequency band has become very crowded and difficult to meet future communication needs. The abundant spectrum resources of the millimeter-wave (mm-Wave) provide potential solutions to this problem, while terahertz (THz) has received increasing attention due to its significant absolute bandwidth advantage in recent years [1]-[3]. The D-band (110-170 GHz), which can provide abundant spectrum resources and is within the frequency range of high-performance devices designed using silicon-based processes, has become more and more attractive in recent years. Communication systems based on amplitude shift keying (ASK) modulation format are widely studied and applied in THz communication systems due to their simple structure.

In order to meet the abundant spectrum resources in the D-band, this paper developed a ultra-wideband ASK transmitter (Tx) operating in the D-band with RF bandwidth of 110-160 GHz, a maximum transmit power of over 6 dBm, and an on-off isolation of over 20 dB. The system is tested using on-off keying (OOK) modulation and achieves an output data rate of over 16 Gb/s with BER Tester. Due to the limitations of testing equipment, the performance beyond this rate cannot be tested. Compared with state-of-the-art ASK transmitters, this work realizes the largest RF bandwidth, providing support for potential higher rates in the future. This work aims at supporting maximum transmission data rates of up to 50 Gb/s using OOK or 100 Gb/s using pulse-amplitude modulation 4 (PAM4) modulation.

Fig. 1 illustrates the block diagram of the proposed transmitter, which includes a signal source with a center frequency of 135 GHz, an ASK modulator, and an ultra-wideband low-noise power amplifier (LNPA). The signal source is composed of a center frequency of 67.5 GHz voltage-controlled oscillator (VCO), a doubler based on the Gilbert core, and a 2-stage common-source driver amplifier (DA). The figure demonstrates two novel designs in this work. Firstly, a dual-mode balun is proposed, which can achieve differential or single-ended output mode by connecting it to a differential output pad, better compatible with different interface antennas or other following devices. Secondly, the usage of LNPA effectively avoids the problem of bandwidth mismatch caused by the respectively design of the power amplifier (PA) and low-noise amplifier (LNA) in the traditional architecture. Due to the overall bandwidth of the transceiver system (BW_{TR}) not being greater than that of the transmitter (BW_{TX}) and receiver (BW_{RX}), some bandwidth loss will occur in the traditional transceiver architecture. However, using the same LNPA both in the transmitter and the receiver can maximize the bandwidth of the amplifier and improve the overall bandwidth of the transceiver.

Due to the lower requirements for carrier frequency in non-coherent modulation, the phase-locked loop (PLL) is unnecessary. A signal source is integrated into the ASK transmitter, eliminating the need for an external carrier recovery and improving system power consumption. The schematic and related test results of the signal source are shown in Fig. 2. To address the difficulty of directly designing the VCO with fundamental frequency near 135 GHz using limited f_{max} devices, a cross-coupled VCO operating near 67.5 GHz is designed, with a pair of small varactors added to handle possible carrier frequency adjustments. Gilbert core is used in the doubler, which offers advantages of full differential structure, high reverse isolation, and low power consumption. A 2-stage common-source amplifier is used as the driver, allowing the output power of the signal source to be adjusted by changing the gate voltage to meet the LO power requirements in different scenarios. For the convenience of performance testing, separate chips of the signal source were fabricated, and the die micrograph is shown Fig. 2. The signal source can be tuned between 130-140 GHz and the fluctuation of output power within the frequency band is within 2 dB. The maximum output power exceeds 3 dBm and can achieve power adjustment between -6 dBm and 3 dBm, with a maximum DC-RF efficiency of 3.3%.

The schematics of the ASK modulator and the LNPA are shown in Fig. 3. A pair of differential transistors (Q16, Q17) is used in the modulator and a pair of transistors (Q13, Q14) with the same size acts as neutralization capacitors to reduce the local oscillator leakage during shutdown which improves isolation. Test results show that the isolation of the modulator is greater than 20 dB and the 3-dB bandwidth covers the entire D-band (110-170 GHz) with the insertion loss less than 5dB. The input 1-dB compression point is 3 dBm at 135 GHz, and corresponding output 1-dB compression point of -1.8 dBm. The LNPA utilizes a 6-stage common-source structure, with transformer peak mismatch matching between stages to achieve ultra-wideband. The first-two stages use low-noise matching to reduce the noise performance of the amplifier, while the post-four stages use maximum power matching to obtain higher output power. Within the bandwidth of 110-160 GHz, the saturation power (P_{sat}) of the LNPA exceeds 4dBm with less than 3dB fluctuation.

The on-chip measurement setup for the transmitter chip is illustrated in Fig. 4. The baseband (BB) signal was inputted into the chip via a Keysight BERT M8041, controlled by a personal computer. The output signal of the transmitter was down converted by the ERAVANT SFB-06-N1 D-Band Mixer and measured using a Keysight 9030 spectrum analyzer. For output power, VDI Erickson PM4 power sensor was used with a WR6.5 to WR10 taper. The results show that the carrier frequency ($f_{carrier}$) of the transmitter can be adjusted between 130-140 GHz, consistent with the measurement results of the signal source. Within the range of carrier frequency changes, the output power of the transmitter exceeds 5 dBm and the isolation degree exceeds 20 dB. The power consumption breakdown of the transmitter chip is also shown in Fig. 4, indicating a working power consumption of 167 mW for the entire transmitter.

By bonding the transmitter chip with an envelope detector (ED) chip onto a PCB board, the modulation function and data rate of the transmitter can be tested. Photographs of the test board and test environment are shown in Fig. 5. The transmitter and ED were both operating in differential states. The baseband signal used for testing was a PRBS signal with a modulation mode of OOK. The baseband signal was supplied by the Keysight BERT M8041, and the baseband signal demodulated by the ED was fed into Keysight DSO Z504A to observe waveforms and eye diagrams, or to return to the BERT for testing bit error rate (BER). The eye diagram at a data rate of 16 Gb/s is shown in the figure, and the BER under this state is less than 10^{-12}. Due to instrument limitations, data rate higher than 16 Gb/s are temporarily not testable, and related tests will be carried out soon.

Fig.6 compares the proposed transmitter with the state-of-the-art, displaying the best RF bandwidth and highest output power. It is worth mentioning that most of the high-performance transmitters currently reported do not include high-performance PAs, or the insufficient bandwidth of the modulator and PAs affects the maximum data transmission rate. The OOK modulator and LNPA in this work have the widest bandwidth among D-band transmitters that include a PA, which provides support for faster data transmission.

References:

[1] X. Yu, S. P. Sah, H. Rashtian, S. Mirabbasi, P. P. Pande and D. Heo, "A 1.2-pJ/bit 16-Gb/s 60-GHz OOK Transmitter in 65-nm CMOS for Wireless Network-On-Chip," TMTT, vol. 62, no. 10, pp. 2357-2369, Oct. 2014.

[2] S. Ooms and P. Reynaert, "A 120-GHz Wireless Link Using 3D-Printed Lens with Flexible Dielectric Fiber Feed and 28-nm CMOS Transceiver," SSCL, vol. 3, pp. 142-145, 2020.

[3] S. Park et al., "A D-Band Low-Power and High-Efficiency Frequency Multiply-by-9 FMCW Radar Transmitter in 28-nm CMOS," JSSC, vol. 57, no. 7, pp. 2114-2129, July 2022.

[4] K. Dens, J. Vaes, S. Ooms, M. Wagner and P. Reynaert, "A PAM4 Dielectric Waveguide Link in 28 nm CMOS," ESSCIRC, Grenoble, France, 2021, pp. 479-482.

[5] B. Hadidian, F. Khoeini, S. M. H. Naghavi, A. Cathelin and E. Afshari, "A 220-GHz Energy-Efficient High-Data-Rate Wireless ASK Transmitter Array," JSSC, vol. 57, no. 6, pp. 1623-1634, June 2022.

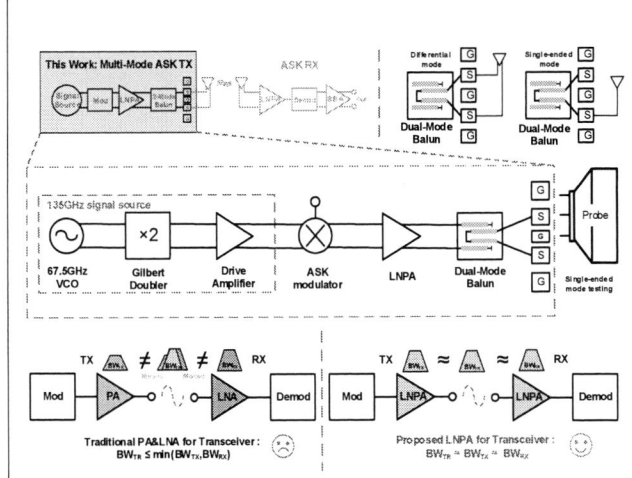

Fig. 1. Block diagram of the 110-160GHz multi-mode ASK transmitter and the advantages of the LNPA.

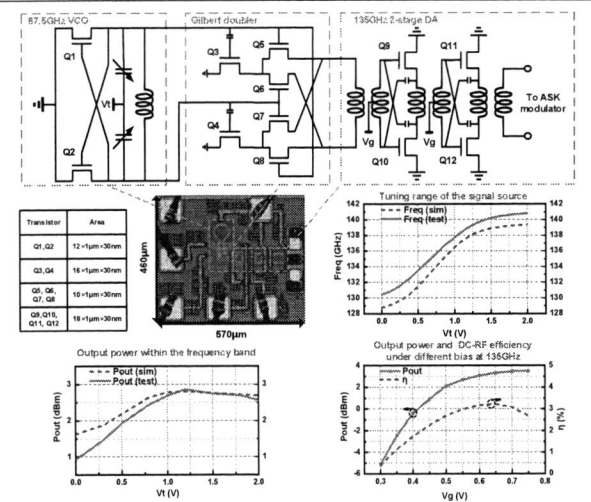

Fig. 2. Schematic and measured results of the signal source.

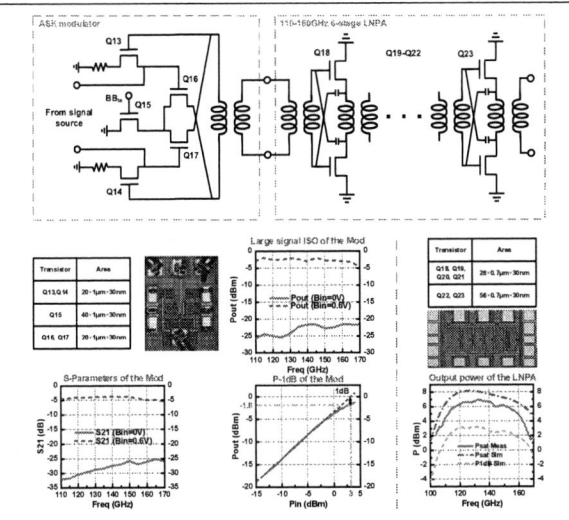

Fig. 3. Schematic and measured results of the Mod and LNPA.

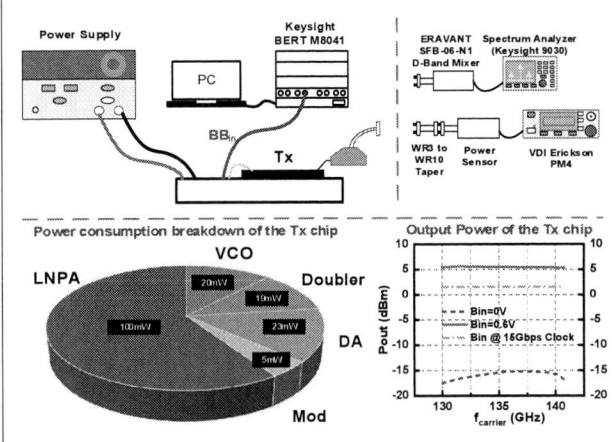

Fig. 4. Measurement setups for the Tx, and power consumption and output power of the Tx chip.

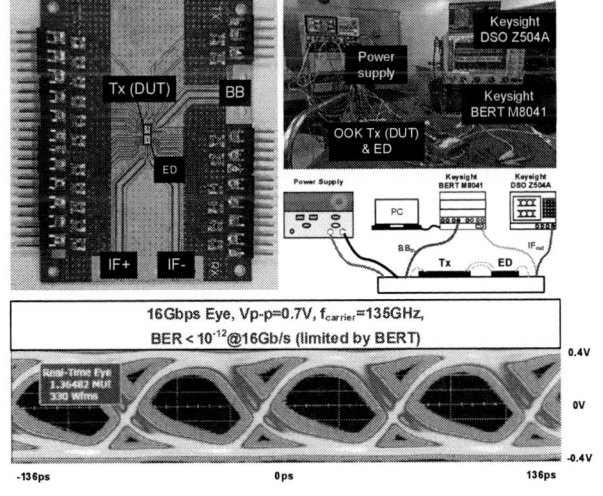

Fig. 5. The measured environment, setup and eye diagram.

Reference	This Work	ESSCIRC'21 [4]	JSSC'21 [5]	JSSC'19 [6]	JSSC'19 [7]
Technology	28nm CMOS	28nm CMOS	55nm SiGe	28nm CMOS	65nm CMOS
Topology	Source+Mod+PA	Osc+Mod+Buffer	Osc+Switch+Ant	DCO+Driver+PA	Modulator+OSC+Mixer
Modulation	ASK	ASK	OOK	CPFSK	PAM-4
Carrier Frequency (GHz)	130-140	135	220	140	127
RF BW (GHz)	110-160 (37%)	N/A	N/A	N/A	N/A
Pout (dBm)	>5	-3	4.2 (EIRP)	≧ 4	0
Symbol Rate (GBd)	> 16 (limited by BERT)	25 (DSP)	20	12	10
BER	10^{-12}	10^{-6}	N/A	10^{-12}	10^{-4}
RF interface	Single / Differential	Single	Single (Ant)	Ant	Differential
Power (mW)	167	43.2	165.2	65	50.8

Fig. 6. Performance comparison with state-of-the-art.

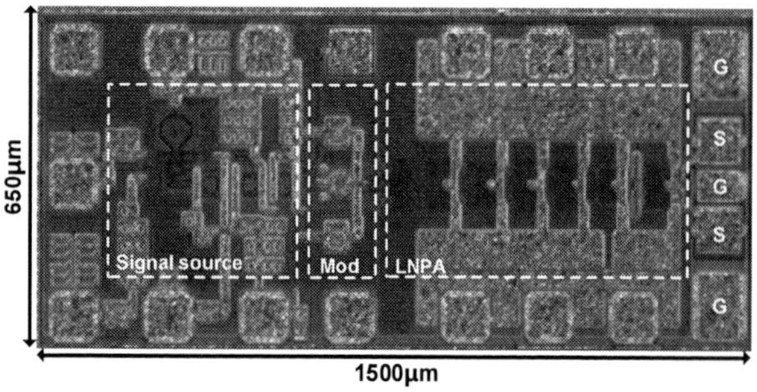

Fig. 7. Die micrograph of the 110-160GHz multi-mode ASK transmitter

Additional References:

[6] M. De Wit, Y. Zhang and P. Reynaert, "Analysis and Design of a Foam-Cladded PMF Link with Phase Tuning in 28-nm CMOS," JSSC, vol. 54, no. 7, pp. 1960-1969, July 2019.

[7] Y. Kim et al., "A Millimeter-Wave CMOS Transceiver With Digitally Pre-Distorted PAM-4 Modulation for Contactless Communications," JSSC, vol. 54, no. 6, pp. 1600-1612, June 2019.

A PAM4 Level Mismatch Adjustment Scheme for 48-Gb/s PAM4 Memory Tester Bridge

Daeho Yun[1,2], Minsu Park[2], Kahyun Kim[1], Kyungmin Baek[1], Eonhui Lee[2], Woo-Seok Choi[1], and Deog-Kyoon Jeong[1]

[1]Seoul National University, Seoul, Korea, [2]SK Hynix, Icheon, Korea

With the ever-increasing bandwidth of memory interfaces, including GDDR, securing link margins is becoming more critical. Starting with the production of memory that employs Pulse Amplitude Modulation-4 (PAM4) such as GDDR6X, each memory manufacturer aims at designing a robust interface with PAM4 signaling. However, direct test equipment is not available to verify the operation of the memory with the controller. For example, T5511 (ADVANTEST) Tester, an existing test solution using Automatic Test Equipment/System Level Test (ATE/SLT), does not support PAM4 signaling and testable data rates are limited to 8 Gbps. Therefore, development of innovative methods such as the Built-Out Self-Test chip (BOST) chip [1] is being designed to ease the high-speed testing as well as reducing the increased cost of new test equipment. Moreover, in the previous work [2], constant impedance matching was achieved by applying the current mode type of the PMOS input to the PAM4 TX operating at high speed. However, testing the memory requires adjustable tuning of the Ratio of Level Mismatch (RLM) as well as impedance matching. In this paper, we propose an optimal bridge architecture with all necessary functions that satisfies the extended function and higher bandwidth in conjunction with existing test equipment. In addition, we describe the method of output level adjustment in the PAM4 driver for improved the RLM. The bridge operates at 48 Gbps per pin and consumes 1.85 pJ/bit and 2.97 pJ/bit for write and read modes of the PAM4 memory, respectively.

Fig. 1 shows the proposed PAM4 Memory/Tester bridge that can test PAM4 signaling memory with a tester that supports only NRZ signaling. The proposed bridge employs a WCK signaling scheme in the GDDR5/6 interface. To support the write operation of the DRAM, the tester transmits NRZ data on 8 pins to the bridge chip with VSS-termination. The received data are serialized into LSB and MSB using 4-phase clocks, which an all-digital PLL (ADPLL) generates. The input level of the main driver, which is the output level of pre-driver, is adjusted in response to the calibration result, and calibration is performed at power-up to improve the RLM of the PAM4 signal. This 4-level adjustable main driver is employed to reduce the impedance variation of the output driver caused by the changes in the output level [3]. The PAM4 main driver is a single-ended voltage mode CMOS driver. The receiver incorporates 3 parallel CTLEs and 1-tap Decision Feedback Equalizer (DFE) along with digital adaptation algorithm. A Cherry-Hooper topology extends bandwidth while bisecting the upper, middle, and lower data levels. A direct-feedback DFE eliminates the first post-cursor, and error samplers are reused for XOR-based eye-opening monitoring (EOM). The low-frequency external WCK is phase- adjusted by a phase interpolator (PI) and a duty cycle corrector (DCC) for maximum frequency control within the bridge.

Fig. 2(a) illustrates the ideal driver's output impedance for each output level, consisting of parallel PMOS and NMOS transistors and three 3Z0 size transistors. The PAM4 binary bridge transmitter requires precise control over each level of the PAM4 signal and impedance matching capability. To produce a 1/2VDDQ output level, all three pull-up transistors are activated while all pull-down transistors are deactivated, creating an output level with an impedance matching that of the termination. For a 1/3VDDQ output level, two pull-up transistors and one pull-down transistor are turned on, resulting in an output impedance of 3Z0||1.5Z0= Z0. The pull-up and pull-down transistors' resistances and the termination resistance determine the output level and impedance. The third picture in Fig. 2(a) shows that even when the output level is 1/6VDDQ, it comprises a pull-up driver of 3Z0 and a pull-down driver of 1.5Z0. However, the implementation using real transistors results in non-linear driver operation that does not resemble the resistor in the figure. To achieve linear operation, corrections can be made by adjusting the width or Vgs according to Vds. Previous work [3] improved linearity

by modifying the width, but in this paper, the gate voltage was adjusted for fine resolution to support output level changes for the DRAM RX testing and the RLM improvement. Fig. 2(b) displays the Fig. 2(a) equation correction values α, β, γ, and δ, which require calibration circuits to fine-tune Vgs values for each output level.

Fig. 3 is the driver input level calibration circuit for the PAM4 TX. The calibration is performed at the beginning of the operation through a training operation of the memory. The calibration circuit consists of 6 types of replica drivers. It operates to adjust the transistor width and gate input level so that each replica driver can reach the main driver's target input level. The first step of the calibration is performed to find the widths of PMOS and NMOS suitable for process variation at 1/6VDD and 1/2VDD, respectively. The first calibrated condition is that PMOS and NMOS are driven without gate level limitation. After determining the transistor width, the calibration circuit determines the input gate level of the remaining four replica drivers. The reference level to adjust the gate input level is generated by the VDAC and selected according to the calibration result. After the calibration, the RLM is optimized. Moreover, the result of calibration code can be read out and it can be re-entered with a different value for evaluating the margin of the DRAM receiver.

Fig. 4 describes the circuit implementation of the PAM4 receiver analog-front-end. A Cherry-Hooper topology consists of a conventional RC-degeneration CTLE, a transimpedance stage (Gm cell), and an additional peaking stage by negative feedback of low-pass-filtered signal. By adding zero in the frequency response, CTLE further extends bandwidth by gain-boosting around the Nyquist frequency. The three threshold voltages, generated from voltage digital-to-analog converters (VDACs), are independently calibrated through the count-based EOM training sequence. This cancels random offset arising from transistor mismatch and data path, thus maximizing eye margin. 2dB to 8dB boosting at Nyquist 12GHz is provided. A 1-tap direct DFE is employed with adaptation algorithm of sign-sign least-mean square (SS-LMS). In each odd and even path, shared summer architecture [4] decreases parasitic capacitance to break bandwidth bottleneck of the CML structure. Furthermore, error samplers, used to track Dlev voltage, can also perform an XOR-based EOM operation [5].

The prototype chip is fabricated in 40nm CMOS technology and uses a total area of $2.13 \times 1.098 mm^2$. Fig. 5(a) and Fig. 5(b) is the measurement result of the RLM before and after calibration, and the RLM of the PAM4 driver is improved from 0.73 to 0.98 at 16 Gbps. Fig. 5(c) shows the data eye diagram obtained from the internal EOM circuit of the bridge. Fig. 5(d) shows the performance of the bridge's PAM4 RX characteristics. With the use of the PAM4 RX, the BER was measured at less than 10^{-12}, and an opened eye is obtained at 48 Gbps. Fig. 6 summarizes the performance and compares it to other works. In this work, the speed is improved from the existing built-out test method, and the PAM4 level can be adjusted, making it possible to test and characterize the memory interface.

References:

[1] H. Jin et al., "A 24Gb/s/pin PAM-4 Built Out Tester chip enabling PAM-4 chips test with NRZ interface ATE," *ASSCC*, pp. 1-3, Nov. 2021.

[2] D. Yun et al., "A 32-Gb/s PAM4-Binary Bridge With Sampler Offset Cancellation for Memory Testing," *IEEE TCAS-2*, vol. 69, no. 9, pp. 3749-3753, Sep. 2022.

[3] Y. –U. Jeong et al., "A 0.64-pJ/Bit 28-Gb/s/Pin High-Linearity Single-Ended PAM-4 Transmitter With an Impedance-Matched Driver and Three-Point ZQ Calibration for Memory Interface," *IEEE JSSC*, vol. 56, pp. 1278-1287, Apr. 2021.

[4] K. Lee et al., "An Adaptive Offset Cancellation Scheme and Shared-Summer Adaptive DFE for 0.068 pJ/b/dB 1.62-to-10 Gb/s Low-Power Receiver in 40 nm CMOS," *IEEE TCAS-2*, vol. 68, no. 2, pp. 622-626, Feb. 2021.

[5] P. -J. Peng et al., "6.1 A 56Gb/s PAM-4/NRZ transceiver in 40nm CMOS," *ISSCC*, pp. 110-111, Feb. 2017.

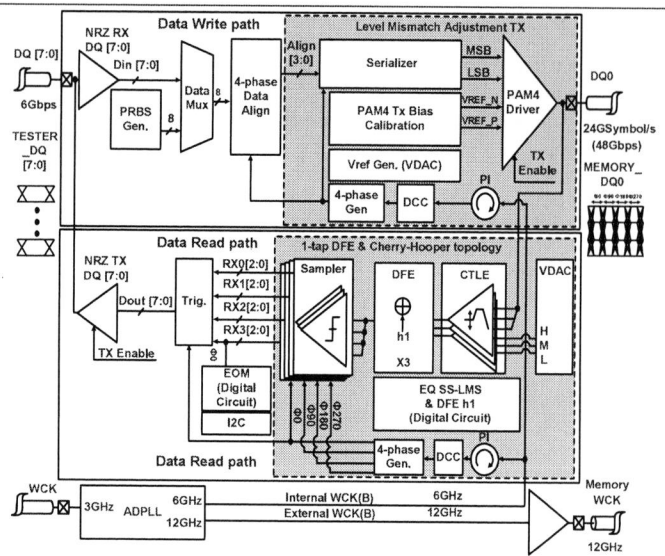

Fig. 1. Overall architecture of the proposed PAM4/Memory tester.

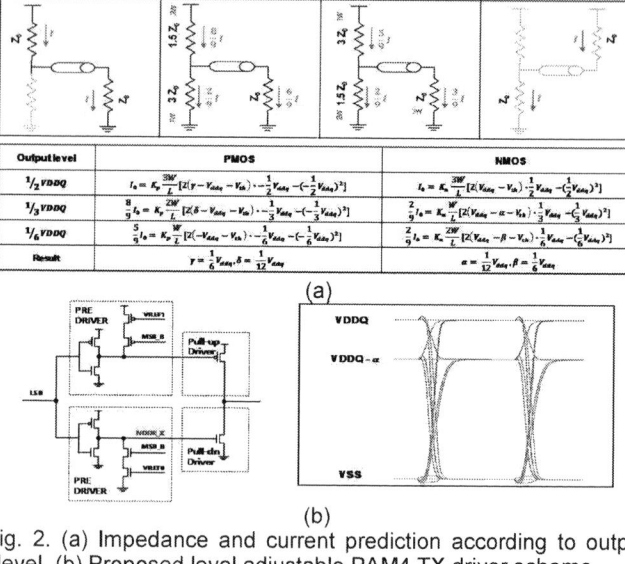

Fig. 2. (a) Impedance and current prediction according to output level. (b) Proposed level adjustable PAM4 TX driver scheme.

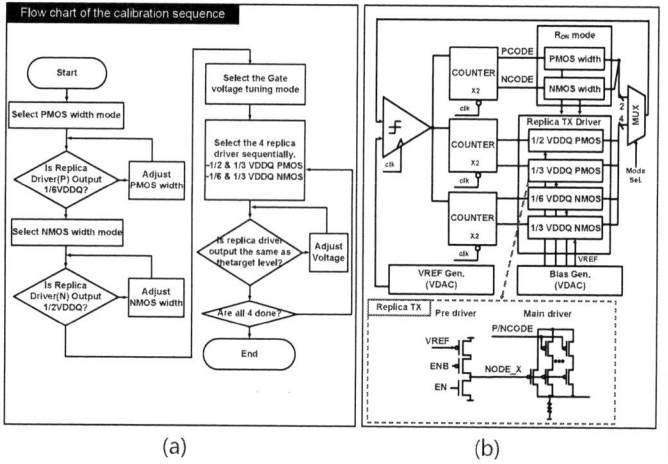

Fig. 3. (a) Flow chart of the calibration sequence and (b) Block diagram of PAM4 driver input level calibration circuit.

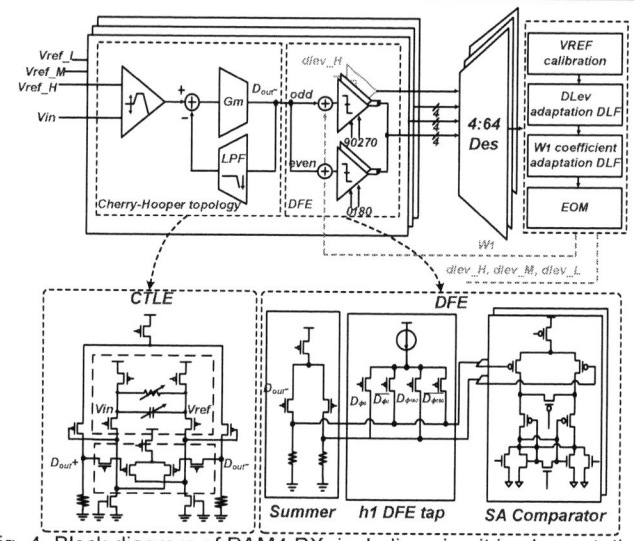

Fig. 4. Block diagram of PAM4 RX, including circuit implementation of analog front end.

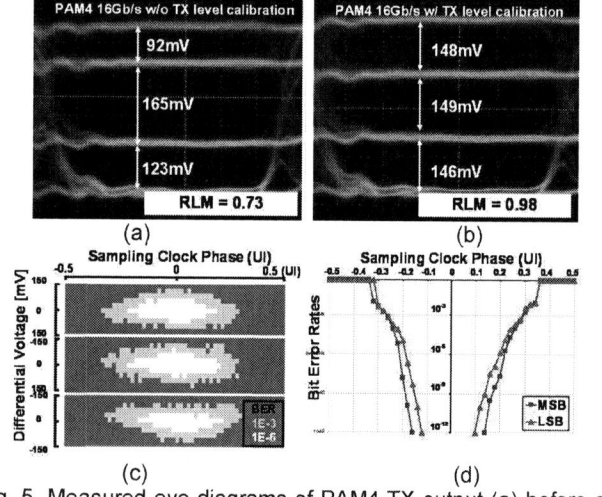

Fig. 5. Measured eye-diagrams of PAM4 TX output (a) before and (b) after calibration, (c) measured EOM, and (d) bathtub curve of PAM4 RX.

	ASSCC'21 [1]	TCAS-II'22 [2]	JSSC'21 [3]	ISSCC'17 [5]	This work
Technology	28nm	40nm	65nm	40nm	40nm
Supply(V)	1.2/1	1.25/0.9	1.0/0.6	1.5/1.0	1.25/0.9
Data rate(Gb/s/pin)	24	32	28	56	48
Tx Driver topology	Voltage-mode	Current-mode	Voltage-mode	Current-mode	Voltage-mode
Tx Equlization	-	2-tap FFE	2-tap FFE	3-tap FFE	-
Rx Slicer topology	1 tap DFE	CTLE	-	CTLE, 3-tap DFE	CTLE, 1-tap DFE
Signaling type	PAM4/NRZ	PAM4/NRZ	PAM4	PAM4/NRZ	PAM4
Clocking type	External	ADPLL	External	PLL	ADPLL
EOM function	-	O	O	O	O
Rx BER(10⁻⁹)	0.36UI @24Gbps	0.5UI @32Gbps	0.5UI @28Gbps	0.35UI @56Gbps	0.3UI @48Gbps
Tx Driver RLM	0.95 @24Gbps	0.95 @32Gbps	0.95 @28Gbps	-	0.96 @48Gbps
Energy effciency (pJ/bit)	-	3.66*	0.65**	3.57@Tx* 6.82@Rx	1.85@Tx* 2.97@Rx

*Includes Rx, PLL and EOM function   **Includes Rx and PLL

Fig. 6. Comparison table: overall performance of the proposed PAM4/Memory tester.

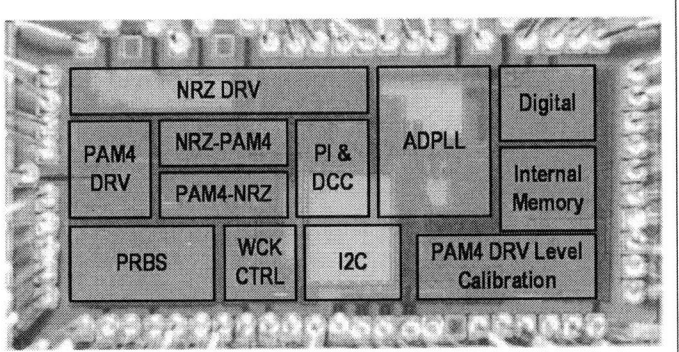

Fig. 7. Chip photomicrograph of the implemented bridge with the detailed area and power consumption

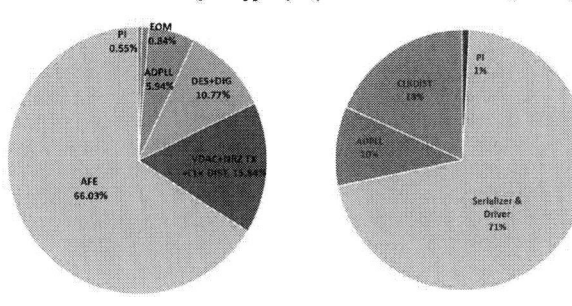

Fig. S1 . Power consumption of the proposed PAM4/Tester bridge.

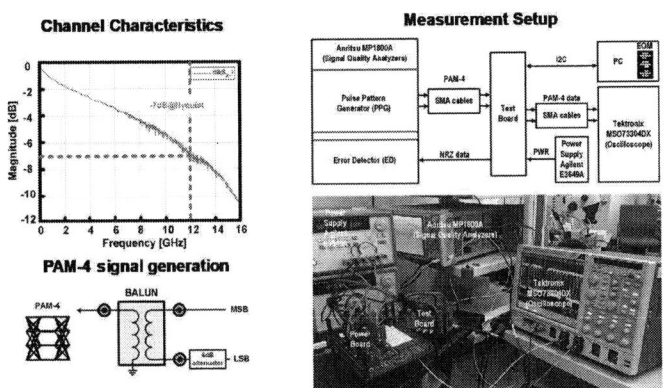

Fig. S2 Measurement setup. PAM-4 signal is made using a passive power divider and 6 dB attenuator. The 6-dB attenuator is utilized for the LSP pattern to lower the swing half. The measurement process is automated by using I2C with Python scripts. The automated measurement performs eye scanning, bathtub plotting, and jitter tolerance graph.

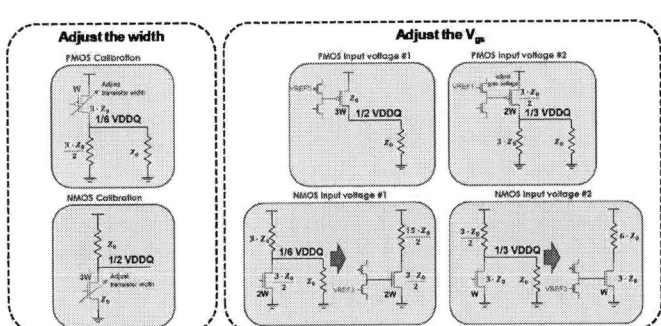

Fig. S3 . Replica driver configuration for Six driver level calibration. Two replica drivers are used to determine the width of the main driver, and four replica drivers are used to determine VREF values for the optimal RLM value.

An 8.73GΩ Input Impedance VCO-Based Neural Front-End with Autonomous Time-Domain Impedance Boosting Technique and Maximum 0.215s Re-Calibration Time

Huaiyu Liu[1], Yang Lin[1], Guoxing Wang[1], and Yan Liu[1]

[1]Department of Micro-Nano Electronics, Shanghai Jiao Tong University, Shanghai, China

Wearable and implantable electrophysiological sensors with integrated readout circuits provide compact solutions for biomedical applications. However, dry electrodes or integrated miniaturized electrodes used in these devices exhibit high interface impedance, therefore require readout circuits with even higher input impedance to prevent severe signal attenuation or CMRR degradation. This is more challenging for capacitively-coupled instrumentation amplifiers (CCIA), which is the most popular structure for power and area constrained designs. Negative capacitance or positive feedback technique has been widely used to compensate the impedance degradation due to coupling capacitors. However, because of the parasitics and random mismatch, precise compensation is challenging and over-compensation can cause instability of the front-ends. Therefore, foreground self-calibration scheme with programmable positive feedback DAC has been introduced recently to tune the positive feedback strength for compensation of the input current [1]-[3], thus the input impedance could be maximumly boosted. However, the calibration procedure usually requires a calibration signal to switch into the signal chain and perform foreground calibration with tens of seconds calibration time. Moreover, the extra switches and calibration signal generator will take extra silicon area and power budget, as well as introducing extra parasitics. A calibration scheme based on the thermal noise and voltage thresholding is proposed without extra calibration signal, but reference generator and comparators are required with extra silicon area and accuracy requirement [4].

To achieve ultra-compact design with minimized power consumption, high dynamic range (DR) direct-digitization front-ends were proposed with CTDSM-based AFEs. With VCO-based loop filter, the area and power consumption can be greatly reduced. However, achieving impedance boosting with direct digital conversion is challenging, since there is no intermediate analog signal output for detection or tunning. The VCO-based CTDSM shares a similar operation with the phase-locked loop (PLL). Therefore, the phase-locking condition can be monitored in real-time for stability detection with fully-digital circuits. In this paper, we propose a self-calibration input impedance boosting scheme to better incorporate with the capacitively-coupled VCO-based CTDSM. A time-domain duty-cycled positive feedback loop is proposed to relax the programmable DAC design. Phase-locking status is used as the stability flag, which can be instantaneously detected by a fully-digital stb-detector, making it possible to perform fast background auto-calibration.

Fig.2 shows the system-level architecture of the VCO-based neural front-end. The front-end is realized by a current-reused transconductor (GM), two current-controlled-oscillators (CCO), 11-stage phase-frequency detectors (PFD) and capacitor DACs. The input signal is converted into differential current outputs by the GM to drive the differential CCOs. The positive feedback loop consists of a 2-bits coarse loop and a 1-bit fine loop, which is modulated by the pulse-width modulation (PWM) wave. The fully-digital stb-detector adopts the frequency-domain level-crossing technique to monitor the phase-locking condition, enabling a fast background calibration process [5]. According to the stability information from the stb-detector, the calibration logic controls the 2-bits coarse positive feedback DAC and generates the PWM wave with specific duty-cycle.

Fig.3 presents the analysis of proposed time-domain impedance boosting technique and illustrates how the PWM modulation controls the positive feedback strength. Fourier analysis of PWM wave shows that the duty-cycle factor 'm' is linearly proportional to the DC component 'C_0'. Since signal modulation is direct multiplication, the DC component 'C_0' determines the positive feedback gain. In digital domain, signal modulation or multiplication can be easily achieved with XNOR operation. Therefore, by changing the duty-cycle of PWM wave, gain of the positive feedback loop can be controlled with relaxed DAC design. Note that it is also convenient to set polarity of the gain. Besides the DC component, PWM wave will bring high frequency harmonic components to the loop, causing stability issue. Considering the CTDSM theory, when calculating the signal transfer function (STF) of the CTDSM, the input signal could be deemed to be filtered by the continuous time filter STF(f) first, then got sampled afterwards. STF(f) has nulls at the frequency of $n \cdot Fs$, thanks to the anti-aliasing nature of CTDSM. This inspires us that if we set the frequency of the PWM wave to $n \cdot Fs$, high frequency harmonic components can be perfectly removed in the loop with no power and hardware overhead.

Fabricated in a 180 nm CMOS process, the prototype occupies an active area of 0.0977 mm². Sampling frequency and PWM frequency are both set to 1.28 MHz. An ELD of 25% Ts is set in the re-timing block to absorb logic delays. Oscillation frequency of CCO is around 300 kHz. Fig.4 shows the measured output power spectral density and dynamic range performance. With 640 kHz chopping frequency, the total power consumption is 24.38 μW. The input referred noise is 1.76 μVrms over 0.5 Hz-1 kHz bandwidth. The overall front-end achieves 72.4 dB SNDR, 87.6 dB SNR, 75.6 dB SFDR and 90.6 dB DR within 1 kHz bandwidth. The fully digital stb-detector only occupies 0.0022 mm² and consumes 0.54 μW.

Fig. 5. (a-b) presents the test setup and measured results of the input impedance measurement. In order to reduce the interference in the environment, the test platform is put into the shielding box, and the platform is powered by battery. The shielding box has a standard ethernet network interface. A FPGA collects the output data from the chip and send the data to the host computer through the router for further processing. The 1GΩ resistor is connected in series at the input, forming a voltage divider with the input impedance. Input impedance can be calculated with the signal power after FFT. Since the large source impedance (the 1GΩ serial resistor) will inevitably raise the noise floor[2], This method can calculate the input impedance more accurately under large noise interference. Measured results shows that the proposed time-domain impedance boosting technique can effectively boost the input impedance. Fig. 5. (c-d) shows the auto-calibration process and measured key signal waveforms. The total calibration time is at maximum of 0.215s from the chip reset.

Fig. 6. summarizes the performance of the chip and compares with state-of-the-art neural-recording IC designs with impedance boosting technique. The proposed method is the first background impedance boosting technique for direct-digitization front-ends. Owning to the CTDSM structure and the time-domain calibration method, the occupied silicon area are saved for both core readout and calibration circuits. With the proposed time-domain background calibration scheme, the tunning resolution can be increased without limitation of the positive feedback DAC LSB, and the calibration tracking time is reduced to a maximum of 0.215s.

Acknowledgements:

This work was supported by National Key Research and Development Program of China (2022ZD0208500), Shanghai Municipal Science and Technology Major Project (2021SHZDZX).

References:

[1] C. -H. Chang et al., "An Analog Front-End Chip With Self-Calibrated Input Impedance for Monitoring of Biosignals via Dry Electrode-Skin Interfaces," IEEE TCAS-I, 64(10), pp. 2666-2678, Oct. 2017.

[2] J. Lee et al., "An Ultra-High Input Impedance Analog Front End Using Self-Calibrated Positive Feedback," IEEE JSSC, 58(8), pp. 2252-2262, Aug. 2018.

[3] Y. Park et al., "A 3.8-μW/Ch, 15-GΩ Total Input Impedance Chopper Stabilized Amplifier with Dual Positive Feedback Loops and Auto-calibration Scheme," VLSI, June, 2021.

[4] P. Wang et al., "A 136 GΩ-Input-Impedance Active Electrode for Non-Contact ECG Using Auto-Calibrated Positive Feedback and Capacitance Scaling in Femtofarad Resolution," ESSCIRC, Sep. 2022.

[5] H. Liu et al., "Analysis and Design of VCO-Based Neural Front-End With Mixed Domain Level-Crossing for Fast Artifact Recovery," IEEE TCAS-I, 70(3), pp. 1214-1227, Mar. 2023.

979-8-3503-3004-5/23 $31.00 © 2023 IEEE

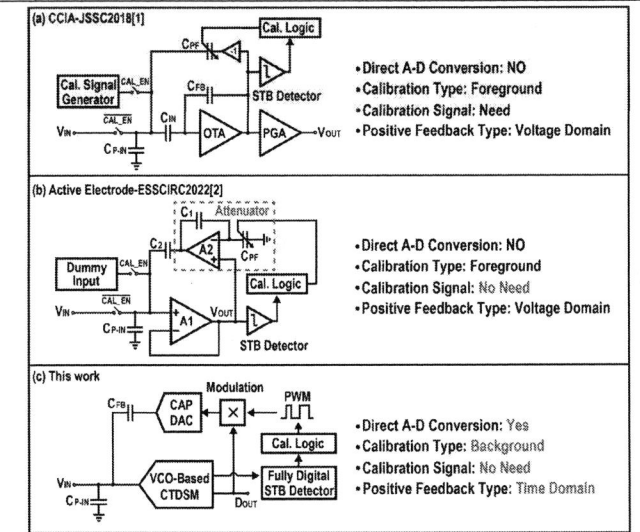

Fig. 1. Auto-calibrated positive feedback impedance boosting technique: prior arts vs. proposed time-domain approach.

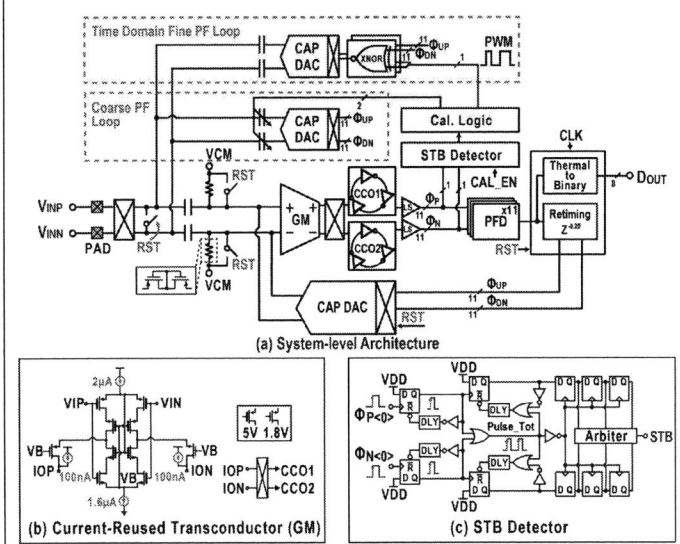

Fig. 2. System-level architecture of the VCO-based neural front-end and circuit implementation of key blocks.

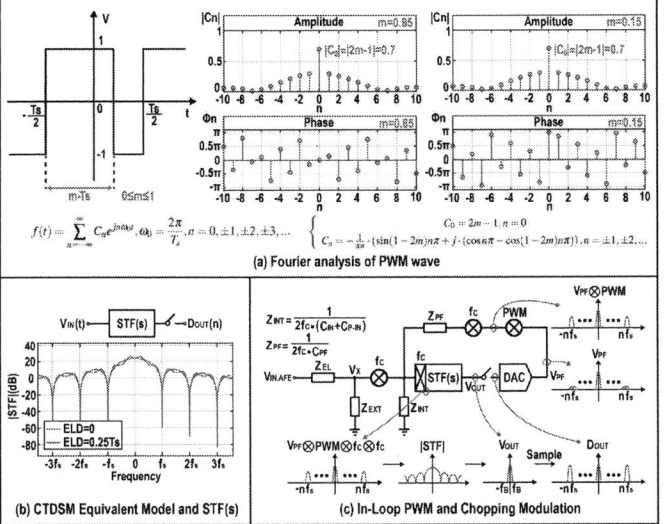

Fig. 3. Analysis of proposed time-domain impedance boosting technique.

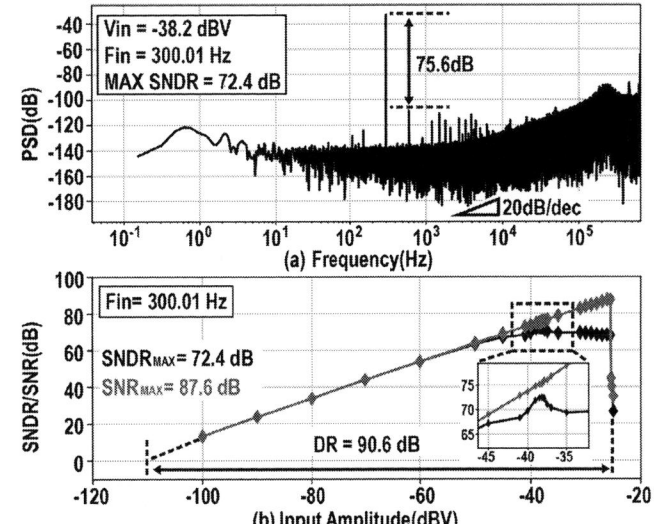

Fig. 4. (a) Measured output power spectral density, (b) dynamic range performance.

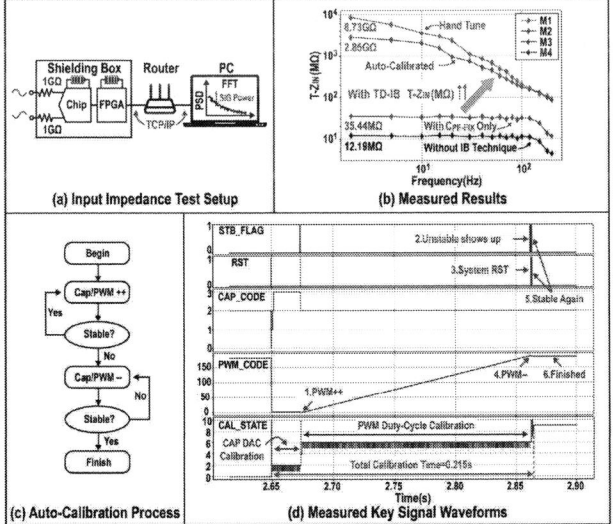

Fig. 5. Input impedance measurement: (a) test setup, (b) result, (c) auto-calibration process, (d) measured key signal waveforms.

Fig. 6. Performance summary and comparison with state-of-the-art neural-recording IC designs with impedance boosting technique.

	This work	VLSI2021[3]	JSSC2018[2]	TCAS-I2017[1]	JSSC2017[6]/ JSSC2018[7]	ISSCC2020[8]	ESSCIRC2022[4]
Technology (nm)	180	110	180	130	40	180	180
Architecture	VCO-Based CTDSM	CCIA+ADC	CCIA+ADC	CCIA	CCIA+ADC	CCIA	Unity-Gain Buffer
Supply(V)	1.8(A)/1.2(D)	1/1.5	0.8	1.255	1.2	1.8	1.2
Power/Ch(μW)	24.38	3.83	0.289	41.6	7.3	3.35	10.46
Bandwidth(Hz)	1k	0.5-300	1-400	0.5-45.5	1-200	0.5-200	0.5-100Hz
IRN (μVrms)	1.76	0.36	8.26	3.75	1.4	1.73	0.72
Input range(Vpp)	150m	~1m[a]	~16m[a]	~0.6m[a]	200m	~9m	700m
Area/Ch (mm2)	0.0977	0.7457	0.58	0.955	0.113	0.2268	0.1855
SNDR(dB)	72.4	59.4	61.97[b]	N/A	86[d]	N/A	N/A
DR(dB)	90.6	~59.8[c]	~56.7	~35.0[c]	90[d]	~65.3	~110.7
CMRR(dB)	>82	92	66	77.6	78	130	>95
PSRR(dB)	>101(A)>73(D)	100	69	74	76	>100	N/A
Input Impedance	8.73GΩ@2Hz	15GΩ@10Hz	200GΩ@1Hz/50GΩ @50Hz	1.42GΩ@20Hz	1.5GΩ@DC	5GΩ@10Hz/1.6G Ω@50Hz	136GΩ@60Hz
Input Impedance Boosting Type	Time-Domain Positive Feedback	Dual-Loop Positive Feedback	Active Shield Positive Feedback	Active Shield Positive Feedback	Pre-Charge	Active Shield Positive Feedback	Positive Feedback Capacitance Scaling
Calibration Type	Background	Foreground	Foreground	Foreground	N/A	N/A	Foreground
Calibration Time	0.215s	32.64s	~11s	~100s	N/A	N/A	N/A
Block Under Consideration	VCO-Based CTDSM	IA+PGA	IA+PGA+ADC	IA+PGA	IA+CTDSM	IA+PGA	Active Electrode

a. Estimated from gain and power supply; b. Performance of the ADC without CCIA; c. Estimated from input range and IRN; d. Performance of the ADC with CCIA.

979-8-3503-3004-5/23 $31.00 © 2023 IEEE

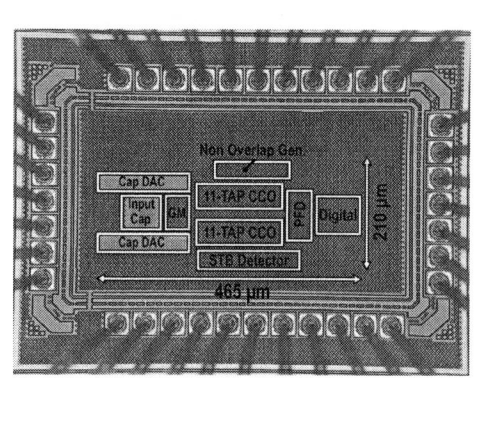

Fig. 7. Die micrograph of the proposed neural front-end.

Additional References:

[6] H. Chandrakumar *et al.*, "An 80-mVpp Linear-Input Range, 1.6-GΩ Input Impedance, Low-Power Chopper Amplifier for Closed-Loop Neural Recording That Is Tolerant to 650-mVpp Common-Mode Interference," *IEEE JSSC*, **52**(11), pp. 2811-2828, Nov. 2017.

[7] H. Chandrakumar et al., "A 15.2-ENOB 5-kHz BW 4.5-μW Chopped CT ΔΣ-ADC for Artifact-Tolerant Neural Recording Front Ends," *IEEE JSSC*, **53**(12), pp. 3470-3483, Dec. 2018.

[8] S. Zhang et al., "A 130dB CMRR Instrumentation Amplifier with Common-Mode Replication," *ISSCC*, Feb. 2020.

GCFP-ACIM: A 40nm 4.74TFLOPS/W General Complex Float-Point Analog Compute-in-Memory with Adaptive Power-Saving for HDR Signal Processing Applications

Zizhao Ma[1], Xianwu Hu[1], Gan Wen[1], Yihao Wang[1], Zeming Wang[1], Yukai Lin[1], Yu Wang[1], Yuhao Guo[2], Yanlei Li[2], Xingdong Liang[2], Xiaoyang Zeng[1], Yufeng Xie[1*]

[1]State Key Laboratory of Integrated Chips and Systems, Fudan University, Shanghai, China

[2]Aerospace Information Research Institute, Chinese Academy of Sciences, Beijing, China

Edge signal processors have difficulty balancing the high throughput and high dynamic range (HDR) requirements of modern digital signal processing (DSP) such as digital beamforming (DBF) and pulse compression due to size, weight and power (SWaP) constraints. Compute-in-Memory (CIM) has proven to be an energy-efficient and high-throughput solution, reducing the data transfer on the bus. However, as shown in Fig. 1, when applied in DSP, the previous CIM macros face new challenges: 1) general-purpose floating-point (FP) CIMs based on logic gates require long pipeline delays[1,2], artificial intelligence (AI) FP CIMs based on mantissa pre-alignment cannot handle HDR data due to quantization loss in pre-alignment[3-5], 2) integer-based analog CIMs require large readout bit widths to satisfy the signal-to-noise ratio (SNR) of DSPs, resulting in larger power consumption and area[6,7], 3) CIMs for real matrix require six rounds of sequential operations to achieve complex matrix multiplication in DSP, resulting in significant data storage and transfer overhead.

To overcome the challenges in CIM designs for DSP applications, this work proposes a general complex floating-point analog CIM (GCFP-ACIM) chip featuring: 1) A general-purpose analog-domain parallel input FP multiplication-in-memory (FMIM) macro for HDR data without quantization loss based on 8T-SRAM bit cells, supporting FP32 floating-point weights (FPW). 2) A power-saving weight-indexed and input-aware adaptive readout array (ARA) that reduces power consumption in analog computation through input peak detection and high-order mantissa block indexing of weights. 3) An energy-efficient and high-throughput complex matrix CIM systolic array that significantly reduces transfer energy by reusing input data and FMIM weights in each processing element (PE). Compared with previous logic-based FP CIMs, the proposed FMIM achieves 5.8x energy efficiency and 95% latency reduction by in-memory multiply-and-accumulate after mantissa unfolding. In addition, ARA reduces 55% of readout energy in the FP32 pulse compression test by dynamically gating the voltage sense amplifier. The proposed GCFP-ACIM chip, fabricated by 40nm technology, measured a maximum DBF processing error of 0.0133% at FP32 inputs and weights.

Fig. 2 illustrates the overall structure and operation of the general-purpose analog-domain parallel FMIM macro for HDR multiplication. The proposed FMIM can compute FP multiplications for any dynamic range in fewer cycles without significant errors, unlike pre-aligned FP CIMs. The FMIM macro contains a multi-precision mantissa analog CIM array (MP-MACA) based on 8T-SRAM CIM bit cells, adaptive readout array, multi-precision timing controller (MP-TC), local adder tree (LAT), partial product accumulator (PDAcc) and exponent adder. The 64x64b MP-MACA is divided into 8x8b blocks for the mantissa of different FP precision. The last row stores the FPW_M from MSB to LSB, shifted one bit for each line up (LSB Dropout). As shown in Fig. 2 (right), during CIM, MP-TC splits the FP input mantissa (FPI_M) into 8 bits and activates MP-MACA by blocks per cycle. Then, ARA and LAT convert the 8 BL voltages into an 11-bit partial product (PD). The higher bits in the high-order block of MP-MACA have fewer activated bit cells, resulting in less fluctuation of the BL voltage. This eliminates the calculation error of the higher-order bits of mantissa, which is critical in HDR DSP algorithms. Considering process variations, the MP-MACA shows less than 0.21% error data with deviations below 1 LSB in Monte Carlo simulations. The mantissa product is obtained by PDAcc through the multi-cycle accumulation of PDs, according to the FP structure with different precision. The floating-point product of FMIM can be obtained by combining the shifted mantissa product with the exponent adder output. Additionally, the proposed FMIM can be programmed to support common data types such as bFP16, FP24, FP32 and extended precision FP40 to cover various DSP algorithms.

Fig. 3 illustrates the circuit structure and power-saving method of MP-MACA and ARA. The ARA consists of an adaptive energy-efficient controller, analog CIM driver & sampling circuits, and 64 voltage sense amplifiers (VSAs). The CIM current generated by the activated cascode transistor in MP-MACA is converted to voltage readout by the sampling circuit. In ARA, this work proposes block-level weight-indexed and bit-level input-aware approaches to dynamically gate partially sensing amplifiers (SAs) to reduce sensing energy. This is because there is almost no sparse data in the DSP's input and weights, unlike machine learning. The block-level weight-indexed controller reduces the enabled VSAs from LSB to MSB for high-order mantissa blocks whose higher bits' effective bit cells are reduced column by column due to the shift unfolding of the mantissa. That is, there is only one enabled VSA in the MSB column. The bit-level input-aware controller limits the maximum number of enabled VSAs in each column by predicting the maximum possible result of the multiply-and-accumulate (MAC) based on the count of the split 8-bit mantissa input. Note that both weight-indexed and input-aware approaches will be valid for the high-order blocks.

Fig. 4 shows the GCFP-ACIM chip's architecture including the proposed complex matrix CIM systolic array consisting of 8x8 complex floating-point multiply-and-accumulate (CFPMAC) PEs that can reduce over 60% of the storage overhead and transfer of complex data. The proposed CFPMAC contains real part FMIM (R-FMIM), imaginary part FMIM (I-FMIM), complex FP multiplication controller (CFMC) and systolic accumulator. The CFMC detects the input FP exceptions first. Then, I-FMIM and R-FMIM perform real and imaginary multiplication in two separate cycles. The CFMC swaps the real and imaginary inputs in the second cycle to reuse the input and weight data. Finally, the systolic accumulator gets the maximum exponent from CFMC and then performs the shift and accumulation. Additionally, the system agent of GCFP-ACIM enables independent operation of each CFPMAC PE via the internal bus, facilitating other requirements like parallel general-purpose FP multiplication.

The prototype of the proposed GCFP-ACIM chip was fabricated in 40nm CMOS technology (Fig. 7). Fig. 5 shows the test system and experimental results of the prototype. The waveform shows the measured critical path delay of the analog CIM in FMIM, where the BL voltage buildup time is 7.02 ns and the ARA readout time is 2.24 ns. The prototype underwent verification for DBF and pulse compression experiments in a DSP system. The measured maximum errors of the chip for DBF processing with bFP16, FP24, and FP32 accuracy are 3.12%, 0.0173% and 0.0133%, respectively. The pulse compression results exhibit good consistency with the software simulation. The weight-indexed and input-aware ARA measured 55% energy saving in the pulse compression experiment compared to the classical voltage readout array. The prototype measured energy efficiency of 4.74 TFLOPS/W for bFP16 and 0.58 TFLOPS/W for FP32 after adding 50% sparse data to the complex signal processing. The prototype operated at 0.89-1.1V and measured a 25 ns latency for bFP16 at 1.1V with an 80 MHz clock.

Fig. 6 presents a comparison of the recently proposed floating-point CIM macros. In the prototype, FP32 throughput still surpasses the logic-based FP CIM despite storing multiple non-sparse, dynamically changing DSP weights and ensuring accuracy through eight-fold redundancy. Since DSP requires high dynamic range floating-point support, the figure of merit (FoM) introduces computational error as a factor. The error is measured using bFP16 pulse compression data.

Acknowledgement:

This work was supported by the National Natural Science Foundation of China under Grant 61874031.

References:

[1] S. Jeong et al., "A 28nm 1.644TFLOPS/W Floating-Point Computation SRAM Macro with Variable Precision for Deep Neural Network Inference and Training," ESSCIRC 2022:145-148.

[2] J. Wang et al., "A 28-nm Compute SRAM With Bit-Serial Logic/Arithmetic Operations for Programmable In-Memory Vector Computing," JSSC 2020, vol. 55, no. 1, pp. 76-86.

[3] P. -C. Wu et al., "A 22nm 832Kb Hybrid-Domain Floating-Point SRAM In-Memory-Compute Macro with 16.2-70.2TFLOPS/W for High-Accuracy AI-Edge Devices," ISSCC 2023:126-128

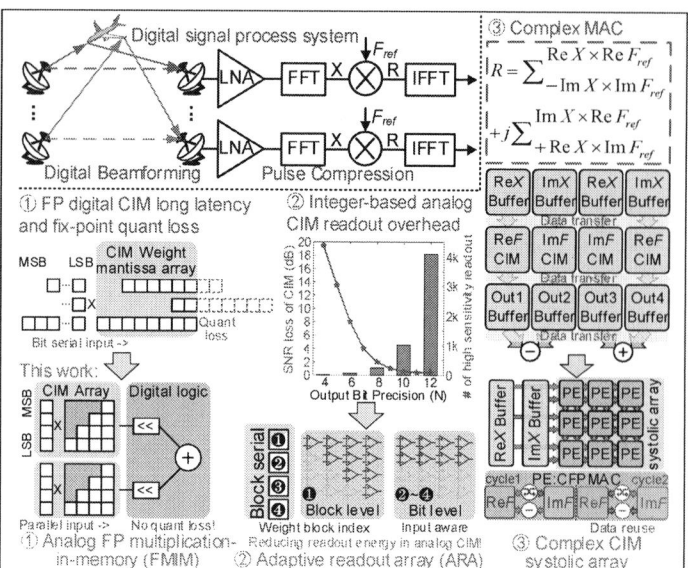

Fig. 1. Challenges of DSP and proposed GCFP-ACIM feature.

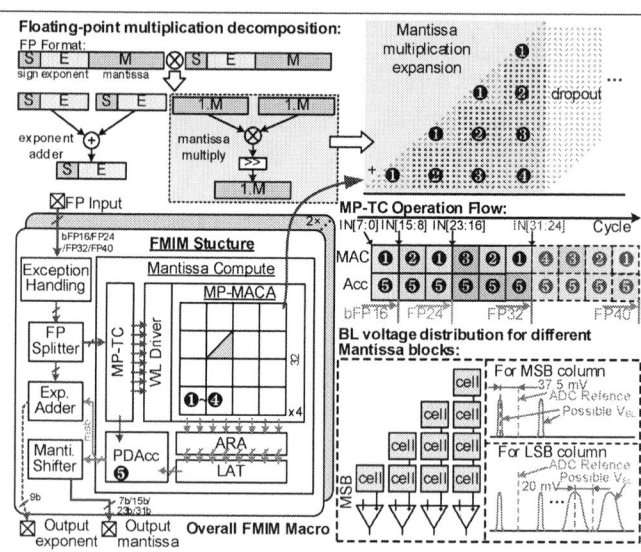

Fig. 2. Structure and operation of general-purpose analog-domain parallel float-point multiply-in-memory (FMIM) macro.

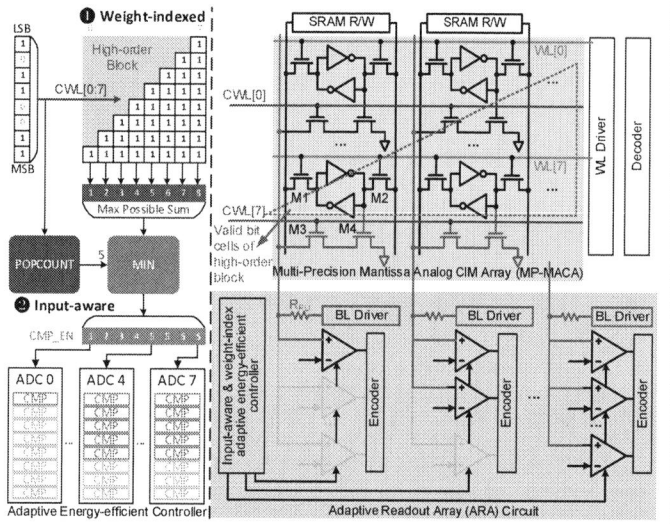

Fig. 3. Structure and power-saving method of MP-MACA and weight-indexed & input-aware adaptive readout array (ARA).

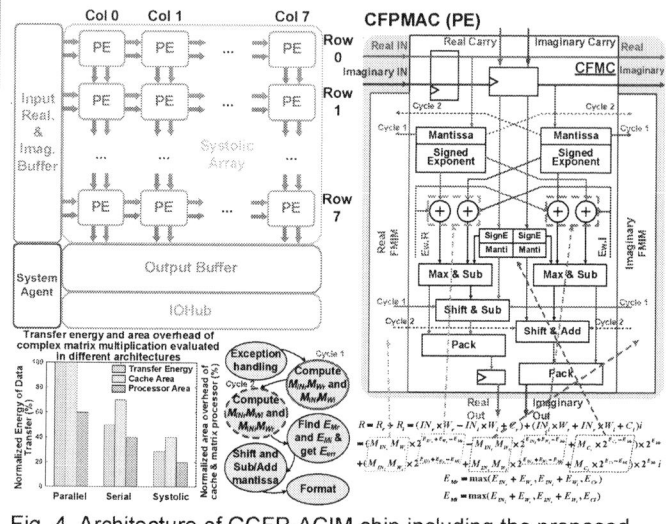

Fig. 4. Architecture of GCFP-ACIM chip including the proposed complex matrix CIM systolic array consisting of CFPMAC PEs.

Fig. 5. Test environment, measured waveform and performance.

	ISSCC '23[3]	ISSCC '23[4]	ESSCIRC'22[1]	ISSCC '22[5]	VLSI' 21[8]	This work
Technology	22nm	28nm	28nm	28nm	28nm	40nm
CIM type	Hybrid	Digital	Logic	Digital	Logic	Analog
Macro size	832Kb	64Kb	16Kb	96Kb	1280Kb	512Kb
Input/Weight precision	BF16	BF16	FP8-32	BF16/ FP32	BF16	BF16-FP32
Output precision	FP32	BF16	FP8-32	BF16/ FP32	BF16	BF16-FP32
Complex support	No	No	No	No	No	Yes
Latency[1](ns)	NA	47.6	446.6	~2390	50	25
Access time(ns)	6.4	6.8	NA	NA	NA	9.28
Throughput[2] (GFLOPS)	1240 @BF16	299 @BF16	4.84	140	119.4 @BF16	5.08
Energy Efficiency (TFLOPS/W)	14.04 @BF16	16.22 @BF16	0.099 @FP32	29.2 @BF16 3.7 @FP32	1.43 @BF16	4.74 @BF16 0.58 @FP32
Error in DSP[3]	3.54%	6.34%	1.27%	7.98%	0.57%	1.07%
FoM[4]	253	164	45	234	161	312

[1] Latency is compared with the accuracy of bFP16
[2] Throughput is measured at the precision of FP32
[3] Measured at BF16 accuracy to ensure consistency, data from pulse compression.
[4] FoM = (input mantissa+1) x (weight mantissa+1) x energy efficiency / (Error x 100).

Fig. 6. Comparison table.

979-8-3503-3004-5/23 $31.00 © 2023 IEEE

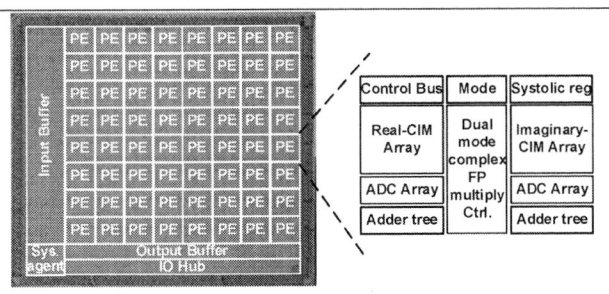

Technology	40nm CMOS			
Macro size	512Kb			
Macro area	17.94mm²			
Frequency	80MHz			
Input / Weight / Output Precision	BF16	FP24	FP32	FP40
Number of systolic	64			
Supply voltage (V)	0.89-1.1			
Access time (ns)	9.28			
Compute latency (ns)	25	50	87.5	137.5
Energy efficiency (TFLOPS/W)	4.74	1.33	0.58	0.38

Fig. 7. Die micrograph and summary table.

Additional References:

[4] A. Guo et al., "A 28nm 64-kb 31.6-TFLOPS/W Digital-Domain Floating-Point-Computing-Unit and Double-Bit 6T-SRAM Computing-in-Memory Macro for Floating-Point CNNs," ISSCC 2023:128-130

[5] F. Tu et al., "A 28nm 29.2TFLOPS/W BF16 and 36.5TOPS/W INT8 Reconfigurable Digital CIM Processor with Unified FP/INT Pipeline and Bitwise In-Memory Booth Multiplication for Cloud Deep Learning Acceleration," ISSCC 2022:1-3.

[6] H. -M. Ou and N. R. Shanbhag, "Enhancing the Accuracy of Resistive In-Memory Architectures using Adaptive Signal Processing," ICASSP 2023:1-5

[7] S. -E. Hsieh et al., "7.6 A 70.85-86.27TOPS/W PVT-Insensitive 8b Word-Wise ACIM with Post-Processing Relaxation," ISSCC 2023:136-138

[8] J. Lee et al., "A 13.7 TFLOPS/W Floating-point DNN Processor using Heterogeneous Computing Architecture with Exponent-Computing-in-Memory," VLSI 2021: 1-2.

A 0.000261 mm² Single-Channel 1 GS/s 8-Bit 3-Stage Capacitor Array-Assisted Charge-Injection DAC-Based SAR ADC in 28 nm CMOS

Chan-Ho Kye[1,2] and Kyojin Choo[1]

[1]EPFL, Neuchâtel, Switzerland

[2]University of Michigan, Ann Arbor, MI, USA

The demand for high-performance and area-efficient SAR ADC has been rising over the past years for ADC-based wireline receiver and mixed-signal computing. The small size of the sub-ADC brings better accuracy and throughput in these array-based applications, because the small implementation mitigates the heavy routing and parasitic effects, thereby improving energy efficiency, signal integrity, and speed. A capacitor-array DAC (CDAC)-based SAR ADC [1] is the de-facto standard architecture for these needs, however, the exponential trade-off between resolution and the CDAC size imposes a limitation on the area-efficiency where more compact designs are desired. CDACs with bridge capacitors alleviate this trade-off, however, the gain error between the stages tolls significant complexity for calibration, i.e., needing a high-speed reference driver or a trimming array. The charge-injection DAC (ciDAC) can achieve higher density by exploiting its cascade-ability as it facilitates the gain error calibration through its (zero-load) bias terminal. Furthermore, hybridizing the CDAC and ciDAC synergistically achieves higher performance density (characterized by information rate density or IRD, TSps·conv-steps/mm² [2]) while taming the artifacts of each architecture (e.g., exponential area, limited input range). This paper demonstrates a 3-stage capacitor array-assisted charge-injection DAC-based SAR ADC (c-ciSAR ADC) achieving the highest IRD. The prototype ADC with 1 GS/s 8-bit resolution achieves a state-of-the-art IRD of 470 TSps·conv-steps/mm² from a 1 V supply in 28 nm CMOS.

The N-stage c-ciDAC structure is illustrated in Fig. 1(a), where the structure is generalized to have one-stage of CDAC followed by multiple stages (N-1) of ciDACs. Based on this structure, we first sought an optimal combination of parameters (i.e., bit allocation, number of stages, and degree of ci-cell reuse) to achieve the highest IRD. Through a parameterized model with all structural variants (exhaustively including signal, noise, distortion, size, and speed effects of the ADC components), it was revealed that there is an optimal construction for IRD at one-stage 3-bit (CDAC) + (2,2)-bit (two-stage ciDACs) for an 8-bit ADC (Fig. 1(b)). An optimal bit allocation for the CDAC is around 2 or 3 bits as the area exponentially suffers when made with a larger bit allocation (although benefiting from a wider input signal range). Also, the ciDAC is designed with the least degree of reuse and is cascaded into two stages (2 ci-cells per 2-bit ciDAC) as reuse imposes an exponential slowdown while a single-stage ciDAC construction requires a larger area of ciDAC output capacitor. This parametrized modeling experiment yields that the maximum IRD achieved through the hybridization of CDAC and ciDACs is roughly 2.4x higher than the optimal CDAC-only design.

The resulting c-ciDAC is shown in Fig. 2(a). The 3-bit differential CDAC is implemented with 1.25 fF unit MOM capacitors (a total of 2 x 13 fF at the sampling node, including the bridge capacitance) followed by the two stages of 2-bit pull-down type ciDACs. The first bridge capacitor (C_{br1}) is set as a relatively large (4.3 fF) portion of the input capacitance to sufficiently couple the ciDAC swing (~ 300 mV) into the overall DAC (~ 2 x 700 mV differentially). However, as a side-effect, the CDAC operation is also reversely coupled into the ciDAC, reducing its usable output linear range. This is a critical threat to achieving high IRD as it demands more bits allocated to the CDAC. For mitigation, we employed two measures: 1) increasing the C_{qv1} to lower the reverse coupling gain, and 2) operating the CDAC to step up-then-down such that reverse coupling helps to operate within the ciDAC output linear range. Accordingly, we used a moderately large C_{qv1} (11.6 fF) carefully chosen to avoid the consequence of needing extremely large C_{br2} and C_{qv2}. Consequently, the DAC area of the cap network is extremely small through hybridization. Also, as the reset state of ciDAC is based at V_{DD} and operates downward, the first CDAC bit operation is designed to step up (and then the rest steps down) which neutralizes the reverse coupling impact on the

ciDAC output linear range while also maintaining the optimal common mode voltage of the comparator for noise critical LSBs (Fig. 2(b)).

The ci-cell in the proposed ciDAC is designed as a differential pull-down type with 2.3 fF of source capacitor while applying the SNR boosting technique [3] to enhance the speed, noise, and DC dependence. Despite this enhancement, the ci-cell in 28 nm CMOS technology still exhibits significant output DC dependence (7.1 % min-max difference across 1 V output change) which could degrade the performance (Fig. 3(a)). To address this issue, we implement the output capacitor for the ciDAC as a combination of MOM capacitor and native MOS capacitor with a 6:4 ratio. Then, the positive voltage coefficient of the ci-cell (in output charge) is effectively canceled out by the positive voltage coefficient of the native MOS capacitance. Introducing a high-density MOS capacitor also improves the density whereas the corner variation can be easily trimmed out using the zero-load bias voltage of ci-cells (Fig. 3(b)). Also, as the DC-dependent ci-cell response manifests only as a slight gain error (rather than INL/DNL) while the ciDAC is differential with a common starting voltage, its impact is further reduced systematically.

A synchronous control is employed to relax the controller complexity. Consequently, the metastability of the comparator becomes a critical design point since the valid decision may not timely arrive at the point of operating the DAC. This is a great concern for the ciDAC as the comparator metastability may cause its activation pulse to disappear resulting in the ADC skipping a DAC cycle. To address this issue, we implement a metastability detector that observes if the comparator's output has not fully transitioned out of the reset state for any cycles. This indicator is then used to update the sticky metastability mask flag that pads the subsequent decisions (down to LSB-1) to a fixed sequence of '0111...' when active (Fig. 4). This way, the ADC can progress without having to wait for the comparator to resolve the decision fully [4], yet still, error due to the metastability is contained within +/- 1 code from the ideal value. The measured output spectrum shows that the ADC sampled at 1 GS/s exhibits SNDR of 44.3 dB at the 132.8125MHz input and SNDR of 43.5 dB at the 492.1875MHz input with decision padding.

The measured results are shown in Fig. 5. INL and DNL are +0.69/-0.75 LSB and +0.8/-1.0 LSB, respectively. The measured input full-scale range is 2 x 700 mV$_{pp,diff}$. The ADC shows < 0.15 bits of ENOB variation across 31.25 to Nyquist input frequencies sampled at 1 GS/s. The SNDR and SFDR are 43.5 dB and 54.1 dB, respectively for the Nyquist input frequency. The power consumption is 2.61 mW from a 1.0 V core supply, and the reference power is not included as the differential ADC does not consume static power from references by construction. A 6.94 ENOB at the Nyquist rate is measured, achieving an IRD of 470 TSps·conv-steps/mm² and FoM$_w$ of 21.3 fJ/conv-steps (Fig. 6). Fig. 7 shows the chip photomicrograph fabricated in 28 nm CMOS. It occupies an active core area of 0.000261 mm², which is 1.2x smaller than the state-of-the-art [5] of 0.000313 mm² (used 5 nm technology). The prototype ADC has 1.92x higher IRD than the state-of-the-art [6] and has the highest IRD among all surveyed area-efficient single-channel Nyquist rate ADCs.

References:

[1] L. Kull et al., "A 3.1mW 8b 1.2GS/s Single-Channel Asynchronous SAR ADC with Alternate Comparators for Enhanced Speed in 32nm Digital SOI CMOS," ISSCC 2013:468-469.

[2] K. D. Choo et al., "Area-Efficient 1GS/s 6b SAR ADC with Charge-Injection-Cell-Based DAC," ISSCC 2016:460-461.

[3] K. Choo et al., "14.1-ENOB 184.9dB-FoM Capacitor-Array-Assisted Cascaded Charge-Injection SAR ADC," ISSCC 2021:372-373.

[4] J. P. Keane et al., "An 8GS/s Time-Interleaved SAR ADC with Unresolved Decision Detection Achieving -58dBFS Noise and 4GHz Bandwidth in 28nm CMOS," ISSCC 2017:284-285.

[5] A. Yonar et al., "An 8b 1.0-to-1.25GS/s 0.7-to-0.8V Single Stage Time-Based Gated-Ring-Oscillator ADC with 2x Interpolating Sense-Amplifier-Latches," ISSCC 2023:266-267.

[6] L. Kull et al., "A 10b 1.5GS/s Pipelined-SAR ADC with Background Second-Stage Common-Mode Regulation and Offset Calibration in 14nm CMOS FinFET," ISSCC 2017:474-475.

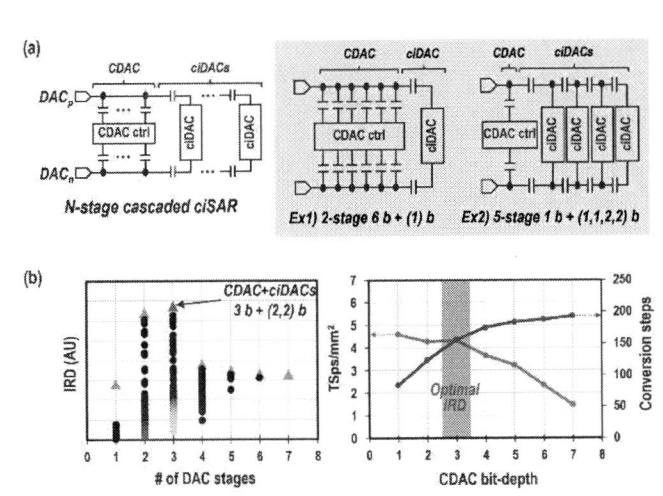

Fig. 1. (a) Hybridization of CDAC and ciDAC, (b) Optimal DAC construction for a 1 GS/s 8-bit ADC.

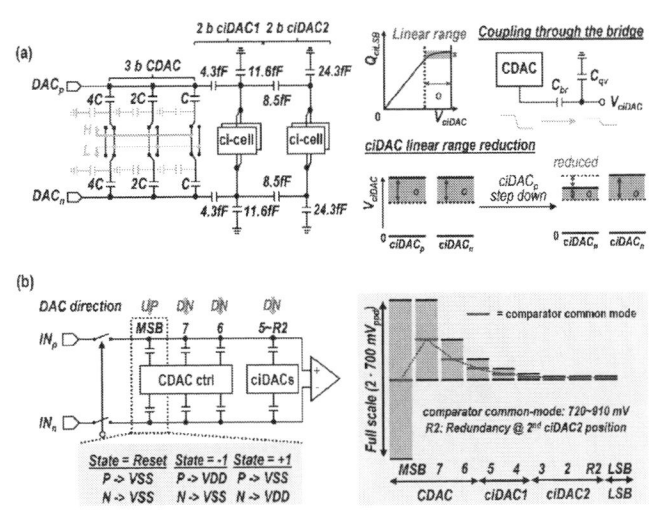

Fig. 2. (a) Proposed c-ciDAC construction and ciDAC linear range reduction, (b) DAC sequence considering ciDAC linear range.

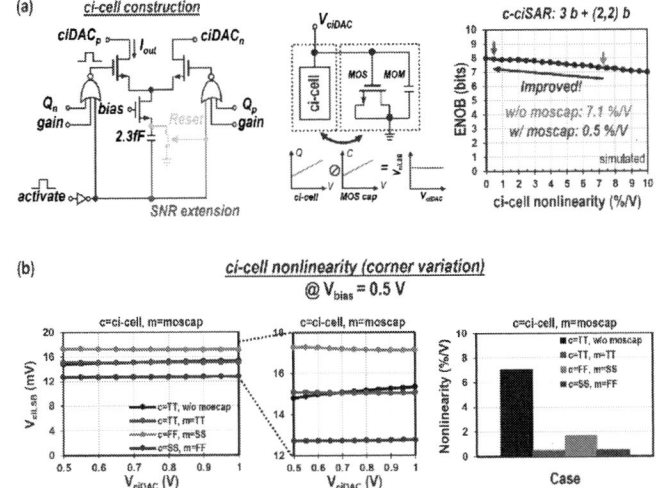

Fig. 3. (a) Proposed ci-cell construction and DC dependence compensation with hybrid ciDAC output capacitance, (b) ci-cell nonlinearity across the corner variation.

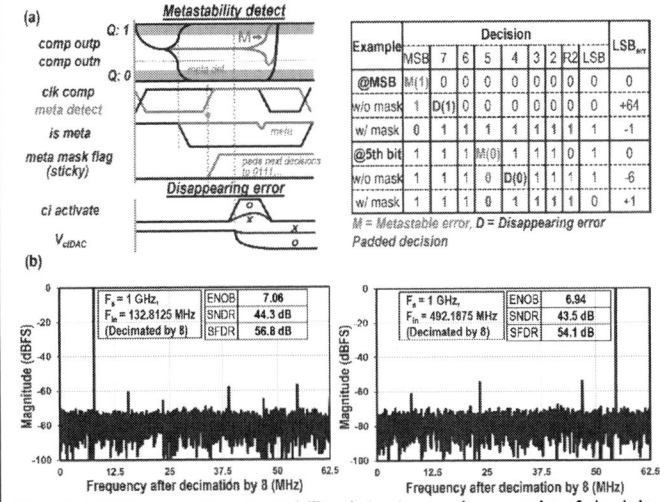

Fig. 4. (a) Proposed metastability detector and example of decision padding, (b) Measured output spectrum and performance with decision padding.

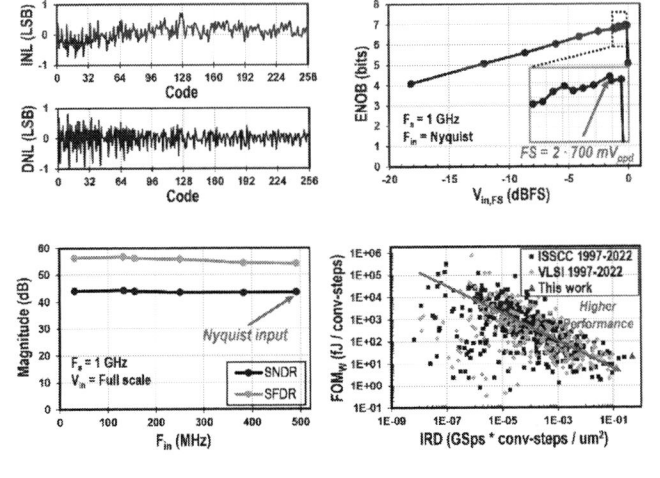

Fig. 5. Measured performance at 1 GS/s of the prototype ADC.

	This work	ISSCC 2017 Kull [6]	ISSCC 2019 Jiang [7]	ISSCC 2016 Choo [2]	ISSCC 2013 Kull [1]	ISSSC 2017 Chan [8]	VLSI 2018 Chen [9]	VLSI 2020 Oh [10]
Structure	c-ciDAC SAR ADC	Pipelined SAR ADC	Pipelined SAR ADC	ciDAC SAR ADC	CDAC SAR ADC	2x TI 1-2b SAR ADC	Two-step SAR ADC	SAR-Flash ADC
Tech nm	28	14	28	40	32	28	40	28
VDD V	1	0.95	1	1	1	0.9	1.2	1.1
Resolution bits	8	10	12	6	8	7	8	8
$F_{s,nyq}$ @channel GSps	1	1.5	1	1	1.2	2.4	1.1	1
SNDR @Nyq. dB	43.5	50.1	60.0	34.6	39.3	40.05	45	45.5
ENOB @Nyq. bits	6.94	8.03	9.68	5.46	6.24	6.36	7.18	7.27
SFDR @Nyq. dB	54.1	58.4	74.6	49.7	50	54.3	66	58
Power mW	2.61	6.92	7.6	1.26	3.1	5	4	2.55
Core area um^2	261	1600	9100	580	1500	4300	1650	2000
FoM$_{walden}$ @Nyq. fJ/conv-steps	21.3	17.7	9.28	28.7	33.8	25.4	25	16.6
IRD @Nyq. TSps* conv-steps/mm²	470	245	89.8	75.6	60.3	45.9	96.7	76.6

Fig. 6. Performance summary and comparison with the state-of-the-art area-efficient single-channel Nyquist rate ADCs.

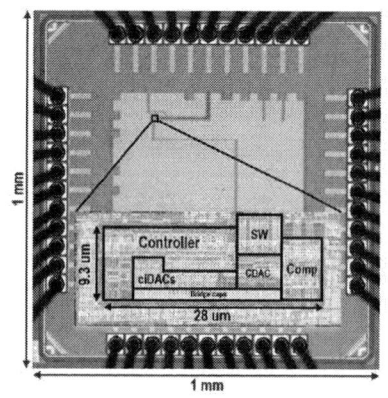

Fig. 7. Chip photomicrograph fabricated in 28 nm CMOS.

Additional References:

[7] W. Jiang *et al.*, "A 7.6mW 1GS/s 60dB SNDR Single-Channel SAR-Assisted Pipelined ADC with Temperature-Compensated Dynamic Gm-R-Based Amplifier," ISSCC 2019:60-61.

[8] C.-H. Chan *et al.*, "A 5mW 7b 2.4GS/s 1-then-2b/cycle SAR ADC with background offset calibration," ISSCC 2017:282-283.

[9] H. Chen *et al.*, "A >3GHz ERBW 1.1GS/s 8b Two-Step SAR ADC with Recursive-Weight DAC," VLSI 2018:97-98.

[10] D.-R. Oh *et al.*, "An 8b 1GS/s 2.55mW SAR-Flash ADC with Complementary Dynamic Amplifiers," VLSI 2020:1-2.

A 1.01V 8.5Gb/s/pin 16Gb LPDDR5x SDRAM with Self-Pre-Emphasized Stacked-Tx, Supply Voltage Insensitive Rx, and Optimized Clock using 4th-Generation 10nm DRAM Process for High-Speed and Low-Power Applications

Hyun-A Ahn, Yoo-Chang Sung, Yong-Hun Kim, Janghoo Kim, Kihan Kim, Donghun Lee, Young-Gil Go, Jae-Woo Lee, Jae-Woo Jung, Yong-Hyun Kim, Ga-Ram Choi, Jun-Seo Park, Bo-Hyeon Lee, Jinhyeok Baek, Daesik Moon, Daihyun Lim, Seung-Jun Bae, Young-soo Sohn, Changsik Yoo, and Tae-Young Oh

Samsung Electronics, Hwaseong, Korea

For the increasing demands in big data processing and cloud computing, today's mobile system has to meet the requirements of higher data bandwidth, lower system cost, and lower power consumption [1]. These requirements force dynamic random access memory (DRAM) to increase its data bandwidth and capacity with low power consumption, and are accelerating the movement from LPDDR5 to LPDDR5x for beyond-5G communications. For low power 8.5Gbps operation, 16Gb LPDDR5x DRAM I/O circuits using the 4th-generation 10nm DRAM process are proposed in this paper. The proposed I/O improves signal integrity (SI) by using a self-pre-emphasized stacked driver in transmitter (Tx), a supply voltage insensitive data receiver in receiver (Rx), and an optimized clock tree in WCK clock-paths. The measured eye-widths of Tx and Rx, which are the indicators of SI, are 0.66 unit interval (UI) and 0.57 UI at 8.5 Gbps, respectively. Also, the measured power consumption is reduced by 20.0% compared to previous LPDDR5 SDRAM products [2].

Fig. 1 shows the block diagram of the proposed LPDDR5x I/O circuits, which are composed of Tx, Rx and WCK path. In the Tx path, a read (RD) duty cycle adjuster (DCA) is adopted as the new feature of LPDDR5x specifications, and a stacked driver with a scalable pre-emphasis is also proposed for reducing power consumption and improving SI. In the Rx path, the per-pin decision feedback equalizer (DFE) and offset cancelation in the data input sense amplifier (DINSA) are adopted for new features of LPDDR5x, and an internal voltage controller (IVC) is employed for supplying stable power to the write (WR) current mode logic (CML) to CMOS converter (C2C). In the WCK path, the load on the WCK buffer is reduced to decrease power consumption.

Fig. 2 (a) shows the schematic diagram of the previous Tx driver, which is composed of main driver and pre-driver. The main driver consists of two n-type transistors (M_{PU} and M_{PD}) for low-voltage swing termination logic (LVSTL) using low supply voltage of VDDQ (0.5 V). Fig. 2 (b) shows the schematic diagram of the proposed stacked driver, which is also composed of pre-driver and main driver. The function of multiplexing data and ZQ codes, which was implemented in pre-driver of previous Tx driver, is moved to ZQ transistors (M_{PUZQ} and M_{PDZQ}) in the main driver. Since D_{PU} and D_{PD} are transferred to the main driver through five-stage repeaters (RPTs) and a single-inverter without any multiplex operation in the pre-driver, the size and the number of RPTs can be decreased. Therefore, the proposed stacked driver decreases power supply induced jitter (PSIJ) and power consumption. Fig.2 (c) illustrates the effect of the self-pre-emphasizing in the proposed stacked driver. When the $PU_{ZQ<5>}$ is logically in a low state and the voltage at node A transits from low to high, the stored charge on C_{P5} at node C helps to increase the slope of DQ. Therefore, the proposed stacked driver has the self-pre-emphasizing effect due to the parasitic capacitors, which are created by the large stacked n-type transistor without any control schemes. The simulation results in Fig. 2 (d) demonstrate the self-pre-emphasis effect of the proposed stacked driver. When the voltage of node A increases, the voltages of node B and node C are bumped up, and then the pre-charged voltage on node C is decreased by discharging operation while the voltage on DQ starts to increase.

Fig. 3 (a) shows the schematic diagram of the proposed IVC, which supplies stable power (IVC_IO) into two WR C2Cs. The IVC converts voltage from VDD1 to IVC_IO using VREF, and supplies IVC_IO for supply voltage insensitive operation of WR C2Cs. Fig. 3 (b) shows the schematic diagram of the proposed WR C2C adopting reverse bias and IVC for reducing the influence of power supply noise of VDD2H. The WR C2C is composed of reverse bias stage, pre-amplifier, and CMOS RPT using resistive feedback inverter. In the reverse bias stage, when VDD2H is increased, BIAS_N is decreased because IVC_IO is stable and the on-resistance of M_{BN} is decreased. And then, since the current flowing through M_{BP} is decreased, BIAS_P is increased. On the other hand, when VDD2H is decreased, BIAS_N is increased because IVC_IO is stable and an on-resistance of M_{BN} is increased. And then, since the current flowing through M_{BP} is increased, BIAS_P is decreased. The delay characteristics of the pre-amplifier can be compensated by BIAS_N and BIAS_P from the reverse bias stage against the movement of VDD2H. In the CMOS RPT, the swing range of clock signals (CLK0 and CLK180) are extended up to rail-to-rail. Because there are no devices using VDD2H in the CMOS RPT, delays of CLK0 and CLK180 can be constant. Therefore, the proposed WR C2C adopting reverse bias and IVC minimizes the delay variation caused by PSIJ in the Rx path. Fig. 3 (c) represents the schematic diagram of the double tail DINSA adopting the proposed per pin offset calibration as a new feature of LPDDR5x. The step range of the offset calibration is set by 7 left steps and 7 right steps (OC_L<2:0> and OC_R<2:0>). As shown in the flow chart of Fig. 3 (d), the offset calibration is started by setting MR15 OP7 to 1. Then, DIN and VREFP are shorted by VSS to decide offset direction. If DIN_O is larger than DINB_O, the left-hand side of offset calibration is executed. On the other hand, if DIN_O is smaller than DINB_O, the right-hand side of offset calibration is executed. Once the direction is selected, the counter repeats up and down to find the optimum code of OC_L<2:0> or OC_R<2:0> until 3 times of 1-bit toggle on DIN_O. When the 3 times of 1-bit toggle are done, the offset calibration code is fixed and the calibration is done.

Figs. 4 (a) and (b) show the block diagrams of the previous and the proposed WCK tree, respectively. In Fig. 4 (a), due to the WCK buffer's heavy line loading of 1520 µm which results in bandwidth limitation, the slope of WCK buffer outputs is decreased. As shown in Fig. 4 (b), because the number of WR C2Cs is decreased from six to two and the length of WCK buffer's line loading is also decreased to 690 µm, the slope of WCK buffer outputs is enhanced and the power consumption is reduced in the proposed WCK tree.

Fig. 5 (a) shows the measured 2-dimensional (2D) shmoo results of RD EYE and WR EYE, which represent 0.66 UI and 0.57 UI, respectively, at 8.5 Gbps. Fig. 5 (b) shows the distribution of the VREF with best timing margin in DINSA when the offset calibration is turned on (red) and off (blue). After the offset calibration, distribution of the VREF is narrowed by 26.5% (from 32mV to 23.5mV). Fig. 5 (c) represents the measurement results of tWCK2DQi_volt, which is defined as the voltage sensitivity of write timing from WCK to DQ. When the proposed DRAM only adopts reverse bias in WR C2C, tWCK2DQi_volt is improved by 40%. Moreover, when the proposed DRAM adopts both of reverse bias and IVC in WR C2C, tWCK2DQi_volt is improved by additional 40%. Also, Fig. 5 (d) shows that the power consumption of the proposed LPDDR5x is reduced by 20.0% compared to the previous LPDDR5.

Various I/O schemes for high-speed low power LPDDR5x are proposed. In Fig. 6, a summary and comparison table shows that the measured eye-widths of Tx and Rx are 0.66 UI and 0.57 UI at 8.5 Gbps, respectively. And, the measured power consumption is reduced by 20.0% compared to the previous LPDDR5 SDRAM [2]. In addition, the input offset of DINSA is reduced by 26.5% by adopting offset calibration, and the supply voltage sensitivity of WR is also decreased by 65.0%.

References:

[1] M. Bucher *et al.*, "A 6.4-Gb/s near-ground single-ended transceiver for dual-rank DIMM memory interface systems," JSSC, vol. 49, no. 1, pp. 127–139, Jan. 2014.

[2] Y. H. Kim *et al.*, "A 16Gb Sub-1V 7.14Gb/s/pin LPDDR5 SDRAM Applying a Mosaic Architecture with a Short-Feedback 1-Tap DFE, an FSS Bus with Low-Level Swing and an Adaptively Controlled

Body Biasing in a 3rd-Generation 10nm DRAM," ISSCC 2021: 346-348.

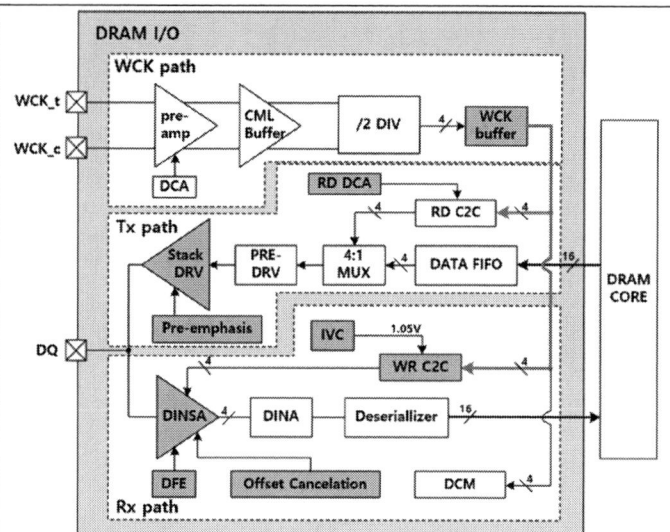

Fig. 1. I/O interfacing functional block diagram of the proposed LPDDR5x. (new scheme: gray)

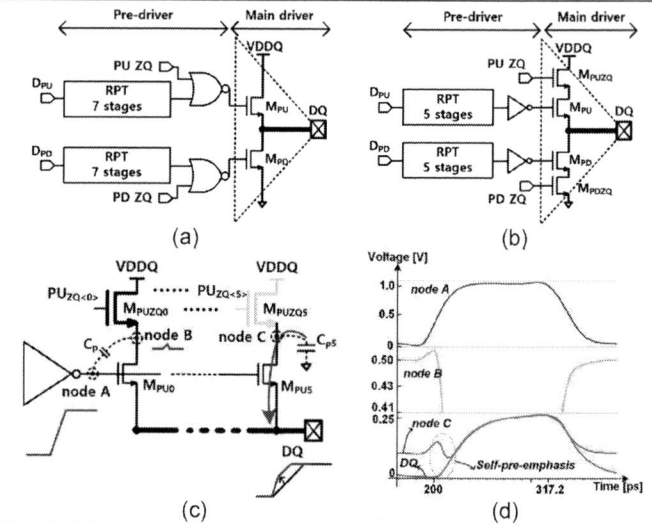

Fig. 2. (a) Schematic diagram of the previous Tx driver, (b) the proposed stacked Tx driver, (c) self-pre-emphasized effect, and (b) the simulation result.

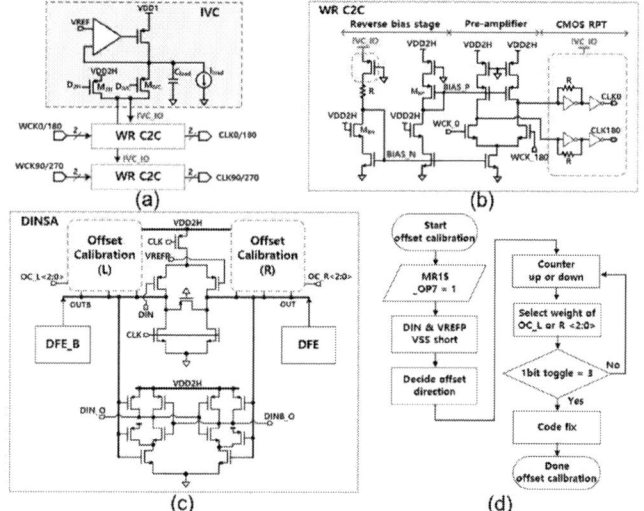

Fig. 3. Schematic diagram of (a) the proposed IVC, (b) WR C2C adopting reverse bias and IVC, (c) DINSA adopting offset calibration, and (d) flow chart of offset calibration.

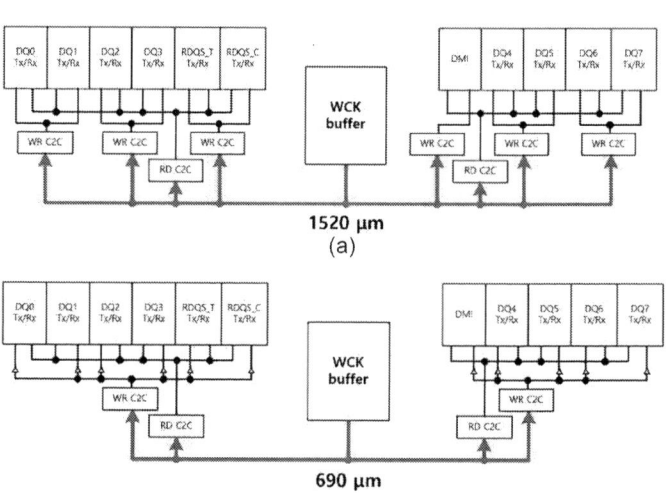

Fig. 4. Block diagrams of (a) the previous WCK tree and (b) the proposed WCK tree.

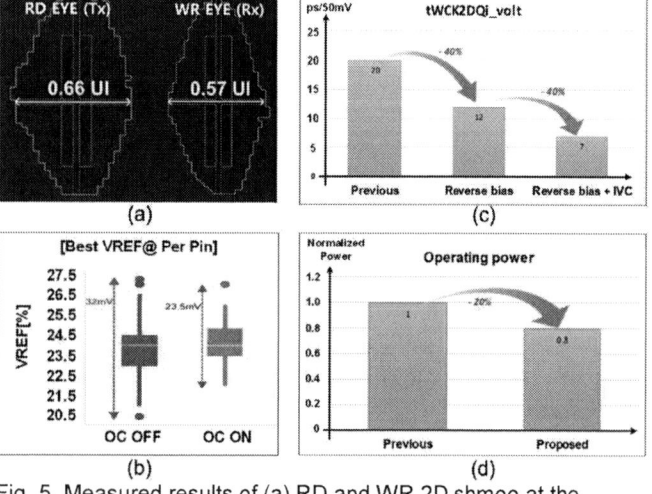

Fig. 5. Measured results of (a) RD and WR 2D shmoo at the 8.5Gbps, (b) VREF variation according to offset calibration on/off, (c) tWCK2DQi_volt, and (d) operating power of the previous LPDDR5[3] and the proposed LPDDR5x (assuming RD:WR = 2:1).

Product		LPDDR5 [2]	This work (LPDDR5x)
Process Technology		3rd-Generation 10nm Process	4th-Generation 10nm Process
Die Density		16Gb	16Gb
EYE Size (at 8.5Gbps)	Read [UI]	0.54	0.66
	Write [UI]	0.49	0.57
Operation Power (Normalized)		1	0.8
tWCK2DQi_volt (Voltage sensitivity)		20ps/50mV	7ps/50mV
Chip Area		60.92 mm²	47.53 mm²

Fig. 6. Summary and comparison table of LPDDR5/5x.

979-8-3503-3004-5/23 $31.00 © 2023 IEEE

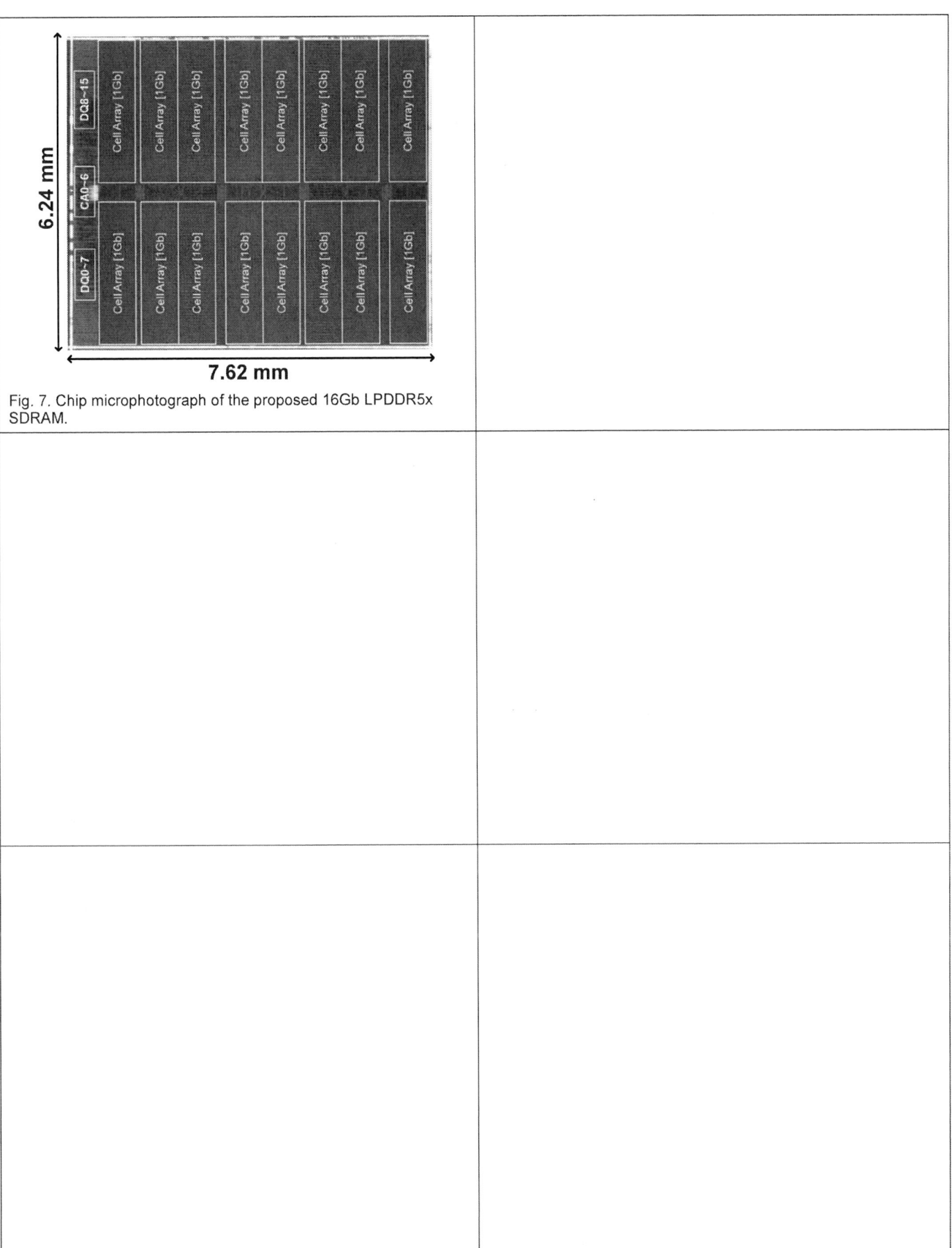

Fig. 7. Chip microphotograph of the proposed 16Gb LPDDR5x SDRAM.

A Low Noise 8Mpixel CMOS Image Sensor with 5.36GHz Global Counter and Dual Latch Skew Canceler for Surveillance AI Camera System

Yoichi Iizuka, Akihide Maezono, Wataru Saito, Atsushi Yamane, Kazuhiko Takami, Fukashi Morishita

Renesas Electronics Corporation, Tokyo, Japan

CMOS image sensors (CIS) are widely used in automotive, industrial, and consumer applications. In recent years, they have become increasingly important in automatic driving, face recognition, and other automatic recognition technologies combined with AI, and are used in surveillance network systems, etc[1-3]. Important performance features for CIS include low noise, high speed, HDR, and high resolution, but in surveillance camera systems used for crime prevention and search purposes, the ability to capture clear images at high frame rates, especially in low light, is required to target fast-moving objects in a variety of environments [4]. Therefore, we propose an image sensor that uses a new high-speed circuit technology to achieve both low noise and high frame rate. We believe that this image sensor will be useful for a wide range of surveillance solutions. A prototype AI camera system combining this image sensor and an AI accelerator has successfully achieved AI recognition even in dark areas by reducing noise while maintaining a high frame rate of 30 fps at 8M pixels.

The architecture of the CIS Test Chip is shown in Figure 1. We used a CML Global Counter with a multi-phase generator(CGCM) and a single slope ADC with a latch section that has a circuit (DLSC) to cancel the transmission time difference (t_{skew_ADC}) between counter signals that occurs during long-distance transmission over an ADC column band. Although multiple sampling is an effective method to suppress noise in order to capture images in dark areas, the following issues arise when sampling multiple times at high frame rates such as 30 fps with a large number of pixels like 8 Mpix. First, in order to perform multiple sampling, it is necessary to increase the counter frequency to a high speed, e.g., 5 GHz or higher to perform four times sampling, but this is difficult to achieve in CMOS logic circuits, The second is that the skew constraint between counter signals transmitted over long distances across thousands of column ADCs becomes more severe with increasing speed. To solve these problems, we propose the following circuit.

The block diagram and operating waveforms of the CGCM are shown in Fig. 2. The multi-phase generator and phase shift counter, which are the high-speed operation part of the counter, are composed of Current Mode Logic (CML) and operate with differential small amplitude, which enables high-speed operation compared to normal CMOS circuits. Since the CGCM itself has a CLK multiplication function, there is no need to change the input CLK specification, and thus, it is possible to increase the speed without affecting other IP such as PLLs. This chip can output a counter signal equivalent to 5.36GHz, which is 4 times faster than the input CLK frequency of 1.34GHz. The counter outputs four Johnson counter codes JC[3:0] for the lower bits consisting of multi-phase CLK at 0.67 GHz and gray codes GR[12:4] for the upper bits at a maximum of 167.5 MHz. The signal itself transmitted over the column band is relatively slow, at a maximum of 0.67 GHz, and can be transmitted over long distances by inserting buffers along the way, achieving an ADC equivalent to 5.36 GHz with a multi-phase CLK phase difference. Simulation waveforms are also described.

When the counter speed is increased, the time t_{lsb} corresponding to 1LSB of ADC becomes shorter, and when t_{lsb} is less than t_{skew_ADC}, an impossible counter code "Missing Code" is output, and a large AD conversion error is generated. Therefore, we propose the Dual Latch timing Skew Canceler (DLSC) to cancel most of the t_{skew_ADC} (Fig. 3). This test chip has 13 counter signals, with 4 multi-phase CLKs which is the Johnson counter codes, and 9 gray codes. There are usually 10 to 20 counter signals depending on the number of bits in the ADC, and it is necessary to match the transmission delay within all these column bands. However, it is difficult to match the transmission delays of many signals with different operating frequencies over a

long distance with high precision. The transmission distance is also increasing due to the increase in the number of pixels, which is one of the reasons for the obstruction of ADC speed-up. In the case of DLSC, if only the transmission delay of the JC[3:0] operating at the same frequency is matched, the AD conversion result is not adversely affected at all even if the transmission delay of the gray codes of the upper bit has a transmission delay difference of 7 times that of t_{lsb}, because the signal of the upper bit is latched at the timing when the signal of the upper bit is stable. Since the JC[3:0] have only four signals that operate at constant velocity and differ only in phase, it is easy to make t_{skew_ADC} small by maintaining the same circuit and layout, matching the parasitic capacitances equivalently, and by keeping the wiring positions close together to suppress the effect of delay differences due to power supply potential differences. In addition, even in the unlikely event that the t_{skew_ADC} of the JC[3:0] exceeds t_{lsb}, it can be restored with a small error because the only relevant bit is the lower bits.

This test chip enables sampling by changing the number of times and timing by register settings. It has an arithmetic circuit that performs the Dark conversion before pixel signal input and the Sig conversion after pixel signal input multiple times each, averages them, and then subtracts the difference. Fig. 4 shows the noise measurement results of multiple sample with different number of sampling times, and example image and timing charts. The ADC reduces the Dark conversion time and Sig conversion time to 1/4 by increasing the counter operating frequency by a factor of 4 and at the same time making the slope of the RAMP wave 4 times steeper in low noise mode. Therefore, it is possible to sample Dark and Sig 4 times each without significantly increasing the time required per AD conversion. When sampling four times each, noise is improved by 31% compared to once. An Image taken with this test chip at an illuminance of 500 lux are also shown.

Figure 5 shows a comparison table. Three Modes are selectable for this test chip. In Normal mode, a normal CMOS counter at 1.34 GHz is used to perform D-CDS (Digital Correlated Double Sampling), which takes the difference between the Dark conversion before pixel signal input and the Sig conversion after pixel signal input, to operate at 30 fps. In Low Noise Mode, by increasing the frequency by 4 times to 5.36GHz using CGCM and DLSC, 4 times CMS and D-CDS are performed, and the frame rate remains the same as in Normal Mode, operating at 30fps. Although the counter frequency is increased by a factor of 4, the Power ratio of the counter in the entire Chip is low, so the power increase is limited to +4%. By increasing the counter frequency, the ADC has the potential to achieve a frame rate of 60 fps or higher if CMS is not performed, but in this test chip, due to IO and other limitations, the counter speed-up is used specifically for low noise High fps mode omits D-CDS and performs only Sig conversion, operating up to 60 fps. Low Noise mode reduced noise to 187uVrms and DNL was 0.68LSB.

A block diagram of the prototype AI camera system is shown in Fig. 6. The camera is equipped with this test chip, an Image Signal Processor (ISP), and an AI-accelerator (RZ-V2M), and has two paths: one to output a Video Stream and the other to output AI recognition results. AI recognition results are output from the AI Codec to RZ-V2M for inference processing. The inference results are output to the PC via Ethernet using the User Datagram Protocol (UDP) protocol. The AI training model is a standard object recognition model (Tiny-Yolo v2) stored on an SD card, and the data received by the PC is overlaid on a Video Stream to depict the AI recognition results. The AI camera equipped with this chip was able to recognize a person even in a low-light environment of 1.3 lux. Test chip micrograph is shown in Fig. 7.

References:

[1] F. Morishita et al., "A CMOS Image Sensor and an AI Accelerator for Realizing Edge-Computing-Based Surveillance Camera Systems," 2021 Symposium on VLSI Circuits, pp. 1-2

[2] T. Kim, C. Kang, Y. Kim and S. Yang, "AI Camera: Real-time License Plate Number Recognition on Device," 2022 IEEE International Conference on Consumer Electronics (ICCE), pp. 1-6

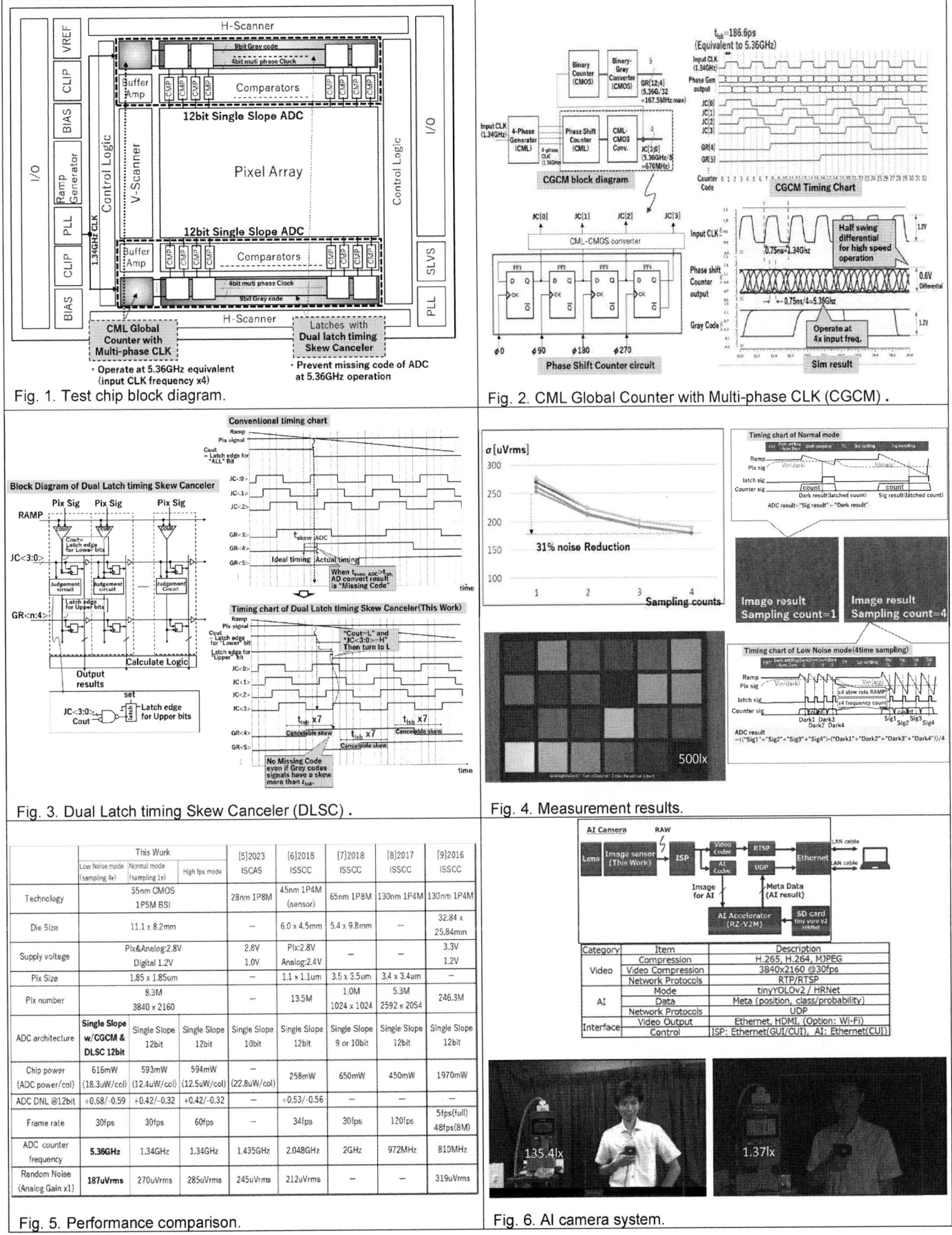

Fig. 1. Test chip block diagram.

Fig. 2. CML Global Counter with Multi-phase CLK (CGCM).

Fig. 3. Dual Latch timing Skew Canceler (DLSC).

Fig. 4. Measurement results.

Fig. 5. Performance comparison.

Fig. 6. AI camera system.

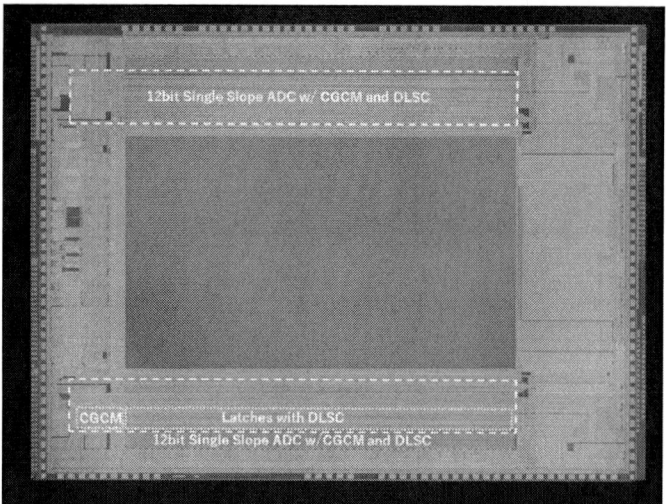

Fig. 7. Test chip micrograph.

Additional References:

[3] J. -J. Lee, U. -j. Gim, J. -H. Kim, K. -H. Yoo, Y. -H. Park and A. Nasridinov, "Identifying Customer Interest from Surveillance Camera Based on Deep Learning," 2020 IEEE International Conference on Big Data and Smart Computing (BigComp), pp. 19-20

[4] F. Morishita, W. Saito, Y. Iizuka, N. Kato, R. Otake and M. Ito, "A 30.2-µ Vrms Horizontal Streak Noise 8.3-Mpixel 60-Frames/s CMOS Image Sensor With Skew-Relaxation ADC and On-Chip Testable Ramp Generator for Surveillance Camera," in IEEE Journal of Solid-State Circuits, vol. 57, no. 10, pp. 3103-3113, Oct. 2022

[5] J. Jun, H. Yang, B. Yoon, Y. Kim and K. Koh, "A Low Power Digitizer with Piecewise- Linear Counting Technique for High Dynamic Range Nonacell-Based 3-D-Stacked CMOS Image Sensor," 2023 IEEE International Symposium on Circuits and Systems (ISCAS) 2023, pp. 1-5

[6] P. -S. Chou et al., "A 1.1µm-Pitch 13.5Mpixel 3D-stacked CMOS image sensor featuring 230fps full-high-definition and 514fps high-definition videos by reading 2 or 3 rows simultaneously using a column-switching matrix," ISSCC 2018, pp. 88-90

[7] C. S. Bamji et al., "1Mpixel 65nm BSI 320MHz demodulated TOF Image sensor with 3µm global shutter pixels and analog binning," ISSCC 2018, pp. 94-96

[8] M. Kobayashi et al., "A 1.8erms− temporal noise over 110dB dynamic range 3.4µm pixel pitch global shutter CMOS image sensor with dual-gain amplifiers, SS-ADC and multiple-accumulation shutter," ISSCC 2017, pp. 74-75

[9] H. Totsuka et al., "6.4 An APS-H-Size 250Mpixel CMOS image sensor using column single-slope ADCs with dual-gain amplifiers," ISSCC 2016, pp. 116-117

Fig. S2 (to be removed for final submission)

Fig. S3 (to be removed for final submission)

Fig. S4 (to be removed for final submission)

A 28-nW Noise-Robust Voice Activity Detector with Background Aware Feature Extraction

Jingsen Yang[1], Liangjian Lyu[2], Zirui Dong[1], Heyu Ren[1], C.-J. Richard Shi[3]

[1]Fudan University, Shanghai, China

[2]East China Normal University, Shanghai, China

[3]University of Washington, Seattle, USA

In light of the increasing number of Internet of Things (IoT) devices, such as intelligent vehicles and smart assistants, it has become imperative to develop low-power Voice Activity Detection (VAD) devices. The always-on VAD devices detect the voice to wake up the target system, thus dominating the standby power consumption of the IoT devices. The typical VAD consists of a feature extractor and a neural-network-based classifier. The algorithm using the frequency-domain features which can be obtained by modulation frequency [1], fast Fourier transform (FFT) [2], and analog filter banks, can achieve high detection accuracy. However, the feature extractor induces high power consumption due to the complex operations. Alternatively, the time-domain analog VADs [5-6] achieve low power consumption, due to the lack of a frequency extractor, but also suffers from reduced accuracy that the audio amplitude is interfered with noise easily, especially in noisy environments with a signal-to-noise ratio (SNR) is lower than 0dB. In summary, achieving high accuracy and low power consumption simultaneously in VAD devices is a critical challenge.

To address the above issues, we present a noise-robust VAD. The low power and high accuracy are guaranteed by processing the following techniques: 1) The VAD record environment noise features to enhance accuracy by comparison. 2) The envelope and zero-crossing rate (ZCR) are extracted as the VAD input features. This method saves the overhead of storage and operation by reducing the input dimension and still retains the information of volume and frequency to avoid accuracy loss.

Figure 1 shows the proposed noise-robust VAD system. The input audio signal of 512 ms length is framed into consecutive frames with a length of 64 ms and a shift of 64 ms. For each frame, the maximum envelope and ZCR are extracted. These features are fed into 4 depth-8 FIFOs. The audio feature FIFOs are updated in each frame, and the background feature FIFOs are only updated when the wake-up FIFO's value is 0, i.e., speaking is not detected in recent times. The first-stage energy VAD works by comparing the envelope FIFO's value sum of audio and background noise. The energy VAD has high accuracy in voice and activates the second stage VAD when the audio volume is getting higher. The second stage neural network VAD uses the features of envelope and ZCR, and operates them in two convolution groups separately, each group has two 1D-Conv layers. Then, the results following the Conv operation are spliced together for a fully connected (FC) layer to obtain the wake-up result. Fig.1 (lower) shows the recognition effect of VAD with different noises. With -5 dB Volvo noise, the ZCRs of the voice and the background are well distinguished even with a large noise due to the frequency stability of Volvo noise.

Figure 2 shows the hardware architecture and data flow of the proposed VAD. The envelope receiver and ZCR receiver receive 500 Hz envelope and 8 KHz zero-crossing, then feed the maximum envelope and ZCR to feature FIFOs in each frame. The energy VAD uses the envelope feature to activate the neural network VAD. The neural network VAD uses the features of envelope and ZCR, and then exports the wake-up signal. The wake-up signal is optimized in the post-process unit and then fed into the wake-up FIFO in the feature FIFOs module to control the update of background feature FIFOs. In Fig.2 (lower left), the deep learning accelerator comprises a control logic, a weight memory, a feature memory, and four PEs (two envelope PEs, and two ZCR PEs). The CNN operation is performed separately on the envelope and ZCR PEs. When executing the FC network, the envelope PEs are utilized for processing the FC layer since the FC network has 2 output channels. In Fig.2 (lower right), the wake-up signal is smoothed first to avoid glitches. Then The frame-extend operation is achieved via a counter. The wake-up signal is output continuously to extend the effect until the counter value reaches zero.

Figure 3 shows the detail of the feature FIFO. The FIFO module consists of four 8-depth FIFOs. When a new frame occurs, the envelope and ZCR from the feature receiver are directly written into the respective voice FIFOs. The update of background information is determined by the wake-up list FIFO, which records the VAD output of the previous 5 frames. The background noise's FIFO is updated only when the wake-up list FIFO's values are all zero. Moreover, each background information FIFO contains a register that stores the FIFO sum. By subtracting the value of background FIFO [0] and adding the current feature value, the energy sum of the new background FIFO is obtained. Then, a shift operation realizes division by 8 to smooth the input of background information. the audio envelope FIFO energy sum is also obtained using a similar approach to the background energy sum calculation. Fig.3 (lower) shows the detail of the energy VAD. The energy-VAD mode can be selected to amplify the background envelope energy by 1-2.5 times, followed by a comparison with the audio envelope energy.

Figure 4 shows the simulated robustness of the proposed VAD system. Fig.4 (upper left) shows the non-speech accuracy with a continuous linear amplification babble noise environment. Fig.4 (upper right) shows the affluence of white noise in different amplitudes added in audio with 10 dB babble noise, which is related to the circuit noise generated in the ADC. Note that, the proposed VAD system maintains consistent accuracy even when the noise has been amplified many times and the white noise has a high amplitude. Fig.4 (lower left) depicts the influence of voice input amplitude variations on the VAD performance. The proposed VAD achieves consistent performance over a wide amplitude range. Fig.4 (lower right) depicts the performance of the multi-feature against the single feature. Using multi-feature can achieve high accuracy with different noise environments.

The proposed VAD is fabricated in a 65nm CMOS process and occupies 0.07mm². The TIMIT dataset and the NoiseX-92 noise set are employed to evaluate the VAD performance. A 50-minute audio file is formed by concatenating 200 audios from the TIMIT test set with a speech-to-silence ratio of 1:4. The measurements are conducted separately using babble noise and Volvo noise. Fig.5 (upper left) shows the measured receiver operating characteristic (ROC) curve with babble noise. Fig.5 (upper right) displays the VAD accuracy versus the different SNRs. The VAD achieves an overall recognition rate of 93.56% in 10 dB babble noise and 91.53% in -5 dB Volvo noise. The digital power breakdown is summarized in Fig.5 (lower left). Fig.5 (lower right) shows the corresponding power when the voltage supply varies from 0.4 to 1.0V, where the power consumption is 28 nW with a 0.4V voltage supply.

Figure 6 provides the performance summary of the proposed noise-robust VAD and the comparison with the state-of-the-art works. The power consumption is 2x less than the digital part of [2], and the core area is 2x less than [4].

Acknowledgment:

This work was supported by the Science and Technology Commission of Shanghai Municipality under Grant No. 2018SHZDZX01. This work was also supported by the National Natural Science Foundation of China under Grant No. 62204085 and the Lingang Laboratory under Grant No. LG-QS-202202-12.

References:

[1] M. Price et al "A Low-Power Speech Recognizer and Voice Activity Detector Using Deep Neural Networks," in IEEE Journal of Solid-State Circuits, Jan. 2018.

[2] M. Cho et al. "A 142nW Voice and Acoustic Activity Detection Chip for mm-Scale Sensor Nodes Using Time Interleaved Mixer-Based Frequency Scanning," ISSCC, Feb. 2019.

[3] M. Yang et al "Design of an Always-On Deep Neural Network-Based 1-μ W Voice Activity Detector Aided with a Customized Software Model for Analog Feature Extraction," in IEEE Journal of Solid-State Circuits, June 2019

[4] M. Croce et al. "A 760nW, 180 nm CMOS Analog Voice Activity Detection System," 2020 IEEE Custom Integrated Circuits Conference (CICC), Boston, MA, USA, 2020

[5] F. Chen et al. "A 108nW 0.8mm2 Analog Voice Activity Detector (VAD) Featuring a Time-Domain CNN as a Programmable Feature Extractor and a Sparsity-Aware Computational Scheme in 28nm CMOS", ISSCC, Feb. 2022.

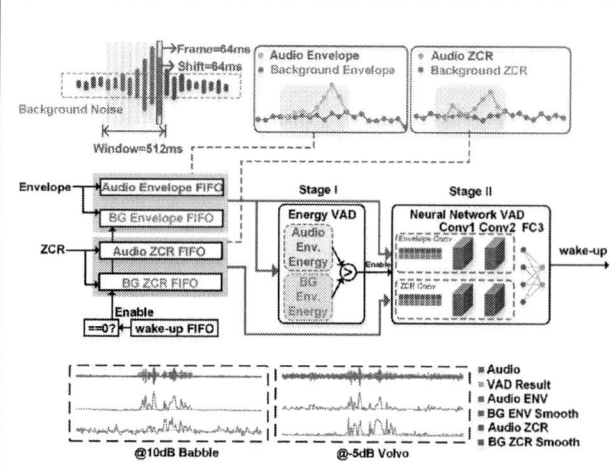

Fig. 1. The noise-robust VAD system architecture (upper). And examples of VAD in different noise environments (lower).

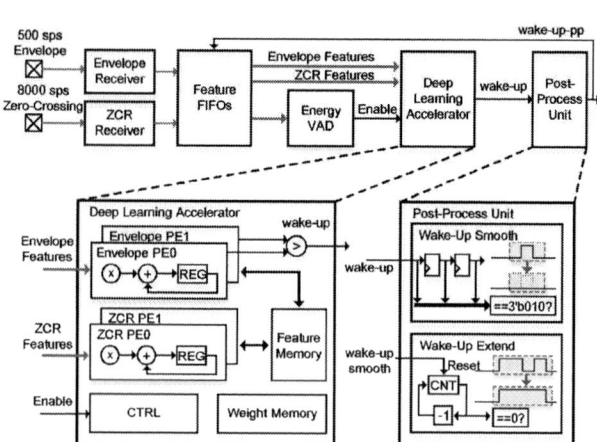

Fig. 2. Hardware architecture and data flow of the proposed VAD. The hardware of the deep learning accelerator (lower left) and the post-process unit (lower right).

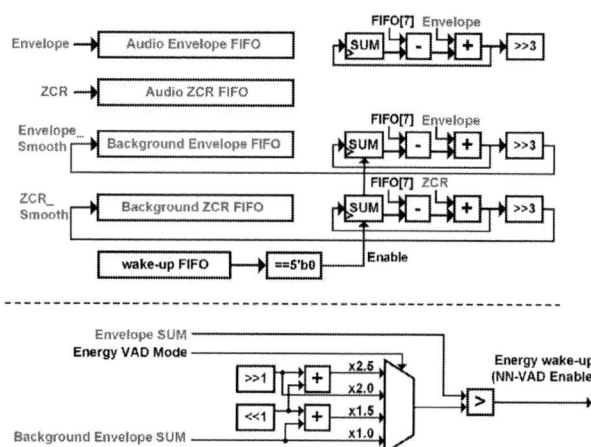

Fig. 3. The architecture of the feature FIFOs module (upper) and the energy VAD module (lower).

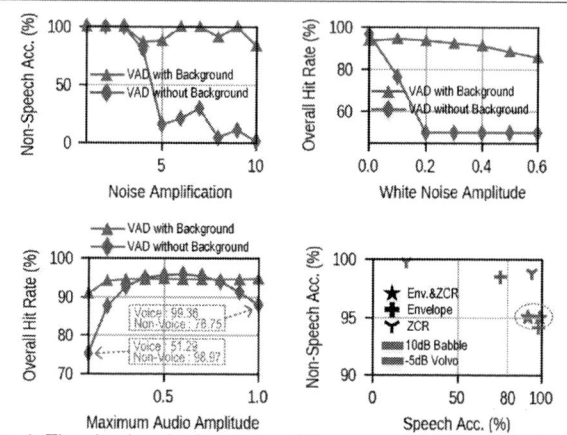

Fig. 4. The simulated robustness of the proposed noise-robust VAD system. The VAD performance analyzes with a changed noise environment (upper left), white noise (upper right), and input audio amplitude (lower left). The comparison between multi-feature and single-feature (lower right) of VAD.

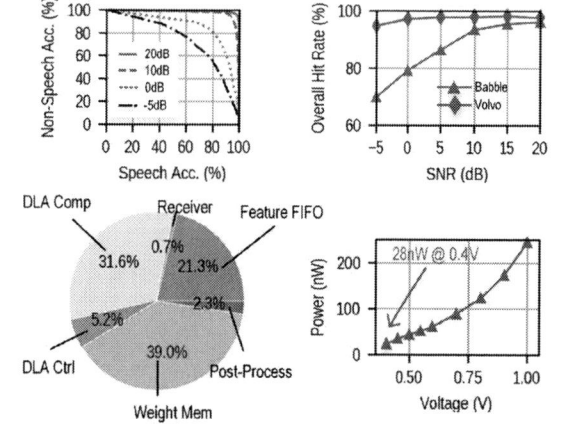

Fig. 5. Chip measurement results. ROC curve with babble noise (upper left). Overall hit rate versus SNR with babble noise and Volvo noise (upper right). Powe breakdown between VAD modules. And the power versus voltage in chips (lower right).

Voice Activity Detector	JSSC'18[1]	ISSCC'19[2]	JSSC'19[3]	JSSC'21[4]	ISSCC'22[5]	ISSCC'23[6]	This Work
Feature Extractor	Modulation Frequency	Down Sample + DCT	Analog to events	Energy	TD-CNN	ST-CNN	Envelope and ZCR
Background Feature	No	No	No	No	No	No	Yes
VAD Stage	1	1	1	1	1	1	2
Window Length (ms)	N/A	16	25	16	10	0.5	512
Latency (ms)	10	512	10	N/A	50	2.5	64
Classifier	NN	FC	BNN	Analog SNR-Based Decision Rule	TD-CNN + BNN	RNN	CNN + FC
Analog Front-End	No	Yes	Yes	Yes	Yes	Yes	No
Parameters	N/A	1.5K-11.9K	N/A	N/A	7.4K (3K1 Bit)	45	98 (8 Bit)
Dataset	Aurora2	LibriSpeech + NOISEX92	Aurora4 + DEMAND	Custom	TIMIT + NOISEX92	TIMIT + NOISEX92	TIMIT + NOISEX92
Speech/non-Speech Hit Rate @ 10dB Babble Noise	N/A	91.5%/90.0%	N/A	N/A	90.1%/94.0%	91% (Overall)	95.14%/91.88%
Speech/non-Speech Hit Rate @ other Noises	90%/90% @7dB Unspecified Context	N/A	84%/65% @ 10dB Restaurant Noise	99.8%/99.7% @ Custom Dataset	N/A	N/A	92.77%/96.69% @ -5dB Volvo Noise
Technology	65	180	180	180	28	65	65
Voltage (V)	0.5 (Digital)	1.4 (Analog)/ 0.6 (Digital)	0.6 (Analog)	1.2 (Analog)	0.9 (Analog)/ 0.6 (Digital)	0.9 (Analog)	0.4 (Digital)
ADC	No	Yes	Yes	Yes	No	No	No
Digital Power (nW)	22300	69.2	626	0	0	0	28
Total Power (nW)	22300	142	1000	760	108	47	28
Chip Area (mm²)	9.51(Core)	17.6 (Core + Pad)	2.5 (Core)	0.14 (Core)	0.8 (Core + Pad)	0.8 (Core + Pad)	0.49 (Core + Pad)/ 0.97 (Core)

Fig. 6. Chip measurement results and comparison with recent VAD works.

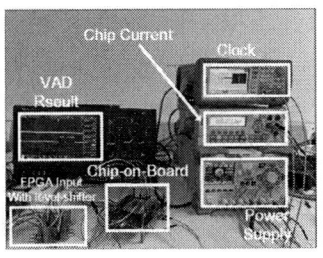

Technology	TSMC 65nm Logic CMOS
Die Area	0.267 x 0.265 mm²
Supply Voltage	0.4V ~ 1.0V
Power	28 nW @ 0.4V
Overall Accuracy (Babble Noise)	93.56 @ 10dB
	79.02 @ 0dB
Overall Accuracy (Volvo Noise)	97.91 @ 10dB
	94.83 @ -5dB

Fig. 7 Chip micrograph of the proposed VAD fabricated in 65nm CMOS (left top). The testing setup (right) and the PCB of the VAD system.

Additional References:

[6] J. Lin et al. "A 47nW Mixed-Signal Voice Activity Detector (VAD) Featuring a Non-Volatile Capacitor-ROM, a Short-Time CNN Feature Extractor and an RNN Classifier," ISSCC, 2023

An 83.6dB-SNDR 101.6dB-SFDR 4th-Order Noise-Shaping SAR with 2nd-Order Nonlinearity Error Shaping

Yanbo Zhang, Xianghui Zhang, Li Tian, Shubin Liu, Zhangming Zhu

Key Laboratory of Analog Integrated Circuits and Systems (Ministry of Education), School of Microelectronics, Xidian University, Xi'an, China

The Noise-Shaping SAR (NS-SAR) is a promising ADC architecture for Internet-of-Everything systems owing to its high resolution and low power [1]. Nevertheless, further precision improvement of NS-SAR ADCs is limited, suffering from nonlinearity caused by capacitor mismatch and voltage-dependent top-plate parasitic capacitance (C_{par}). Data-weighted averaging (DWA) is used to address the mismatch issue [2]. However, the DWA is only applied to MSB-DACs, and the LSB-DACs mismatch still restricts SFDR. Meanwhile, the DWA could not be easily extended to the entire DAC due to its hardware complexity exponentially growing with the DAC resolution. The segmentation method employing DWA on MSB-DACs and LSB-DACs individually has the gain error issue [3]. Ref. [4-6] separate the DAC into an M-bit thermometer-coded MSB-DAC with DWA and an N-bit binary-weighted LSB-DAC with mismatch error shaping (MES). The hardware complexity of MES grows linearly with the DAC resolution. However, the low quantization noise calls for a high DAC resolution resulting in a small unit capacitor while suffering from the C_{par} error. On the other hand, the mismatch error is inversely proportional to the size of the capacitor. NS is an appropriate way to shrink DAC resolution and increase the unit capacitor size under the same kT/C noise budget. However, high NS efficiency needs active amplifiers for compensation. The large-sized devices required for large gain connect on the top plate, which enlarges the C_{par} and worsens the linearity. To overcome the constraints above, instead of adopting the dedicated effort to handle the nonlinearity error, this work presents a multi-stage-noise-shaping (MASH) SAR architecture that cascades two 2nd-order quantization NS (QNS) stages to incorporate together the 4th-order QNS and the 2nd-order nonlinearity error shaping (NES). Our prototype achieves >100dB-SFDR over a 62.5kHz bandwidth (BW).

Fig. 1 shows the presented MASH-SAR signal flow. We separate the SAR ADC into two NS stages, MSB NS and LSB NS, each consisting of an error-feedback (EF) Loop. The quantization noise is shaped by both the MSB and LSB NS, resulting in a high-order overall NTF. Besides that, this new approach has four other important advantages: 1) The mismatch error from LSB (E_L) and the gain error (Δ) between the two stages can be shaped by NTF_M. Meanwhile, DWA can shape the mismatch error from MSB (E_M) with low hardware overhead. As E_M, E_L and Δ are shaped, the harmonic distortion from C_{par} is dominant. To alleviate it, 2) the bottom-plate sample technology can be adopted due to the eliminated presetting operation. Moreover, 3) the errors caused by C_{par} during LSB switching ($\mathcal{E}_{par}$) can be further shaped by NTF_M, resulting in a kT/C-limited DAC. Finally, 4) this architecture efficiently renders a high-order QNS to decrease the DAC resolution, fundamentally reducing the mismatch error in physical processing. The simulated spectrums reveal the harmonic distortions caused by E_L and C_{par} are shaped to high frequencies by NTF_M and the gain error of up to ±1.5% may be tolerated while still achieving >100-dB SNDR.

Fig. 2 illustrates the architecture of the MASH-SAR ADC. Owing to the high order QNS, this work adopts a 6-bit fully differential SAR ADC, leading to a 28fF unit capacitor. For simplicity, the single-ended equivalent is shown. The DAC is segmented into a 3b thermometer MSB-DAC and a 4b binary LSB-DAC with 1b redundancy. During Φ_S, the ADC processes the bottom-plate sampling. And then the comparator starts to work during Φ_C. Once the 3b MSB conversion is finished, the residue on CDAC, $-Q_M$, is amplified during Φ_{AM} and passed through FIR_M. The filtered voltage is feedback in the next cycle during Φ_{nsM}, resulting in an $NTF_M=1-G_M A_{CSM}\cdot(2z^{-1}-z^{-2})$, where G_M is the amplifier gain and A_{CSM} is the charge sharing attenuation factor of residue feedback. After the amplification, the ADC quantizes the remaining 4b LSB sequentially. In this design, the NS in LSB conversion is identical to MSB but with an overall time shift to enable a split-over-time amplifier-reused operation. When Φ_{AL} goes high,

the residue on CDAC, $-Q_L$, is amplified and feedback in the next cycle during Φ_{nsL}, resulting in an $NTF_L=1-G_L A_{CSL}\cdot(2z^{-1}-z^{-2})$. $G_M A_{CSM} = G_L A_{CSL} =1$ is designed to achieve the standard 4th-order overall $NTF=(1-z^{-1})^4$, boosting the SQNR to 112dB at OSR=16. To correct the gain error of G_M, a dither-based LMS calibration [7] is adopted by injecting a 1b pseudo-random number (PRN) in the dummy unit capacitor after the MSB-DAC conversion.

One challenge of this architecture is the implementation of relatively high linearity of the amplification in the MSB NS. The other constraint is a stable output common mode for subsequent quantization in LSB. Fig. 3 shows the schematic of the NS filter with an open-loop gm-R-based amplifier. To resolve the challenges above, 1) the gm cell is implemented by a differential flipped voltage follower for high linearity. 2) the loading capacitor is placed between the outputs of the differential amplifier, which achieves the output common-mode voltage free and obtains an inherent 2x gain owing to differential operation. Hence, the gm-R only provides half the value of the required gain. 3) owning to the EF NS, the amplifier's input can be a normal voltage ($-Q_M$) instead of an amplified voltage ($-NTF_M\cdot Q_M$) in feed-forward NS. Combined with the attenuation factor ($1-3A_{CSM}$), the amplifier can see a small input swing, which benefits the linearity. Except for the linearity during amplification, the tail nodes of the differential input pairs are pull-up to VDD during the sampling and conversion phase to alleviate the C_{par} error. The chopping is used to address the amplifier's flicker noise and offset.

Fig. 3 also details the NS operations with timing control. The $FIR_{M(L)}$ filter consists of $C_{resM(L)1,a-b}$ and $C_{resM(L)2,a-c}$, creating one-cycle and two-cycle delays, respectively. The MSB NS is taken as an example to explain the overall operation. In the (i+1)-th cycle, $C_{resM1,a}$ and $C_{resM2,a}$ load the corresponding residues. During the (i+2)-th cycle, $C_{resM1,a}$ and $C_{resM2,a}$ process the one-cycle and two-cycle delay. They are then merged with CDAC to realize MSB NS. The ratio among C_{resM1}, C_{resM2} and C_{DAC} is 2:1:1, forming $NTF_M=(1-z^{-1})^2$ with $A_{CSM}=1/4$ and $G_M=4$ (where $g_m\cdot R=2$). The $C_{resM1,a}$ and $C_{resM2,a}$ are connected with DAC until LSB conversion finishes to maintain a matching between signal and reference. The operation in the FIR_L filter is identical, with a larger gain to reduce the capacitance. By connecting another identical resister in parallel, $G_L=2G_M$ is achieved. The ratio between C_{resL1}, C_{resL2} and ($C_{DAC} + C_{resM1} + C_{resM2}$) is 2:1:5 to achieve $NTF_L = (1-z^{-1})^2$.

Fig. 7 displays the micrograph. Fig. 4 shows the FFT result with an 11kHz input. The measured SNDR, SNR and SFDR are 83.6dB, 84dB and 101.6dB, respectively, with 4th-order NS. It obtains a 62.5kHz BW at a 2MHz sampling frequency with an OSR of 16. Under MASH 0-2 mode, the SFDR can achieve 92dB SFDR without calibration, earning the advantage of large unit capacitor size. The MSB NS improves SNDR, SNR and SFDR by 5.6dB, 5.5dB and 9.1dB, respectively, which shows that the thermal noise is dominant. It also shows that the SNDR and SFDR variation across five samples in two modes, respectively. Fig. 5 plots SNR and SNDR versus input amplitude for an 11kHz input, indicating a measured DR of 85.3dB. The ADC consumes 68μW from a 1.2V supply voltage. The resultant FoM$_S$ is 173.2dB. Fig.6 summarizes the performance compared with state-of-the-art ADCs. The proposed MASH-SAR architecture leverages aggressive QNS to achieve the smallest DAC resolution. It enlarges the size of the capacitor, fundamentally alleviating the mismatch error. Meanwhile, it simultaneously shapes the nonlinearity errors from the capacitor mismatch, voltage-dependent capacitor, and segment gain error, achieving >100dB SFDR.

Acknowledgements:

This work was funded by the National Natural Science Foundation of China (62204192, 92164301, 62261160649, 62021004). Corresponding author: Shubin Liu and Zhangming Zhu.

References:

[1] C. -C. Chen, et al., "A 12-ENOB Second-Order Noise Shaping SAR ADC with PVT-insensitive Voltage-Time-Voltage Converter," IEEE A-SSCC, pp. 1-3, 2021.

[2] C. C. Liu, et al., "A 0.46mW 5MHz-BW 79.7dB-SNDR Noise-Shaping SAR ADC with Dynamic-Amplifier-Based FIR-IIR Filter, " ISSCC, pp. 466–467, Feb. 2017.

979-8-3503-3004-5/23 $31.00 © 2023 IEEE

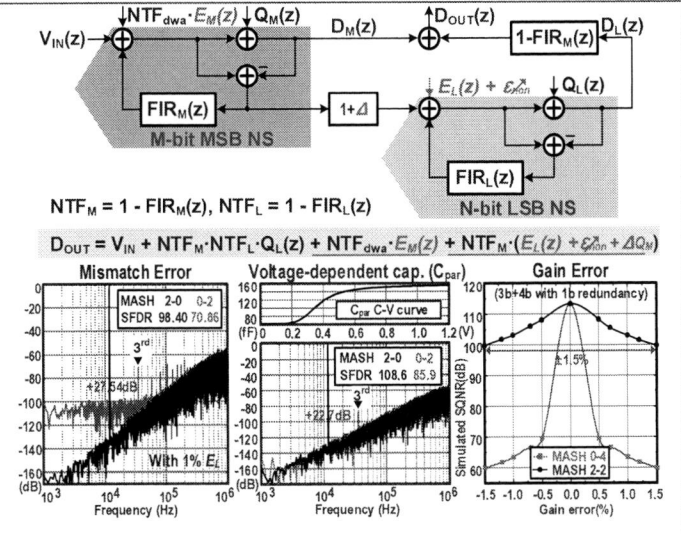

$$NTF_M = 1 - FIR_M(z), \quad NTF_L = 1 - FIR_L(z)$$

$$D_{OUT} = V_{IN} + NTF_M \cdot NTF_L \cdot Q_L(z) + NTF_{dwa} \cdot E_M(z) + NTF_M \cdot (E_L(z) + \varepsilon_{non}^2 + \Delta Q_M)$$

Fig. 1. The signal-flow diagram of the MASH-SAR ADC and the simulation results of error shaping in mismatch, C_{par} and gain.

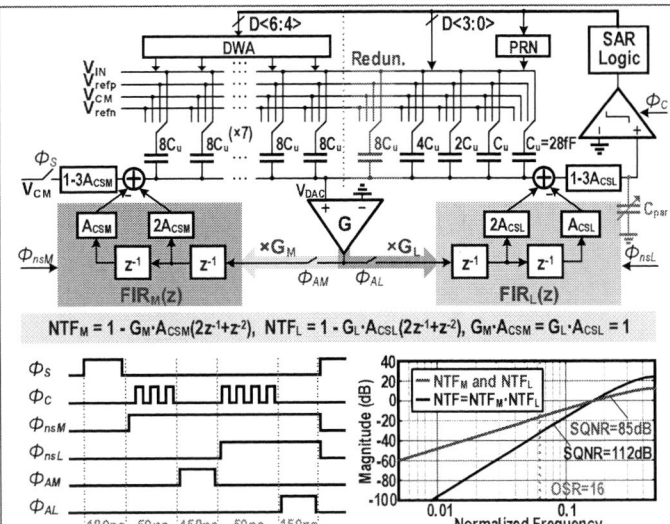

$$NTF_M = 1 - G_M \cdot A_{CSM}(2z^{-1} + z^{-2}), \quad NTF_L = 1 - G_L \cdot A_{CSL}(2z^{-1} + z^{-2}), \quad G_M \cdot A_{CSM} = G_L \cdot A_{CSL} = 1$$

Fig. 2. The overall architecture and the timing diagram of MASH-SAR ADC, reusing the single amplifier in two stages NS.

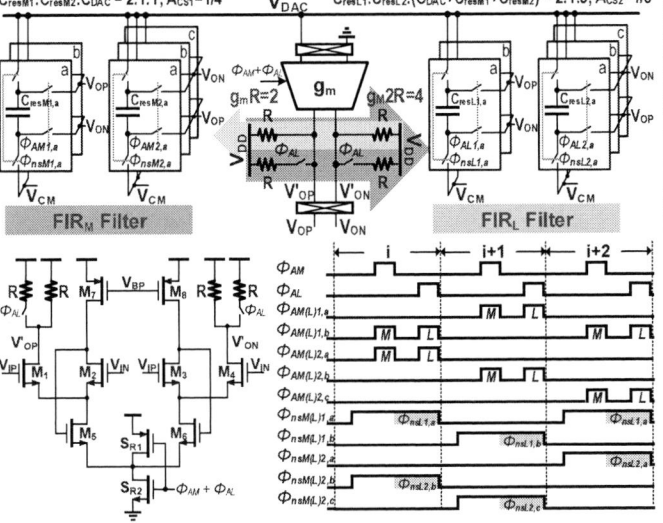

Fig. 3. The schematic, detailed operation and the corresponding

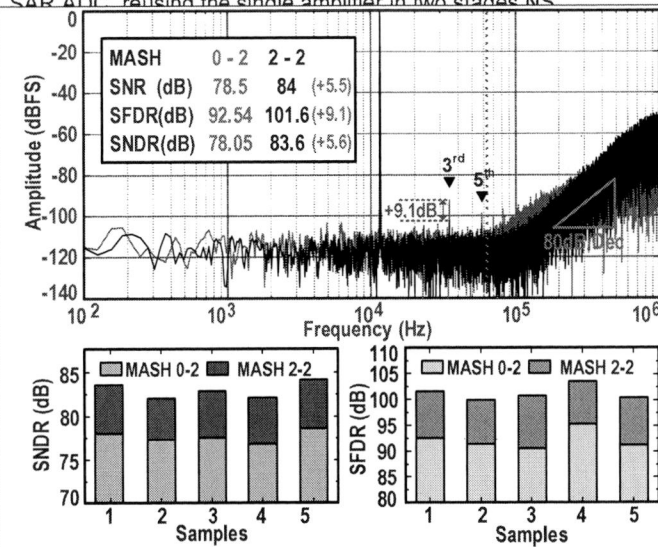

Fig. 4. Measured FFT spectra under different modes of sample #1 and SNDR/SFDR variation over 5 samples in two modes.

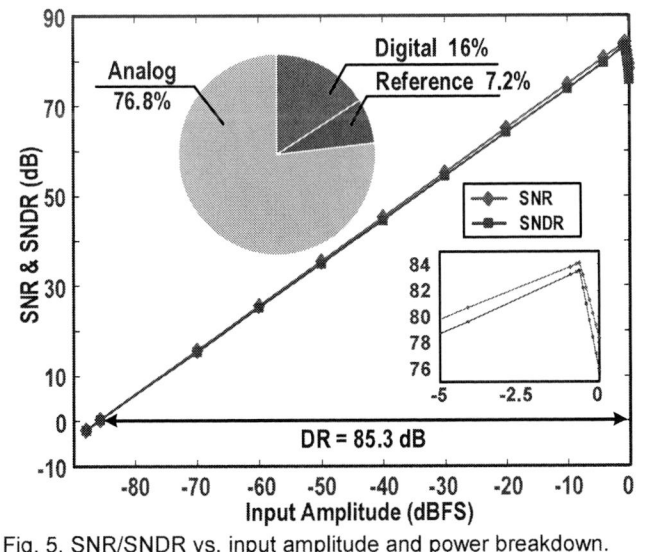

Fig. 5. SNR/SNDR vs. input amplitude and power breakdown.

	ISSCC 16 Shu[4]	ISSCC 20 Liu[5]	CICC 21 Shen[6]	ASSCC 21 Chen[1]	This Work
Technology[nm]	55	40	65	90	65
Architecture	MES+QNS	MES+QNS	MES	QNS	NES+QNS
C_{par} error shaping	×	×	×	×	√
Sharp NTF	×	×	×	√	√
NS Order	1	1	-	2	4
MES Order	1	2	1	0	2
DAC resolution	12	13	14	9	6
Fs[MHz]	1	2	0.128	10	2
OSR	500	25	16	8	16
BW[kHz]	1	40	4	625	62.5
SNDR[dB]	101	90.5	84.5	73.8	83.6
SFDR[dB]	105.1	102.2	103	88.13	101.6
DR[dB]	101.7	94.3	85	-	85.3
Power[μW]	15.7	67.4	0.98	71	68
FoMs[dB]	178.9	178.2	180.1	173.2	173.2

Fig. 6. Performance summary and comparison with state of the art.

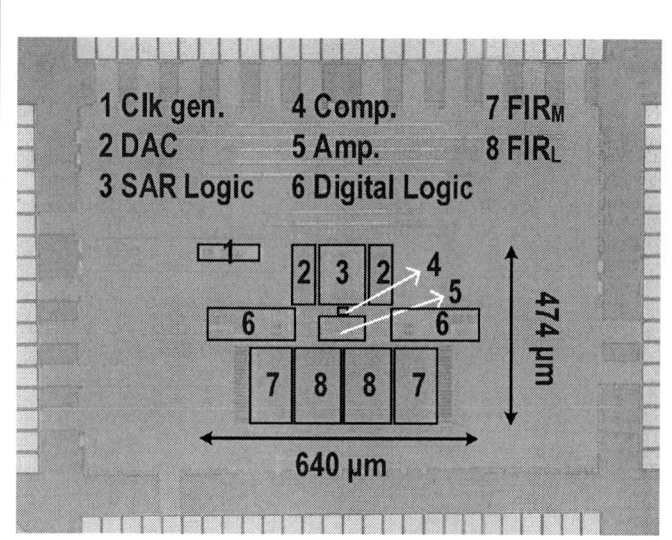

Fig. 7. Die micrograph.

Additional References:

[3] R. Adams and K. Q. Nguyen, "A 113-dB SNR oversampling DAC with segmented noise-shaped scrambling," in IEEE Journal of Solid-State Circuits, vol. 33, no. 12, pp. 1871-1878, Dec. 1998.

[4] Y. Shu, et al., "An oversampling SAR ADC with DAC mismatch error shaping achieving 105dB SFDR and 101dB SNDR over 1kHz BW in 55nm CMOS," ISSCC, pp. 458-459, Feb. 2016.

[5] J. Liu, et al., "A 40kHz-BW 90dB-SNDR Noise-Shaping SAR with 4× Passive Gain and 2nd-Order Mismatch Error Shaping," ISSCC, pp. 158-160, Feb. 2020.

[6] Y. Shen, et al., "A low-cost 0.98μW 0.8V oversampled SAR ADC with pre-comparison and mismatch error shaping achieving 84.5dB SNDR and 103dB SFDR," IEEE CICC, pp. 1-2, Apr. 2021.

[7] S. Li, et al., "A 13-ENOB 2nd-Order Noise-Shaping SAR ADC Realizing Optimized NTF Zeros Using an Error-Feedback Structure," ISSCC, pp.234-236, Feb. 2018.

A Compact 1,257-Gbps/W Byte-Serial AES Accelerator for IoT Applications in 22 nm

Shutao Zhang, Malte Wabnitz and Tobias Gemmeke

RWTH Aachen University, Aachen, Germany

Security is becoming even more critical for the digital society such as IoT. Especially quantum computers pose serious threats to the currently deployed crypto-systems [1], which requires double key size for symmetric cryptographic algorithms to preserve security. As the de-facto standard block cipher, Advanced Encryption Standard (AES) acts as the mainstay to secure versatile applications like data communication and storage. Therefore, an even more compact and energy-efficient AES accelerator is in urgent demand for billions of miniaturized and battery-supplied devices in IoT field considering quantum security. In contrast to the traditional 128-bit datapath designs, several byte-serial architectures have been proposed obtaining smaller area [2-7]. On one hand, single SBox is exploited for both data encryption and key expander for further area reduction [2-5]. However, significant redundant data movements result in large energy overhead, and additional cycles for ShiftRow lower the throughput [2-3]. The additional registers for the intermediate storage add to control complexity as well as power consumption [4-5]. The long latency (336/337 cycles per encryption) negatively impacts the hardware utilization and energy efficiency. On the other hand, double SBoxes boost the throughput with a corresponding area penalty [6-7]. Compared to the separate SBox for data encryption and key expander in [7], one of the two Sboxes is shared between data encryption and key expander in [6], which reduces the latency from 160 cycles/encryption to 113 but is still above the algorithmic lower bound of 100 cycles for 2 SBoxes. In this paper, the proposed AES accelerator reaches a throughput equivalent to the corresponding lower bound of 200 cycles/encryption using a single SBox at 100% utilization and achieves the lowest number of registers (32 bytes) for data storage. The redundant data movement is significantly reduced with efficient data structure while preserving high clock frequency. Through the multi-objective optimization, the proposed AES-128 encryption accelerator is well suited for embedded systems with tight area, throughput and energy constraints.

Fig. 1 shows the conventional byte-serial architecture of AES encryption with area and power breakdown and the motivation of this design targeting IoT field. Registers consume a large portion (over 50%) in both area and power. As the most compute-intensive block, SBox dominates the critical path and the power consumption in the combinational logic. Further, the utilization rate of SBox determines the throughput. Thus, this work targets minimizing the number of registers and maximizing the utilization of the SBox. By exploiting efficient memory organization, register renaming and pipelining techniques, both high clock frequency and energy efficiency are achieved.

Fig. 2 presents the overall architecture of the proposed 8-bit datapath AES accelerator. The whole architecture can be partitioned into two parts: data encryption (left) and key expander (right) with a shared SBox. The data encryption part includes 4 scratch pads each composed of 4 bytes, with each byte implemented with clock gated multi-bit FFs. Data shift between adjacent scratch pads happen once in each cycle of data encryption maintaining data locality. This arrangement keeps the data going to the SBox locally stored in the uppermost scratch pad, which reduces the critical path. Furthermore, this memory structure ensures that the 4 bytes of data going to MixColumn are read out from 4 distinct scratch pads. Next, the results are in-place written back to distinct scratch pads without any memory access conflict. An optimized trade-off between write and read complexity is achieved with such a hybrid shift-memory structure. Instead of the sequential order of SBox and MixColumn in [4-7], SBox and MixColumn are reordered to be processed in parallel, which further reduces the critical path and avoids the serial accumulator inside the MixColumn to save energy. The MixColumn is bypassed in the 10th round. The key expander part is composed of 2 scratch pads (4 bytes and 12 bytes), round constant generator and XORs for key expansion. The key byte read out from the 4-byte scratch pad has three destinations: AddRoundKey for data encryption, the 12-byte scratch pad as intermediate storage and XOR for key expansion in the current round. The read port is reused

for multiple purposes for timing, area and energy efficiency. While the final read-out key byte from the 12-byte scratch pad is sent to XOR for key expansion and further written into the 4-byte scratch pad, the read-out key byte from part of the 12-byte scratch pad (K12-K15) propagates to SBox for key expansion. This data locality not only reuses the existing read port for area savings, but also reduces the timing critical path. No read conflict happens because any simultaneous read operation accesses different parts of scratch pad in this architecture. Fully utilizing the XOR for key expansion realizes a compact key expander. To reduce area, the round constant is generated on-the-fly based on the Linear Feedback Shift Register (LFSR) as opposed to LUT. Overall results are very positively affected by the efficient dataflow and novel data organization in both data encryption and key expander.

Fig. 3 describes the hierarchical dataflow schedule through message-, round- and cycle-levels. For clearness, the one additional cycle due to the pipelining in the SBox is not shown. Each message has a 216-cycle latency, including 10 normal rounds of 20 cycles and one additional round of 16 cycles to perform final AddRoundKey and output the ciphertext. The accelerator overlaps the first round of the next message and the last round of the current one resulting in a throughput corresponding to the inverse of 200 cycles per encryption. In the normal round, the 100% utilized SBox is deployed for data encryption in the first 16 cycles and is reconfigured for key expansion in the last 4 cycles. During the last 4 cycles, ShiftRow and MixColumn are performed simultaneously with the adaptive read and write pattern. The in-place write back with register renaming technique eliminates the need for additional registers and redundant data movement for both area and energy efficiency. For example, four-byte data {D0, D5, D10, D15} are sent to MixColumn and the results are written back to DataReg in the next cycle with the updated name {D0, D4, D8, D12}. By distributing the expansion of the 16 key bytes into 16 cycles, the related XORs are reused to their largest extent. The expanded key bytes are stored in the 4-byte scratch pad in key expander, which are transferred to AddRoundKey in 4 cycles.

Fig. 4 presents the pipelined SBox, parallel MixColumn and area/energy comparison between the proposed design and baseline. Compared to the straightforward LUT-based and extension-field $GF(2^8)$ implementations, the Sbox in composite field $GF(2^4)^2$ attains a more compact area. The pipelining in the SBox balances the datapath for an ultra-high clock frequency (> 2 GHz at 0.8V). The parallel MixColumn reuses multiplication by 2 as part of the multiplication by 3 in $GF(2^8)$ gaining 25% area reduction. By eliminating additional registers, preserving 100% utilization of SBox and developing efficient data storage as well as dataflow, more than 50% reduction in area and energy are obtained as compared to the baseline.

Fig. 7 shows a micrograph of the test chip (1,704 μm^2) fabricated in a 22-nm FDSOI CMOS technology. Fig. 5 summarizes the chip measurements results. Under the supply voltage range of 0.45-0.8 V, the clock frequency increases by more than 8.9X from 251 to 2,237 MHz, which thereby boosts the throughput from 161 to 1,432 Mbps and area efficiency from 94 to 840 Gbps/μm^2. The voltage scaling from 0.8 to 0.45 V achieves 3.1X increase in energy efficiency from 400 to 1,257 Gbps/W. Fig. 6 presents the comparison with prior art. Our proposed design outperforms previous designs across different metrics including throughput, energy and area efficiency. In comparison with [5], this design achieves 1.8X better energy efficiency, 1.8X better area efficiency and 2.4X higher throughput at iso-technology at nominal supply voltage. Even compared to the 2-SBox design in [6], our proposal achieves 1.21X better energy efficiency, 1.65X better area efficiency and 1.48X higher throughput at iso-technology at nominal supply voltage. Overall, our proposed high-throughput, area- and energy-efficient AES-128 accelerator is well suited for versatile lightweight applications. All proposed techniques are applicable to AES-256 with double key size for quantum security and other AES-like ciphers.

Acknowledgment: This work was partially funded by the German BMBF project NEUROTEC II under Grant no. 16ME0399.

References:

[1] S. Zhang et al., "A 22-nm 1,287-MOPS/W Structured Data-Path Array for Binary Ring-LWE PQC," ESSCIRC, 2023.

979-8-3503-3004-5/23 $31.00 © 2023 IEEE

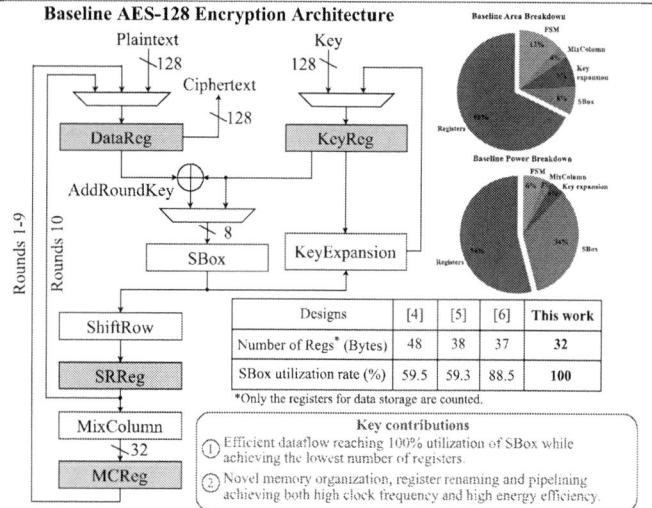

Fig. 1. Byte-serial baseline implementation of AES-128 encryption with area/power breakdown (post-layout simulation) and design motivation.

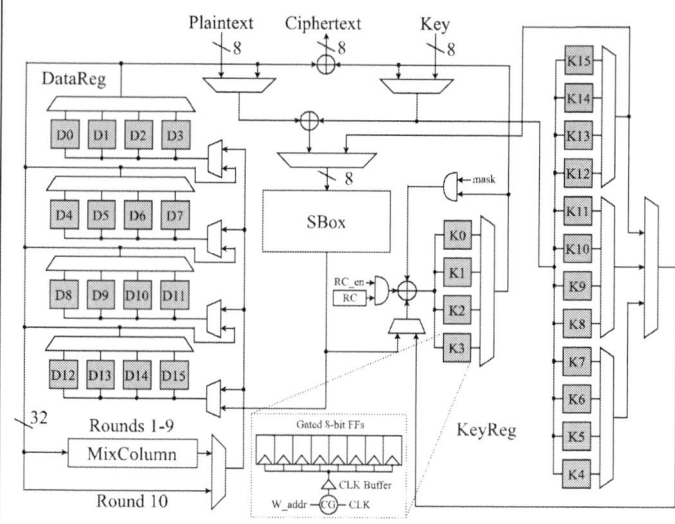

Fig. 2. Overall architecture of the proposed 8-bit datapath AES accelerator.

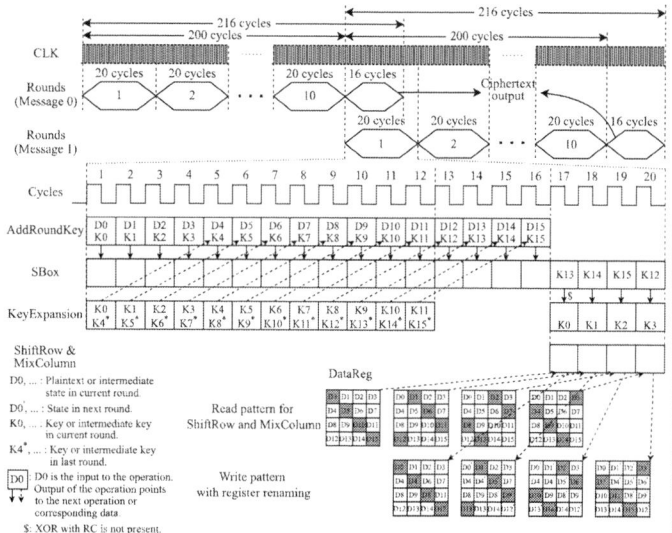

Fig. 3. Hierarchical dataflow schedule and memory access pattern.

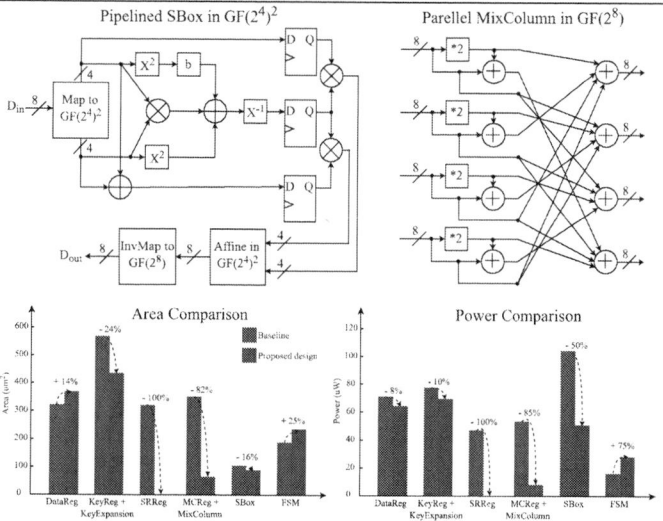

Fig. 4. Pipelined SBox in $GF(2^4)^2$, parallel MixColumn in $GF(2^8)$ with logic reuse and area/power comparison between baseline and proposed design (post-layout simulation).

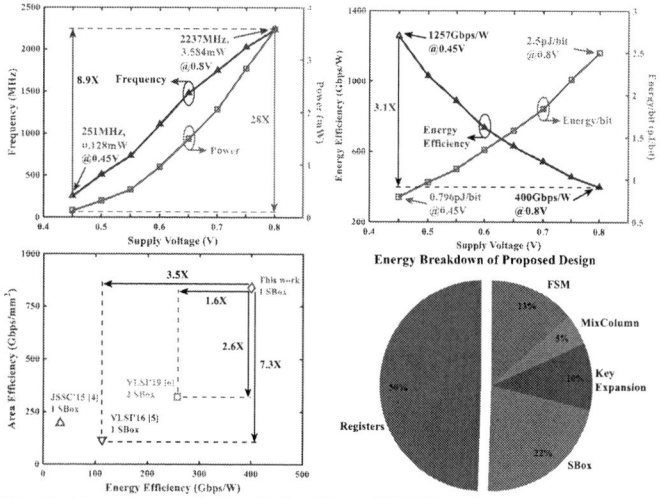

Fig. 5. Measurement results in silicon (25°C) and energy breakdown of the proposed design in post-layout simulation.

	JSSC'15 [4]	VLSI'16 [5]	VLSI'19 [6]	**This work**
Process (nm)	22	40	28	22
No. Registers (Bytes)	48	38	37	32
No. SBox	1	1	2	1
Encryption Cycles	336	337	113	200
SBox Utilization (%)	59.5	59.3	88.5	100
Supply Voltage (V)	$0.43^£$	$0.47^£$	$0.41^£$	$0.45^£$
	$0.9^\$$	$0.9^\$$	$0.9^\$$	$0.8^\$$
Power (mW)	$0.45^£$	$0.1^£$	$0.0306^£$	$0.128^£$
	$13^\$$	$4.39^\$$	$3.85^\$$	$3.584^\$$
Frequency (MHz)	$220^£$	$122^£$	$25^£$	$251^£$
	$1100^\$$	$1300^\$ (1596)^†$	$875^\$ (856)^†$	$2237^\$ (2.61X)^*$
Throughput (Mbps)	$83.6^£$	$46.2^£$	$28^£$	$161^£$
	$432^\$$	$494^\$ (606)^†$	$991^\$ (969)^†$	$1432^\$ (1.48X)^*$
Area (μm²)	2200	$4290 (1298)^†$	$3090 (1908)^†$	$1704 (-11\%)^*$
Energy Efficiency (Gbps/W)	$186^£$	$446^£$	$923^£$	$1257^£$
	$33^\$$	$113^\$ (225)^†$	$257^\$ (330)^†$	$400^\$ (1.21X)^*$
Area Efficiency (Gbps/mm²)	$38^£$	$11^£$	$9^£$	$94^£$
	$196^\$$	$115^\$ (467)^†$	$321^\$ (508)^†$	$840^\$ (1.65X)^*$

£ Measurement at minimum energy point   \$ Measurement at nominal supply voltage

† Scaled to 22 nm technology and 0.8 V supply voltage

* Comparison results to [6] at iso-technology

Fig. 6. PPA summary and comparison with prior art.

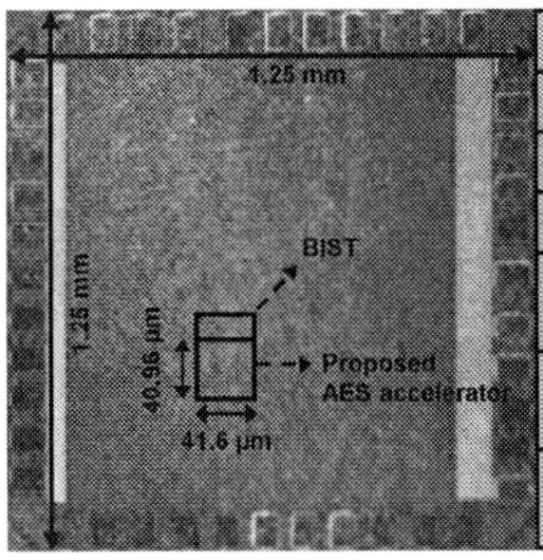

AES Accelerator Specification	
Technology	22 nm FDSOI
Supply Voltage	0.45-0.8 V
Core Area	1,704 μm^2
Max. Frequency	2237 MHz@0.8 V 251 MHz@0.45 V
Throughput	1432 Mbps@0.8 V 161 Mbps@0.45 V
Energy Efficiency	400 Gbps/W@0.8 V 1257 Gbps/W@0.45 V

Fig. 7. Chip micrograph and specification.

Additional References:

[2] A. Moradi *et al.*, "Pushing the limits: A very compact and a threshold implementation of AES, " EUROCRYPT, 2011.
[3] S. Banik *et al.*, "Atomic-AES: A Compact Implementation of the AES Encryption/Decryption Core," INDOCRYPT, 2016.
[4] S. Mathew *et al.*, "340 mV–1.1 V, 289 Gbps/W, 2090-Gate NanoAES Hardware Accelerator With Area-Optimized Encrypt/Decrypt GF(2 4) 2 Polynomials in 22 nm Tri-Gate CMOS," JSSC, 2015.
[5] Y. Zhang *et al.*, "A compact 446 Gbps/W AES accelerator for mobile SoC and IoT in 40nm," VLSI, 2016.
[6] W. Shan *et al.*, "A 923 Gbps/W, 113-Cycle, 2-Sbox Energy-efficient AES Accelerator in 28nm CMOS," VLSI, 2019.
[7] P. Hamalainen *et al.*, "Design and Implementation of Low-Area and Low-Power AES Encryption Hardware Core," EUROMICRO DSD, 2006.

A 200MHz Bandwidth 84% Peak Efficiency AC-Coupled Envelope Tracking Supply Modulator with 10MHz Constant Switching Frequency

Peng Xu, Xueli Zhang, Tao Wang, Peng Cao, Jiawei Xu, Zhiliang Hong.

State Key Laboratory of Integrated Chips and Systems, Fudan University

Envelope Tracking (ET) is widely used to improve the efficiency of linear power amplifiers (PAs) in mobile handsets. Power dissipation caused by the large peak-to-average power ratio (PAPR) can be reduced by adjusting the RF PA supply voltage according to the transmitted signal's envelope. The conventional analog ET supply modulators (SMs) are usually composed of a high-efficiency switching power modulator (SPM) and a wide-bandwidth linear amplifier (LA). The ET bandwidth is mainly determined by LA and has been extended from 20MHz to 130MHz [1-3]. To accommodate 5G New Radio (NR), wide-bandwidth ET SMs are required, but the large PA decoupling capacitor and output voltage exceeding 5V remain bottlenecks [4]. Digital ET SM [5] provides a solution for NR 200MHz transmission, but the generated stair-case voltage is more like a fast average power tracking (APT) SM rather than an ET SM, resulting in limited power efficiency. In addition, the switched-capacitor voltage divider has more than 11 off-chip capacitors and 20 power transistors.

This work proposes a 200MHz 84% peak-efficiency ET SM with less number of external components by applying the following techniques: (1) the LA with rail-to-rail input to accommodate a low supply voltage for improved efficiency; (2) the switching frequency is constant, which is preferable in the wireless communication system for the known electromagnetic interference noise spectrum; (3) Using actual PA as the load instead of a PA model in [4].

Figure.1 shows the architecture of the proposed ET SM. The unit-gain feedback of the linear amplifier plays a key role in tracking the 200MHz bandwidth envelope. The pole at the negative input of LA is removed due to the unity-gain feedback to prevent deteriorating the loop response [2-4]. The required bandwidth of the LA is also relaxed since the close-loop gain of LA at 200MHz is 0dB rather than 6dB [4]. Furthermore, the ac-coupling capacitor C_C isolates the output of the LA (V_{LA}) from the output of ET SM (V_{ET}), thereby enabling a V_{ET} beyond the supply voltage of SPM (VDD_{SPM}). The voltage on C_C (V_{CAP}) is regulated to be equal to a reference voltage (V_{REF}) by SPM, thus V_{ET} is the sum of V_{REF} and V_{ENV}. Moreover, the proposed SPM utilizes a pulse-width-modulation (PWM) control strategy with a 10MHz constant switching frequency. To improve the transient response speed of the SPM, as illustrated at the bottom of Fig.1, the descending slope (k_{SAW1} & k_{SAW2}) of the sawtooth (V_{SAW}) is variable according to the direction of the output current of LA (i_{LA}). When the output power increases, the error voltage (V_{EA}) rises to increase Duty. Meanwhile, i_{LA} accurately tracks the variations of output power, this information is then added to adjust the swing range of V_{SAW} via D_{LA}. Under the combined actions of V_{EA} and V_{SAW}, the adjustment speed of Duty is improved.

The simplified LA schematic is shown in Fig.2. It consists of an OTA with rail-to-rail input and class-AB output stages. The dominant poles are located at the gate of the power transistors (M_{PP} and M_{PN}) while the non-dominant pole appears at the output node (V_{out}). Since the bandwidth of the open-loop LA is more than 200MHz, any other poles below 2GHz may deteriorate the phase margin. The structure of the LA is simplified to the greatest extent to reduce the number of poles. Four poles located at the gates of the current mirrors are pushed to higher frequencies by the compensation resistors (R_{g1}-R_{g4}) [4]. Also, for simplicity, the constant transconductance design in conventional operational amplifiers is not adopted in the proposed rail-to-rail input stage, but the frequency response still needs to meet the bandwidth and phase margin requirements at any common mode input voltage. The ac-coupling capacitor C_C is included in a load of LA to isolate the DC path between LA and PA model. The simulated frequency responses are shown in the table in Fig.2, when V_{ip} is 0.5V or 2.8V, either PMOS (M_{P1} & M_{P2}) or NMOS (M_{N1} & M_{N2}) input pairs are active to amplify the input signal. When V_{ip}=1.7V, as shown in the bottom of Fig.2, the frequency response shows that the UGB is 198MHz and

the phase margin is 48°. When the LA is set to the unity-gain mode, the UGB is extended to 280MHz and the phase margin is improved to 72°.

The proposed descending slope variable sawtooth generator (VSG) is shown in Fig.3. V_{RAMP} is a 10MHz ramp waveform with a swing range of 0-V_H. Controlled by the same CLK signal, V_{SAW} has the same frequency as V_{RAMP}. The descending slope of V_{SAW} varies according to the logic voltage of D_{LA}. To illustrate the principle that VSG improves the transient speed of SPM, we assume that the input of LA (V_{ENV}) and V_{ET} are constant. As the waveforms shown in Fig.3, at the steady state, LA sinks the inductor ripple current (i_{SPM}) to keep i_{PA} constant. When there is a step in V_{ENV}, the current i_{PA} and i_{LA} immediately follow this step and change Δi_{PA}. Ignoring the change of V_{EA} in this period, the duty cycle varies by Δd due to the slope change of V_{SAW}. As a result, i_{SPM} changes Δi_L. According to the geometric relationship, when (k_{SAW2}/k_{SAW1}) > (1-2V_{ET}/VDD_{SPM}), Δi_L<2Δi_{PA}, thus i_{SPM} always tracks i_{PA} and the system is stable. In this design, k_{SAW1}:k_{SAW2}=1:3 is used to ensure V_{ET}>1.55V (output power > 600mW).

Figure 4 shows the measured ET waveforms for a 200MHz envelope NR signal with an 8.84dB PAPR, and 4Ω//600pF is used as the PA model. The measured output voltage V_{ET} can well track the desired envelope V_{ENV}. When applying a 700mV step to the input of LA, with V_{SAW}, i_{SPM} can complete the response in 500ns. To visualize the tracking ability of i_{SPM}, we enabled a 20MHz low-pass filter function in the oscilloscope to filter V_{ET}. The filtered waveform represents the low-frequency power of V_{ET}, which is well-tracked by i_{SPM}. Figure 5 (left) shows the measured performance of SPM. When tracking the NR 200MHz envelope, the spectrum of V_{SW} indicates that the switching frequency stays 10MHz. The efficiency is measured under different peak-to-peak envelope voltages. The ET SM is capable of delivering up to 2.8W average output power at 2V_{pp} and 3V_{pp} envelope voltages with efficiencies of 84% and 77.6%, respectively.

To evaluate the performance of the ET PA in practice, we also used a PA prototype as the load. The first step is to align the outputs of the ET and the PA, as shown in the top-right of Fig.5. The PAE of a stand-alone PA and an ET PA, and the power saving under different PA output power scenarios are shown in the right-bottom of Fig.5. At 26dBm output power, the ET PA improves PAE by 1.5% and saves 228mW, accounting for 8.9% of stand-alone PA power consumption. It should be noted that when calculating the PAE of PA, the loss of its power supply is not considered. Figure 6 (top) compares the output spectrum. The ACLR of PA is -29dBc and -33.7dBc at 26dBm. Without using the pre-distortion technique, the ACLR of the ET PA is reduced by 1.2dBc and 3.1dBc, respectively. The comparison table in Fig.6 benchmarks this work with the state-of-the-art ET SMs. The proposed analog ET SM achieves the largest bandwidth of 200MHz with excellent power efficiency. The chip was fabricated in a 0.18μm BCD 5V process with an area of 1.5mm×1.2mm (Fig.7).

References:

[1] X. Liu et al., "A Multi-Loop-Controlled AC-Coupling Supply Modulator with a Mode-Switching CMOS PA in an EER System with Envelope Shaping," IEEE J. Solid-State Circuits, vol. 54, no. 6, pp. 1553-1563, June 2019.

[2] Y. -C. Lin et al., "A CMOS Envelope Tracking Supply Converter for RF Power Amplifiers of 5G NR Mobile Terminals," IEEE Trans. Power Electronics, vol. 36, no. 6, pp. 6814-6823, June 2021.

[3] D. Kim et al., "33.9 A Hybrid Switching Supply Modulator Achieving 130MHz Envelope-Tracking Bandwidth and 10W Output Power for 2G/3G/LTE/NR RF Power Amplifiers," ISSCC 2021:476-478.

[4] P. Xu et al., "A 2.7W AC-coupled hybrid supply modulator achieving 200MHz envelope-tracking bandwidth for 5G new radio power amplifier," A-SSCC 2021.

[5] J. -S. Bang et al., "14.5 2-Tx Digital Envelope-Tracking Supply Modulator Achieving 200MHz Channel Bandwidth and 93.6% Efficiency for 2G/3G/LTE/NR RF Power Amplifiers," ISSCC 2022: 236-237.

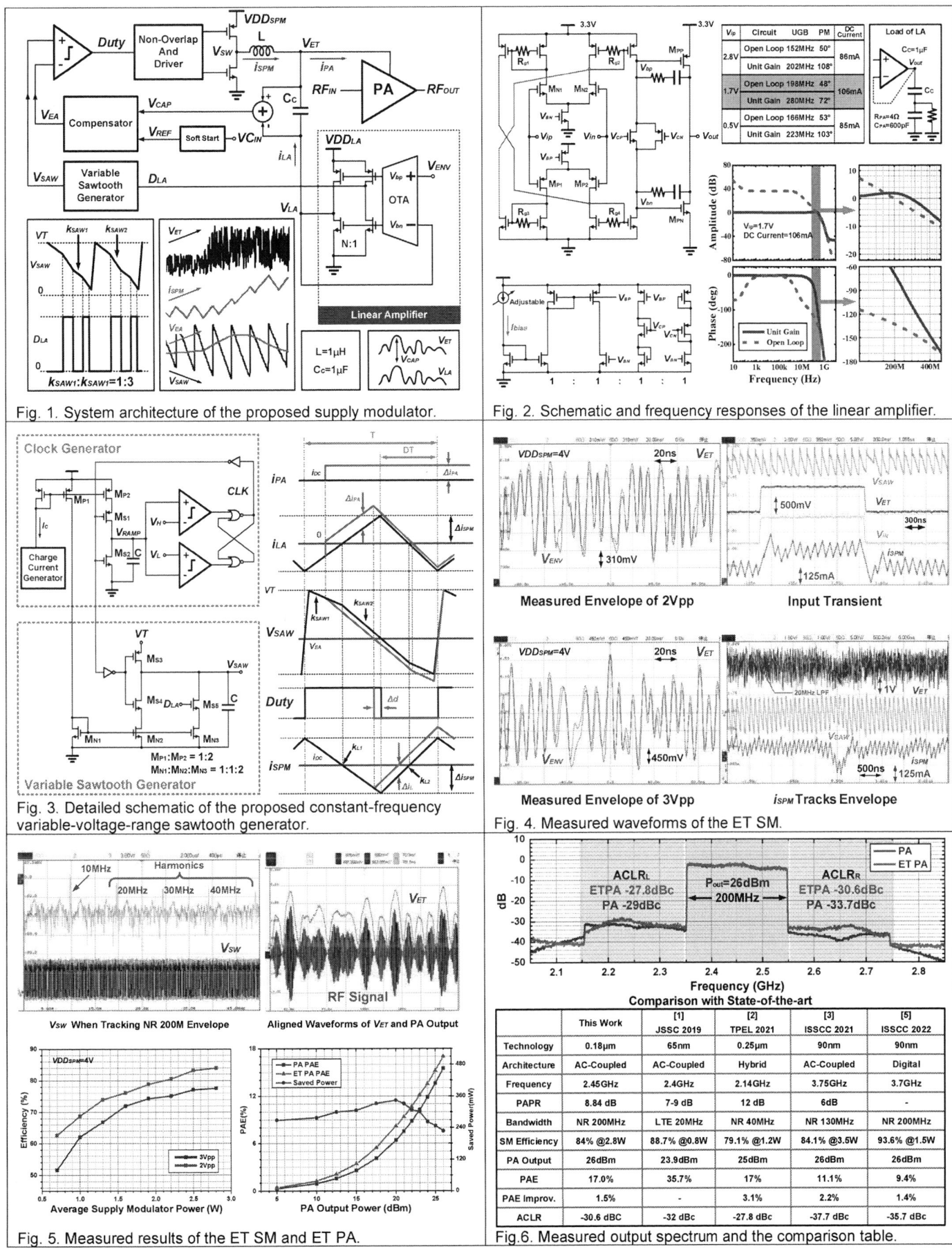

Fig. 1. System architecture of the proposed supply modulator.

Fig. 2. Schematic and frequency responses of the linear amplifier.

Fig. 3. Detailed schematic of the proposed constant-frequency variable-voltage-range sawtooth generator.

Fig. 4. Measured waveforms of the ET SM.

Fig. 5. Measured results of the ET SM and ET PA.

Fig.6. Measured output spectrum and the comparison table.

Comparison with State-of-the-art

	This Work	[1] JSSC 2019	[2] TPEL 2021	[3] ISSCC 2021	[5] ISSCC 2022
Technology	0.18μm	65nm	0.25μm	90nm	90nm
Architecture	AC-Coupled	AC-Coupled	Hybrid	AC-Coupled	Digital
Frequency	2.45GHz	2.4GHz	2.14GHz	3.75GHz	3.7GHz
PAPR	8.84 dB	7-9 dB	12 dB	6dB	-
Bandwidth	NR 200MHz	LTE 20MHz	NR 40MHz	NR 130MHz	NR 200MHz
SM Efficiency	84% @2.8W	88.7% @0.8W	79.1% @1.2W	84.1% @3.5W	93.6% @1.5W
PA Output	26dBm	23.9dBm	25dBm	26dBm	26dBm
PAE	17.0%	35.7%	17%	11.1%	9.4%
PAE Improv.	1.5%	-	3.1%	2.2%	1.4%
ACLR	-30.6 dBC	-32 dBc	-27.8 dBc	-37.7 dBc	-35.7 dBc

979-8-3503-3004-5/23 $31.00 © 2023 IEEE

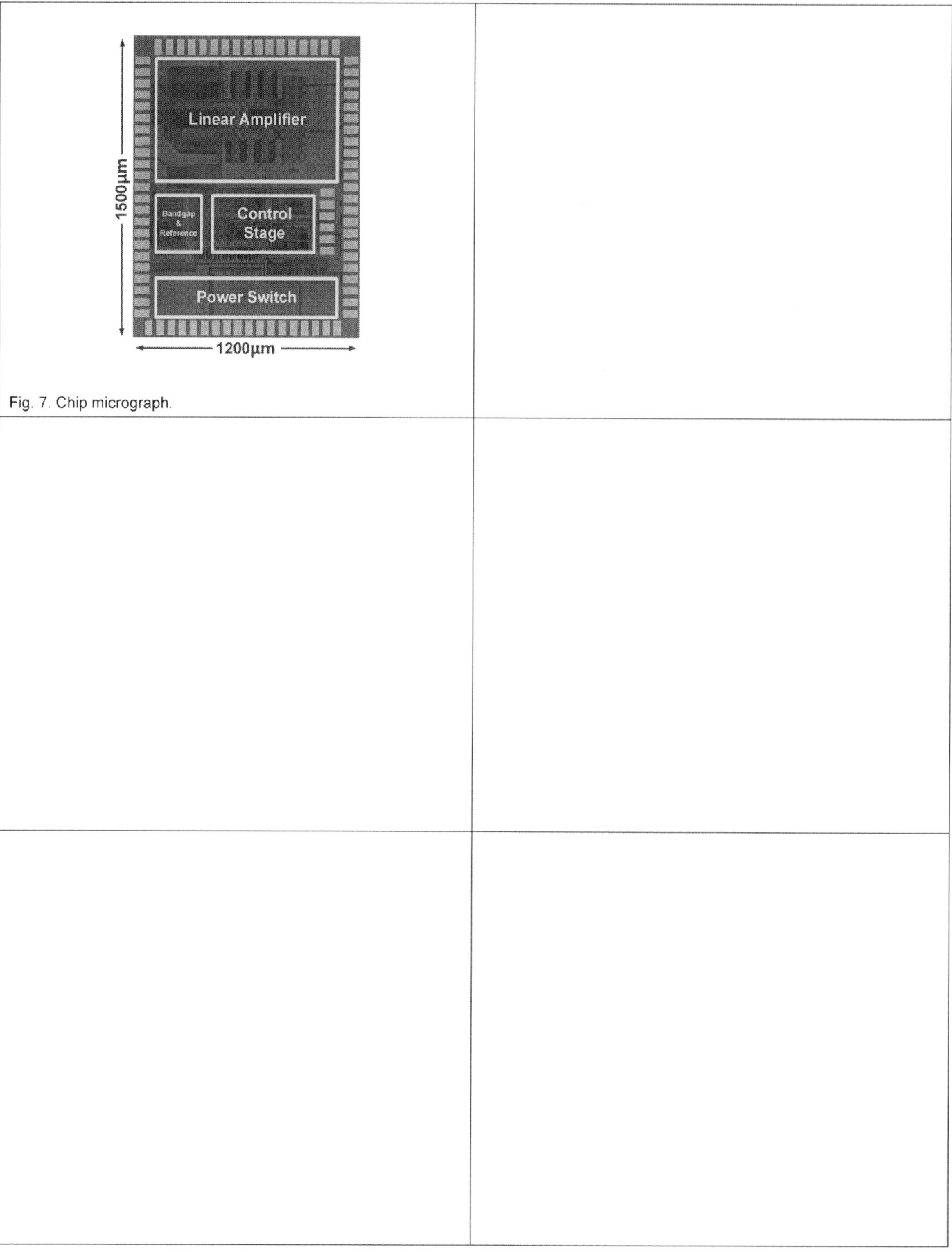

Fig. 7. Chip micrograph.

An Energy-Efficient Heterogeneous Fourier Transform-based Transformer Accelerator with Frequency-wise Dynamic Bit-Precision

Jingu Lee, Sangjin Kim, Wooyoung Jo and Hoi-Jun Yoo

Korea Advanced Institute of Science and Technology (KAIST)

Recently, transformer models [1] have achieved human-level accuracy and become the mainstream algorithm for diverse Natural Language Processing (NLP) tasks with the self-attention mechanism. The transformer models comprise a series of repeated self-attention layers that find the relationships between tokens and mixing information (TokenMix) and the subsequent feed-forward network (FFN) for capturing complex patterns. The self-attention layer calculates the relevance weights for each input token and employs these weights to a weighted-sum operation for calculating the output token. Therefore, the model can capture contextual relationships by mixing the information in multiple input tokens into output tokens. The Fourier transform-based TokenMix (FT-TokenMix) algorithm [2] replaces the computationally-heavy self-attention algorithm with a light-weight Fourier transform while achieving nearly the same performance. Since Fourier transform also finds the correlation between a different data point and mixes the information of the input token, it can capture the contextual relationships. Also, because the computation complexity of the Fourier transform is much lower than the self-attention layer with the Fast Fourier transform (FFT) algorithm, 99.6% of token mixing computation can be reduced.

However, the FT-TokenMix-based model encounters several challenges in hardware acceleration as shown in Fig. 1. First, unlike the traditional transformer model, the computation characteristics of TokenMix and FFN are different. Therefore, this paper proposes a heterogeneous architecture for FFT and FFN. Next, since FT-TokenMix only reduces the computation of TokenMix, the FFN layer accounts for 99.7% of the total computational requirements. As a result, FFN layer optimization is required for high-performance inference.

In this paper, a heterogeneous transformer accelerator is proposed for fast and energy-efficient inference with 2 key features: 1) Frequency-wise Dynamic Bit-precision (FDB) reduces 24.3% of overall computations in FFN by changing 60% of 8-bit operation into 4-bit operation. 2) Heterogeneous FFN core cluster (HeNCoC) improves throughput 21% by maintaining high utilization and efficiency even on the varying workload characteristics with different layers and tasks. Finally, the proposed processor achieves 268.94mJ/Inf. on the CoLA dataset of the GLUE benchmark which is 18% lower than the previous state-of-the-art [3, 4].

Fig.2 shows the motivation and operation of the proposed FDB. FFN requires a large computation amount for neural network operations and it also requires at least 8-bit precision for weight (W) and input activation (IA) to maintain accuracy. To alleviate its high computation cost, FDB is proposed for reducing the average computation precision of FFN with a property of FT-TokenMix. After the FFT layer, each vector at a different frequency represents the contextual relationships between different tokens because the FFT layer calculates correlation with different periods. However, some relationships are not crucial and their importance can be easily defined by the amplitude of FFT layer output as in the previous self-attention-based transformer [5]. Since the vectors at the low-frequency (LF) range represent a global relationship with its long period and vectors at the high-frequency (HF) represents a less-important local relationship with its short period, the amplitude of data is large in LF and small in HF. By this, FDB aggressively quantizes HF data to low bit-precision(4-bit) and uses the high bit-precision(8-bit) for LF data. Moreover, since the frequency range for important data are varying depending on the layers and tasks, FDB adaptively determines the frequency range for quantization to minimize average computing precision without an accuracy drop. In conclusion, FDB shows a negligible accuracy drop (<1%) on the GLUE benchmark, while replacing 60% of 8-bit operations with 4-bit operations.

Fig.3 shows the overall architecture of the proposed heterogeneous transformer accelerator with FT-TokenMix. It consists of pipelined FFT core for FT-TokenMix and HeNCoC for FFN. FFT core is made up of six 1D FFT lanes which support 256-point pipelined 1D FFT by five Radix-2^2 single-path delay feedback (SDF) units [6]. Also, a single Radix-2^2 SDF unit consists of 2 butterfly units, 2 delay buffers, and a complex multiplier with read-only memory for the FFT twiddle factor. Two or three lanes can be combined using a 2/3 radix signal flow unit for supporting 512-point and 768-point FFT respectively. As a result, pipelined FFT core can support 2D FFT operation without utilization drop on the input matrix with a size of 512×768 by the reconfiguration of 1D 256-point FFT lane. HeNCoC consists of a high-bit core that supports 8-bit operation and an adaptive bit-core that supports both 4-bit and 8-bit operation. High-bit core integrates a total of 84KB block memory(BMEM) and 64×16 8-bit processing element (PE) array. Adaptive-bit core integrates a total of 264KB BMEM and 64×64 4-bit PE array. Both cores integrate the same input-stationary PE array with different dimensions, partial-sum accumulator, and GeLU and quantization unit.

Fig.4 described the architecture and operation of the proposed HeNCoC to maintain high hardware utilization and improve throughput. FDB replaced the part of an 8-bit operation with a 4-bit operation but remained 8-bit operation portion is varying from 12% to 100% across different layers and tasks. However, there is a static 8-bit workload (<20%), and the remaining dynamic workload (>80%) is changed between 4-bit and 8-bit operations, HeNCoC utilizes both dedicated high-bit core and adaptive-bit core. For detail, a high-bit core computes only a static 8-bit workload and an adaptive-bit core computes either a 4-bit workload and an 8-bit workload depending on their ratio. Despite the adaptive-bit core just using a 4-bit PE array, it can support 8-bit operation by sequentially processing MSB and LSB of IA and W with shift and accumulation logic. Also, the 5-bit multiplier is used for unit PE to support sign extension. Lastly, by adaptively reconfiguring an adaptive-bit core for dynamic workload, HeNCoC achieves a 21% throughput increase.

Fig.5 explains the detailed implementation of HeNCoC and performance on the benchmark. Since three 8-bit multipliers can be packed into 1DSP, a high-bit core with a 64×16 PE array can be synthesized with 77% of the DSP. Also, unit PE of adaptive-bit core consumes 15~18LUT without DSP and 66% of logic resources are used for adaptive-bit core. Finally, by integrating the heterogeneous core of HeNCoC, a total of 100% of DSP and 82% of logic are used and most of the FPGA resources are utilized. Since both high-bit core and adaptive-bit core uses the same weight for different IA vector, the weight memory can be shared and their memory access can be synchronized. Therefore, 37.93% memory footprint of HeNCoC is reduced in the CoLA dataset. Using FDB and HeNCoC, latency was reduced by 71.58% compared to self-attention and 54.84% compare to FT-TokenMix without FDB. Also, on the GLUE benchmark, the proposed processor achieves ×2.5~×4.0 higher FPS compare to self-attention and ×1.6~×2.2 compare to FT-TokenMix without FDB.

Fig.6 shows the implementation results with the demonstration system and the comparison table. The proposed accelerator is implemented on Intel's Cyclone V, which operates at a maximum 100MHz clock frequency as shown in Fig. 7. It achieves 76.62ms latency with 3.51W power consumption with 12-layer FT-TokenMix-based transformer inference on the CoLA dataset. Also, its energy efficiency is 268.94mJ/Inf. which is ×1.2 to ×2.4 more efficient than the previous state-of-the-art FPGA accelerator for transformer [3, 4]. In conclusion, the proposed accelerator optimizes TokenMix and FFN layers together and achieves state-of-the-art energy efficiency for transformer inference on NLP tasks.

References:

[1] T. Lin *et al.*, "A Survey of Transformers," AI Open 2022:111-132.

[2] J. Lee-Thorp *et al.*, "FNet: Mixing Tokens with Fourier Transforms," NAACL 2022:4296-4313.

[3] B. Li *et al.*, "FTRANS: Energy-Efficient Acceleration of Transformers using FPGA," ISLPED 2020:175-180.

[4] H. Fan *et al.*, " Adaptable Butterfly Accelerator for Attention-based NNs via Hardware and Algorithm Co-design," MICRO 2022:599-615.

[5] T. Yang *et al.*, "DTQAtten: Leveraging Dynamic Token-based Quantization for Efficient Attention Architecture," DATE 2022:700-705.

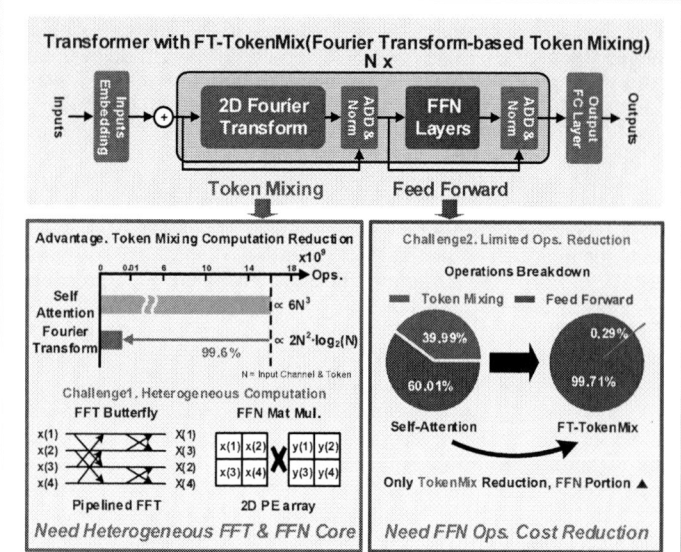

Fig. 1. Advantage and Challenge of Transformer with FT-TokenMix

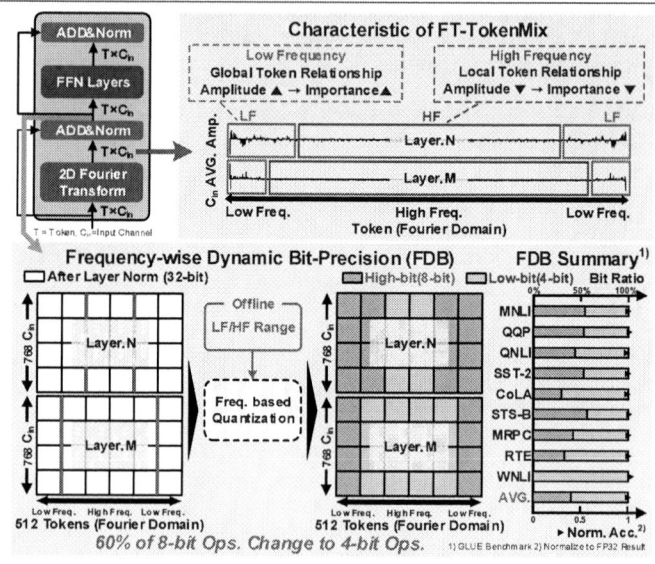

Fig. 2. Frequency-wise Dynamic Bit-Precision (FDB)

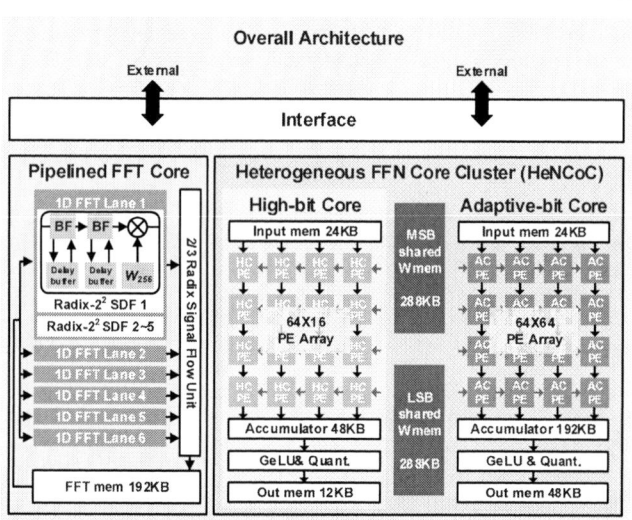

Fig. 3. Overall Architecture

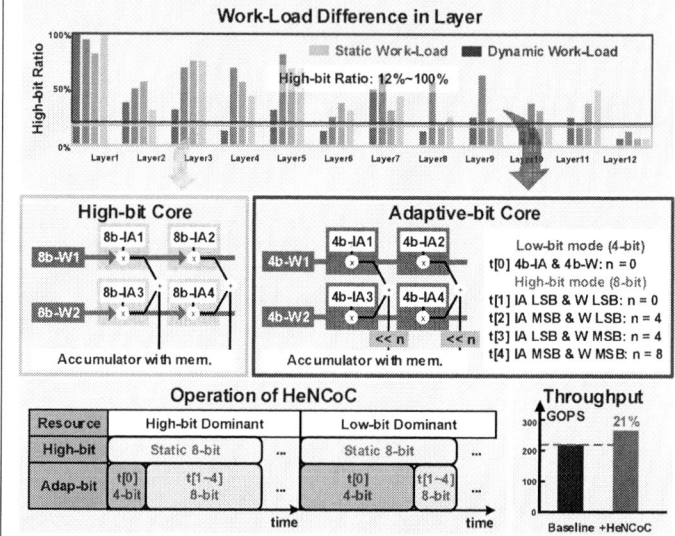

Fig. 4. Heterogeneous FFN Core Cluster (HeNCoC)

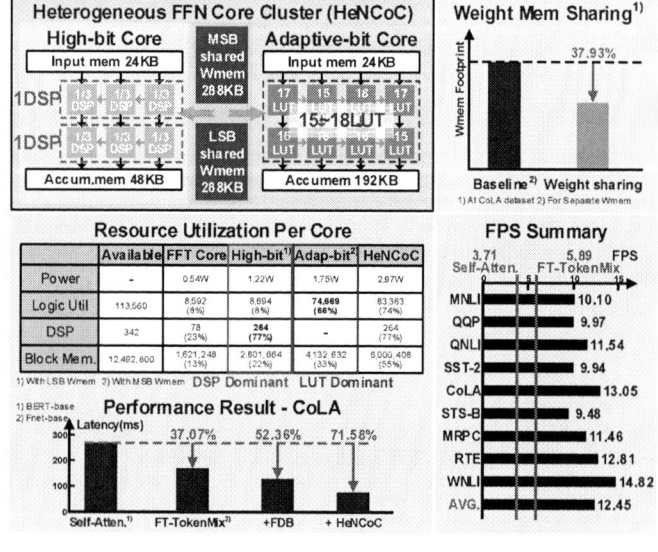

Fig. 5. HenCoC's Detail Implementation and Performance Summary

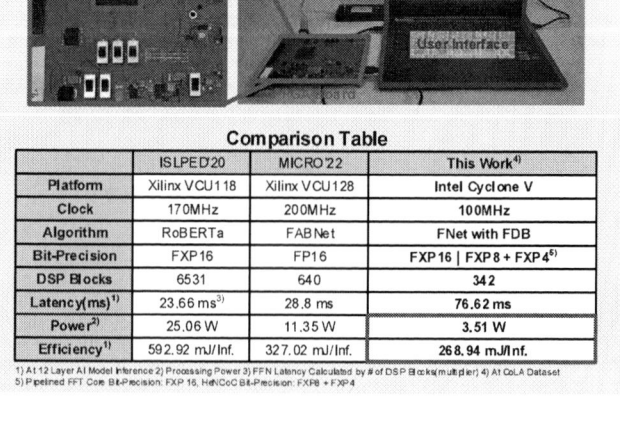

Fig. 6. Demonstration System and Comparison Table

Comparison Table

	ISLPED'20	MICRO'22	This Work[4]
Platform	Xilinx VCU118	Xilinx VCU128	Intel Cyclone V
Clock	170MHz	200MHz	100MHz
Algorithm	RoBERTa	FABNet	FNet with FDB
Bit-Precision	FXP16	FP16	FXP16 | FXP8 + FXP4[5]
DSP Blocks	6531	640	342
Latency(ms)[1]	23.66 ms[3]	28.8 ms	76.62 ms
Power[2]	25.06 W	11.35 W	3.51 W
Efficiency[1]	592.92 mJ/Inf.	327.02 mJ/Inf.	268.94 mJ/Inf.

1) At 12 Layer AI Model Inference 2) Processing Power 3) FFN Latency Calculated by # of DSP Blocks(multiplier) 4) At CoLA Dataset
5) Pipelined FFT Core Bit-Precision: FXP 16, HeNCoC Bit-Precision: FXP8 + FXP4

979-8-3503-3004-5/23 $31.00 © 2023 IEEE

FPGA Layout

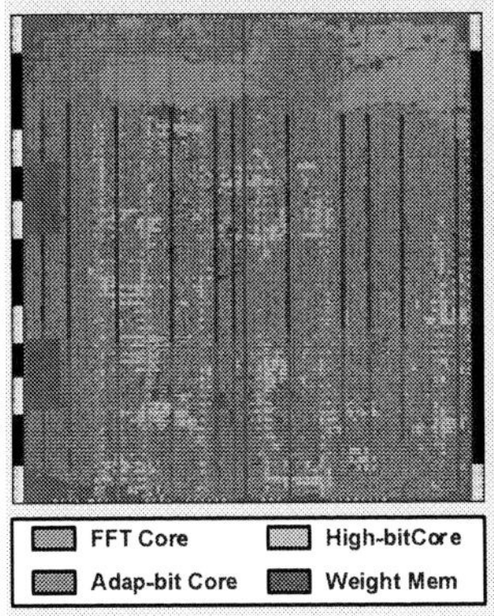

FPGA Specifications	
Platform	Intel Cyclone V
Clock	100MHz
LUT	91,995 (82%)
DSP	342 (100%)
Block Mem	7,621,656 (68%)

Implementation Results	
Algorithm	FNet with FDB
Bit Precision	FXP16 | FXP8 + FXP4
FFT Support	O
Latency	76.62 ms
Power	3.51 W
Efficiency	268.94 mJ/Inf.

FFT Core High-bitCore
Adap-bit Core Weight Mem

Fig. 7. FPGA Layout and Specifications.

Additional References:
[6] M. Garrido *et al.*, "A Survey on Pipelined FFT Hardware Architectures," J Sign Process Syst 2022:1345-1364.

An 8.5ps Resolution, 2000μm^2 Phase-Domain Delta-Sigma TDC for Lidar Applications

Yoondeok Na[1], Seokho Yun[1], Myung-Jae Lee[2], and Youngcheol Chae[1]

[1]Yonsei University, Seoul, Korea

[2]Korea Institute of Science and Technology, Seoul, Korea

A direct time-of-flight (dToF) lidar is used for distance measurement by detecting the round-trip time of a laser pulse [1-4]. Single-photon avalanche diodes (SPADs) have been successfully realized in a CMOS process, and the timing circuit and memory are implemented to store multiple timestamps over conversion cycles. An effective way to detect the timestamps is to detect the arrival time via a time-to-digital converter (TDC), and to sort arrival times into histogram bins. However, as the number of pixels increases, so does the data capacity. This leads to a direct trade-off between distance resolution and the maximum distance range. To address this, partial histogram approaches have been investigated, which successively change the histogram bins by zooming or sliding [3, 4]. However, it requires a significant increase in conversion time or limits distance resolution.

This paper proposes a phase-domain delta-sigma (PDΔΣ) TDC for dToF lidar, detecting timestamps of the SPAD in a ΔΣ manner within a reference time window. By exploiting highly digital gated-ring oscillator (GRO)-based architecture, the TDC can be implemented in a compact area of 2000μm^2. It utilizes an extended counting and hardware reuse, achieving a higher resolution without area penalty. A proposed TDC implemented in a 65nm CMOS achieves a time resolution of 8.5ps in an integration time of 0.5ms and the maximum INL of 80ps over 200ns time range. Compared to state-of-the-art [3], it achieves an 11.7× improvement in time resolution with a 66× reduction in integration time, while occupying a 2.25× smaller area.

The histogram method is aiming for *peak value* detection in histogram bins, as shown in Fig. 1(top). When a photon reaches the SPAD, a digital pulse is generated that is measured via a TDC, often realized with a GRO and a counter. A timestamp is stored in memory and this process is repeated acquisition times N_{acq} to generate a histogram, and the most frequently recorded value becomes the output. It requires a large number of histogram bins per pixel. For instance, each pixel requires 2048 bins to cover the 100m distance range for cm-range resolution. In contrast, the phase shift of the SPAD output can be digitized by a PDΔΣ TDC. The block diagram of a first-order PDΔΣ TDC is shown in Fig. 1(middle). It accumulates the digital pulses by adding or subtracting them relative to a reference time window, and the PDΔΣ loop converges the phase difference compared to laser pulses. The modulator's bitstream average is accompanied with a counter as a decimation filter and represents the *mean value* of the timestamps. Since the difference between the two results is only a constant offset as shown in Fig. 1(bottom) [5], the *mean value* of the timestamps can be well approximated with the actual distance.

Fig. 2 shows the simplified block and timing diagram of the proposed PDΔΣ TDC. The use of a GRO and an up/down counter can replace the analog loop filter of the modulator as in [6]. When a photon reaches the SPAD within the time window (TW), its output signal *TRG* is triggered. When the feedback signal *Sign* is 0, the up/down counter performs down-counting in region B. Conversely, when the *Sign* is 1, up-counting is carried out in region A. For example, if the GRO performs a total of 9 counts within the TW, with the rising edge of the *TRG* at 2/3 point. As shown in Fig. 2, during the first cycle, the up/down counter has a value of -3, and the MSB becomes 1. The *Sign* is then changed based on the MSB by the DFF's output. During the next TW, by counting up +6 in region A, the up/down counter value becomes +3. During the conversion time of N_{cycle}, the averaged *Sign* can be achieved with a ripple counter, resulting in a digital output D_O. The time interval T_M between SPAD trigger and laser pulses can be calculated according to the equations in Fig. 2(bottom).

The first challenge associated with interfacing a SPAD is errors caused by the absence of the SPAD trigger during the conversion. This often happens depending on the photon detection probability (PDP) of the SPADs. To address this, the proposed PDΔΣ TDC uses two equally assigned time windows (TW$_1$ and TW$_2$), but with opposite counting directions. This enables that there are no changes in the result if the SPAD's trigger is missing. The second challenge is the long conversion time of a 1st-order PDΔΣ TDC. With the proposed TDC, on the other hand, the residue value at the end of conversion is stored in the digital up/down counter and can be used directly for the extended counting (EC) without any additional effort. By using the quantized residue value, a higher resolution can be achieved with a significantly lower oversampling ratio (OSR).

Fig. 3(top) presents the schematic and timing diagram of the proposed TDC. Active quenching and recharging circuits are implemented via *SET* and TW$_2$ signals. A counter is used at the SPAD output that counts the total number of triggers C$_O$. A level shifter is used for realizing the low-power operation of the TDC while maintaining the high excess bias of the SPAD. The use of dual GROs (GRO$_1$ and GRO$_2$) can avoid the time-domain Q-noise of the counter as in [6], resulting in an improved resolution. A selector operates GRO$_1$ or GRO$_2$ based on TW$_1$, TW$_2$, *Sign*, and *TRG* signals. To count dual GROs, two counters and a subtractor are used for the up/down counter operation. The up/down counter outputs (I$_O$) are used for the EC operation. As shown in Fig. 3(middle), assume that a GRO with a frequency of 16 counts during two time windows and the edge of *TRG* is positioned at 17/32 of TW$_1$. During the first cycle, there are 4 up-counting, increasing GRO$_1$ phase by +0.25, and 11 down-counting, decreasing GRO$_2$ phase by -0.75. In the second cycle, the *Sign* is 1, leading to 12.25 up-counting and -3.75 down-counting. During the third cycle with no trigger, up-counting is performed during TW$_1$, and the same down-counting occurs during TW$_2$, resulting in no change in the net value. By repeating this process, the ratio of *Sign* signal values 0 and 1 becomes 8.5:7.5. The final output of 17/32 by combining the EC output is realized, as shown in Fig. 3(bottom).

The prototype TDC was fabricated using a 65nm CMOS process, occupying only 2000μm^2. It consumes 136μW from supply voltages of 1.5 / 3V at GRO frequency of 562MHz. Fig. 4(a) shows the measured time error over the time interval of 200ns. With an OSR of 32, the TDC achieves the measured time error of 80ps(+53ps/-27ps). After including a SPAD frontend, the error is increased to 192ps, as shown in Fig. 4(b). Fig. 5(a) shows the measured PSD for a fixed time input. Fig. 5(b) shows the maximum INL according to the number of cycles. The output with EC has a much lower max INL in fewer cycles than without EC. Depending on the tunable GRO frequencies of 325MHz, 445MHz, and 562MHz, the maximum INL is also measured as in Fig. 5(c). After the OSR of 32, the results are well preserved with a GRO frequency of 562MHz. Fig. 5(d) shows the measured noise versus the number of cycles. At OSR of 32, the TDC resolution of 27.6ps is achieved. This can be improved to 8.5ps at an OSR of 1000, corresponding to the integration time of 0.5ms. Fig. 6 presents the performance summary and comparison with other state-of-the-art TDCs for lidar sensors [1-3, 7]. In comparison to [1, 2], the use of the ΔΣ-based mean detection method significantly reduces the area overhead. In contrast to [7], this work can be applied to direct ToFs. Compared to state-of-the-art TDC [3], this work achieved an 11.7× improvement in time resolution with a 66× reduction in integration time and a 2.25× smaller size.

Acknowledgement:

This work was supported by the Technology Innovation Program (or Industrial Strategic Technology Development Program) (20019301) funded by the Ministry of Trade, Industry & Energy (Korea) and the authors are thankful to the Integrated Circuits Design Education Center (IDEC).

References:

[1] A. T. Erdogan *et al.*, "A CMOS SPAD Line Sensor With Per-Pixel Histogramming TDC for Time-Resolved Multispectral Imaging," JSSC 2019:1705-1719.

[2] C. Zhang *et al.*, "A 30-frames/s, 252 × 144 SPAD Flash LiDAR With 1728 Dual-Clock 48.8-ps TDCs, and Pixel-Wise Integrated Histogramming," JSSC 2019:1137–1151.

[3] S. Park *et al.*, "An 80×60 Flash LiDAR Sensor with In-Pixel Histogramming TDC Based on Quaternary Search and Time-Gated Δ-Intensity Phase Detection for 45m Detectable Range and Background Light Cancellation," ISSCC 2022:98-99.

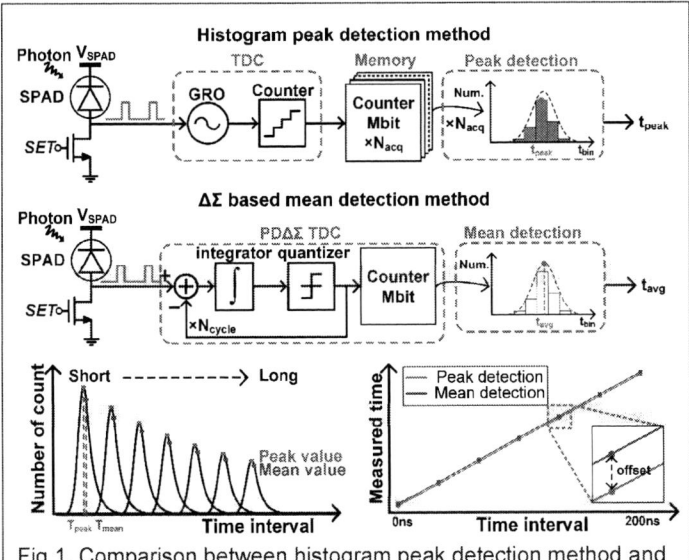

Fig 1. Comparison between histogram peak detection method and ΔΣ-based mean detection method.

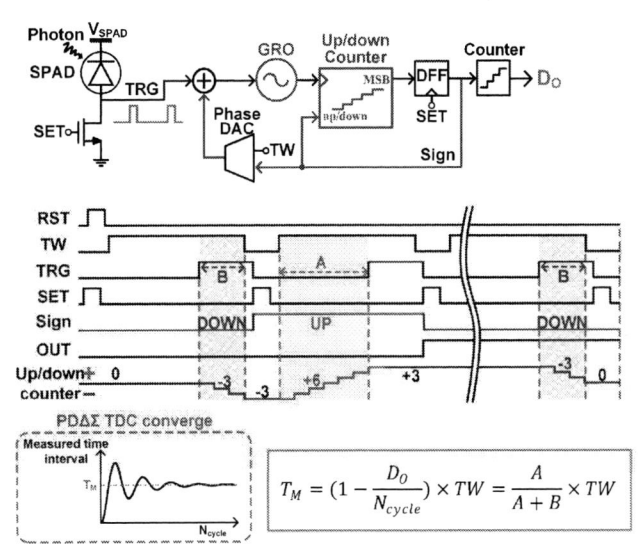

$$T_M = (1 - \frac{D_O}{N_{cycle}}) \times TW = \frac{A}{A + B} \times TW$$

Fig 2. Simplified PDΔΣ TDC block diagram and timing diagram.

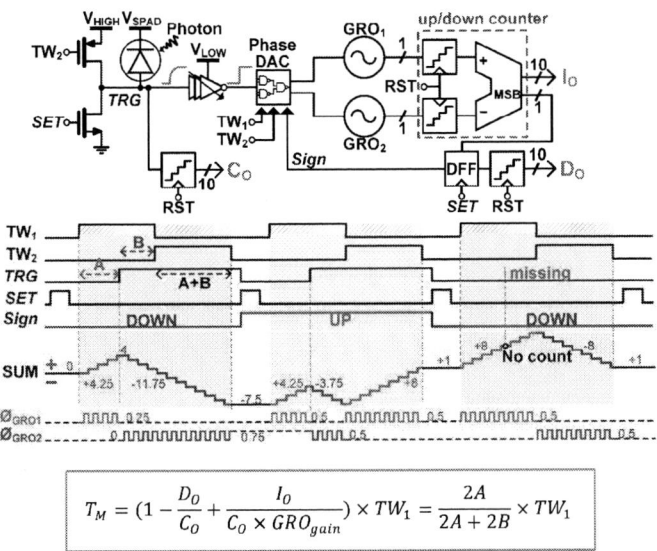

$$T_M = (1 - \frac{D_O}{C_O} + \frac{I_O}{C_O \times GRO_{gain}}) \times TW_1 = \frac{2A}{2A + 2B} \times TW_1$$

Fig 3. Schematic and timing diagram of the proposed PDΔΣ TDC.

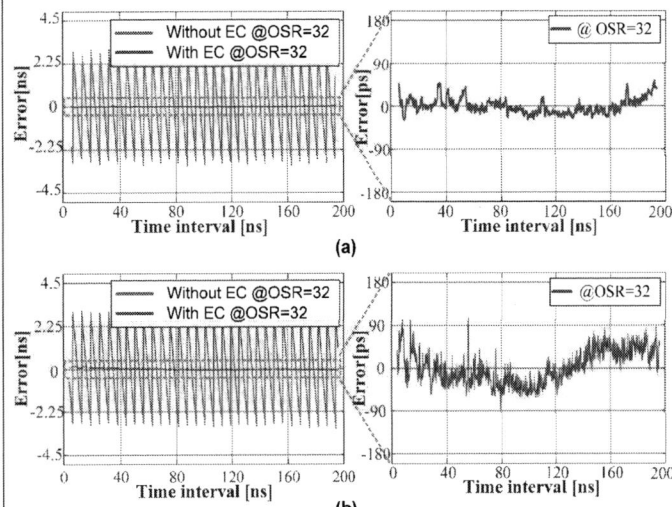

Fig 4. (a) Measured time error over time interval of 200ns, and (b) measured time error with SPAD frontend.

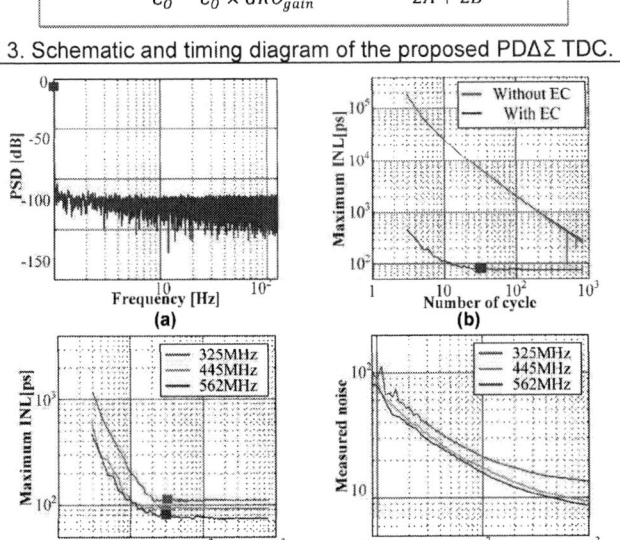

Fig 5. (a) Measured PSD, (b) measured maximum INL, (c) maximum INL according to the GRO frequency, and (d) measured noise according to the number of cycle.

Features	This work	ISSCC'11 [7]	VLSI'17 [2]	JSSC'19 [1]	ISSCC'22 [3]
Technology[nm]	65	130	130	180	110
ToF type	Diect	In-direct	Diect	Diect	Diect
Readout type	In pixel	In pixel	scanning	Column parralel	In pixel
Scalability	High	High	High	LOW	High
Readout architecture	ΔΣ	ΔΣ	Histogram	Histogram	Zoom
[1)]Area [μm²/ch]	2,000	1613	[1)]47,320	[1)]1,226,200	4,500
Maximum INL [ps]	80	333	361	125	167
TDC depth [bit]	20	-	11	12	13
TDC resolution [ps]	[2)]27.6 [3)]8.5	-	51.2	631.5	100
Integration time [ms]	[2)]0.016 [3)]0.5	50	-	33.3	33.3

[1)]Area: TDC area + histogram area for each channel [2)]When OSR is 32 [3)]When OSR is 1000

Fig 6. Performance summary and comparison with other state-of-the-art TDCs.

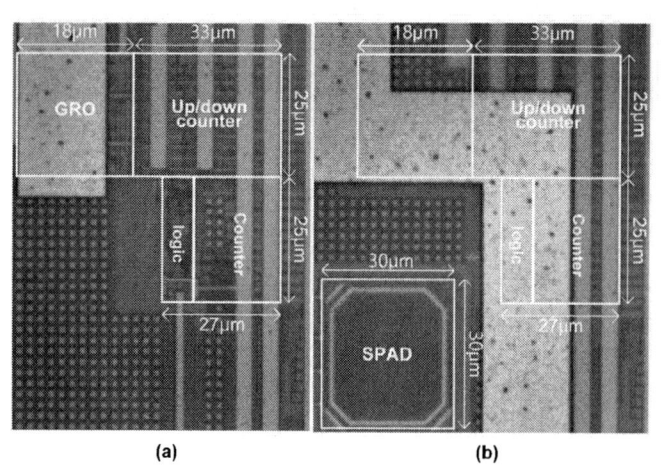

Fig 7. Chip micrograph: (a) Pixel without SPAD and (b) pixel with SPAD.

Additional References:

[4] D. Stoppa *et al.*, "A Reconfigurable QVGA/Q3VGA Direct Time-of-Flight 3D Imaging System with On-chip Depth-map Computation in 45/40nm 3D-stacked BSI SPAD CMOS," IISW 2021:53–56.

[5] F. Sun *et al.*, "A Simple Analytic Modeling Method for SPAD Timing Jitter Prediction," JEDS 2019:261-267.

[6] J. Angevare *et al.*, "A Highly Digital 2210μm^2 Resistor-Based Temperature Sensor with a 1-Point Trimmed Inaccuracy of ±1.3°C (3σ) from -55°C to 125°C in 65nm CMOS," ISSCC 2021:76-77.

[7] R. Walker *et al.*, "A 128×96 Pixel Event-Driven Phase-Domain ΔΣ-Based Fully Digital 3D Camera in 0.13μm CMOS Imaging Technology," ISSCC 2011:410-411.

3D-ISC: A 65nm 3D Compatible In-Sensor Computing Accelerator with Reconfigurable Tile Architecture for Real-time DVS Data Compression

Gokul Krishnan[1], Gopikrishnan Raveendran Nair[1], Jonghyun Oh[2], Anupreetham Anupreetham[1], Pragnya Sudershan Nalla[1], Ahmed Hassan[1], Injune Yeo[1], Kishore Kasichainula[1], Jae-sun Seo[1], Mingoo Seok[2], Yu Cao[1]

[1]Arizona State University, [2]Columbia University

A dynamic vision sensor (DVS) outputs a sequence of digital events, each representing a change in brightness at a particular time. As DVS keeps scaling down the pixel size and increasing the resolution, a larger volume of data is generated at limited bandwidth, resulting in reduced system-level throughput, as shown in Fig. 1(a). Hence, conventional interconnect solutions cannot meet the requirement of real-time data output from a high-speed DVS.

Targeting the above limitations, we design 3D-ISC, an all-digital 6T-SRAM-based In-Memory Computing (IMC) accelerator in 65nm for real-time DVS data compression. Fig. 1(b) shows the envisioned 3D stacking. 3D-ISC has three main features: 1) 3D compatibility (emulated in this work by saving data in multiple banks of SRAM) to integrate the accelerator with the DVS sensor using through-silicon-vias (TSVs), 2) an all-digital IMC (DIMC)-based processing element (PE), and 3) a load balanced compute partition with custom gather and scatter modules, minimum buffering, and a maximum hardware reuse dataflow. Moreover, we design an AutoEncoder (AE) algorithm (Fig. 1(c)) to demonstrate compression efficacy. 3D-ISC accelerates the encoder part of the AE algorithm.

Fig. 2(top) shows the overall architecture of the 3D-ISC accelerator. It consists of 4 tiles that operate independently and parallelly on different quadrants of the DVS sensor. The sensor is envisioned to stack with the accelerator using TSVs. Each tile has inputs from the assigned set of pixels. The input is first collected in an SRAM that consists of five banks. A custom gather and scatter (GS) module is designed to fetch the data from the SRAM banks and move it to the PEs. Each tile consists of 48 custom-designed DIMC-based PEs that perform a large number of vector-matrix multiplications (VMMs). The tile also consists of an activation function (AF) unit, a max pooling engine (MPE), a scaler unit for activation quantization, and a feature map accumulator (FMA) unit. A custom control logic is implemented using self-timed blocks to perform pipelined execution in the tiles. The accelerator has a deep pipeline, generating outputs periodically after the initial pipeline delay. Finally, to optimize power efficiency and adaptability, a custom power switch is incorporated into each tile controlled by the control logic to activate tiles when necessary.

Fig. 2(bottom) shows the overall data flow at the system level. Each tile follows the same dataflow while operating on different quadrants of the DVS sensor. The tile computes two layers of the AE encoder, with the first 16 DIMC PEs in the first compute cluster (CC1) for the first layer and the remaining 32 DIMCs in compute cluster two (CC2) for the second layer. Such a partition maximizes the benefit of the deep pipeline and enables a weight stationary dataflow. We design two GS units for each of the CCs within the tile. The GS unit consists of five shallow SRAM banks to match the maximum size of the kernel window supported (5×5). The GS unit for CC1 provides 16 adjacent 5×5 windows such that the scatter module slices the gathered data to account for the stride. 3D-ISC can support different kernel sizes and strides. Furthermore, the GS unit performs both gather and scatter simultaneously through an additional row for the pre-fetch of pixels, enabling the pipeline to achieve high throughput.

The CC1 output is moved to the AF unit for the ReLU function. The 12-bit ReLU output is scaled down to 4-bits using the scaler unit. The scaler performs a first-degree compression from 12 to 6-bits and then scales and rounds to 4-bits. The output activations are collected in a shallow buffer within the MPE to reuse the hardware across time and consume all the data from CC1. The MPE output is collected in the gather and scatter unit 2 (GS-2) and moved to CC2 once we get an entire kernel worth of pixels. The same set of operations are performed to the CC2 output with an addition of the FMA. The FMA accumulates the pixels across different feature maps and then generates the final output pixel. We perform a similar hardware reuse

for the FMA to that of the MPE to reduce the total memory requirement within the 3D-ISC.

We design an all-digital IMC macro to maximize the robustness while reducing the area overhead of digital circuits (Fig. 3). We adopt and improve a time-multiplexing architecture [1], where the macro employs 25×64 compact 6T bitcells. Every twelve bitcells time-share one multiplier, and every 25×4 bitcells time-share a set of 5b ripple carry adder and a shift-accumulator. The overall floorplan of the 65nm 25×64 DIMC macro takes 0.023 mm^2. The DIMC macro supports up to 4b×4b MAC operations, providing 16 different 12-bit outputs. The macro takes 52 cycles to generate the VMM output, while achieving an excellent compute density of 7.3 TOPS/mm^2.

Fig. 4 (left) shows system-level dynamic current at various activation and weight sparsity. Sparsity is generated in an IMC-friendly structured manner [6]. We measure the current draw of the system at 10MHz for three different voltages and across three different levels of activation/weight sparsity, 0%, 60%, and 80%. We use 4-bit precision for both activation and weight. For the weight sparsity, 3D-ISC achieves up to 43% reduction in dynamic current. Fig. 4(right) shows the dynamic power consumed for 2-bit and 4-bit weight precision at different voltage and frequency configurations. A lower bitwidth reduces power consumption depicting the adaptability and reconfigurability of 3D-ISC.

Fig. 5(a) shows the overall 3D-ISC inference power at different voltage and frequency values. 3D-ISC achieves a sub-6mW power profile across all configurations at 4-bit activation and weights with no sparsity. Fig. 5(b) shows the AE algorithm compression ratios (CR) and accuracy for the DVS-MNIST, N-Cars, and DVS-IBM-Gesture datasets. The algorithm achieves 92%, 98%, and 90% CR with a minimal drop of 2.0%, 3.0%, and 3.8% in accuracy. Fig. 5(c) shows an example DVS input image and the corresponding encoded image output obtained from the 3D-ISC accelerator, achieving more than 80% sparsity.

DVS only outputs pixels with an activity, or in other words, whose intensity value changes. Hence, the entire sensor will not light up all the time. To support this, we partition 3D-ISC into four identical tiles, with each tile operating on a different quadrant of the image. Each of these tiles can be dynamically enabled or disabled depending on the input frame. Fig. 6(top) shows the system's power consumption under different tile configurations. With selective enabling of tiles, we achieve up to 76% reduction (one tile enabled) as compared to the baseline that enables all tiles.

The test chip is fabricated using TSMC 65nm technology and occupies 20.52 mm^2. Fig. 7(a) shows the package photograph of the 3D-ISC accelerator, Fig. 7(b) shows the test setup used for the measurement of the accelerator, and Fig. 7(c) shows the die photograph showing the 3D-ISC accelerator with each tile occupying 2.2×1.6 mm^2. Fig. 6(bottom) table shows the comparison of our work with prior works. 3D-ISC achieves up to 14.35× improvement in TOPS/W (normalized to 8-bit MAC OPS) as compared to prior works. For the MINST dataset inference, 3D-ISC achieves 2.6× and 4.7× improvement in latency and energy per frame compared to [5], respectively. We achieve a peak system efficiency of 14.07 TOPS/W at 0.8V and 10MHz normalized to 8-bit precision, and no sparsity.

Acknowledgement

This material is based upon work supported by the Defense Advanced Research Projects Agency (DARPA) under Contract No. HR001121C0134.

References

[1] B. Yan et al., "A 1.041-Mb/mm2 27.38-TOPS/W signed-INT8 dynamic-logic-based ADC-less SRAM compute-in-memory macro in 28nm with reconfigurable bitwise operation for AI and embedded applications," ISSCC, 2022.
[2] Q. Dong et al., "A 351TOPS/W and 372.4GOPS compute-in-memory SRAM macro in 7nm FinFET CMOS for machine-learning applications," ISSCC, 2020.
[3] H. Jia et al., "A programmable neural-network inference accelerator based on scalable in-memory computing," ISSCC, 2021.
[4] J. Yue et al., "A 2.75-to-75.9TOPS/W computing-in-memory NN processor supporting set-associate block-wise zero skipping and ping-pong CIM with simultaneous computation and weight updating," ISSCC, 2021.
[5] J. Yang, et al. "A 65nm 8-bit all-digital stochastic-compute-in-memory deep learning processor," A-SSCC, 2022.
[6] G. Krishnan, et al. "Hybrid RRAM/SRAM in-memory computing for robust DNN acceleration," TCAD, 2022.

979-8-3503-3004-5/23 $31.00 © 2023 IEEE

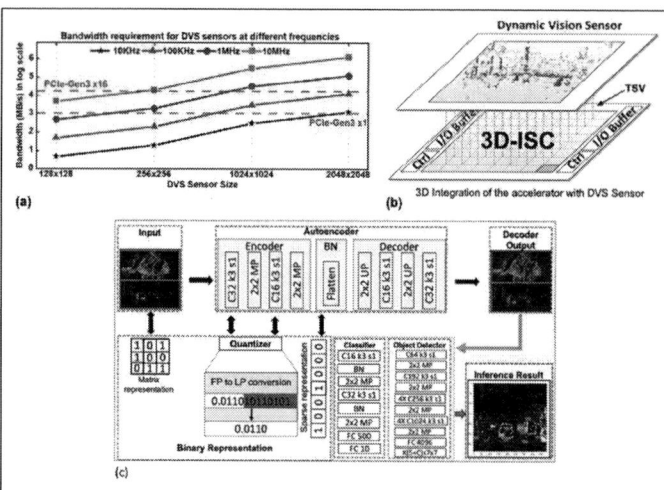

Fig. 1. (a) Bandwidth requirements for different DVS sensor resolutions and frame rates, (b) 3-D stacking of the proposed 3D-ISC accelerator with a DVS sensor, and (c) the proposed autoencoder algorithm for DVS data compression.

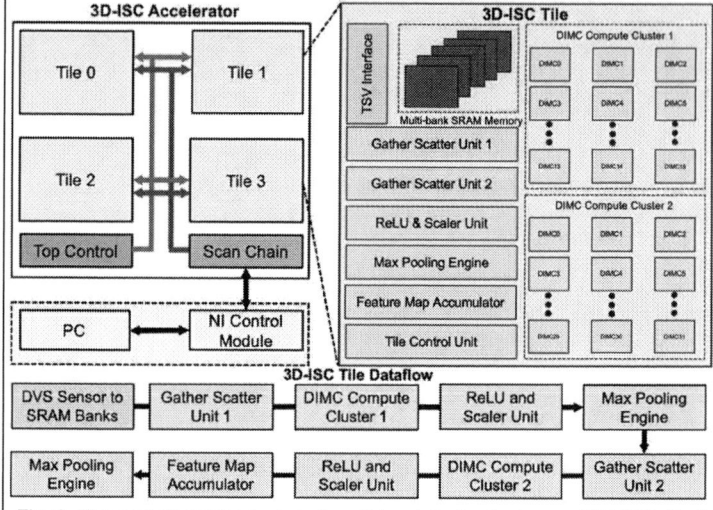

Fig. 2. The overall architecture and multi-level design features of 3D-ISC. Each tile consists of two compute clusters, a TSV-compatible interface, and associated control logic to process the pixels from the DVS sensor.

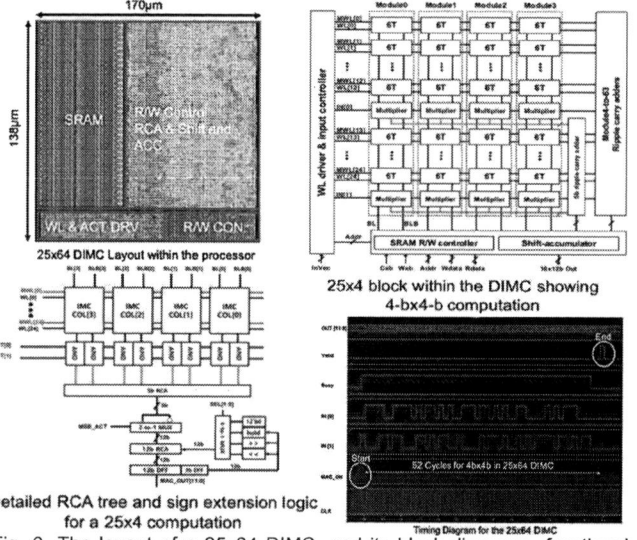

Fig. 3. The layout of a 25x64 DIMC, and its block diagrams, functional diagram, and timing diagram.

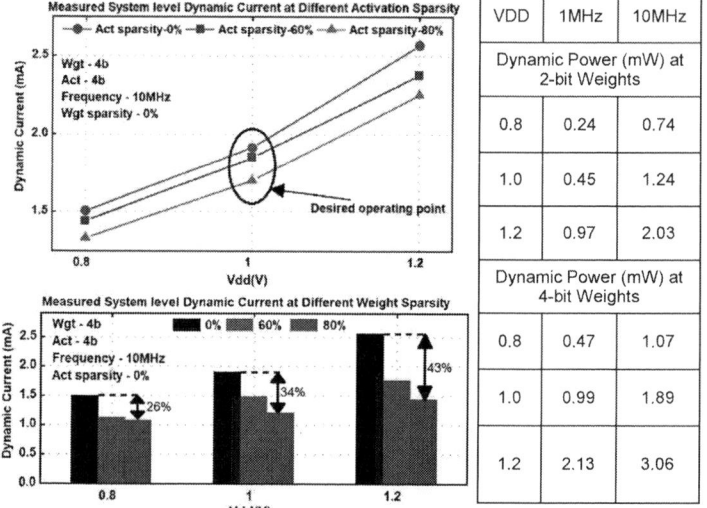

Fig. 4. (left) System-level dynamic current at different activation and weight sparsity, and (right) dynamic power at different weight precisions.

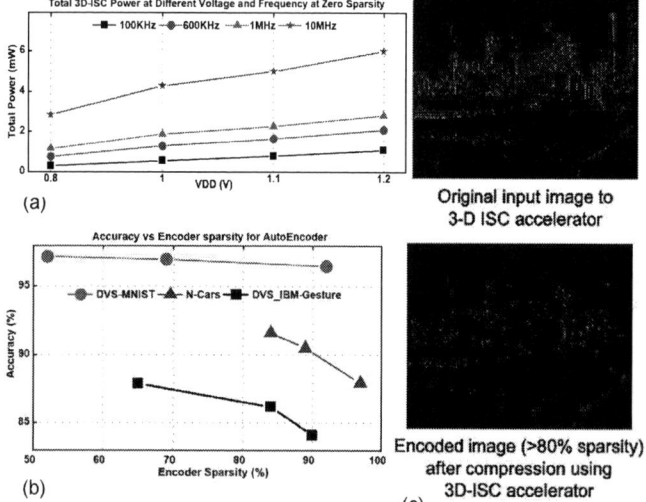

Fig. 5. (a) Total power of 3D-ISC across different voltages and frequencies, (b) accuracy vs. output sparsity for different datasets with AutoEncoder for 3D-ISC, and (c) DVS images before and after compression using the 3D-ISC accelerator.

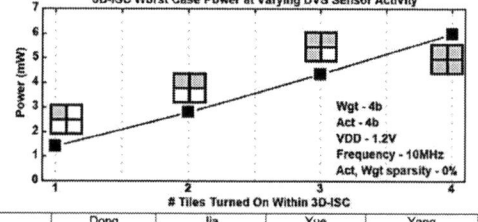

	Dong ISSCC 20 [2]	Jia ISSCC 21 [3]	Yue ISSCC 21 [4]	Yang ASSCC 22 [5]	This Work
Tech	IMC, 7nm	IMC, 16nm	IMC, 65nm	SIMC, 65nm	DIMC, 65nm
Size	0.032 mm²	25 mm²	12 mm²	9.4 mm²	20.52 mm²
Voltage	1V	0.8V	0.65V (Digital), 1V (IMC)	0.7-1.05V(IMC), 0.8V (SRAM)	0.8-1.2V(DIMC), 0.8V (SRAM)
On-Chip IMC Capacity	4Kb	4.5Mb	64Kb	520Kb	300Kb
IMC Macro Density[1]	16.6 Kb/mm²	22 Kb/mm²	16 Kb/mm²	130 Kb/mm²	66 Kb/mm²
Precision	4b Fxt Pt	8b Fxt Pt	8b Fxt Pt	8b Fxt Pt	4b Fxt Pt
(TOPS/W) [2, 3]	0.98 (Macro)	1.88 (System)	2.3 (System)	7.96 (System)	14.07 (System)
NN Performance (DVS)	N/A	N/A	N/A	N/A	3.23 uJ/fr, 1.05 ms/fr
NN Performance (MNIST)[4]	N/A	N/A	N/A	5.85 ms/fr 8.3 uJ/fr @0.8V, 4MHz	2.25 ms/fr 1.73 uJ/fr @0.8V, 1MHz

[1] Scaled to 65nm and [2] # OP scaled with bit precision same as in [5]
[3] Peak System Energy-efficiency scaled to 8-bit, [4]Used same network structure as [5]
For this work TOPS/ uses DIMC MAC OPS (multiplication and addition as 2 OPS) and the system power

Fig. 6. (top) Worst case power consumption for 3D-ISC across different DVS sensor activity and (bottom) comparison table with other state-of-the-art works.

979-8-3503-3004-5/23 $31.00 © 2023 IEEE

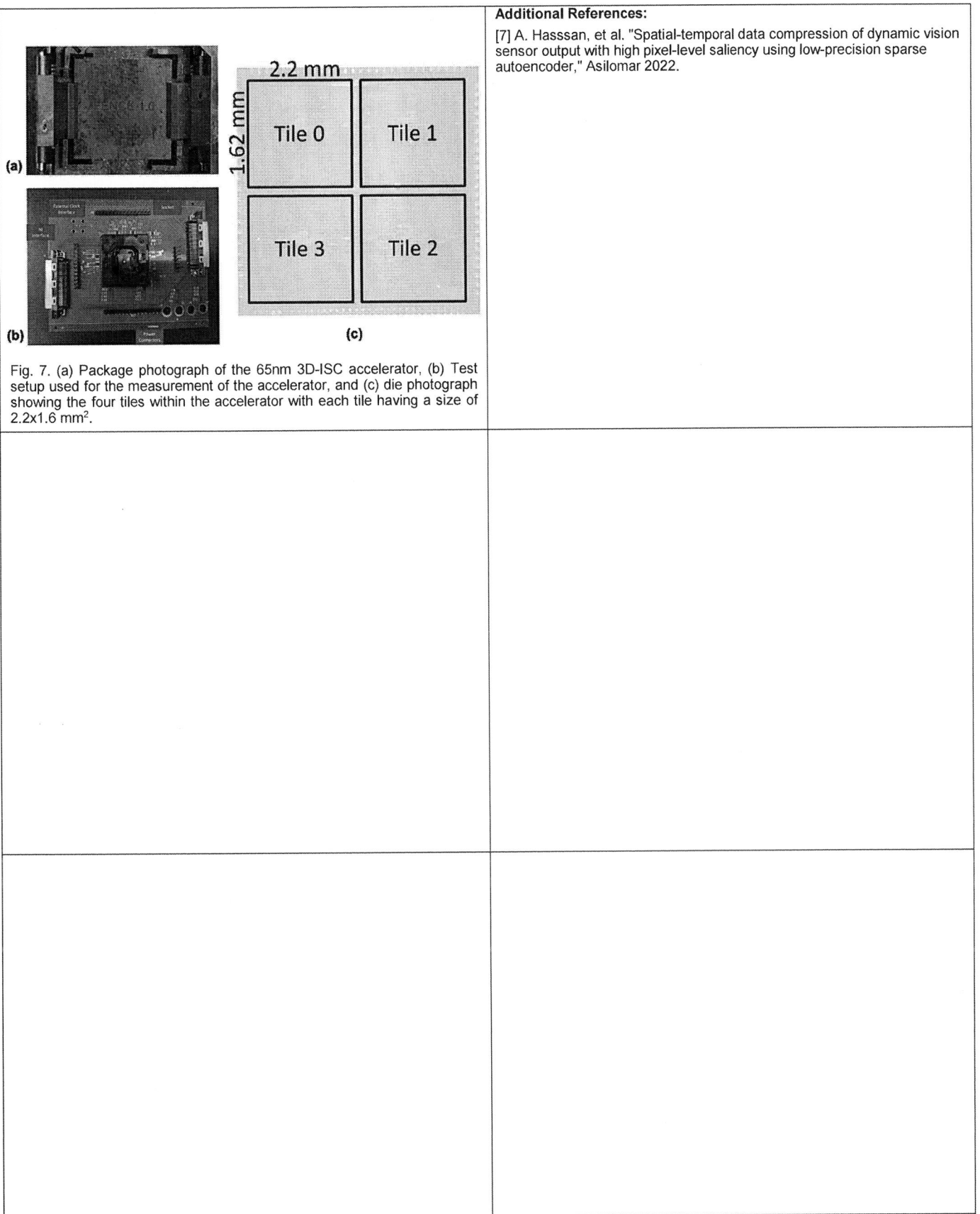

Fig. 7. (a) Package photograph of the 65nm 3D-ISC accelerator, (b) Test setup used for the measurement of the accelerator, and (c) die photograph showing the four tiles within the accelerator with each tile having a size of 2.2x1.6 mm².

Additional References:

[7] A. Hasssan, et al. "Spatial-temporal data compression of dynamic vision sensor output with high pixel-level saliency using low-precision sparse autoencoder," Asilomar 2022.

A 23.9 µW 13.6-bit Period Modulation-Based Capacitance-to-Digital Converter with Dynamic Current Mirror Front-end Achieving Capacitor Range of 1 to 68 pF

Hyeeyeon Lee[1], Donguk Seo[2], Young-Jin Woo[3], Yoonmyung Lee[2], Inhee Lee[4], and Youngcheol Chae[1]

Yonsei University, Seoul, Korea[1]

Sungkyunkwan University, Suwon, Korea[2]

LX Semicon, Seoul, Korea[3]

University of Pittsburgh, PA, USA[4]

A multi-purpose capacitance-to-digital converter (CDC) needs to cover a wide input range (C_{IN} > 50 pF) for various sensor applications such as a pressure sensor and a strain sensor. Typical CDCs using SAR or delta-sigma modulation have limitations in maintaining their energy-efficiency for such a wide input range, as they require to drive a large reference capacitor (C_{REF} > 10 pF) with a reference voltage buffer, resulting in high power consumption. On the other hand, CDCs using period-modulation (PM) are good alternative to achieve a large input range [1, 2]. The work in [1] digitizes an input capacitance (C_{IN}) using an iterative delay-chain discharge method. It achieves an excellent energy-efficiency of 0.14 pJ/step, but the effective resolution is limited to 7.9-bit since the charge captured on the C_{IN} is discharged nonlinearly. The work in [2] generates D_{OUT} proportional to a period ratio of the C_{IN} to the C_{REF} and achieves a high resolution of 114 aF at the cost of power and area consumption due to the use of a reference oscillator (140 µW and 0.175 mm²).

This paper presents a PM-based CDC with a dynamic current mirror (DCM) front-end whose output generates a ratio-metric (C_{IN}/C_{REF}) current. This current is then converted into a PM output that is digitized by a digital counter. Without the use of a power-hungry reference voltage buffer, the DCM allows the CDC to achieve the C_{IN} range of up to 68 pF, which is >2.3× larger than the prior art CDCs. In addition, it achieves a resolution of 13.6-bit (versus 7.9-bit in [1]), with 23.9 µW and 0.05 mm² (5.9× and 3.5× smaller than [2]).

Fig. 1 shows the simplified diagram of the proposed CDC. The DCM, consisting of M_{P1}, M_{P2}, C_{REF} (= 50 pF), and C_{IN}, generates a mirroring current (I_{DCM}) with a multiplication factor of C_{IN}/C_{REF} to a reference current (I_B). The mirroring accuracy between two transistors, M_{P1} and M_{P2}, is improved by an auxiliary amplifier A_1 that sets the same voltage on the gate and drain of M_{P2}. Simulation results show a gain error of < 1% at the C_{IN} from 5 pF to 68 pF. The I_{DCM} is integrated into an integration capacitor C_{INT}, and its voltage V_{INT} is compared with a reference voltage V_{REF}. The CDC employs a self-biased reference current source I_B as in [3] to ensure a good tolerance of 0.38% on a ±10% variation from a 1 V supply. The V_{REF} also utilizes the same reference current, resulting in a good supply immunity for the DCM front-end. When the comparator output (f_{INT}) goes high, M_{NR} resets the C_{INT} and the integration is repeated. A two-stage continuous-time comparator is used for current-to-frequency conversion, where a pMOS input pair with cross-coupled nMOS loads is used at the first stage. It achieves a DC gain of 54 dB, while drawing 8 µA and guarantees its operation under 0.9 V to 1.1 V supply range. The f_{INT} is proportional to the settled I_{DCM}, and thus also proportional to C_{IN}/C_{REF}. The C_{INT} and R_{REF} are set to 800 fF and 20 kΩ, respectively, considering the min-max code of the D_{OUT} corresponding to the target C_{IN} range since the DCM's dynamic property predefines the available timing window for the D_{OUT}.

Fig. 2 shows the detailed operation of the DCM front-end, consisting of two phases: reset phase (RST=1) and operating phase (RST=0). When RST=1, two capacitors C_{REF} and C_{IN} are reset before conversion, and the source voltages of M_{P1} and M_{P2} ($V_{S,REF}$ and $V_{S,IN}$) become V_{DD}. Simultaneously, the initial current for the I_{DCM} is set to I_B. Once RST=0, the DCM starts copying I_{REF} to I_{DCM} by the factor of C_{IN}/C_{REF}. Note that the sizes of M_{P1} and M_{P2} are identical and constant. The I_{DCM} maintains a fixed value in a target timing window (Δt). Here, the charge variations on the reference and mirror branches are $C_{REF} \cdot \Delta V_{S,REF}$ ($= I_B \cdot \Delta t$) and $C_{IN} \cdot \Delta V_{S,IN}$ ($= I_{DCM} \cdot \Delta t$), respectively. Since $V_{S,IN}$ chases $V_{S,REF}$ to maintain the source-gate voltage V_{SG} of M_{P2} ($\Delta V_{S,REF} = \Delta V_{S,IN}$), I_{DCM} becomes $I_B \cdot C_{IN}/C_{REF}$ with a time constant of C_{IN}/g_{mP}, where the g_{mP} is the transconductance of the M_{P2}. As shown in Fig. 2, $V_{S,IN}$ follows $V_{S,REF}$ when $C_{IN} = C_{REF}$ (middle), since M_{P1} and M_{P2} have the same degeneration

capacitance. Even though the C_{IN} value is different from that of C_{REF} (top and bottom), $\Delta V_{S,IN}$ follows $\Delta V_{S,REF}$ after some predetermined time, where PRE changes from 1 to 0.

Fig. 3 shows the noise calculation of the proposed CDC, including three main noise sources such as the DCM, the comparator, and the V_{REF} generator. The 1/f noise of the DCM is directly shown at the DCM output, so the sizes of M_{P1} and M_{P2} are carefully chosen. Here, the g_{mP} is one of the key parameters of the proposed front-end because not only the transient response but also the noise performance is determined by the g_{mP}. The 1/f corner frequency is about 3 kHz and the thermal noise level is 2.07×10^{-25} A²/Hz with the C_{IN} of 1 pF, where g_{mP} and g_{mN} are 20 and 26 µS, respectively. The output thermal noise of the DCM increases as the C_{IN} increases because the mirror ratio ($\beta = C_{IN}/C_{REF}$) is applied to the thermal noise of the reference current branch. The output current noise of the V_{REF} generator consists of the R_{REF} thermal noise and the $10 \cdot I_B$ noise. All the noise sources are analyzed at the input of the comparator. At the C_{IN} of 1 pF, the overall noise is dominated by the DCM and the comparator. On the other hand, as the C_{IN} increases, the DCM noise portion increases up to 98% at the C_{IN} of 68 pF.

The prototype CDC is implemented in a 65 nm CMOS process, occupying only 0.05 mm² (Fig. 7). It consumes 23.9 µA from a 1 V supply at the C_{IN} of 47 pF. Fig. 4 shows the measured resolution versus integration time, achieving 13.6-bit effective number of bits (ENOB) with the C_{IN} of 1 pF and an integration time of 4 ms. With the C_{IN} of 68 pF, the capacitor resolution is measured to 7.6 fF, where the dominant noise source is the DCM with the increased C_{IN}. Fig. 4 shows the measured capacitance with the integration time of 4 ms, resulting in a good linearity of 0.17% even for a very wide C_{IN} range. The linearity is obtained by $\log_2(1/\lambda)$, where $\lambda = (M_{2+3}-M_{1+3})/(M_2-M_1)-1$ and the measured digital output M_X for the capacitance C_X [6]. A combination of four capacitors (10, 20, 33, and 43 pF) is used for the calculation. The f_{INT} variation over different supply voltages from 0.9 V to 1.1 V is also shown in Fig. 4. With the C_{IN} of 20 pF, a supply-dependent shift is only 3%. The CDC is also verified with a commercial pressure sensor (SCB-10H), where the sensor is put on the through-hole of the board as shown in Fig. 5. The D_{OUT} properly captures the pressure-dependent capacitor changes over the pressure range of 525 to 840 mmHg. The nonlinear relationship between the pressure input and the measured capacitance output comes from the sensor itself. With the pressure input of 525 mmHg, the CDC successfully measures a sensor capacitance of 20.04 pF with a resolution of 5.8 fF.

Fig. 6 compares the proposed CDC with the state-of-the-art (SOTA) PM-based CDCs. It achieves >2.3× input range than the prior art CDCs (except the work on [1]), with a competitive FoM of 4.1 pJ/step (at the integration time of 1 ms) and a small area of 0.05 mm². Compared to the work on [1], this work achieves a significantly improved resolution. This work offers a promising solution for CDCs with multi-sensor applications.

Acknowledgements:

The authors thank the IDEC for support of IC fabrication and this work is supported by the LX Semicon Future Research Center.

References:

[1] W. Jung et al., "A 0.7pF-to-10nF Fully Digital Capacitance-to-Digital Converter Using Iterative Delay-Chain Discharge," IEEE ISSCC, 2015:484-485.

[2] A. K. George et al., "A 114-aF$_{rms}$-Resolution 46-nF/10-MΩ-Range Digital-Intensive Reconfigurable RC-to-Digital Converter with Parasitic-Insensitive Femto-Farad Baseline Sensing," IEEE Symp. VLSI circuits 2018:157-158.

[3] M. S. E. Sendi et al., "Temperature Compensation in CMOS Peaking Current References," IEEE TCAS-II, 2018:1139-1143.

[4] Y. He et al., "A 0.05mm² 1V Capacitance-to-Digital Converter Based on Period Modulation," IEEE ISSCC, 2015:486-487.

[5] O. Aiello et al., "Capacitance-to-Digital Converter for Operation Under Uncertain Harvested Voltage down to 0.3V with No Trimming, Reference and Voltage Regulation," IEEE ISSCC, 2021:74-75.

[6] Z. Tan et al., "An Energy-Efficient 15-Bit Capacitive-Sensor Interface Based on Period Modulation," IEEE JSSC, 2012:1703-1711.

979-8-3503-3004-5/23 $31.00 © 2023 IEEE

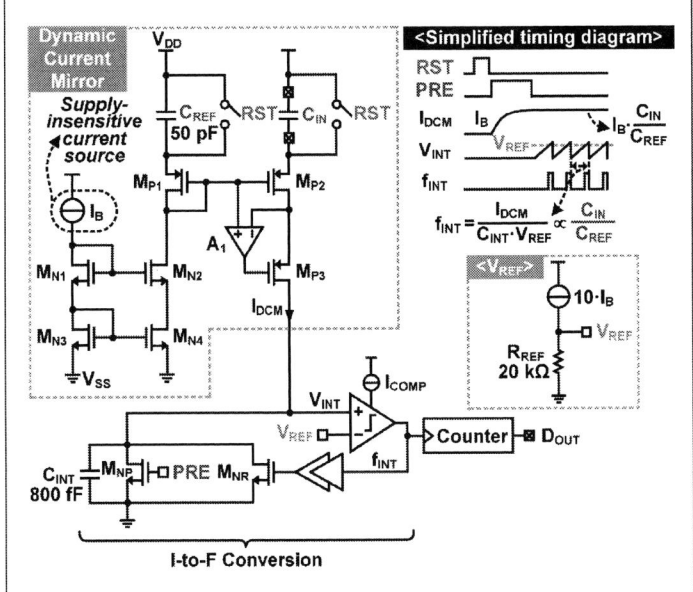

Fig. 1. Proposed PM-based CDC with DCM front-end.

Fig. 2. Detailed operation of the proposed DCM.

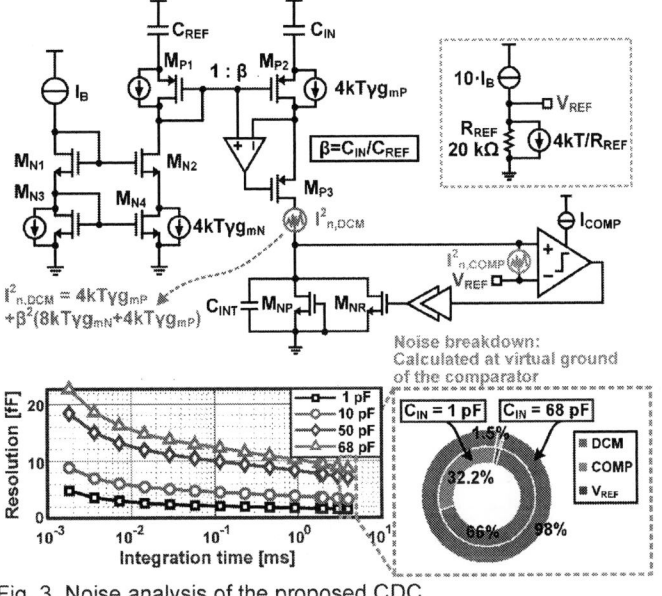

Fig. 3. Noise analysis of the proposed CDC.

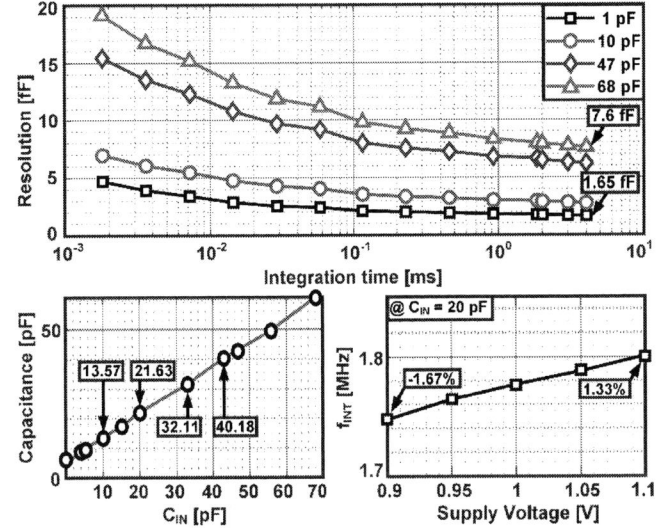

Fig. 4. Measured resolution (top), D_{OUT} (bottom-left), and f_{INT} over different supply voltage (bottom-right).

Fig. 5. Measurement setup with a SCB-10H pressure sensor (top) and its measurement results (bottom).

	This Work	ISSCC'15 [1]	VLSI'18 [2]	ISSCC'15 [4]	ISSCC'21 [5]
Process [nm]	**65**	40	180	160	180
Method	**PM**	PM	PM	PM	PM
Supply Voltage [V]	**0.9 ~ 1.1**	1	1	1	1.6 ~ 2
Supply-Dependent Drift [%]	**3**	N/A	N/A	N/A	N/A
Area [mm²]	**0.05**	0.00168	0.175	0.05	0.2
Power [µW]	**23.9**	1.84	140	14	0.00137
Resolution [fF]	**1.65 / 3.4**	12.3	0.114	0.255	67
Meas. Time [ms]	**4 / 1**	0.019	2.93	6.86	1040
Input Range [pF]	**68**	9999.3	11.6	8	30
ENOB[1] [bit]	**13.6 / 12.5**	12.8*	16.3	13.1	7.02
FoM[2] [pJ/step]	**7.78 / 4.1**	0.14	4.04	10.6	11.1

$\text{ENOB}^{1} = \frac{\text{RES}-1.76}{6.02}$, where $\text{RES} = 20 \times \log(\frac{\text{Input range}/2\sqrt{2}}{\text{Resolution}})$. $\text{FoM}^{2} = \text{Energy}/2^{\text{ENOB}}$.

*Input range 10.6 pF for the calculation as given in [1].

Fig. 6. Performance summary and comparison with SOTA works.

Fig. 7. Chip Micrograph.

MorphBungee: A 65nm 7.2mm² 27μJ/image Digital Edge Neuromorphic Chip with On-Chip 802 frame/s Multi-layer Spiking Neural Network Learning

Tengxiao Wang[1], Min Tian[1], Zhengqing Zhong[1], Haibing Wang[1], Junxian He[1], Fang Tang[1], Xichuan Zhou[1], Shuangming Yu[2], Nanjian Wu[2], Liyuan Liu[2], Cong Shi[1]*

[1]School of Microelectronics and Communication Engineering, Chongqing University, Chongqing 400044, China

[2]Institute of Semiconductors, Chinese Academy of Sciences, Beijing 100083, China

The spiking neural network (SNN) and event-driven neuromorphic architectures have gaining ever increasing popularity in low-cost and energy-efficient edge intelligent systems, as they closely mimic the human brain mechanism by utilizing spatiotemporally sparse binary spikes to represent and process sensory information in the neurons [1]. Especially, the small-scale digital ones attract more interests for their wide deployment potential in versatile edge-node applications such as wearables, drones and mobile platforms where cost, energy and latency budgets are strictly constrained [2]-[6]. However, many edge neuromorphic chips can only run shallow SNNs, leading to low object recognition accuracies. Moreover, they provide no or just single-layer on-chip learning function, failing to smartly and quickly adapt to the ever changing environment.

In this work, we propose a digital edge neuromorphic processor with on-chip multi-layer fully-connected (FC) SNN learning for fast and accurate object recognition at low hardware costs. A pre-mature sub-optimized version of this chip was earlier released in [7], and the final formal design is given in this work. The proposed dual-level multi-core chip architecture is depicted in Fig. 1. One macro-core (MCore) maps to part of or a whole FC layer, wherein each micro-core (μCore) executes the operations of multiple Leaky Integrate-and-Fire (LIF) spiking neurons in that FC layer, via a time-multiplexing fashion. An output error unit computes SNN output errors, and broadcasts them to all MCores to realize the DeepTempo training [7] on the chip.

Hybrid Parallelism: In the SNN feedforward pass, the MCores are event-driven and they work independently. Once the input spike FIFO of an MCore is not empty, the core is triggered to fetch and process the spike data. The spikes are encoded in the address-event representation (AER) format [1]. In the backward pass during on-chip learning, in contrast, all the MCores operate synchronously. The classification errors at the output layer are simultaneously projected back to each layer for the calculations of synaptic weight updates in every neuron. In both passes, the μCores in each MCore work in a naïve array parallelism. By adopting such hybrid core parallelism (the lower part of Fig.1), we can maximize the processing throughputs of the neuromorphic chip for both on-chip learning and inference stages.

Meta-Crossbar for Spike Communication: For the feedforward phase, we propose a meta-crossbar routing fabric as in Fig. 2 to efficiently deliver spike events among the cores. A spike AER issued by any μCore in an MCore is injected into the meta-crossbar, and arrives at all μCores in all the MCores that are configurably linked to the source MCore. This AER packet is implicitly used by all LIF neurons in the destination cores, realizing full connections between SNN layers. Via such meta-crossbar paradigm, the huge routing overheads for spike delivery between individual neurons as commonly via a 2D mesh network-on-chip (NoC) [4] are mitigated. The numbers of neurons and synapses (inputs) per neuron in a μCore are programable under the constraint that total synapses per μCore cannot exceed 2^{14} (16K).

Event-Driven Neuronal Processing: In an MCore, an input spike AER activates the entirely event-driven exponential unit and the μCores' neuronal processing units to execute operations sequentially for their LIF neurons $j = 1, 2, ...,$ as well as some variable updates if during the learning stage, as depicted in Fig. 3. To minimize computational overheads and energy consumption, this work permits each neuron to fire at most once, and converts each input image pixel into only one spike (this conversion is done by software off-chip), whose time is inversely related to the pixel brightness. In Fig. 3, the operations of the neurons that have fired are skipped for successive input AERs.

On-chip Stochastic Learning: At the end of one input training sample presentation, the output error unit computes the 2b ternary error of every output neuron, and broadcasts the error vector simultaneously to all MCores (i.e. all SNN layers) via fixed random binary Matrices **B** to obtain local neuron errors for synchronous and serial synaptic weight updates in μCores' learning units over each synapse i of each neuron j, as shown in Fig. 4. Refer to [7] for details of the DeepTempo learning rule. The lowest 8 bits of the high-precision 16b weight update amount Δw is stochastically rounded up by comparing the 8 bits with an 8b unsigned random number from a linear feedback shift register (LFSR), before Δw adds to the low-precision 8b weight. Compared to storing high-precision 16b weights, such stochastic learning technique saves 1.27 mm² (17.6%) active silicon area.

A prototype chip was fabricated using a 65-nm 1P9M CMOS process, and it occupied only 7.2 mm² active region area. The measurement system is exhibited in Fig. 5. An FPGA communicates encoded input spike AERs and processing results between the host computer and the neuromorphic chip. Several popular image datasets were fed to the chip for on-chip learning and classification (the category whose output neurons have fired the most spikes is the classification result [7]), and relatively higher recognition accuracies were obtained. Particularly, our chip attained 96.06% accuracy on MNIST images, at a high speed of 802 (2270) frame/s for learning (inference) and an energy efficiency of 10.32 (3.04) pJ/SOP (SOP = synaptic operation) or equivalently, 76.1 (26.9) μJ/image for learning (inference).

Our neuromorphic chip is compared with prior state-of-the-art digital edge neuromorphic processors in Fig. 6. It indicates that our chip is the only one owning high-speed and high-accuracy on-chip multi-layer SNN learning capability, at reasonable silicon area and power consumptions. Although the work in [6] exhibits very high processing throughput and high energy efficiency in inference, it has no on-chip learning ability. This drawback compromises its wide deployments in many edge and mobile application scenarios, where system cost and latency for off-chip learning might be unaffordable. Figure 7. gives our chip microphotograph and summarizes the chip specifications.

In conclusion, our edge neuromorphic chip is characteristic of hybrid dual-level multi-core parallelism, meta-crossbar for efficient spike communication between cores dedicated for SNN FC layers, entirely event-driven spike-based neuronal processing, and effective on-chip multi-layer SNN learning. The fabricated prototype chip was small and obtained comparably high on-chip learning accuracy at relatively high processing speed and moderate energy efficiency.

Acknowledgment:

This work was funded in part by the National Key Research and Development Program of China (Grant No. 2019YFB2204303), in part by the National Natural Science Foundation of China (Grant No. U20A20205), in part by the Key Project of Chongqing Science and Technology Foundation (Grant No. cstc2019jcyj-zdxmX0017, cstc2021ycjh-bgzxm0031), and in part by the Pilot Research Project from Chongqing Xianfeng Electronic Institute Co., Ltd (Grant No. H20201100).

References:

[1] K. Roy et al., "Towards spike-based machine intelligence with neuromorphic computing," Nature 2019, 575(7784): 607-617.

[2] C. Frenkel et al., "A 0.086-mm² 12.7-pJ/SOP 64k-Synapse 256-Neuron Online-Learning Digital Spiking Neuromorphic Processor in 28-nm CMOS," TBioCAS 2019, 13(1): 145–158.

[3] H. Tang et al., "Spike Counts Based Low Complexity SNN Architecture With Binary Synapse," TBioCAS 2019, 13(6): 1664-1677.

[4] J. Pu et al., "A 5.28-mm² 4.5-pJ/SOP Energy-Efficient Spiking Neural Network Hardware With Reconfigurable High Processing Speed Neuron Core and Congestion-Aware Router," TCAS-I 2021, 68(12): 5081-5094.

[5] Y. Zhong et al., "A spike-event-based neuromorphic processor with enhanced on-chip STDP learning in 28nm CMOS," ISCAS 2021: 1-5.

[6] Q. Chen et al., "A 67.5 μJ/prediction accelerator for spiking neural networks in image segmentation," TCAS-II 2021, 69(2): 574-578.

[7] T. Wang et al., "MorphBungee: An Edge Neuromorphic Chip for High-Accuracy On-Chip Learning of Multiple-Layer Spiking Neural Networks," BioCAS 2022: 255-259.

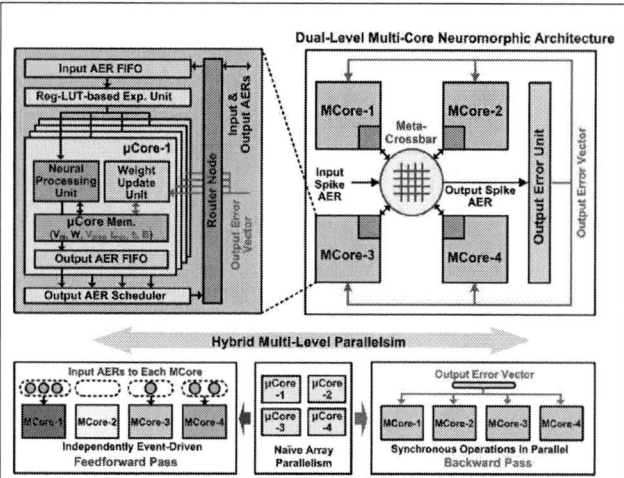

Fig. 1. The proposed dual-level multi-core edge neuromorphic chip architecture with hybrid core processing parallelism.

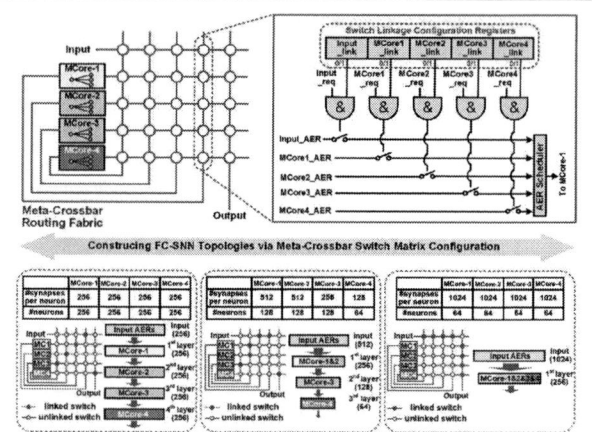

Fig. 2. The proposed meta-crossbar for efficient spike routing. By configuring individual switches in the crossbar, every MCore can have its neurons to be fully-connected to those neurons in different MCores so as to flexibly construct various FC-SNN topologies.

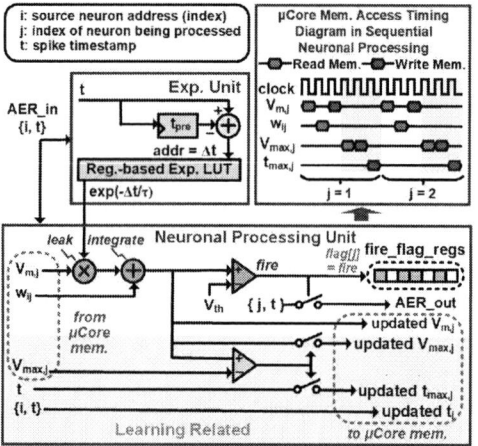

Fig. 3. The event-driven spike processing units. Necessary register buffers along the datapaths are not drawn for visual clarity.

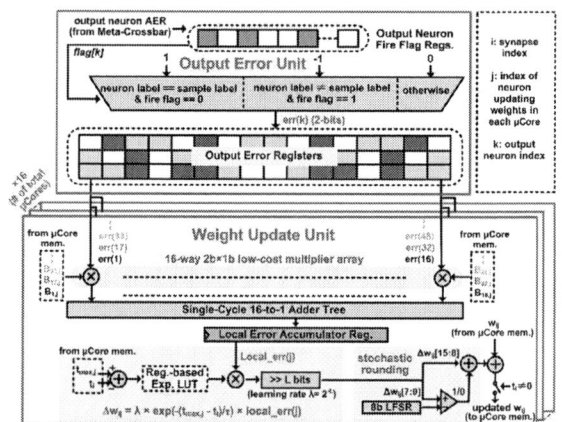

Fig. 4. The circuits of blocks implementing on-chip learning. The register-based exponential LUT in Fig. 3 is reused here as it has multiple access ports. Those synapses that have received no spike (i.e., $t_i = 0$) during the training sample do not update their weights.

Fig. 5. Prototype chip measurement system.

	ODIN [2]	Tang's [3][a]	Pu's [4][a]	Zhong's [5]	Chen's [6][a]	ours
Process	28 nm	65 nm	40 nm	28 nm	28 nm	65 nm
Chip area (mm²)	0.086 (active)	0.39 (active)	5.28 (total)	1.41 (active) 2.44 (total)	0.89 (active)	7.2 (active) 10.65 (total)
Scaled area* (mm²)	0.086 (active)	0.072 (active)	0.98 (total)	1.41 (active) 2.44 (total)	0.89 (active)	1.34 (active) 1.98 (total)
On-chip learning	SDSP	Spike-counti-based	Supervised STDP+	STDP	N/A	DeepTempo
On-chip learning acc.	84.5%	87.4%	89.6%	N/A	N/A	96.06%
Clock freq.	100 MHz	384 MHz	100 MHz	250 MHz	500 MHz	100 MHz
Core power supply	0.55-1.0 V	1.2 V	0.85 V	0.9 V	0.81 V	1.0 V
Chip power	477 μW	N/A	8.2 mW	3.348 mW	149.3 mW	61 mW
Throughput on MNIST	N/A	N/A	N/A	N/A	346k fps (infer.)	802 fps (learn.) 2270 fps (infer.)
Energy efficiency on MNIST	8.4 pJ/SOP (learn.)[△]	1.42 pJ/SOP (learn.)[△] 0.26 pJ/SOP (infer.)[△]	4.5 pJ/SOP (infer.)[▲]	4.1 pJ/SOP (learn.)[▲] 0.97 pJ/SOP (infer.)[▲]	0.43 μJ/image (infer.)[▲]	10.32 pJ/SOP (learn.)[△] 3.04 pJ/SOP (infer.)[△] 76.1 μJ/image (learn.)[▲] 26.9 μJ/image (infer.)[▲]

[a] Not fabricated. All results were simulated by software programs or estimated by EDA tools on circuit layouts.
[△] Incremental energy efficiency, the extra energy paid per SOP other than the consumed energy when no SOP occurs [2].
[▲] Global energy efficiency, calculated as the total energy consumption divided by the number of SOPs [2].
* Scaled area = Reported area / (Process / 28nm)².

Fig. 6. Comparison with state-of-the-art edge neuromorphic chips.

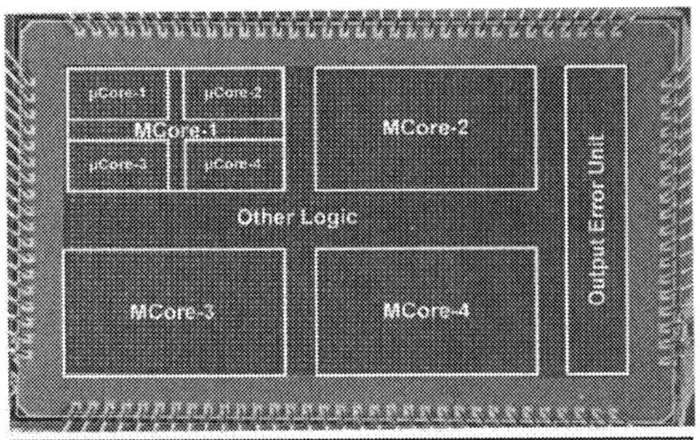

Technology	65 nm 1P9M CMOS
Chip area (active/total)	7.2/10.65 mm²
max #neurons/synapses	4K/256K
Online learning	DeepTempo
Clock frequency	100MHz
Supply voltages	1.0V for core, 3.3V for I/O
Power dissipation	61 mW
Throughput on MNIST	802/2270 fps @learn./infer.
Energy efficiency on MNIST	10.32/3.04 pJ/SOP @learn./infer. 76.1/26.9 μJ/image @learn./infer.

Fig. 7. Prototype chip microphotograph and measured results.

Additional References:

[8] C. Shi et al., "DeepTempo: A Hardware-Friendly Direct Feedback Alignment Multi-Layer Tempotron Learning Rule for Deep Spiking Neural Networks," TCAS-II 2021, 68(5): 1581-1585.

SESOMP: A Scalable and Energy-Efficient Self-Organizing Map Processor with Computing-In-Memory and Dead Neuron Pruning

Yuncheng Lu*[1], Xin Zhang*[1], Zehao Li[1], Bo Wang[2], and Tony Tae-Hyoung Kim[1] (* Equally contributed authors)

[1]Nanyang Technological University, Singapore

[2]Singapore University of Technology and Design

Self-organizing map (SOM), as a versatile artificial neural network, has been used in extensive fields, such as data clustering, pattern recognition, image quantization, etc. Massive parallel vector computations in the SOM necessitate specific hardware accelerators [1]-[5] for real-time data processing. However, the existing endeavors face the following limitations (Fig. 1(top)): 1) high power consumption caused by frequent memory access for massive vector processing, 2) energy waste from redundant computations in seldomly updated neurons (called "dead neurons"), and 3) mapping difficulty for various network sizes due to the unscalable architecture. To overcome the above limitations, this paper proposes a Scalable Energy-efficient SOM Processor (SESOMP) (Fig. 1(bottom)) with three key features: 1) computing-in-memory (CIM) macros for reducing memory access during SOM operation, 2) adaptive pruning to avoid redundant computation caused by dead neurons, and 3) a scalable architecture to support various network sizes.

The SOM includes the recall and training modes. In the recall mode, for each input vector ($\mathbf{X}$), the neuron with the minimum Manhattan distance (MD) to $\mathbf{X}$ is selected as the global winner (GW). The GW represents the input space that closely matches the current input, which can be used for data clustering and classification tasks. As shown in Fig. 1(bottom), the data scheduler broadcasts the negation of the input vector (-$\mathbf{X}$) to the four CIM macros, computing $\Delta\mathbf{W}_i$ by summing -$\mathbf{X}$ with the each neuron weight ($\mathbf{W}_i$). Then, the winner search engine (WSE) derives the MD of each neuron by accumulating the absolute value of the elements (Δw_{ij}) in each $\Delta\mathbf{W}_i$, and selects the local winner neuron (LW) on each chip with the minimum MD. When the SOM network is mapped on multiple chips, the LWs on different chips are compared through the chip-to-chip link (C2CL) to determine the GW. In the training mode, after winner selection, the neuron weights are updated according to their spatial distance to the GW, allowing the SOM to gradually adapt to the input data distribution. The neuron update engine (NUE) computes the update coefficient (α) for each neuron, and the CIM macros update the weight of each neuron following $\mathbf{W}_i = \mathbf{W}_i + \Delta\mathbf{W}_i \times 2^{-\alpha}$.

The CIM macros in the SESOMP can compute the $\Delta\mathbf{W}_i$ and update the $\mathbf{W}_i$ of the neurons in-situ, which reduces the memory access. Fig. 2 illustrates that each CIM macro comprises four subarrays and bidirectional peripheral circuits. Each subarray consists of 32 CIM local arrays (CLA), with each CLA handling a weight element (w_{ij}). The CLA incorporates an in-memory full adder (IMFA) for bit-serial addition and thirty-six 8T SRAM bit-cells storing w_{ij} and Δw_{ij}. The 8T bit-cell performs the CIM operation using the horizontal wordline (HWL) and the local bit-lines (LBL & LBLB), while also enabling column-wise data read-out of each CLA through the vertical wordline (VWL) and the horizontal bitlines (HBL & HBLB). The IMFA consists of sum logic and carry logic to conduct the bit-wise in-memory addition, long with a 6T bit-cell storing the carry. During the in-memory computation, the vertical bit-lines (VBL & VBLB) precharge the local bit-lines of the CLAs in the same column to a certain voltage level simultaneously. Then, with the switch SW in each CLA turned off, the CLAs sharing the same VBL(B) can conduct the bit-wise addition on their LBL(B) in parallel, resulting in a 4× improvement in throughput.

Fig. 3 illustrates the operation details of the CLA. During $\Delta\mathbf{W}$ computation, the pre-stored w_{ij} in the CLA, is added with the externally applied -x_j bit serially. In each clock cycle, one bit of -x_j and w_{ij} are added following the pre-charging, computing, and saving stages. In the pre-charging stage, the -$x_j[k]$ ($\overline{-x_j[k]}$) on the VBL (VBLB) is loaded to the LBL (LBLB) with the SW turned on. In the computing stage, with a 2.5 ns pulse applied on the HWL of the $w_{ij}[k]$ bit-cell, the pre-charged LBL(B) will be discharged by the zero value at the Q (QB) of the bit-cell. Therefore, the AND (NOR) between -$x_j[k]$ and $w_{ij}[k]$ is achieved on the LBL (LBLB). Using the results on the local

bit-lines, the IMFA computes the bit-wise sum (S_o) and the carry-out (C_o) following the truth table in Fig. 3. In the saving stage, when the EN in the IMFA switches off dividing LBL(B) into two separate parts, the S_o and C_o are saved in the $\Delta w_{ij}[k]$ and carry bit-cells concurrently. During weight update, different from the $\Delta\mathbf{W}$ computation, the operands (w_{ij} and Δw_{ij}) are both pre-stored in the CLA. In the pre-charging stage, the LBL(B) is pre-charged to VDD by VBL(B). During the computing stage, after activating the $w_{ij}[k]$ and $\Delta w_{ij}[k]$ bit-cells simultaneously, the AND and the NOR results on the local bit-lines are used to compute the S_o and C_o by the IMFA. The bit-wise addition results are then saved in the $w_{ij}[k]$ and the carry bit-cells in the saving stage. Note that a lower HWL voltage (0.8 V) is utilized to avoid data corruption when two bit-cells are activated simultaneously in each CLA. The proposed CLA can compute $\Delta\mathbf{W}$ and update weight, which reduces the memory access during the recall and the training mode of SOM by 66.7% and 83.3%, respectively.

During the training process of SOM, the dead neurons exhibit extremely low update rates due to the significant deviation between their initial weights and the training dataset. However, their related sub-arrays in the macro keep computing $\Delta\mathbf{W}$, which wastes energy. Therefore, the NUE monitors the update frequency of each neuron and prunes the dead neuron adaptively. As shown in Fig. 4, the NUE consists of 16 update signal generators (USG) to compute the α values of neurons in 16 memory subarrays in parallel. Each USG includes a neuron monitor to track the status of a specific neuron. The neuron monitor comprises a Neuron Update Frequency register (NUF) to record the number of times a neuron is updated within a certain period and a Dead Neuron Flag register (DNF) to indicate whether a neuron is dead. The pruning process begins after the neurons have been trained by T1 epochs to avoid mistakenly pruning the normal neurons. T1 is set to 50% of the total training epochs (T_{max}), assuming that normal neurons have begun to converge to the training set after T1 epochs. Then, the NUF counts the number of times each neuron is updated within every T2 interval, where T2 is 12.5% of T_{max}. The neurons updated less than 10 times within the T2 interval are considered as dead neurons, and their DNF register is set to one. Based on the DNF, the top controller deactivates the hardware blocks related to the dead neurons, such as the USG and the corresponding subarrays in the CIM macros. Experimental results on the iris dataset show that the proposed dead neuron pruning reduces the power by up to 12.5%. Additionally, compared with the original SOM, the quantization error only increases by 0.2% after pruning the dead neurons.

When the SOM network is mapped on multiple chips, the GW needs to be selected among the LWs and broadcasted to all chips. As shown in Fig. 5, the C2CL supports two inter-chip connection modes: the centralized mode and the systolic mode. In the centralized mode, a host is used to collect the LWs from all chips and determine the GW. In the systolic mode, the LW with a smaller MD between two adjacent chips will be selected and forwarded to the next chip in sequence and the GW is determined in the last chip. The centralized mode consumes less latency for the winner searching, while the systolic mode is free of the host. The data clustering and image quantization experiments are conducted on four chips with each being allocated a 4×4 sub-network. Thanks to the scalable architecture, the proposed SESOMP can adapt to SOM networks with different sizes flexibly.

Fig. 6 presents the comparison with the state-of-the-art SOM processors [1]-[5]. Featuring the CIM-based architecture and the dead neuron pruning, the proposed SESOMP achieves the highest power efficiency. Moreover, the SESOMP supports different network sizes and variable weight precision (1-16 bit) thanks to the scalable architecture and the bit-serial computation in the CIM macro. Fig. 7 shows the test platform, chip micrograph, and measurement results of the prototype chip.

Acknowledgement:

This work was supported by a RIE2020 ASTAR AME IAF-ICP under Grant I1801E0030.

References:

[1] C. Shi et al., "A 1000 fps Vision Chip based on a Dynamically Reconfigurable Hybrid Architecture Comprising a PE Array Processor and Self-Organizing Map Neural Network," JSSC, 2014.

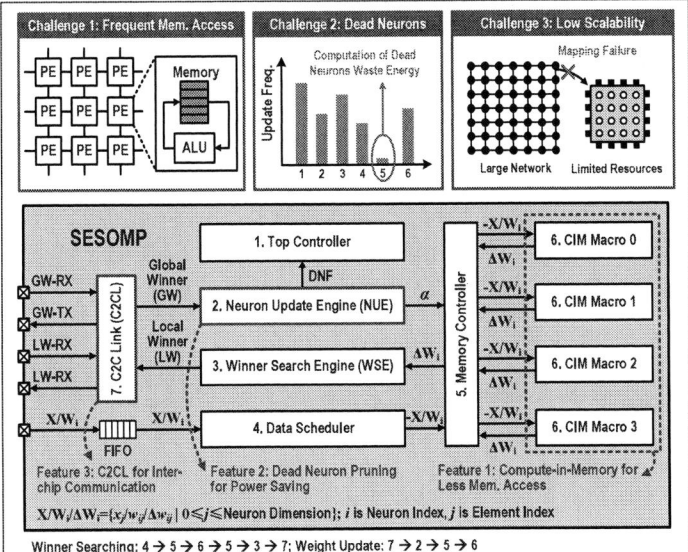

Fig. 1. The motivation and system architecture.

Fig. 2. Overall architecture of the CIM macro and detailed circuits.

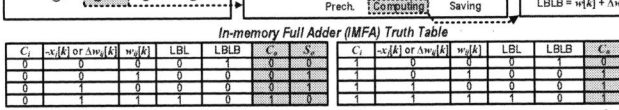

Fig. 3. Operation principles of each CLA during ΔW computation and weight update and the truth table of the IMFA.

Fig. 4. Dead neuron pruning flow, architecture of neuron update engine, and measurement results after dead neuron pruning.

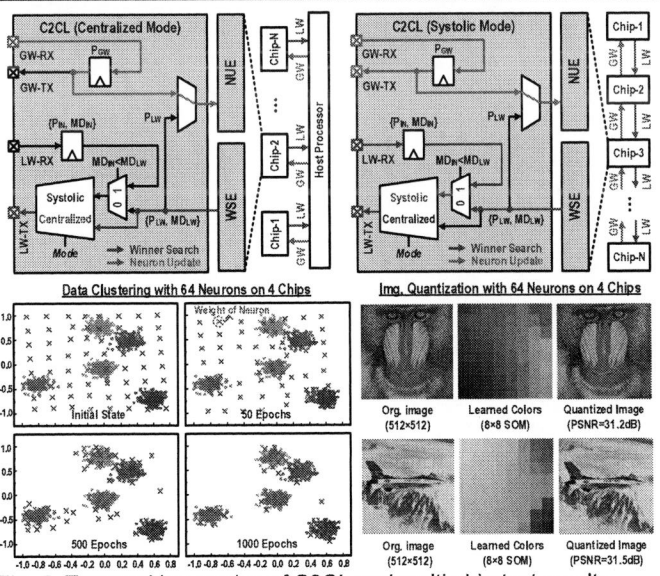

Fig. 5. Two working modes of C2CL and multi-chip test results.

		JSSC'14 [1]	TMSC'16 [2]	TCASI'18 [3]	TCASII'19 [4]	NN'20 [5]	This Work
Technology		COMS 180nm	CMOS 65nm	CMOS 65nm	FPGA	FPGA	CMOS 65nm
Architecture		Distributed	Distributed	Distributed	Distributed	Distributed	CIM
Frequency (MHz)		50	100	150	100	28.57	100
Voltage (V)		1.8	0.8	1	NA	NA	Digital: 0.96 Macro: 0.8
Paralism		Element-parallel	Sub-neuron-parallel	Sub-neuron-parallel	Neuron-parallel	Element-parallel	Element-parallel
Power		630 mW	9.4 mW	21.5 mW	204 mW	4.267 W	2 mW
Peak Throughput	Recall (MCPS)	258	3091	4753.6	NA	NA	948.1 ~ 1113*
	Learning (MCUPS)	NA	3080	4730.7	7263	37620	701.4 ~ 898.2*
Peak Power Efficiency	Recall (GCPS/W)	0.409	328.8	221.1	NA	NA	474 ~ 556.5*
	Learning (GCUPS/W)	NA	327.7	220	35.6	8.8	350.7 ~ 449.1*
Scalability		NO	NO	NO	NO	NO	YES
Dead Neuron Pruning		NO	NO	NO	NO	NO	YES
Neuron Number (per Chip)		256	256	256	16-256	100	Recall: 1-512 Learning: 1-64
Neuron Dimension		1-64	32	128	3	79	1-128
Precision (bit)		16	16	16	16	16	1-16
Application		Image Classification	Data Clustering	Data Clustering	Data Clustering Image Quantization	Pose Recognition	Data Clustering Image Quantization Image Classification
Accuracy		Gesture: 92% Face: 86%	NA	NA	NA	53%	Gesture: 94.3%

*The peak throughput and power efficiency is measured with weight precision ranging within 8-16 bit

Fig. 6. Comparison with the state-of-the-art SOM accelerators.

979-8-3503-3004-5/23 $31.00 © 2023 IEEE

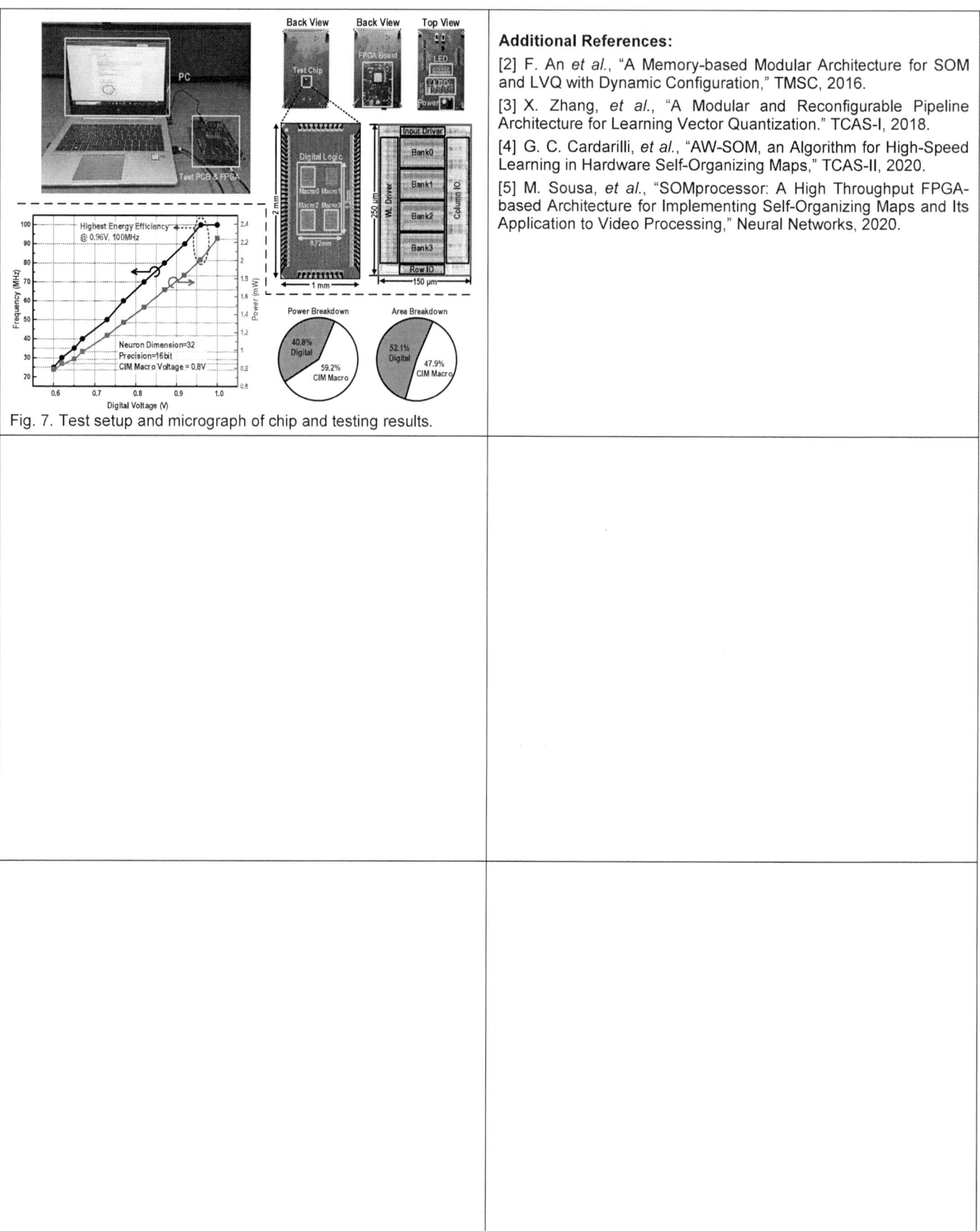

Fig. 7. Test setup and micrograph of chip and testing results.

Additional References:

[2] F. An *et al.*, "A Memory-based Modular Architecture for SOM and LVQ with Dynamic Configuration," TMSC, 2016.

[3] X. Zhang, *et al.*, "A Modular and Reconfigurable Pipeline Architecture for Learning Vector Quantization." TCAS-I, 2018.

[4] G. C. Cardarilli, *et al.*, "AW-SOM, an Algorithm for High-Speed Learning in Hardware Self-Organizing Maps," TCAS-II, 2020.

[5] M. Sousa, *et al.*, "SOMprocessor: A High Throughput FPGA-based Architecture for Implementing Self-Organizing Maps and Its Application to Video Processing," Neural Networks, 2020.

A Time-based PAM-4 Transceiver using Single Path Decoder and Fast-Stochastic Calibration Techniques

Dong-Hyun Yoon[1], He Junsen[1], Kwang-Hyun Baek[2], Youngdon Choi[3], Jung-Hwan Choi[3], and Tony Tae-Hyoung Kim[1]

[1]Nanyang Technological University, [2]Chung-Ang University, and [3]Samsung Electronics

The demand for high-speed transceivers for memory interfaces is steadily increasing. Recently, time-based (TB) transceivers [1]-[4] are obtaining more attention because of the implementation using simple digital logic gates assisted with a voltage-to-time converter (VTC) for power reduction. Since the VTC gain is inversely proportional to the supply voltage, the TB transceivers show higher power efficiency. However, VTC and logic gate delay are highly susceptible to PVT variations. This paper proposes a single path decoding scheme and linear VTC for PAM-4 signaling to enhance power efficiency and signal integrity (SI). We also present calibration techniques to improve the variation tolerance in the VTC gain, the single-to-differential amplifier (S2D), the time threshold (T_{TH}), and the decision feedback equalizer (DFE).

The top-left of Fig. 1 shows the conventional TB PAM-4 decoding. The input signal (T_{VTC}) is compared with three-time thresholds ($T_{TH,L/M/H}$) [2]. Therefore, twelve delay cells and three time-comparators (T-Comp) are required per channel. On the other hand, the proposed PAM-4 decoding scheme uses only a single path. As shown in the top right of Fig. 1, M_P and M_N are compared to recover MSB. The first T-Comp outputs are fed to the delay cells, and a delay of T_{TH} is added to the "preceding signal" to decode LSB. The bottom of Fig. 1 depicts the timing diagram of the proposed decoding scheme when T_{TH} is set to 2D (D is the unit time difference). If the input is "00" or "01", because M_P leads M_N, M_P is the preceding signal. Thus, MSB becomes "0" and M_P is delayed more than M_N by 2D. If the input is "00", the preceding signal is still L_P even though M_P has a longer delay, and thus LSB also becomes "0". However, in the case of "01", the time difference of D makes L_N the preceding signal after delaying M_P by 2D, decoding LSB as "1". The same principle is applied to the input "10" and "11". The proposed decoding scheme needs only four delay logics and two T-Comps, it saves thirty two delay cells, four T-Comps, and PAM-4 decoders compared with [2].

However, T_{TH} should be placed between D and 3D. As shown at the top of Fig. 2, if T_{TH} is shorter than D, the preceding signal is unchanged for all input patterns, which makes MSB and LSB always identical. On the other hand, if T_{TH} is longer than 3D, the preceding signal is always changed, and MSB and LSB become complementary. The proposed T_{TH} calibration detects the D and 3D. First, T_{TH} is set to the minimum value and increases step by step. Then, the point where MSB and LSB become complementary is D. Next, T_{TH} is set to the maximum value and decreases until finding 3D where MSB and LSB are identical. After that, an optimal T_{TH}, 2D, can be obtained by averaging D and 3D.

The nonlinearity of VTC causes a poor ratio of level mismatch (RLM). The bottom left of Fig. 2 shows the conventional VTC using transistor stacking for delay adjustment [1]. The delay is proportional to the transistor's on-resistance but inversely proportional to the input voltage. However, the proposed linear VTC controls the delay using a feedback transistor. Because the delay is inversely proportional to the on-resistance, it shows a linear relationship with the input voltage. Consequently, the proposed VTC demonstrates a higher RLM of 0.96 compared with the conventional VTC with an RLM of 0.18, as presented in the bottom right of Fig. 2.

Even though TB transceivers achieve high power efficiency because of digital delay cells, the delay is sensitive to PVT variations, degrading the operating frequency and SI. For example, a high VTC gain provides a significant time difference but leads to a large delay. Conversely, a small VTC gain decreases the timing margin between the PAM signal levels. However, the VTC gain and the delay are highly susceptible to the PVT variation, as shown in the top left of Fig. 3. Therefore, time-domain calibration is essential to maximize the time difference without operating frequency degradation. Also, since V_{REF} errors in S2D reduce the margin, V_{REF} needs to be calibrated for variation-tolerant operation. As illustrated in the top

right of Fig. 3, the PAM-4 signaling follows the normal distribution at every signal level. If M_P and M_N are compared with the center of the "00" level, the T-comp's output ratio is 0.875 to 0.125. The conventional stochastic method accumulates the comparator output and compares it with a reference ratio [5]. Then, it updates the parameters and repeats the calibration cycles until the ratio reaches the reference value at the cost of many cycles. However, the proposed method finds the target parameters using only a single cycle. If the time reference (T_{REF}) departs 0.84σ from the "00" level, its ratio will be 0.8:0.2, which simplifies the circuit implementation. The bottom left of Fig. 3 shows the block diagram of the proposed calibration techniques. For VTC gain calibration, M_P is compared with T_{REF}. The comparator output is accumulated having 1/4 weight of the negative output to meet the 0.8:0.2 ratio. The 8-bit DAC generates VTC_{Gain} to control the VTC gain. As a result, the M_P level of "11" is aligned with T_{REF}. V_{REF} is also calibrated in the same principle using M_N. In Fig. 4, (1) and (2) show the simulation results of the VTC gain and V_{REF} calibration, respectively. When the VTC gain is smaller than the target, M_P-M_N increase during the calibration. After 4,096 clock cycles, the calibration circuit is disabled. On the other hand, the magnitude of M_P-M_N is maintained but shifted upward during the V_{REF} calibration. The above calibration stages are repeated until a final state is found. After "(1) gain calibration", M_P is aligned with T_{REF} by increasing the VTC gain, but M_N is not aligned yet. Therefore, V_{REF} is calibrated using M_N in the same way. However, M_P and M_N are shifted upward together. As a result, M_P and M_N are aligned with T_{REF} in "Final State", as depicted in the top of Fig. 4. The proposed DFE calibration is also performed in the time domain. The bottom right of Fig. 3 shows the voltage and time domain waveforms of the proposed DFE calibration. In the ideal case, the signal levels are well-defined. However, the signal levels are disturbed if the signal is under or over-equalized. The proposed DFE calibration extracts the "00" to "11" transition and compares M_N with T_{REF}. Because M_N is aligned during the V_{REF} calibration, the comparison results indicate either under-equalization or over-equalization.

Fig. 4 illustrates the top block diagram of the proposed transceiver. The transmitter adopts the half-rate structure, 3-tap FFE, and a source series terminated driver. The transmitter's output (V_{TX}) is fed to the receiver through the off-chip channel, and the single-ended input (V_{RX}) is converted to the differential signal. Also, a sample and hold circuit (S&H) is adopted to block the input voltage fluctuation of VTC during the clock transition [2]. Then, the input voltage is decoded through the proposed single-path decoder. The proposed calibration logic provides the V_{REF}, VTC_{Gain}, $T_{TH}<4:0>$, and DFE_tap<3:0> to compensate for the PVT variations.

The proposed transceiver was fabricated in a 65 nm CMOS process and tested under 0.75V TX supply and 0.85 V RX supply at 6 Gbps. The eye diagrams of the TX are shown in Fig. 5. In the top left of Fig. 5, the peak-to-peak error and the root mean square error (RMSE) of the VTC gain and the V_{REF} calibration are 13/15 LSB and 3.8/4.0 LSB, respectively. The top-right of Fig. 5 shows that the VTC gain and V_{REF} calibration tracks the target VTC gain and V_{REF} after ten cycles. After that, the DFE calibration logic detects the optimal DFE weight. The measured bathtub curve is presented in Fig. 5. The measured eye margin is 0.2 UI which is improved by 0.02 UI utilizing the DFE adaptation. BERs are also measured over supply voltages in Fig. 5. The proposed calibration improves the supply margin from 80 mV to 120 mV. The top of Fig. 6 shows the time domain eye diagram with different V_{REF} and T_{TH} values. In the V_{REF} eye diagram, the average of the V_{REF} calibration is marked, achieving BER lower than 10^{-12}. T_{TH} is also calibrated by averaging D and 3D, as shown in the T_{TH} eye diagram. Because the minimum value of the fabricated chip is larger than D, the minimum code is regarded as D. This work is the first paper that proposed the time-domain calibration technique in the table of Fig. 6. The proposed transceiver achieves the 10^{-12} BER at 6 Gbps with 7.59 mW power consumption.

References:

[1] P.-W. Chiu et al., *ISSCC*, 2020. [2] H. Park et al., *JSSC*, 2021. [3] M. -K. Chae et al., *TCAS-II*, 2021. [4] I.-M. Yi, et al., *JSSC*, 2018. [5] H. Won et al., *TCAS-I*, 2017.

979-8-3503-3004-5/23 $31.00 © 2023 IEEE

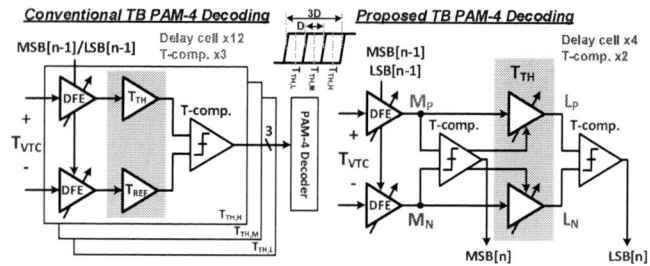

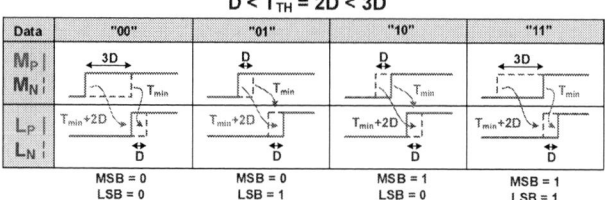

Fig. 1. Proposed single path PAM-4 decoding scheme and timing diagram

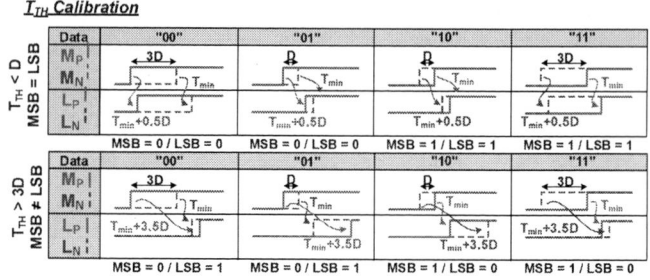

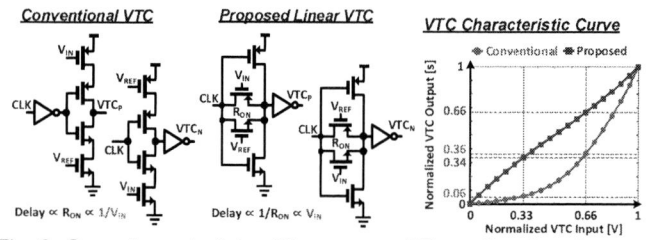

Fig. 2. Operating principle of the proposed T_{TH} calibration, the conventional VTC, and the proposed linear VTC

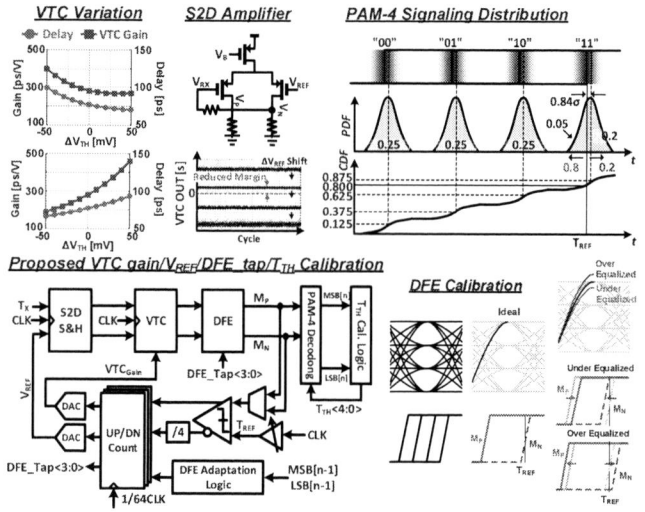

Fig. 3. VTC/S2D/PAM-4 signaling characteristics, block diagram of proposed of calibration, and operation principle of DFE calibration

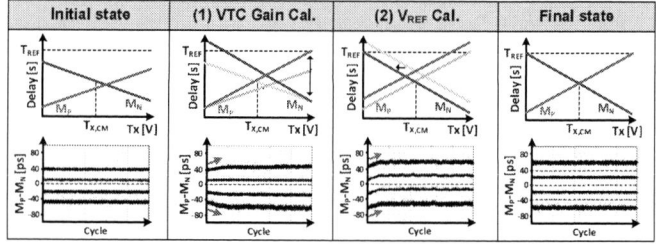

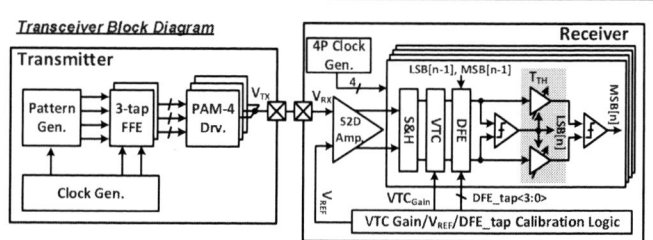

Fig. 4. Operating principle of the proposed VTC gain/V_{REF} calibration and block diagram of proposed transceiver

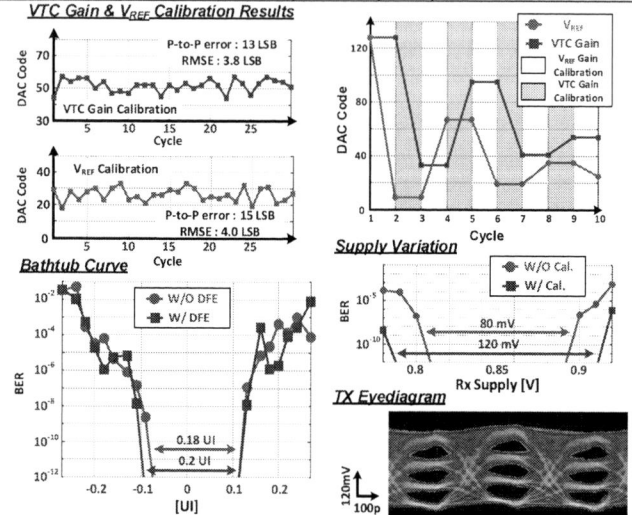

Fig. 5. Measured VTC gain/V_{REF} calibration, bathtub curve, BER under supply variation, and eye diagram

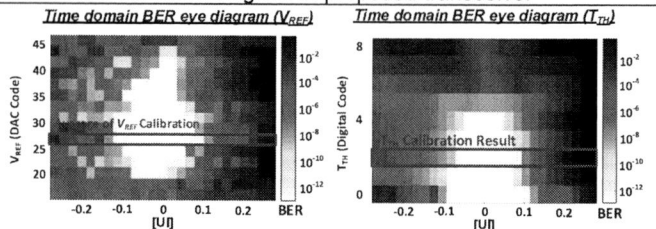

Fig. 6. Time domain BER eye diagram and performance comparison

Performance Comparison Table

	This Work	[1] ISSCC '20	[2] JSSC '21	[3] TCAS-II '17	[4] JSSC '18
Signaling	PAM-4	PAM-4	PAM-4	Duo Binary	NRZ
RX Type	Time-based	Time-based	Hybrid	Time-based	Time-based
Decoding	Fully Time-based / Single Path	Fully Time-based / Conven. Decoding	MSB - Voltage / LSB - Time	Fully Time-based / Look-ahead	
Equalizer	3-tap FFE / 1-tap DFE		3-tap FFE / 2-tap DFE	1-tap FFE / 1-tap DFE	2-tap DFE
Date Rate	6 Gbps	32 Gbps	56 Gbps	15 Gbps	12.5 Gbps
Process	65 nm	65 nm	28 nm	28 nm	65nm
Calibration	VTC Gain / V_{REF} T_{TH} / DFE	-	V_{TH}	-	-
Power	7.59 mW	31.04 mW	54.6 mW	12 mW	6.125 mW
Area	0.033 mm² *(0.019 mm²)	0.009 mm²	0.024 mm²	0.004 mm²	0.009 mm²

*Without calibration circuit

979-8-3503-3004-5/23 $31.00 © 2023 IEEE

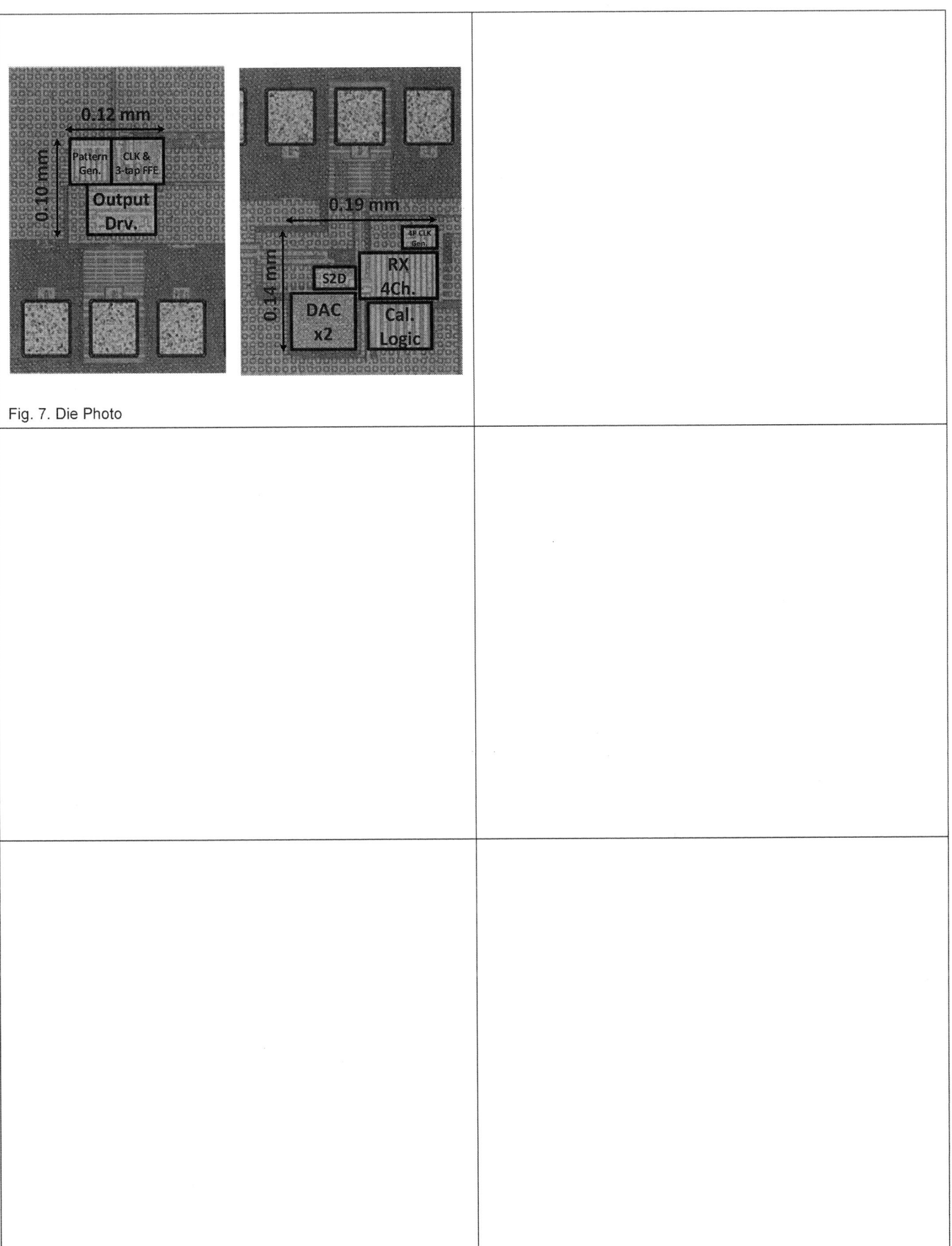

Fig. 7. Die Photo

A 160×120 Indirect Time-of-Flight Sensor with Pixel-level Adaptive $\Delta\Sigma$-operations for Background Light Cancellation

Yongjae Park[1], Jubin Kang[1], Dahwan Park[2], Insang Son[1], Jung-Hye Hwang[1], and Seong-Jin Kim[1]

[1]Ulsan National Institute of Science and Technology, Ulsan, Korea
[2]SK hynix Inc., Icheon, Korea

An indirect time-of-flight (iToF) sensor has been spotlighted for its high-quality depth images within a few meter ranges, making it a strong candidate for realizing metaverse applications in mobile devices [1]. However, one of the unresolved challenges in iToF sensors is low background light (BGL) resilience, which can saturate floating diffusion (FD) nodes and deteriorate a demodulation contrast (DC), causing large depth distortion and noise. Even though in-pixel auto-zeroing and chopping techniques have achieved BGL robustness up to 200 klx, it still has large noise of over 3% due to noise contribution from the BGL cancellation technique [2]. The in-pixel voltage subtraction scheme has been reported for compensating BGL up to 180 klx, but it requires tons of switching operations, generating large noise as well as charge injections [3]. The column-parallel $\Delta\Sigma$ BGL canceling technique has been introduced to subtract a common-mode voltage of each FD node and accumulate differential signals [4]. However, the column-parallel architecture has limited scalability and still suffers from additional reset noise due to the dedicated number of $\Delta\Sigma$-operations. Another big challenge of iToF sensor is the degradation of depth non-linearity and precision caused by FD mismatches [5]. Although common-mode voltage has been successfully compensated in [4], FD mismatches can still convert residual BGL to phase information.

This paper presents a 160×120 iToF sensor with a pixel-level adaptive $\Delta\Sigma$-scheme that automatically controls the optimal number of $\Delta\Sigma$-operation to suppress large BGL with minimum switching noise. It also increases depth dynamic range at a short distance in a condition without large BGL. In addition, in-pixel FD swapping further reduces residual BGL while enhancing depth linearity.

Fig.1 shows the conceptual operating principles of the proposed adaptive $\Delta\Sigma$-scheme and potential diagrams of the two mismatched FD nodes for two different cases. A total integration time is divided into eight sub-int-times followed by eight $\Delta\Sigma$-operations. The number of $\Delta\Sigma$-operations is decided automatically by comparing the potential of the FD-A node with the monitoring voltage (V_{MON}) after the first and second sub-int-times. After the first sub-int-time, if the potential of the FD-A is lower than V_{MON}, the number of $\Delta\Sigma$-operations is selected to the largest number, eight. Then, the first $\Delta\Sigma$-operation is operated, removing the common-mode signal while leaving residual BGL charges due to FD mismatches. The residual BGL and depth non-linearity can be further reduced by the FD swapping after the second $\Delta\Sigma$-operation. For a low BGL environment or a long-range target, each pixel integrates ToF signals without adding switching noise by skipping $\Delta\Sigma$-operations.

Fig. 2 shows the proposed in-pixel charge amplifier with its timing diagram. Unlike previous works [2, 3], PPDs with transfer gates (TX-A, B) demodulate reflected light so that the amplifier bandwidth can be much lower than the modulation frequency (50-MHz). Each FD node can be accessed by bridge switches (BR-A, B), and four feedback switches (In-ph, Out-ph) are implemented to transfer the FD charges and flip the C_{INT}. The C_{INT} is designed with two 50-fF capacitors reconfigurable for an analog gain of two. During the $\Delta\Sigma$-operations, two capacitors are connected in parallel to minimize the noise, while they are reconfigured in series to double the output voltage for the final readout time. A delta-double sampling subtracts $V_{180°} - V_{0°}$ from $V_{0°} - V_{180°}$, achieving an additional gain and removing the charge amplifier offset. Charge injection from BR-A and -B switches is one of the candidates that degrade the depth linearity. The Q_{OFF-FD} and Q_{OFF-CP}, charges injected when they are turned off, are signal-independent so that they can be compensated by the $\Delta\Sigma$-operation. Other injected charges in the on-state, the Q_{ON-A} and Q_{ON-B}, are signal-dependent on each FD node voltage. However, they can only affect transfer voltage gain as shown in Fig. 2, which can also be canceled out when calculating the distance.

Fig. 3 shows an in-pixel logic circuitry for realizing the adaptive $\Delta\Sigma$-scheme and its timing diagram. The V_{MON}, the threshold voltage of the inverter-based comparator, is variable by tuning the gate voltage (V_{B-TH}) of an additional PMOS. Two flag latches memorize the comparison results to select the ΦRST among $\Phi RSTx2$, x4, and x8 for governing the number of $\Delta\Sigma$-operations, and the latches are not reset until obtaining quadrature phase value for maintaining the number of $\Delta\Sigma$-operations and the transfer gain. The comparator is only turned on twice during the whole integration time to capture in- and quadrature-phase. The T-FF governed by TX-Swap-M and BR-Swap-M is turned on at every $\Delta\Sigma$-operation time to update the previous and present states and control FD swapping operations.

A proposed iToF sensor prototype was fabricated in a 110-nm BSI process. Fig. 4 describes the BGL experiment setup and measured depth images with the different number of $\Delta\Sigma$-operations and the adaptive $\Delta\Sigma$-operation mode for 20-ms int-time. A halogen lamp is adopted to validate the BGL cancellation technique, and its optical power is measured as 3.3 mW/cm^2 at 940 nm with a 30-nm bandwidth on a target surface. Considering the wavelength and bandwidth, equivalent sunlight is estimated to be 300 klx. Our 940-nm IR emitter with a single VCSEL supplies 100 times lower optical power of 31 µW/cm^2 measured on the target with the same optical power meter, offering the BGL-to-signal ratio (BSR) of over 100. The more $\Delta\Sigma$-operations are applied, the less distorted depth image can be acquired as expected. The adaptive $\Delta\Sigma$-scheme can also recover depth without distortion successfully. The last map illustrates the decided number of $\Delta\Sigma$-operations in each pixel, demonstrating that the adaptive $\Delta\Sigma$-logic successfully selects optimal points relying on the BGL strength.

Fig. 5(a) shows measured depth linearity and precision, which are less than 2 % and 2.7 % from 0.3 m to 2.9 m, respectively, with a 10-ms int-time. In this measurement, the adaptive $\Delta\Sigma$-scheme selects eight, four, and two $\Delta\Sigma$-operations for 0.3 m to 0.4 m, 0.45 m to 0.55 m, and longer ranges, respectively. Fig. 5(b) shows the capability of short-range depth acquisition and the depth precisions with different numbers of $\Delta\Sigma$-operations. The pixels are saturated with two and four $\Delta\Sigma$-operations in a long int-time of 40 ms, whereas eight and adaptive $\Delta\Sigma$-operations provide the exact distance. In contrast, the more operations are applied, the worse depth precision can be acquired in the long range. The adaptive $\Delta\Sigma$-scheme automatically adjusts the number of operations to two, achieving better noise performance.

Fig. 6 summarizes the chip characteristics and performance comparison The proposed prototype achieves largest BGL cancellation, and it can extend the minimum detectable depth to 0.3 m with comparable depth non-linearity. We believe a relatively large pixel area can be alleviated by utilizing the 3D stacking process. The chip photograph and pixel layout are disclosed in Fig. 7.

Acknowledgment:

This work was supported in part by SK hynix Inc., Co. Ltd, and in part by the COMPA grant funded by the Korea government(MSIT) (No. 20211100).

References:

[1] C. S. Bamji, *et al.*, "1Mpixel 65nm BSI 320MHz demodulated TOF Image sensor with 3µm global shutter pixels and analog binning," in *IEEE ISSCC Dig. Tech. Papers*, Feb. 2018, pp. 94-96.

[2] B. Park, *et al.*, "A 64×64 APD-based ToF image sensor with background light suppression up to 200 klx using in-pixel auto-zeroing and chopping," in *Proc. IEEE Symp. VLSI Circuits*, Jun. 2019, pp. C256-C257.

[3] T.-H. Hsu, *et al.*, "A CMOS Time-of-Flight depth image sensor with in-pixel background light cancellation and phase shifting readout technique," in *IEEE J. Solid-State Circuits*, vol. 53, no. 10, pp. 2898-2905, Oct. 2018.

[4] D. Kim, *et al.*, "Indirect Time-of-Flight CMOS Image Sensor With On-Chip Background Light Cancelling and Pseudo-Four-Tap/Two-Tap Hybrid Imaging for Motion Artifact Suppression," in *IEEE J. Solid-State Circuits*, vol. 55, no. 11, pp. 2849-2865, Nov. 2020.

[5] J. Kang, *et al.*, "A 640×480 Indirect Time-of-Flight Image Sensor with Tetra Pixel Architecture for Tap Mismatch Calibration and Motion Artifact Suppression," in *Proc. IEEE Symp. VLSI Circuits*, Jun. 2022, pp. 44-45.

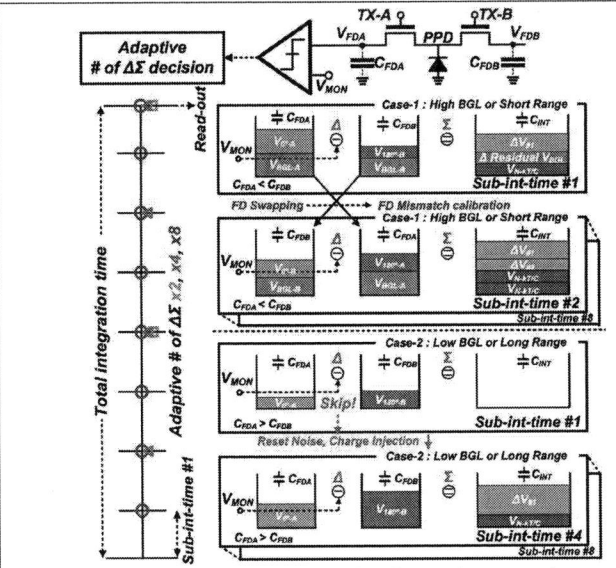

Fig. 1. Conceptual operating principle of adaptive ΔΣ-scheme.

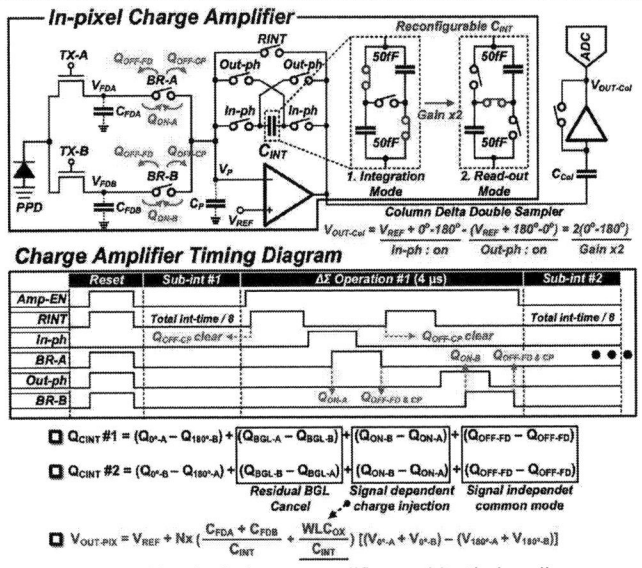

Fig. 2. Proposed in-pixel charge amplifier and its timing diagram.

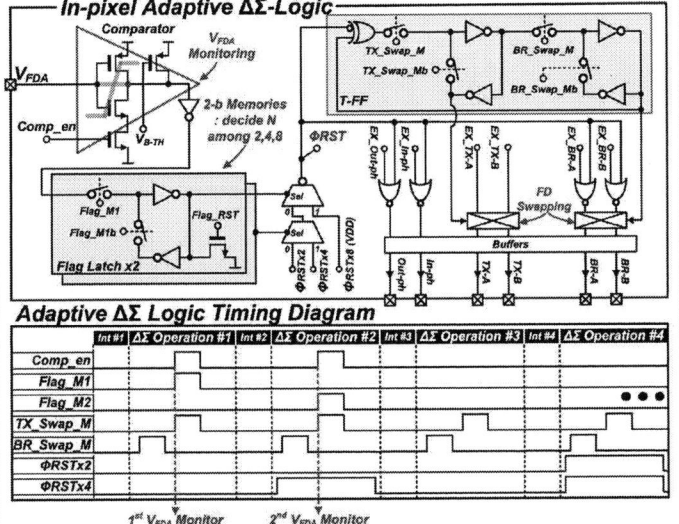

Fig. 3. Proposed in-pixel adaptive ΔΣ logic and its timing diagram.

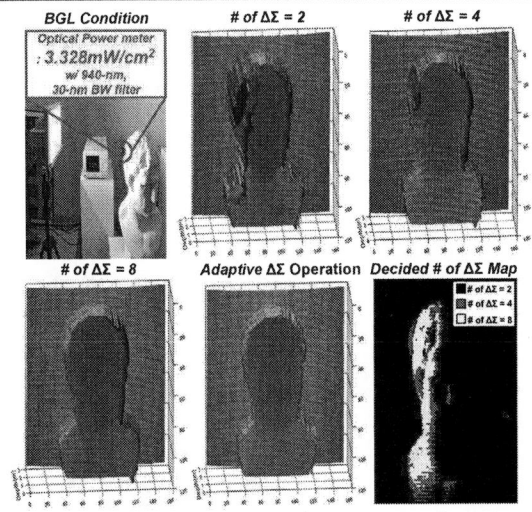

Fig. 4. BGL cancellation demonstration setup and captured depth image with different numbers of ΔΣ-operations and adaptive ΔΣ-operation with a corresponding selected number of ΔΣ image.

Fig. 5. (a) Measured depth with its non-linearity and precision from 0.3 m to 2.9 m and (b) measured depth of a 60-cm target with different integration times and depth precision comparison with different numbers of ΔΣ-operations.

	JSSC'18 [3]	VLSI'19 [2]	JSSC'20 [4]	JSSC'23 [6]	This Work
Technology	180 nm / FSI	110 nm / FSI	90 nm / BSI	110 nm / BSI	110 nm / BSI
Pixel Array	64 x 64	64 x 64	320 x 240	640 x 480	160 x 120
BGL Cancellation Technique	In-pixel voltage subtraction	In-pixel auto-zeroing & chopping	Column level Δ-INT	Short pulse w/ drain gate	In-pixel adaptive ΔΣ
	180 klux	200 klux	120 klux	100 klux	300 klux
BGL Power	N/A	N/A	142 μW / cm² @ 855nm	N/A	3.3mW / cm² @ 940nm, 30nm BW
BGL to Signal Raito	N/A	N/A	N/A	N/A	>100 (3.3 mW / 31 μW)
Pixel Pitch	20 μm	32 μm	8 μm	5.6 μm	21.6 x 25.2 μm
Fill Factor	33 %	21 %	43 %	N/A	33 %
Modulation Frequency	2.5 MHz	25 / 1.56 MHz	10 – 100 MHz	17-19 ns pulse width	50 MHz
Frame Rate	26 fps	96 fps	10 – 60 fps	15 – 120 fps	30 fps
Depth Range	0.75 – 7.5 m	0.5 – 2 m / 2 – 20 m	0.75 – 4 m	1 – 20 m	0.3 – 2.9 m
Depth Non-linearity	< 1.1 % @ 0.75 – 7.5 m	< 0.8 % @ 0.5 – 2 m < 0.75 % @ 2 – 20 m	< 0.84 % @ 0.75 – 4 m	< 0.75 %(A) @ 1- 20m	< 2 % @ 0.3 – 2.9 m
Depth Noise	3.3 %(A) @ 3 m	3.4 %(A) @ 2 m	0.54 % @ 0.75 – 4 m	< 1.3 % @ 1 – 20 m	2.7% @ 2.9m

(A) : Estimated from given information.

Fig. 6. Chip characteristics and performance comparison.

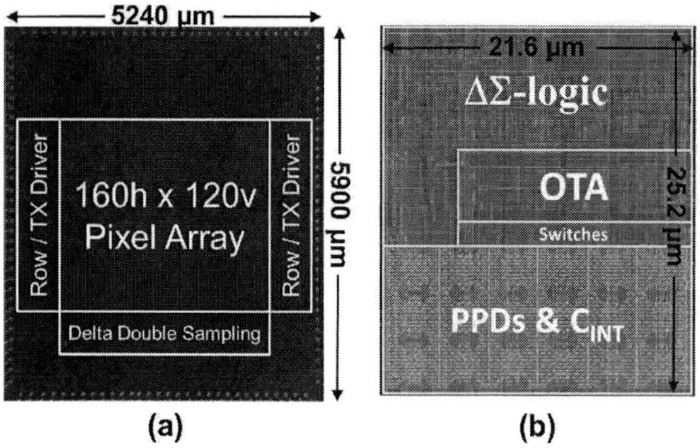

Fig. 7. (a) Chip photograph and (b) pixel layout.

Additional References:

[6] K. Hatakeyama, *et al.*, "A Hybrid ToF Image Sensor for Long-Range 3D Depth Measurement Under High Ambient Light Conditions," in *IEEE J. Solid-State Circuits*, vol. 58, no. 4, pp. 983-992, April. 2023.

An 11bit 360MS/s Pipelined SAR ADC with Dynamic Negative-C Assisted Residue Amplifier

Yigi Kwon[1,2], Sangwoo Lee[1], Changuk Lee[1], Hyunchul Yoon[1], Byounghan Min[2], and Youngcheol Chae[1],

[1]Yonsei University, Seoul, Korea

[2]LG Electronics, Seoul, Korea

Pipelined SAR ADCs are widely used for moderate-to-high resolution applications. In this architecture, a residue amplifier plays a key role in speed and linearity, so it is inevitable that the amplifier consumes a large amount of power to achieve the target performance. Several amplifier topologies such as G_m-based amplifier, ring amplifier, and floating inverter amplifier (FIA) [1-3] have been developed to improve energy efficiency and represent a successful alternative to conventional OTAs. However, achieving high bandwidth while maintaining good energy efficiency of ADCs is still very challenging.

Alternatively, a negative-C assisted amplifier has been proposed in [4], [5], and applied to a pipeline ADC [6]. By exploiting a negative-C (NC) on the virtual ground, the sampling capacitor is cancelled during amplification, which increases the feedback factor and relaxes the requirements of the residue amplifier. However, mainly due to the use of a static NC implementation [5], [6], they can reduce the gain error, but not improve the amplifier's speed. This work improves the speed of the NC-assisted residue amplifier by introducing a new dynamic NC circuit. Due to combinations between static and dynamic amplifiers, it can achieve both PVT robustness and high energy efficiency. The proposed technique is applied to an 11bit 360MS/s pipelined SAR ADC on 28nm CMOS. It achieves 58 dB SNDR and 78 dB SFDR, while consuming 3.9mW. This results in a competitive energy-efficiency FoM: $FoM_W = 14fJ/step$ and $FoM_S = 164.6dB$.

Fig. 1 shows a schematic of the residue amplifier during the amplification phase. A closed-loop amplification provides robust interstage gain against PVT variations. However, the amplifier's finite gain and bandwidth make the virtual ground V_G non-ideal and cause errors in the feedback current I_{FB}, resulting in a gain error G_E and a bandwidth limitation determined by the feedback factor β ($=C_F / (C_S+C_F)$). Therefore, considering small β due to large interstage gain, the amplifier should have large open-loop gain (A_{OL}) and sufficient transconductance G_m to achieve target accuracy and speed, resulting in large power consumption. On the other hand, when NC ($=-kC_S$) is applied to V_G, the required I_{FB} is reduced by a compensation current I_{NC} generated by the NC. Since the NC cancels the C_S due to its parallel connection, the feedback factor β_{NC} ($=C_F/(C_S+C_F-kC_S)$) at an ideal compensation ($k=1$) becomes 1. Note that the NC does not change the closed-loop gain of the amplifier ($=C_S/C_F$). Assuming $k=1$ and $C_S/C_F=16$, β_{NC} is changed to 1 instead of 1/17. Accordingly, the NC enables the gain error $G_{E,NC}$ of the amplifier to be $1/A_{OL}$. If the NC fully compensates for the C_S, the gain error of the NC-assisted amplifier $G_{E,NC}$ can be reduced to 5.8% compared to the amplifier without NC. Furthermore, the bandwidth of the NC-assisted amplifier increases up to the amplifier's UGB, which means that the amplifier's G_m can be designed to be 17x smaller to achieve the same bandwidth.

There are three design challenges in implementing the NC. First, the speed of NC should be much higher than that of the main amplifier to achieve the bandwidth extension, which should also be done in an energy-efficient manner. Second, the NC value should be accurate enough to ensure its stability. Third, the NC's operating point should be decided immediately after its connection. Otherwise, the NC slow down the overall speed of the amplifier.

The proposed NC can address these challenges. Fig. 2(a) shows a schematic of the proposed dynamic NC, consisting of a dynamic bias and an NC core implemented with a capacitively-degenerated cross-coupled latch. During the sampling phase, the reservoir capacitor C_B is charged to supply voltage. During the amplification phase, the C_B is connected to the NC core, and its outputs are connected to V_G (V_{GP} and V_{GN}). The speed of the NC-assisted amplifier is determined by both G_m of the main amplifier and $G_{m,NC}$ of the NC. If $G_{m,NC} >> G_m$, both BW extension and gain compensation can be achieved, but when $G_{m,NC} \approx G_m$ or $G_{m,NC} < G_m$, only the gain compensation is possible. Assuming that a static NC is used, the required $G_{m,NC}$ is about 40mS to reach the target speed (<1ns), as shown in Fig. 2(b).

consuming significant power. However, by using the dynamic bias with C_B, the proposed NC can achieve a large $G_{m,NC}$ at the start of amplification (①) and a small one at steady-state (②). Simulation results show that the average $G_{m,NC}$ takes only 10mS to achieve the same speed, meaning the dynamic NC is about 4x more efficient than its static counterpart. In addition, thanks to C_B, the NC achieves a good CMRR without an explicit CMFB, and the NC's operating point is set very quickly after connection.

The NC's output impedance Z_N is represented as a series connection of two negative resistors ($-R_{NC}/2$) and a negative capacitor ($-2C$) (Fig.2). A matched NC value ($-2C$) to C_S ensures the stability of the amplifier. The parasitic negative resistors ($-R_{NC}/2$), however, reduce the bandwidth of the amplifier and deteriorate stability. An effective way to eliminate this is to insert matched resistors in signal path, as shown in Fig. 3(a). This can be realized by using on-resistance R_{ON} of connection switch with C_S. Fig. 3(b) shows the step response of the amplifier versus the degree of compensation α. For $\alpha=1$, the amplifier can reach the maximum bandwidth and is also stable. For $\alpha >1$, however, the amplifier becomes unstable. To maintain the stability, α is chosen to be 0.9 (Fig. 3(c)). The robustness against R_{ON} mismatch is confirmed by Monte-Carlo simulations (Fig. 3(d)).

The proposed amplifier is used for an 11bit 360MS/s pipelined SAR ADC, consisting of a 16x residue amplifier between a 5-bit coarse SAR and a 7-bit fine SAR with 1-bit redundancy (Fig. 4). The first sampling capacitor C_{S1} is chosen to be 0.5pF and the residue amplification is done within 1ns. The dynamic NC is implemented with design parameters: $C_B=8pF$, $k=0.93$, and $C=120fF$. A clock boost switch is used to realize the R_{ON}. The main amplifier has an A_{OL} of 50dB and a G_m of 4mS. The step response of the proposed amplifier shows clear improvements in both DC gain and bandwidth.

The prototype ADC is fabricated in a 28-nm CMOS process and occupies only 0.016mm^2 (Fig. 7). It consumes only 3.9mW from a 0.95V supply and the dynamic NC overhead is 13% of the total power consumption. Fig. 5(a) shows the static performance. The DNL is -0.47/+0.72 LSB and the INL is -0.96/+1.37 LSB. Fig. 5(b) shows the PSDs at a sampling frequency of 360MS/s. For a 19MHz input, the SNDR and SFDR are 59.5dB and 76.9dB, respectively. With a 163MHz input, the SNDR and SFDR are 58dB and 77.9dB, respectively. As shown in Fig. 5(c), both SNDR and SFDR are well maintained for different sampling rates from 90 to 360MS/s and different input frequencies up to 200MHz.

Fig. 6 shows the performance summary and comparison with prior works [1-3], [6]. Compared to SOTA works [1-3], this work achieves a very competitive energy-efficiency FoM (FoM$_W$ of 14fJ/step and FoM$_S$ of 164.6dB), without gain calibration while occupying one of the smallest area. Compared to [6], this work achieves a better FoM$_S$ by 8.6dB with an increased sampling rate.

Acknowledgement:

The authors thank the IDEC for support of IC fabrication.

References:

[1] K. Moon, et al., "A 9.1-ENOB 6-mW 10-Bit 500-MS/s Pipelined-SAR ADC With Current-Mode Residue Processing in 28-nm CMOS," in *IEEE JSSC*, vol. 54, no. 9, pp. 2532-2542, Sep. 2019.

[2] W. Jiang, et al., "A Single-Channel 14b 500 MS/s Pipelined-SAR ADC with Reference Ripple Mitigation Techniques and Adaptive-Biased Floating Inverter Amplifier," in *IEEE A-SSCC*, Nov. 2022, pp. 1-3.

[3] J. Lagos, et al., "A Single-Channel, 600-MS/s, 12-b, Ringamp-Based Pipelined ADC in 28-nm CMOS," in *IEEE JSSC*, vol. 54, no. 2, pp. 403-416, Feb. 2019.

[4] Y. Chae, et al., "A Negative R-Assisted Amplifier on the Virtual Ground and Its Applications," in *IEEE CICC*, April 2021, pp. 1-6.

[5] S. Jeon, et al., "A Parasitic Insensitive Catheter-Based Capacitive Force Sensor for Cardiovascular Diagnosis," in *IEEE TBCAS*, vol. 12, no. 4, pp. 812-823, Aug. 2018.

[6] W. Jang, et al., "A Pipeline ADC with Negative C-assisted SC Amplifier Canceling Gain Error and Nonlinearity," in *IEEE ESSCIRC*, Sep. 2022, pp. 317-320.

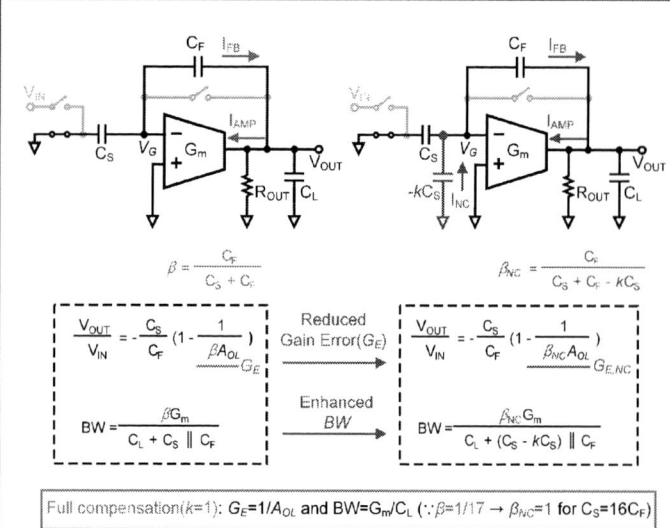

Fig. 1. Conventional residue amplifier vs. Proposed NC assisted residue amplifier.

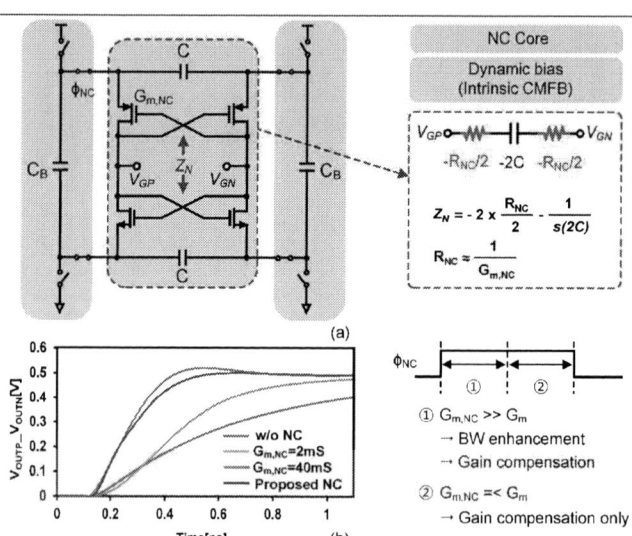

Fig. 2. (a) Schematic of the dynamic NC, (b) Transient response for different $G_{m,NC}$ (at G_m = 4mS).

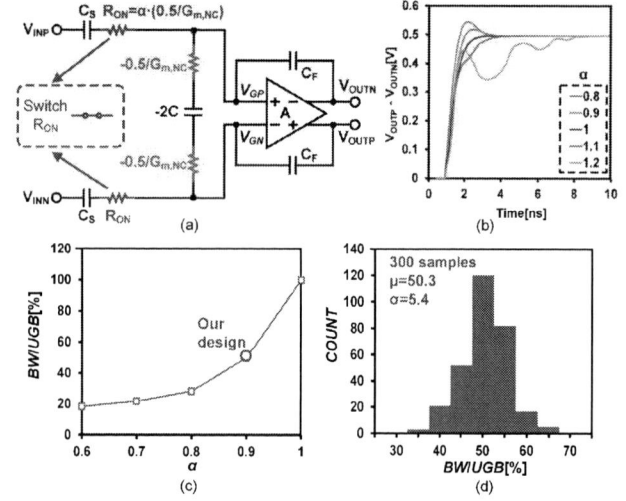

Fig. 3. (a) Schematic of residue amplifier width dynamic NC, (b) Step response with different α, (c) Simulated BW/UGB vs. α, and (d) Monte-Carlo simulation results (BW/UGB).

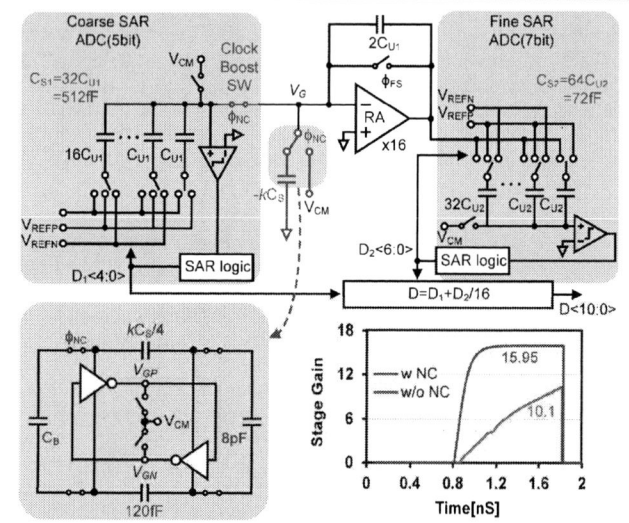

Fig. 4. Proposed pipelined SAR ADC with dynamic NC-assisted residue amplifier.

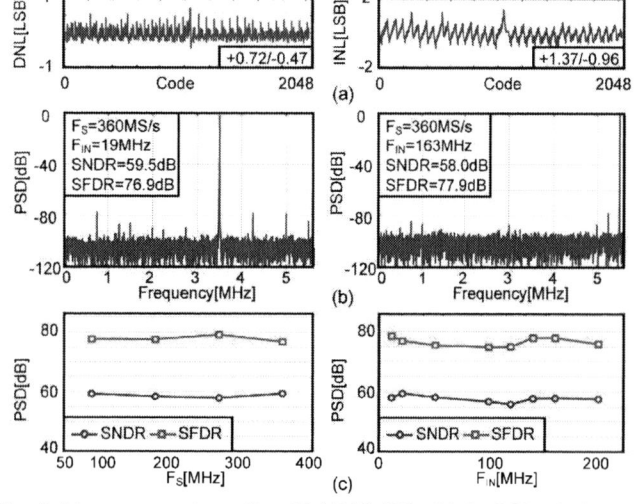

Fig. 5. Measurement results of (a) DNL/INL, (b) the PSDs decimated by 32, and (c) SNDR/SFDR vs F_S and F_{IN}.

	This Work	JSSC19 Moon[1]	A-SSCC22 Jiang[2]	JSSC19 Lagos[3]	ESSCIRC22 Jang[6]
Architecture	Pipelined SAR	Pipelined SAR	Pipelined SAR	Pipeline	Pipeline
Residue amplification	Closed-loop	Open-loop	Closed-loop	Closed-loop	Closed-loop
Amplification technique	Assisted with dynamic NC	G_m based amplifier	Adaptive Bias FIA	Ring amplifier	Assisted with static NC
Gain calibration	Not required	On-chip	Not required	Off-chip	Not required
Process [nm]	28	28	28	28	28
Sampling rate [MS/s]	360	500	500	600	320
Area [mm²]	0.016	0.015	0.018	0.621	0.302
Supply [V]	0.95	1	1.2/1	0.9	1
Power [mW]	3.9	6.0	6.3	14.5	11.3
SNDR [dB]	58.0	56.6	64.2	58.7	54.7
SFDR [dB]	77.9	69.2	80.6	72.4	71.8
FoM$_W$ [fJ/step]	14	21.7	9.6	34.4	79.6
FoM$_S$ [dB]	164.6	162.8	170.2	161.9	156

Fig. 6. Performance summary and comparison with prior works.

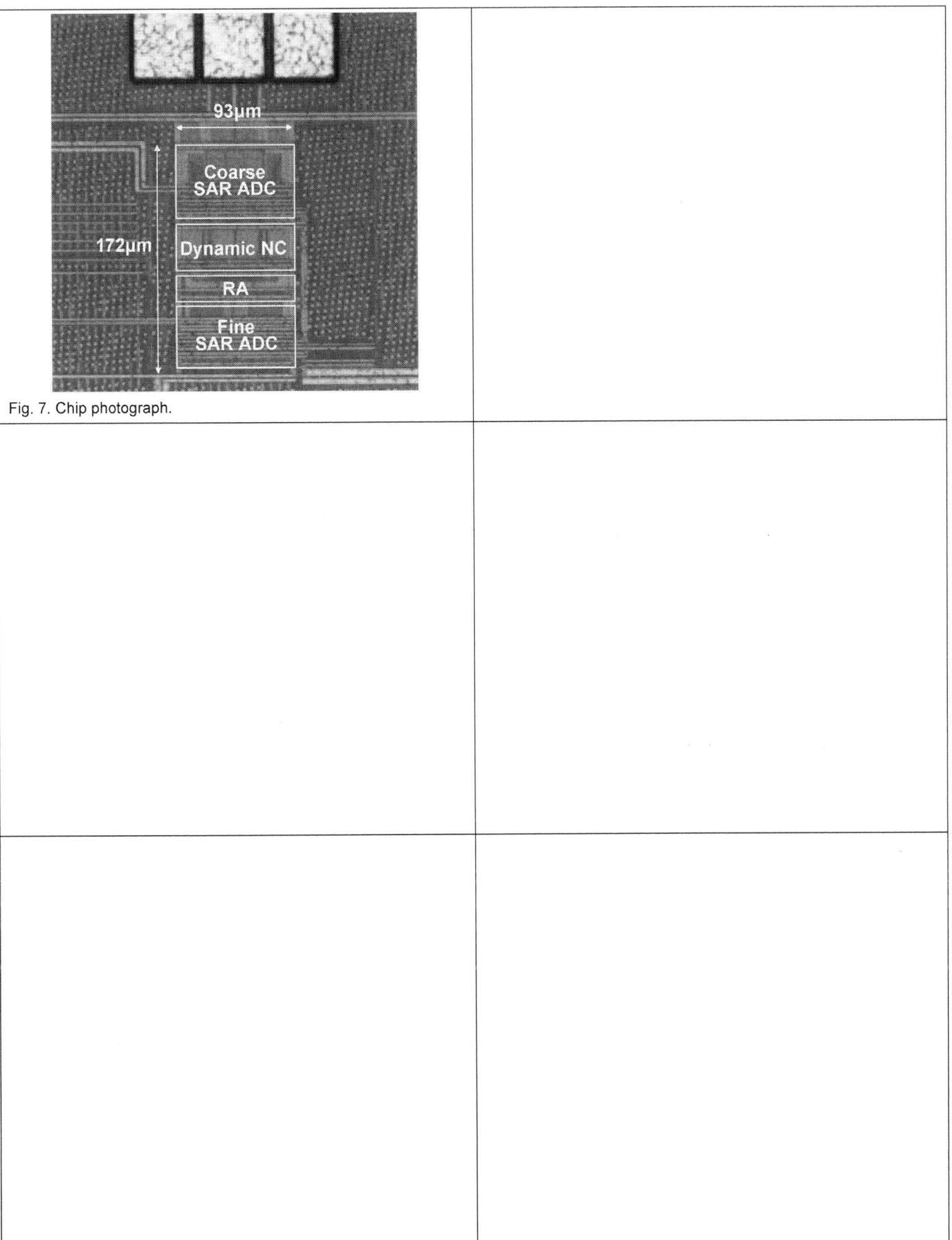

Fig. 7. Chip photograph.

A Jitter-Programmable Bang-Bang Phase-Locked Loop using PVT Invariant Stochastic Jitter Monitor

Yong-Jo Kim[1], Taekwang Jang[2], SeongHwan Cho[1]

[1]KAIST, Daejeon, Korea, [2]ETH Zurich, Zurich, Switzerland, Email: kimmuse35@kaist.ac.kr, chosta@kaist.ac.kr

Ring-oscillator based BBPLL is widely used in clock generators for its simplicity. However, satisfying the jitter requirement under various PVT conditions is not easy and thus worst case design is usually employed with many iterations of fabrication and testing, which leads to waste of power in normal conditions and delayed time-to-market. Although several works have tried to minimize the variation of jitter or PLL loop bandwidth [1, 2], the output jitter cannot be set as desired and thus worst case design and iterations are still necessary. In this work, we propose a BBPLL whose jitter can be set to a desired target value. It employs a stochastic jitter monitoring circuit (JMC) and automatic loop gain control (ALGC) that allows the PLL to optimize its power and bandwidth to achieve the target rms jitter.

The proposed BBPLL shown in Fig. 1 employs an ALGC and an on-chip PVT invariant stochastic JMC. The ALGC sets the optimum loop bandwidth so that the output jitter distribution is Gaussian [6] and the JMC controls the VCO power so that the BBPLL achieves the desired target jitter. Hence by running both the ALGC and JMC, the PLL can achieve the target rms jitter with optimal loop bandwidth and minimum power consumption. Note that unlike previous jitter monitors [3-5], the proposed JMC does not require foreground calibration or off-chip processing to obtain the output rms jitter.

When the jitter distribution is Gaussian, 16% of the jitter outcome is above the rms jitter. Hence if the occurrence of the output jitter (τ_{out}) that exceeds the target rms jitter (τ_{target}) is equal to 16%, then it means that the rms of the output jitter is equal to the target jitter. To find the occurrence of τ_{out} that is larger than τ_{target}, we use a time-to-voltage converter (TVC) with a comparator. In concept, pulses corresponding to τ_{out}, τ_{target} are created and converted to voltage for comparison. To implement this concept, a couple of challenges must be addressed. First, a PVT invariant τ_{target} must be available. To achieve this, we propose using the VCO period (T_{VCO}) which is PVT invariant when the PLL is in lock. By setting τ_{target} as a fractional value of the T_{VCO} (i.e., $\tau_{target} = D_{target}/M \times T_{VCO}$, where D_{target} and M are integers), a PVT invariant τ_{target} can be obtained. Second, a pulse corresponding to this fractional T_{VCO} must be created and converted to voltage. As the pulse width is very narrow, we exploit the following equivalence for implementation. Suppose that the TVC creates voltage by charging a capacitor (C_0) using a current source (I_0) during a certain input time. If the inputs to the TVC are $T_{VCO} + \tau_{out}$ and $T_{VCO} + \tau_{target}$, respectively, then the outputs can be expressed as,

$$V_{out} = (T_{VCO} + \tau_{out}) \times \frac{I_C}{C_0}, V_{target} = \left(T_{VCO} + \frac{D_{target}}{M}T_{VCO}\right) \times \frac{I_C}{C_0} \quad (1)$$

Note that an offset time of T_{VCO} is added to both τ_{out} and τ_{target} since jitter is typically in picoseconds and is too short for the circuits to handle. Since these two voltages will be compared, we can rearrange the terms and rewrite the equation $V_{out} \lessgtr V_{target}$ as,

$$(T_{VCO} + \tau_{out}) \times \frac{I_C}{(M+D_{target})C_0} \lessgtr T_{VCO} \times \frac{I_C}{MC_0} \quad (2)$$

Therefore, comparing τ_{target} to τ_{out} is equivalent to comparing the output voltages of the TVC that use input pulse and capacitor as ($T_{VCO} + \tau_{out}$, $(M+D_{target}) \cdot C_0$) and (T_{VCO}, $M \cdot C_0$). Thus, we can define τ_{target} by simply changing D_{target}/M, where the resolution of the τ_{target} is set by M. For example, to achieve a target jitter of 1ps for a 2GHz clock with a 0.2ps programmable jitter resolution, one should have D_{target} =5 and M=2500. Note that the proposed scheme is not affected by VTC non-linearity or PVT variation, as long as the transfer curve of the VTC is monotonic.

The schematic of the JMC is shown in Fig. 2. The pulse corresponding to $\tau_{out} + T_{VCO}$ is generated using phase detector and

T_{VCO} is obtained by windowing the output of the VCO. Next, they are alternately converted to voltages by pre-charging the load capacitors ($C_{out} = (M+D_{target}) \cdot C_0$, $C_{target} = M \cdot C_0$) and discharging them when the input pulses are high via M1. The converted voltages are compared and the output is counted to obtain the rms jitter. The stochastic JMC adjusts the bias current of the VCO shown in Fig. 2 and thus its power consumption to improve or degrade the output jitter.

The proposed JMC can suffer from comparator offset and capacitor mismatch. To solve this issue, an offset self-calibration loop is proposed as shown in Fig. 3. The basic idea is to run the JMC with the same input (T_{VCO}) and the same capacitance of $M \cdot C_0$ (i.e. $C_{out} = C_{target} = M \cdot C_0$). The self-calibration loop adjusts the capacitance C_{cal} so that the average comparator output becomes zero, hence removing the offset time error. Note that mismatch error can exist in the capacitor $D_{target} \cdot C_0$. However, its effect is small since M is much larger than D_{target}. The noise from the VCO, specifically its period jitter (i.e. variation in T_{VCO}) can deteriorate the accuracy of the JMC. However, since period jitter is much smaller than the instantaneous jitter (i.e. absolute jitter), it does not hurt the accuracy of the JMC.

The proposed jitter programmable BBPLL is fabricated in 28nm CMOS. Measured results are shown in Fig. 4, where it can be seen that the PLL achieves Gaussian distribution with the ALGC. The rms jitter changes from 1.85ps to 3.01ps when the power consumption of the VCO changes from 9.1mW to 5.6mW. The difference between the output rms jitter and the programmed target rms jitter is measured and the average difference is 0.13ps (4.9%) and the maximum is 0.26ps (11.3%), which is limited by the quantization levels of VCO's bias control code. The performance under different corner (TT, FF, SS), supply (1.1V ± 0.1V) and temperature (-20°C~85°C) is shown in Fig. 5 for different τ_{target}. It can be seen that the JMC reduces the variation of the output jitter so that those that do not meet the target jitter are within the target spec, and those that overachieve the target has its power reduced, thereby moving the output jitter closer to the target. For the overachievers whose jitter is lower than the target by more than 10%, the average power saving is 7.2% and the maximum is 15.5%. Among the 90 measurements, 33 results which did not meet the target all achieved the desired target by using the proposed JMC.

TABLE I compares the proposed work with the state-of-the-art ring oscillator-based BBPLLs. It achieves state-of-the-art FoM over PVT variations and it is the only work whose output jitter can be user-defined. Fig. 6 shows the die photo and power breakdown.

Acknowledgment

This work was supported by Institute of Information & Communications Technology Planning & Evaluation(IITP) grant funded by the Korea government(MSIT), (No. 2023-00262634, An Ultra-low Power On-chip Training AI Semiconductor for Keyword Spotting). Chip fabrication and EDA tool were supported by IDEC of Korea.

References:

[1] W. Wu, et al., "A 14nm Analog Sampling Fractional-N PLL with a Digital-to-Time Converter Range-Reduction Technique Achieving 80fs Integrated Jitter and 93fs at Near-Integer Channels," ISSCC, pp. 444-446, 2021.

[2] T. Jang, et al., "A 2.5ps 0.8-to-3.2GHz Bang-Bang Phase- and Frequency-Detector-Based All-Digital PLL with Noise Self-Adjustment," ISSCC, pp. 148-150, 2017.

[3] K. Niitsu, et al., "CMOS Circuits to Measure Timing Jitter Using a Self-Referenced Clock and a Cascaded Time Difference Amplifier with Duty-Cycle Compensation," IEEE JSSC, vol. 47, no. 11, pp. 2701-2710, 2012.

[4] M. -H. Chou, et al., "Embedded PLL Phase Noise Measurement Based on a PFD/CP MASH 1-1-1 ΔΣ Time-to-Digital Converter in 7nm CMOS," Symp. VLSI Circuits, pp. 1-2, June, 2020.

[5] K. Choo, et al., "A 0.02mm2 Fully Synthesizable Period-Jitter Sensor Using Stochastic TDC Without Reference Clock and Calibration in 10nm CMOS Technology," ISSCC, pp. 120-122, 2018.

[6] T. -K. Kuan, et al., "A Digital Bang-Bang Phase-Locked Loop with Automatic Loop Gain Control and Loop Latency Reduction," Symp. VLSI Circuits, pp. 1-2, June, 2020.

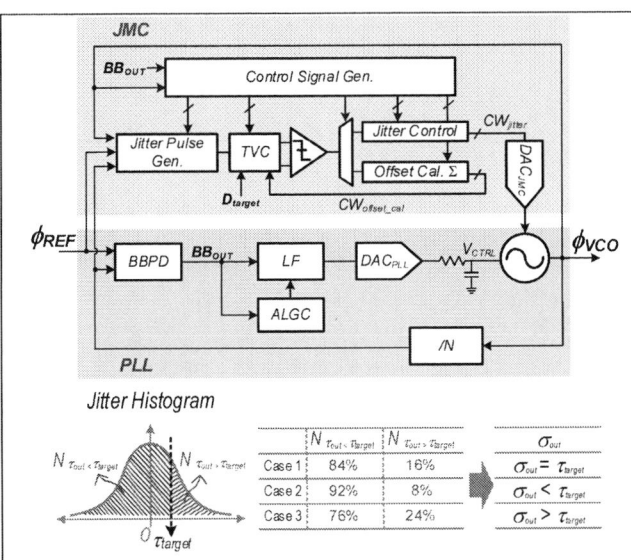

Fig. 1. Jitter programmable PLL and principle of the stochastic JMC.

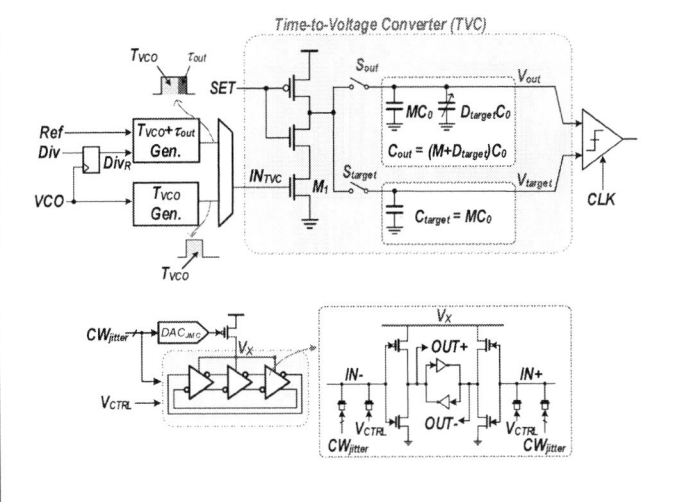

Fig. 2. Schematic of the JMC and ring-oscillator.

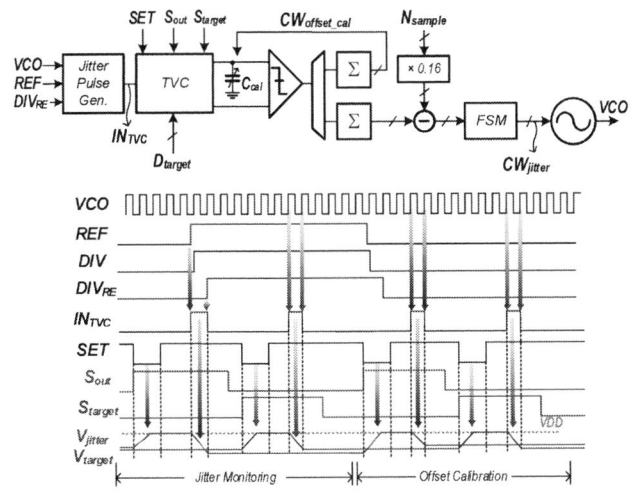

Fig. 3. Block diagram and timing of JMC and offset calibration.

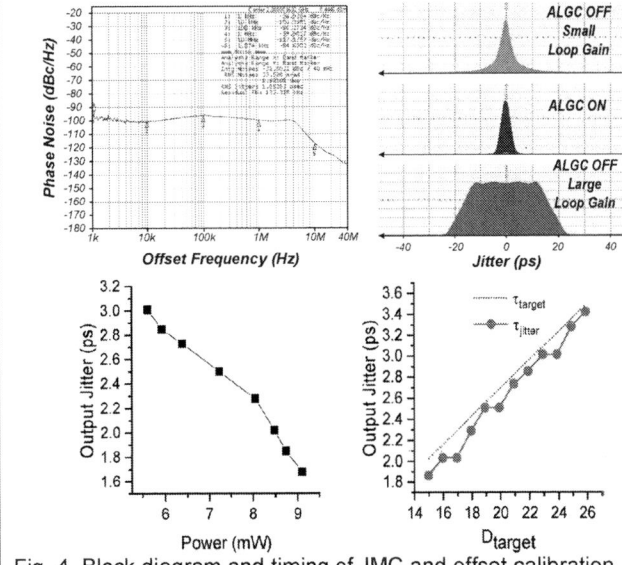

Fig. 4. Block diagram and timing of JMC and offset calibration.

Fig. 5. Measured jitter difference ($\sigma_{out} - \tau_{target}$) over PVT variations.

	Y-C. Huang ISSCC'14	M. Song ISSCC'15	C-W. Yeh ISSCC'16	T. Jang ISSCC'17	This work
Process	40nm	14nm FinFET	40nm	28nm	28nm
Architecture	BBPLL	BBPLL	BBPLL	BBPLL	BBPLL
Oscillator Type	Ring	Ring	Ring	Ring	Ring
Reference Frequency	26MHz	50MHz	200MHz	50MHz	36MHz
Output Frequency	2.4GHz	2GHz	3.2GHz	2.4GHz	2.88GHz
Division Value	92.3	48	16	48	80
Real-time Jitter Monitor	No	No	No	Yes	Yes
Jitter Programmability	No	No	No	No	Yes
Corner Sweep	SS, TT, FF	No	No	No	SS, TT, FF
Supply Sweep	0.95V ~1.45V	No	No	No	1.0V ~ 1.2V
Temp. Sweep	-20°C ~ 100°C	No	0°C ~ 80°C	No	-20°C ~ 85°C
Integrated RMS Jitter (Integrated Range)	3.29ps (10k~40MHz)	18.8ps (100k~100MHz)	3.54ps (N/A)	2.5 ~ 15ps (100k~100MHz)	1.68 ~ 3.01ps (1k~40MHz)
Power [mW]	6.4	2.06	2.92	1.7 ~ 5.0	7.7 ~ 11.2
Figure of Merit* [dB]	-221.6	-211.4	-224.4	-214.2 ~ -225.1	-221.6 ~ -225.0
Area	0.013mm²	0.009mm²	0.0216mm²	0.049mm²	0.108mm² (0.035mm² **)

*FoM=20log(jitter/1s)+10log(P/1mW)   ** Active Area

Fig. 6. Summary and comparison of Ring-osc. based BBPLL

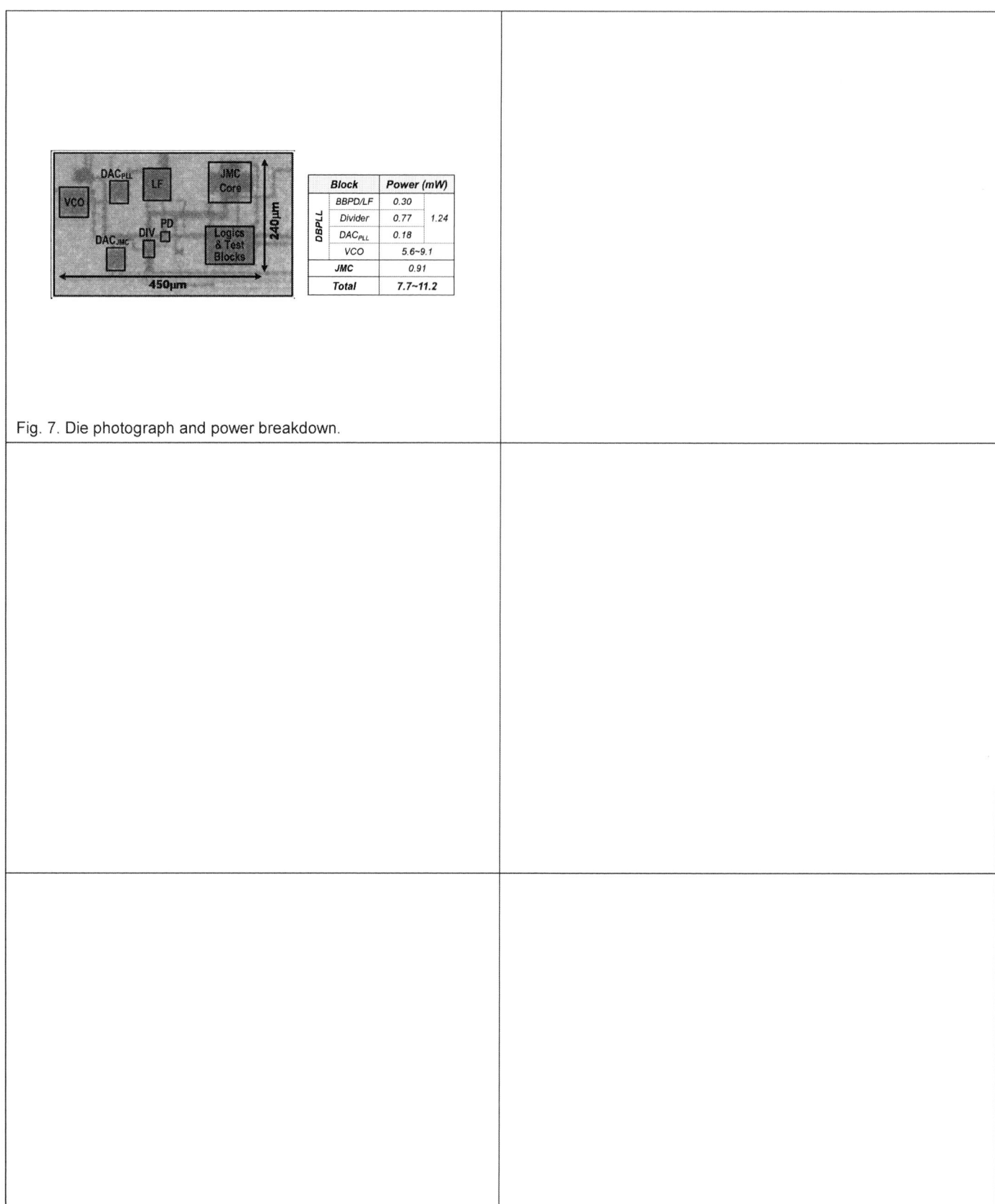

Fig. 7. Die photograph and power breakdown.

TEPD: A Compound Timing Detection of Both Data-Transition and Path-Activation for Reliable In-Situ Timing Error Detection and Correction in 28nm CMOS

Zhengguo Shen, Junyi Qian, Keran Li, Ziyu Li, Lishuo Deng, Weiwei Shan*

Southeast University, Nanjing, China

Conservative timing margins are reserved in digital IC to resist process, voltage, and temperature (PVT) variations, causing large power waste. It can be mitigated or eliminated by in-situ error detection and correction (EDAC) [1-3] based adaptive voltage/frequency scaling (AVFS), which monitors some selected critical paths and reduces the supply voltage until reaching the point of first failure (PoFF) with error-correction (Fig. 1 top). However, it endures inactive-path risk due to the limited monitoring of critical paths because illegal voltage scaling may be applied when none of the monitored paths are inactivated (Fig. 1 bottom left) so that there is no timing error for a certain time. Analyzing a keyword spotting (KWS) task in a neural network (NN) accelerator as an example, the majority data of the MAC output is quite small (Fig. 1 bottom right) that its most significant bits could be inactivated for hundreds of cycles before the accumulated data reaches a large value to activate the critical paths. Thus, reducing VDD is at risk on this occasion. Therefore, we propose a compound timing detection of both data transition and critical path activation together, ensuring reliable voltage scaling in EDAC systems. It can confirm whether the timing is violated and whether the path is activated at the same time, thus providing reliable information on both timing and activation status (Fig. 1 middle).

To detect the timing error and path-activation status at the same time, we propose a timing error and path-activation detector (TEPD) with only 15 transistors, with its schematic and layout shown in Fig. 2 (top). It consists of two XOR gates, an inverter, and a charge and discharge path composed of 3 transistors of M1, M2, and M3, where signals of "D" and "Q" are input/output of the endpoint flip-flop, and "mq" is the output of the intermediate master latch. Its working principle and timing diagrams are shown in Fig. 2 (bottom):

(1) Mode 1 in the negative clock phase, TEPD monitors the activation status. If there is a data D flipping before setup time, Q stays unchanged (and is different from D) until after the clock rising edge, thus generating a positive pulse in the "Active" signal by performing the XOR operation on D and Q, indicating the path is activated. Moreover, here M2 is a high-threshold transistor to minimize the leakage power.

(2) Mode 2 in the positive clock phase is to detect whether the data arrives late. Here node N1 is first pre-charged to VDD and the output "Error" is reset to "0" during the negative clock phase. Then, in the positive clock phase for error detection, when D arrives late (e.g., from "0" to "1"), while "mq" from the main latch of the flip-flop remains low as being latched, the output of XOR1 turns high. Thus, M2 is turned on to generate a high-level "Error" signal through the inverter. Monte-Carlo simulation shows its working reliability at a near-threshold voltage (NTV). The dynamic power and leakage power are 25.66nW and 0.185nW respectively at TT, 0.55V, and 25°C, which are comparably low as compared to the state-of-the-art (SoTA) work (Fig. 2 bottom).

An in-situ error-correcting flip-flop (ECFF) is proposed to correct the occasional timing, based on a DFF and a MUX inserted between the master and slave latches, as shown in Fig. 3 (top). The error signal from TEPD controls the MUX to make the ECFF to be either a regular flip-flop or a high-voltage-transparent latch. Here the MUX is designed based on transmission gates with only 6 transistors to minimize its circuit area. The working principles of ECFF are illustrated in Fig. 3 (bottom left) with two stages. At stage 1 when there is no timing violation, the MUX selects the "mq" of the master latch to pass it to the slave latch. At stage 2, when a timing error occurs that "Error" equals "1", the MUX passes the D data straight to the output Q. At this time, it is equivalent to a latch to utilize the time-borrowing feature to allow the late-arriving signal to pass. Compared with the standard flip-flop, only 6 transistors are added, and its delay, dynamic power, and static power only increase by 15%, 11%, and 14% respectively, as shown in Fig. 3 (bottom right). Compared with latch-based EDACs [1], our circuit is reliable and compatible with the digital circuit design flow by avoiding the use of end-point latches. Compared with other dynamic flip-flop conversion circuits using two clocks and complex logic in flip-flop [4], our circuit avoids the extra clock frequency in the slave latch and alleviates the design complexity. Note that our ECFF introduces the short path (SP) issue, and it is solved by using clock-controlled transmission gate (CTG) based short-path padding [5], with 1654 CTGs inserted in the short paths, corresponding to only 1.2% area overhead.

Fig. 4 (top) shows our AVFS system applied on a NN accelerator for KWS [6], where the critical paths lie in the 48 Processing Elements (PEs) which are reused in different NN layers. Here 289 flip-flops are replaced by TEPDs and ECFFs with an insertion rate of 12.7% among a total of 2269 endpoint flip-flops. As shown in Fig. 4 (middle), the generated "Active" and "Error" signals are clustered by two dynamic-OR trees respectively to accelerate the collection of all timing alarm signals. The hybrid AVFS strategy mainly copes with two conditions (Fig. 4 middle left). First, it counts the total timing errors in a certain time to tune the voltage up if there are continuous/frequent errors, otherwise, the occasional error is corrected by the ECFFs directly. Second, it counts the cycles of no-error conditions and tunes the voltage down by an off-chip DC-DC converter. Therefore, this AVFS system pushes the operating point down to V_{min} reliably for either lower power or higher frequency, avoiding the risk of unauthentic VDD scaling down caused by inactivated paths.

Fabricated in a 28nm CMOS process, our TEPD-based AVFS chip operates at a wide voltage down to NTV with a core size of 0.276 mm^2. Its die photo and test platform are shown in Fig. 6. Here the test platform includes a PCB with a socket for testing chips and FPGA for data/weight load, a power supply and measurement unit, an oscilloscope, and a high/low-temperature chamber. We test 40 chips for the TEPD function and then AVFS gains. Their operating conditions range from 0.54 V ~ 0.9 V at 10 MHz ~ 525 MHz when the function is right. Fig. 5 shows the measured waveforms of the "clk", "Error", "Active" and the related Voltage-tuning signals. Here "Votage_Up" is generated when the continuous error occurs, regardless of the activation status since a detected timing error ensures the path is activated. But a reliable "Voltage_Down" is generated only when no-error for a long period and path-activated in this period. As for the AVFS gains (Fig. 5 bottom), our chip reduces 37~52% power consumption as compared to a margined baseline chip with fixed voltages ranging from 0.65~0.9V at room temperature. On the other hand, it may also obtain a frequency gain of up to 56~123% as compared to the fixed frequency baselines over a wide voltage range.

Compared with SoTA EDAC methods [1][5][7][8] illustrated in Figure 6, our TEPD is the first approach that takes into account path activation detection for ensuring reliable AVFS tuning in EDAC systems, with the best power gains of up to 52% at NTV. It operates effectively across a wide-voltage-range, and the ECFF ensures the compatibility of regular digital design flow and endures no accuracy loss for NN tasks. Our method effectively addresses the risk of unauthentic VDD scaling resulting from the monitoring of inactivated paths, thus enabling safe AVFS tuning for resilient circuits.

Acknowledgment:

This work was supported in part by the National Key R&D Program of China under Grant 2022YFB4400600.

References:

[1] Y. Zhang et al., "iRazor: Current-based error detection and correction scheme for PVT variation in 40-nm ARM Cortex-R4 processor," JSSC, 2018:619-631.

[2] Z. Shen et al., "Beyond eliminating timing margin: an efficient and reliable negative margin timing error detection for neural network accelerator without accuracy loss," JSSC, 2022:1462-1471.

[3] H. Zhang et al., "A DFT-compatible in-situ timing error detection and correction structure featuring low area and test overhead," TCAD, 2023:1015-1028.

[4] M. Ahmadi et al., "A dynamic timing error avoidance technique using prediction logic in high-performance designs," TVLSI, 2019:734-737.

979-8-3503-3004-5/23 $31.00 © 2023 IEEE

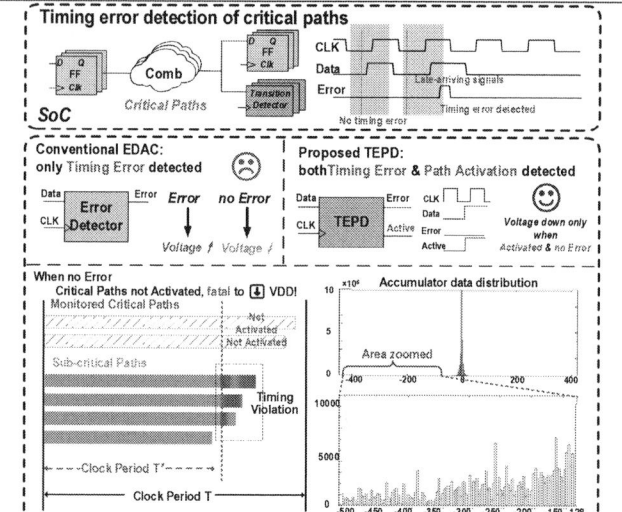

Fig. 1. Problem of conventional EDAC critical path monitoring method and our proposed TEPD to detect both timing error detection and path activation for safe AVFS adjustment.

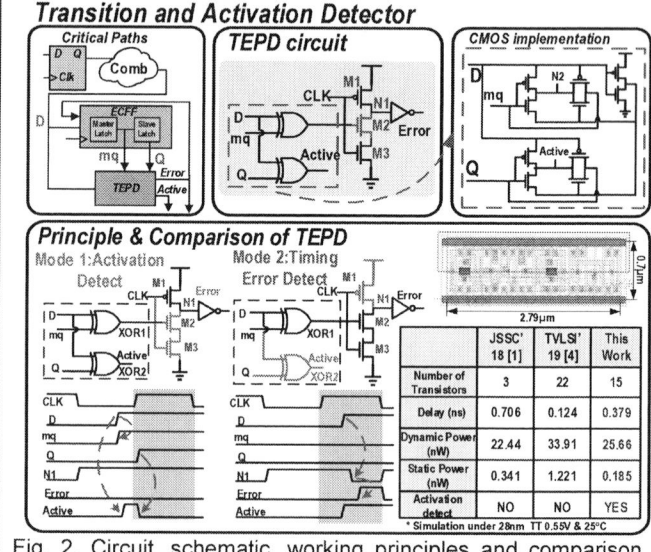

Fig. 2. Circuit, schematic, working principles and comparison of proposed Transition and Activation Detector (TEPD).

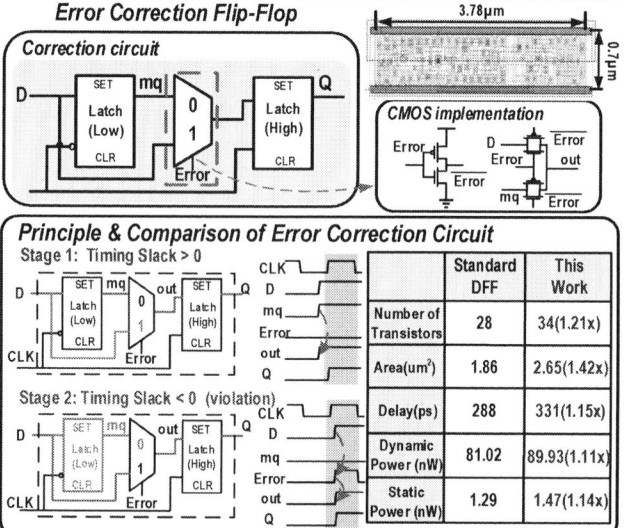

Fig. 3. Circuit, schematic, working principles, and comparison of proposed Error Correction Flip-flop (ECFF) at TT, 25°C.

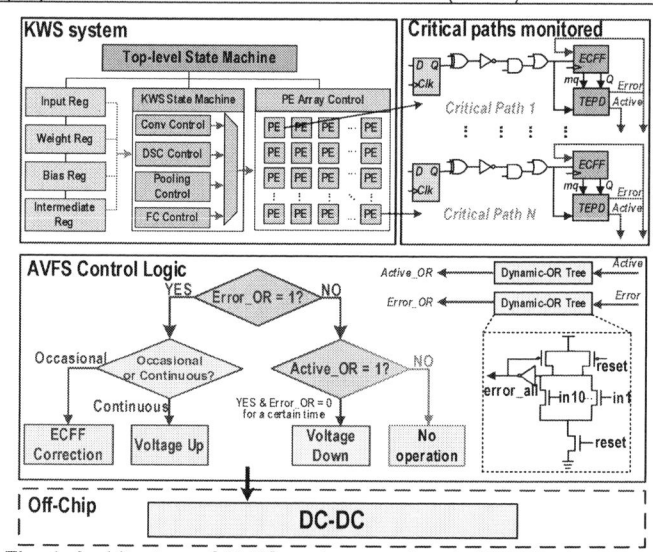

Fig. 4. Architecture of AVFS system consisted of TEPD, the NN accelerator, ECFF, Dynamic-OR, and the AVFS control logic.

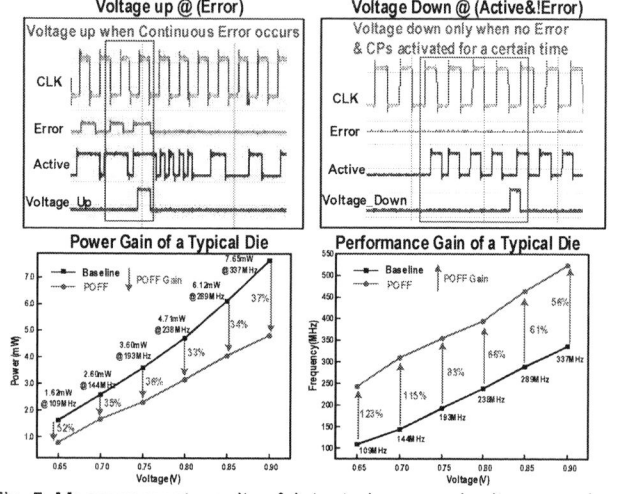

Fig. 5. Measurement results of detected error and voltage-up signals (top left); Long time no error with critical paths activated and voltage-up (top right); Power and frequency gain of a typical die as compared to baseline across 0.65V~0.9V.

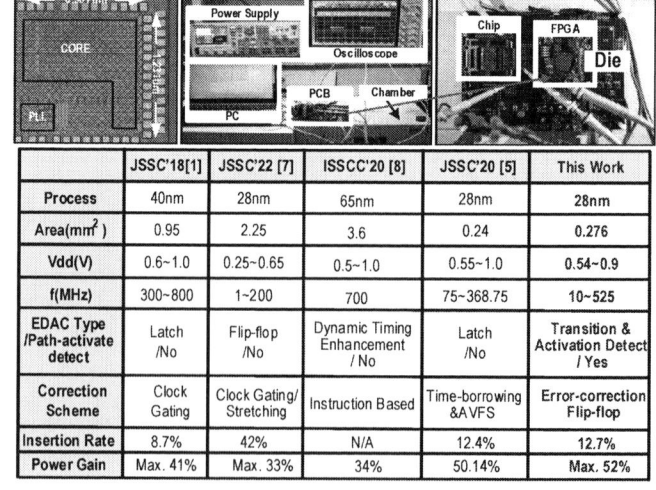

Fig. 6. Testing platform, PCB board, measurement results, and comparisons with the state-of-the-art work.

Additional References:

[5] W. Shan et al., "TG-SPP: A one-transmission-gate short-path padding for wide-voltage-range resilient circuits in 28-nm CMOS," JSSC, 2020:1422-1436.

[6] W. Shan et al., "A 510nW 0.41V low-memory low-computation keyword-spotting chip using serial FFT-based MFCC and binarized depthwise separable convolutional neural network in 28nm CMOS," ISSCC, 2020:230-232.

[7] R. Uytterhoeven and W. Dehaene, "Design margin reduction through completion detection in a 28-nm near-threshold DSP processor," JSSC, 2022:651-660.

[8] T. Jia et al., "A compute-adaptive elastic clock-chain technique with dynamic timing enhancement for 2D PE-array-based accelerators," ISSCC, 2020:482-484.

Flexible and Efficient Implementation of CRYSTALS-KYBER SIMD RISC-V Coprocessor Based on Customized Vector Instruction-Set Extension

Jiaming Zhang, Jiahao Lu, Dongsheng Liu, Aobo Li, Xiang Li, Shuo Yang, Ang Hu, Xuecheng Zou

Huazhong University of Science and Technology

With the development of quantum computers in recent years, the security of traditional public-key encryption algorithms is facing serious threats, and post-quantum cryptography (PQC) algorithms that can resist quantum computer attacks are urgently needed. CRYSTALS-KYBER as the finalized NIST key-encapsulation scheme, is continuously advancing the standardization process. The existing hardware implementations of Kyber mostly use compact architectures to pursue high speed and high performance with the cost of programmability, while most hardware-software co-designs suffer from low parallelism and performance. Aiming at flexibly and efficiently implementing the key encapsulation mechanism (KEM) of Kyber, this work presents a single instruction multiple data (SIMD) Kyber coprocessor that supports the RISC-V instruction-set. A reconfigurable polynomial and logic unit (PLU) is designed, which can accelerate all types of polynomial vector instruction operations, and a dynamic hardware scheduling strategy is proposed to enable different types of instructions to be executed parallelly, improving the coprocessor pipeline throughput. Implemented on the Ultrascale+ FPGA platform and evaluated under SMIC 40nm technology, the proposed coprocessor achieves the fastest computing speed with the lowest power consumption and 3.5×/6.2× improvement in FPGA/ASIC AT product efficiency.

Figure 1 shows the entire Kyber coprocessor with a scalar RISC-V core. The coprocessor can be tightly coupled into the RISC-V core pipeline alongside the load/store unit (LSU), sharing tightly coupled memory (TCM). Vector instruction-set extension are used in instruction fetching unit (IFU) to make the coprocessor suitable for fine-grained operations while providing programmability. The sample unit uses a single Keccak core and implements a 4-parallel rejection sampler and a 16-parallel central binomial distribution (CBD) sampler, requiring only about 80 and 16 clock cycles to complete a polynomial sampling of **A** and **s/e/r**. The poly unit consists of four groups of parity PLU as the core calculators to implement the polynomial vector operations. Finally, three dual-port RAMs are utilized to store the coefficients during the computation and one ROM to store the number theoretic transform (NTT) twiddle factors.

Figure 2 shows the designed hardware architecture of the PLU. The modular multiplier needs to consume one DSP resource to realize the multiplication of two coefficients and subsequently uses the divide-and-conquer method to modulo the product of coefficients. Compression and decompression of polynomials can be achieved by multiplexing DSPs at the same time. The proposed modular adder, modular subtractor and modular divider are implemented using a combination of carry save adder (CSA) and carry propagate adder (CPA). Three levels of registers are inserted into the modular multiplier and designs the PLU as a pipeline architecture, which can significantly improve the performance of the PLU and make it more suitable for vector instruction operations.

As shown in Figure 3, PLU_OE consists of a pair of PLUs that process odd and even vectors respectively. By controlling the selector, the PLU_OE can be reconfigured and implement different polynomial operations. The polynomial pointwise multiplication (PWM) operation uses the Karatsuba algorithm, which reduces the number of multiplications from 5 to 4 at the cost of extra two additions and one subtraction. Benefiting from the hardware structure of PLU_OE array, the clock cycles of vector PWM are reduced from 640 to 72, which is decreased by 88.75% compared to original time.

As shown in Figure 4, coefficients of the Kyber polynomial vector are stored in dual-port RAM with a parity sequential storage strategy, and the PLU_OE array reads two rows of coefficients at a time. To ensure pipelined operations, the VLEN in customized vector instruction-set is set to 96 to match the bandwidth of PLU_OE array. Taking the most time-consuming NTT as an example, by using this memory operation strategy, the delay of a 256-point NTT operation can be reduced by 87.5%. In order to improve the computational efficiency, an out-of-order instruction flow is designed in this work. The coprocessor fetches the next instruction immediately after the previous instruction is dispatched. This method allows the poly unit to be activated while the sample unit starts the next round of vector sampling. In order to avoid memory access conflict, this work proposes a dynamic hardware scheduling strategy: the poly unit uses two poly RAMs for ping-pong operation after completing the first round of fetching and computing the vectors from the sample RAM, and the sample unit will wait for the end of the first round before performing absorb and squeeze operations on the vectors in the sample RAM. This strategy improves the parallelism and throughput of the entire coprocessor while reducing the storage capacity.

Figure 5 shows the three-stage pipeline structure of the whole SIMD coprocessor. The RISC-V core fetches the instruction and sends scalar R-type instructions sequentially, then the coprocessor will customize these instructions into extended vector instruction and check the Read After Write (RAW) hazard between different types of instructions. Since Kyber is a Lattice-based algorithm with 4×4 as the largest operation matrix, and the number of basic polynomial coefficients is 256, the RAM of depth 256 can be divided into 8 vector addresses, and the vs1/vs2/vd operands of instructions point to these vector spaces, enabling the coprocessor to complete a polynomial vector operation with a single instruction.

Figure 6 compares this work with a series of Kyber1024 security-level implementations in Table I. This work is implemented on Xilinx UltraScale+ platform. Benefiting from the parallelism of vector instructions and the reconfigurability of PLU_OE array, this work achieves 5.5×, 3.5× and 3.3× speedups respectively compared to [1], [2] and [3], and realizes the highest operating frequency. A maximum 3.5× improvement in FPGA AT product efficiency is obtained in this work. Compared with other works' DC synthesis results, this work achieves the highest operating frequency with the lowest power consumption under SMIC 40nm technology, due to the dynamic hardware scheduling strategy. Since [4] only implements a scalar PQC accelerator with poor performance, this work improves 250× in ASIC AT product efficiency. Compared to [5], which is implemented based on scalar RISC-V processor, the improvement in AT reaches 6.2×, and to [1] is 7.5×. This application customized instruction-set coprocessor is programmable, and has a high degree of operational parallelism, allowing flexible and efficient implementation of KEM operations of Kyber.

Acknowledgements:

This work is supported by the National Key Research and Development Program of China (No. 2021YFA0715502), the National Natural Science Foundation of China (No. 61874163, 62104076, 6213400 2), the National Key Analog Integrated Circuit Laboratory Project of China (No. JCKY2021210C004), the Introduced Innovative R&D Team of Dongguan (No.201760712600139), and the Laboratory Open Fund of Beijing Smart-chip Microelectronics Technology Co. Ltd,. The corresponding author is Jiahao Lu. E-mail: lujiahaohust@foxmail.com

References:

[1] M. Bisheh-Niasar et al., "Instruction-Set Accelerated Implementation of CRYSTALS-Kyber." TCAS-I: Regular Papers, vol. 68, no. 11, pp. 4648-4659, Nov. 2021.

[2] W. Guo et al., "An Efficient Implementation of KYBER." TCAS-II: Express Briefs, vol. 69, no. 3, pp. 1562-1566, March. 2022.

[3] A. Li et al., "A Flexible Instruction-based Post-quantum Cryptographic Processor with Modulus Reconfigurable Arithmetic Unit for Module LWR&E," A-SSCC 2022:1-3.

[4] F. Tim et al. "RISQ-V: Tightly Coupled RISC-V Accelerators for Post-Quantum Cryptography. " TCHES 2020:446.

[5] Y. Zhao et al., "A High-Performance Domain-Specific Processor With Matrix Extension of RISC-V for Module-LWE Applications," TCAS-I: Regular Papers, vol. 69, no. 7, pp. 2871-2884, July. 2022.

[6] D. Kundi et al., "Ultra High-Speed Polynomial Multiplications for Lattice-Based Cryptography on FPGAs," Transactions on Emerging Topics in Computing, vol. 10, no. 4, pp. 1993-2005, Oct.-Dec. 2022.

979-8-3503-3004-5/23 $31.00 © 2023 IEEE

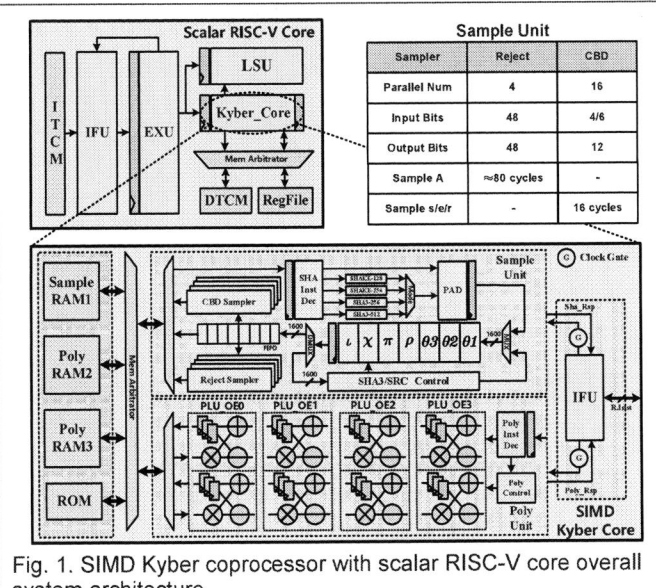

Fig. 1. SIMD Kyber coprocessor with scalar RISC-V core overall system architecture

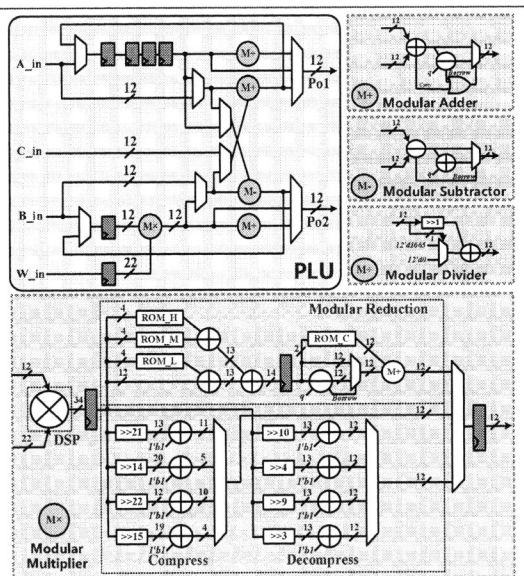

Fig. 2. Hardware architecture of polynomial and logic unit

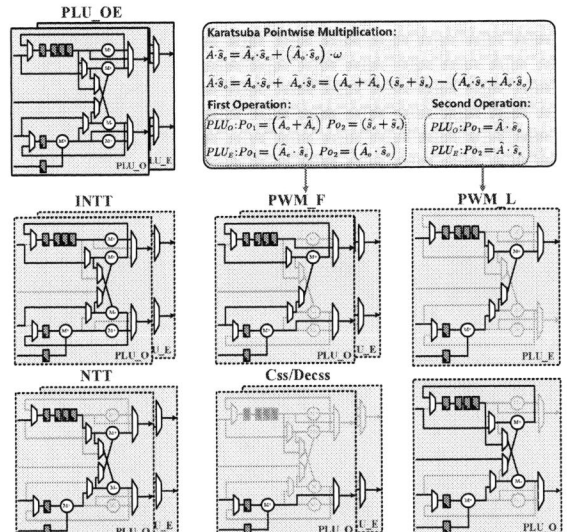

Fig. 3. PLU functional reconfiguration and Karatsuba polynomial pointwise multiplication

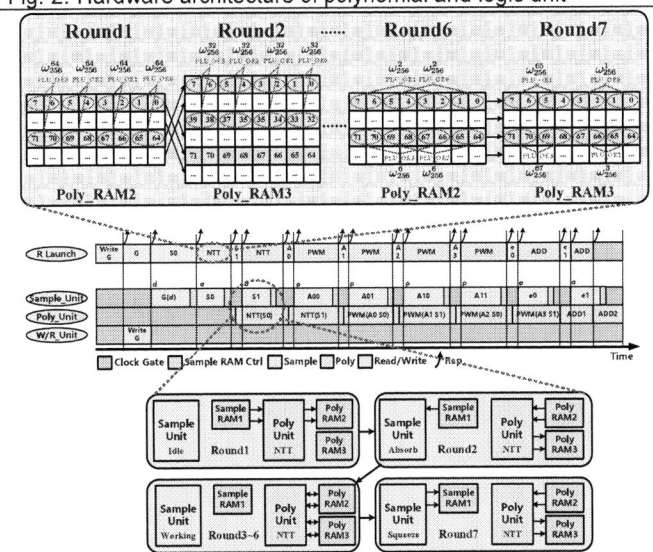

Fig. 4. Coefficient storage method and coprocessor dynamic hardware scheduling strategy

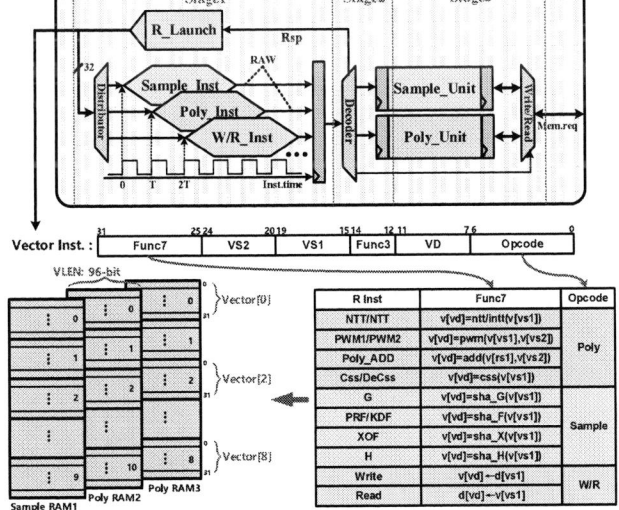

Fig. 5. Three-stage pipeline SIMD coprocessor and customized vector instruction set extension

Table I Results Comparison of FPGA Implementation for Kyber1024

	TCASI'21 [1]	TCASII'22 [2]	ASSCC'22 [3]	This work
Platform	Virtex-7	Artix-7	UltraScale+	UltraScale+
Frequency	156MHz	159MHz	250MHz	360MHz
Slices	5000	2300	3997	3202
DSPs	12	4	36	8
BRAMS	17	16	5	9
Keygen/ Encaps/ Decaps(µs)	64.1/ 89.7/ 115.4	49.1/ 52.8/ 66.0	-/ 70.4/ 87.3	11.1/ 17.5/ 20.3
AT*	1.00	0.98	1.35	0.28
Flexibility	Fixed Inst.	Fixed Inst.	Config. Inst.	Programmable Inst.

*FPGA area and time production (AT) = (DSP×100+BRAM×196+Slices) × Total Time (s) [6]

Table II Results Comparison of ASIC Synthesis for Kyber1024

	CHES'20 [4]	TCASI'21 [1]	TCASI'22 [5]	This work
Tech	65nm	65nm	28nm	40nm
Frequency	45MHz	200MHz	540MHz	660MHz
Logic(kGE)	21	104	623	83
SRAM(kB)	465	190	36.75	8
Total Time(µs)	26222.2	210.0	29.4	27.1
Total Energy(µJ)*	307.68	-	16.24	0.97
AT**	550.6	16.6	13.7	2.2

*Post-synthesis energy consumption is performed through PTPX
**ASIC area and time production (AT) = Logic Gate (kGE) × Total Time (ms) [5]

Fig. 6. Results comparison

979-8-3503-3004-5/23 $31.00 © 2023 IEEE

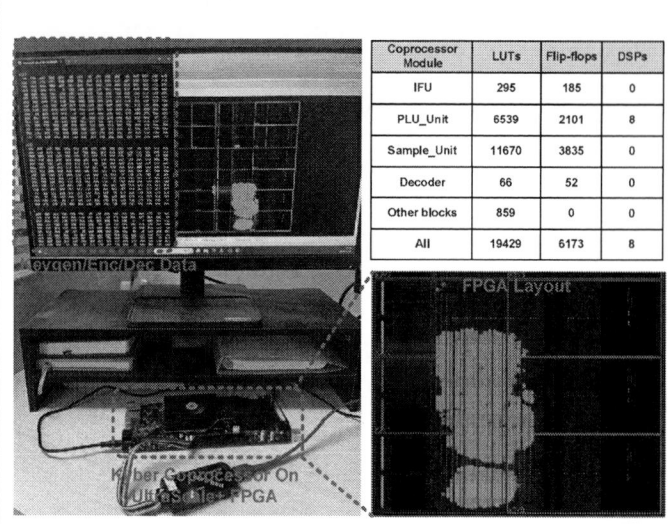

Coprocessor Module	LUTs	Flip-flops	DSPs
IFU	295	185	0
PLU_Unit	6539	2101	8
Sample_Unit	11670	3835	0
Decoder	66	52	0
Other blocks	859	0	0
All	19429	6173	8

Fig. 7. Experiment on UltraScale+ FPGA platform

A 1.2 V 2.3 µW 94.7 dB DR Delta-Sigma Modulator With Dynamic-Range Enhancement and Tri-Level CDAC

Cong Wei[1], Rongshan Wei[1*], Lijie Huang[1], Gongxing Huang[1], Jinze Lai[1], Zhichao Tan[2]

[1]Fuzhou University, Fuzhou, China

[2]Zhejiang University, Hangzhou, China

Email: wrs08@fzu.edu.cn

Discrete-time delta-sigma modulators (DTDSM) have excellent energy efficiency in low bandwidth, high dynamic range, and high precision applications, making them ideal for battery-powered IoT sensor devices with bandwidths within the kHz range. They achieve high signal-to-noise ratio (SNR) and dynamic range (DR) primarily through higher-order loops, higher OSR, multi-bit quantizers, and multi-stage noise-shaping (MASH). However, higher-order loops can decrease the maximum stable amplitude (MSA), higher OSR increases system power consumption, multi-bit quantizers bring additional dynamic element matching (DEM) overhead, and MASH leads to noise leakage problems. Tri-level quantization schemes combine the advantages of single-bit and multi-bit approaches [1], making them more suitable for high-energy efficiency applications. SAR quantizers have gained popularity in low-bandwidth DTDSM designs in recent years [2-4]. This paper presents a third-order single-loop DTDSM based on an asynchronous 1.5-bit SAR quantizer, tri-level feedback CDAC, and dynamic range enhancement (DRE) technique, eliminating the need for DEM overhead. With the assistance of DRE technology, the loop gain for small input amplitudes is enhanced. This allows the DTDSM to achieve 2.3 µW power consumption while achieving 94.7 dB DR and 92.2 dB SNDR at a 700 Hz bandwidth. As a result, the Schreier FoM for SNDR and DR is 177 dB and 179.5 dB, respectively.

Fig. 1 illustrates the block diagram of the proposed DTDSM, which is based on an asynchronous 1.5-bit SAR quantizer, tri-level feedback CDAC, and DRE technique. Compared to [1], the tri-level feedback CDAC employing common-mode voltage (V_{CM}) is multiplexed with the input sampling capacitor C_{S1}, which ensures absolute feedback linearity and eliminates the need for DEM overhead. When the input amplitude is less than -50 dBFS, choosing a smaller threshold voltage (V_{TH}) to increase the loop gain of the modulator improves its noise-shaping capability. Using a larger V_{TH} for large input amplitudes yields higher benefits, as it provides lower quantization noise and a larger MSA. Compared to a single-bit quantizer, it brings an approximate 6 dB improvement in peak SQNR. Hence, this paper proposes a DRE technique, which utilizes a digital decimation filter to monitor the amplitude of the input signal and provide feedback in the form of a digital code to control the dynamic switching of the 1.5-bit quantizer's V_{TH}. The normalized V_{TH} is set to 0.1 and 0.4, considering system linearity and coefficient-matching trade-offs. It should be noted that if the input sampling capacitor dominates the system noise, the enhancing capability of DRE technology will be attenuated. To simplify implementation, this DRE technique is realized off-chip.

The simplified circuit and timing diagram using a half-cycle modulator are shown in Fig. 2. In this circuit, it mainly consists of integrators, an asynchronous SAR quantizer based on variable threshold V_{TH}, a compensation circuit, DR enhancement section, and a tri-level feedback CDAC. Reference [5] presents the design of a cascode floating inverter amplifier (FIA) with low noise and high gain advantages. Although cascode FIA limits the output swing of the integrator, the reduced swing does not impose severe restrictions on DTDSM due to the embedded input feedforward path and scaling techniques. Therefore, it is suitable for the target application in this paper. To reduce chip area, the cascode FIA does not utilize a self-zeroing scheme [6] but instead adopts chopping techniques to suppress flicker noise and offset.

Passive summation is necessary to avoid the active summation overhead [4]. Compared to Flash ADC, the proposed DTDSM chooses asynchronous SAR ADC as the quantizer, which makes it

easier and more cost-effective to implement DRE techniques. The simplified circuit and timing are illustrated in Fig.3. During ∅2， the feedforward voltage is sampled to the bottom plate of the feedforward capacitor. During ∅S1, the feedforward summation operation is completed, and simultaneously, the bottom plate potential of capacitor C_1 is adjusted to set the threshold voltage to -V_{TH}. When the rising edge of ∅S11 arrives, it triggers the SAR logic to generate the comparator clock ∅C for the first comparison. After completing the first comparison, the asynchronous logic generates the ∅S clock to complete the +V_{TH} setting, followed by the second comparison. Therefore, the DRE technique based on variable V_{TH} only requires a scaling capacitor C_2 for C_1, with sizes of 15.5 fF and 62 fF, respectively, making the overhead of the DRE technique negligible. When the digital decimation filter detects an input signal amplitude lower than -50 dBFS, it generates an enable signal (DR_{EN}) to switch the working capacitor C_1 to C_2. It is worth noting that this quantization is insensitive to parasitic capacitance since the attenuation coefficients of the V_{TH} and the feedforward summation voltage are strictly consistent. Although a kickback voltage drop is introduced at V_{FP}, this can be compensated for by using a mirrored feedforward capacitor and CDAC array driven by opposite SAR logic. Since the compensation circuit only consumes switching power during SAR quantization, it is negligible compared to the overall power consumption of the DTDSM system. Furthermore, the impact of quantizer offset voltage can be ignored in the design due to the large margin of the 1.5-bit quantizer. The comparator proposed in [7] is utilized here to reduce the power consumption of the quantizer further.

This prototype DTDSM is fabricated in a 180 nm CMOS process and occupies a chip area of 0.27 mm² (Fig. 7). As shown in Fig. 4 (top), it achieves a DR of 94.7 dB with the DRE technique enabled. Fig.4 (bottom) shows the measured output spectrum with the input shorted. After opening the DRE, the loop gain increases, and the contribution of quantization noise is reduced. The measured spectrum of a -2.9 dBFS input at 197.3 Hz is shown in Fig 5 (top). With the chopper on and DRE off, the measured SNR, SNDR, and SFDR are 92.4 dB, 92.2 dB, and 105.8 dB, respectively, in a 700 Hz bandwidth. The power consumption breakdown of this DTDSM is shown in Fig. 5 (bottom). At a supply voltage of 1.2 V and a sampling rate of 128 kHz, the designed modulator consumes 2.3 µW (reference 0.17 µW, analog circuits 0.54 µW, digital circuits 1.6 µW). Fig. 6 summarizes the measured performance of this work and compares it with state-of-the-art micro-power high-resolution ADCs. The presented DTDSM achieves a FoM$_{S\_SNDR}$ of 177 dB and a FoM$_{S\_DR}$ of 179.5 dB.

References:

[1] M. G. Kim et al., "A 0.9 V 92 dB Double-Sampled Switched-RC Delta-Sigma Audio ADC," JSSC 2008:1195-1206.

[2] Y. Hu et al., "A 2.87µW 1kHz-BW 94.0dB-SNDR 2-0 MASH ADC Using FIA with Dynamic-Body-Biasing Assisted CLS Technique," I-SSCC 2022:410-412.

[3] S. Mehrotra et al., "A 590 µW, 106.6 dB SNDR, 24 kHz BW Continuous-Time Zoom ADC with a Noise-Shaping 4-bit SAR ADC," E-SSCIRC 2022:253-256.

[4] J. Han et al., "A 96dB Dynamic Range 2kHz Bandwidth 2nd Order Delta-Sigma Modulator Using Modified Feed-Forward Architecture with Delayed Feedback," TCAS-II 2021:1645-1649.

[5] R. S. A. Kumar et al., "Analysis and Design of a Discrete-Time Delta-Sigma Modulator Using a Cascoded Floating-Inverter-Based Dynamic Amplifier," JSSC 2022:3384-3395.

[6] Y. Zhao et al., "A 94.1 dB DR 4.1 nW/Hz Bandwidth/Power Scalable DTDSM for IoT Sensing Applications Based on Swing-Enhanced Floating Inverter Amplifiers," CICC 2021:1-2.

[7] H. S. Bindra et al., "A 1.2-V Dynamic Bias Latch-Type Comparator in 65-nm CMOS With 0.4-mV Input Noise," JSSC 2018:1902-1912.

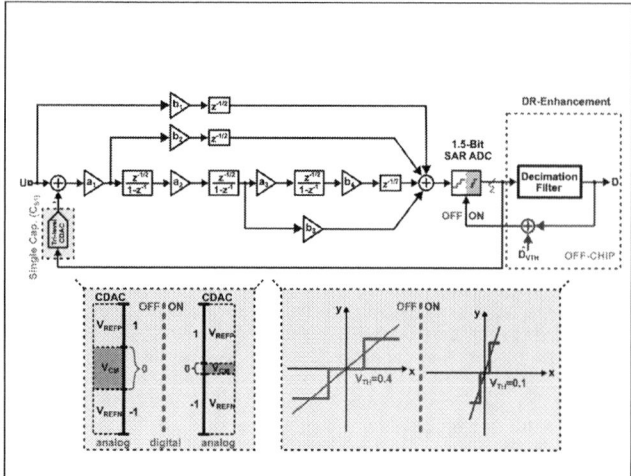

Fig. 1. The block diagram of the proposed DTDSM.

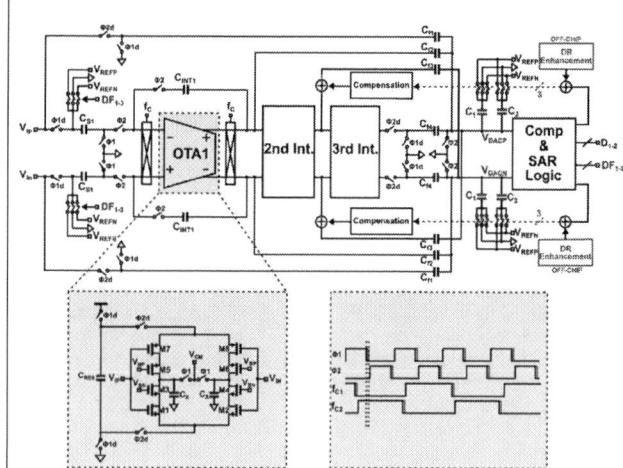

Fig. 2. Simpled circuit and timing diagram of the proposed DTDSM.

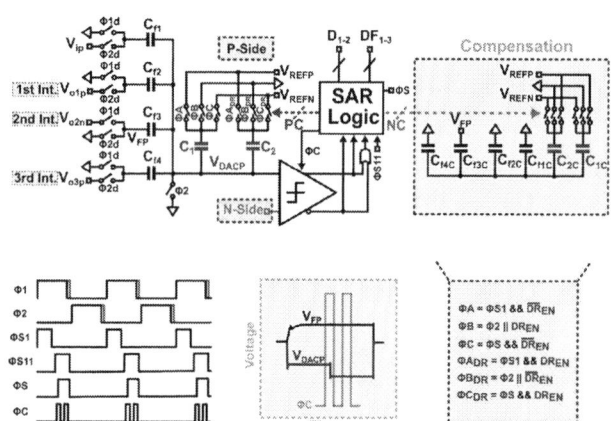

Fig. 3. Simplified circuit diagram of the asynchronous SAR quantizer and DRE technique.

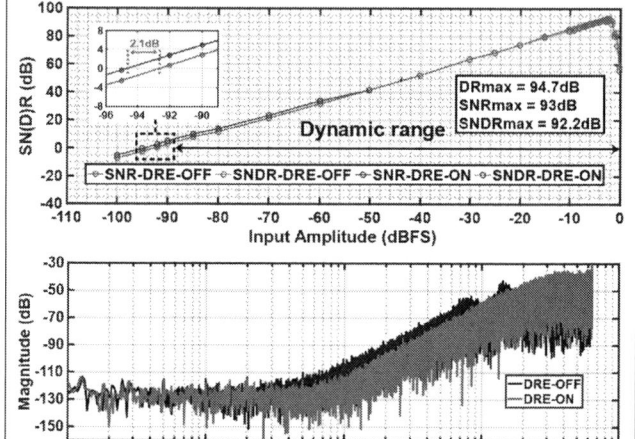

Fig. 4. Measured SNDR/SNR vs. input amplitudes (top); measured output spectrum with the input shorted (250k points) (bottom).

Fig. 5.Measure out spectrum (2^{16} FFT points) (top); measured power consumption when VDD=1.2 V (bottom).

	ISSCC-19 M.Konijnenburg	ISSCC-23 G.Kim	ISSCC-22 Y. Hu	ISSCC-21 C.Pochet	ISSCC-21 J. Huang	This work	
Architecture	NS-SAR	NS-SAR	Zoom	VCO	VCO	SDDT	
Process [nm]	55	180	55	65	65	180	
Fs [kHz]	32	128	250	200	64	128	
OSR	107	64	125	100	64	91	
BW [kHz]	0.15	1	1	1	0.5	0.7	
Supply [V]	1.3	1	1.2	1.2	0.8	1.2	
Power [μW]	0.9	9.29	2.87	5.8	1.68	2.3	
SNDR [dB]	78.8	95.3	94	92.3	94.2	92.2	
DR [dB]	NA	98.5	96.9	92.3	95.1	92.6 (DRE-OFF)	94.7 (DRE-ON)
FoM$_{S1}$ [dB]	161.1	179.4	179.4	174.7	178.9	177	
FoM$_{S2}$ [dB]	NA	181.9	182.3	174.7	179.8	177.4 (DRE-OFF)	179.5 (DRE-ON)

FoM$_{S1}$=SNDR+10*log10(BW/Power)
FoM$_{S2}$=DR+10*log10(BW/Power)

Fig. 6. Performance comparison with previous works.

979-8-3503-3004-5/23 $31.00 © 2023 IEEE   134

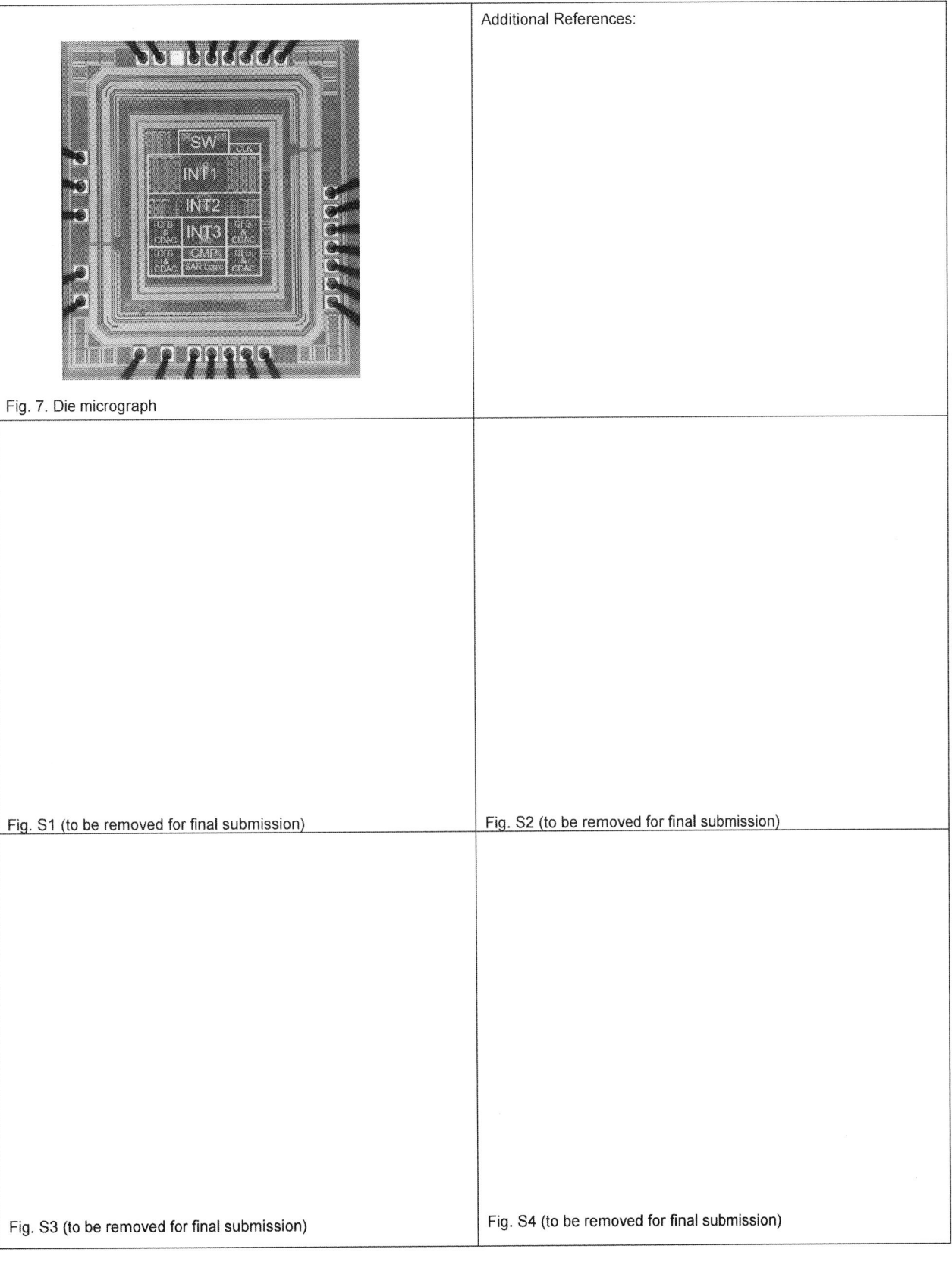

Fig. 7. Die micrograph

Additional References:

Fig. S1 (to be removed for final submission)

Fig. S2 (to be removed for final submission)

Fig. S3 (to be removed for final submission)

Fig. S4 (to be removed for final submission)

Integrated Circuit to Compensate Parasitic Leakage Component for WL Leakage Current in NAND Flash Memory

Bu-il Nam, Jayang Yoon, Kyunghea Lee, Sol Kim, Junhong Park, Chi-Weon Yoon, Eunkyoung Kim

Samsung Electronics, Hwasung, Korea

As high-tech industries such as 5G, artificial intelligence (AI), and high-performance computers develop, the importance of processing and managing large amounts of data is increasing. Accordingly, large-capacity and high-performance products are required for VNAND Flash Memory, which records and stores data. In order to satisfy these market demands, flash memory is increasing the number of layers of word lines (WL), which serve as data storage, every generation[1-2]. Due to the increased number of WL stacked stages to increase cell density, the space between adjacent WLs is reduced and the difficulty of the channel hole etch process is increasing. WL defects are increasing due to increased process difficulty and structural weakness. These defects cause minute leakage current on the WL and cause deterioration in product quality. In order to improve quality, electric die sorting (EDS) evaluation detects micro-leakage current on WL and screens defects, but it is difficult to accurately detect defects because it cannot reflect the influence of parasitic leakage current components of peripheral circuits. For accurate defect detection, parasitic leakage current components in the peripheral area must be removed. In this paper, we propose a new method to independently detect leakage current in the main cell area by removing parasitic components in the peripheral area that affect defect detection accuracy. In addition, Gm mismatch calibration technology is introduced to increase the accuracy of the defect detection circuit.

To detect WL defects, a detecting circuit using a comparator is used (Fig.1). First, a certain level is pre-charged using a generator, and then the switch is opened and left for a certain period of time (development time, t_{DEV}). After t_{DEV}, Sensing is performed to determine Good/Defective. If there is a WL defect, a voltage drop occurs due to leakage current and falls below the reference voltage (V_{REF}) that determines pass/fail, and a fail signal is generated in the comparator. Conversely, if there is no WL defect, the voltage drop hardly occurs and stays above the V_{REF} level, and the comparator generates a pass signal.

However, this method is not perfect and causes two kinds of errors (Fig.2(a)). The first error is that in the case of a chip with a large peripheral circuit leakage current, the input voltage (V_{IN}) becomes lower than the V_{REF} at the sensing timing even though there is no WL defect, and the good chip is judged as a defective chip. The second error is that in the case of chips with too small leakage current of the peripheral circuit, WL defects with small leakage current cannot be detected. To improve this, we propose a new method that can calculate and remove the parasitic leakage current components of different peripheral circuit areas for each chip (Fig.2(b)). This proposal method measures the current in the peripheral circuit area before determining the WL defect in the main cell area and corrects it when making the final decision. The first step separates and operates only the peripheral circuit and enables the counter when developing. Then, it updates the new t_{DEV} by compensating for the parasitic leakage of the peripheral circuit of the chip to the existing t_{DEV}. After that, if the entire area circuit operation is performed, the peripheral parasitic leakage current component is removed and the detection accuracy is improved.

Fig.3 shows the circuit diagram and flow chart. First, by using the first transistor of the part where main cell area and peripheral circuit area come into contact, independent control of only peripheral circuit area is possible. And by adding counter, resistor, and arithmetic and logical unit (ALU), t_{DEV} can be measured, calculated, and saved. The proposal method follows the following flow. First, turn off the transistor to disconnect the node on the main cell side so that only the peripheral circuit area operates. And after pre-charge, the counter operates in the developing section. And record the number of counting at the time when V_{OUT} of the comparator becomes '1' (the first time when V_{IN} is lower than V_{REF}). Then, by calculating the counted t_{DEV} value, an appropriate t_{DEV}_screen value is set. By turning on the transistor, all paths are opened and measured again to determine pass/fail.

Normally, the comparator input stage uses two NMOS, but the current amplification factor (g_m) of the comparator input stage NMOS can be affected by MOSFET mismatch characteristics such as variability in the manufacturing process and layout influence. In this paper, these factors were removed in the design process to increase the accuracy of the defect detection circuit. To this end, the g_m mismatch calibration technology that reduces the offset voltage by adjusting the g_m of the comparator input stage was introduced (Fig.4). As shown in Fig.4(a), a calibration circuit was implemented using the selected MOSFET and 8-bit Cal.code. Offset Calibration operation shorts the two inputs of the compactor, then sweeps Cal.code and monitors the V_{OUT} value. If the V_{OUT} value changes as the code is swept, it is determined that the offset is minimized, and after saving the Cal.code as a latch, the main operation proceeds using this Cal.code value. This g_m mismatch calibration technology can reduce the offset voltage from 56mV to 6.5mV(800nA →100nA).

Fig. 5(a) shows a method of independently extracting only the current in the peripheral area. Peripheral current can be calculated using t_{DEV}. In Fig. 5(b), the conventional method performs a batch screen by summing the WL and the peripheral area. However, since the proposed method can extract only the peripheral current, it can be calibrated to detect only the WL defects.

In Fig. 6, As a result of comparing the detection power of the WL defective of the proposal method, it was confirmed that it is perfectly possible to screen without two errors of over-kill and under-kill. WL defects were verified with physical failure analysis. After applying proposal method, a program-erase cycle evaluation was conducted to represent customer quality. As a result, the WL defects that occurred at 4000ppm when the conventional method was applied became 0ppm when the proposal method was applied (Fig. 7). The area slightly increased to 1.0006-A.U. A circuit that compensates for parasitic leakage and detects WL defects was applied to the 8th generation V-NAND, and Fig.8 is a photo micrograph of the die [3].

References:

[1] S. Choi et al., "A cell current compensation scheme for 3D NAND FLASH memory," 2015 IEEE Asian Solid-State Circuits Conference (A-SSCC), 2015, pp. 1-4.

[2] S. Kim et al., "Review of Semiconductor Flash Memory Devices for Material and Process Issues," Adv. Mater. 2022, 2200659.

[3] M. Kim et al., "A 1Tb 3b/Cell 8th-Generation 3D-NAND Flash Memory with 164MB/s Write Throughput and a 2.4Gb/s Interface," 2022 IEEE International Solid- State Circuits Conference (ISSCC), 2022, pp. 136-137.

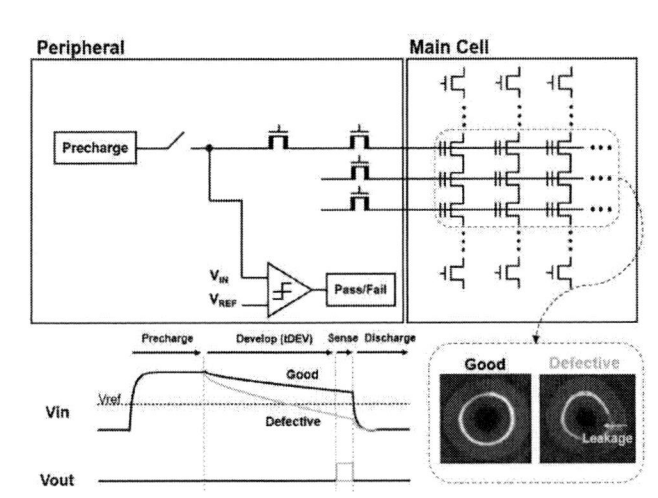

Fig.1. Conventional WL leakage current detector

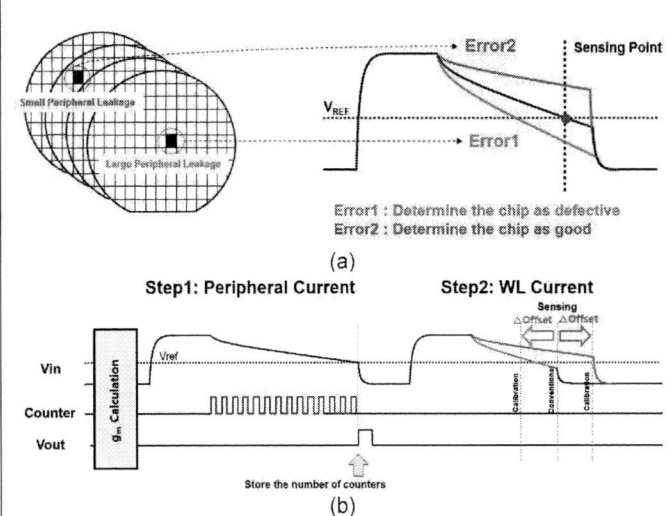

Fig.2. (a) Error case of conventional WL leakage current detector. (b) The concept of WL leakage current detector compensating for parasitic components

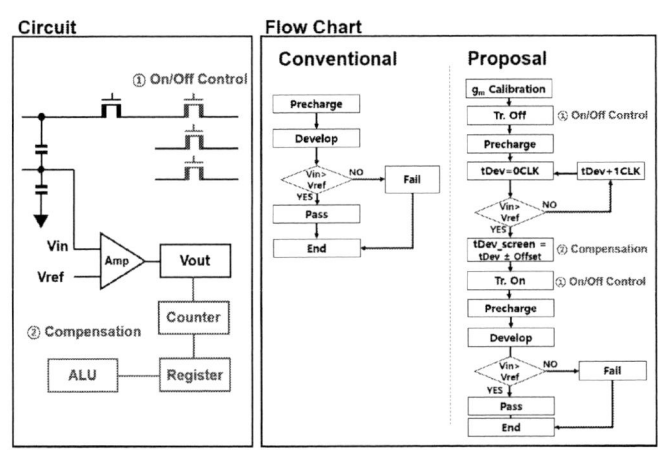

Fig.3. The schematic of WL leakage current detector compensating for parasitic components

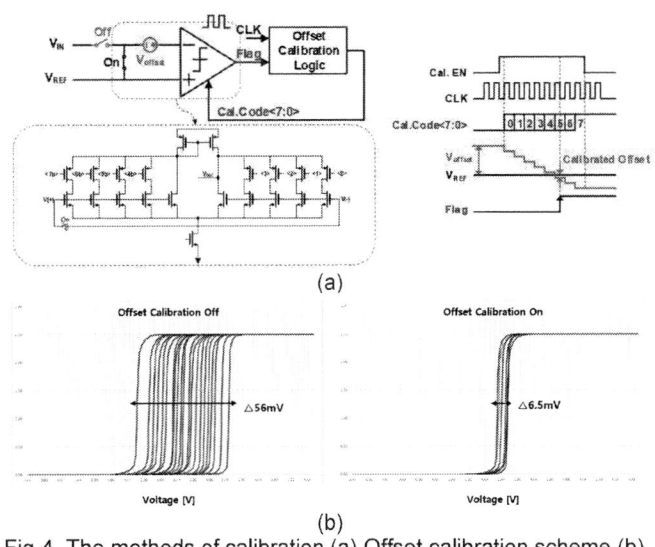

Fig.4. The methods of calibration (a) Offset calibration scheme (b) Results of offset calibration

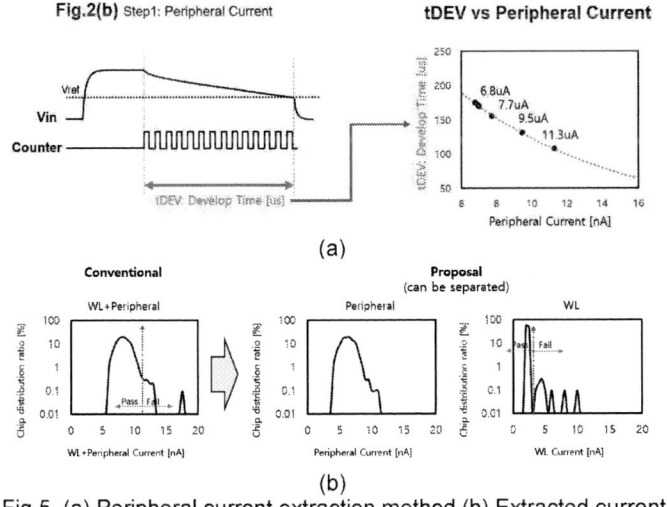

Fig.5. (a) Peripheral current extraction method (b) Extracted current distribution of conventional and proposal methods

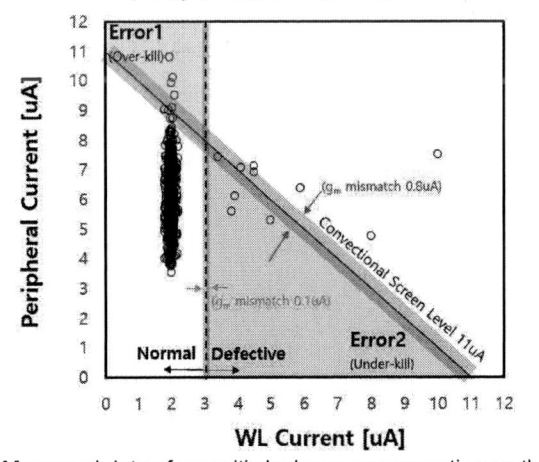

Fig.6. Measured data of parasitic leakage compensation methods

979-8-3503-3004-5/23 $31.00 © 2023 IEEE

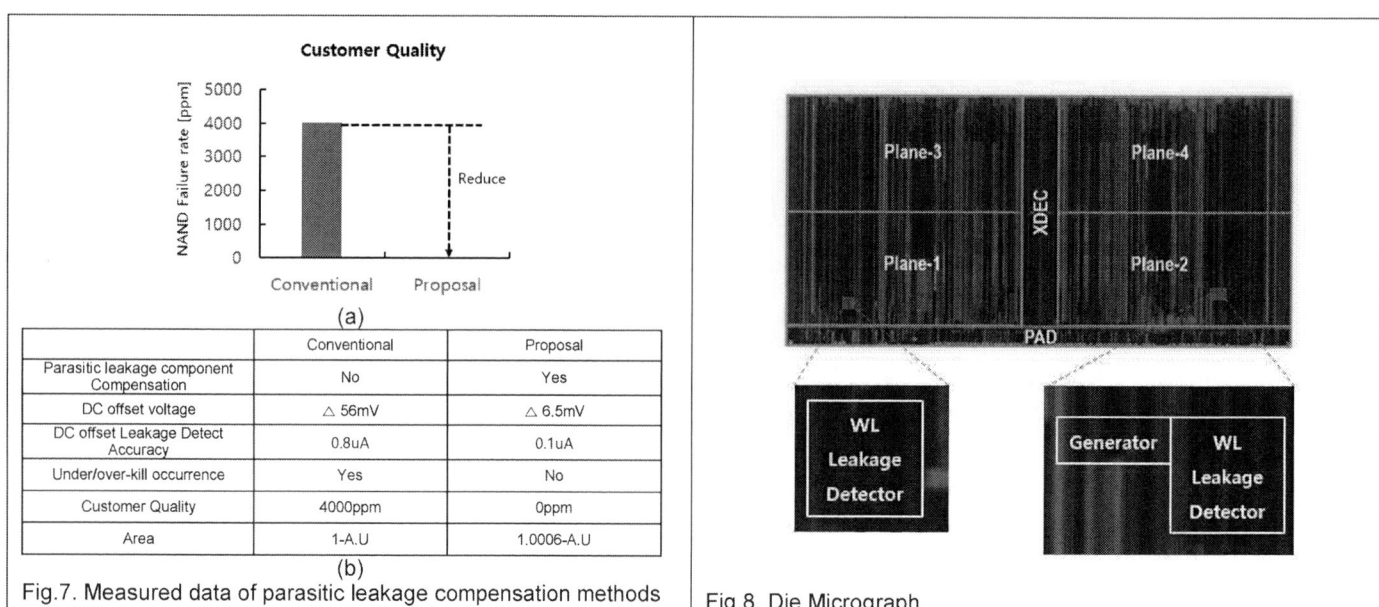

Customer Quality

	Conventional	Proposal
Parasitic leakage component Compensation	No	Yes
DC offset voltage	△ 56mV	△ 6.5mV
DC offset Leakage Detect Accuracy	0.8uA	0.1uA
Under/over-kill occurrence	Yes	No
Customer Quality	4000ppm	0ppm
Area	1-A.U	1.0006-A.U

(b)

Fig.7. Measured data of parasitic leakage compensation methods

Fig.8. Die Micrograph

A 12.75-to-16-GHz Spur-Jitter-Joint-Optimization SS-PLL Achieving -94.55-dBc Reference Spur, 31.9-fs Integrated Jitter and -260.1-dB FoM

Yixi Li[1, 2], Zhao Zhang[1, 2]*, Yong Chen[3], Xinyu Shen[1, 2], Zhaoyu Zhang[1, 2], Nan Qi[1, 2], Jian Liu[1, 2], Nanjian Wu[1, 2], Liyuan Liu[1, 2]

[1]Institute of Semiconductors, Chinese Academy of Sciences, China

[2]University of Chinese Academy of Sciences, China

[3]University of Macau, Macau, China

*E-mail: zhangzhao11@semi.ac.cn

The remote-radio-head transceivers have started digitalizing radio frequency technology, thanks to its unparalleled bandwidth and flexibility, by employing giga-samples-per-second data converters [1]. Direct synthesizing and sampling of these converters must demand excellent phase noise (PN) performance, namely, the simultaneous realization of low spur and low jitter. Sub-sampling phase-locked loop (SS-PLL) is the most promising candidate to realize this target under low-power consumption due to high phase-detection gain [2], as shown in Fig. 1(top). Yet, the periodically sub-sampling operation induces several non-idealities, e.g., binary-frequency-shift-keying (BFSK) effect, charge sharing, and clock feedthrough, seriously deteriorating the reference spur. In [2] and [3], placing an isolated buffer between the voltage-controlled oscillator (VCO) and the sub-sampling phase detector (SS-PD) can effectively alleviate the above issues, but the high-frequency buffer consumes extra power and adds noise contribution. Interestingly, the isolated SS-PDs (iSS-PDs) have been devised to improve spur and save power simultaneously [4], [5]. Yet, as the VCO frequency increases, the transistor size of the iSS-PD becomes a remarkable consideration, as illustrated in Fig. 1(bottom). The large size of isolated NMOS (M_S) and sampling switches can ensure enough iSS-PD gain to suppress the in-band noise, but the large parasitic capacitance of M_S will seriously degrade the reference spur. On the other hand, the small size of the iSS-PD is beneficial to enhance the isolation for the improved reference spur, but it is hard to lower the in-band noise due to the undersized iSS-PD gain.

This paper reports the joint-optimization technique of the spur-jitter performance for the SS-PLL. We develop a low-power low-jitter regulated narrow pulse generator (R-NPG) for supplying the stable narrow sampling pulse. High-swing Class-F VCO and low-spur iSS-PD are combined to co-optimize the low disturbance and high iSS-PD gain. Our SS-PLL prototype scores -94.55-dBc reference spur, 31.9-fs$_{rms}$ integrated jitter, and -260.1-dB figure of merit (FoM).

Fig. 2 outlines the top architecture of the proposed low-spur low-noise SS-PLL. We feed the clean reference signal into a low-power low-jitter R-NPG to create an ultra-narrow pulse with a low-jitter sampling edge, which samples the large-swing output of the Class-F VCO [6] directly by our low-spur iSS-PD. Without the usage of the VCO buffer, the extra noise contribution and power can be avoided. Fig. 2(bottom) details the proposed ultra-low-spur co-optimization scheme. Firstly, the iSS-PD [4] enhances the isolation between the VCO tank and the sampling capacitor (C_S). However, the BFSK effect still exists since the equivalent parasitic capacitors of the isolated M_S are periodically modulated at the reference frequency, called the drain voltage modulation. Thus, the auxiliary path with the dummy switches and dummy capacitor (C_S) [7] are merged into the iSS-PD to ensure the constant load of M_S during sampling and holding modes, greatly alleviating the BFSK effect. Specifically, the size of M_S and switches should be set small enough to reduce the kickback to the VCO tank. That is, the small size of the iSS-PD has a small parasitic capacitor, indicating the great suppression of the BFSK effect, clock feedthrough, and charge sharing. Paradoxically, this small-size active-buffer-based iSS-PD will seriously degrade its gain, thus worsening in-band PN. For the sake of high iSS-PD gain, we design the Class-F VCO to deliver a ~4xV_{DD} voltage swing at the gate node of M_S and provide a suitable voltage bias for M_S. Besides, we narrow the width (Δt) of the sampling pulse (V_{PUL}) to significantly reduce the impact of the periodic sub-sampling operation on the reference spur. Meanwhile, the clock feedthrough will also be eased because of lower fundamental frequency energy.

Fig. 3(top) outlines the conventional NPG. The sampling edge is disturbed by INBUF, inverter, delay cell, and and-gate in sequence, indicating that the in-band PN will go up due to the increased reference PN. To reduce this PN contribution, the size of the NPG should be large enough, thus consuming more power. Also, the width of the sampling pulse (V_{PUL}) is sensitive to the process, voltage, and temperature (PVT) variation, whose uncertainty will limit the improvement of the spur-jitter performance.

To surmount the above issues, we devise the low-power low-jitter regulated NPG (R-NPG), as detailed in Fig. 3(bottom). Herein, we use a large-size NMOS (M_N) to create the stable sampling edge, meaning that only M_N contributes to the final in-band PN. This large size can directly lower the flicker and thermal noise of M_N. On the other hand, the final width of the V_{PUL} is determined by the delay path of the R-NPG. Since the noise of the delay path doesn't degrade the jitter of the sampling edge, the delay path can be optimized in a low-power way. Note that the delay t_1 is still sensitive to PVT variations. Hence, the Δt regulation loop is proposed to keep a constant pulse width. By setting the static discharging current N times larger than the charging current, the average discharging current in one reference cycle is controlled by the width of the sampling pulse. The width of the V_{PUL} is finally determined upon the average discharging current being equal to the charging current, thus obtaining the delay t_1. The final Δt is only dependent on the ratio between the charging and discharging currents. The PVT issues can be effectively eschewed by carefully optimizing the layout matching. Unlike the conventional NPG, the ultra-narrow pulse t_2 of $<\Delta t$ on the delay path is output by the DFF-reset-delay for reliability consideration.

Fig. 4 shows the measured spectrum and PN profile under F_{OUT}=14 GHz when inputting a 125-MHz sinusoidal reference. It is seen that the reference spur is -94.55 dBc, and the RMS jitter integrated from 1 kHz to 100 MHz is 31.9 fs, corresponding to -260.1 dB. To further verify the robustness of our design, first, we perform the multi-chip assessments. Fig. 5(top) plots the jitter of <36 fs and the spur of <-90 dBc over 5 samples. Next, we measure the impact of the supply voltage (V_{DD}) variation of the R-NPG and iSS-PD on the jitter-spur performance, as shown in Fig. 5(bottom-left). The reference spur is always below -90 dBc due to our joint optimization. The jitter gradually decreases as the V_{DD} increases, since high V_{DD} enhances the iSS-PD gain and reduces the reference noise simultaneously. The jitter is <37 fs across the 200-mV V_{DD} variation. Fig. 5(bottom-right) plots <37-fs jitter and <-90-dBc spur over 12.75-to-16 GHz.

Fig. 6 benchmarks our design with recent high-performance PLLs. We first lower the spur of SS-PLL to -90 dBc, and meanwhile score competitive jitter and FoM compared with recent various PLLs. Fig. 7(top) shows the 40-nm SS-PLL prototype occupying a core area of 0.134 mm^2 and consuming a 9.5-mW power.

Acknowledgments:

This work was supported by the National Natural Science Foundation of China (No. 62222409 and 62174153) and Key Research Program of Frontier Sciences, CAS (No. ZDBS-LY-JSC008).

References:

[1] D. Turker, et al., "A 7.4-to-14GHz PLL with 54fs$_{rms}$ jitter in 16nm FinFET for integrated RF-data-converter SoCs," *ISSCC*, pp. 378-380, Feb. 2018.

[2] X. Gao, et al., "A Low Noise Sub-Sampling PLL in Which Divider Noise is Eliminated and PD/CP Noise is Not Multiplied by N2," *IEEE JSSC*, vol. 44, no. 12, pp. 3253-3263, Dec. 2009.

[3] Z. Zhang, *et al.*, "A 0.65-V 12–16-GHz Sub-Sampling PLL With 56.4-fs$_{rms}$ Integrated Jitter and −256.4-dB FoM," *IEEE JSSC*, vol. 55, no. 6, pp. 1665-1683, Jun. 2020.

[4] Z. Yang *et al.*, "16.8 A 25.4-to-29.5GHz 10.2mW Isolated Sub-Sampling PLL Achieving -252.9dB Jitter-Power FoM and -63dBc Reference Spur," *ISSCC*, pp. 270-272, Feb. 2019.

[5] D. -G. Lee, *et al.*, "A Sub-mW 2.4-GHz Active-Mixer-Adopted Sub-Sampling PLL Achieving an FoM of −256 dB," *IEEE JSSC*, vol. 55, no. 6, pp. 1542-1552, Jun. 2020.

[6] Y. Chen, *et al.*, "High-performance harmonic-rich single-core VCO with multi-LC tank: A tutorial," *IEEE Trans. Circuits Syst. II, Exp. Briefs*, vol. 69, no. 7, pp. 3115–3121, Jul. 2022.

979-8-3503-3004-5/23 $31.00 © 2023 IEEE

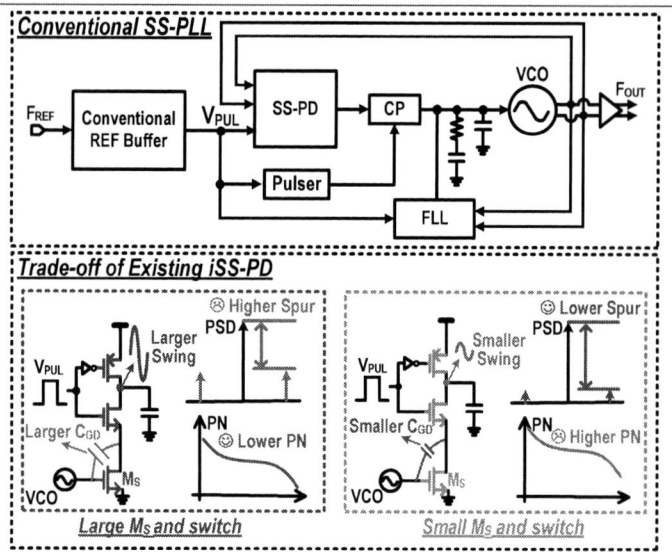

Fig. 1. Generic architecture of the SS-PLL (top) and the obvious tradeoff of the existing iSS-PD.

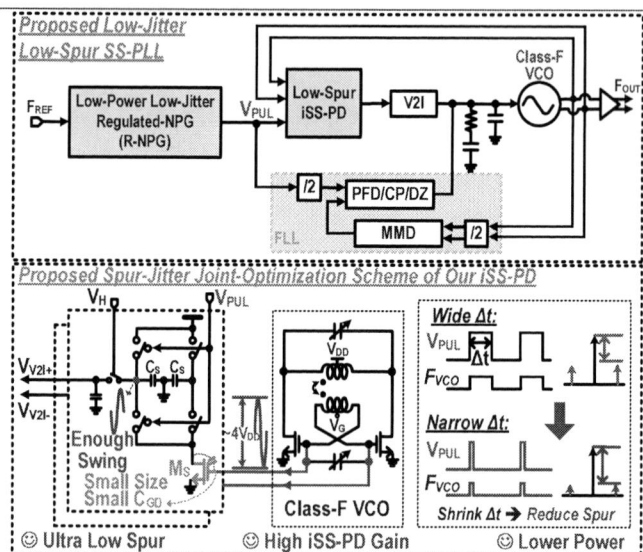

Fig. 2. Block diagram of the proposed low-spur low-noise SS-PLL (top) and the proposed ultra-low-spur co-optimization (bottom).

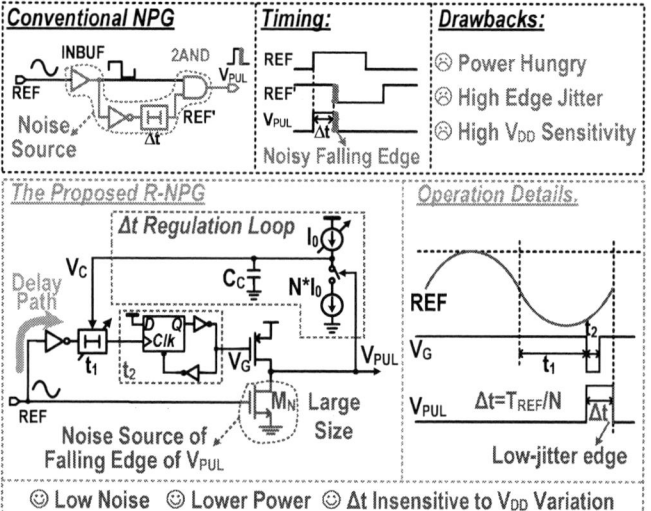

Fig. 3. Operation details of the conventional NPG (top) and the proposed (bottom) R-NPG.

Fig. 4. Measured spectrum (top) and PN profile (bottom) under f_{VCO}=14 GHz.

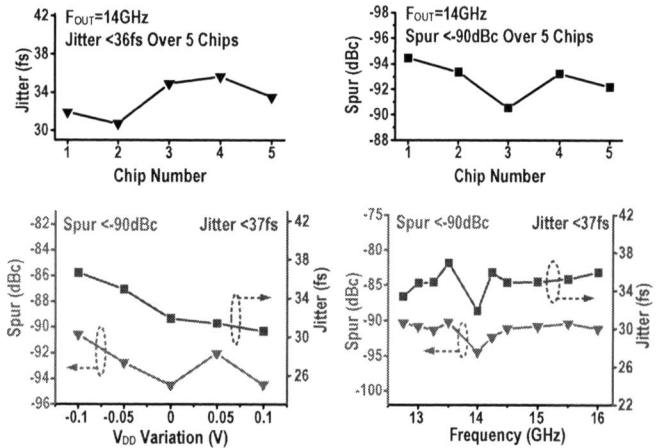

Fig. 5. Measured jitter-spur performance considering five chips, V_{DD} variation of the PG, and tuning range.

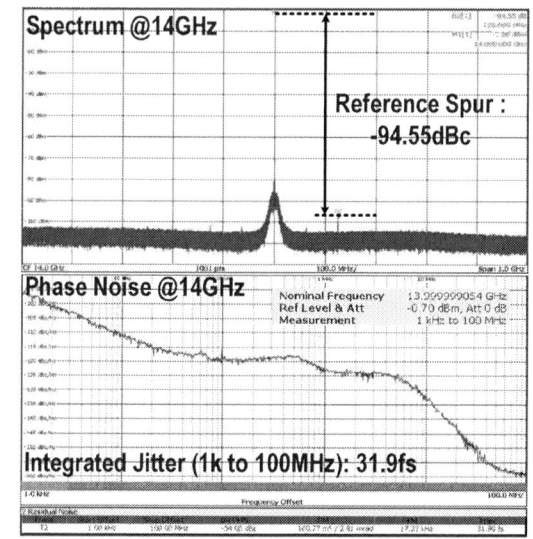

	This work	Zhao [8] JSSC'23	Gong [9] JSSC'22	Huang [10] TCAS-1'23	Turker [1] ISSCC' 18	Zhang [11] CICC'23	Yang [7] TVLSI' 22
Architecture	SS-PLL	DS-PLL	CS-PLL	S-PLL	CP-PLL	S-PLL	SS-PLL
CMOS (nm)	40	28	40	65	16	65	65
f_{REF} (MHz)	125	250	100	105	500	100	103
f_{OUT} (GHz) /FTR (GHz)	14 /(12.75–16)	20 /NA	11.2 /(9.6-12)	3.78 /NA	12.5 /(7.4-14)	10.6 /(10.3-11.1)	3.286 /NA
RMS Jitter (fs) /Integrated Range (MHz)	31.9 /(0.001-100)	20.9 /(0.01-100)	48.6 /(0.001-100)	39.6 /(0.001-100)	53.6 /(0.01-10)	33.7 /(0.01-40)	64.9 /(0.001-40)
Reference spur (dBc)	-94.55	-66	-77.3	-70.96	-75.5	-83.9	-82.2
Power (mW)	9.5	12	5	6.15	45	242	7.53
FoM* (dB)	-260.1	-262.8	-259.2	-260.2	-246.8	-245.6	-255
FoM$_N$** (dB)	-280.2	-282.1	-279.7	-275.8	-260.8	-265.8	-270
Core Area (mm²)	0.134	0.06	0.13	0.397	0.35	0.7	0.146

*FoM = 20log(σ_{rms})+10log(P_{DC}/1mW)
** FoM$_N$ = FoM -10log(N), the lower, the better

Fig. 6. Performance summary and comparison with recently published high-performance PLLs.

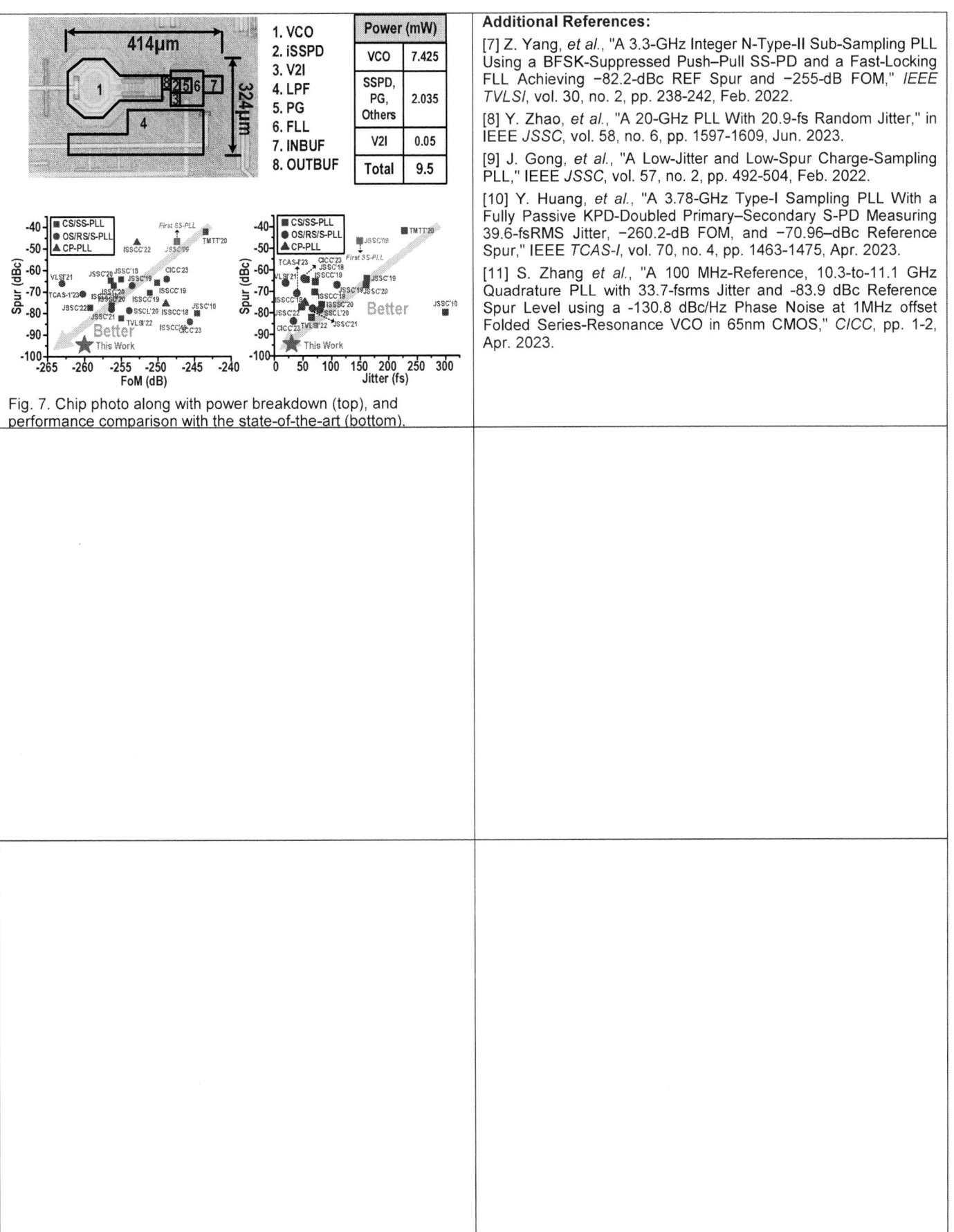

Fig. 7. Chip photo along with power breakdown (top), and performance comparison with the state-of-the-art (bottom).

Additional References:

[7] Z. Yang, *et al.*, "A 3.3-GHz Integer N-Type-II Sub-Sampling PLL Using a BFSK-Suppressed Push–Pull SS-PD and a Fast-Locking FLL Achieving −82.2-dBc REF Spur and −255-dB FOM," *IEEE TVLSI*, vol. 30, no. 2, pp. 238-242, Feb. 2022.

[8] Y. Zhao, *et al.*, "A 20-GHz PLL With 20.9-fs Random Jitter," in IEEE *JSSC*, vol. 58, no. 6, pp. 1597-1609, Jun. 2023.

[9] J. Gong, *et al.*, "A Low-Jitter and Low-Spur Charge-Sampling PLL," IEEE *JSSC*, vol. 57, no. 2, pp. 492-504, Feb. 2022.

[10] Y. Huang, *et al.*, "A 3.78-GHz Type-I Sampling PLL With a Fully Passive KPD-Doubled Primary–Secondary S-PD Measuring 39.6-fsRMS Jitter, −260.2-dB FOM, and −70.96–dBc Reference Spur," IEEE *TCAS-I*, vol. 70, no. 4, pp. 1463-1475, Apr. 2023.

[11] S. Zhang *et al.*, "A 100 MHz-Reference, 10.3-to-11.1 GHz Quadrature PLL with 33.7-fsrms Jitter and -83.9 dBc Reference Spur Level using a -130.8 dBc/Hz Phase Noise at 1MHz offset Folded Series-Resonance VCO in 65nm CMOS," *CICC*, pp. 1-2, Apr. 2023.

A Wideband Low-Noise Linear LiDAR Analog Front-End Achieving 1.6 GHz bandwidth, 2.7 pA/Hz$^{0.5}$ Input Referred Noise, and 103 dBΩ Transimpedance Gain

Yidan Zhang[1,2], Zhao Zhang[1,2]*, Yiqing Xu[1,2], Xinyu Shen[1,2], Nan Qi[1,2], Nanjian Wu[1,2], Jian Liu[1,2], Liyuan Liu[1,2]

[1]Institute of Semiconductors, Chinese Academy of Sciences, China

[2]University of Chinese Academy of Sciences, China

*zhangzhao11@semi.ac.cn

Linear LiDAR can obtain distance by measuring time of flight (TOF), which is widely used in unmanned, geomorphic mapping and other fields. In recent years, the demand of long-distance LiDAR is getting increased. The maximum detection range of pulsed LiDAR is mainly dictated by the signal-to-noise ratio (SNR) [1]. Fig.1 (top) indicates that improving the SNR relies on enhancing the peak emission power of the laser pulse and reducing the noise of the analog front-end (AFE) circuit. Moreover, to comply with safety regulations for eye-safe laser energy [1], narrow high-power laser pulse is required. Since the pulse width of high-power laser has already reached sub-nanosecond range [2], an AFE with a wide bandwidth of over-1 GHz and lower noise is required, thus significantly challenging the circuit design. In the AFE, the input referred noise (IRN) is mainly dominated by the transimpedance amplifier (TIA). Prior shunt-feedback TIA (SF-TIA) [3] performs well with a narrow bandwidth around 100 MHz. But it suffers from strong trade-off between IRN, gain, bandwidth and stability. Thus, the IRN is higher than 4.5 pA/Hz$^{0.5}$ with over-200-MHz bandwidth [4, 5], indicating unsuitable for the design of low-noise TIA with over-1-GHz bandwidth and high gain. A low IRN of 2.6 pA/Hz$^{0.5}$ was achieved by using a narrow-band TIA along with an equalizer for bandwidth enhancement (190 MHz) [6]. But this is still not sufficient for the design of over-1-GHz TIA.

This paper presents an AFE circuit for linear LiDAR scoring a wide bandwidth of 1.6 GHz and a low IRN of 2.7 pA/Hz$^{0.5}$ with a gain of 103 dBΩ concurrently, as the block diagram illustrated in Fig. 1 (bottom). The NMOS feedforward TIA with inner feedback resistor (NFFR-TIA) is devised to mitigate the design trade-offs of prior SF-TIAs, thus achieving wide bandwidth around 400MHz with low IRN and high gain. Two stage continuous-time linear equalizers (CTLE) are utilized to further boost the bandwidth to 1.6 GHz with a flatten in-band AC response and without IRN degradation.

The widely used SF-TIA is depicted in Fig. 2 (top), which suffers from strong trade-offs between bandwidth, IRN, stability and gain: using a large feedback resistor R_f reduces IRN but comes with low bandwidth; a TIA with high open-loop gain A_0 helps to ensure sufficient bandwidth, yet, it may degrades stability; a feedback capacitor C_f improves the stability [3] whereas it increases equivalent input capacitance of the TIA, thus degrading bandwidth.

To surmount these issues, we propose the NFFR-TIA, as the detailed schematic shown in Fig. 2 (bottom). The three-stage cascade topology is used to enhance open-loop gain for wideband design consideration. To solve the stability issue of the conventional SF-TIA with high open-loop gain, we introduce an NMOS feedforward stage, which consists of a NMOS M_L and a resistor R_L (R_L can avoid oscillation caused by the inductance of the input bonding wire). This reduces the output resistance by paralleling M_L to increase the non-dominant pole, thus improving the phase margin for better stability. Meanwhile, using the feedforward M_L also reduces the input resistance of the TIA, leading to higher bandwidth with the same feedback resistance R_{F1}. This indicates that both lower IRN and higher bandwidth can be achieved concurrently by properly setting R_{F1}, compared to the conventional SF-TIA. To further suppress IRN, the first stage of the NFFR-TIA core amplifier is designed with a high transconductance.

The gain of the TIA should be adjustable with sufficient tuning range to meet the requirement of measuring different distances. The is performed by tuning the resistance of feedback resistor R_f in the conventional SF-TIA. However, in our case of three-stage TIA, if reducing R_{f1} at low-gain mode, the bandwidth gets increased, thus causing stability issue due to the fixed value of the non-dominant poles at the inner nodes of the TIA (see node A and B in Fig. 2 (bottom)). So, to mitigate this issue, we introduce a tunable feedback

R_{F2} in the second stage of this NFFR-TIA, and R_{F1} and R_{F2} are tuned simultaneously. At low gain mode, both R_{F1} and R_{F2} are adjusted to a small value; thus, the poles at node A and B are increased due to the reduced resistances of node A and B with lower R_{F2} to ensure the stability. At high gain mode, both R_{F1} and R_{F2} are increased to obtain a sufficient open-loop gain.

The NFFR-TIA can improve the bandwidth to near 400 MHz with a low IRN of 2.0 pA/Hz$^{0.5}$ according to the post-layout simulation results. Yet, this is still far away from our design target of over-1-GHz bandwidth. Therefore, we employ two stage CTLEs to further improve the bandwidth, as illustrated in Fig. 3 (top). The CTLE boosts the high-frequency gain by using the source-degenerated resistor (R_S) and capacitor (C_S) array to compensate for the high-frequency attenuation due to the limited bandwidth of the TIA, and the boosted AFE bandwidth can be adjusted by tuning the R_S and C_S, respectively. Since the IRN of the whole AFE circuit is mainly dictated by the relatively high gain of NFFR-TIA, the CTLE contributes negligible noise to the total IRN, thus suitable for wideband low-noise AFE design. Note that one stage CTLE only contains one peak, thus showing limited capability of bandwidth enhancement, as illustrated in Fig. 3 (bottom). Hence, two stage CTLEs are introduced, providing two peaks with two different frequencies, as shown in Fig. 3 (top left). The first stage CTLE boosts the low-frequency gain with a low-frequency peak, while the second stage's higher peak frequency further boosts the bandwidth to over-1 GHz. According to the post-layout simulation results, by properly tuning the peak frequencies of the two stage CTLEs, the bandwidth can be boosted to 1.8 GHz with a flatten in-band frequency response and an over-100 dBΩ gain.

The measured AFE AC frequency responses are present Fig. 4 (top), exhibits a 1.6-GHz bandwidth with the highest gain of 103 dBΩ, and a gain range of 73-103 dBΩ with flatten AC frequency response. The CTLE boosts the bandwidth of the AFE from 460 MHz to 1.6 GHz, as shown in Fig.4 (bottom). Fig. 5 (top) presents the measured AFE output RMS noise using an oscilloscope. The AFE is configured with 103-dBΩ gain and 1.6-GHz bandwidth. The left figure displays the combined noise of the oscilloscope and AFE, while the right figure illustrates the oscilloscope's noise. Thus, we get the integrated IRN of 109 nA and an equivalent average IRN of 2.7 pA/Hz$^{0.5}$ of our AFE accordingly. This AFE also shows low average IRN of 2-2.7 pA/Hz$^{0.5}$ within the tunable bandwidth range, as shown in Fig.5 (bottom left). Fig. 5 (bottom right) presents the 3-Gb/s output eye mask of the AFE with CTLE compensation, also demonstrating that the AFE bandwidth can be boosted to over-1 GHz. Fig. 6 depicts that our AFE achieves almost 1-order higher bandwidth of 1.6 GHz with lower average IRN, comparable gain and integrated IRN, compared to other recently published works. The 40-nm CMOS prototype occupies 0.15 mm^2 and consumes 10 mW excluding output testing buffer (OB), as shown in Fig. 7.

Acknowledgments:

This work was supported by the National Natural Science Foundation of China (No. 62222409) and Key Research Program of Frontier Sciences, CAS (No. ZDBS-LY-JSC008).

References:

[1] M. Kashmiri et al., "23.1 a 4GS/s 80dB DR current-domain analog front-end for phase-coded pulse-compression direct Time-of-Flight automotive LiDAR," *ISSCC*, pp. 344–345, Feb. 2020.

[2] VARYDISK E150 RPMC Laser Datasheet. Accessed: Jun. 2023. [Online]. Available: https://www.rpmclasers.com/product/varydisk-e150-thin-disk-laser/

[3] H. Cho et al., "A high-sensitivity and low-walk error LADAR receiver for military application," *IEEE TCAS-I*, vol. 61, no. 10, pp. 3007-3015, Oct. 2014.

[4] H. Zheng et al., "A linear-array receiver AFE circuit embedded 8-to-1 multiplexer for direct ToF imaging LiDAR applications," *IEEE TCAS-I*, vol. 69, no. 12, pp. 5050-5058, Dec. 2022.

[5] X. Wang et al., "A low walk error analog front-end circuit with intensity compensation for direct ToF LiDAR," *IEEE TCAS-I*, vol. 67, no. 12, pp. 4309-4321, Dec. 2020.

[6] P. Wang et al., "A multi-Channel low-noise analog front end circuit for linear LADAR," *IEEE TCAS-II*, vol. 67, no. 7, pp. 1209-1213, Jul. 2020.

979-8-3503-3004-5/23 $31.00 © 2023 IEEE

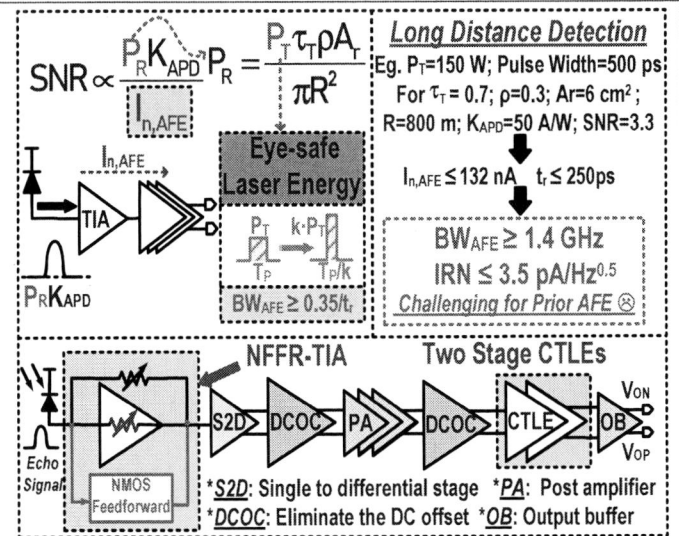

Fig. 1. Design challenges of AFE for long-distance LiDAR (top), and the proposed LiDAR AFE (bottom).

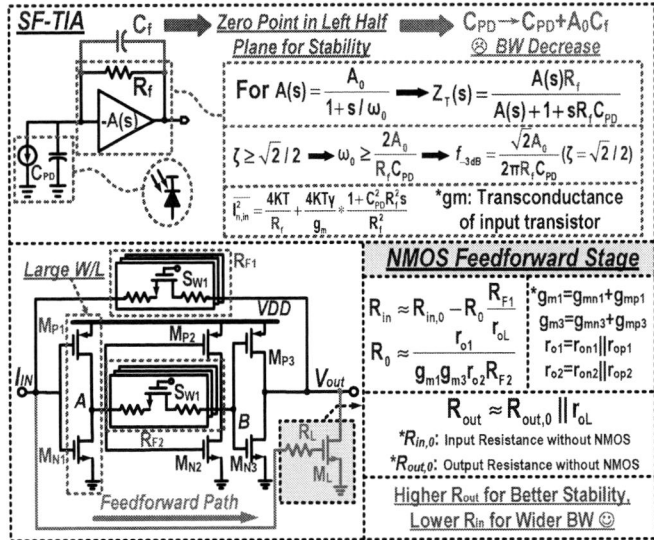

Fig. 2. Conventional capacitive compensated SF-TIA (top) and the proposed NFFR-TIA (bottom).

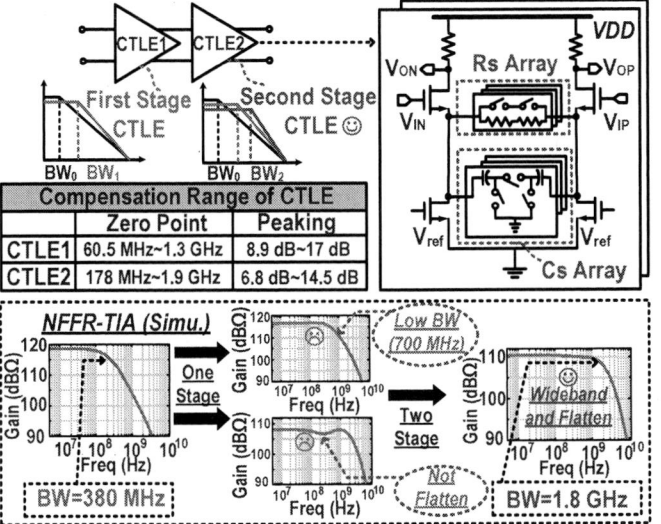

Fig. 3. Schematic (top) and simulation results (bottom) of two-stage CTLEs.

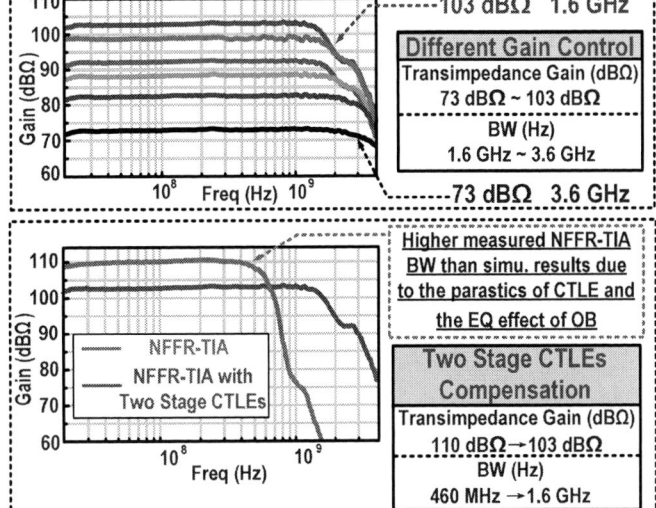

Fig. 4. Measured frequency response of AFE (top), and the measurement result of the CTLE compensation (bottom).

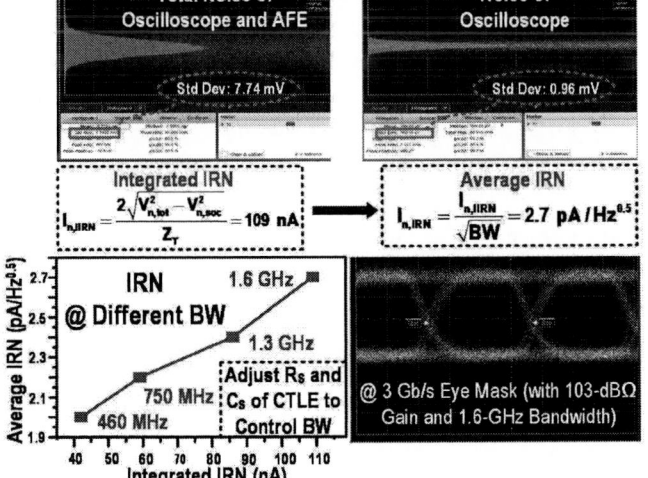

Fig. 5. Measured AFE IRN (top) at high-gain mode, IRN at different AFE bandwidths (bottom left), and output eye mask using a 3-Gb/s PRBS-31 string with CTLE compensation (bottom right).

	This work	ISSCC'20 [1]	TCASI'22 [4]	TCASI' 20 [5]	TCASI'17 [7]	TCASII'20 [6]
CMOS (nm)	40	130	65	180	350	180
Detector	APD	APD	APD	APD	APD	APD
C_{PD} (pF)	1.2	2.0	1.0	1.2	/	2
Architecture	TIA+CTLE+AMP	CURRENT AMP	TIA+AMP	TIA+AMP	TIA+AMP	TIA+AMP+EQ
Transimpedance Gain (dBΩ)	103	/	106	86	~100	101
BW (Hz)	1.6 G	200 M	220 M	281 M	230 M	190
Integrated IRN (nA)	42 (460 MHz)	70	75	78	~100	36
	109 (1.6 GHz)					
Average IRN (pA/Hz$^{0.5}$)	2.0 (400 MHz)	5.0	5.1	4.68	6.6	2.6
	2.7 (1.6 GHz)					
Power Consumption (mW)	10 (wo OB)	119 (AFE+ADC)	17 (AFE+COM)	200*	180*	37
Core Area (mm²)	0.15	0.56	1.44**	2.2**	4**	3.41 (32 ch)

*Power consumption of the entire receiver circuit
**The area of the whole chip

Fig. 6. Performance summary and comparison.

979-8-3503-3004-5/23 $31.00 © 2023 IEEE

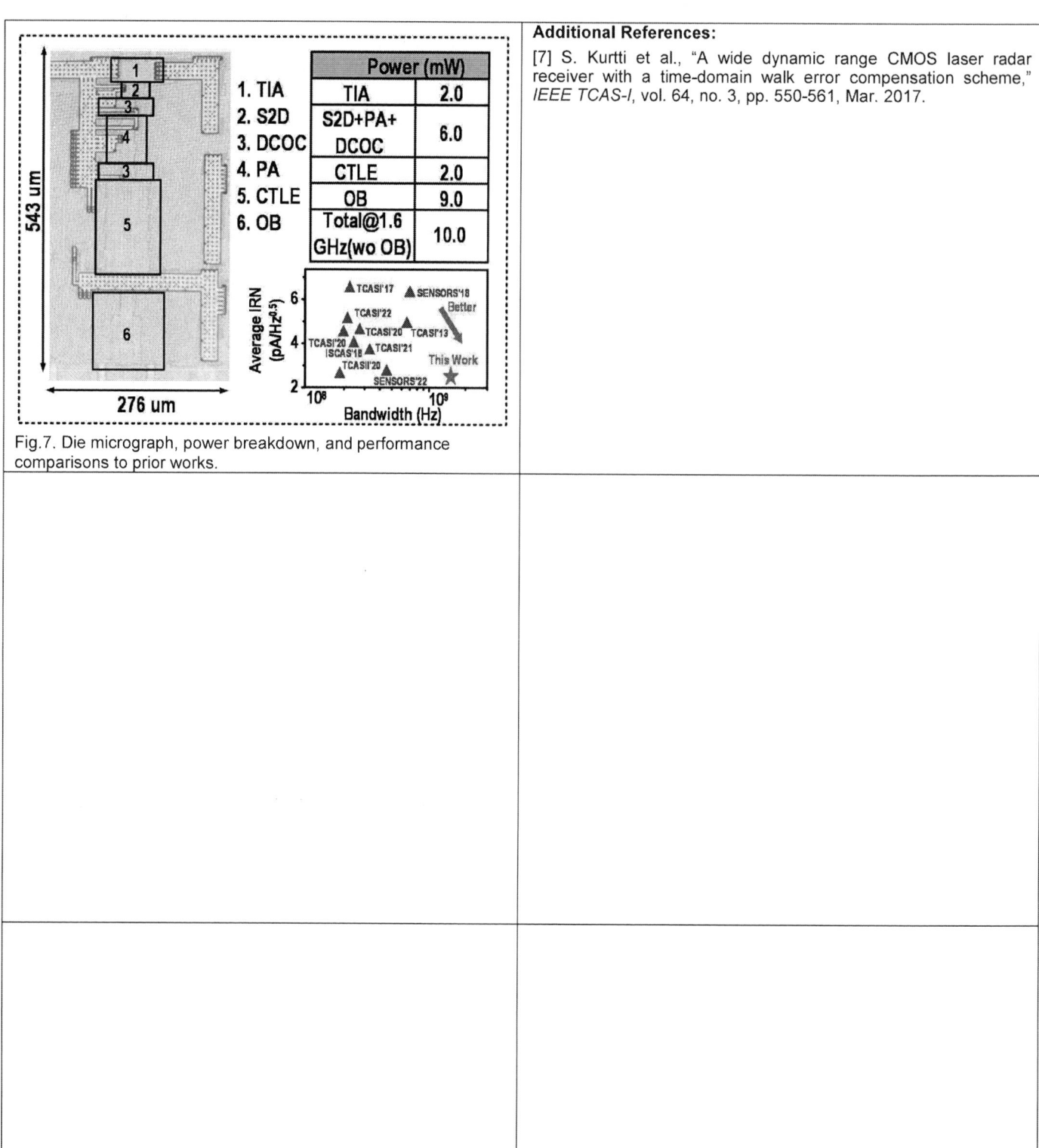

Power (mW)	
TIA	2.0
S2D+PA+ DCOC	6.0
CTLE	2.0
OB	9.0
Total@1.6 GHz(wo OB)	10.0

1. TIA
2. S2D
3. DCOC
4. PA
5. CTLE
6. OB

Fig.7. Die micrograph, power breakdown, and performance comparisons to prior works.

Additional References:

[7] S. Kurtti et al., "A wide dynamic range CMOS laser radar receiver with a time-domain walk error compensation scheme," *IEEE TCAS-I*, vol. 64, no. 3, pp. 550-561, Mar. 2017.

An Adaptive-sampling Digital LDO With Statistical Comparator Selection Achieving 99.99% Maximum Current Efficiency and 0.25ps FoM in 65nm

Shun Yamaguchi, Takashi Hisakado, Osami Wada, Mahfuzul Islam
Kyoto University, Kyoto, Japan

Attention to digital LDO has been increasing due to their low-power and wide-voltage-range operation. For fine-grain power control of an SoC (System-on-a-Chip), LDOs with small output capacitance and low quiescent current are required. A typical synchronous Digital LDO with a fixed clock frequency suffers from power and transient response trade-off. To achieve a fast transient response with low quiescent current, techniques such as adaptive sampling, event-driven control, and level-triggered asynchronous loop are proposed [1-7]. Asynchronous LDOs often utilize delay or logical threshold voltage for droop detection. However, the droop detection voltage and the quantization resolution deviate largely under different voltages. Furthermore, delay and logical threshold voltages are highly susceptible to PVT variation, and the control algorithm tends to become complex [2]. In the case of synchronous LDOs, adaptive sampling with the help of a built-in VCO (Voltage Controlled Oscillator) can achieve both low power and fast transient response. Several droop detectors are proposed to boost a synchronous LDO's transient response [3-7]. These droop detectors require additional reference voltages that poses a significant challenge in integrating such LDOs with small areas and power. Furthermore, comparators with upsized transistors or preamplifiers are employed to realize low offset voltage, increasing kick-back noise or quiescent current. Because of high kickback noise, low output impedance reference generators are required causing power overhead [4]. Thus, we identify stable reference voltages as the key issue in realizing low-power LDOs with variable output voltage.

To solve the reference problem, we propose to utilize the intrinsic threshold voltage variation. The threshold voltage difference is stable across voltage and temperature and thus becomes an attractive choice to realize stable references using this voltage difference. As comparator offset voltage is mainly determined by threshold voltage difference, desired references can be obtained from a comparator bank with statistical selection. Although typical statistical selection techniques require DACs for calibration, we propose a rank-based calibration where a one-time sweep of the input terminal by a current source can select the comparators [8]. The offset voltage distribution can be estimated with the foundry-provided statistical models. Thus, we can choose the rank values during design time. As offset voltage variation is utilized here, near minimum-sized transistors can be used for the comparators saving power and reducing kick-back noise. In this paper, we integrate 512 comparators and choose two whose offset voltage ranks correspond to $-\Delta$ and $+\Delta$. As a result, stable references of $V_{REF}-\Delta$ and $V_{REF}+\Delta$ are obtained. The comparators detect overshoot and undershoot without additional external references. We demonstrate the offset voltage-based droop detector using an adaptive-sampling digital LDO fabricated in a 65nm low-power process. With >200x frequency scaling and droop detectors, the LDO achieves a maximum current efficiency of 99.99% and a transient FoM of 0.25ps.

Fig. 1 shows the architecture of our proposed LDO. The LDO consists of a 14-bit binary-sized power MOSFET, a preamplifier, a VCO to generate the adaptive clock signal, an inverter-based dead-zone comparator, a comparator bank, a calibration circuit, and an adaptive-gain controller. The preamplifier generates a differential signal which the VCO utilizes for generating the clock signal having symmetric frequency against V_{OUT} change [9]. The amplified signals are also used to detect V_{OUT} deviation. Symmetric frequency realizes fast transient response for both the undershoot and overshoot.

Fig. 2(a) shows the preamplifier with two inverters to detect the output V_{OUT} change with a small dead-zone window [9]. The amplified differential signal is also used in the VCO shown in Fig. 2(b) to achieve wide frequency scaling. Figs. 2(c) and (d) show the comparator input voltage V_{INP} waveform, the count value that tracks offset voltage rank, and the comparator enable signal during the calibration. Two comparators for overshoot and undershoot detection are chosen from pre-defined rank values. When calibration starts, the input node V_{INP} is separated from V_{OUT}, and the LDO

operates without the droop detectors. The VCO generates the clock signal with the maximum frequency to reduce calibration time and control delay. Comparators are enabled one by one repeatedly while the V_{INP} node discharges. A counter counts the number of flipped comparators. Once a comparator output is flipped, the counter won't count up this comparator again to avoid duplicate count. After the comparators with the target ranks are found, droop detection is enabled, and the LDO enters the adaptive sampling mode. For wide-frequency scaling, VCO utilizes the amplified error signal and capacitance-canceling technique [10].

Fig. 3(a) shows the structure of the droop detector. During calibration, SW1 is turned OFF. The input node is first charged to V_{IN} by closing SW2 and opening SW3. SW2 is then opened and SW3 is closed to discharge V_{INP} by the current source, I_b. As only ranks are measured in our calibration, the requirement on the I_b is relaxed. Thus, a moderate design suffices here. Fig. 3(b) shows the strongARM comparator. M_{tail} is inserted to increase the offset voltage stability across a wide V_{IN} and V_{REF} range. V_{NBIAS} is generated internally. Fig. 3(c) shows the Monte Carlo simulation result of offset voltage selection for three emulated chips under different V_{IN} and V_{REF}. Stable references are realized over a wide V_{REF} range.

We implemented our LDO in a commercial 65nm low-power CMOS process. Fig. 7 shows the layout of our LDO. C_{load} is only 0.1nF. The active area is 0.067mm^2. Only a single reference is used here. Two comparators are active during normal operation; the others are clock gated. The inactive comparators act as capacitors for the V_{REF} node, relaxing the output impedance requirement for the V_{REF} generator.

Fig. 4 shows the measured load transient response at V_{IN} = 1.0V. A binary search algorithm is adopted when large fluctuations are detected. Overshoot and undershoot of V_{OUT} are 59mV and 145mV under a load step between 9.7mA and 41.3mA.

Fig. 5(a) shows the steady-state output voltage, V_{OUT}, against different I_{load} for V_{IN} from 0.6 to 1.2V. V_{OUT} is regulated to V_{REF} over I_{load} range from 0.1mA to 62mA. Fig. 5(b) shows V_{OUT} against V_{IN} for I_{load} of 15mA. Good stability is observed over the V_{IN} range from 0.6 to 1.2V. Figs. 5(c) and 5(d) show the measured quiescent current and current efficiency. The quiescent current, I_Q, is 17.4uA at V_{IN} = 1.0V, and the maximum current efficiency is 99.99% thanks to its wide frequency scaling and low-power droop detectors. Fig. 5(e) shows the waveform of comparator input during calibration. During calibration, the maximum clock frequency is generated. After calibration, the LDO enters the adaptive-sampling mode.

Table I compares our LDO with state-of-the-art synchronous LDOs [3-7]. The proposed LDO achieves a high maximum current efficiency of 99.99% using only one external reference voltage. The transient FoM of 0.25ps is achieved with fully synchronous control. More than 200x frequency scaling is achieved thanks to the preamplifier and capacitance-canceling technique.

Acknowledgment:
This work was supported by JSPS KAKENHI 22K11953, a Rohm joint research program, and VDEC, The University of Tokyo, in collaboration with Synopsys, Cadence, and Mentor Graphics.

References:
[1] J. Oh et al., "An output-capacitor-free synthesizable digital LDO using CMP-triggered oscillator and droop detector," JSSC 2022:1-13.

[2] M. A. Akram et al., "Output-Capacitorless Tri-Loop Digital Low Dropout Regulator Achieving 99.91% Current Efficiency and 2.87 fs FOM," in IEEE Trans. on Power Electronics 2021:2044-2058.

[3] M. Huang et al., "A fully integrated digital LDO with coarse-fine-tuning and burst-mode operation," TCASII 2016:683–687.

[4] X. Liu et al., "A digital LDO in 22nm CMOS with a 4b self-triggered binary search windowed flash ADC featuring automatic analog layout generator framework," ASSCC 2022:2–4.

[5] J. Bang et al., "A 0.0084- mV-FOM, fast-transient and low-power external-clock-less digital LDO using a gear-shifting comparator for the wide-range adaptive sampling frequency," ESSCIRC 2021: 351–354.

[6] S. Kundu et al., "A Fully Integrated Digital LDO With Built-In Adaptive Sampling and Active Voltage Positioning Using a Beat-Frequency Quantizer," JSSC 2019:109-120.

979-8-3503-3004-5/23 $31.00 © 2023 IEEE

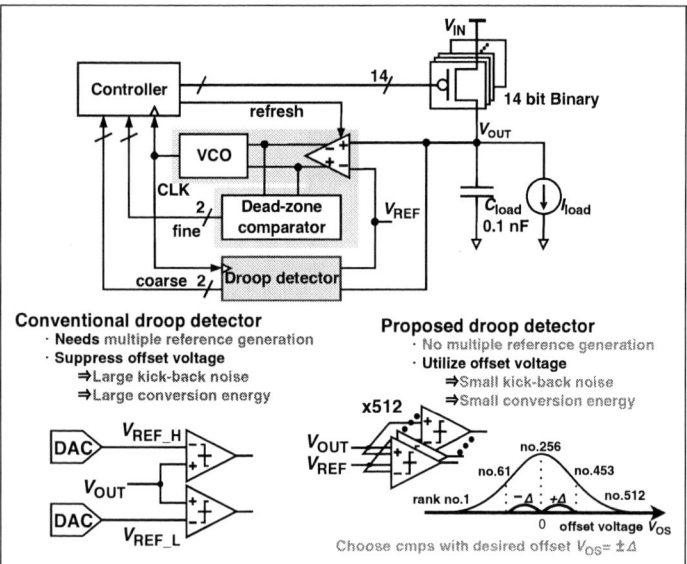

Conventional droop detector
· **Needs** multiple reference generation
· Suppress offset voltage
⇒Large kick-back noise
⇒Large conversion energy

Proposed droop detector
· No multiple reference generation
· Utilize offset voltage
⇒Small kick-back noise
⇒Small conversion energy

Choose cmps with desired offset $V_{OS} = \pm\Delta$

Fig. 1. Proposed digital LDO architecture with droop detectors.

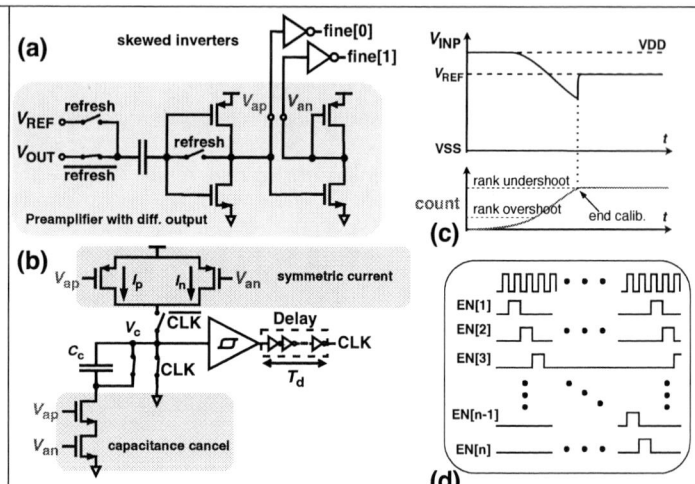

Fig. 2. (a) Schematic of dead-zone comparator with preamplifier, (b) schematic of VCO, (c) comparator input voltage V_{INP} and the count value that tracks offset voltage rank during calibration, and (d) the comparator enable signal during the calibration.

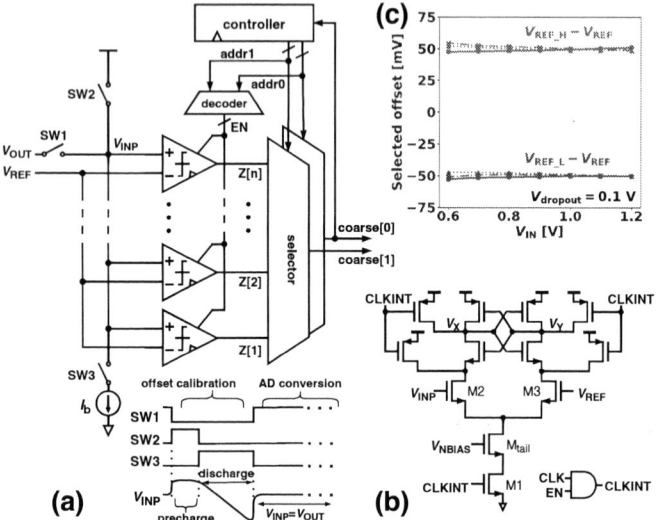

Fig. 3. (a) Droop detector utilizing offset voltage, (b) schematic of comparator, and (c) simulated offset voltage.

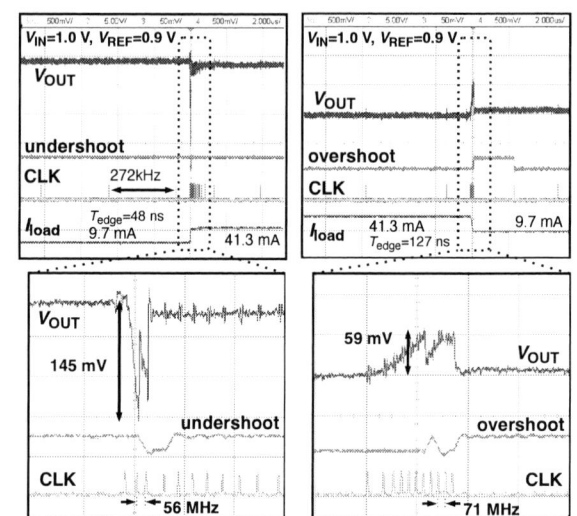

Fig. 4. Measured transient response under a load step between 9.7mA and 41.3mA at V_{IN}=1.0V.

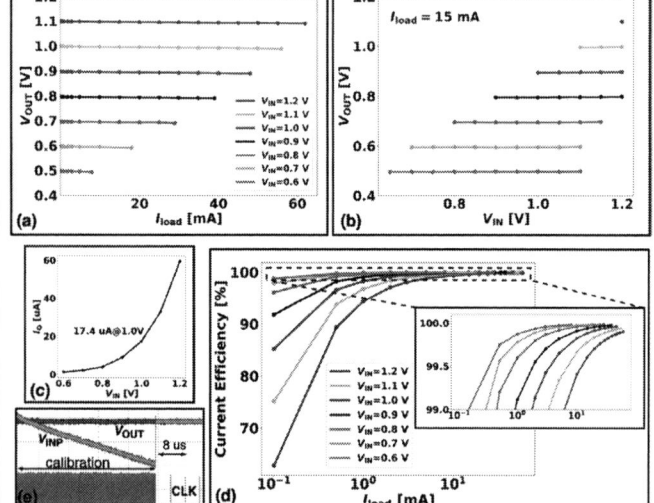

Fig. 5. (a) Measured load regulation, (b) line regulation, (c) quiescent current, (d) current efficiency, and (e) offset calibration.

	This work	[3] TCASII'16	[4] ASSCC'22	[5] ESSCIRC'21	[6] JSSC'19	[7] ASSCC'22
Process	65 nm	65nm	22nm	65nm	65nm	65nm
Number of external reference voltage	1	3	15	1	1	1
V_{IN} [V]	0.6-1.2	0.6-1.1	0.75-1.2	0.5-1.0	0.6-1.2	0.8-1.2
V_{OUT} [V]	0.5-1.1	0.4-1	0.7-1.15	0.45-0.95	0.4-1.1	0.7-1.1
Max I_{load} [mA]	62	100	75	30	100	105
C_{load} [nF]	0.1	1	1.17	0.1	0.04	0.1
I_Q [uA]	1.2-59.4 17.4@1.0V	82	400	10-169	100-1070	80-323
Max current efficiency [%]	99.99	99.92	N/A	99.93	99.5	99.88
$\Delta V_{OUT}@\Delta I_{load}$	145 mV @31.6mA	55mV @98mA	42.7mV @75mA	101mV @12mA	108mV @50mA	102mV @50mA
FoM [ps]	0.25	0.47	3.55	0.70	1.38	0.76

$FoM = C_{load}\Delta V_{OUT} I_Q/(\Delta I_{load})^2$

Fig. 6. Performance summary and comparison with state-of-the-art synchronous LDOs.

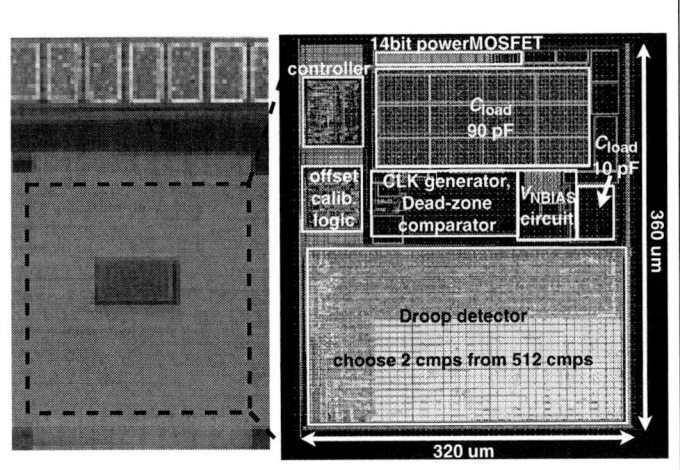

Fig. 7. Chip micrograph and LDO layout. Active area is 0.067mm².

Additional References:

[7] X. Wang *et al.,* "A Self-Clocked TDC-Based Unified Clock and Voltage Regulator with Replica Frequency-Locked Loop and Hysteresis Switching in 65nm CMOS," A-SSCC 2022:14-16.

[8] T. Kitamura *et al.,* "Order statistics based low-power flash ADC with on-chip comparator selection," IEICE 2022:1450–1457.

[9] S. Yamaguchi *et al.,* "Low-power design of digital LDO with nonlinear symmetric frequency generation," TCASII 2022:4644–4648.

[10] S. Yamaguchi *et al.,* "A fully synchronous digital LDO with built-in adaptive frequency modulation and implicit dead-zone control," A SPDAC 2023:186–187.

A Resource-Efficient Super-Resolution FPGA Processor with Heterogeneous CNN and SNN Core Architecture

Jiwon Choi, Sangyeob Kim, Wonhoon Park, Wooyoung Jo, and Hoi-Jun Yoo

Korea Advanced Institute of Science and Technology (KAIST)

Recently, super-resolution (SR) has become widely employed to improve the resolution of images and videos in a variety of fields, including medical diagnosis [1]. Among deep learning-based methods, generative adversarial network (GAN) is commonly used in SR due to its ability to generate high perceptual quality images. Fig.1 shows the requirements for GAN to generate high-quality images at high speed. Firstly, due to the high computational complexity (>500 GOPs), high throughput is required. Secondly, high precision (>8 bit) is necessary for high quality. Thirdly, PReLU is required to generate both positive and negative values for each pixel. However, previous FPGA processors struggled to meet these requirements. First of all, previous CNN processors suffered from low resource utilization. In FPGA, multiplier for high-precision (>8 bit) operands tends to be synthesized into DSP. Therefore, previous works dominantly used DSPs, while underutilizing LUTs, resulting in low resource efficiency [2-4]. On the other hand, spike-neural-network (SNN) processors utilized only LUTs because SNN requires integrate-and-fire instead of MAC [5-7]. Hence, in this paper, we utilize both CNN and SNN, with different resource characteristics, to increase resource efficiency and achieve high throughput.

However, there are still 2 challenges in utilizing them together. In SNN, spike encoding converts the input into a spike train with a length of the time window (TW). At each time step within the TW, the weight accumulates on the membrane potential when a pre-spike occurs, and a post-spike is generated when the potential exceeds the threshold. For high input precision, the amount of information increases, resulting in longer TW, which hinders achieving high throughput due to the increased latency. Another challenge is that previous SNN processors cannot support PReLU. Although PReLU in the SNN domain with positive/negative thresholds was proposed [8,9], its hardware implementation is lacking. To solve these issues, we propose a resource-efficient super-resolution FPGA processor with 3 key features: 1) Heterogeneous CNN-SNN core architecture (HECSA), 2) Mixed precision-based workload allocation (MPWA), 3) Operand-switchable SNN core (OSSC) with the signed spike encoder (SISE) and the dual-thresholding unit (DTU).

Fig.2 illustrates the overall system. It consists of a host CPU system and FPGA board which integrates EZ-USB FX3, SDRAM, flash memory, and proposed processor. The host processor exchanges input and output image with the board via USB communication. The EZ-USB FX3 is integrated for this, and SDRAM is used to store the intermediate feature map of each layer. The flash memory is integrated to store the thresholds from ANN-to-SNN conversion [10], network configurations, and weights. In addition, the proposed processor consists of a top controller, workload allocator, and the HECSA. The HECSA includes a CNN core and the OSSC, and the workload allocator divides the workloads to CNN and SNN.

Fig.3 shows the HECSA with CNN core and OSSC. The CNN core consists of 16×16 PE array with FXP16 MAC, 32KB IMEM, 128KB WMEM, and 32KB OMEM. The PE array employs a weight stationary structure, where input is broadcasted and weights are unicasted. The CNN PE has FXP16 multiplier which is synthesized using DSP, resulting in the DSP-dominant characteristic. The OSSC consists of 12 SNN units, SNN counter, 128KB high precision memory (HMEM), and 32KB low precision memory (LMEM). Each unit includes a 16×16 PE array for FXP16 addition, the DTU, and 32KB of output memory. Only when the pre-spike is transferred to each PE, the weight is transferred to the 16-way adder tree to generate membrane potential. These adder trees are synthesized using LUTs, resulting in the LUT-dominant characteristic. By utilizing a DSP-dominant CNN core and a LUT-dominant OSSC in a heterogeneous manner, resources are effectively utilized with 76.0% of DSPs and 85.7% of LUTs in the HECSA. As a result, HECSA increases the throughput by 93.3% compared with using only the CNN core.

Fig.4 shows the MPWA for effective workload allocation. In the OSSC, spike encoding for high-precision input leads to long latency. The MPWA is proposed to maximize throughput by allocating a low-precision workload to the OSSC. It is applied before each layer's computation and consists of two steps. The first step involves generating an input mask to indicate the processing domain of each tile. The input is spatially divided into tiles of size 8x8, and the sum of absolute differences is computed for each tile. If the result exceeds a predefined threshold, indicating a significant impact on the final PSNR, the tile is identified as a high-precision tile (HP-Tile) with FXP16. If the value is below the threshold, indicating a lesser impact, the tile is identified as a low-precision tile (LP-Tile) and quantized to FXP6. Despite utilizing the MPWA, when the ratio of HP-Tiles is high, the throughput improvement is limited due to the long latency involved in task offloading from the CNN core to the OSSC. To solve this problem, weights are divided into high-precision (HP-W) with FXP16 and low-precision (LP-W) with FXP6. Among the workloads for HP-Tile, only the workloads involving LP-W are allocated to the OSSC, while workloads involving HP-W are allocated to the CNN core. Then, by encoding LP-W instead of HP-Tile in the OSSC, the TW becomes shorter, leading to throughput improvement. As a result, we increase throughput by 83.6% compared to the case where only the HECSA was used without the MPWA. Furthermore, we obtain a 3.55× higher throughput than the baseline.

Fig.5 illustrates the proposed OSSC. It supports spike encoding for low-precision operands (input or weight) selected by the MPWA. The OSSC operates in two different modes: input spike encoding (ISE) mode for workloads with LP-Tile and weight spike encoding (WSE) mode for HP-Tile with LP-W. In both modes, low-precision (FXP6) operands are stored in LMEM, while HMEM stores high-precision (FXP16) operands. In this regard, the OSSC should encode negative values in input and weight due to the MPWA and PReLU. The SISE enables the spike encoding for both positive and negative values, which was not possible in previous works. For positive data, the 5-bit magnitude, excluding the sign bit, is stored in a 5-bit register. For negative data, the signed magnitude conversion is performed, and the resulting 5-bit data is stored in the register. Then, the SISE generates a 1-bit spike pattern by selecting different bits at the predetermined time step by the counter which operates during the TW, and the spike and sign-bit are broadcast to the PE row. When a pre-spike occurs, if the sign bit is '0', the weight is just sent to the adder tree to compute membrane potential. However, if the sign bit is '1', the 2's complement of the weight is sent to the adder tree to process negative spikes from negative values. After the adder tree operation, the result is transferred to the DTU to compare the membrane potential with both positive and negative thresholds for PReLU. If it exceeds one of the thresholds, a soft reset is applied by decrementing the potential by the threshold. Thanks to the SISE and DTU, 27.34 dB PSNR and 0.732 SSIM on average are achieved for SRGAN with x4 scaling factor on Set5, Set14, and BSDS100.

Fig.6 shows the demonstration system and comparison table. By utilizing HECSA, throughput is increased by 93.3% compared to using only the CNN core. Additionally, MPWA, OSSC, and SISE further increased the throughput by 83.6%. The SISE can also be jointly used with the DTU to support PReLU in the SNN domain, which can broaden the applicability of SNN. The proposed processor is implemented on Intel's Cyclone V with a maximum clock frequency of 200 MHz. It consumes 3.1 W and achieves a throughput of 363.5 GOPS. Compared to previous FPGA processors, we achieve a state-of-the-art resource efficiency of 1.40 GOPS/DSP by leveraging both CNN and SNN. Fig.7 shows the layout of the proposed processor. In conclusion, an SR processor with a heterogeneous CNN-SNN core is proposed to achieve state-of-the-art resource efficiency.

References:

[1] J. S. Isaac et al., "Super resolution techniques for medical image processing," ICTSD 2015:1-6.

[2] S. Lee et al., "CNN acceleration with hardware-efficient dataflow for super-resolution," IEEE Access 2020, 8:187754-187765.

[3] J. Chang et al., "An Energy-Efficient FPGA-Based Deconvolutional Neural Networks Accelerator for Single Image Super-Resolution." IEEE TCSVT 2020, 30(1):281-295.

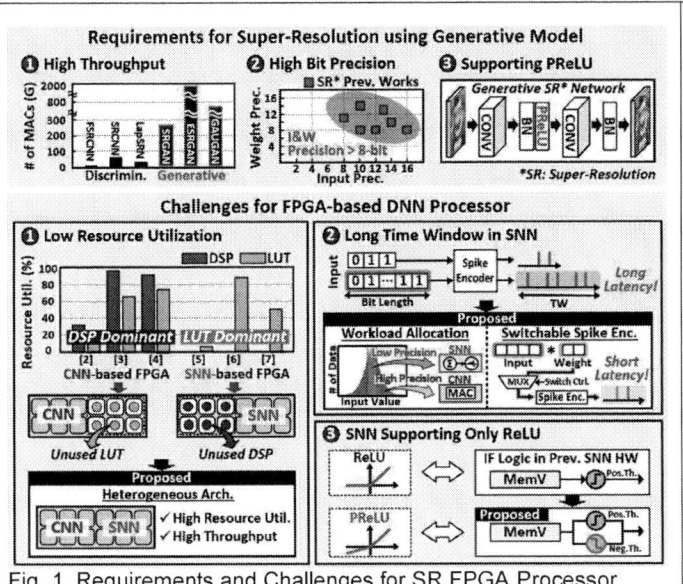

Fig. 1. Requirements and Challenges for SR FPGA Processor

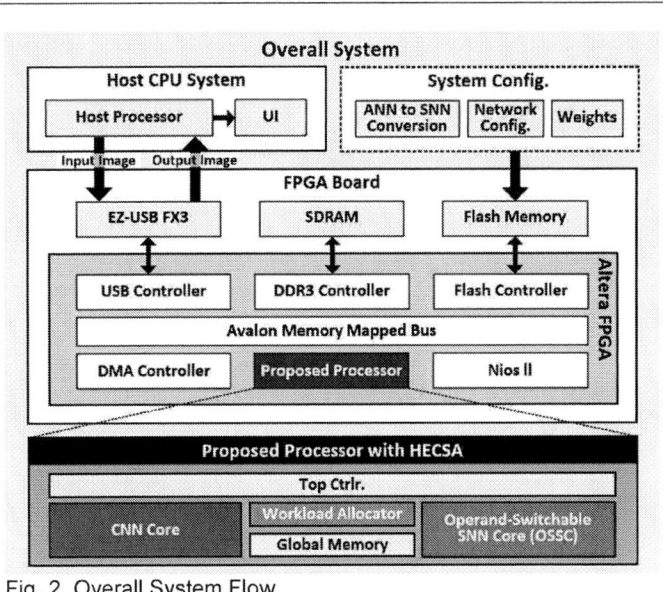

Fig. 2. Overall System Flow

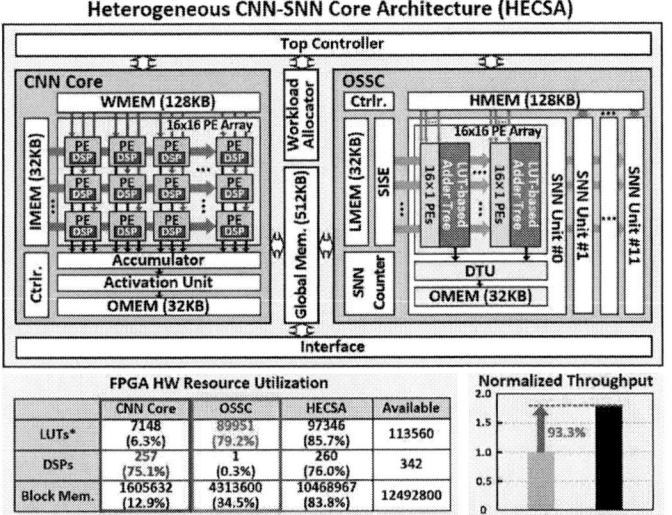

Fig. 3. Heterogeneous CNN-SNN Core Architecture (HECSA)

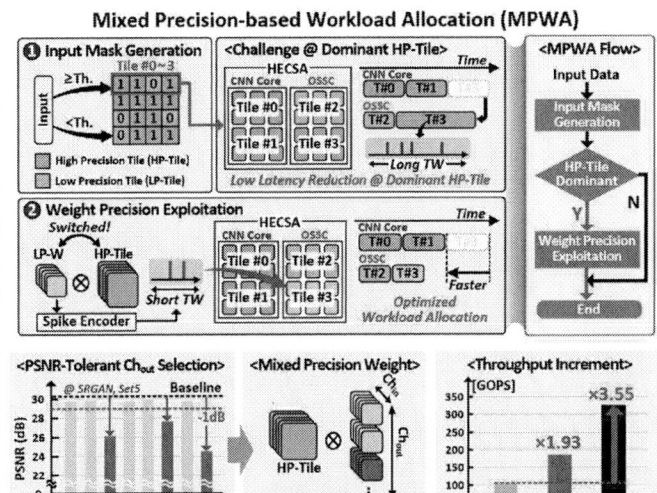

Fig. 4. Mixed Precision-based Workload Allocation (MPWA)

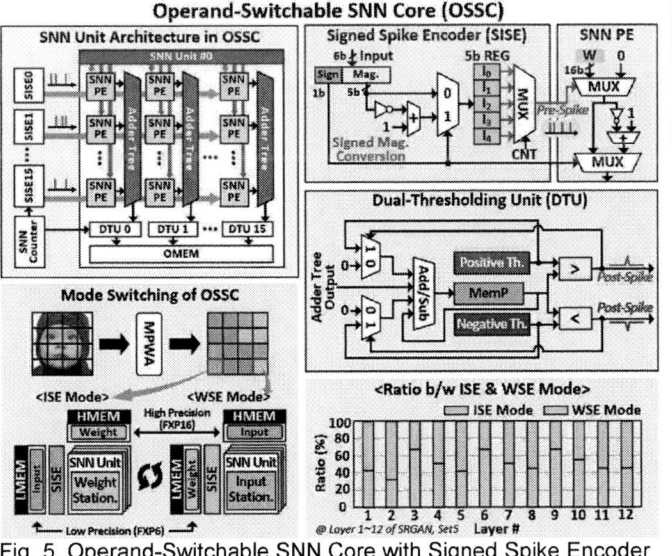

Fig. 5. Operand-Switchable SNN Core with Signed Spike Encoder

Fig. 6. Demonstration System and Comparison Table

Comparison Table

	[2]	[3]	[4]	[5]	This Work
Platform	Xilinx XCVU9P	Xilinx ZCU102	Intel Arria 10 SoC	Alveo U250	Intel Cyclone V
Frequency (MHz)	200	200	200	75	200
Target Domain	CNN	CNN	CNN	SNN	CNN/SNN
Input Precision	FXP 14	FXP 13	FXP 16	FXP 8	FXP 6[1]/FXP 16[2]
Weight Precision	FXP 10	FXP 13	FXP 16	FXP 8	FXP 16[1]/FXP 6[2]
Support SR	O	O	X	X	O
Generative Model	X	X	O	X	O
DSPs / kLUTs	2146/94.0	1512/167.0	1536/189.0[3]	0/21.6	260/97.3[3]
Throughput (GOPS)	624.0	780.0	670.0	-	363.5
Resource Efficiency (GOPS/DSP)	0.29	0.52	0.44	-	1.40
Power (W)	6.9	5.4	5.9	0.8	3.1
Energy Efficiency (GOPS/W or GSOPS/W)	90.3	144.9	113.5	0.22[4]	117.3/218.0[4]

1) ISE Mode   2) WSE Mode   3) ALMs for Intel FPGA   4) SOPS/W = (# of Spikes/Time) / (Power)

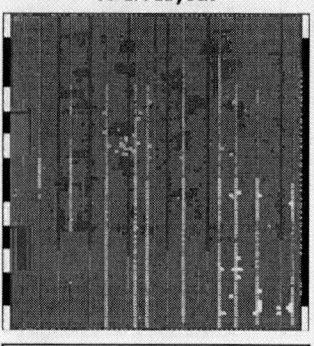

FPGA Layout

CNN Core	CNN Core Mem
OSSC	OSSC Mem

FPGA Specifications	
Platform	Intel Cyclone V
Clock	200 MHz
Logic Util.	97346
DSP	260
Block Mem.	10468967

Implementation Results	
Application	Super-Resolution
Network	SRGAN
Bit Precision	FXP16/FXP6
Peak Perf.	363.5 GOPS
Resource Eff.	1.40 GOPS/DSP
Power	3.1 W
Energy Eff.	117.3 GOPS/W

Fig. 7. FPGA Layout and Specifications

Additional References:

[4] S. Kim *et al.*, "A CNN Inference Accelerator on FPGA With Compression and Layer-Chaining Techniques for Style Transfer Applications." TCAS-I 2023.

[5] C. Fang *et al.*, "A 217.8 MSOPs/W FPGA-based Online Learning SNN Processor Using Unified Event-Driven Structure and Topology Aware Data Reuse Strategies." A-SSCC 2022.

[6] Y. Kuang *et al.*, "ESSA: Design of a Programmable Efficient Sparse Spiking Neural Network Accelerator," TVLSI 2022

[7] X. Ju *et al.*, "An FPGA implementation of deep spiking neural networks for low-power and fast classification," Neural Computation 2020, 32(1):182-204.

[8] S. Kim *et al.*, "Spiking-yolo: spiking neural network for energy-efficient object detection," AAAI 2020, 34(7):11270-11277.

[9] Y. Qiang *et al.*, "Constructing accurate and efficient deep spiking neural networks with double-threshold and augmented schemes." IEEE TNNLS 2021, 33(4):1714-1726.

[10] N. Rathi *et al.*, "Diet-snn: A low-latency spiking neural network with direct input encoding and leakage and threshold optimization." IEEE TNNLS 2021.

A 142.8-µW 98.1dB-SNDR Power/Bandwidth Configurable Fully Dynamic Discrete-Time Zoom ADC with Interstage Leakage Shaping

Yuke Shen, Shubin Liu, Kui Wen, Yanbo Zhang, Yi Shen,

Ruixue Ding, Zhangming Zhu

Key Laboratory of Analog Integrated Circuits and Systems (Ministry of Education), School of Microelectronics, Xidian University, Xi'an, China

In different application scenarios, high-performance Internet-of-Things (IoT) devices powered by batteries or energy harvesters demand analog-to-digital converters (ADCs) that possess a broad dynamic range (DR>100dB), adjustable power and bandwidth (BW), and relatively high energy efficiency (FoM>180dB). Continuous-time delta-sigma modulators (CTDSM) [1-2] are quite suitable for low-power designs due to its inherent anti-aliasing characteristics and relaxed settling requirements. However, loop filters based on RC time-constants cannot scale with sampling frequency. In addition, the design of traditional OTAs must accommodate the maximum BW, resulting in a trade-off between static power and dynamic performance. This paper presents a power/BW configurable fully dynamic discrete-time zoom ADC with 1st-order interstage leakage shaping (ILS). A low-cost 5-bit NS-SAR is used as the coarse ADC, which can substantially suppress the in-band quantization noise leakage in the zoom architecture due to non-unity STF of the fine DSM stage. As an added advantage, the quantization error leakage to the output due to gain error between the two stages can also be shaped out of band simultaneously. Fabricated in a 65nm CMOS process, the prototype ADC achieves 98.1dB SNDR over a 20kHz BW with an OSR of 128 and consumes 142.8µW, resulting in a DR-based Schreier FoM of 182.9dB. Measurement results show that the ADC can support 8x power/BW configurable with SNDR> 98dB.

The classic zoom ADC consists of a coarse Nyquist ADC and a fine $\Delta\Sigma$ ADC, as shown in Fig. 1 (top). The coarse ADC dynamically updates the DAC reference levels and transfers the residue to the next stage. The fine stage only needs to process the small residue signal, which significantly reduces the requirement for the first integrator. The two-stage zoom architecture improves the DR performance of the whole system. However, when targeting applications with different BW requirements, OTA based loop filters cannot be scalable according to the sampling frequency. In addition, due to the absence of the direct feedforward path for the signal, the STF of the second stage is not equal to 1. This, combined with the gain error of the coarse ADC, results in quantization noise leakage in the final output, expressed as $[1-(1-\Delta)STF]Q_1(z)$, impacting the in-band linearity and increasing the requirement of the subsequent digital decimation filter. Several methods have been proposed [3-4] to address this issue by compensating the STF, but they all require additional hardware budget.

Fig. 1 (bottom) shows the proposed fully dynamic discrete-time zoom ADC with ILS. The conventional OTA is replaced by the dynamic amplifier (DA) to implement the fine loop filter. Due to the fully dynamic operations in both stages without static currents, power and BW configurability can be achieved by simply changing the sampling frequency. Furthermore, by incorporating an error feedback (EF) path consisting of a gain stage and an FIR filter in the coarse ADC, noise shaping capability is achieved, modulating most of the first stage's quantization error leakage to higher frequency, expressed as $[1-(1-\Delta)STF][1-GH_{FIR}]Q_1$. Compared to the previous works, employing signal modulation instead of signal compensation, there is no need for an additional DAC or reconstruct the STF off-chip.

Fig. 2 (top) shows a simplified system-level block diagram of the proposed fully dynamic zoom ADC with ILS. A 5-bit SAR ADC with a EF path serves as the coarse NS quantizer while a single-bit 3rd-order CIFF discrete time $\Delta\Sigma$ modulator is employed as the fine quantizer. Both stages operate at the same sampling frequency, enabling real-time tracking and processing of the input signal. The coarse and fine conversion results are then processed in digital domain and calibrated by the following DWA to improve the linearity before sent to control the N-bit fine DAC. The proposed fully dynamic zoom ADC operates at a maximum sampling rate of 5.12 MHz with

an OSR of 128 and an over-ranging factor R=2. DAC levels versus time with and without the EF path is shown in Fig. 2 (bottom). For the same over-ranging factor, the intermediate DAC levels unused before can be utilized by employing NS quantizer in the coarse stage.

Fig. 3 shows the detailed circuit implementation and the timing diagram of the proposed zoom ADC. The two-stage FIA architecture [5] with class-AB biased output stage is employed in each integrator for improved output swing without dedicated CMFB circuits. The cascoded FIA is used as the second stage in the first integrator for higher DC gain. Closed-loop amplification operation based on dynamic FIA is enabled in each stage with UGB and PM scaled dynamically. The first integrator is chopped at Fs/2 to suppress the flicker noise and DC offset. In this design, the input sampling capacitors are reused as the feedback capacitors with a unit value of 400fF, making the total sampling capacitor C_S equal to 12.4pF. The bootstrap switches in each branch share the same peripheral bootstrapping circuit and the level-shift capacitor to save layout area. Bottom-plate sampling is used in the coarse ADC for better linearity and CM voltage keeps stable during the conversion by employing split capacitor array. The NS capability is realized by adding an EF path, where the loop filter is implemented by using capacitor stacking and a low-cost open-loop unity-gain buffer. The 1st-order FIR filter is achieved by residual sampling capacitors operating in a ping-pong manner. An inverter-based amplifier is used as the unity-gain buffer [6]. The open-loop gain is essentially determined by the gm ratio of the input and load transistors, which is relatively PVT-insensitive. Simple CMFB circuit is used for the fully differential application. The deterioration of gain accuracy can be negligible since the NS capability is intended to eliminate the leakage of the first stage's quantization noise. The overall noise transfer function (NTF) is well defined by the second fine stage.

The prototype ADC core in 65nm CMOS occupies an active area of 0.22mm² (Fig. 7). When clocked at 5.12MHz for a 20kHz BW, it only consumes 142.8µW from a 1.2V supply ($\Delta\Sigma$ modulator 90µW; NS-SAR 44.2µW; digital calibration 8.6µW). Fig. 4 (top) shows the measured output spectrum with and without DWA for a -1.6 dBFS input signal at 2.8kHz. Nearly 45.6dB improvement of the SFDR is substantially realized when the DWA is enabled, from 63.4 to 109dB. Fig. 4 (bottom) shows the measured output spectrum with ILS turned "off" and "on". It can be seen that the out of band tones due to the leakage of first stage's quantization noise are significantly attenuated and the peak SNDR and SFDR are improved by 1.5 and 1.4dB, respectively. The measured SNR and SNDR versus different input amplitudes are shown in Fig. 5 (top), illustrating a DR of 101.5dB. Performance and power versus different sampling frequency is shown in Fig. 5 (bottom). Measurement results show that the ADC can support 8x power/BW configurable with SNDR> 98dB just by changing the sampling frequency (OSR=128). Fig. 6 shows the performance summary and comparisons with state-of-the-art works having similar specifications. The proposed fully dynamic zoom ADC achieves competitive FoM of 182.9dB and can configure the power and BW flexibly, which is quite suitable for smart sensors and event-driven IoT devices with various requirements.

Acknowledgements:

This work was funded by the National Natural Science Foundation of China (92164301, 62021004, 62261160649, 62204192, 62022065, 62004052). Corresponding author: Shubin Liu and Zhangming Zhu.

References:

[1] Y. Zhong et al., "A 78.6 dB-SNDR 520mVpp-full-scale 620MΩ-Zin 105dB-CMRR VCO-based Sensor Readout Circuit Using FVF-Based Gm-Input Structure," A-SSCC 2022:1-3.

[2] M. Jang et al., "A 134-µW 99.4-dB SNDR Audio Continuous-Time Delta-Sigma Modulator With Chopped Negative-R and Tri-Level FIR-DAC," JSSC 2021:1761-1771.

[3] E. Eland et al., "A 440µW, 109.8dB DR, 106.5dB SNDR Discrete-Time Zoom ADC with a 20kHz BW," VLSI 2020:1-2.

[4] S. Karmakar et al., "A 280 µW Dynamic Zoom ADC With 120 dB DR and 118 dB SNDR in 1 kHz BW," JSSC 2018:3497-3507.

[5] X. Tang et al., " A 13.5b-ENOB Second-Order Noise-Shaping SAR with PVT-Robust Closed-Loop Dynamic Amplifier," ISSCC 2020:162-164.

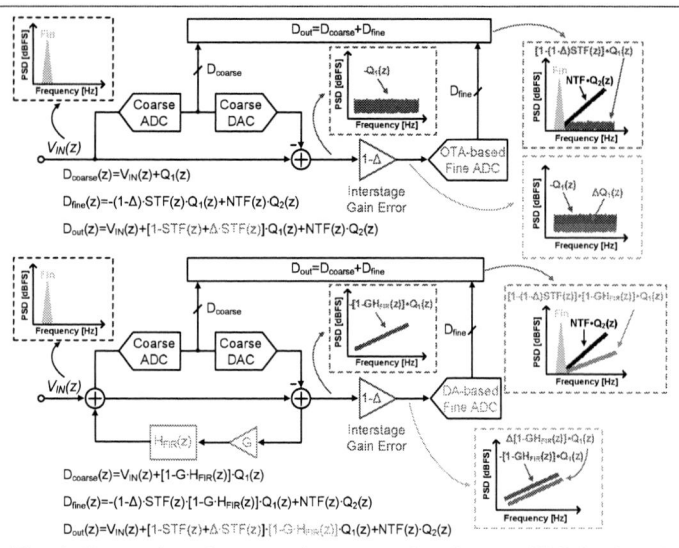

Fig. 1. Comparison between the conventional zoom ADC (top) and the proposed zoom ADC with interstage leakage shaping (bottom).

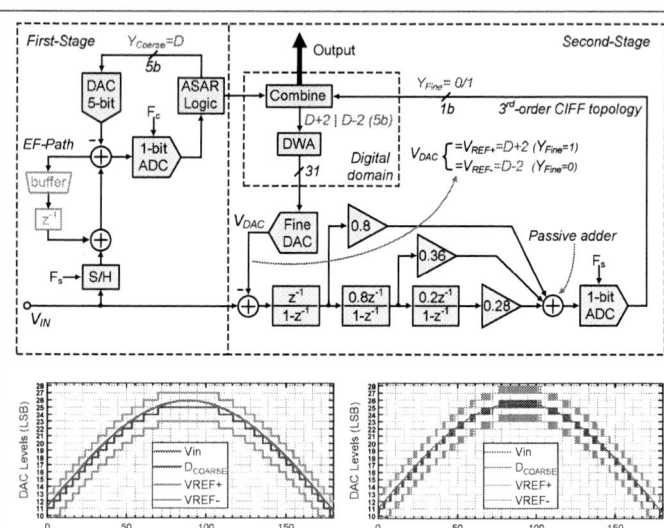

Fig. 2. Block diagram of the DT zoom ADC (top), waveforms of the reference zooming without and with the EF path (bottom).

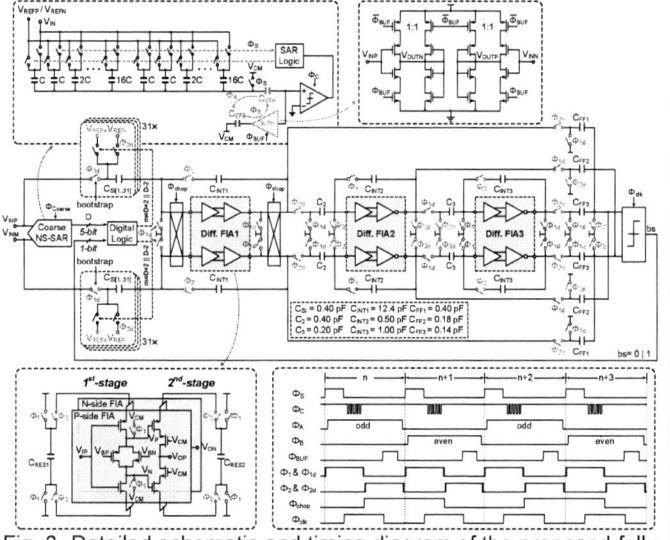

Fig. 3. Detailed schematic and timing diagram of the proposed fully dynamic zoom ADC with interstage leakage shaping.

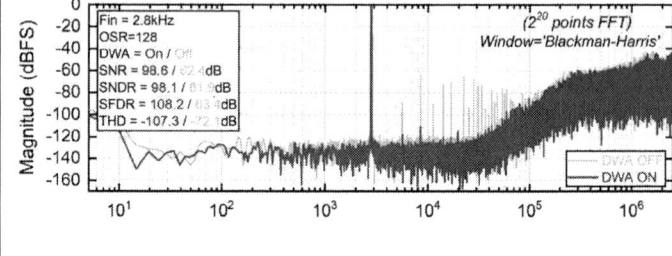

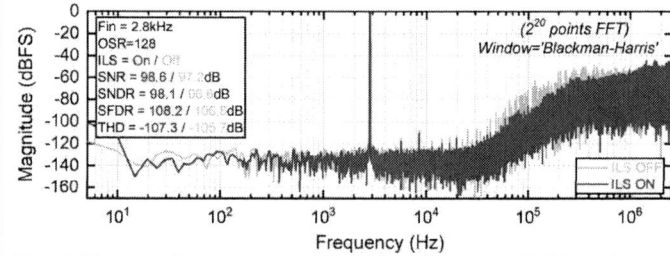

Fig. 4. Measured output spectrum with and without DWA (top), measured output spectrum with ILS turned "off" and "on" (bottom).

Fig. 5. Measured SN(D)R vs. input amplitude (top), SN(D)R vs. sampling rate (bottom left), and power vs. BW (bottom right).

Reference	ISSCC'22 Hu	ASSCC'22 Zhong	VLSI'20 Eland	ISSCC'20 Jang	ISSCC'18 Karmakar	This work
Architecture	MASH 2-0 DTDSM	VCO-based CTDSM	DT Zoom ADC	CTDSM	DT Zoom ADC	DA-based Zoom ADC
Power/BW Configurable	Yes	No	No	No	No	Yes
Technology [nm]	55	28	160	65	160	65
Area [mm²]	0.136	0.095	0.27	0.28	0.25	0.22
Supply Voltage [V]	1.2	1.1/0.55	1.8	1.2	1.8	1.2
Bandwidth [kHz]	1	10	20	24	1	20
Power [µW]	2.87	7.1	440	134	280	142.8
SNR [dB]	95.8	78.7	107.5	101	119.1	98.6
SNDR [dB]	94	78.6	106.5	99.4	118.1	98.1
DR [dB]	96.9	79.3	109.8	103.5	120.3	101.5
NDSD [dBFS/Hz]	-124	-118.6	-149.5	-143.2	-148.1	-141.1
FoM$_{SNDR}$ [dB]	179.4	170.1	183.1	181.9	183.6	179.5
FoM$_{DR}$ [dB]	182.3	170.8	186.4	186	185.8	182.9

NDSD=-[SNDR+10log(BW)]; FoM$_{SNDR}$=SNDR+10log(BW/Power); FoM$_{DR}$=DR+10log(BW/Power);

Fig. 6. Performance summary and comparisons.

Fig. 7. Chip micrograph.

Additional References:

[6] X. GUO *et al.*, "A Calibration-Free 13b 625MS/s Tri-State Pipelined-SAR ADC with PVT-Insensitive Inverter-Based Residue Amplifier," CICC 2022:01-02.

A 71-to-86GHz, 20.4dBm P_out, 6.0dB NF Transceiver with Quadrature Direct-Modulated Transmitter and Reflectionless Heterodyne Receiver in 40nm CMOS

Jie Zhou[1,2], Changxuan Han[1,2], Wen Chen[1,2], Bingzheng Yang[1,2], Xun Luo[1,2]

[1]Shenzhen Institute for Advanced Study, University of Electronic Science and Technology of China (UESTC), Shenzhen, China
[2]Center for Advanced Semiconductor and Integrated Micro-System, UESTC, Chengdu, China

In recent years, E-band transceivers (TRXs) with multi-Gb/s data-rate and high sensing accuracy are dramatically demanded for communication and radar systems [1,2]. To improve the wireless system integration level, transmitter (TX), receiver (RX), and phase-locked loop (PLL) are usually designed in a single chipset. Besides, the super-heterodyne architecture is widely adopted in E-band TRX, due to its simple implementation [3–5]. Whereas, the direct conversion approach requires a high isolation between TX and RX terminals. Then, the dual-mixing sliding-IF architecture is introduced [6,7]. Nevertheless, increased mixing blocks limit the entire system efficiency and introduce extra in-band mixing products. To overcome these issues, a TRX architecture with quadrature direct-modulated TX and reflectionless heterodyne RX is proposed for E-band applications. Note that, different signal modulation and demodulation methods are used in TX and RX paths to avoid the self-interferences.

The block diagram of the proposed E-band TRX is shown in Fig. 1. In TX path, the mm-wave digital-to-analog converters (mm-DACs) with quadrature architecture are introduced to convert the baseband signals into modulated signals directly. Then, to improve the output power, a two-way power-combined analog power amplifier (PA) is cascaded. In RX path, the common-gate common-source (CG-CS) noise-cancelling technique is utilized to achieve the low noise amplifier (LNA) circuit. Then, the received RX signals is down-converted to IF by a mixer. Besides, to reduce out-of-band reflection signals, a signal absorption network is inserted between the mixer and IF amplifier. The out-of-band reflection signals are absorbed by the resistors. The LO signals for TX and RX paths are generated by a sub-sampling PLL (SSPLL). A quadrature sub-sampling phase detector (QSSPD)-based dead zone automatic controller (DZAC) is presented in the SSPLL for fast-locking. The output signal of SSPLL is divided into two paths. Meanwhile, signal in each path is doubled by the injection-locked frequency multiplier (ILFM) and amplified by the LO driver. Note that, since the IF frequency of RX path is selected at 12GHz, the TRX can be applied in a frequency division duplexing (FDD) system. The TX and RX can also works on the same frequency by time-sharing switching operation of PLL.

Figure 2 (left) shows the schematic of proposed quadrature direct-modulated TX. Four 7-bit mm-DACs are utilized to construct the quadrature direct-modulator. Each mm-DAC is implemented by a Gilbert-type cell with digital-controlled current-DAC sub-arrays. Besides, each current-DAC sub-array consists of 4-bit MSB and 3-bit LSB unit cells, which are controlled by the thermometer-code-type baseband signals (i.e., BBI_n and BBQ_n). In addition, the quadrants of the output signals are selected according to the sign-bits (i.e., SI and SQ). Then, the corresponding two mm-DACs are worked on, while the rest are worked off. Extra sign-map circuits are not required. To further improve output power, an analog PA with three driver stages and one amplifier stage is introduced. The cascode topology with 2.2V supply is utilized in the driver and amplifier stages. Besides, a 4-to-1 power-combining transformer is adopted to combine two-way PAs for high output power. The schematic of proposed reflectionless heterodyne RX is shown in Fig. 2 (right). A three-stage CS and CG path are combined by an asymmetrical power-combing transformer to construct the LNA. The phase and amplitude imbalances between CG and CS paths are reduced by the asymmetrical transformer, which can improve the noise-cancelling ratio. Meanwhile, a positive feedback is introduced in the CG path to improve gain and partially reduce noise voltage simultaneously. A signal absorptive network, composed of a CG stage and two resistors, is placed between mixer and IF amplifier. Such network with a reflectionless bandpass filtering response can dissipate the out-of-band reflection signals to decrease the unwanted high-order mixing products.

Figure 3 (top) shows the block diagram of the proposed wideband fast-locking SSPLL, which is composed of two feedback loops (i.e., sub-sampling loop (SLL) and frequency-locked loop (FLL)) and a QSSPD-based DZAC. Besides, a mm-wave quad-mode oscillator is integrated in the SSPLL to achieve a wide frequency range. The SLL is used for phase locking. However, when the phase error of SSPD in SLL exceeds the dead zone, a long lock time is required. Then, the QSSPD-based DZAC automatically switch on the FLL for fast-locking. Note that, during the locking state, only the SSL is active. The jitter from FLL has no contribution on the entire jitter of the proposed PLL. Thus, long locking time caused by the dead zone of the FLL is avoided without any jitter degradation. The measured locking time is better than 1.3μs, when the QSSPD-based DZAC is enable. The LO chain connecting SSPLL and TX/RX is shown in Fig. 3 (bottom), which is composed of a VCO buffer, an ILFM, and an output driver. The ILFM is implemented by a differential-in-differential-out topology with a transformer-based LC tank. The cross-coupled pair is connected to the transformer's primary coil and the input current is directly injected into the cross-coupled pair. Besides, differential CS stages are adopted to implement the buffer and the driver for high drive capability. Note that, in order to achieve TX and RX working on the same frequency, the LO chains are optimized at different frequencies.

The presented E-band TRX with quadrature direct-modulated TX and reflectionless heterodyne RX is fabricated in a conventional 40nm CMOS technology (Fig. 7) and occupies a 2.8mm² chip area. The whole TRX consumes 768.8mW, where TX consumes 581.5mW, RX consumes 50.7mW, SSPLL consumes 23.6mW, LO chain consumes 50.8mW for RX and 62.2mW for TX. Figure 4 (top left) shows the measured output power (P_out) and system efficiency (SE) of the TX. The TX achieves a peak P_out of 20.4dBm at 77GHz. Besides, the calculated peak SE is 19.2%. Note that, all the power consumptions in the TX path are considered in the calculation of SE. The measured conversion gain and noise figure (NF) of the RX is shown in Fig. 4 (bottom left). The RX achieves a peak gain of 22.0dB at 72GHz. Besides, the measured NF is from 6.0 to 8.5dB within 71 to 86GHz, while the minimal NF is 6.0dB at 71GHz. Meanwhile, the proposed E-band TRX supports 3Gb/s 64-QAM signals (Fig. 4 right). Fig. 5 shows the measured typical phase noises (PNs), RMS jitter, and reference spur. The measured output integrated jitter is less than 86fs within the operation frequency range. The measured results are summarized and compared with the prior-art E-band TRXs in Fig. 6. The fabricated E-band TRX achieves a competitive NF and P_out with high SE in a conventional CMOS technology.

Acknowledgement:

Authors thank the support of National Key R&D Program of China under Grant 2021YFE0205600, National Natural Science Foundation of China under Grant 61934001, 62161160310, and Shenzhen Science and Technology Program under Grant JCYJ20210324120004013.

References:

[1] N. Ebrahimi et al., "A 71-76/81-86 GHz, E-band, 16-Element Phased-Array Transceiver Module with Image Selection Architecture for Low EVM Variation," RFIC 2020:95–98.

[2] T. Arai et al., "A 77-GHz 8RX3TX Transceiver for 250-m Long-Range Automotive Radar in 40-nm CMOS Technology," JSSC 2021:1332–1344.

[3] B. Ginsburg et al., "A Multimode 76-to-81GHz Automotive Radar Transceiver with Autonomous Monitoring," ISSCC 2018:158–159.

[4] M. Kucharski et al., "A Scalable 79-GHz Radar Platform Based on Single-Channel Transceivers," TMTT 2019:3882–3896.

[5] T. Ma et al., "A CMOS 76-81-GHz 2-TX 3-RX FMCW Radar Transceiver Based on Mixed-Mode PLL Chirp Generator," JSSC 2020:233–248.

[6] R. Levinger et al., "High-Performance E-Band Transceiver Chipset for Point-to-Point Communication in SiGe BiCMOS Technology," TMTT 2016:1078–1087.

[7] N. Ebrahimi et al., "A High-Fractional-Bandwidth, Millimeter-Wave Bidirectional Image-Selection Architecture with Narrowband LO Tuning Requirements," JSSC 2018:2164–2176.

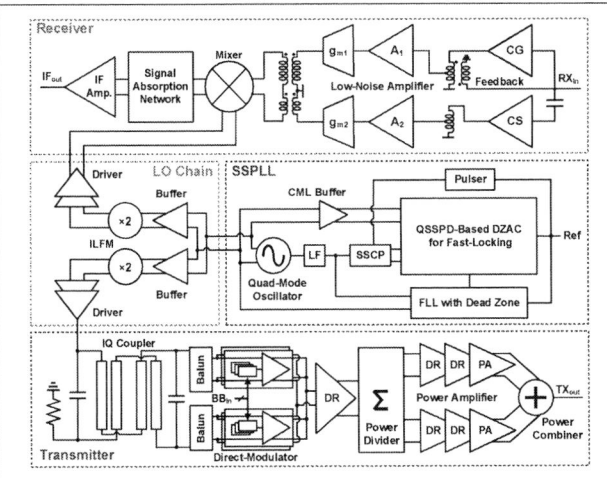

Fig. 1. Architecture of the proposed E-band transceiver.

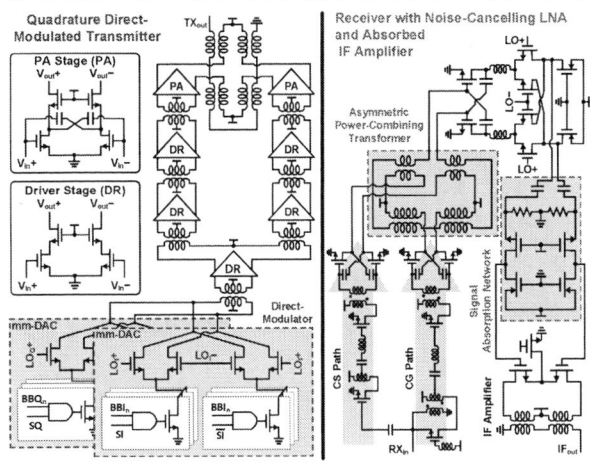

Fig. 2. Schematic of the quadrature modulated transmitter and reflectionless heterodyne receiver.

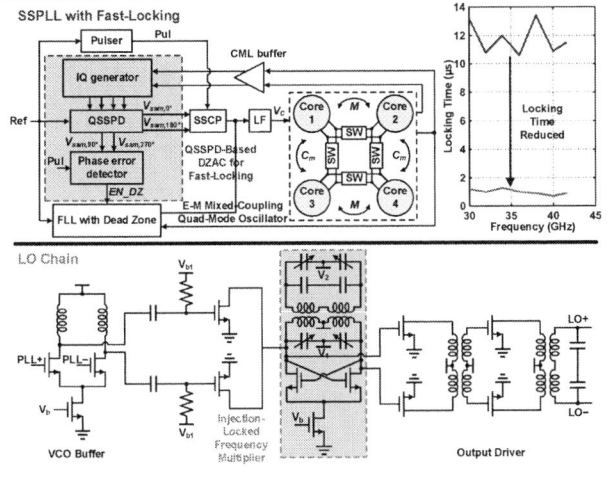

Fig. 3. Block diagram of the fast-locking SSPLL and schematic of the LO chain.

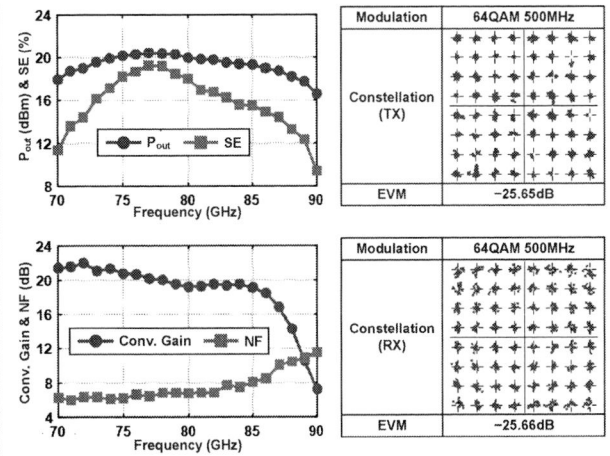

Fig. 4. Measured output power and system efficiency of transmitter. Measured conversion gain and noise figure of receiver. Measured constellation of 3Gb/s 64-QAM signals at 73GHz.

Fig. 5. Measured phase noise, RMS jitter, and reference spur of the SSPLL.

Ref.	This Work	RFIC 2020 [1]	JSSC 2021 [2]	TMTT 2019 [4]
Architecture	FL-SSPLL + Direct-Modulated TX + Reflectionless RX	4-Element Weaver Sliding IF Image-Selection TRX	PLL + Time-Division-Multiplexing MIMO 3TX/8RX	PLL + Quadrature Direct-Conversion TRX
Technology	40nm CMOS	90nm SiGe	40nm CMOS	130nm SiGe
Frequency (GHz)	71–86	71–76/81–86	76–77	79
PN (dBc/Hz@1MHz)	−114@16.2GHz	N.A.	−91@76GHz	−80@39.2GHz
TX P_{out} (dBm)	18.7–20.4	8	14.1	19.2
TX SE (%)	13.6–19.2	N.A.	N.A.	N.A.
RX CG (dB)	18.5–22.0	32	17	27.5
RX NF (dB)	6.0–8.5	<9	<14	7.2
Modulation	64QAM 3Gb/s	64QAM 9Gb/s	N.A.	N.A.
EVM	−25.65dB	−24dB	N.A.	N.A.
P_{DC} (W)	0.77(TRX+PLL)	0.25(TX), 0.16(RX)	3.2(8RX+3TX+PLL)	1.02(TRX+PLL)
Chip Area (mm²)	2.8(TRX+PLL)	12.8(4TRX)	48.2(8RX+3TX+PLL)	6.68(TRX+PLL)

Fig. 6. Comparison with the state-of-the arts.

979-8-3503-3004-5/23 $31.00 © 2023 IEEE

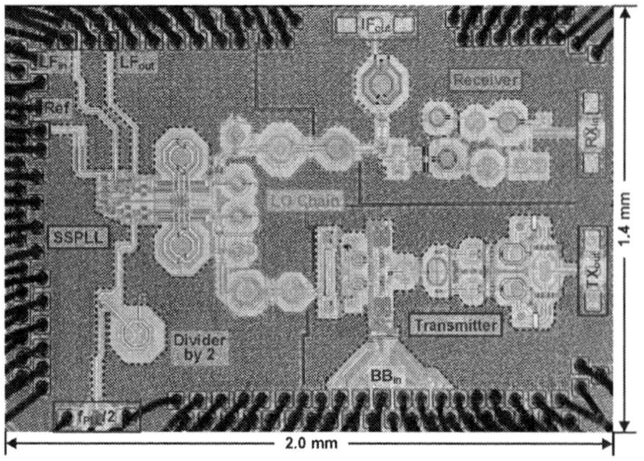

Fig. 7. Chip Microphotograph.

K/Ka-Band 4-Element 4-Beam Hybrid Phased-Array Transmitter and Receiver Front-Ends with Compact Layout Floor-Plans and Fault-Tolerant Digital Circuits

Mengru Yang[1,2], Chenyu Xu[1,2], Peng Gu[1,2], Yongran Yi[1,2], Mohan Guo[1,2], Liangliang Liu[2], Xiangxi Yan[2], Xiaohu You[1,2], Dixian Zhao[1,2]

[1]National Mobile Communication Research Laboratory, Southeast

University, Nanjing, China

[2]Purple Mountain Laboratories, Nanjing, China

Space-terrestrial networks utilizing low-Earth-orbit (LEO) satellite constellations have emerged as promising 6G applications. The K-/Ka-bands are allocated with frequencies of 17.7-20.2 GHz and 27.5-30 GHz as transmit and receive bands in the space segment. Due to the limited payload of satellites, fully-connected multi-beam phased-arrays [1] are exploited to improve data rates and spectral efficiency. However, realizing such architectures faces three challenging issues. First, the layout floor-plans are often in complexity, resulting in long interconnects, serious couplings and low area utilization. Second, LEO satellites have limited space and cannot afford large size of the antenna board. To achieve better EIRP and G/T metrics, high output power in the transmitter (TX) and low NF in the receiver (RX) are required. The third issue is that single-event effects (SEE) induced by energetic particles can potentially produce register flips in digital circuits when the ICs are used in space. To overcome these issues, this paper presents K/Ka-band 4-element 4-beam hybrid phased-array TX and RX front-ends. The proposed TX and RX achieve compact and simplified floor-plans, enhanced output power and NF performance, and fault-tolerant serial peripheral interface (SPI).

The block diagram of the TX is illustrated in Fig. 1, consisting of one CMOS beamformer and four GaAs PAs. The four beams are fed from the top and bottom, and the four antenna elements will be interfaced on the left and right sides. By splitting the channels into two columns, coherent signals can be combined in two distinct areas, which thus shortens the length of interconnects by almost half. The design of the tree-type beam-combination network becomes simplified. In addition, the channels from one beam are staggered on both sides, lending themselves to symmetrical placement via the rectangular beam-distribution network (RBDN). The detailed schematic of the RBDN is illustrated in Fig. 2, comprising two 1:4 differential power dividers (PD) and coplanar waveguides with isolation-enhanced shielded walls. The PD utilizes the capacitor-free folded transmission-line (T-line) structure. A distributed model is constructed to account for the parasitic effects of folded T-lines for better optimization. By arranging multiple PD cells into two rows, area utilization is greatly improved. To guarantee signal consistency, special attention has been paid to equalizing the length of interconnects. The electro-magnetic (EM) simulation reveals that phase and amplitude mismatches are less than 3.9° and 0.2 dB across different routings.

Minimizing power dissipation of one channel is significant for such 4-element 4-beam architecture as the total power will increase 16 times. Given that fully-passive beamforming blocks introduce intolerable insertion loss, a passive vector-modulated phase shifter (VMPS) along with a variable-gain amplifier (VGA) is employed, resulting in a total dc power of only 5 mW. In the VMPS, transistor arrays from the I- and Q-path are independently controlled, ensuring an invariable impedance among different phase states. The VGA provides a 31.5-dB tuning range by switching 8-bit weighted transistor arrays. Since the tuning step will gradually increase with the gain decreased, an additional 4-bit attenuator is followed to maintain 0.5-dB resolution in low-gain mode. Besides, the CMOS power amplifier (PA) adopts a differential common-source structure and the second harmonic trap technique, providing a highly-linear power gain to drive the GaAs PA. Series RC feedback networks are added to extend the bandwidth.

The RX configuration in Fig. 1 is similar to the TX while paying more attention to low power due to the relaxed linearity requirements. Therefore, the LNA adopts a single-ended cascode structure with small size of transistors. To achieve broadband, the inter-stage of the LNA introduces a Π-type dual-resonant network. Additionally, the input and output matchings use the inductive source degeneration and L-type inductor-enhanced network to reduce the return losses. The L-type network is built by two magnetically-coupled inductors

with opposite current flows. As depicted in Fig. 2, such transformer-like topology leads to additional design freedoms. In the Smith chart, its trajectory moves from a high impedance to 50 Ω. Meanwhile, it combines the advantages of broadband, compactness and robust to the process variations. The added GaAs LNA can suppress considerable noise from the CMOS beamformer.

The fault-tolerant SPI is illustrated in Fig. 3. The implementation of triple redundancy for the central control logic module is proposed. The three sets of modules operate in parallel and employ a majority voting mechanism to determine the final output. The majority voting is realized through combinational logic blocks with optimized routing connections. If there is a discrepancy between two modules, the third one can be used to identify and correct the single register flip. An example of the proposed SPI subjected to radiation is depicted in Fig. 3, proving superior failure tolerance than traditional designs. To further minimize the probability of simultaneous errors occurring in two registers, these modules are independently routed and keep at a distance from each other. The SPI operates at 100 MHz, ensuring a fast and flexible configuration to multiple beams.

The TX and RX beamformers are fabricated in 65-nm bulk CMOS with areas of 4x5 mm^2 (Fig. 7), while consuming a maximal power of 240 and 144 mW, respectively. The measurements are performed by probing one input and one output while leaving all other ports open-circuited. The measured performance of the TX is shown in Fig. 4. It achieves a typical gain of 11 ± 0.8 dB across 17.7-20.2 GHz with <1.5 dB variations among channels. The measured OP$_{1dB}$ reaches 9 dBm accompanied by a PAE$_{1dB}$ of 16%. The RMS phase and gain errors are less than 2.5° and 0.2 dB. Besides, the channel coupling (e.g., CH$_4$ to the others) is investigated by evaluating changes of phase and amplitude when nearby channels' states are swept. The measured couplings are below -40 dB, indicating good isolation among beams. Similar to the TX, the measured performance of the RX is depicted in Fig. 5. It achieves a typical gain of 8 dB across 27.5-30 GHz with <1.7 dB variations among channels. Thanks to the proposed L-type network, S$_{22}$ is consistently below -12 dB. The measured NF is about 13 dB, while the IP$_{1dB}$ is -20 dBm. The RMS phase and gain errors are less than 3° and 0.28 dB, respectively. The measured couplings remain below -40 dB across all cases.

Fabricated in 0.1μm GaAs pHEMT, the GaAs PA and LNA occupy 4 and 2 mm^2 (Fig. 7). The measurement results are depicted in Fig. 4 and Fig. 5, respectively. The GaAs PA delivers 24.5-dBm OP$_{1dB}$ and the GaAs LNA achieves 1.9-dB NF. Benefit from them, the overall TX PAE$_{1dB}$ achieves 40% and the overall RX NF is about 3.3 dB. Figure 6 compares this work with the state-of-the-art multi-beam phased-array ICs [1-5]. This work demonstrates the first-reported 4-element 4-beam hybrid phased-array front-ends. It supports flexible main-lobe and nulling steering for simultaneous four beams. Besides, high TX efficiency and low RX RF power consumption are achieved, which satisfies the LEO satellite requirements.

Acknowledgement:

This work was supported in part by the National Key Research and Development Program of China under Grant 2019YFB1803000 and in part by the Major Key Project of Peng Cheng Laboratory, Shenzhen, China, under Project PCL2021A01-2.

References:

[1] N. Li et al., "A 4-Element 7.5-9 GHz Phased Array Receiver with 8 Simultaneously Reconfigurable Beams in 65 nm CMOS Technology." *IEEE RFIC*, pp. 83-86, Aug. 2020.

[2] S. Mondal et al., "A 25–30 GHz Fully-Connected Hybrid Beamforming Receiver for MIMO Communication," *IEEE JSSC*, vol. 53, no. 5, pp. 1275-1287, May 2018.

[3] H. Mohammadnezhad et al., "A 64–67GHz partially-overlapped phase-amplitude-controlled 4-element beamforming-MIMO receiver," *IEEE CICC*, pp. 1-4, 2018.

[4] Y. -S. Yeh et al., "Multibeam Phased-Arrays Using Dual-Vector Distributed Beamforming: Architecture Overview and 28 GHz Transceiver Prototypes," *IEEE TCAS-I*, vol. 67, no. 12, pp. 5496-5509, Dec. 2020.

[5] N. Peng et al, "A Ka-Band CMOS 4-Beam Phased-Array Receiver With Symmetrical Beam-Distribution Network," *IEEE SSC-L*, vol. 3, pp. 410-413, 2020.

979-8-3503-3004-5/23 $31.00 © 2023 IEEE

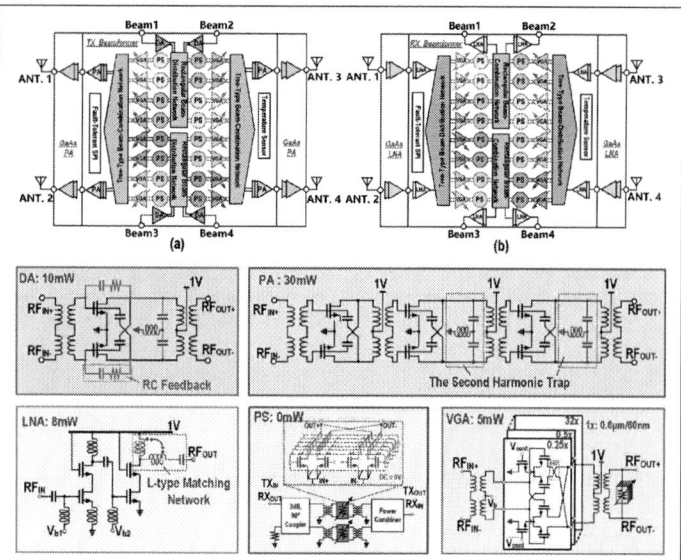

Fig. 1. Block diagrams of the 4-element 4-beam phased-array (a) TX, (b) RX front-ends, and schematics of key building blocks.

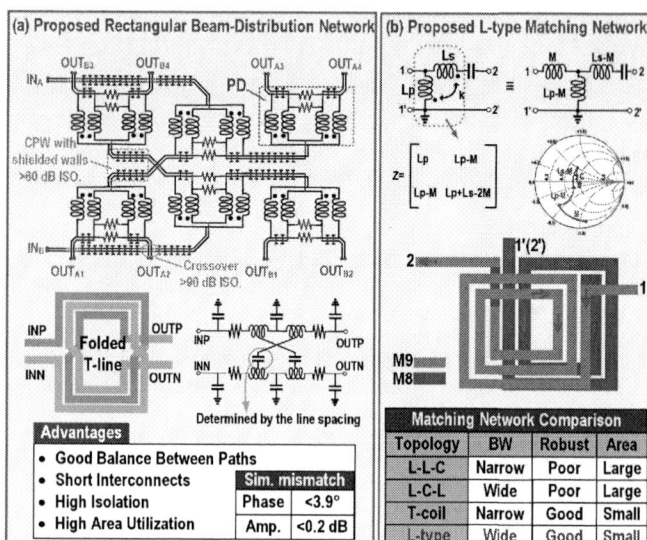

Fig. 2. Details of (a) proposed rectangular beam-distribution network and (b) proposed L-type matching network.

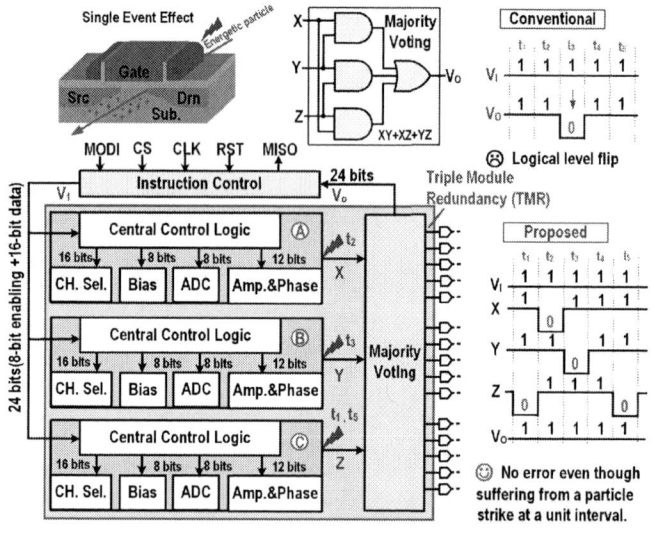

Fig. 3. Fault-tolerant SPI based on triple module redundancy.

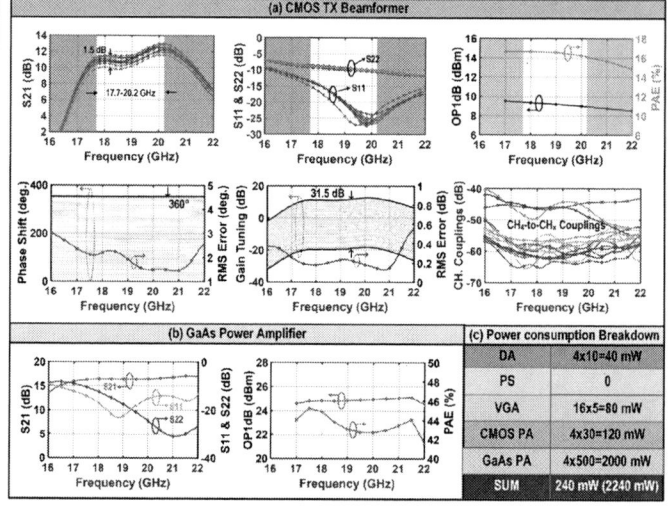

Fig. 4. Measured performances of (a) CMOS TX beamformer and (b) GaAs PA, as well as (c) power consumption breakdown.

Fig. 5. Measured performances of (a) CMOS RX beamformer and (b) GaAs LNA, as well as (c) power consumption breakdown.

References	This Work		RFIC[1]	JSSC[2]	CICC[3]	TCAS-I[4]	SSC-L[5]
Process	65-nm CMOS 0.1μm GaAs pHEMT		65-nm CMOS	65-nm CMOS	130-nm BiCMOS	120-nm BiCMOS	65-nm CMOS
Beamforming Architecture	Fully-con. TX FE	Fully-con. RX FE	Part.-con. RX FE	Fully-con. Full RX	Fully-con. TX FE	DVDB RX FE	Fully-con. RX FE
Frequency (GHz)	17.7-20.2	27.5-30	7.5-9	25-30	57-66	26-30	27-31
Phase Step (°)	5.625	5.625	5.625	5.625	22.5	11.25	5.625
RMS Phase Error (°)	<2.5	<3	<2		<8.8	<2.3	<4
Gain Tuning Range/Step (dB)	31.5/0.5	31.5/0.5	15.75/0.25	-	7/1	-	15.5/0.5
RMS Gain Error (dB)	<0.2	<0.28	<0.32		<0.53		<0.32
Normal Gain (dB)	11 (26a)	8 (23a)	20	34b	15	7.3	2
TX OP1dB (dBm)	9 (23a)	-	-		7.1		
RX NFmin (dB)	-	13 (3.3a)	8	7.3b		5	12
RX IP1dB	-	-20		-29		-16	-25
No. of Elements x No. of beams	4x4		8x4	8x2	3x2	4x4##	2x4
Total Power (mW)	240(2240a)	144(264a)	864	340b	404	500	40
PDC/CH. (mW)	15(140a)	9(16.5a)	27	27.5b	101	125.7	5
Chip Area (mm²)	20	20	19.6#	6.16	2.88	8.91	10.4

a Adding GaAs PA/LNA b Including down-conversion
Excluding pads ## Supported by DVDB architecture

Fig. 6. Performance comparison with state-of-the-art works.

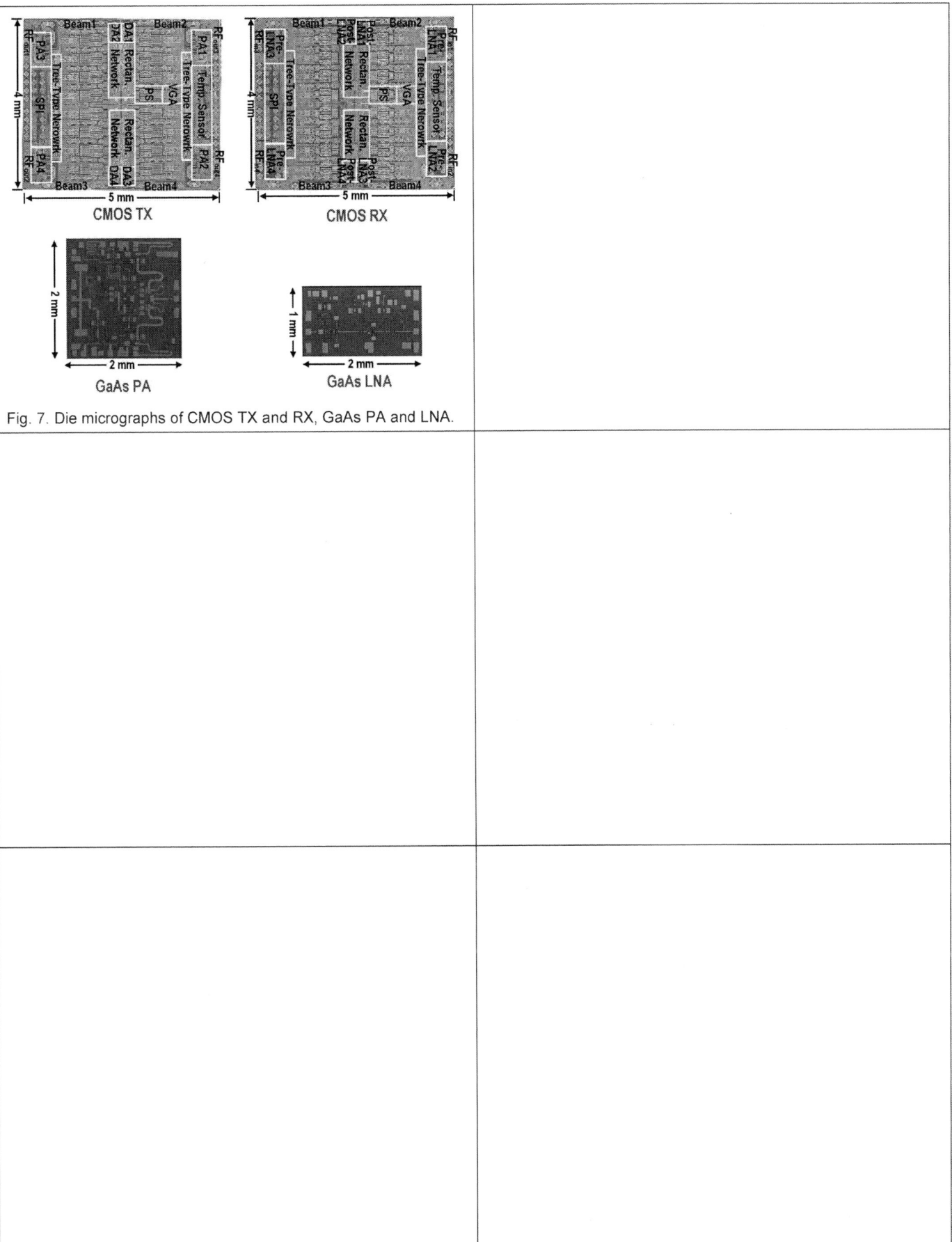

Fig. 7. Die micrographs of CMOS TX and RX, GaAs PA and LNA.

A 54-to-69-GHz Wideband 2T2R FMCW Radar Transceiver Employing Cascaded-PLL Topology and PTAT-Enhanced Temperature Compensation in 40-nm CMOS

Chenyu Xu[1,2], Xiaofei Liao[1,2], Feifan Hong[1,2], Mengru Yang[1,2], Peijuan Ju[1,2], Wendi Chen[1,2], Pengfei Diao[1,2], Hao Gong[1,2], Xiang Liu[2], Xiaohu You[1,2] and Dixian Zhao[1,2]

[1]National Mobile Communications Research Laboratory, Southeast University, Nanjing, China

[2]Purple Mountain Laboratories, Nanjing, China

The unlicensed frequency band around 60 GHz (i.e., 57 to 66 GHz) is always the hot topic of researches due to its abundant spectrum resources. In addition to the high-data-rate wireless communications, current research efforts have also been focused on the short-range frequency-modulated continuous wave (FMCW) radar systems [1-3]. However, there are still some challenges to address. First, realizing a fully-integrated V-band FMCW radar transceiver (TRX) is nontrivial. Prior arts either exclude the integration of critical blocks (e.g. FMCW synthesizer, analog baseband (ABB)) or contain with limited chirp linearity and slope [1-3]. Second, due to the uncertain operating environment of radar systems, maintaining constant performance is required over a wide temperature range. Finally, it remains not easy to design the TRX that covers the entire band. In this paper, we present a 54-to-69-GHz fully-integrated wideband FMCW radar TRX with two transmitter and receiver channels (2T2R). To simplify the off-chip components, a cascaded-PLL topology is employed, achieving a maximum chirp bandwidth of 7.2 GHz with a fast slope up to 468.5 MHz/μs. Thanks to the proposed proportional-to-absolute-temperature (PTAT)-enhanced temperature compensation (TC) technique, the average conversion gain (CG) varies less than 1.1 dB from -55 to 105 °C.

Two typical applications are shown in Fig. 1. Thanks to the 2T2R architecture, not only the traditional forward detection but a full 360° detection is possible with proper antenna configurations. Two receiver (RX) antennas and two transmitter (TX) antennas are placed alternatively every 90° with radiation direction parallel to the chip's front face. Each antenna has a radiation angle exceeding 180°. Since the minimum antenna gain position of TX exactly aligns with the maximum position of RX, an approximate unit antenna gain of the whole TRX is achieved. The block diagram of the radar TRX is also depicted. The FMCW signal is generated by two cascaded PLLs and fed to the frequency quadrupler, multiplying the signal to V-band and then delivering it to all RF channels. The entire RF signal chain is fully differential connected by magnetically coupled resonators (MCRs). Each MCR is deliberately designed for wideband operation.

The diagram of the cascaded-PLL FMCW generator is illustrated in Fig. 2. PLL1 is an integer-N structure with a narrow loop bandwidth (BW), providing a low-noise 600-MHz reference signal for PLL2. An active and on-chip 2-order low-pass filter is employed to reduce the required capacitance. PLL2 adopts a fractional-N structure, serving as a FMCW modulator with a large loop BW for quick frequency tracking. A wideband VCO with flat gain sensitivity is implemented in PLL2 to extend the continuous sweeping band and guarantee the chirp linearity. It confers three key merits of this topology. 1) Low phase noise (PN) and fast frequency sweeping can be achieved simultaneously by setting the two PLLs with different loop BWs [4]. 2) As the reference signal is multiplied by PLL1, the frequency and cost of the off-chip crystal are decreased. 3) An extra 200-MHz signal generated from PLL1 can serve as a homologous clock for the whole radar system, simplifying the off-chip clock management.

For wideband operation, the performance of quadrupler is crucial as it should drive the whole TRX with a flat frequency response. The schematic of the quadrupler is depicted in Fig. 2, which consists of three amplifiers and two frequency doublers cascaded alternately. However, the unbalance caused by single-ended-to-differential conversions and high inserting loss introduced by wideband MCRs degrade the performance severely. To address them, a capacitor is inserted into the center tap of each transformer and resonates with the connected coil. This resonant tank absorbs the undesirable in-band common-mode signal and enhances the CG. As each stage provides sufficient in-band linearity, the final output power of the quadrupler is flat.

The PTAT current is a popular TC technique used in RF circuits to mitigate the temperature sensitivity [5]. However, two crucial issues hamper its employment in this design. 1) The traditional PTAT slope is incapable to compensate the gain variation versus temperature of V-band RF building blocks, which are usually biased in the deep-saturation region for high power gain. 2) The constant slope of PTAT cannot satisfy diverse temperature characteristics of different RF modules. To overcome these limitations, this work proposes a PTAT-enhanced TC technique, demonstrated in Fig. 3. An amplifier (A_2) magnifies the difference between the PTAT voltage (V_P) and the reference voltage (V_R), outputting a PTAT-enhanced voltage (V_{P-E}). The normalized V_{P-E} can be expressed as $((1 + R_5/R_4) \times (T/300K) - 2))$, with an enhancement factor R_5/R_4. V_R and V_{P-E} are used to generate zero-to-absolute-temperature currents (I_{ZTATS}) and PTAT-enhanced currents (I_{P-ES}), respectively. The TC current (I_{TC}) is the generated by adding up a 4-bit I_{ZTAT} and a 4-bit I_{P-E} with complementary control words to ensure consistency at room temperature. The slope of I_{TC} can be configured according to different RF modules' requirements.

Fabricated in 40-nm bulk CMOS, the TRX occupies 3×3.3 mm^2. The chip micrograph is shown in Fig. 7. Powered by a 1.1-V and a 3.3-V supply voltages, the total power consumption of the TRX is 695 mW, with all channels turned on. Figure 4 summarizes the measurement results of the TRX. The CG of the RX can be tuned from 20 to 87 dB with a 3-dB bandwidth over 15 GHz (i.e., 54-69 GHz). The total 67-dB tuning range is achieved consisting of 60 dB by ABB at IF and 7 dB by LNA at RF, respectively. The 7-dB RF gain tuning also adjusts the IP1dB from -21 to -12 dBm. The measured IF bandwidth covers the range from 10 kHz to 20 MHz with configurable high-pass and low-pass cut-off frequencies. The average variation of RX CG and NF are less than 1.1 dB and 2.3 dB from -55 to 105 °C, respectively. The P_{SAT} of TX drifts about 2.1 dBm across this temperature span.

Figure 5 illustrates the performance of the cascaded-PLL FMCW synthesizer. Measured directly at the output of PLL2, the PN is -105.3 dBc/Hz at 1-MHz offset, which is equivalent to -93.3 dBc/Hz when normalized to 60 GHz. The chirp results are measured at the output of the divider-16 and normalized to 60 GHz with a multiplication factor of 64. A 7.2-GHz modulation chirp bandwidth is achieved with a slope rate up to 468.5 MHz/μs. The RMS FM errors at slow and fast chirp slopes are 3.2 kHz and 1.4 MHz (i.e., 0.0004‰ and 0.019%), respectively. Under the sawtooth waveform mode, the flyback time is less than 1 μs.

Figure 6 summarizes the comparison with the state-of-the-art FMCW radar TRXs. This work achieves the widest bandwidth and RX CG tuning range. The chirp slope and linearity are superior among them. Other parameters such as NF and integration are also competitive.

Acknowledgement:

This work was supported in part by the National Key Research and Development Program of China under Grant 2019YFB1803000 and in part by the Major Key Project of Pengcheng Laboratory (PCL) under Grant PCL2021A01-2.

References:

[1] J. Rimmelspacher et al., "Low Power Low Phase Noise 60 GHz Multichannel Transceiver in 28 nm CMOS for Radar Applications," IEEE RFIC 2020: 19-22.

[2] H. Kuo et al., "A Fully Integrated (60-GHz CMOS Direct-Conversion Doppler Radar RF Sensor With Clutter Canceller for Single-Antenna Noncontact Human Vital-Signs Detection," IEEE T-MTT 2016: 1018-1028.

[3] K. Dandu et al., "High-Performance and Small Form-Factor Mm-Wave CMOS Radars for Automotive and Industrial Sensing in 76-to-81GHz and 57-to-64GHz Bands," IEEE ISSCC 2021: 39-41.

[4] X. Liao et al., "An 18.8-to-20.3-GHz Wide-Ramping-Range Cascaded-PLL-Based FMCW Generator with 44.1-kHz RMS Frequency Error and -105.6-dBc/Hz Phase Noise in 40-nm CMOS," IEEE ASSCC 2022: 1-3.

[5] D. Zhao et al., "A K-Band Hybrid-Packaged Temperature-Compensated Phased-Array Receiver and Integrated Antenna Array," IEEE T-MTT 2022: 1-15.

[6] Z. Duan et al., "A 76-to-81GHz 2 × 8 FMCW MIMO Radar Transceiver with Fast Chirp Generation and Multi-Feed Antenna-in-Package Array," IEEE ISSCC 2021: 228-230.

979-8-3503-3004-5/23 $31.00 © 2023 IEEE

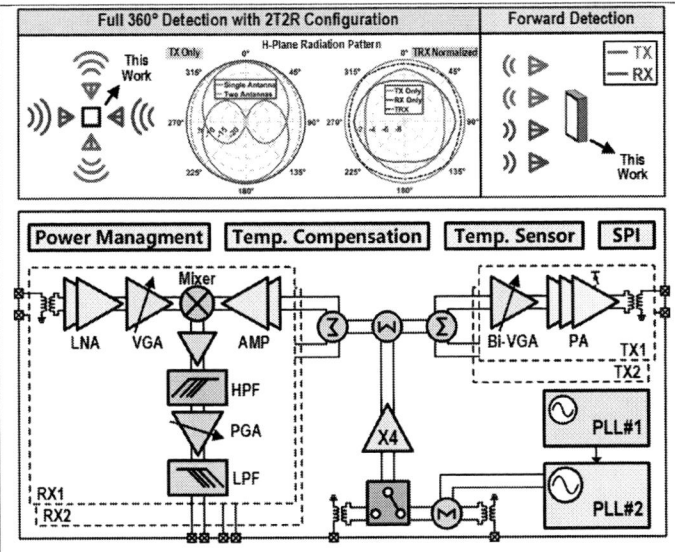

Fig. 1. Two typical applications of this work with corresponding antenna configurations and the diagram of the FMCW radar TRX.

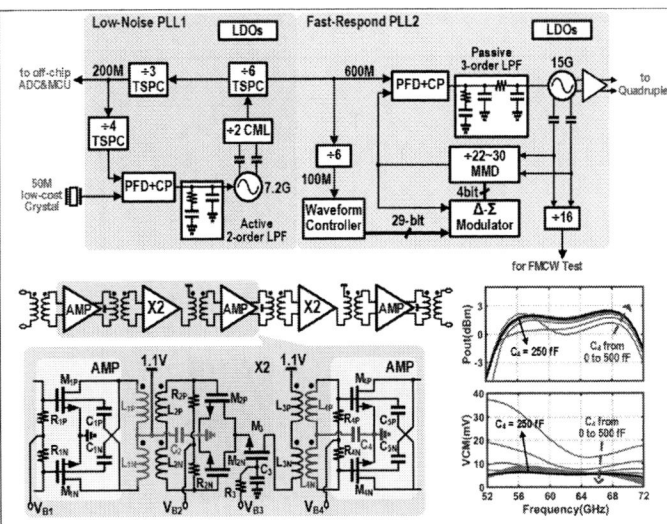

Fig. 2. The diagram of the cascaded-PLL FMCW synthesizer and the schematic of the frequency quadrupler with the common-mode resonant tank.

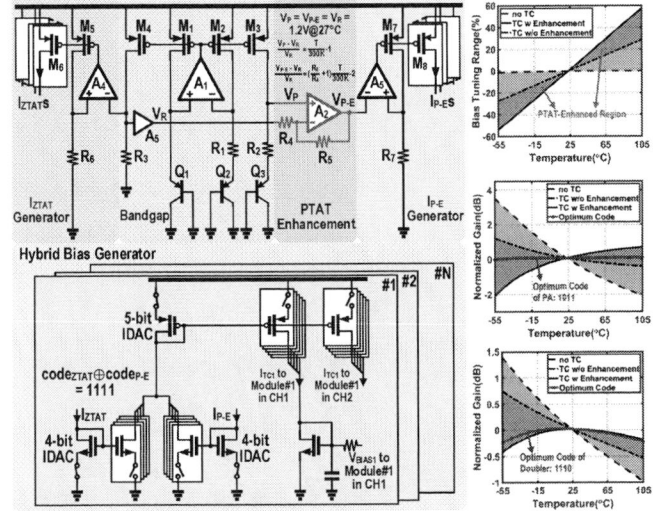

Fig. 3. The schematic and simulated results of the PTAT-enhanced temperature compensation technique.

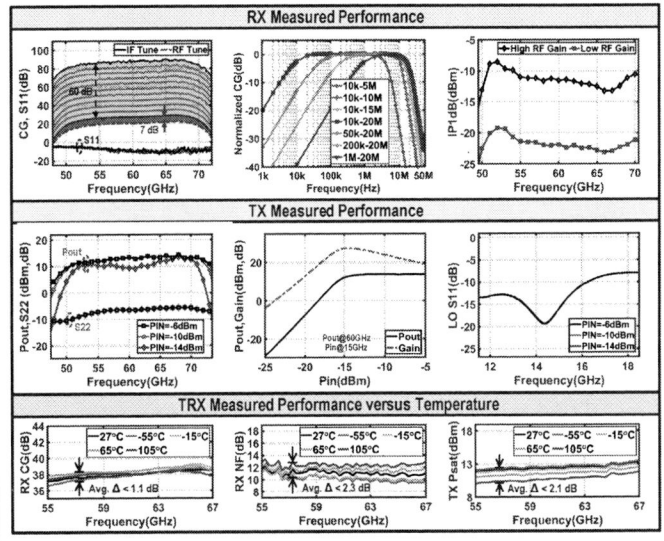

Fig. 4. The measurement results of the TRX parts.

Fig. 5. The measurement results of the FMCW generater parts.

Fig. 6. Performance comparison with the state-of-art works.

	[1]	[2]	[3]	[6]	This Work
Technology	28-nm CMOS	350-nm SiGe	45-nm CMOS	65-nm CMOS	40-nm CMOS
Integration	Front-ends ABB VCO Only	Front-ends ABB VCO Only AiP	Front-ends ABB FMCW Gen. On-chip LDO AiP Digital Baseband	Front-ends ABB FMCW Gen. On-chip LDO AiP	Front-ends ABB FMCW Gen. On-chip LDO
Frequency (GHz)	57-64	57-64	57-64	76-81	54-69
Channels	2T3R	2T4R	3T4R	2T8R	2T2R
RX Gain (dB)	77	19	N/A	18-66	20-87
RX NF (dB)	15	9.5	14	12	11
RX IP1dB (dBm)	-12	-8.5	-10	-7	-12
IF Band (Hz)	N/A	N/A	20M	100k-20M	10k-20M
TX P$_{SAT}$ (dBm)	10	4	12	14	13.5
LO PN @ 1MHz (dBc/Hz)	-99.4	-105	-93	-90	-93.3
Max Chirp BW (GHz)	N/A	N/A	4	7.2	7.2
Max Chirp Rate (MHz/μs)	N/A	N/A	250	340	468.5
Chirp Linearity[#1] (% @ MHz/μs)	N/A	N/A	0.04 @ 250	0.03 @ 18 0.66 @ 340	0.00004 @ 3.7 0.019 @ 468.5
Temperature	N/A	N/A	-55-125	-55-125	-55-105
Area (mm²)	7.45	196[#2]	225[#2]	576[#2]	9.9
Power (mW)	478	990	3500	3150	695

[#1] Chirp Linearity = $\frac{RMS\ Chirp\ Error}{Chirp\ BW}$

[#2] Packaged with AiP

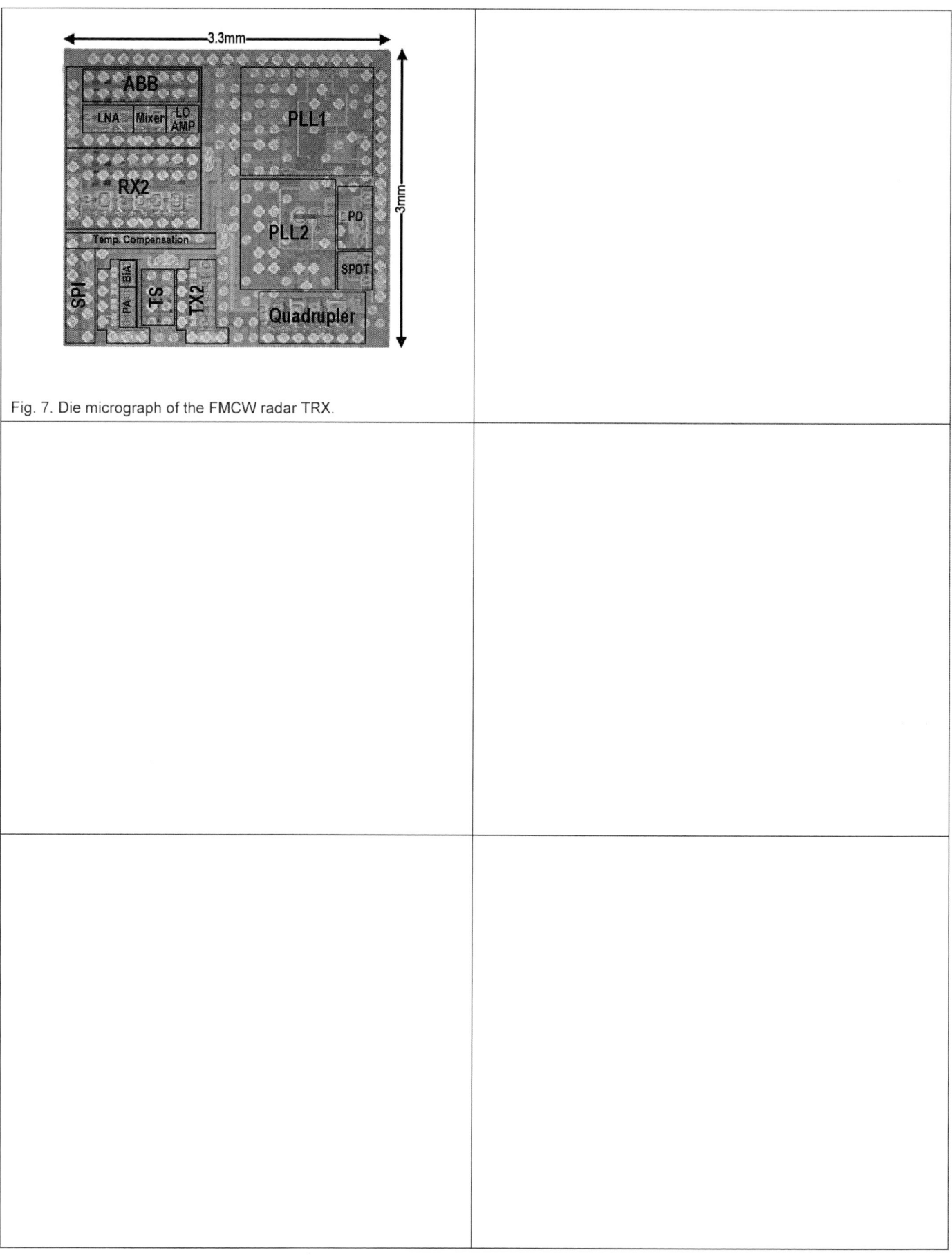

Fig. 7. Die micrograph of the FMCW radar TRX.

A Low-Voltage Bias-Current-Free Pseudo-Differential Hybrid PLL Using a Time-Interleaving Flip-Flop Phase Detector

Liqun Feng, Qianxian Liao, Woogeun Rhee, Zhihua Wang

Tsinghua University, Beijing, China

Low-voltage clock generation is critical for dynamic voltage scaling in modern microprocessors and SoC designs. The performance of several main PLL architectures are limited under low supply voltage as illustrated in Fig. 1. It is difficult for a CP-PLL to maintain good matching with a wide output voltage swing under low supply. In the case of a sub-sampling phase detector (SSPD), the on-resistance of sampling switches is increased, significantly degrading phase noise performance. A digital-intensive PLL architecture would be desirable, but a low-voltage TDC suffers from the substantial variation of time resolution over process and temperature [1]. As a result, a large decoupling capacitor with good supply regulation is required in practice to achieve robust in-band noise performance.

Given the fact that the use of a large decoupling capacitor with supply regulation is a must for low-voltage design, a PLL using a voltage-mode PD such as an XOR-based PD or a flip-flop PD (FFPD) could be an alternative architecture for low-voltage design as the in-band noise and the loop bandwidth do not vary with the small fluctuation of a supply voltage. As shown at the bottom figures in Fig. 1, the voltage-mode PDs followed by a passive loop filter can be implemented in type-I or hybrid PLLs [2]–[5]. However, those PDs output a square wave, resulting in a large reference spur. To reduce voltage ripples, a master-slave sampling filter with harmonic traps and tunable duty cycle are utilized in [2] and [3], respectively, which leads to additional noise and power. Programmable notch filters based on RC network are used in [4] and [5], but they occupy a large area and degrade phase margin. A dual-path quadrature XOR technique is proposed to remove the notch filter, but it requires quadrature signals and an OTA for differential-to-single-ended conversion [6]. In this work, we propose a time-interleaving FFPD (TI-FFPD) that does not require additional filtering circuitry in the loop, while achieving good reference spur performance. With the proposed TI-FFPD, a bias-current-free hybrid PLL with a pseudo-differential loop control is implemented in 65nm CMOS, showing that the proposed PLL is promising for ultra-low-voltage clock generation with advanced CMOS technology.

As depicted in Fig. 2, in the conventional HPLL with the FFPD, the voltage ripple in two paths is differential. Thus, it modulates the VCO frequency and generates large spur. In the proposed architecture, the rising edges of clock signals at the PD input are first distributed to two paths by a time interleaver. The time interleaver is controlled by a simple built-in controller which switches the two paths alternatively. The distributed signals are then fed into to two RS-latch-based FFPDs with appropriate polarity. Thanks to the TI-FFPD structure, the voltage ripple in two differential paths becomes a common-mode voltage and can be cancelled by differential varactors in the ring VCO (RVCO). Therefore, the RVCO frequency is not much disturbed by the ripple. The RVCO is designed based on a bias-current-free 4-stage pseudo-differential structure with differential varactors for positive and negative frequency tuning. An integral path that provides frequency-tracking capability is also important for reference spur performance. A background-calibrated frequency doubler (FD) with duty-cycle correction (DCC) [7] is designed to further improve the spur performance by doubling the reference frequency. The error information for DCC calibration logic can be directly found at the BBPD input of the integral path without having a dedicated comparator. Compared with previous spur reduction techniques relaying on the loop filter, the proposed method suppresses the reference spur based on the PD structure, making the spur performance less dependent on circuit mismatch and nonlinearity over PVT variations. And as it does not introduce poles in the loop, the loop stability is not degraded. Since two parallel feedforward paths operate concurrently, it is also immune to supply noise and achieves a good power-supply rejection.

The detailed structure and operation principle of the proposed TI-FFPD are shown in Fig.3. The proposed PD structure is simple, thus offering a small area and low power. The time interleaver is basically a 1-to-2 demultiplexer (DEMUX) controlled by a DFF-based digital controller. By connecting the negative output to the data input, the controller generates 0 and 1 alternately to switch the selected paths for edge distribution. The TI-FFPD is realized by combing two TIs and two FFPDs with correct polarity. The operating waveforms are shown in Fig. 3. In a locked state, the output waveform at FFPD outputs $PDOP$ and $PDON$ are square waves with aligned phases. After passing the LPF with reduced a common-mode variation amplitude, they are suppressed by differential varactors in the RVCO which minimizes the RVCO modulation with common-mode rejection. If DIV lags REF, the phase error can be found at the difference of $PDOP$ and $PDON$, which drives the RVCO to increase frequency and reduce the phase error. The PD gain is $V_{DD}/2\pi$, which is the same as the conventional FFPD. Since the TI-FFPD provides two parallel paths with a common-mode ripple and a difference-mode signal, this pseudo-differential operation can provide good noise immunity while maintaining phase error detection functionality.

The proposed bias-current-free HPLL is implemented in 65nm CMOS. A chip micrograph is shown in Fig. 7. The RVCO supports an output frequency of 1.6GHz and 0.9GHz under 0.9V and 0.6V supply voltages, respectively. The PD can be reconfigured to the conventional PD structure for performance comparison. A reference frequency of 50MHz is used for all measurements. Figure 4 shows the measured reference spur performance with a wide loop bandwidth of about 5MHz. With the proposed TI-FFPD structure and the DCC, the reference spur is reduced from −46.1dBc to −73.4dBc with 0.9V supply and from −44.4dBc to −72.1dBc with 0.6V supply, showing that the proposed technique is highly effective in suppressing the reference spur over different supply voltages. Figure 5 shows the measured phase noise performance. With 0.9V supply, the HPLL achieves an in-band phase noise of −113.1dBc/Hz and an integrated RMS jitter of 0.86ps. Under 0.6V supply, an in-band noise of −110.4dBc/Hz and an integrated RMS jitter of 1.98ps are measured. The in-band noise variation is less than 3dB for PD supply over 0.6–0.9V, which shows its advantage over than TDC. Figure 6 summarizes and compares the measured performance with state-of-art RVCO-based integer-N PLLs. Without requiring a notch filter and bias current, the proposed HPLL supports low-voltage operation with less than −70dBc spur and good noise performance. It is worth noting that over 90% power is consumed by the RVCO in this HPLL, and significant power reduction is expected if the proposed HPLL is implemented with advanced CMOS technology.

Acknowledgement:

This work is supported by National Key Research and Development Program of China (No. 2020YFB2205602).

References:

[1] N. Pourmousavian et al., "A 0.5-V 1.6-mW 2.4-GHz fractional-N all-digital PLL for Bluetooth LE with PVT-insensitive TDC using switched-capacitor doubler in 28-nm CMOS," IEEE JSSC, vol. 53, no. 9, pp. 2572–2583, Sep. 2018

[2] L. Kong et al., "A 2.4 GHz 4 mW integer-N inductorless RF synthesizer," IEEE JSSC, vol. 51, no. 3, pp. 626–635, Mar. 2016.

[3] S. Yang et al., "A 0.2-V energy-harvesting BLE transmitter with a micropower manager achieving 25% system efficiency at 0-dBm output and 5.2-nW sleep power in 28-nm CMOS," IEEE JSSC, vol. 54, no. 5, pp. 1351–1362, May 2019.

[4] C.-C. Lin et al., "Spur minimization techniques for ultra-low-power injection-locked transmitters," IEEE TCAS-I, vol. 67, no. 11, pp. 3643–3655, Nov. 2020.

[5] X. Xu et al., "A bias-current-free fractional-N hybrid PLL for low-voltage clock generation," IEEE TCAS-I, vol. 68, no. 9, pp. 3611-3620, Sept. 2021.

[6] Y. Liang et al., "A low-jitter and low-reference-spur 320 GHz signal source with an 80 GHz integer-N phase-locked loop using a quadrature XOR technique," IEEE T-MTT, vol. 70, no. 5, pp. 2642-2657, May 2022.

[7] A. Elkholy et al., "A 2.5–5.75-GHz ring-based injection-locked clock multiplier with background-calibrated reference frequency doubler," IEEE JSSC, vol. 54, no. 7, pp. 2049-2058, July 2019.

979-8-3503-3004-5/23 $31.00 © 2023 IEEE

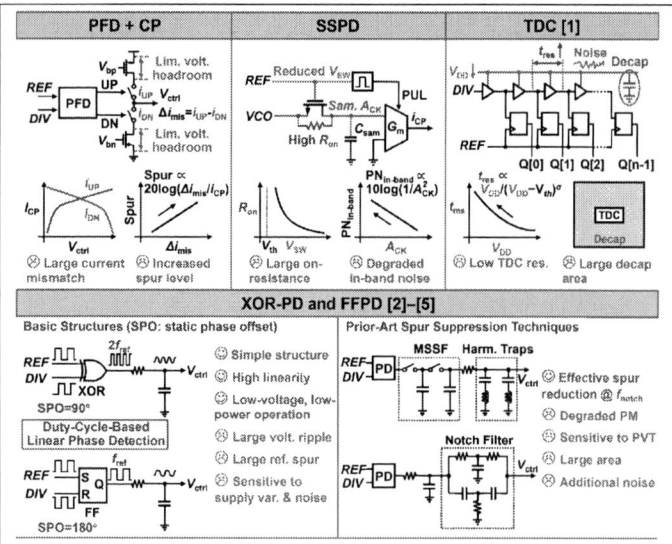

Fig. 1. PD structures under low-voltage design and prior-art reference spur reduction techniques for voltage-mode PDs.

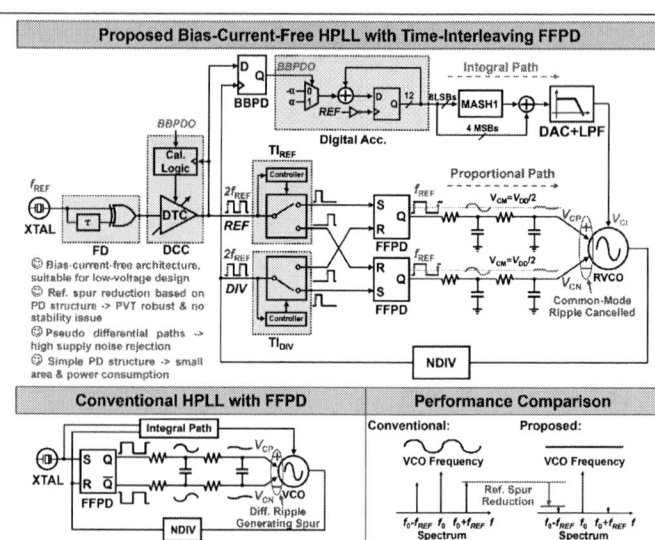

Fig. 2. Architecture of proposed bias-current-free HPLL with TI-FFPD and comparison with conventional FFPD.

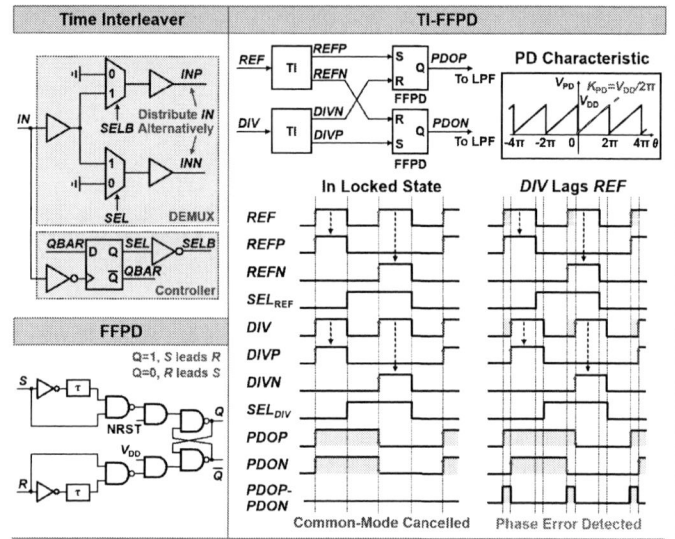

Fig. 3. Detailed structure and operation principle of TI-FFPD.

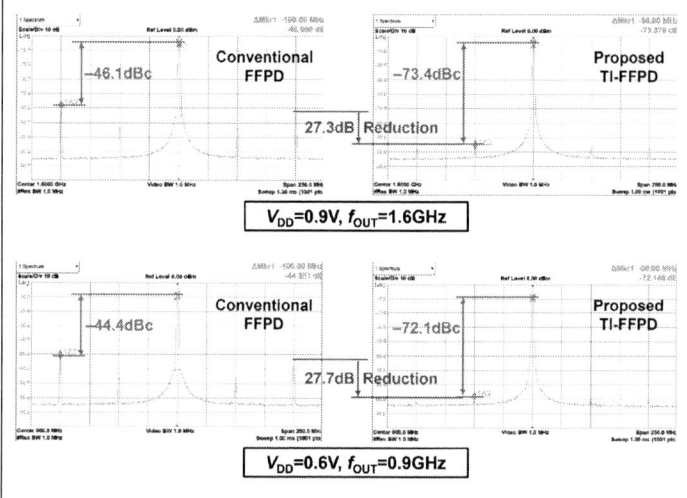

Fig. 4. Measured reference spur performance.

Fig. 5. Measured phase noise performance.

Fig. 6. Performance comparison with state-of-art ring-oscillator-based integer-N PLLs.

	This Work		X. Xu TCAS-I'21	L. Kong JSSC'16	B. Xiang CICC'20	S. Yang TCAS-II'21	T. Tsai VLSI'18	J. Zhu JSSC'17
Architecture	Hybrid PLL		Hybrid PLL	Type-I PLL	CP-PLL	Hybrid PLL	Hybrid PLL	Time-based PLL
PLL Type	Integer-N		Fractional-N*	Integer-N	Integer-N	Integer-N	Integer-N	Integer-N
Oscillator Type	Ring		Ring	Ring	Ring	Ring	Ring	Ring
PD Type	Flip-Flop		Flip-Flop	XOR	PFD	BBPD	PFD	PFD
Notch Filter	No		Yes	Yes	No	No	No	No
Bias-Current-Free	Yes		Yes	No	No	Yes	No	No
Reference (MHz)	50		60	22.6	100	250	250	275
Supply Voltage (V)	0.9	0.6	0.6 (0.9 for VCO)	1	0.8	0.8	0.65	1
Output (GHz)	1.6	0.9	1.26	2.4	3.2	2.0	3.0	2.2
PLL BW (MHz)	5	4.5	2.5	10	10	69	20	10
In-band PN (dBc/Hz)	−113.1	−110.4	−97.7	−113.8	−110	−117.4	−120 (w/i realign.)	−102.8
RMS Jitter (ps)	0.86	1.98	3.5	0.97	2.3	1.71	0.62	3.73
RMS Degree (°)	0.50	0.64	1.59	0.84	2.65	1.23	0.67	2.95
Int. Range (Hz)	10k–100M		1M–1G	1k–200M	100k–100M	10k–300M	100k–1G	10k–300M
Ref. Spur (dBc)	−73.4	−72.1	−76 (−40 w/o notch)	−65	N/A	−60	−52.3	−40.5
Power (mW)	7.97	1.92	1.85	4	0.68	0.38	2.3	1.82
FoM_JIT (dB) **	−232.3	−231.2	−226.4	−234.1	−234.4	−239.5	−240.5	−226.0
FoM_JIT,N (dB) ***	−247.3	−243.8	−239.7	−254.5	−249.5	−248.6	−251.3	−235.0
Active Area (mm²)	0.046		0.23	0.015	0.015	0.0006	0.012	0.0021
Tech. (nm)	65		65	45	22	28	7	65

* Use measurement results in integer-N mode.

** FoM_JIT = 20log(Jitter_rms/1s) + 10log(P_DC/1mW)

*** FoM_JIT,N = FoM_JIT − 10log(f_out/f_ref)

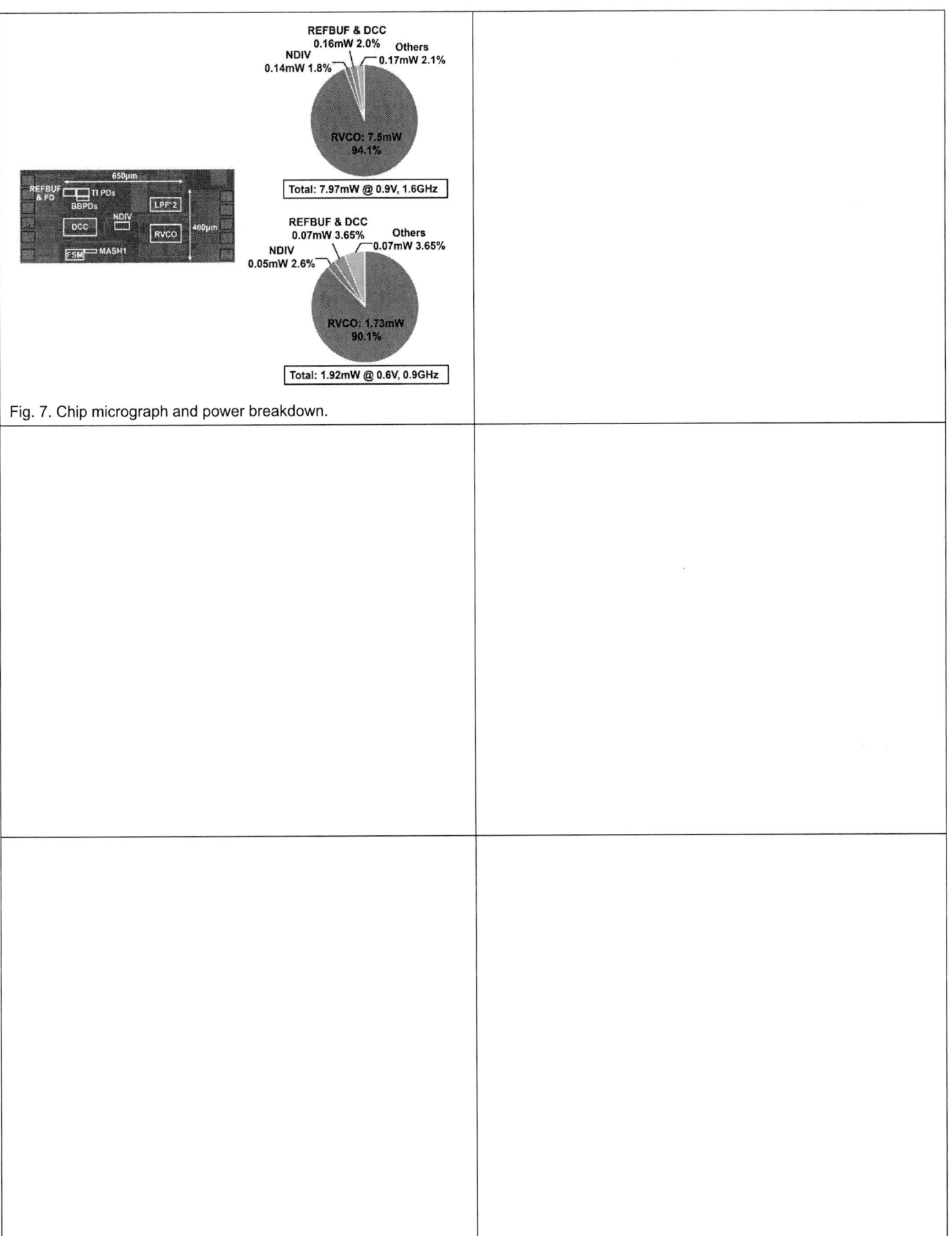

Fig. 7. Chip micrograph and power breakdown.

A Cryo-CMOS 4.5~7GHz Dual-Qubit Homodyne Reflectometer Array with High Q Degenerate Parametric Amplifier Through Dynamic Mode Coupling

Yujie Geng[1], Haichuan Lin[2], Bo Wang[3], Cheng Wang[1*]

[1]University of Electronic Science and Technology of China (UESTC), Chengdu, China, [2]Chengdu Data Automation System Technologies, Chengdu, China, [3]Hamad Bin Khalifa University, Doha, Qatar

Error-corrected quantum computing does not only require large-scale (10^6-10^9) physical Qubit arrays but also relies on high-fidelity, scalable quantum-to-classic interface electronics. Quantum control and readout systems based on cryogenic CMOS integrated circuits mitigate the error of Qubit manipulation, reduce the cabling complexity, and facilitate the quantum feedback, which pave the way towards practical quantum error correction (QEC) and logic quantum gates. RF reflectometry is widely adopted for qubit states readout. Reflectometry conducts Qubit states discrimination (collapsed to "0" or "1") by detecting the resonant frequency through RF reflection of a high-Q LC tank associated with the sensor Qubit (Fig. 1). The quantum states are encoded in either the amplitude or the phase of the reflected RF signal. Compared with its DC counterpart, the reflectometry achieves unprecedented charge sensitivity, reduces the integration time, and avoids long-term drifting. However, conventional heterodyne reflectometer schemes [1][4][5] adopt sophisticated up/down mixers, ADC, DAC, and PLL for the Qubit excitation and down-mixing, respectively. They suffer from poor TX-RX isolation, inter-modulation, and high DC power, which deteriorate the scalability and readout fidelity. To address these challenges, a highly scalable dual-Qubit homodyne reflectometer array is proposed in this paper. It drives the LC tank by a single PLL and detects the reflection by a high-gain degenerate parametric amplifier (DPA). Along with fast modulation and lock-in detection, it mitigates the interference, complex cabling, and high DC power of conventional heterodyne reflectometers.

The proposed dual-qubit homodyne reflectometer array diagram is shown in Fig.1. It sends fast-modulated TX pulses (f_m = 10~100MHz) generated by a PLL-driven CML divider by 2 to the sensor Qubit pair. After RF reflection, the HF and LF RX channels amplify the reflected TX pulses with the DPA, pumped by the same TX PLL. The amplitude and phase of TX pulses are carefully controlled by a vector modulation (VM) to achieve the maximum parametric gain. Next, the envelope detection and baseband noise filtering are performed by a square-law detector and a lock-in amplifier successively. The proposed single PLL architecture leads to significant circuit simplification. Thanks to the frequency selective parametric amplification, fast TX modulation, and lock-in detection, the flicker noise and intermodulation components are suppressed.

The Josephson junction parametric amplifiers (JPA) at 20mK exhibit superior cryogenic performance by traveling-wave amplification. However, since the inductors of the traveling-wave scheme are chip area-hungry on CMOS, conventional Cryo-CMOS parametric amplifiers (PA) [2][3], adopting low Q, single resonant tank (Q=5~15), which is resistively coupled with the 50Ω RF input source, exhibit low RF gain and high noise. In this paper, a degenerate parametric amplifier (DPA) is proposed, as shown in Fig. 2. It delivers high parametric RF gain by coupling the common-mode (CM) RF input signal dynamically into a differential-mode (DM), unloaded, high Q resonant tank (Q=30~40). The proposed DPA consists of a pair of anti-parallel non-linear transmission lines (APNLTL) with accumulation mode PMOS varactors, inductors, and a 50Ω load resistor. On the upper side of APNLTL, the CM RF input at f_{RF} seeing the varactor's anode is parametrically amplified if it is phase-aligned with the pumping signal at $f_{pump}=2f_{RF}$. However, On the lower side, the CM RF input signal is attenuated under the same phase alignment by seeing the varactor cathode instead. Thus, a DM signal at f_{RF} is created by the parametric amplification under the pumping of f_{pump} =2f_{RF} CM input. Then, APNLTL is half wavelength at f_{RF}, which works as an unloaded, high Q λ/2 resonator for high gain parametric amplification, since the DM signal shorts at both ends of APNLTL. Next, the DM signal is coupled from the APNLTL center to maximize the CM to DM conversion and further amplified by the

degenerated NMOS pairs buffer to improve CM rejection. Meanwhile, the above-mentioned dynamic mode coupling technique achieves effective input-to-output isolation and reverse leakage reduction due to CM-to-DM mode isolation. The CM, traveling-wave RF input signals at f_{RF} and f_{pump} are finally absorbed by the 50Ω load for impedance matching, which stays invariant under the dynamic mode coupling. Since the effective Q of DM tanks is improved to 30~40, the DPA not only provides higher RF gain (>20dB) but also performs frequency selective filtering for input interference reduction. The equivalent noise voltage model of DPA is also presented in Fig.3 (left). The simulation shows that 48.57% of noise comes from the varactors' parasitic substrate resistance R_{sub}, which can be further improved by using isolated varactors in the future. Lastly, in the TX, a CML divider by 2 with off-state interlocking is implemented, which reduces the modulation delay to <100ps. The digital delay line (DDL) is adopted to avoid off-state uncertainty.

The chip is fabricated on a commercial 65nm bulk CMOS process. Fig. 2 (top middle) shows the measured varactor C-V curve using the LHE-6H-6 cryogenic probe station from ZEPTOOLS. The varactor capacitance tunability and quality factor exhibit decent improvement from 298K to 10K. The chip is further tested in a 4.2K refrigerator KDE415SA from CSIC Pride (Nanjing) Cryogenic Technology Co., LTD. Fig.3 (right) shows the chip photo, packaging, and cryogenic test setup. At 4.2K, shown in Fig. 4, a 4.5~7GHz (43.4%) bandwidth and a peak RF gain of 52dB at 6.2GHz are measured. The associated min. noise temp. T_{noise} is 78K, measured by the classic hot-cold method. The saturation of T_{noise} is observed when T_{env} < 30K (Fig. 4). It is believed that the noise-temp. scale is deteriorated by the device self-heating and the parasitic substrate resistance of the varactor cathode on the lower TL, which has not been depleted by the substrate freezing out. The DPA presents competitive noise temperature with other Cryo-CMOS PAs [2][3]. Thus, further efforts are still necessary to break the noise bottleneck/problem of Cryo-CMOS PAs. The measured TX RF output range is 3~8.5GHz at 4.2K with an output power of -22dBm and a 30dB attenuation range (Fig. 5 (left)). A TX output signal modulated at 10MHz is also shown as expected. Fig. 5 (right) presents the measured phase noise (PN) of TX input, TX divided output, and RX output by looping TX output back to RX input. The PN measurement is limited by the low RF input power of the PN analyzer. The 4.2K loop-back test also indicates the input P_{1dB} of -75dBm and -73dBm for HF and LF RX channels, respectively.

In conclusion, a dual-Qubit homodyne reflectometer array with a DPA for scalable quantum sensing is presented in this paper, which works from 4.5~7GHz with a 52dB peak RF gain, a 78K noise temperature and a -22dBm modulated TX pulses with 30dB tunability at 4.2K. Compared with the prior works (Fig. 6), with the lowest 16.5mW DC power/channel, this chip achieves broadband, high gain, and dual-Qubit readout.

Acknowledgements:

The authors appreciate the help of Dr. Yu He, and Dr. Guangchong Hu at International Quantum Academy, Shenzhen, China, Prof. Guangwei Deng at UESTC, Chengdu, China, and Prof. Xufeng Kou in Shanghai Tech, Shanghai, China on the cryogenic measurement.

References:

[1] K. Kang, et al., "A Cryo-CMOS Controller IC with Fully Integrated Frequency Generators for Superconducting Qubits," ISSCC 2022, pp.362-363.

[2] G. Baek, et al., "13-K T_{noise} Cryo-CMOS Parametric Amplifier at 80 mK for Quantum Computers," VLSI 2022, pp.236-237.

[3] M. Mehrpoo, et al., "A Cryogenic CMOS Parametric Amplifier" SSCL 2020, pp.5-8.

[4] A. Ruffino, et al., "A Fully-Integrated 40-nm 5-6.5 GHz Cryo-CMOS System-on-Chip with I/Q Receiver and Frequency Synthesizer for Scalable Multiplexed Readout of Quantum Dots" ISSCC 2021, pp.210-211.

[5] B. Prabowo, et al., "A 6-to-8GHz 0.17mW/Qubit Cryo-CMOS Receiver for Multiple Spin Qubit Readout in 40nm CMOS Technology" ISSCC 2021, pp.212-213.

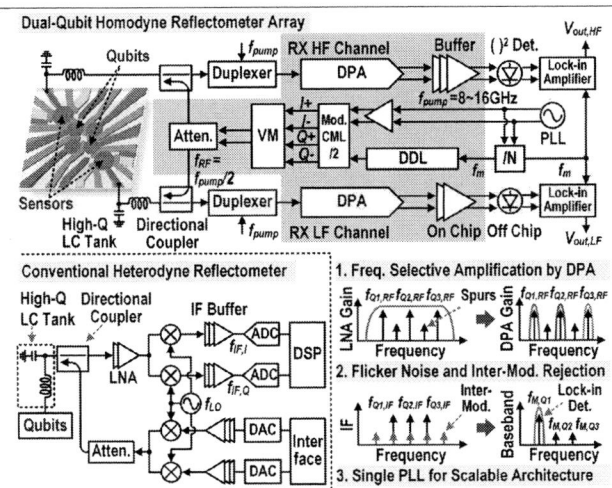

Fig. 1. The proposed dual-Qubit homodyne reflectometer array and the conventional heterodyne reflectometer.

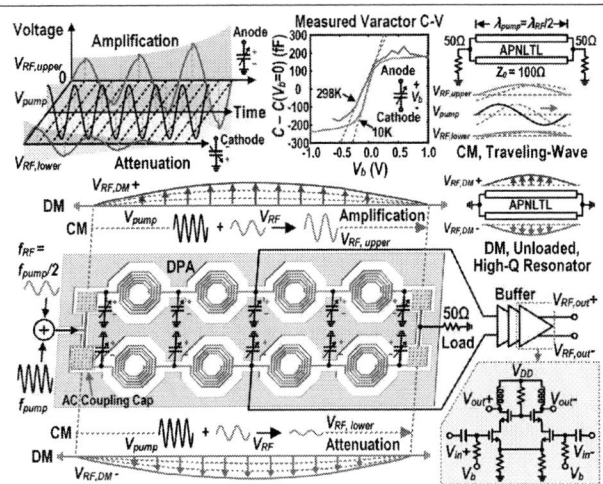

Fig. 2. The operation principle, schematic, and EM structure of degenerate parametric amplifier (DPA).

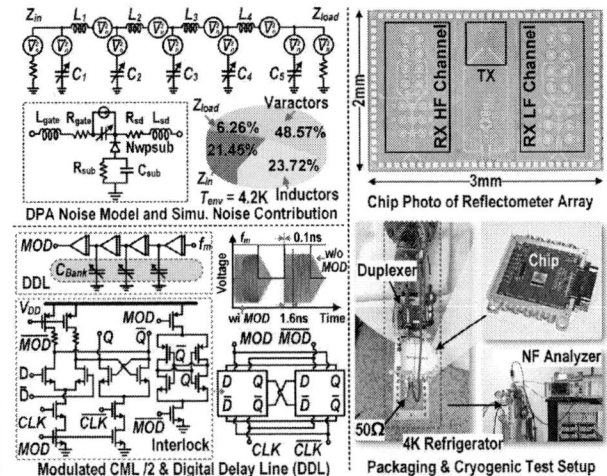

Fig. 3. The noise model of DPA, the fast-modulated CML divider with digital delay line (DDL), the chip photo, and 4.2K meas. setup.

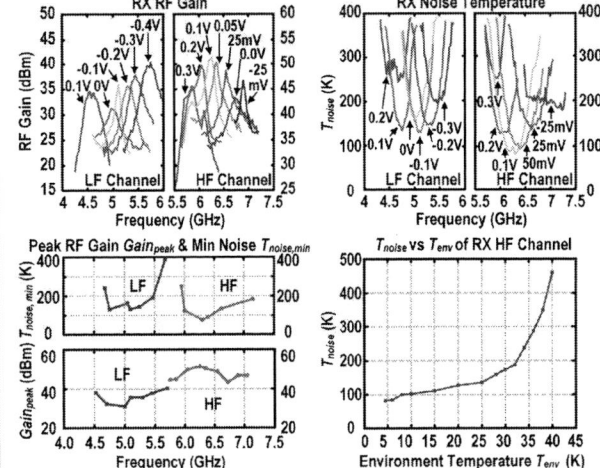

Fig. 4. 4.2K Meas. results: the RX Gain & noise temp., the peak RF gain & min. noise temp., and the HF channel T_{noise} vs. T_{env}.

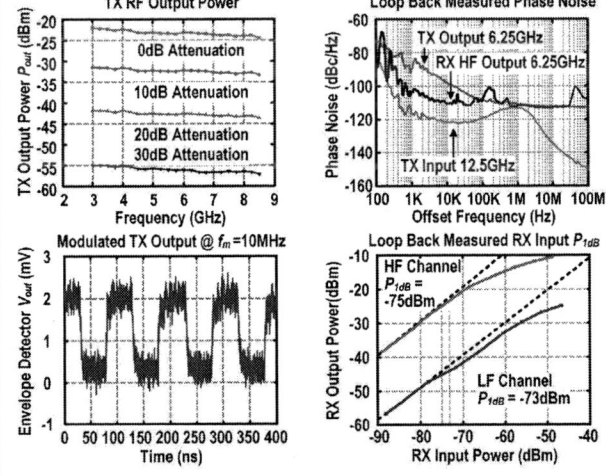

Fig. 5. 4.2K TX-to-RX loop-back meas. results: the TX RF power, the TX modulation, the phase noise, and the RX input P1dB.

	This work	ISSCC 2022 [1]	VLSI 2022 [2]	SSCL 2020 [3]	ISSCC 2021 [4]	ISSCC 2021 [5]
Test Temp.	4.2K	3.5K	80mK	8.5K	3.5K	4.2K
Frequency	4.5~7GHz	5~7.25GHz	5.3~5.5GHz	6.8~8.6GHz	5~6.5GHz	6~8GHz
Operation Bandwidth	2.5GHz (43.4%)	2.25GHz (36.7%)	0.2GHz (3.7%)	1.8GHz (23.4%)	1.5GHz (26.1%)	2GHz (28.6%)
Architecture	Homodyne 1 TX, 2 RX	Heterodyne 4 TX, 2 RX	N/A	N/A	Heterodyne 1 TX, 1 RX	Heterodyne 0 TX, 1 RX
Low Noise Amplifier	Degenerate Parametric Amplifier	3-Stage Cascode LNA	Classic Parametric Amplifier	Classic Parametric Amplifier	3-Stage Cascode LNA	Cascaded Common Source LNA
RX Peak RF $Gain_{peak}$	52dB	47dB[1]	17dB	9dB	40dB	30dB
RX $T_{noise,min}$	78K	84K	13K	100K	39K	44K
RX $P_{1dB,input}$	-75dBm	-49dBm	N/A	N/A	-85dBm	-58.4dBm
TX P_{out}	-57~-22dBm	-10dBm	N/A	N/A	N/A	N/A
TX Mod.	Amp./Phase	Amp.	N/A	N/A	N/A	N/A
Technology	65nm CMOS	40nm CMOS	65nm CMOS	40nm CMOS	40nm CMOS	40nm CMOS
DC Power	33mW	60mW[2]	N/A	33mW	108mW	66mW

1. RF + Mixer + IF Gain; 2. Estimated by 20mW / RX Channel and 5mW / TX Channel

Fig. 6. The performance comparison table.

979-8-3503-3004-5/23 $31.00 © 2023 IEEE

A 2-W, 90%-Efficiency Single-Stage Dual-Output Wireless Power Receiver with 0.1 to 700-mA Output Current Range Through Dynamic Delay Compensation and Bootstrap Adaptive Body Biasing Circuit

Yutang Chen[1], Yuxuan Luo[1], Jianping Guo[1*], Xian Tang[2], and Dihu Chen[1]

[1]Sun Yat-sen University, Guangzhou, China

[2]Shenzhen International Graduate School, Tsinghua University, Shenzhen, China

Wireless power transfer with multiple-output is very helpful for portable devices like wireless solid-state drives, implantable biomedical devices, etc [1-5]. The conventional way to achieve multiple-output is using multiple-stage receiver (RX) for different outputs [1], which would degrade the power efficiency due to the cascade stages. A single-stage dual-output (SSDO) RX was implemented by using a regulated rectifier [2]. However, ten power transistors are needed. An SSDO RX proposed in [3] only required six power transistors, but the two outputs are quite close due to the cross-regulation problem. A full-bridge-separated SSDO RX was proposed to solve the cross-regulation problem [4], thus providing two completely independent outputs. However, the separated rectifier halves the maximum output power. The SSDO RX proposed in [5] overcame the cross-regulation problem through synchronous control and maintained a full-bridge rectifier at high-side output (V_{OH}). However, the output ripple of low-side output (V_{OL}) becomes worse if the output current I_{OL} at V_{OL} is much larger than I_{OH} at V_{OH}, reducing the output current range.

The portion of power consumption at V_{OH} increases with more ways of NAND in solid-state drives [6]. Thus, the current I_{OH} with more proportion of the input current is desirable, meanwhile a wide-range I_{OL} is necessary for the stand-by and operating modes. In this work, a 6.78-MHz SSDO RX with bootstrap adaptive body biasing (B-ABB) circuit and dynamic delay compensation is presented. The adopted full bridge maintains the full-cycle–rectified output current I_{OH} at 3.3-V V_{OH}. The B-ABB circuit overcomes the cross-regulation problem at 1.8-V V_{OL}, providing a wide-range output current I_{OL} without limitation. The dynamic on/off-delay compensations improve the output current capability and system efficiency.

The proposed SSDO RX architecture is shown in Fig. 1, including a reconfigurable regulated rectifier, a B-ABB circuit, a zero-current-detecting (ZCD) based dynamic delay compensation circuit, and a control circuit consisted of three comparators, three series resistor chains, and three basic logic gates. Fig. 1 also shows the conceptual regulated output voltage waveform of the proposed RX. The output capacitors C_{OH} and C_{OL} will be charged during 1X mode meanwhile be discharged during 0X mode, respectively. If both of V_{OH} and V_{OL} are under 1X mode, the input current will be given priority to V_{OL}.

Fig. 2(a) shows the topology and operation of the proposed RX. In stage I, the proposed RX works as two half-bridge rectifiers, both V_{OH} and V_{OL} are allocated half the input current. In stage II, V_{OH} side freewheels and V_{OL} is still as a half-bridge rectifier. In stage III, V_{OL} freewheels and RX works as a full-bridge rectifier at V_{OH}. In stage IV, both two sides are under freewheeling mode.

The cross-regulation problem at low output V_{OL} cannot be ignored in SSDO RX. As shown in Fig. 2(b), the body diode of transistor M_{P2} will conduct when V_{IN2} is greater than ($V_{OL} + |V_{THP}|$), if its body potential is only connected to the source. An ABB circuit is used to tie the body to the highest voltage [3] for solving the cross-regulation problem. However, the conduction resistance of the transistors M_{S1} and M_{S2} will lead to an unexcepted body effect, and turn on M_{P2} eventually. The traditional way is to increase the size of M_{S1} and M_{S2}, but the increased parasitic capacitance brings a great challenge to the driving capability and area cost. A bootstrap circuit is combined with the ABB circuit in this work. By boosting the supply voltage of the buffer, M_{P2} will be turned off completely. Furthermore, the capacitor C_{BST} is just about 50 fF because M_{P2} is biased at the subthreshold region, nearing the threshold voltage $|V_{THP}|$. Thus, the parasitic device capacitance of M_{P2} is rather small. The proposed B-ABB circuit only covers an area of 16 μm × 62 μm. Fig. 2(b) shows the simulated waveform of the general ABB circuit and proposed B-

ABB circuit. The simulated average leakage current in a cycle decreases greatly from 2.2 mA to 90 μA, which only remains parasitic capacitance leakage current, expanding the minimum output current at V_{OL}.

To improve the output current capability and system efficiency, a ZCD-based dynamic delay compensation scheme is proposed. As shown in Fig. 3, the ZCD signal V_{OFFEN} is obtained by a dual-side current sensor and a comparator for enabling the off-delay compensation. However, a longer propagation delay occurs in the current sensor under high-level input current. To obtain a more accurate zero-current switching point, an adaptive threshold voltage generator is adopted. The adaptive voltage generator works like an envelope detector and the voltage V_{TH} increases with the input current, decreasing the propagation delay. Furthermore, the signal V_{ONEN} is generated by two different circuits for on-delay compensation. The above signals V_{ONEN} and V_{OFFEN} connect to the switches, controlling the offset currents and enabling the on/off-delay compensation in two types of push-pull comparators, respectively. As shown in the Monte Carlo simulated results, the reverse current and body-diode conduction time are reduced considerably with the proposed scheme. Because the proposed delay compensation is only available in ns-level time, there is no multiple-pulsing thus over-compensation problem.

The proposed SSDO RX has been fabricated in 0.18-μm CMOS technology. Fig. 4 shows the measured result of the proposed delay compensation. Both short body-diode conduction time and zero-current switch are achieved. Measured startup waveform shows the successful self-startup under two different coil distances of 6 mm and 10 mm, respectively. As shown in the measured steady-state waveforms, the output voltage V_{OH} and V_{OL} are regulated at 3.3 V and 1.8 V, respectively. It also indicates the proposed four-stage operation regulating the two outputs successfully.

Fig. 5 shows three types of measured load transient response waveforms. The output ripples are 100 mV at both two outputs under different output current. Instant response recovery time and unnoticeable cross regulation are observed. Fig. 5 also depicts the measured system efficiency of the proposed RX. The maximum output current of I_{OH} reaches 702 mA at V_{OH} due to the full-bridge rectifier. The proposed B-ABB circuit overcomes the cross-regulation problem, obtaining a 0.1 to 380-mA I_{OL} at V_{OL}. A peak efficiency of 90.1% is achieved when I_{OL} equals to 1 mA and I_{OH} equals to 702 mA, respectively. Compared with the prior arts, this work generates two completely independent outputs with competitive peak system efficiency and excellent load transient performance. Due to the proposed B-ABB circuit and dynamic delay compensation, wide output current range and high maximum output power for both outputs are achieved, which helps to broaden the application. Fig. 7 shows the chip micrograph and measurement setup. Two coils are the same with a 550-nH inductor and a 300-mΩ parasitic resistance. A wireless power transmitter board is adopted with a 5.5-V DC supply voltage.

Acknowledgement:

This work was supported by the Natural Science Foundation of Guangdong Province, China, under Grant 2022A1515011054.

References:

[1] Y. Lu, *et al.*, "A Dual-Output Wireless Power Transfer System with Active Rectifier and Three-Level Operation," TPEL 2017: 927-930.

[2] Z. Xue *et al.*, "Single-Stage Dual-Output AC-DC Converter for Wireless Power Transmission," CICC 2018: 1-4.

[3] Q. W. Low, *et al.*, "A Single-Stage Dual-Output Tri-Mode AC-DC Regulator for Inductively Powered Application," TCAS-I 2019: 3620-3630.

[4] J. Lin, *et al.*, "A Single-Stage Dual-Output Regulating Rectifier with Hysteretic Current-Wave Modulation," JSSC 2021: 2770-2780.

[5] Z. Luo *et al.*, "A 90%-Efficiency 40.68MHz Single-Stage Dual-Output Regulating Rectifier with ZVS and Synchronous PFM Control for Wireless Powering," ISSCC 2023: 454-456.

[6] K.-T. Park *et al.*, "Three-Dimensional 128 Gb MLC Vertical NAND Flash Memory With 24-WL Stacked Layers and 50 MB/s High-Speed Programming," JSSC 2015: 204-213.

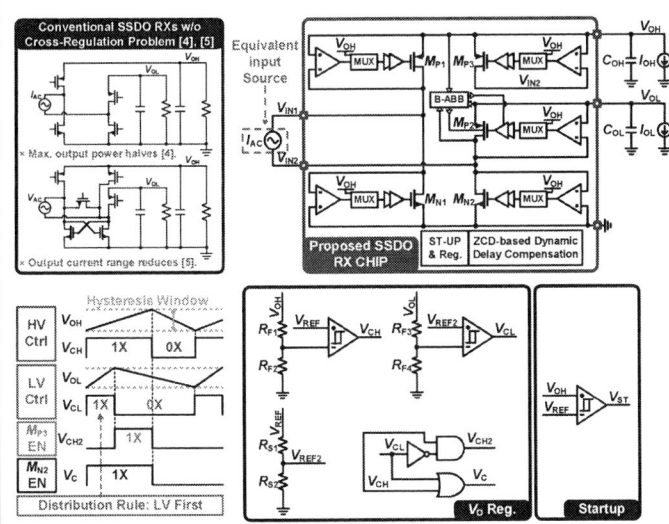

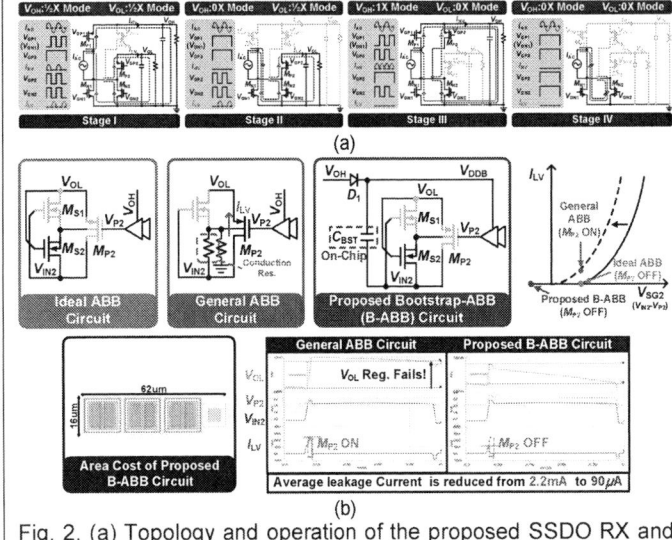

Fig. 1. Block diagram of the proposed SSDO RX and the conceptual regulated output voltages waveforms.

Fig. 2. (a) Topology and operation of the proposed SSDO RX and (b) the structure and simulated waveforms of the proposed B-ABB circuit.

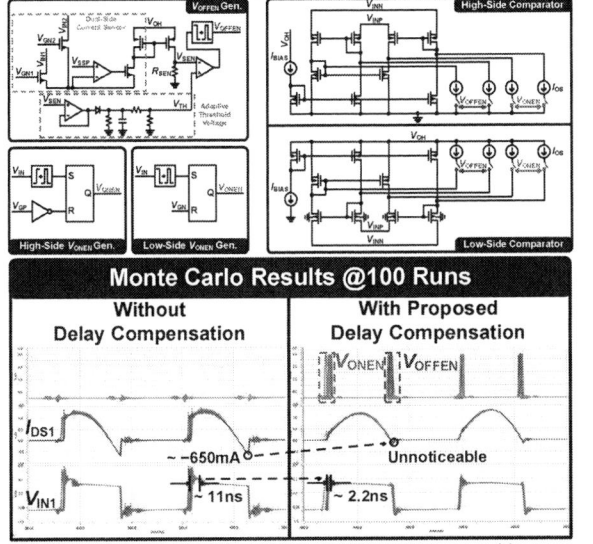

Fig. 3. Block diagram of the push-pull comparators with proposed on/off-delay compensation and relevant Monte Carlo results.

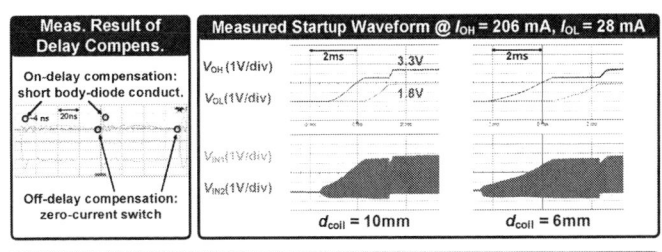

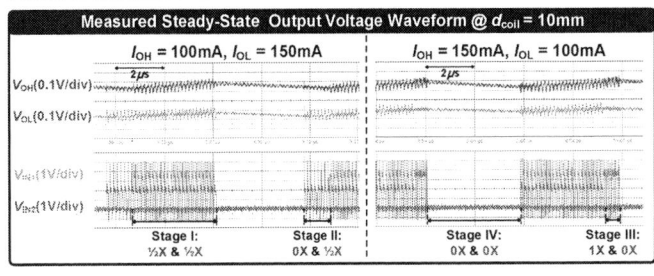

Fig. 4. Measurement results of the delay compensation, startup, and steady-state waveforms of the proposed SSDO RX.

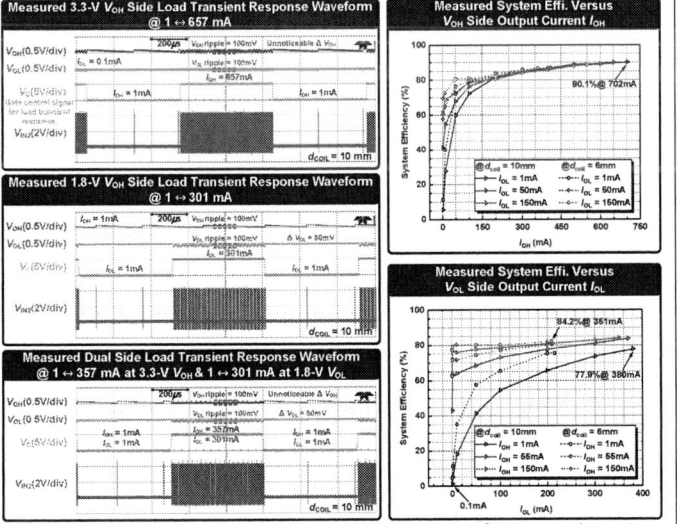

Fig. 5. Measured load transient response waveforms and system efficiency of the proposed SSDO RX.

Performance comparison with state-of-the-art dual-output RX

	[1] TPE 2017	[2] CICC 2018	[3] TCAS-I 2019	[4] JSSC 2021	[5] ISSCC 2023	This work
Technology	0.35 μm CMOS	0.18 μm CMOS	0.18 μm CMOS	0.18 μm CMOS	65nm CMOS	0.18 μm CMOS
Topology	Rectifier + SIMO	SSDO Rectifier	SSDO Rectifier	SSDO Rectifier	SSDO Rectifier	SSDO Rectifier
LC tank	Parallel	Parallel	Parallel	Series	Parallel	Series
Freq. (MHz)	6.78	13.56	0.125	6.78	40.68	6.78
Power switches	8	10	6	4	6	5
Power Stage	2	1	1	1	1	1
V_O	1.8 V, 3.8 V	0.9-1.8 V, 1.3-2.75 V	1.6-2 V, 1.6-2 V	1.8 V, 3.3 V	1.1 V, 3.3 V	1.8 V, 3.3 V
Output Current Range (mA)	N/A	N/A	N/A	20-200@1.8 V 20-200@3.3 V	1-5@1.1 V 2.5-25@2.2 V	0.1-380@1.8 V 1-702@3.3 V
I_{O_MAX}/I_{O_MIN}	N/A	N/A	N/A	10	25	7020
Max. Output Power (mW)	600	10	114	1020	60.5	2316
Peak Effi.	80.5%	79%	91.7%	91.9%	90.1%	90.1%
Off-chip Components	4 cap. & 1 ind.	8 cap.	2 cap.	2 cap.	2 cap.	2 cap.
Load Transient Response	N/A	N/A	540 μs	Unnoticeable	Unnoticeable	Unnoticeable
Cross Regulation	N/A	N/A	N/A	Unnoticeable	Unnoticeable	Unnoticeable

Fig. 6. Performance comparison with state-of-the-art dual-output RXs.

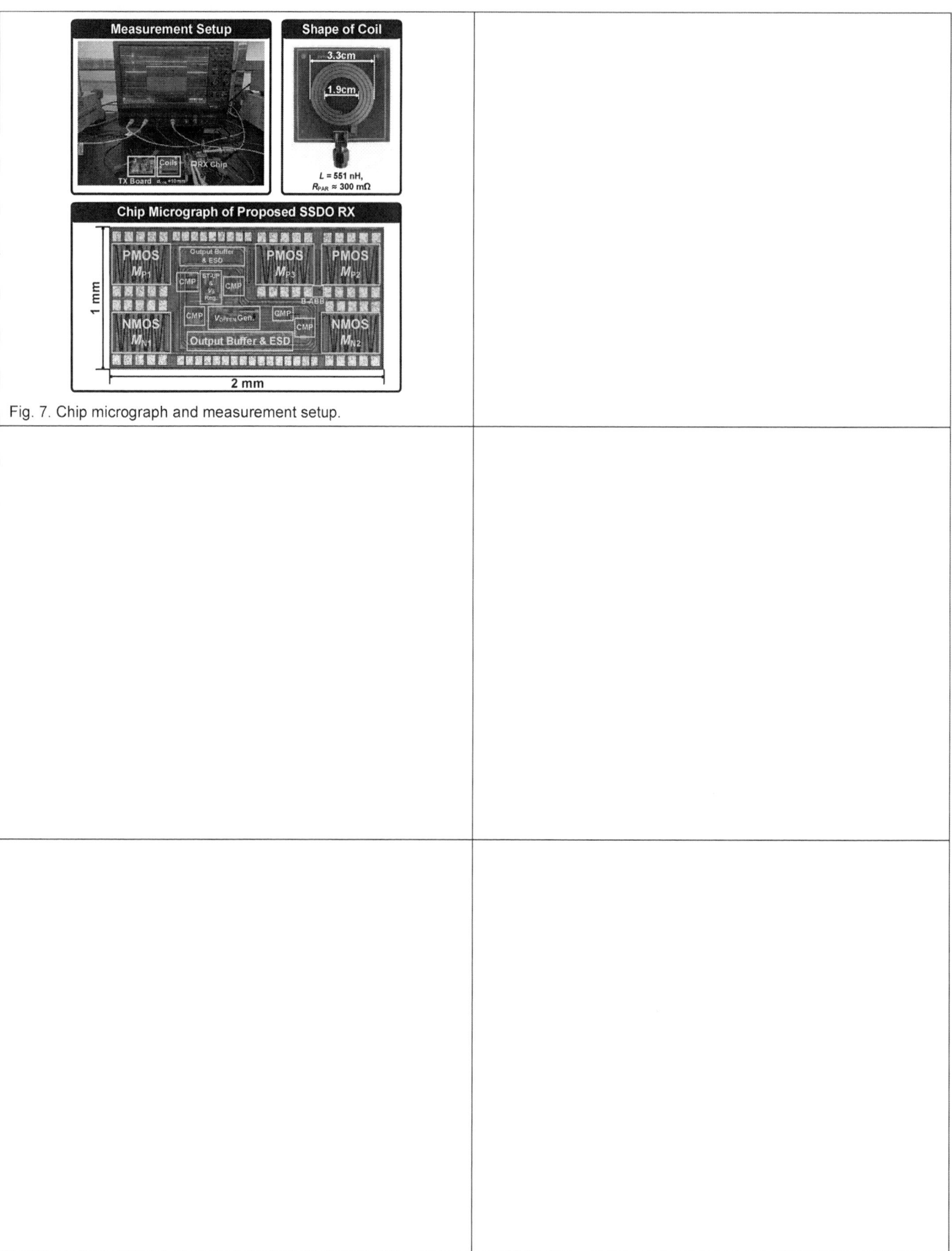

Fig. 7. Chip micrograph and measurement setup.

An 18-nW, 170°C Temperature Range, Voltage and Current Reference Circuit with Low Line Sensitivity

I-Fan Lin[1], Yu-Chu Tsai[2], Heng-Li Lin[2], and Yu-Te Liao[1]

Institute of Electrical and Computer Engineering, National Yang Ming Chiao Tung University[1], uPI Semiconductor Corp.[2]

In Internet-of-Things applications, there is an increasing focus on power efficiency and small form factors in electronics. Designing for low power consumption involves using a small volume battery while maintaining a long operational time. The reliability of the electrical circuits is highly dependent on having a stable voltage and current reference (VCR). For a reference circuit to be well-designed, it needs to be low power, supply-independent, stable over a wide temperature range, and occupy a small chip area. Integrated VCRs are appealing because of the strong correlation between the local and global distributions of voltage and current.

Several VCR types have recently been published [1]-[6]. The conventional ΔV_{BE}-based VCRs tend to be large in size and are difficult to operate with a sub-1V supply due to the use of BJTs [1][2]. For sub-1V applications, some VCRs divide the V_{BE} of a BJT by a current mirror to reduce the number of BJTs [3]. Others use MOSFETs instead of BJTs for reference generation [4]. Unfortunately, the CMOS-only reference current (I_{REF}) exhibits high variations due to the process discrepancy of MOSFETs. Moreover, the type of I_{REF} that generates the reference voltage (V_{REF}) has poor line sensitivity (LS), and vice versa. To mitigate process dependency, one VCR utilizes a stacked PMOS to create V_{REF}. It employs a resistor for defining I_{REF} with V_{REF} and an amplifier to protect the generation of I_{REF} from the channel length modulation effect [5]. However, without a stable current source in this circuit, V_{REF} becomes vulnerable to the supply voltage, leading to poor LS of I_{REF}. Additionally, the high leakage current at high temperatures limits the usability of the VCR to a small temperature range. Finally, a CMOS-only V_{REF} with stacked transistors can generate I_{REF} via a feedback amplifier [6]. The self-feedback architecture improves the LS. However, to compensate for the temperature coefficient (TC) of the resistors, the VCR requires multiple types of resistors, introducing additional variations in I_{REF}. The LS of this VCR can be enhanced by designing a current source with better LS for V_{REF}, instead of directly using I_{REF}. This paper proposes a low-power VCR with excellent LS and TC across a broad temperature range.

Fig. 1 shows the schematic of the proposed VCR. The V_{REF} is generated by a two-stage stacked-diode metal oxide semiconductor transistor (SDMT) [7]. The first-stage output is designed to be complementary to the absolute temperature (CTAT), as a bias for I_{REF}, which corresponds to the TC of the resistor, R_1; the second-stage output is the V_{REF}, and can be derived as follows:

$$V_{REF} \approx m_{n1}V_T ln\left(\frac{2K_{n1}}{K_{n2}}\right) + \left(\frac{m_{n1}}{m_{n2}}V_{TH,n2} - V_{TH,n1}\right)$$
$$+ m_{n3}V_T ln\left(\frac{K_{n3}}{K_{n4}}\right) + \left(1 - \frac{m_{n1}}{m_{n2}}\right)V_{G,n2}, \quad (1)$$

where $K_x = \mu_x C_{ox,x} W_x/L_x$, μ_x, $C_{ox,x}$, W_x, L_x, V_T, V_{THx}, m_{nx} represent mobility, unity oxide capacitance, width, length, thermal voltage, threshold voltage, and slope factor of a transistor M_{nx}, respectively. Note that m_{n3} and $V_{TH,n3}$ are nearly equal to m_{n4} and $V_{TH,n4}$, respectively, since M_{n3} and M_{n4} are identical devices. The $V_{TH,n2}$ and $V_{G,n2}$ terms are neglected. By adjusting the aspect ratios of the four transistors, a CTAT bias and V_{REF} are generated. Using a thick-oxide transistor (M_{n2}) increases the voltage difference $(m_{n1}/m_{n2})V_{TH2}-V_{TH1}$, resulting in a higher output voltage.

A compensation scheme is employed to compensate for the high leakage current, which constitutes a significant portion at high temperatures [7]. A reversely biased transistor M_{p7} is used to mitigate the leakage loss of V_{REF}. Fig. 2 illustrates the simplified model of the compensation and the result of the VCR in simulation. To achieve compensation, M_{p7} is placed in parallel with M_{p2}, with a current set to α times the current in the current mirror. Consequently, the current of transistors $M_{n2, n3, n4}$ is increased by $(1+\alpha)$. The current ratio of M_{n4}/M_{n3} remains unchanged, while the ratio of M_{n2}/M_{n1} is increased by $(1+0.5\alpha)$. Additionally, an α-controlled proportional to absolute temperature (PTAT) term is added to V_{REF}, compensating for the leakage at high temperatures. The temperature compensation result

of V_{REF} based on the parameter α is presented in Fig 2. To improve the power supply rejection ratio (PSRR), M_{p7} is placed in parallel with M_{p2} instead of M_{p5}. This approach weakens the V_{DD} interference through the cascode structure. Furthermore, the second stage of the SDMT functions as a high-pass filter (HPF) by subtracting the two signal paths at the output. In addition, I_{REF} is generated through a feedback amplifier with compensated V_{BIAS}. A V_{REF}-biased transistor M_{N8} is placed in parallel with R_1 to compensate for the TC of R_1, which follows a second-order curve (Fig. 2). Finally, $M_{p3,p6,n5,n6}$ generate the bias current with high LS to the core circuit and the two-stage amplifier. The bias current is set to 1% of I_{REF} at room temperature for power reduction.

To further analyze the LS at low frequency, we first simplify the output resistance of the cascode current mirror (M_{p1}, M_{p4}) and (M_{p2}, M_{p5}) to R_P. The LS of V_{REF} is derived as follows:

$$\frac{\Delta V_{REF}}{\Delta V_{DD}} \approx \frac{4}{R_P}\left[\left(\frac{1}{g_{m,n4}} - \frac{1}{g_{m,n3}}\right) - \frac{1}{2}\left(\frac{1}{g_{m,n2}} - \frac{1}{2g_{m,n1}}\right)\right], \quad (2)$$

where $g_{m,x}$ represents the transconductance of transistor M_x. From (2), the LS can be minimized by setting $g_{m,n2} = 2\,g_{m,n1}$ for the first stage and $g_{m,n4} = g_{m,n3}$ for the second stage. Thus, adequately sizing the SDMTs and increasing the output resistance of the current mirror can improve the LS. Under this condition, the LS of I_{REF} created by V_{REF} can be derived as

$$\frac{\Delta I_{REF}}{\Delta V_{DD}} \approx \frac{\Delta V_{Bias}}{\Delta V_{DD}R_1} + g_{m,n8}\frac{\Delta V_{REF}}{\Delta V_{DD}}, \quad (3)$$

Note that the amplifier is designed with a 60-dB gain over 100 Hz, making the LS in (3) barely affected by the amplifier's output. When the temperature is low, the bias current is small, causing the second term of the expression to be much smaller than the first term. Hence, the LS is primarily determined by the LS of V_{BIAS}, which can be minimized by setting $g_{m,n2} = 2\,g_{m,n1}$; the second term can also be minimized according to (2). For the startup of the circuit, C_1 is added [7]. Fig. 3 shows the Monte-Carlo simulation result of V_{REF} and I_{REF} for 1000 runs. A good correlation (0.987) between voltage and current reference and a good LS (<0.1%/V) can be achieved in the proposed VCR design.

The VCR circuit was fabricated using a 0.18 μm CMOS process. At 25 °C and a voltage supply of 0.8 V, the average power consumption of the 13 samples is 18.51 nW, where the V_{REF}, amplifier, and I_{REF} consume 504 pW, 1.66 nW, and 16.34 nW, respectively, which are 2.7 %, 9.0 %, and 88.3 % of the total. Across the 13 samples, the μ and σ/μ of V_{REF} are 270.3 mV and 0.81 %, respectively, while the μ and σ/μ of I_{REF} are 20.4 nA and 1.03 %, respectively. Fig. 4 illustrates the PSRR of V_{REF} versus frequency and the LS of V_{REF} and I_{REF}. A -60.9 dB PSRR at 100 Hz is achieved in this design. The average LS of V_{REF} is 0.011%/V and the average LS of I_{REF} is 0.094% over a supply voltage range of 0.8 to 1.8 V. To determine the optimal trimming code, a sample was measured from -40 to 130 °C in a temperature chamber. Subsequently, 13 samples were measured using the same code. Fig. 5 shows the measured V_{REF} and I_{REF} versus temperature, respectively, from -40 to 130 °C. The average TC of V_{REF} is 124 ppm/°C, while the average TC of I_{REF} is 264 ppm/°C. Finally, Table I presents a comparison with recent VCR designs. Among these designs, the proposed VCR stands out with its good LS and PSRR of V_{REF} and I_{REF}, σ/μ of I_{REF}, and comparable TCs over a wider range of temperatures. Fig. 7 shows the chip micrograph, and the core area is 25000 μm².

Acknowledgments: The authors thank the Taiwan Semiconductor Research Institute for chip fabrication, the National Science and Technology Council, Taiwan for financial support (grant 112-2628-E-A49-015), and uPI Semiconductor Corp. for contributions.

References:

[1] Y. Wenger and B. Meinerzhagen, "Low-voltage current and voltage reference design based on the MOSFET ZTC effect," *IEEE Transactions Circuits Systems I: Regular Papers,* vol. 66, pp. 3445–3456, Sep. 2019.

[2] W. Huang, L. Liu, and Z. Zhu, "A Sub-200nW All-in-One Bandgap Voltage and Current Reference Without Amplifiers," *IEEE Transactions on Circuits and Systems II: Express Briefs,* vol. 68, pp. 121-125, Jan. 2021.

979-8-3503-3004-5/23 $31.00 © 2023 IEEE

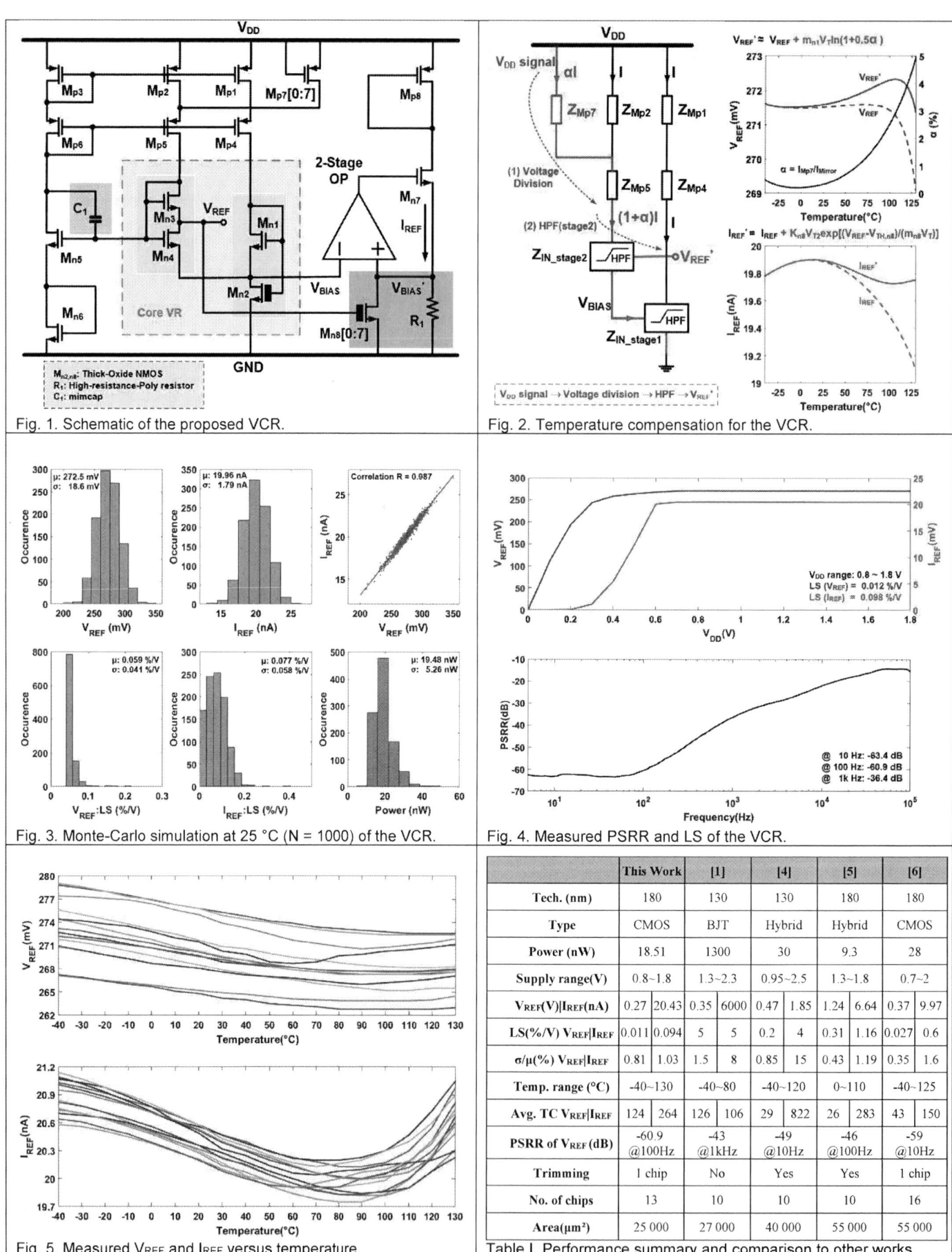

Fig. 1. Schematic of the proposed VCR.

Fig. 2. Temperature compensation for the VCR.

Fig. 3. Monte-Carlo simulation at 25 °C (N = 1000) of the VCR.

Fig. 4. Measured PSRR and LS of the VCR.

Fig. 5. Measured V_{REF} and I_{REF} versus temperature.

Table I. Performance summary and comparison to other works

	This Work		[1]		[4]		[5]		[6]	
Tech. (nm)	180		130		130		180		180	
Type	CMOS		BJT		Hybrid		Hybrid		CMOS	
Power (nW)	18.51		1300		30		9.3		28	
Supply range(V)	0.8~1.8		1.3~2.3		0.95~2.5		1.3~1.8		0.7~2	
V_{REF}(V)|I_{REF}(nA)	0.27	20.43	0.35	6000	0.47	1.85	1.24	6.64	0.37	9.97
LS(%/V) V_{REF}|I_{REF}	0.011	0.094	5	5	0.2	4	0.31	1.16	0.027	0.6
σ/μ(%) V_{REF}|I_{REF}	0.81	1.03	1.5	8	0.85	15	0.43	1.19	0.35	1.6
Temp. range (°C)	-40~130		-40~80		-40~120		0~110		-40~125	
Avg. TC V_{REF}|I_{REF}	124	264	126	106	29	822	26	283	43	150
PSRR of V_{REF} (dB)	-60.9 @100Hz		-43 @1kHz		-49 @10Hz		-46 @100Hz		-59 @10Hz	
Trimming	1 chip		No		Yes		Yes		1 chip	
No. of chips	13		10		10		10		16	
Area(µm²)	25 000		27 000		40 000		55 000		55 000	

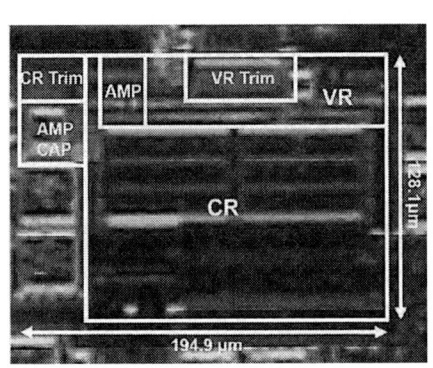

Fig. 7. Chip micrograph

Additional References:

[3] G. Chowdary, et al, "A 1-nW 95-ppm/°C 260-mV Startup-Less Bandgap-Based Voltage Reference," *IEEE International Symposium on Circuits and Systems (ISCAS)*, Oct. 2020.

[4] D. Shetty, et al, "Ultra-Low-Power Sub-1 V 29 ppm/°C Voltage Reference and Shared-Resistive Current Reference," *IEEE Transactions on Circuits and Systems I: Regular Papers,* vol. 70, pp. 1030-1042, Mar. 2023.

[5] Y. Ji, et al, "A 9.3nW all-in-one bandgap voltage and current reference circuit," *IEEE International Solid-State Circuits Conference (ISSCC),* Feb. 2017.

[6] L. Wang, and C. Zhan, "A 0.7-V 28-nW CMOS Subthreshold Voltage and Current Reference in One Simple Circuit," *IEEE Transactions on Circuits and Systems I: Regular Papers,* vol. 66, pp. 3457-3466, Sep. 2019.

[7] C-Z Shao, S-C Kuo, and Y-T Liao, "A 1.8-nW, -73.5-dB PSRR, 0.2-ms startup time, CMOS voltage reference with self-biased feedback and capacitively coupled schemes," IEEE Journal of Solid-State Circuits, vol. 56, no. 6, pp. 1795–1804, Jun. 2021.

A 400MHz 249.1TOPS/W 64Kb Fully-Reconfigurable SRAM-Based Digital Compute-In-Memory Macro for Accelerating CNNs

Xin Zhang*[1,2], Vishal Sharma*[1], Yuncheng Lu[1], Yongjun Jo[1], and Tony Tae-Hyoung Kim[1] (*Equally contributed authors)

[1]Nanyang Technological University, Singapore

[2]MediaTek, Singapore

Compute-in-memory (CIM) is a promising solution for neural networks performing vector-matrix multiplication with limited power by processing data inside a memory and minimizing energy-hungry data transfer between memory and processing elements. However, analog-based designs face several challenges such as non-ideal performance from PVT variations, ADC area and energy overhead, and conversion precision loss as illustrated in Fig. 1(Top) [1-2]. Digital CIMs have no accuracy degradation but show relatively low performance and energy efficiency [3-5]. The limited bit precision reconfigurability is also a big challenge as it restricts the use of CIM macro only for fixed neural network topologies.

This work proposes an SRAM-based digital CIM macro, highlighted with three key features: 1) fully reconfigurable weight precision from 1b to 8b, 2) improved CIM operation speed for higher performance, and 3) SA-less SRAM read operation. The overall architecture of the proposed CIM macro is shown in Fig. 1(Bottom). The 512×128 array is divided into sixteen blocks (BLKs), each containing sixty-four 8×8 sub-blocks (SBLKs) and one binary adder tree. The macro can process fully reconfigurable weight precision and operate at 400 MHz, which shows the highest energy efficiency among the state-of-the-art digital CIMs reported. The SRAM read operation can be executed through the adder tree, which saves the silicon area for sense amplifiers.

Each SBLK includes eight digital processing units (DPUs) for computing reconfigurable multiplication. Each DPU consists of 8×1 SRAM bit-cells and one shared control logic, where the multiplication of 1b input and 1b weight is conducted as shown in Fig. 2. An n-bit weight could be stored in the n bit-cells from the same row in different DPUs. For example, an 8b weight can be stored in the bit-cells in the first row from eight DPUs. In each CIM cycle, 1b input is applied to one of the eight RWLs in the SBLK and multiplied by the bit-cells storing n-bit weight. The n-bit dot product result is generated at CBLs, which is accumulated in the binary adder tree. For different applications, the size of neural networks can vary. Input precision can be reconfigured through bit-serial computation. However, weight reconfiguration has not been well addressed in the reported works [4-5]. This work utilizes the weight-select (WS) signal to provide the full-precision weight reconfiguration. For example, for 8b weight precision, WS<7:0> will be "1" so that all the eight control logic blocks can participate in the CIM operation. Higher bit precision (more than 8b) can also be realized by employing multiple CIM macros with additional shift-and-add operation. Thus, the proposed digital CIM macro can be easily employed by various neural networks with arbitrary input and weight precisions.

Fig. 3(a) shows the detailed circuit for the CIM operation. This work utilizes a decoupled 8T SRAM bit-cell for dot-product without disturbance issue, which also improves the discharging speed at the bit-lines (BLXs). N1 and N2 are also implemented with LVT devices for faster discharging. The 6T control logic circuit is shared with eight SRAM bit-cells to mitigate the area overhead. During standby, P1 and P2 pre-charge BLX and BLI to VDD and the inverter formed by P3, N4, and N5 is disabled. During CIM operation, a multiplication result is generated at BLX, depending on the input (RWL) and the weight (Q). BLX will be discharged only when RWL=Q="1". Otherwise, BLX will stay at floating VDD. The selected BLX voltage is transferred to BLI through the enabled N3 (WS="1"). Once EN goes high, the BLI node voltage is converted to a digital multiplication value at CBL. The voltage at CBL is reserved till the next evaluation cycle with EN = "0". The high operating frequency prevents the voltage at CBL from being affected by any potential leakage. The proposed CIM macro operates at 400MHz because of the combined effect of smaller loading at BLX, and multiplication through a dynamic circuit with LVT devices. Consequently, the improved operation speed enhances the energy efficiency of the CIM macro. Fig. 3(b) summarizes the critical signals at different operation modes.

Fig. 3(c) shows the operation of one BLK in the proposed CIM macro. Each BLK consists of sixty-four 8×8 SBLKs and one binary adder tree for accomplishing the MAC operation of one neuron. Each SBLK produces an 8b product result (1b input × 8b weight) at CBLs for the adder tree to accumulate. Overall, sixty-four SBLKs will provide sixty-four 8b dot product results and each binary adder tree generates a 14b partial sum result. The 14b adder tree output will be shifted and accumulated for calculating the final output (Y<21:0>) as illustrated in Fig. 3(d).

The adder tree in Fig. 4(a) accumulates the sixty-four 8b dot product results and generates a 14b partial sum (Pi<13:0>). The accumulator receives the LSB of the multi-bit input first and receives the subsequent input bits in a bit-serial manner in each CIM cycle. To preserve the positional values, the 14b partial sums from more significant input bits are left-shifted in each cycle and added to the partial sum over the subsequent cycles until all the partial sums from the adder tree are processed. Fig. 4(b) shows an example of accumulating eight 14b partial sums through the accumulator. P0<13:0>, generated from the adder tree using the LSB of the 8b input, is added with the 1b left-shifted P1<13:0> from the adder tree using the second LSB of the input. Note that P7<13:0> is left-shifted by 7b before being added to other Pi<13:0>. Eventually, the accumulator produces a 22b result (S<21:0>) as the MAC result of one neuron, which is from the sixty-four 8b weights and the 8b input.

In general, SRAM read requires sense amplifiers, which occupies significant area. This work improves area efficiency by removing SAs and conducts read operation through the adder tree. For instance, if the selected bit-cell stores "1", the CBL of the corresponding control logic outputs "1" while the other CBLs become "0". Then, the adder tree output leads to Y<13:0> = "00000000000001" ~ "00000010000000", depending on the location of the selected column by WS<n>. Since only WS<i> of the selected weight bit will be enabled, the final read data can be obtained by a simple OR logic to Y<7:0>, as indicated in Fig. 4(c). Note that Y<13:8> is not used for the SRAM read operation.

The macro can operate with fully reconfigurable input and weight precision from 1b to 8b. Fig. 5(a) summarizes the throughput of the CIM macro at various bit precisions. The throughput of the macro is 819.2 GOPS with 8b weight and 1b input. The proposed digital CIM macro was fabricated in 65nm technology. The area and power breakdown (CIM mode, W = IN = 8b) of the macro are depicted in Fig. 5(b). The measured energy efficiency with different input and weight bit precisions is presented in Fig. 5(c). The proposed CIM macro achieves the energy efficiency of 249.1TOPS/W with 1b input and 1b weight. Fig. 6 compares the proposed work with other state-of-the-art digital CIM macros. The proposed digital CIM macro achieves higher operation frequency, higher energy efficiency, and area efficiency than [3]. The macro supports better weight precision reconfigurability than [4-5]. The overall performance and reconfigurability makes the proposed macro an attractive choice for neural network accelerators requiring various network sizes over multiple layers.

Acknowledgement:

This work was supported by a RIE2020 ASTAR AME IAF-ICP under Grant I1801E0030.

References:

[1] C. Yu et al., "A 16K Current-Based 8T SRAM Compute-In-Memory Macro with Decoupled Read/Write and 1-5bit Column ADC," CICC 2020:1-4.

[2] A. et al., "CONV-SRAM: An Energy-Efficient SRAM With In-Memory Dot-Product Computation for Low-Power Convolutional Neural Networks," JSSC 2019:217-230.

[3] H. Kim et al., "A 1-16b Precision Reconfigurable Digital In-Memory Computing Macro Featuring Column-MAC Architecture and Bit-Serial Computation," ESSCIRC 2019: 345-348.

[4] Y. -D. Chih et al., "An 89TOPS/W and 16.3TOPS/mm2 All-Digital SRAM-Based Full-Precision Compute-In Memory Macro in 22nm for Machine-Learning Edge Applications," ISSCC 2021: 252-254.

[5] B. Yan"A et al., "1.041-Mb/mm2 27.38-TOPS/W Signed-INT8 Dynamic-Logic-Based ADC-less SRAM Compute-in-Memory Macro in 28nm with Reconfigurable Bitwise Operation for AI and Embedded Applications," ISSCC 2022: 188-190

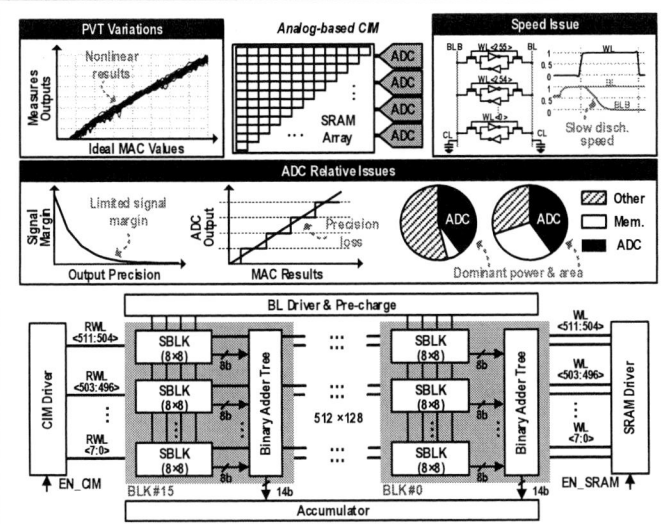

Fig. 1. Design challenges in analog-based CIM macro and the overall structure of the proposed 64Kb CIM macro.

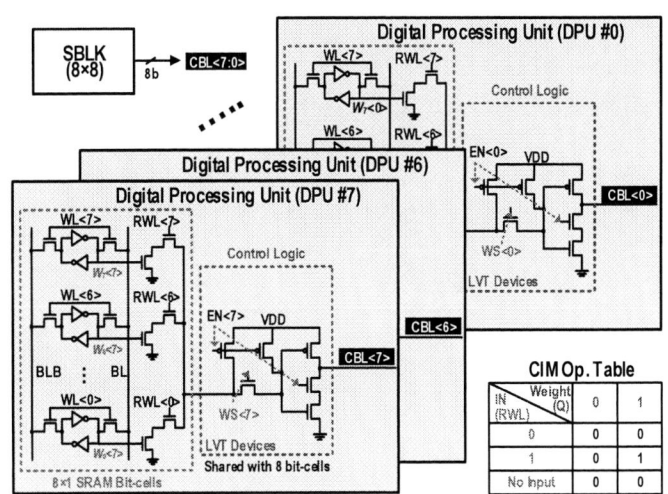

Fig. 2. Structure of one 8×8 sub-block (SBLK) and Digital processing unit (DPU), and CIM operation table for 1b multiplication.

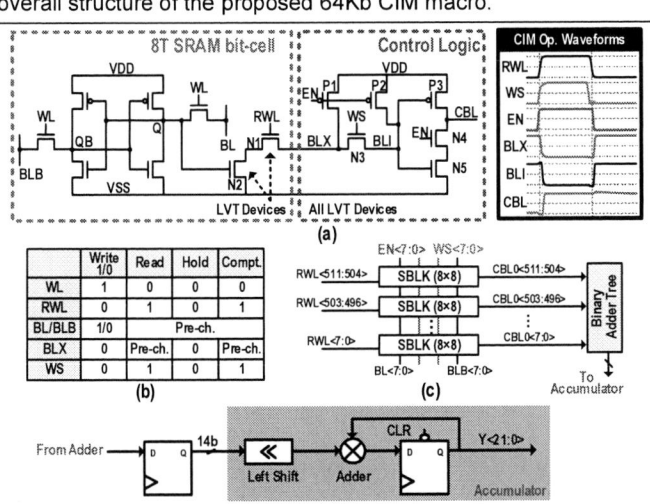

Fig. 3. (a) Schematic of DPU and operation waveforms, (b) control signals for various operation modes, (c) structure of one block (BLK#i), and (d) simplified circuit diagram of the accumulator.

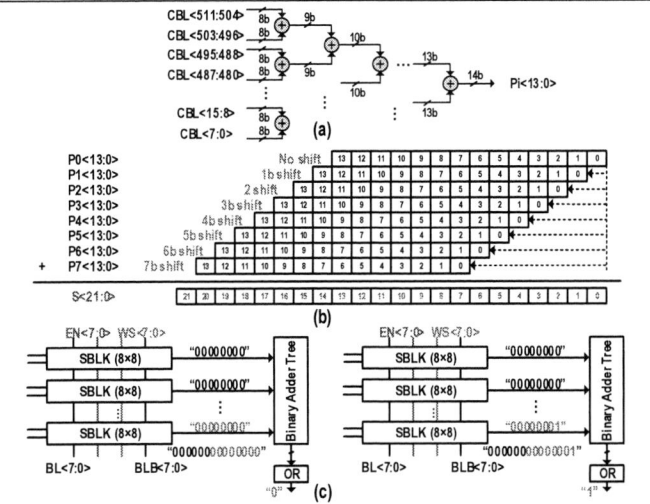

Fig. 4. (a) Structure of one binary adder tree, (b) accumulation scheme using the adder tree with 8b input over eight CIM cycles, and (c) SA-less read operation.

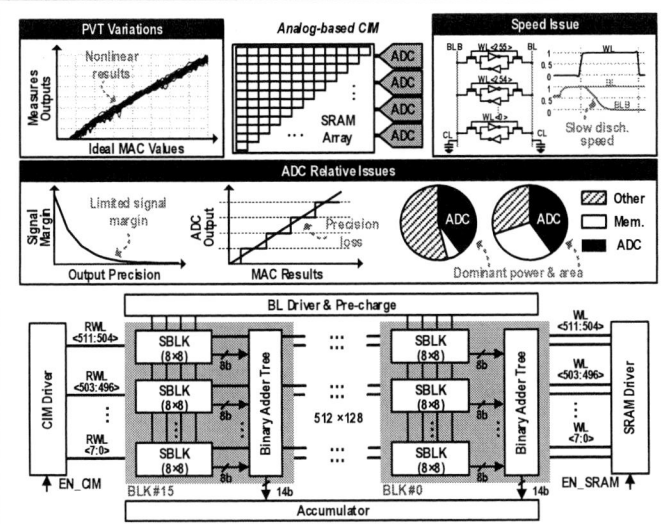

Fig. 5. (a) Performance summary table, (b) area and power breakdown, and (c) energy efficiency of the macro at different input and weight precisions.

	ESSCIRC' 19 [3]	ISSCC' 21 [4]	ISSCC' 22 [5]	This Work
Technology	65nm	22nm	28nm	65nm
Supply Voltage	0.6-0.8 V	0.72 V	0.8 V	1 V
Array Size	16 Kb	64 Kb	32 Kb	64 Kb
Bit-cell	6T	6T	6T	8T
Bit-cell Area	10.53 um²	0.379 um²	0.257 um²	2.3452 um²
Macro Area	0.2272 mm²	0.202 mm²	0.030 mm²	0.3136 mm²
Precision Fully-Reconfigurable	Yes	No	No	Yes
Input Precision	1-16b	1-8b	1-8b	1-8b
Weight Precision	1-16b	4/8/12/16b	1/4/8b	1-8b
Output Precision	8-23b	24b(8b/8b)	21b	14b
Energy Efficiency	117.3 TOPS/W (1b/1b)	24.7 TOPS/W (8b/8b)	27.38 TOPS/W (8b/8b)	249.1 TOPS/W (1b/1b)
Frequency	138 MHz	400 MHz (4b/1b)	333 MHz	400 MHz (8b/1b)

Fig. 6. Comparison with state-of-the-art digital CIMs.

979-8-3503-3004-5/23 $31.00 © 2023 IEEE

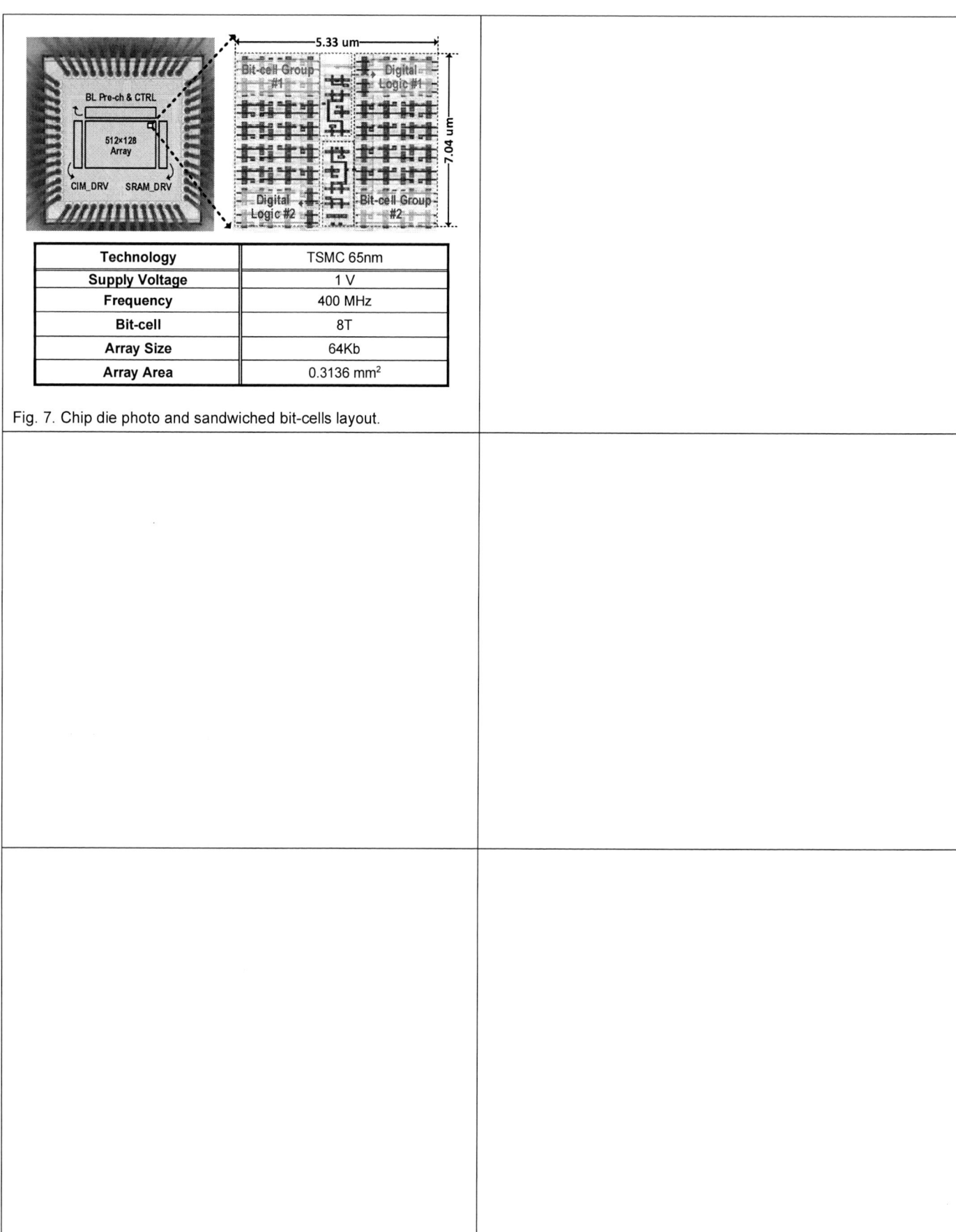

Technology	TSMC 65nm
Supply Voltage	1 V
Frequency	400 MHz
Bit-cell	8T
Array Size	64Kb
Array Area	0.3136 mm²

Fig. 7. Chip die photo and sandwiched bit-cells layout.

A 73.8K inference/mJ SVM Learning Accelerator for Brain Pattern Recognition

Tzu-Wei Tong, Yi-Yen Hsieh, Tai-Jung Chen,

Chia-Hsiang Yang

National Taiwan University, Taipei, Taiwan

Sensing brain dynamics for neural disease control and brain-computer interface has become feasible through brain pattern recognition. The applications include seizure detection/prediction for epilepsy control, neural signal sensing for neuroprostheses, and brainwave processing for motor-assistive devices [1]. Fig. 1 shows the workflow for brain activity recognition, in which the acquired neural signals are sent to a machine learning (ML) kernel for pattern recognition. The outputs could be used to drive a stimulator or an actuator for control/interfacing. Of the ML kernels, support vector machine (SVM) has been widely used for neural signal classification [2-3] due to the effectiveness given the limited training samples. Online SVM adaptation (through training) also becomes essential to combat time-varying neural signals [2]. As signal dimension increases with the electrode array size for more sophisticated tasks [4], conventional SVM becomes inefficient due to the inherent sequential update characteristic. In this work, cluster-partitioning (CP)-SVM [5] is adopted to reduce the latency by processing data in parallel. Energy dissipation is another issue for energy-constrained devices. This work presents an energy-efficient SVM learning accelerator that supports a high throughput for neural sensing devices with a massive electrode array.

Fig. 2 illustrates the CP-SVM workflow for training and inference. The training data are first partitioned into N clusters in the clustering stage by leveraging k-nearest neighbors (KNN) clustering. In the group assignment stage, the N clusters are split into two groups: Group 1 includes clusters with the same label, while Group 2 includes clusters with different labels. For Group 1, training is not required since there is no need for separating the clusters through a hyperplane. For Group 2, a hyperplane to separate data with different labels needs to be found for each cluster. The hyperplane of each cluster within Group 2 can be updated iteratively in parallel [2]. A training pair comprising one misclassified data with the largest evaluation error and one randomly selected data is chosen, and the hyperplane is updated accordingly. The evaluation error is computed based on the distance between the data and the updated hyperplane. For inference, input data are assigned to clusters according to the distance to each centroid, which is determined through KNN. Data in Group 1 are classified into the cluster with the nearest centroid, and only data in Group 2 need to be evaluated. According to our experiments, on average, 18% of the training data and 16% of the inference data for the seizure prediction task in [3] need to be processed. This enables a 93x faster convergence for training. The inference latency is also reduced by 91% in comparison to the SVM without cluster-partitioning.

Fig. 3 shows the architecture of the proposed CP-SVM learning processor, which includes a scheduler array, a processing element (PE) array, and a classifier. The scheduler array including four schedulers distributes the clustering computations to multiple PEs in a hierarchical way. Each scheduler is assigned to process the data of one cluster in CP-SVM. A pipelined sorter is shared by all schedulers to perform sorting for finding the data pair with the most significant evaluation error. The PE array computes the evaluation for inference and training. In each PE, an L1-norm calculator computes the SVM kernel function by employing the coordinate rotation digital compute (CORDIC) units. Multiple PEs need to access the same input data in a cluster to perform evaluation. Therefore, data exchangers are designed to facilitate data sharing with adjacent PEs for the same cluster. The classifier comprises a cluster assigner and an evaluator. The cluster assigner performs cluster partitioning to determine which cluster the data belongs to, and sends the data to the corresponding PE. The evaluator receives both the cluster labels from the cluster assigner and the evaluation results from the PE array to perform inference.

Fig. 4 illustrates algorithm-architecture co-optimizations. The level of parallelism is explored to achieve an optimal design. For the PE array with 16 PEs, the area-time product (ATP) is the lowest, reducing the latency by 89% compared to the architecture with only one PE. To facilitate effective data flow control, the PEs can be reconfigured to support up to 4 clusters. The data of each cluster in Group 2 is partitioned and stored in the local memory of the corresponding PE. Sparsity-aware skipping is proposed to eliminate redundant computations for the evaluation error. For the case that the Lagrange multiplier $\alpha = 0$ (i.e., the training evaluation error is zero), the corresponding data pair in Group 2 make no contribution to training. Statistically, only 19% of α are nonzero for the seizure prediction task in [3], saving the latency by 24%. Parallel processing and scheduling are utilized to increase the utilization of PEs, which reduces the latency by 46%. The overall processing latency for PE array is reduced by 96%.

Fig. 5 shows the details of the sub-modules. Each scheduler sorts the evaluation results from the PE array for finding the data pairs. A configurable mask is used to find the most significant evaluation error. All schedulers share a cross-cluster sorter to perform sorting before applying the mask, which reduces the area of the sorter by 52%. In the direct-mapped architecture, a fully-connected data exchanger can be used to connect all PEs at a cost of high hardware complexity. In this work, a chained data exchanger is proposed to connect a PE with adjacent PEs, reducing the data exchanger area by 93%. In addition, architecture exploration is conducted for optimizing the kernel calculator. Given the same throughput, the design with 4 interleaved CORDIC units achieves the lowest ATP and is selected. For SVM kernel transformation, the base e in the Gaussian kernel is replaced by the base 2 to eliminate multiplications. The L1-norm, instead of the L2-norm, is utilized to facilitate computations on CORDIC units. Overall, the PE array area is reduced by 42% compared to the baseline design.

Fabricated in 40nm CMOS, the proposed SVM learning accelerator integrates 2.1M logic gates in core area of 1.4mm^2. Fig. 6 shows the measurement results. The chip dissipates 9.68mW at a clock frequency of 40MHz from a 0.85V supply, achieving a maximum energy efficiency of 73.8K inference/mJ for inference and 811 training/mJ for training. The chip delivers a throughput of 714.3K inference/s and a 7.8K training/s. It achieves an area efficiency of 510K inference/s/mm^2 and 5.6K training/s/mm^2. The performance of the chip is compared with the SVM kernel with a similar configuration in the prior neural signal processor [3]. As shown in Fig. 6, this work achieves 3.4x and 6.9x higher energy efficiency for inference and training, respectively. The chip also delivers 19.3x and 40.9x higher area efficiency for inference and training, respectively. This work demonstrates a promising energy-efficient brain pattern recognition kernel for massive-array neural sensing systems. Fig. 7 shows the chip micrograph and performance summary.

Acknowledgment:

This work is supported by NSTC of Taiwan and Intelligent & Sustainable Medical Electronics Research Fund in NTU. The authors thank TSRI for technical support on chip design and fabrication.

References:

[1] A. Gupta et al., "Neuroprosthetics: from sensorimotor to cognitive disorders," Communications Biology, vol. 6, no. 1, pp. 14, 2023.
[2] S.-A. Huang et al., "A 1.9mW SVM processor with on-chip active learning for epileptic seizure control," IEEE J. Solid-State Circuits, vol. 55, no. 2, pp. 452-464, Feb. 2020.
[3] Y.-Y. Hsieh et al., "A 96.2nJ/class neural signal processor with adaptable intelligence for seizure prediction," Int. Solid-State Circuits Conf. (ISSCC), pp. 506-508, Feb. 2022.
[4] A. Gunduz et al., "Differential roles of high gamma and local motor potentials for movement preparation and execution," Brain-Computer Interfaces, vol. 3, no. 2, pp. 88-102, 2016.
[5] Y. You et al., "Design and implementation of a communication-optimal classifier for distributed kernel support vector machines," IEEE Trans. Parallel Distrib. Syst., vol. 28, no. 4, pp. 974-988, Apr. 2017.

979-8-3503-3004-5/23 $31.00 © 2023 IEEE

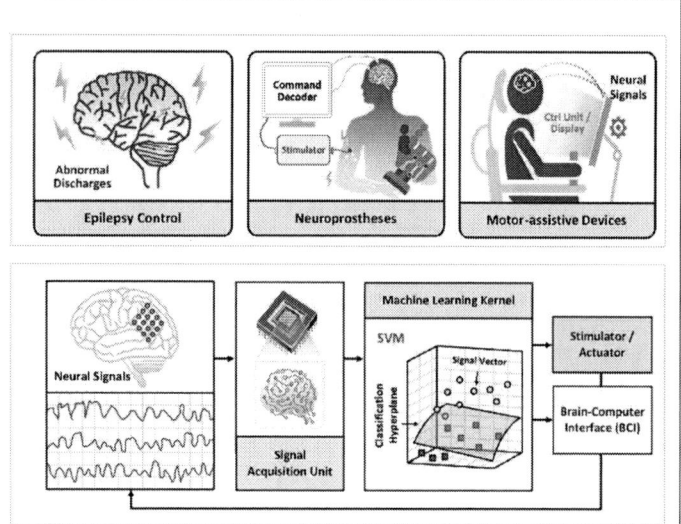

Fig. 1. Applications of neural disease control and brain activity recognition workflow.

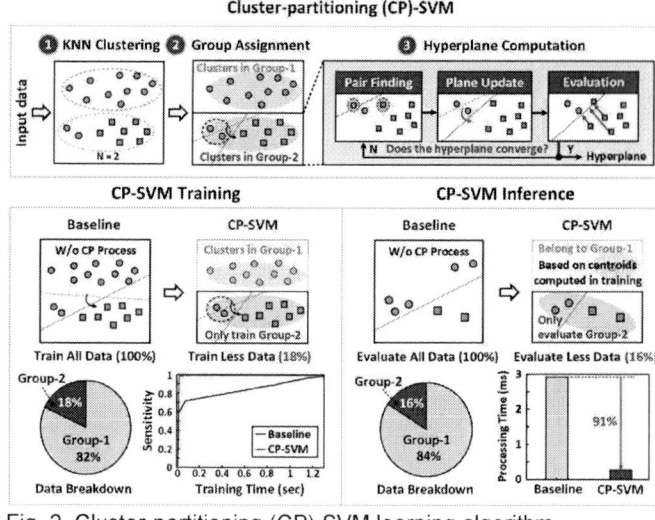

Fig. 2. Cluster-partitioning (CP)-SVM learning algorithm.

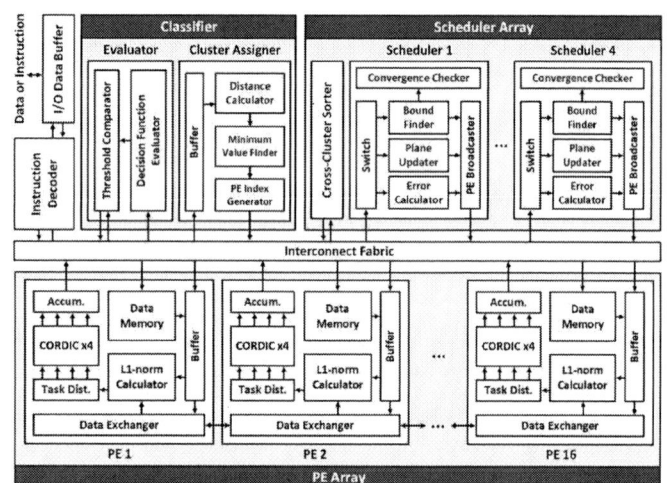

Fig. 3. System architecture of the proposed SVM Learning Accelerator.

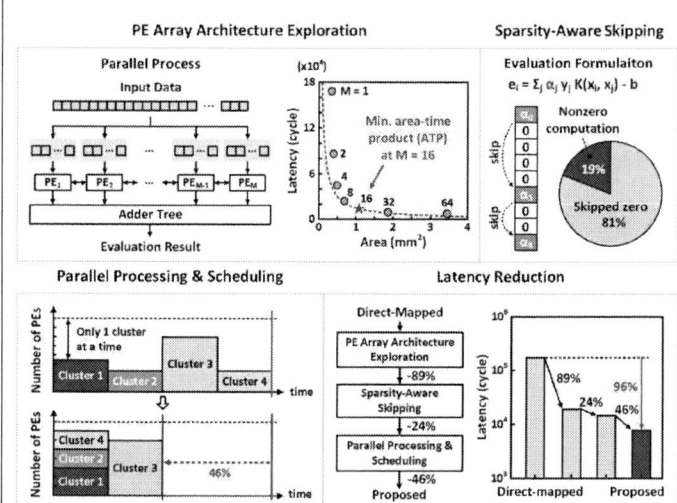

Fig. 4. Algorithm-architecture co-optimizations.

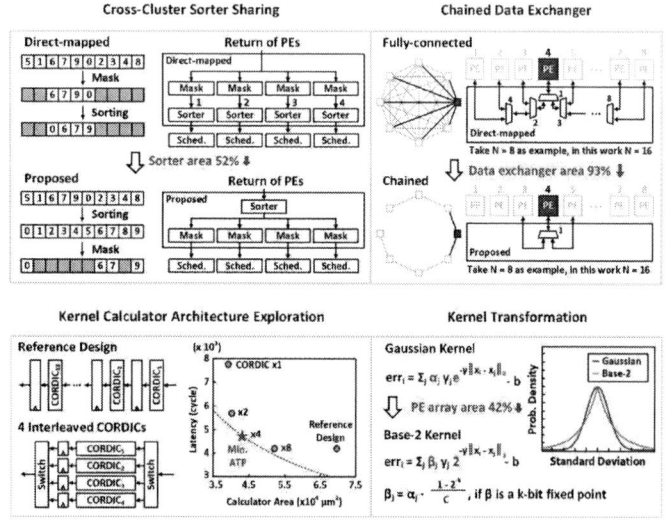

Fig. 5. Design details of the sub-modules.

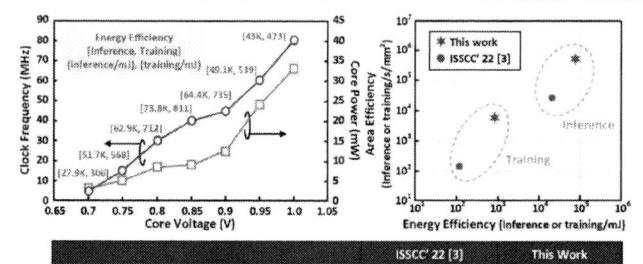

Fig. 6. Experimental verification and performance comparison.

	ISSCC' 22 [3]	This Work	
Technology	40nm	40nm	
SVM Algorithm	ADMM	CP-SVM	
# of Training Data	512	512	
Core Area (mm²)	0.9*	1.4	
Clock Frequency (Hz)	6M	5M − 80M	
Supply Voltage (V)	0.49 − 0.9	0.7 − 1.0	
Power (mW)	1.1*	9.68⁺	
Inference Throughput (inference/s)	23.8K§	714.3K§	
Training Throughput (training/s)	126.6⁺	7.8K⁺	
Energy Efficiency (inference/mJ) / (training/mJ)	21.6K	117	73.8K⁺	811⁺
Area Efficiency (inference/s/mm²) / (training/s/mm²)	26.4K	137	510K⁺	5.6K⁺

* Estimated from the chip micrograph   ⁺ Estimated from [3] '@0.85V, 40MHz

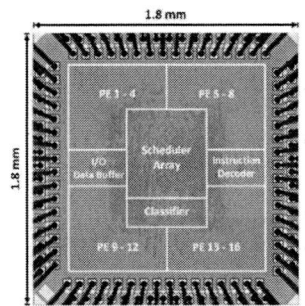

Technology		TSMC 40nm 1P9M
Area (mm²)	Chip	1.8 x 1.8
	Core	1.2 x 1.2
Gate Count		2.1M
On-chip SRAM (KB)		19
Clock Frequency (MHz)		5 - 80
Supply Voltage (V)		0.7 - 1.0
Power (mW)		9.68 @0.85V, 40MHz
Training Throughput (training/s)		7.8K
Inference Throughput (inference/s)		714.3K

Fig. 7. Chip micrograph and summary.

An Instant-Setup On-Delay Compensation Scheme Based on Phase Detection for Series-Resonant Wireless Power Receiver With Up-to-4.41% Efficiency Enhancement

Yuxuan Luo[1], Yutang Chen[1], Jianping Guo[1*], Xian Tang[2], and Dihu Chen[1]

[1]Sun Yat-sen University, Guangzhou, China

[2]Shenzhen International Graduate School, Tsinghua University, Shenzhen, China

Wireless power transfer (WPT) based on series-resonant secondary (SS) topology is becoming increasingly popular for portable electronic applications [1-3]. To enhance the power conversion efficiency (PCE) of the WPT receiver (RX), active diodes with lower forward voltage drop are adopted in rectifier circuit. However, when operating at high frequencies, the propagation delays of the active diodes prevent the power MOSFETs from being turned on promptly (on-delay). As shown in Fig. 1, in SS topology, this on-delay leads to body-diodes conduction (point A to point B), which reduces the PCE.

Some delay compensation schemes have been proposed to minimize the on-delay time for PCE improvement, which can be mainly classified into two types: adaptive offset [3] and adaptive delay line [4], as shown in Fig. 1. Both of them require accurate source and drain voltages when power MOSFETs are turning-on. However, due to the voltage fluctuation caused by the current noise (point B in Fig. 1 when switching from body-diodes conduction to power MOSFETs conduction) and the inductance of the off-chip bond-wires, these voltages are difficult to be sampled [1]. Moreover, the voltage fluctuation will become more severely when higher current is transferred [1-2]. In [3], an on-delay compensation scheme with 8-bits offset calibration was proposed. However, because the on-delay time will change in different operating conditions, the offset for on-delay compensation should be calibrated bit by bit once RX enters a new operating state. As a result, it takes slow setup-time to reach a steady-state offset, and the PCE enhancement due to the on-delay compensation will be reduced before optimal offset is obtained. Moreover, the on-delay compensation accuracy is limited by the least significant bit in [3]. In [4], a delay-line-based on-delay compensation scheme with cycle-based timing control was proposed to achieve an accurate on-delay compensation in parallel-resonant secondary (PS) topology, where the setup time is around 28 μs.

To address above issues, this paper presents an instant-setup accurate on-delay compensation scheme based on phase detection for SS-WPT reconfigurable resonant regulating (R^3) rectifier RX to enhance the PCE. The proposed on-delay compensation scheme is employed in a hysteretic controlled dual-mode (1X: full-wave rectifier mode, 0X: freewheeling mode) R^3 RX with low design complexity. Meanwhile, active diodes with offset voltage for off-delay compensation are adopted to enhance the PCE further as well as boost the power-delivering capability. Because the proposed on-delay compensation is free of cross-compensation problem, the off-delay compensation works all the time. The whole RX chip diagram is shown in Fig. 2 which mainly consists of four parts: rectifier core, push-pull common-gate comparators with offset voltage, on-delay compensation and 0X/1X hysteretic controller.

As indicated by the gate voltage signals (V_{GP1} and V_{GN1}) shown in Fig. 1, in SS topology, the power MOSFETs (M_{P1} and M_{N1}) in the same power-input-ground branch (PV_{DD}-V_{AC1}-PV_{SS}) will be off at the same time for a period of time (on-delay time) while the input current is continuous. Therefore, this on-delay time can be obtained through a phase detection logic. As shown in Fig. 3, the proposed on-delay compensation scheme can be divided into three steps. Firstly, the on-delay time (V_{ON1}) is obtained from the outputs of comparators CMP_{P1} and CMP_{N1} (V_{CP1} and V_{CN1}) on the same power-input-ground branch (PV_{DD}-V_{AC1}-PV_{SS}). Secondly, the edge trigger signals of the outputs of comparators CMP_{P1} and CMP_{N1} (V_{CP1_RD} and V_{CN1_FD}) are used to obtain the on-delay time of the specific power MOSFETs (V_{ON1_P} and V_{ON1_N}, M_{P1} and M_{N1}). The last step is to compensate the specific on-delay time signals (V_{ON1_P} and V_{ON1_N}) to the original comparator output signals (V_{CP1} and V_{CN1}), so as to realize the on-delay compensation (V_{CP1_2} and V_{CN1_2}).

As shown in Fig. 3, the regulated output voltage and off-delay compensation effect will affect the PCE enhancement when using the on-delay compensation scheme. The lower the output voltage, the better the PCE enhancement of the on-delay compensation is. Unfortunately, with suitable off-delay compensation, the PCE enhancement because of the on-delay compensation will be worse.

Some qualitative simulations have been taken to verify the effect of the proposed on-delay compensation. The input of the R^3 RX is simulated by a sine current source with a peak value of 500 mA and the power distribution is estimated by measuring ten 0X-1X cycles in regulating stage. As shown in Fig. 3, the proposed on-delay compensation can decrease the body-diode conduction loss from 0.0441 W to 0.0282 W, thus enhancing the PCE from 90.8% to 92.93%.

The detailed circuit schematic of the delay compensation controller is shown in the top of Fig. 4. Traditional push-pull common-gate comparator with current-mode offset voltage introduction is used to generate the initial comparator output signals. The offset voltage is introduced all the time to compensate the off-delay. The proposed on-delay compensation circuit is realized by simple phase detection circuits which can achieve instant-setup accuracy on-delay compensation with negligible power consumption. Meanwhile, a short delay is introduced to ensure adequate dead time. The measurement setup of PCE is also shown in Fig. 4.

The measured waveforms, including startup, steady state, and load transient, are shown in Fig. 5. The resonant frequency is set in 6.78 MHz. Based on the time and amplitude of body-diode conduction in V_{AC1} and V_{AC2} from the measured steady-state waveform under the heaviest load condition ($d = 34$ mm and $I_O = 0.6$ A), it can be concluded that the proposed on-delay compensation circuit can effectively compensate for the on-delay, and the compensation setup-time is instant (for better visually observing the effect of on-delay compensation, this design specially disables the on-delay compensation at the beginning and ending of 1X mode through a certain logic operation). The bottom of Fig. 5 shows the measured load transient waveform with I_O steps between 100 mA and 600 mA in which on-delay compensation still works well.

The top of the Fig. 6 shows the PCE measurement graphs. The best PCE improvement of 4.41% with on-delay compensation ($d = 34$ mm, $I_O = 0.6$ A) is achieved when off-delay compensation is not enough. Besides, the peak PCE of 92.28% ($d = 14$ mm, $I_O = 0.2$ A) with 1.82% improvement is achieved when the off-delay compensation is suitable.

The bottom of the Fig. 6 exhibits the performance comparisons. Compared with [1] and [2], the proposed on-delay compensation overcomes the difficulty of sampling the source and drain voltages during power MOSFETs turning-on. Therefore, the proposed on-delay compensation still works well when transferring higher current. Meanwhile, compared with [3] and [4], the proposed on-delay compensation scheme is instant-setup and the circuit implementation is simple. Moreover, compared with [3], the proposed on-delay compensation is free of cross-compensation problem. The RX with the proposed on-delay compensation is fabricated in 0.18-μm CMOS technology. Fig. 7 shows the chip micrograph, coil, and measurement setup.

Acknowledgement:

This work was supported by Natural Science Foundation of Guangdong Province, China, under Grant 2022A1515011054.

References:

[1] J. Lin et al., "A 6.78-MHz Single-Stage Wireless Power Receiver With Ultrafast Transient Response Using Hysteretic Control and Multilevel Current-Wave Modulation," *TPEL*, Sept. 2021.

[2] L. Cheng et al., "A 6.78-MHz Single-Stage Wireless Charger With Constant-Current Constant-Voltage Charging Technique," *JSSC*, April 2020.

[3] F.-B. Yang et al., "Structure-Reconfigurable Power Amplifier (SR-PA) and 0X/1X Regulating Rectifier for Adaptive Power Control in Wireless Power Transfer System," *JSSC*, July 2021.

[4] Z. Luo et al., " A 40.68-MHz Active Rectifier With Cycle-Based On-/Off-Delay Compensation for High-Current Biomedical Implants," *JSSC*, Feb. 2023.

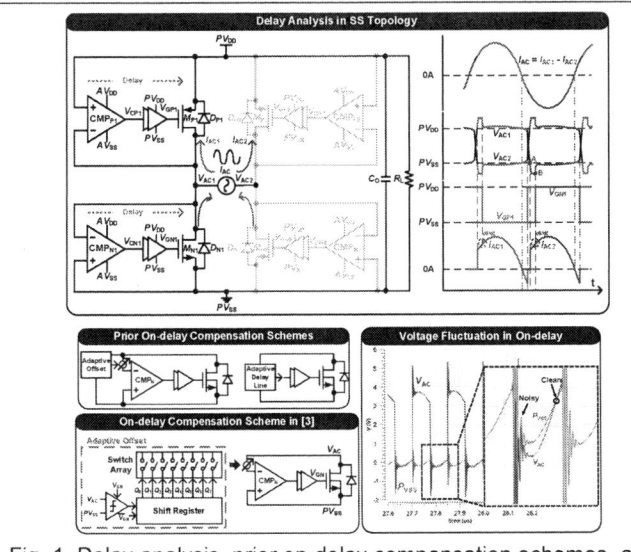

Fig. 1. Delay analysis, prior on-delay compensation schemes, and voltage fluctuation in on-delay.

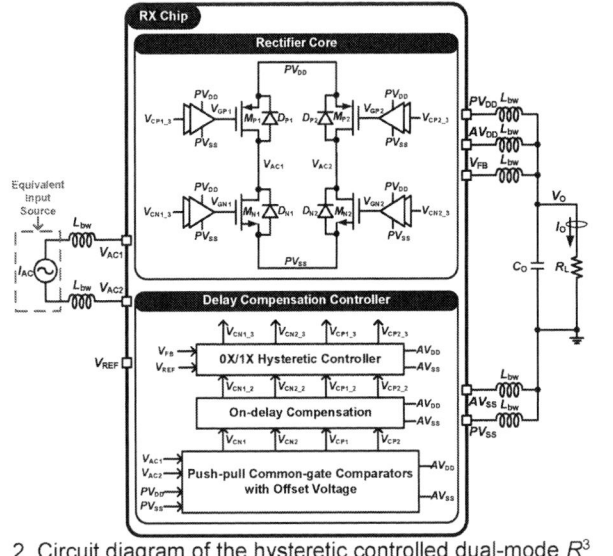

Fig. 2. Circuit diagram of the hysteretic controlled dual-mode R^3 RX with the proposed on-delay compensation scheme.

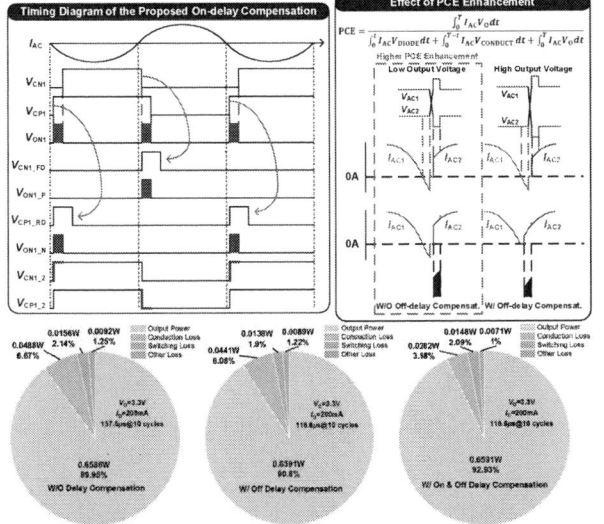

Fig. 3. Operation timing diagram, the effect of PCE enhancement, and power distribution.

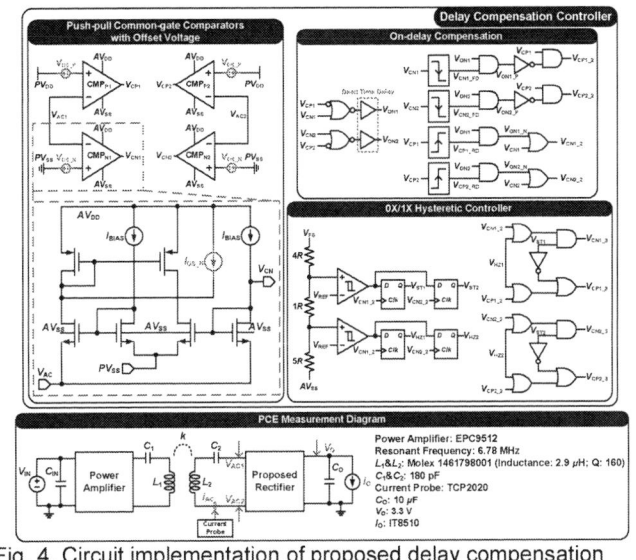

Fig. 4. Circuit implementation of proposed delay compensation controller and PCE measurement diagram.

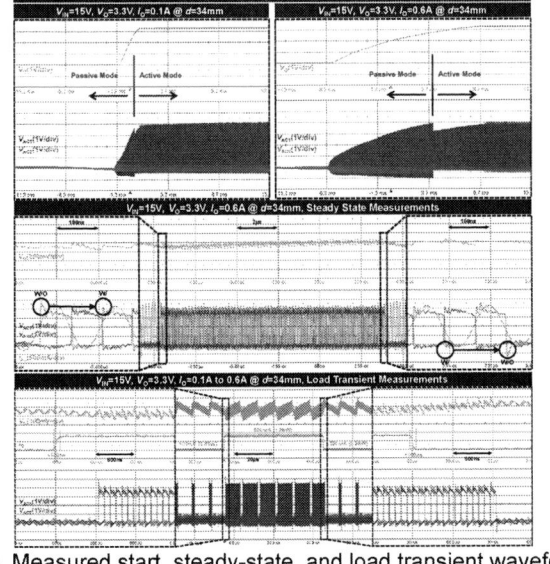

Fig. 5. Measured start, steady-state, and load transient waveforms.

	[2] JSSC'20	[1] TPEL'21	[3] JSSC'21	[4] JSSC'23	This Work
Technology	0.35-μm CMOS	0.18-μm CMOS	0.25-μm CMOS	0.18-μm CMOS	**0.18-μm CMOS**
Resonant Frequency	6.78 MHz	6.78 MHz	6.78 MHz	40.68 MHz	**6.78 MHz**
Secondary LC Topology	SS	SS	PS	—	**SS**
RX Structure	3-Mode R^3	3-Mode R^0	2-Mode R^0	Active Rectifier	**2-Mode R^0**
On-delay Compensat.	NO	NO	Adaptive Offset	Adaptive Delay-line	**Phase Detection**
On-delay Compensat. Setup-time	N/A	N/A	1.18 μs[1]	28 μs[2]	**Instant**
On & Off Delay Cross-Compensat.[3]	N/A	N/A	Yes	No	**No**
RX V_O (V)	2.8 - 4.2	3.3	5	1.65 - 3.66[2]	**3.3**
Max. P_O (W)	6.3	1.98	0.4	0.207	**1.98**
I_O Range (A)	1 - 1.5[4]	0.06 - 0.6	0.001 - 0.08	0.0036 - 0.035[2]	**0.05 - 0.6**
Coil Distance (mm)	20[5]	7 - 16	3	4	**14 - 34**
Peak Eff. With Off Delay Compensat.	92.3% (4.2 V & 1 A)	92.14% (0.6 A & 12 mm)	N/A	N/A	**90.46%** **(0.2 A & 14 mm)**
Peak Eff. With On & Off Delay Compensat.	N/A	N/A	92.9% (0.08 A)	86.0%	**92.28%** **(0.2 A & 14 mm)**
Max. PCE Enhanc. With On-delay Compensat.	N/A	N/A	N/A	N/A	**4.41%** **(0.6 A & 34 mm)**
Chip Area (mm²)	8	0.652	2.07	1.488	**2**

1. Maximum value calculated according to the circuit structure.
2. Estimated from the testing results.
3. Whether the on-delay and off-delay compensation can exist simultaneously.
4. Constant current mode.
5. One side is 20 mm and the other side is more than 20 mm.

Fig. 6. Measured PCE and performance comparisons.

Fig. 7. Chip micrograph, coil, and measurement setup.

A 4 × 112-Gb/s PAM-4 Silicon-Photonic Transceiver Front-End for Linear-Drive Co-Packaged Optics

Han Liu[1,2], Nan Qi[1,2*], Donglai Lu[1,2], Zizheng Dong[1], Zhihan Zhang[1,2], Jian He[1], Guike Li[1,2], Leliang Li[1,2], Ye Liu[3], Ziyue Dang[3], Daigao Chen[3,4], Zhao Zhang[1,2], Jian Liu[1,2], Nanjian Wu[1,2], Xi Xiao[3,4*], Liyuan Liu[1,2*]

[1]Institute of Semiconductors, Chinese Academy of Sciences
[2]University of Chinese Academy of Sciences
[3]National Information Optoelectronics Innovation Center
[4]Wuhan Research Institute of Posts & Telecommunications
*Corresponding authors

The explosive growth of the data-intensive AI computing demands for datacenter interconnects with ever-increasing bandwidth (BW). Traditional pluggable optical modules can hardly meet the DC power and form-factor requirements by the new 51.2T ethernet switch. By tightly integrating the compact optical engine (OE) into the switch package, on-board metal traces are mostly replaced by optical fiber, which thereby saves the signaling power and improves the front-panel BW-density. Depicted in Fig. 1(a), each 3.2T OE employs 4 quad-lane optical transceivers (O-TRX) in the package, enabling 16×112 Gb/s fiber channels directly attached to the chip edge. Meanwhile, the pluggable optical module has to employ a power-hungry DSP-based O-TRX to compensate for the channel-loss beyond 100 Gb/s per lane. The integrated co-packaged optics (CPO) in contrast, simplifies the transceiver by replacing the DSP with linear driver and amplifier with embedded continuous-time linear equalizers (CTLE). The so-called linear-drive (LD) CPO draws increasing attentions, since it significantly reduces power consumption, cost, and latency. The embedded CTLE needs to equalize 10-12 dB channel loss between the switch and each OE.

To achieve arrayed integration and improved performances, a pair of customized Silicon-Photonic (SiPh) modulator and photodetector (PD) are proposed (Fig. 1b). The Mach-Zehnder modulator (MZM) is designed with 2-mm-long traveling-wave (TW) phase shifters. Cascaded P-N-N-P junctions are configured in a push-pull manner and distributed along the TW electrode. By removing the substrate beneath, its -3-dB E-to-O BW is extended to above 40 GHz under -3 V_{DC} junction bias [1]. The Ge-on-Si PD uses double-sided silicon nitride (Si_3N_4) waveguides. Light is injected into the lateral Si_3N_4 waveguides and then be evanescently coupled to the Ge absorption region, which enables high responsivity with small junction capacitance. The PD features 0.52 A/W responsivity and less than 30 fF capacitance at the TIA input [2].

Fig. 2 shows the proposed optical transmitter (OTX), consisting of a SiGe BiCMOS driver hybrid integrated with the SiPh MZM. The driver is designed with above 30-GHz BW, delivering 3-V_{ppd} voltage swing to the modulator. To keep linear under large-swing input (max. 800 mV_{ppd}), an wideband attenuator with adjustable resistor-ladder (R_2 and R_3) and shunt-peaking inductor is employed as the 1st stage. An emitter-follower decouples the input termination from the following variable CTLE. To accommodate the input channel loss, an active feedback (AFB) 2nd-order CTLE is proposed with more flexible frequency response. Three zero-pole pairs are constructed by the two forward stages accompanied with one G_m-feedback stage, which cover 5-to-30GHz frequency tunable range and up to 15dB EQ capability. The EQ strength can be adjusted by the degeneration resistor R_F in the G_m-feedback path. Considering the different input swing due to various channel loss, a variable-gain amplifier (VGA) is employed to maintain a same voltage swing at the driver stage input. An open-collector (OC) driver stage is adopted to generate 3 V_{ppd} voltage swing on the 40-Ω MZM termination. Cascode transistors are added to avoid overstress and improve the linearity. It also helps to prevent the Miller Effect from over-loading the VGA stage. Degeneration R and C are inserted for higher linearity and equalization capability. Compared to CML, the OC driver gets rid of the local termination and thereby saves half of the current for a fixed swing. However, the output node must be carefully designed, since it is more sensitive to signal reflections. In this work, the output trace is deliberately designed with ~80pH inductance to distribute the parasitic capacitance from the ESD and bond PADs for higher BW.

Fig. 3 shows the proposed optical receiver (ORX), comprising of an arrayed SiPh PD wire-bonded to the quad-channel transimpedance amplifier (TIA). The photo-current is firstly converted to voltage in a trans-impedance stage (TIS), which is built in a CE-EF topology with resistive feedback. A T-type LC network is built at the input node, consisting of the bond-wire inductor, C_{PAD} & C_{ESD}, as well as a series-peaking inductor L_s, which is utilized to distribute parasitic cap and extend the BW up to 37GHz. A dummy stage defines the bias current I_C and thus the transistor f_T, duplicating it to the main-path TIS through a wideband LDO. It helps match the differential inputs at the VGA stage and improve the common-mode noise rejection. Three stages of voltage amplifiers (VAS) are implemented following the TIS to provide additional 16dB voltage gain and 20dB dynamic range. RC degenerated CTLE is distributed into each stage, not only relieving the BW stress, but also compensating for the output channel loss of the ORX. In this work, a high-order CTLE is formed, proving up to 12-dB tunable EQ. The AC simulation results are included to show the distributed EQ capability.

The proposed driver and TIA are implemented in a 290 GHz f_T SiGe BiCMOS process, occupying 4×1.7 mm^2 and 3.2×1.1 mm^2 area respectively. The SiPh MZM and PD are fabricated in a 180-nm SOI-CMOS process, and co-designed with matched pitch footprints. The OTX and ORX are realized by COB wire-bonding, with the fiber array (FA) directly attached to the chip edges (shown in Fig. 7).

Fig 4 depicts the measured driver's frequency responses and OTX time-domain eye diagrams. The S_{21} shows 35-GHz analog BW and 20dB voltage gain of the driver. The S_{11} keeps below -10 dB across the driver BW, which shows good impedance matching with the help of wideband passive attenuator. The linear driver supports both 56 Gbaud NRZ and PAM-4 modulation. Fig. 4(c) shows the complete E-to-O transmitter characteristics, in which up to 112 Gb/s eye-diagram with 4.9 dB optical extinction ratio (ER) is observed without any oscilloscope EQ.

Fig. 5 shows the measured TIA AC responses and ORX time-domain eye diagrams. Only one output of TIA is connected to the vector network analyzer, with the other end 50-Ω terminated. Measurement errors are also calibrated in the curves. The TIA features 37 GHz analog BW and 12 dB EQ capability at its default mid-gain settings. Impedance is well-matched across the entire BW. The trans-impedance gain T_z is deduced from above S-parameters, which is measured at 67dBΩ at maximum gain. The 112-Gb/s PAM-4 O-to-E output eye diagrams are demonstrated in Fig. 5(c) (on left), and 5-tap EQ from the oscilloscope is applied to emulate the XSR SerDes EQ in the switch (on right).

Fig. 6 shows the performance comparison of this work with prior published OTXs and ORXs. Complete optoelectronic integration is realized. Driver demonstrates sufficient analog BW with large output swing and good power efficiency. For TIA, a 112-Gb/s data rate is attained with a large T_z.

To emulate the application scenario in the CPO, a compact-size evaluation board (EVB) is developed, in which the electrical channels are designed to present the same loss to the O-TRX. All 4 channels are measured when running at 112Gb/s PAM-4, and all the corresponding eye diagrams are captured to prove the validity.

Acknowledgement:

This work is supported in part by the National Key Research and Development Program of China (Grant 2021YFB2206504), in part by the National Natural Science Foundation of China (Grant 62235017), and in part by the Beijing Municipal Science and Technology Project (Grant Z221100007722024). The authors would like to thank Mr. Yan Tong and SuZhou ToPhotonics Co., Ltd. for the packaging.

References:

[1] M. Li et al., Photon. Res. 6, pp. 109-116, 2018.
[2] X. Hu et al, Photon. Res. 9, pp. 749-756, 2021.
[3] A. H. Ahmed et al., ISSCC, pp. 484-486, 2019.
[4] C. Loi et al., ISSCC, pp. 120-122, 2019.

979-8-3503-3004-5/23 $31.00 © 2023 IEEE

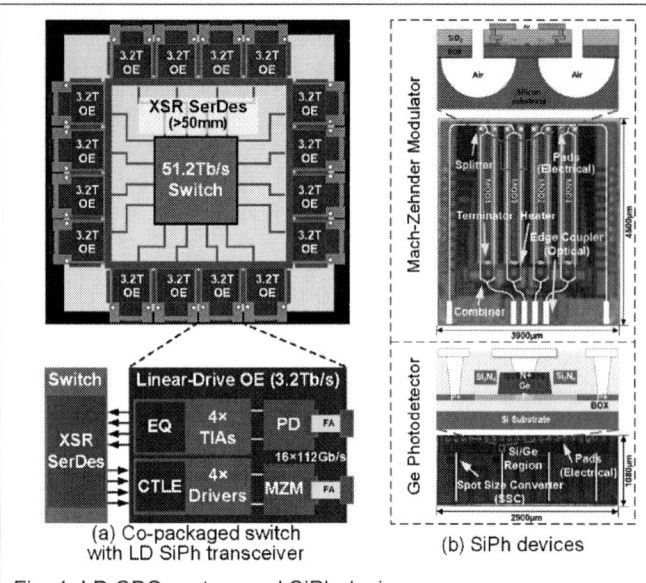

(a) Co-packaged switch with LD SiPh transceiver

(b) SiPh devices

Fig. 1. LD CPO system and SiPh devices

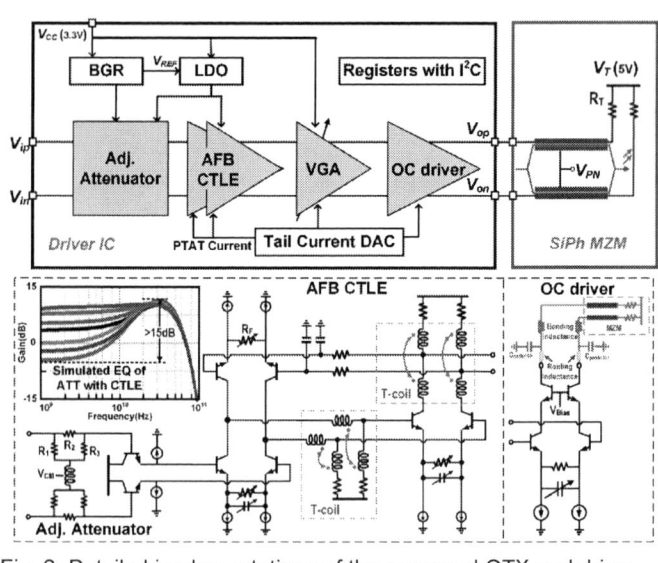

Fig. 2. Detailed implementations of the proposed OTX and driver

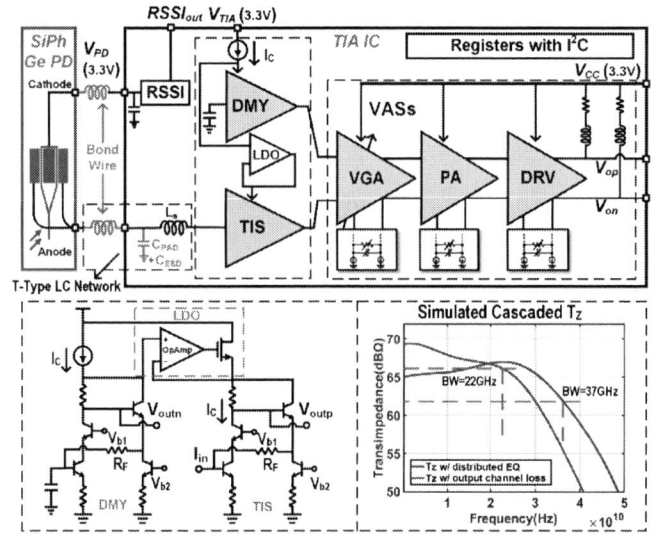

Fig. 3. Detailed implementations of the proposed ORX and TIA

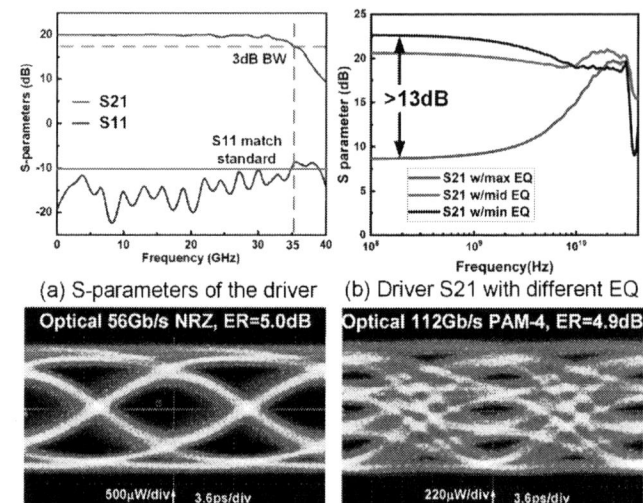

(a) S-parameters of the driver   (b) Driver S21 with different EQ

(c) Output eye diagrams of the OTX

Fig. 4. Measurement results of the OTX

(a) S-parameters of the TIA   (b) TIA S21 with different EQ

(c) Output eye diagrams of the ORX

Fig. 5. Measurement results of the ORX

OTX	This Work	ISSCC'19 [3]	ISSCC'19 [4]	ISSCC'20 [5]
Data Rate (Gb/s)	**112**	272*	106	53.15
Bandwidth (GHz)	35	40	>30	-
E/O Integration	**Hybrid**	Hybrid	No	Hybrid
Output Swing (V_{ppd})	3	6	1.1	-
Power Efficiency (pJ/bit)	4.8	3.68	8.5	5.2
Technology	**180nm BiCMOS**	130nm BiCMOS	16nm FinFET	55nm BiCMOS
ORX	This Work	JSSC'18 [6]	SSCL'21 [7]	PTL'22 [8]
Data Rate (Gb/s)	**112**	100	112	112
E/O Integration	**Hybrid**	No	No	Hybrid
Max. T_z (dBΩ)	67	65	71	60
Power Efficiency (pJ/bit)	**2.95**	1.5	3.1	2.8
Technology	**180nm BiCMOS**	130nm BiCMOS	250nm BiCMOS	180nm BiCMOS

*34Gbaud DP-16QAM

Fig. 6. Performance comparison of the OTX and ORX

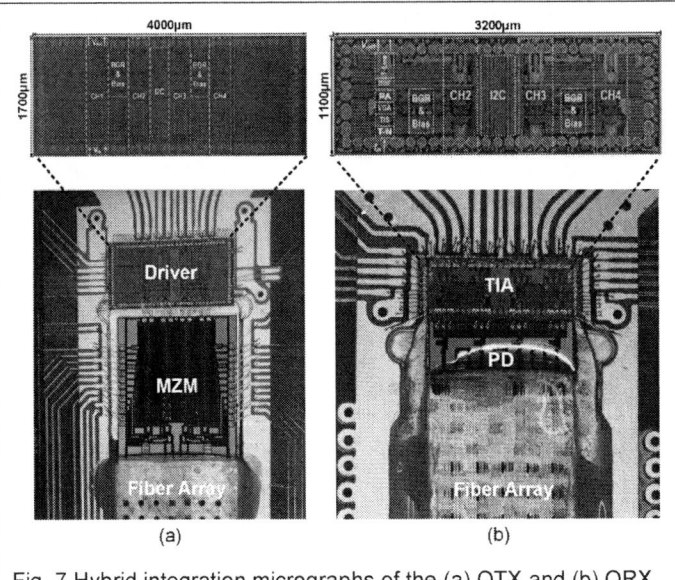

Fig. 7 Hybrid integration micrographs of the (a) OTX and (b) ORX

Additional References:

[5] E. Sentieri et al., ISSCC, pp. 210-212, 2020.

[6] I. García López et al., JSSC, pp. 458-469, 2018.

[7] M. M. Khafaji et al., SSCL, pp. 76-79, 2021.

[8] D. Okamoto et al., PTL, pp. 189-192, 2022.

A 33.6 FPS Embedding based Real-time Neural Rendering Accelerator with Switchable Computation Skipping Architecture on Edge Device

Jongjun Park, Donghyeon Han, Junha Ryu, Dongseok Im, Gwangtae Park and Hoi-jun Yoo

Korea Advanced Institute of Science and Technology (KAIST)

A neural radiance field (NeRF) algorithm which utilizes a deep neural network (DNN) is rising as a new realistic 3D rendering technology. Recently, explicit NeRF [1] that utilizes interpolation on 3D embedding grids benefits 22% more compact memory footprint compared with implicit NeRF which is fully based on multi-layer perceptron (MLP) operations. Fig.1 shows embedding-based explicit NeRF. It consists of 3 computing stages: 1) sampling, 2) density computation, and 3) color computation. The sampling stage casts rays from each image pixel and determines whether the sample is valid or not. The density computation stage performs interpolation between the two nearest embedding values from pre-decomposed X, Y, and Z density embedding vectors and sums up the interpolated values along ranks. If the calculated density is above the threshold, the color computation is performed. The color computation stage computes interpolation and multiplying basis matrix.

Explicit NeRF has two main challenges to achieve real-time (> 30 FPS) rendering on edge computing. First, it requires an excessive computational amount even applying temporal pixel reuse [2]. The 3D interpolation accounts for 70.6% of the density computation and matrix multiplication accounts for 83.2% of the color computation. Second, different sample rates of density computation and color computation across target models bring about the workload imbalance between the two computations, resulting in performance degradation.

To solve the challenges, efficient computation skipping architecture is proposed. The proposed architecture has 3 key features. First, the pruning aware skipping (PAS) and the switchable pruning aware skipping core (SPASC) are suggested to maximize throughput. Second, sample-centric workload balancing (SCWB) is proposed to schedule the workload for preventing a decline in throughput. Third, embedding accumulation early stopping (EAES) is suggested for a throughput increase to achieve real-time rendering.

Fig. 2 illustrates the overall architecture of the proposed accelerator. It consists of a main DNN core, a skipping control unit, and a workload controller. The main core contains embedding buffers and the mode switchable PE array. The 0.42-KB embedding buffers which store 4-bit quantized embedding values consist of 9 embedding blocks. The mode switchable PE array consists of 3×3 switchable PEs (SPEs) and 3 switchable accumulators (SACCs). The SPE contains 96 multi-way switchable multipliers and 6-stage adder trees that are shared across the density computation and color computation. The SACC accumulates the SPE results in different ways depending on modes. The skipping control unit fetches sample coordinates from the buffer and calculates the virtual volume that decides the skipping operation. The core workload controller controls the operation mode and provides input data to the SPE array column by column.

Fig 3. describes the proposed SPASC and PAS. SPASC supports both density and color computations by exploiting the switchable PE array. In density mode, each SPE handles 96 ranks of a sample and accumulates the density values in the SACC. In color mode, 3 SPEs process 288 ranks of a single sample and sum them up in the SACC. SPASC accelerates density and color computations by exploiting the

PAS. Interpolation of 1D pre-decomposed vectors can be replaced as following computation : $\sum_{i=0}^{7} E_i V_i$ (E_i: virtual embedding, V_i: virtual volume). This representation provides additional skipping chances by pruning. The virtual volume ranges between 0 and 1, and the most of values are closer to 0 compared to the uniform distribution of weights in linear interpolation. Therefore, the SPASC utilizes 0.125 as a pruning threshold and achieves an average 68.75% skipping rate with only < 1.2 dB minimal PSNR drop. To maximize computing efficiency, the SPASC exploits 2-bit quantized weights. The skipping control unit checks only MSB of quantized weight values to compute virtual volumes and decides the skipping operations by using a look-up table (LUT). Only valid virtual embeddings are pushed to the FIFO and processed in the PE array. In summary, the SPASC achieves an ideal 2.66× higher throughput with the PSNR an average 30.18 dB.

Fig 4. shows the operation of the SCWB. Different sample numbers between density and color cause the utilization drop of the SPEs and increase the complexity of managing workloads. For example, if the skipping rates of each sample in an SPE column are different, a low skipping rate SPE should take additional time to complete the task. To solve the problem, the SCWB suggests workload balancing and dynamic buffer allocation based on sample coordinates. First, the core workload controller allocates the workloads to the SPE column based on the density skipping rate of the sample. Since the samples with the same density skipping rate take the same time for processing, each SPE column can process corresponding density and color computation without stall. Second, the dynamic buffer allocation controls the embedding values stored in embedding buffers and determines the target address space corresponding to the SPE column. The buffer address generator allocates the address space for each axis according to the range of samples along each axis from the ratio calculator. As a result, the SWCB increases the throughput of the accelerator about 37%.

Fig 5. describes the EAES. Since the calculated density value's 73% is negative value and it goes through the ReLU layer, density computation could be early stopped in SACC if the summation of virtual volumetric embedding is below the threshold after processing the first 3 values. It results in throughput enhancement. Moreover, a threshold value determines the skipping rate and PSNR accuracy. To achieve reasonable PSNR result and throughput increase, the threshold is set to be 0, achieving a 45.86% skipping rate and 97.83% of true positive rate. As a consequence, the EAES achieves 17% higher throughput compared with the PAS result.

Fig 6. illustrates the measurement results and implementation of the FPGA demonstration system for explicit NeRF. The proposed accelerator is implemented on Intel's Cyclone V, which operates at a maximum 200-MHz clock frequency. Peak performance is 242.9 GOPS while consuming 5.8 W power. Finally, it achieves 33.6 FPS of rendering speed which is higher than the latest works [3-4]. Fig. 7 shows an FPGA layout of the proposed accelerator and rendering results in the Synthetic NeRF dataset. In conclusion, the proposed accelerator successfully demonstrates the real-time neural rendering system.

References:

[1] C., Anpei, et al., "Tensorf: Tensorial radiance fields." ECCV 2022.

[2] D. Han et al., "2.7 MetaVRain: A 133mW Real-Time Hyper-Realistic 3D-NeRF Processor with 1D-2D Hybrid-Neural Engines for Metaverse on Mobile Devices," ISSCC, 2023

[3] C. Rao et al., "ICARUS: A Specialized Architecture for Neural Radiance Fields Rendering, " ACM Trans. Graph. 2022.

979-8-3503-3004-5/23 $31.00 © 2023 IEEE

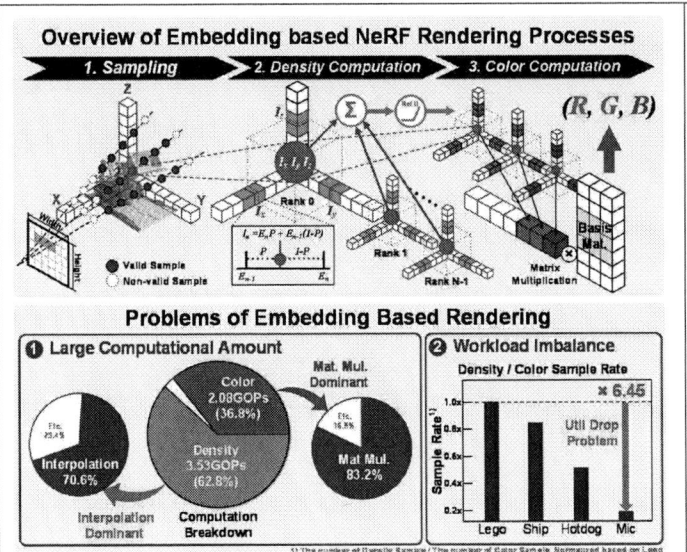

Fig. 1. Overview of Explicit NeRF

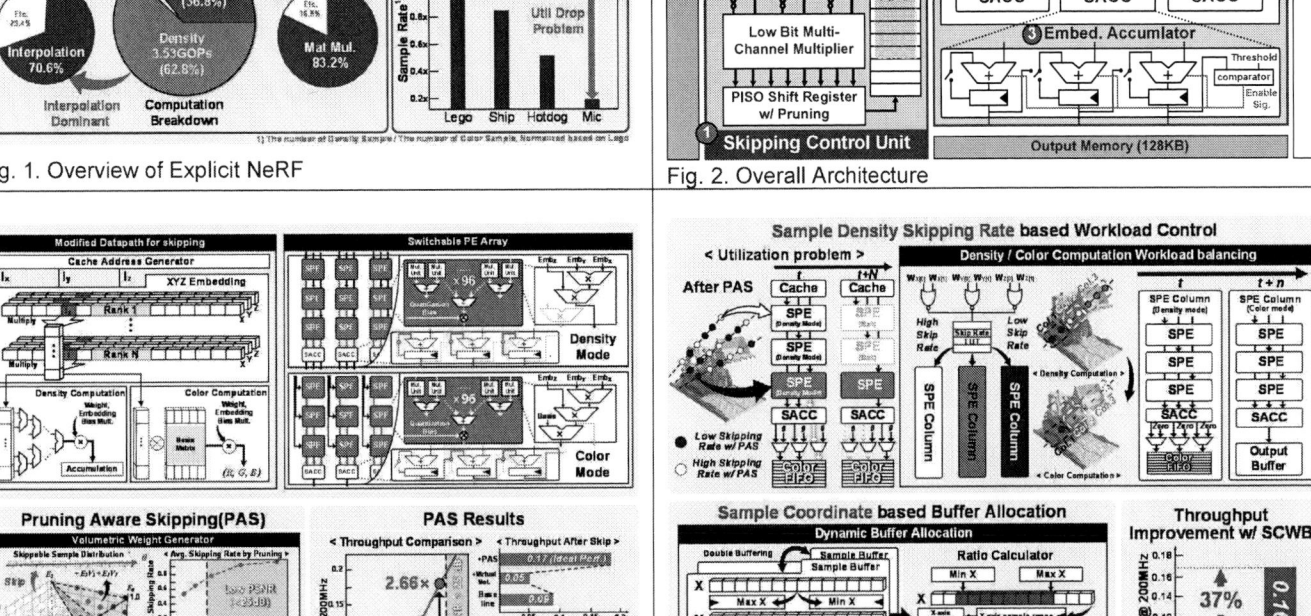

Fig. 2. Overall Architecture

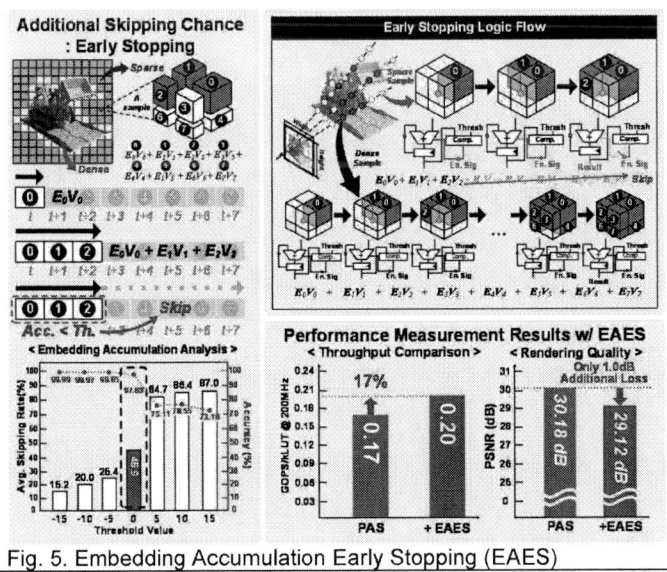

Fig. 3. Switchable Pruning Aware Skipping Core (SPASC)

Fig. 4. Sample Centric Workload Balancing (SCWB)

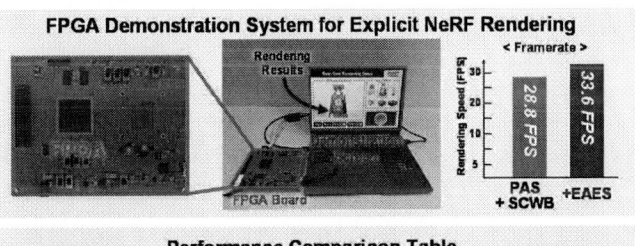

Fig. 5. Embedding Accumulation Early Stopping (EAES)

Performance Comparison Table

	ACM TOG'22	SOCC'22	This Work
Platform	ASIC (Simul.)	ASIC	Cyclone V [5CEBA9F31C7]
Application	Implicit NeRF	Implicit NeRF	Explicit NeRF
PSNR	28.86dB[1]	24.71dB	29.12dB[1]
Clock	400MHz	100MHz	200MHz
Power	282.8mW	69.1W	5.8W
Bit Precision	FXP9 (Weight)	FXP8/9 (Act./Weight)	FXP4/2 (Embed./Weight)
FPS	0.02FPS	15.4FPS[2]	33.6FPS[3]
Energy Efficiency	0.0773 FPS/W	0.2226 FPS/W	5.79FPS/W

1) Avg. Synthetic NeRF dataset
2) 800×800 resolution, 16 samples per ray 3) 800×800 resolution, Avg. 13% samples per ray

Fig. 6. Demonstration system and Comparison Table

979-8-3503-3004-5/23 $31.00 © 2023 IEEE 187

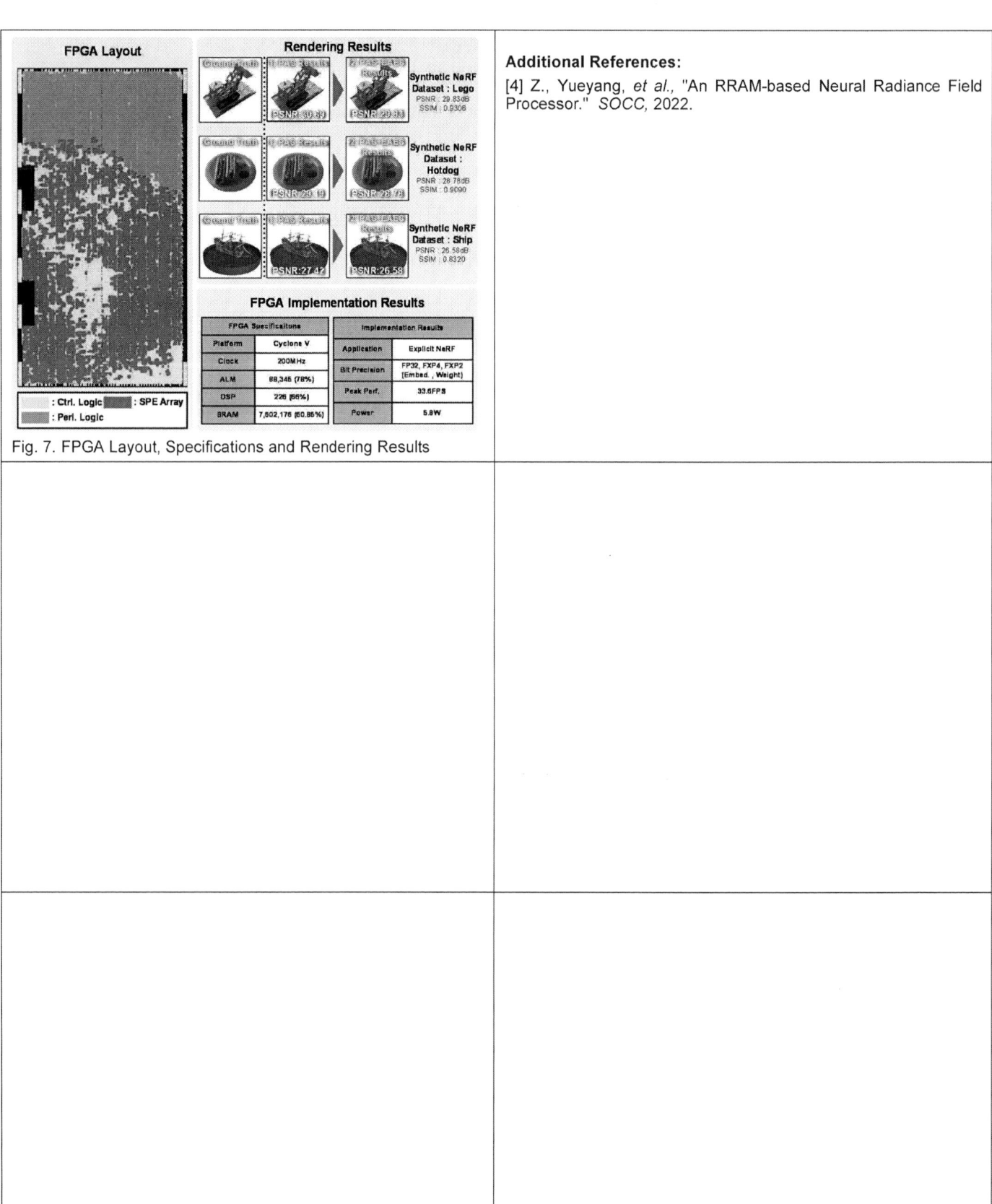

Fig. 7. FPGA Layout, Specifications and Rendering Results

Additional References:

[4] Z., Yueyang, *et al.,* "An RRAM-based Neural Radiance Field Processor." *SOCC,* 2022.

A 26.9-GHz 4-Element Code-Domain Hybrid Beamforming Phased-Array Receiver

Ziyi Lin, Haikun Jia, Chuanming Zhu, Wei Deng, Huabing Liao, Bao Shi, Lujie Hao, Xiangrong Huang, and Baoyong Chi[1]
[1]School of Integrated Circuits, BNRist, Tsinghua University
Email: jiahaikun@tsinghua.edu.cn

Millimeter-wave (mm-wave) phased-arrays have received much attention for richer spectrum resources and for enabling higher data rates. Phased-array systems of large antenna arrays are developed to provide high spatial multiplexing gain to overcome the higher propagation path loss caused by the higher carrier frequency.

Fig. 1. shows the different structures of beamformers. Analog beamforming (ABF) provides a large hardware simplicity. Working as a phased-array, the beamforming is implemented in the RF domain. Thus, a single RF and IF chain is needed. However, ABF has a limited ability to process beams from different directions, which means it cannot provide spatial multiplexing gain. In contrast, a fully digital beamforming (DBF) structure is adopted to preserve the maximum flexibility of beamforming and obtain the maximum spatial multiplexing gain. However, the number of RF and IF chains is equal to the number of transceiver antennas, which makes DBF much unrealistic because of the huge system complexity and power consumption. Requiring only a small number of RF chains, a hybrid beamformer (HBF) is proposed as a trade-off structure between hardware cost and spatial multiplexing gain and can be classified into a sub-connected structure and a fully-connected structure. For a sub-connected structure, the hardware cost is further reduced, but fewer streams can be processed and less gain can be obtained. And for a fully connected structure, more flexibility is preserved to recover more streams at the expense of high complexity both in the RF domain and algorithm design.

Code-division multiplexing method is proposed to share the IF and LO interface in the DBF receiver [1]. The hardware cost is largely reduced but the flexibility in the RF domain is eliminated. We proposed a configurable code-domain hybrid beamforming (CDHBF) structure in this work. Code-division multiplexing is used to reduce the RF chain and ensure hardware simplicity and power efficiency. Flexibility in the digital domain is fully preserved by code multiplexing. Flexibility in the RF domain is also preserved in this work by implementing an RF phased-array. The phased-array and code modulator can be turned on or off according to applications, which means this structure can be reconfigured into ABF, HBF, or DBF structure.

Figure 2 illustrates the architecture of the proposed 26.9-GHz 4-element phased-array receiver with CDHBF. An antenna receives signals for each element, and a three-coil transformer based LNA proposed in [2] is followed to provide a high gain and good noise figure. Gm of the LNA is boosted by the source-to-gate magnetic feedback using folded three-coil transformer and dual-feedforward techniques. A 2-stage common-source buffer is used to provide enough gain. An attenuator with a 6-bit resolution and 31.5 dB tuning range is implemented to adjust the gain of each element. 8-shape inductors are used in the attenuator to reduce coupling effects between different inductors and also improve area efficiency. A phase shifter of vector modulation type with a 360° tuning range is followed. A hybrid coupler is used to generate four quadrant signals and two VGAs with 6-bit resolution are used to adjust the amplitude of I/Q signals, respectively. A passive modulator is used for code-domain multiplexing in the RF domain. Walsh codes are selected in this design because of their orthogonality and simplicity. After code-domain modulation, signals of four elements are combined through a 4-to-1 Wilkinson combiner and sent to the output. After off-chip quadrature down-conversion and quantization, the output RF signal is fed to DSP, including code-demodulation, integration, and digital beamforming, to recover the input data streams.

Figure 3 shows the schematic of the proposed modulator and Walsh code generator. The differential Walsh generator is based on inverter buffers and digital dividers. Output mux is used to power down the code-division multiplexing. To avoid the phase ambiguity after the digital dividers, a phase select circuit is implemented, ensuring the arrival sequence of two clocks when powering on the circuit. The proposed modulator consists of a common-source buffer and a passive switch. Transformer-based inner-stage matching is adopted. The modulator proposed in [1] is based on active Gilbert cells and integrated with the LO path, which means that a mixer is needed for each element. In this work, the modulator is implemented independently, therefore, only one mixer is needed for all elements to reduce power. Modulator based on passive switches improves linearity, whose IP1dB is over 2.5 dBm and has little impact on the overall system linearity. The good linearity of this scheme enables a small switch size, 8 μm in width, which also improves the modulator bandwidth and reduces power in baseband Walsh generator circuits. The waveform after modulation is also shown in Fig. 3.

The proposed 4-element phased-array receiver has been fabricated in 65-nm CMOS technology and occupies 2.7 mm×1.6 mm including pads, whose die micrograph is shown in Fig. 7. The entire chip consumes 326 mA in phased-array mode and 338 mA in CDHBF mode from a 1.0 V supply voltage.

The RF parameters of the receiver operating in phased-array mode are measured through an on-chip probing system, as shown in Fig. 4. The measured peak gain of a single channel is 15.3 dB, and the 3-dB bandwidth is 2.6 GHz from 25.7~28.3 GHz. The measured minimum value of NF is 5.9 dB and the peak OP1dB is -14.6 dBm. The 6-bit attenuator has a measured attenuation tuning range of 31.5 dB and the RMS amplitude error is less than 0.22 dB in-band. The phase shifter has a measured attenuation tuning range of 360° and the phase resolution is 2.8°. The RMS phase error is less than 1.4°, while the RMS gain error is less than 0.14 dB in-band.

Figure 5 shows the over-the-air wireless measurement setup for the CDHBF mode test. Two modulated 16-QAM signals, whose carrier frequencies are 26.9 GHz and symbol rates are 100 Ms/s and 200 Ms/s, respectively, are generated independently by two arbitrary signal generators. Two beams are then transmitted through two antennas on PCBs from different directions. Two beams are received simultaneously by a four-antenna-array on PCB, in which each antenna is connected to each on-chip element through a trace on PCB and a bonding wire. The transmit antennas are more than 80 cm far from the receiver antennas, which meet far-field conditions. Clock signals are generated with Keysight E8257D and the output RF signal of the chip is recorded by oscilloscope Keysight DSAZ594A. The waveforms are captured and processed offline with MATLAB. And constellation diagram and EVM can be measured after DSP. The measured EVMs are -22.7 dB and -20.5 dB for 100 Ms/s and 200 Ms/s streams, respectively, without any equalization. When operating in the phased-array mode and processing one stream of signal, the measured EVM is -25.8 dB without equalization for 200Ms/s stream. Performance comparison with relative work is shown in Fig. 6.

Acknowledgment:

This work was supported by the National Natural Science Foundation of China (No. 62074090) and the Beijing Innovation Center for Future Chips (ICFC).

References:

[1] M. Johnson et al., "Code-Domain Multiplexing for Shared IF/LO Interfaces in Millimeter-Wave MIMO Arrays," JSSC, vol. 55, no. 5, pp. 1270-1281, May 2020.

[2] X. Huang et al., "28 GHz Compact LNAs with 1.9 dB NF Using Folded Three-Coil Transformer and Dual-Feedforward Techniques in 65nm CMOS," RFIC 2022: 223-226.

[3] R. Garg et al., "A 28-GHz Beam-Space MIMO RX With Spatial Filtering and Frequency-Division Multiplexing-Based Single-Wire IF Interface," JSSC, vol. 56, no. 8, pp. 2295-2307, Aug. 2021.

[4] S. Shakib et al., "A Wideband 28-GHz Transmit–Receive Front-End for 5G Handset Phased Arrays in 40-nm CMOS," IEEE TMTT., vol. 67, no. 7, pp. 2946-2963, July 2019.

[5] P. Guan et al. "A 33.5-37.5 GHz 4-Element Phased-Array Transceiver Front-End with High-Accuracy Low-Variation 6-bit Resolution 360° Phase Shift and 0~31.5 dB Gain Control in 65 nm CMOS," A-SSCC 2021: 1-3.

979-8-3503-3004-5/23 $31.00 © 2023 IEEE

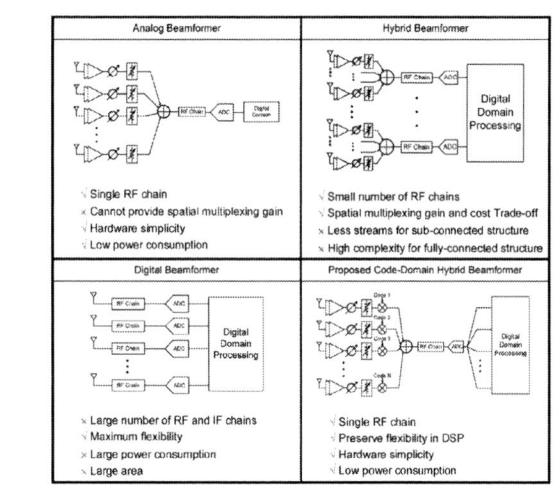

Fig. 1. Different structures of beamformers in mm-wave systems.

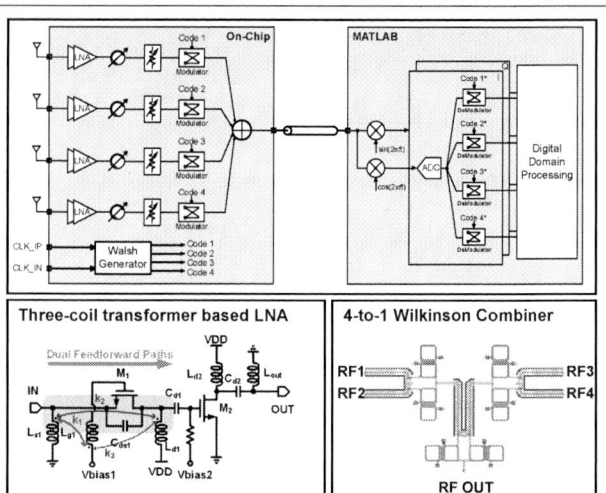

Fig. 2. The architecture of the proposed 4-element phased-array receiver with CDHBF.

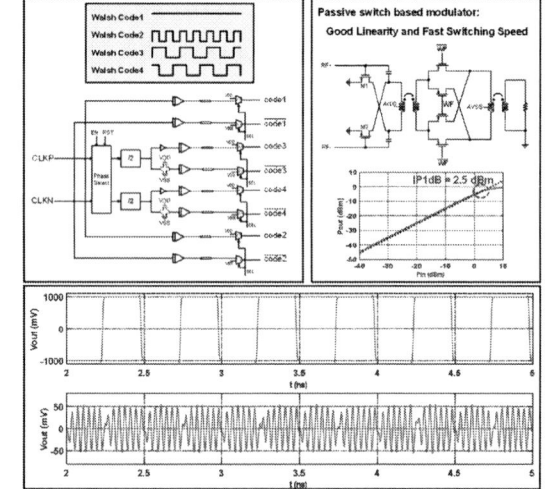

Fig. 3. The schematic and simulation results of the proposed passive modulator and Walsh code generator.

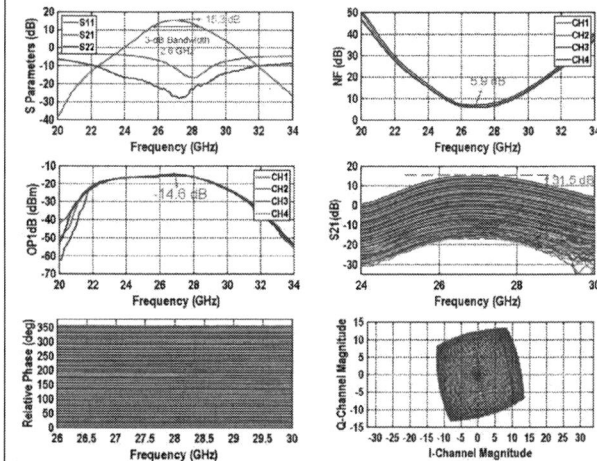

Fig. 4. Measured results of S-parameters, NF, OP1dB, and attenuation and phase shift characteristics of the receiver operating at phased-array mode.

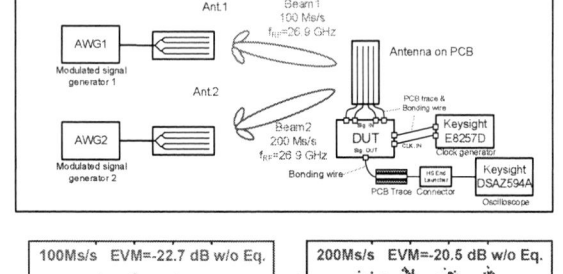

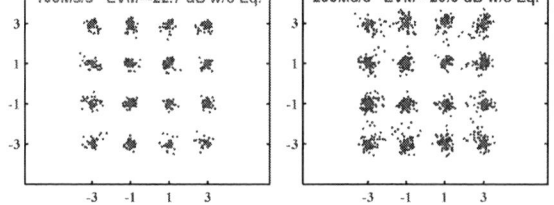

Fig. 5. The dual-stream over-the-air measurement setup, and the demodulated constellation.

	This work	JSSC 2020 [1]	JSSC 2021 [3]	TMTT 2018 [4]	A-SSCC 2021 [5]
Technology	65nm CMOS	65nm CMOS	65nm CMOS	65nm CMOS	65nm CMOS
Architecture	Phased-array & Code Domain Multiplexing	Code Domain Multiplexing	Frequency Domain Multiplexing	Phased-array	Phased-array
RF Frequency (GHz)	26.9	29.2	28	28	35.5
No. of elements	4	4	4	1	4
No. of Beams	2	NA	4	1	1
Supply Voltage (V)	1.0	1.2	1.2	1.1	1.2
Power (mW)	338	NR	450	32	656
Chip Area (mm²)	4.32	5.75	28.1	0.99	25.9
NF (dB)	5.9	10	6-7.8	5.5	4.2
IP1dB (dBm)	-31	-20	-36.9	-16	NA
RX BW (MHz)	2600	200	≥450	3000	4000
Gain Range (dB)	31.5	NA	NA	19	31.5
Gain RMS err. (dB)	0.22	NA	NA	<0.5	0.16
Eff. Phase bit	7	NA	NA	3	6
Phase RMS err. (deg)	<1.4	NA	NA	<10	2
EVM (dB)	-22.7 @400Mbps -20.5 @800Mbps	NA	-16.55 @400Mbps	NA	-28.2 @800Mbps

Fig. 6. Performance comparison.

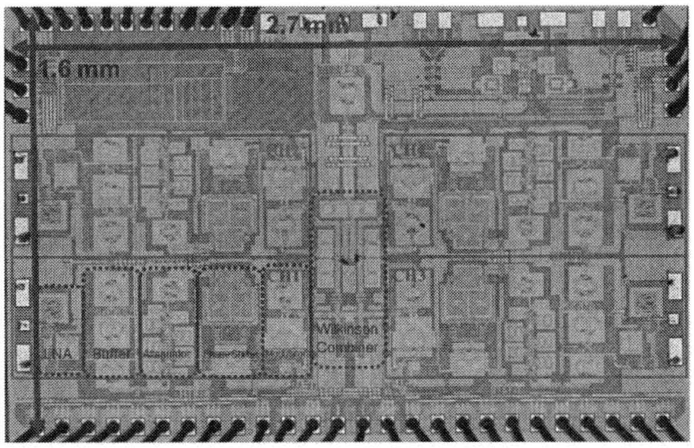

Fig. 7. Die micrograph of the proposed 4-element receiver.

Additional References:

A Triple-band radio for WLAN 11b/g/n/ax in 45nm CMOS

D Sahu, V Srinivas, R Chatterjee, M Agrawal, P Agrawal, R Juluri, M. Mukherjee, Vimal E, A. Yerramsetty, G Bakalzuk, O Rahmanony, K Rajmohan, A Sancheti, R Anand

Texas Instruments (India), Texas Instruments (Israel)

A significant increase in the deployment of internet-of-things has led to many battery-operated smart connections. WLAN, due to its wide installation base, is the technology of choice for majority of such applications. A long battery life and simplified use-case enables many new smart products featuring WLAN. This paper describes the RF core of an 802.11a/b/g/n/ax radio implemented in 45nm CMOS technology. A dynamic bias technique for the power amplifier and a novel local oscillator scheme significantly reduced the transceiver active current consumption, while innovations in the PLL and front-end architectures improved the performance at no additional cost.

The transmitter (fig. 2) consists of 12-bit current-steering DAC working at 400MHz sample rate. DAC replicas and noise are attenuated by operational-amplifier/resister-capacitor based filter. This filter output is upconverted to 2.4GHz or 5-to-7GHz bands by two dedicated active mixers and driven to dedicated power amplifiers (PA) for the two bands. The power amplifier has a pre-amplifier followed by final amplifier (fig. 2). Traditionally, PAs in WLAN radio suffer heavily in efficiency due to high peak-to-average-ratio of the modulated signal and linearity requirements. Efficiency of a linear PA drops very quickly as output power is lowered from the saturated power. In this PA, the output of pre-amplifier is rectified by an envelope detector and used to optimize the bias and DC current of the PA to improve the efficiency at higher back-off. The output of envelope detector is filtered to attenuate RF harmonics and yet be very quick in response to avoid any memory effect which could introduce signal asymmetry and degrade the transmit-EVM.

An analog fractional-N PLL generates the local oscillation (LO) (fig. 3). It uses a delta-sigma modulated feedback divider. The fractional quantization noise is minimized by using a higher compare frequency. The crystal clock is multiplied by a delay-locked-loop (DLL) for this. The DLL uses a single delay element as a relaxation oscillator to avoid any spurious due to mismatch. The noise added by the DLL is kept well below the phase noise of reference crystal oscillator. The DLL keeps it flexible to multiply the reference clock by any integer to choose the optimum compare frequency. The PFD-charge-pump combination of the PLL delivers the sum of up and down currents using a pair of co-related PFDs. The scheme is immune to mismatch and avoids folding of fractional quantization noise. The use of the sum of up and down currents also reduces the detector's noise floor. A major contribution of noise from the stabilizing zero-setting resistor of the loop filter is avoided by a new topology of the loop-filter. The proportional correction is derived directly by averaging the up and down pulses of the PFD and adding directly to the output of the integrator. The voltage levels of up/down pulses and the scaling provided by R_1, R_2 and R_3 establishes the equivalence between this new scheme and that of a regular stabilizing resistor, normally added in series with the capacitor C_1.

The transceivers employ direct conversion with an offset frequency plan such that the fundamental of the RF channel or its harmonics do not align with VCO frequency or its harmonics. This is to avoid any RF coupling between signal chain and VCO so that there is no amplitude dependent frequency pulling of the VCO, that could result in phase-distorted EVM. In this radio the VCO frequency is converted to 2.4GHz channel by a digital offset mixer (fig. 3) where 1.6GHz and 800MHz clocks are generated by dividing the VCO output, and then mixed by XOR logic. This generates unwanted 800MHz and 4GHz components along with the desired 2.4GHz channel. The L-C tuned band-pass stages of the signal chain provide the required rejection of these components. A third-harmonic selector amplifier generates the LO for 5-to-7GHz bands. The output frequency of this tuned amplifier is divided to generate in-phase and quadrature components of LO. The amplifier's tuning range is widened by a coupled inductor switching arrangement.

A single RF port is shared between the PA and low noise amplifier (LNA). The effective transmit/receive (Tx/Rx) switching is implemented as shown in fig. 4. C_c functions as a shunt capacitor, resonating with PA's balun when the switches S_1 and S_2 are closed in Tx mode. The switches are opened in Rx mode and C_c couples the RF signal from the pin to the LNA. The LNAs have single-ended input and differential output. LNA is biased through a shunt inductor L_g. L_g and source side inductor L_s provide the matching of impedance at the LNA input. The combined RF pin adds significantly to the ease of use and compactness of the final solution. The RF impedance matching at the PA-LNA interface is more complex at the 5-to-7GHz pin. A cascaded T-match has been introduced in a single coupled metal-spiral coil to achieve this. To support 5-to-7GHz band, the LNA uses single RF pin for input but two sets of outputs for 5-to-6 and 6-to-7GHz bands. The LNA is followed by a trans-conductor amplifier and passive mixers for in-phase and quadrature paths. The passive mixers drive the down-converted current signals into the virtual ground inputs of the active base-band filters.

The base-band consists of a second order Butterworth filter with variable gain, and drives into the ADC. The Butterworth filter can also be configured as a cascade of real filters for a narrow filtering bandwidth in Bluetooth (BLE) operation. The ADC is a second order sigma-delta modulator operating at 1.6GHz sample rate.

Another very critical part of the transceiver is the auxiliary receiver (AuxRx) to calibrate the impairments of the transmitter, like amplitude and phase non-linearities of the PA, DC offset due to LO-leakage and mismatch between in-phase and quadrature paths. The AuxRx down-converts the RF output using passive components to maintain linearity. Data from multiple configurations of the AuxRx are analyzed to calibrate out its own gain and phase mismatch between in-phase and quadrature paths so that transmitter's impairments can be calibrated and compensated accurately during transmission. The different configurations of the AuxRx introduce programmable phase delays in looping the RF signal back through it.

This WLAN radio has been implemented in 45nm CMOS process and uses an QFN package. Figure 5 shows the transmitter performance at 5600MHz channel for MCS7. An EVM of -28.6dB has been achieved in 5600MHz channels while delivering 16dBm of output power. In 2.4GHz band an EVM of -31.6dB was achieved. Both the RF pins could be matched to 50ohms without any external components. A noise figure of 3dB has been achieved in 2.4GHz bands. The noise figure of a-band receiver is 6dB. This WLAN radio also supports BLE 5.0 1MHz and 2MHz modes. A low IF frequency has been chosen for BLE operations to avoid problems due to 1/f noise and phase noise at very low offset frequencies. The radio can work with any reference clock frequency between 26MHz to 80MHz.

References:

[1] F. Wang A. Yang, et al., "Design of Wide-Bandwidth Envelope Tracking Power Amplifier for OFDM Applications," IEEE Transactions on Microwave Theory & Techniques, Vol. 53, Issue. 4, April 2005, pp. 1244-1255.

[2] K. Cho, R. Gharpurey, "A 25.6dBm wireless transmitter using RF-PWM with carrier switching in 130-nm CMOS," IEEE radio frequency Integrated Circuits Symposium, May 2015, pp. 139-142.

[3] S. Tam, Y. Lu, et al., "A dual band (2G/5G) IEEE 802.11b/g/n/ac 80MHz bandwidth AMAM envelope feedback power amplifier with digital pre-distortion," IEEE radio Frequency Integrated Circuits Symposium, May 2015, pp. 123-126.

[4] N. Ryu, B. Park, Y. Jeong, "A Fully Integrated High Efficiency RF Power Amplifier for WLAN Application in 40nm Standard CMOS Process," IEEE Microware & Wireless Components Letters, Vol. 25, Issue. 6, June 2015, pp. 382-384.

[5] R. Kumar, T. Krishnaswamy, G. Rajendran, D. Sahu, et al., "A Fully Integrated 2X2 b/g and 1X2 a-band MIMO WLan SoC in 45nm CMOS for Multi-radio IC," ISSCC 2013, pp. 328-329.

[6] E. Lu, W. Li, et al., "A 4X4 Dual-Band Dual-Concurrent WiFi 802.11ax Transceiver with integrated LNA, PA and T/R switch Achieving +20dBm 1014-QAM MCS11 Pout and -43dB EVM Floor in 55nm CMOS," ISSCC 2020, pp. 178-179.

[7] C. Wu, C. Hunter, et al., "A 28nm CMOS Wireless Connectivity Combo IC with a Reconfigurable 2x2 MIMO WiFi supporting 80+80MHz 256-QAM, and BT 5.0," IEEE Radio Frequency Integrated Circuits Symposium 2018, pp. 300-303.

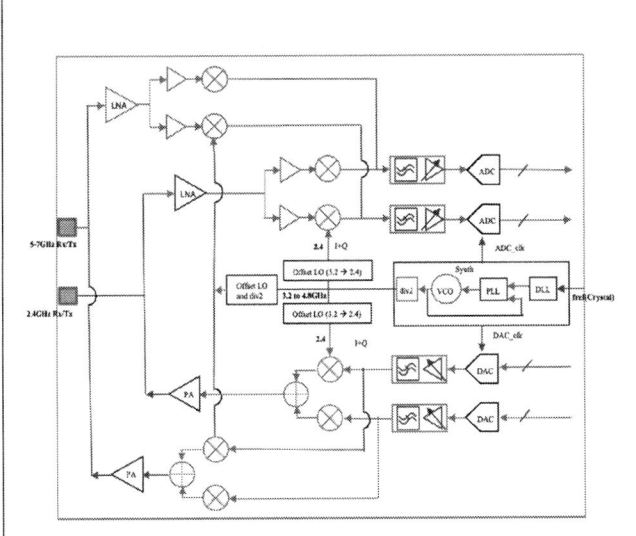

Fig. 1. Block diagram of the radio transceiver

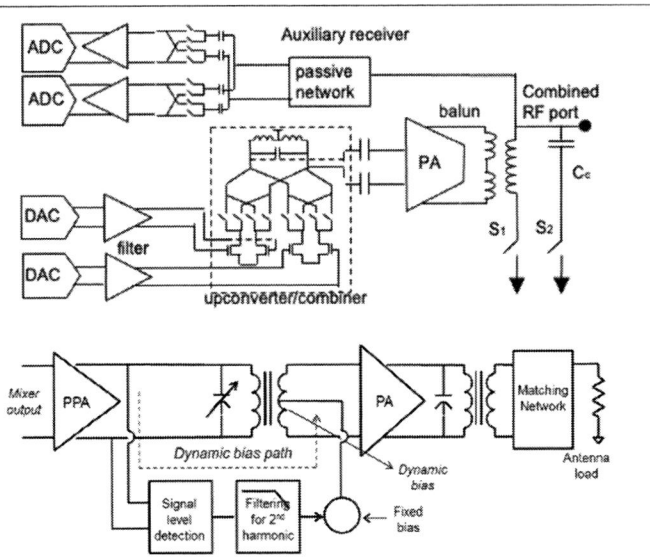

Fig. 2.Transmitter architecture and the Power Amplifier

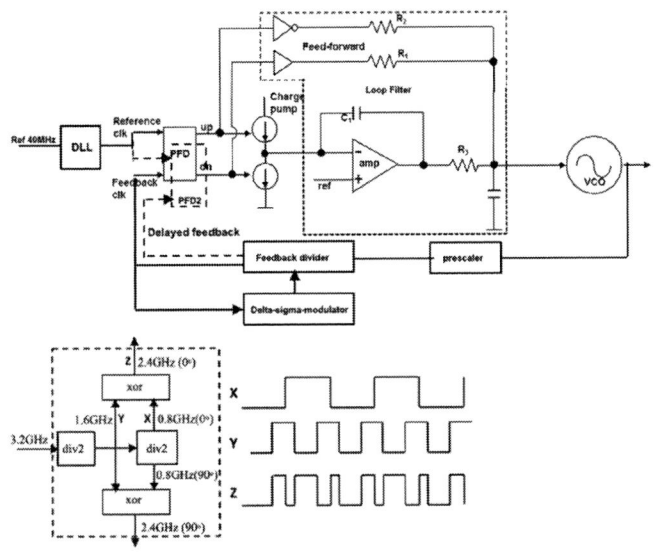

Fig. 3. PLL and the Digital Offset LO generator

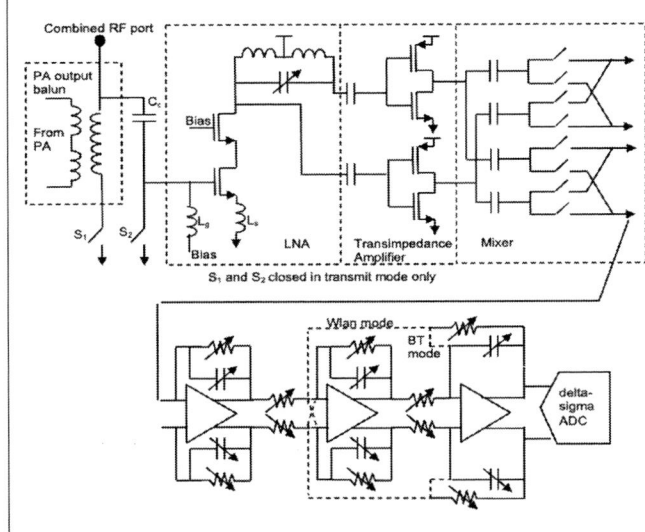

Fig. 4. LNA-PA interface and the receiver

Fig. 5. Transmitter performance at 5600MHz band

	This work	Ref(5)	Ref(6)	Ref(7)
Integrated T/R Switch[b/g, a]	yes, yes	yes, no	yes, yes	no, no
a-band range [GHz]	5 - 7.125	5 - 5.9	5 – 5.9	5-5.9
Process	45nm	45nm	55nm	28nm
PA Psat b/g (a) band [dBm]	28 (23)	29 (26)	28 (26.5)	26.5 (25.5)
NF b/g (a) [dB]	3 (6)	5 (6)	4.2 (4.8)	3.7 (3.8)
Area[mm²]	2.5	3.8	22.9 (4X4 radio)	11.7 (*2)
Core supply	1.35	1.35	1.35	-
Core receiver current b/g (a) [mA]	32 (65)	65 (68)	322 (420)mW (*1)	-
Core Tx current excluding PA b/g (a) [mA]	55 (100)	82 (91)	4324 (4850)mW (*1)	-
PA current at 15dBm b/g (a). [mA]	120 (140)	200 (225)	230 (260) (20dBm)	-

Fig. 6. Performance summary table

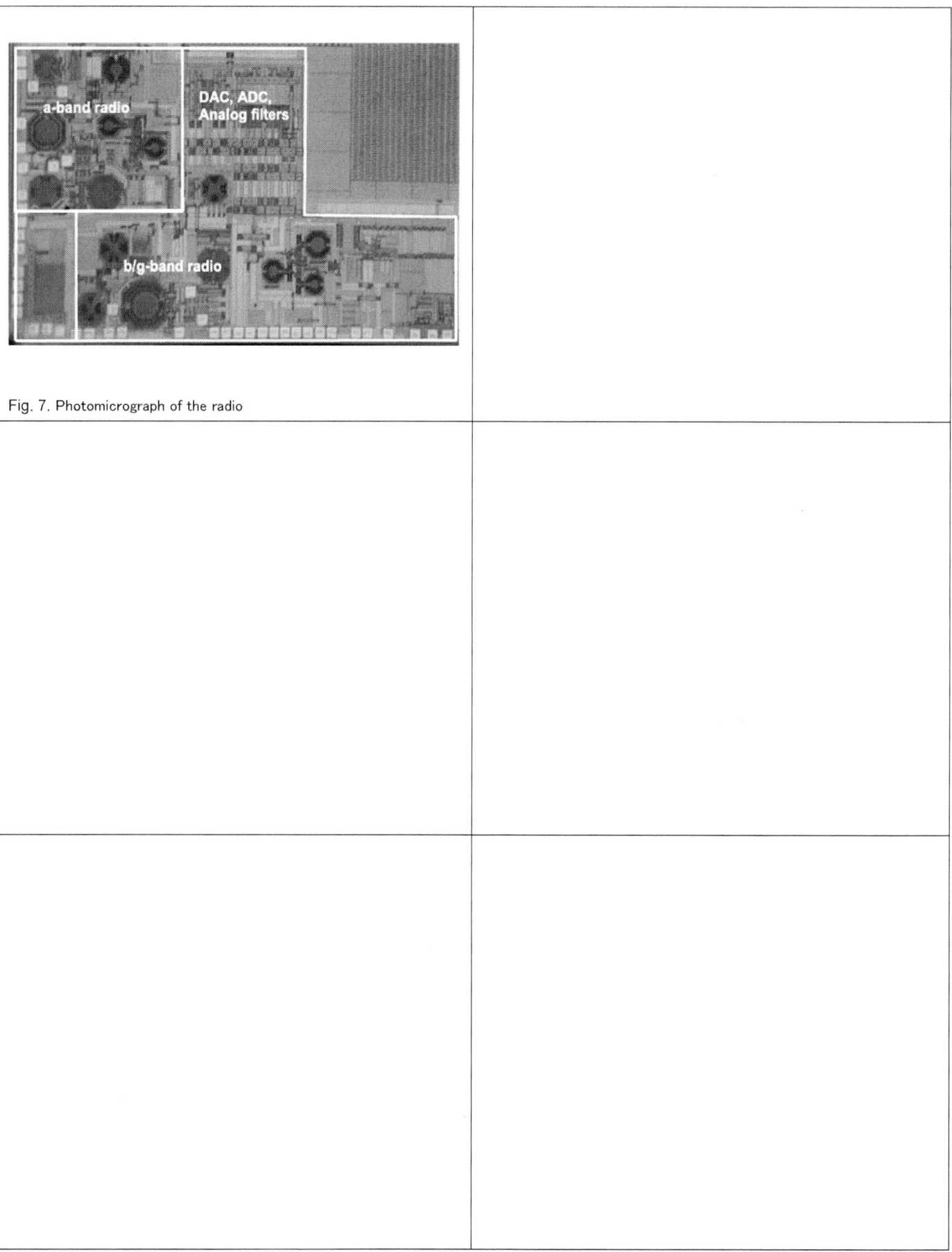

Fig. 7. Photomicrograph of the radio

An Area and Power Efficient Fully Nonlinear 10-bit Column Driver with Time-Shared Multi-Gamma-Slope DAC and Time-Interleaved Sampling Buffer for Mobile AMOLEDs

Seung Hun Choi[1], Jeongmin Kim[1], Jaewoong Ahn[1], Junyeol An[1], Jiwoong Kim[2], Ohjo Kwon[2], Ki-Duk Kim[3], and Hyung-Min Lee[1]

[1] Korea University, Seoul, Korea, [2] Samsung Display, Hwaseong, Korea, [3] C&Tech, Seoul, Korea

A color depth of 10 bits or higher is required to provide more natural colors in active-matrix organic light-emitting diode (AMOLED) display panels. The color depth or gamma voltage means the stages or the number of colors that the display can reproduce while it is determined by the bit resolution of the digital-to-analog converter (DAC) in the column driver IC. Fig. 1 (top left) shows a conventional column driver IC including a 10-bit resistor DAC (RDAC) and a buffer amplifier. This architecture suffers from a critical problem in that the number of analog switches is exponentially increased according to the bits, which also makes the RDAC area increase exponentially. Various DACs have been studied to overcome this trade-off between area and bit resolution. One of the promising techniques is the piecewise linear interpolation as shown in Fig. 1 (top right) [1-3, 5-6]. The interpolation column driver includes an upper M-bit RDAC and a lower N-bit interpolation sub DAC, while the sub DAC can adopt a DAC-embedded buffer, a current DAC, or a capacitor DAC, decreasing the area of column drivers. However, these interpolation techniques are still limited due to "piecewise-linear" output voltages, which cannot cover the whole nonlinear gamma levels for driving the panel. For example, a 10-bit column driver with a 6-bit RDAC and a 4-bit piecewise-linear sub DAC suffers from the lower effective bits (e.g., 7 or 8-bit) for providing nonlinear gamma voltages.

For area and power efficient column drivers that can provide the fully nonlinear gamma levels, the proposed time-shared column driver adopts three key ideas as shown in Fig. 1 (bottom): 1) a multi-slope time-shared DAC, 2) a rail-to-rail interleaving sample/hold (S/H) with high-slew push-pull source followers, and 3) a display-data-dependent adaptive biasing buffer amplifier. The 10-bit nonlinear voltages from the gamma resistor string pass through a global slope generator that provides 64 (=2^6) time-shared gamma slopes, and the shape of each slope depends on corresponding gamma levels. Each slope includes 16 gamma steps corresponding to lower 4-bits, so each column driver channel only requires an upper 6-bit RDAC without any additional sub DACs. The upper 6-bit RDAC selects one of the multi-gamma slopes, and the sample and hold (S/H) circuit samples the corresponding gamma voltage among the gamma slopes based on the lower 4-bits. Then, the buffer amplifier drives the sampled voltage from the time-shared gamma slope through the interleaving feedback loop. The buffer amplifier also adopts the adaptive biasing that depends on the difference between previous and current upper 3-bit display data, further reducing the power consumption.

Fig. 2 (top) shows the overall configuration of column driver channels with the multi-slope DAC and parallel gamma lines. The generated slope has a 1-H time period and 16 (=2^4) gamma voltage steps. Each step has the same step period, T_{step}. Each slope has an initialization period that maintains the same voltage level as the voltage of the first step for twice the T_{step}. As the distance from the slope generator to the channel increases, the loading effect caused by parasitic components such as routing resistance (R_R) and channel capacitance (C_{CH}) affects more to the gamma slopes. To ensure the settling of each gamma step within T_{step} across a number of channels, parallel gamma lines and the S/H with push-pull source followers (SF) were utilized. The parallel gamma lines can reduce the effect of C_{CH} by reducing the number of connected channels per gamma line. Fig. 2 (bottom left) also shows how to combine S/H with push-pull SF to reduce C_{CH} and its loading effect. When the conventional S/H is used, its input capacitance becomes as large as the storage capacitance (C_s), which is about several hundred fF, severely increasing C_{CH}. To overcome this, the push-pull SF with much smaller gate capacitance can be used to not only isolate the gamma slope from each channel S/H but also charge/discharge C_s rapidly within a short sampling period. Fig. 2 (bottom right) also includes post-layout simulation results showing the settling time of each step in the gamma slope when RC models for 1920 channels were connected. The RC distributed models were derived by extracting parasitic RC components through post-layout simulations. Settling time indicates the time when the gamma step voltage reaches within 0.1 LSB difference from the target voltage. The

result shows that each gamma step can settle at the 1920th channel within 230.4ns, which is less than the target T_{step} of 400ns (with a target 1-H time of 7.2µs).

Fig. 3. shows the rail-to-rail interleaving S/H with high-slew push-pull SF and its operational phases. The push-pull operation is determined by the voltage stored in C_s (V_{CS}) and V_{DAC}. To cover the whole range of gamma voltages, either PMOS push-pull SF or NMOS push-pull SF is activated depending on the MSB. When V_{DAC} is connected to PMOS SF and lower than its previous value, M_1 turns on and discharges C_s while M_2 turns off. On the other hand, when V_{DAC} becomes higher than its previous value, the drain of M_4 and the gate of M_3 increase, turning on M_2 and charging C_s, while M_1 turns off. Through this complementary push-pull operation, the SF can charge and discharge C_s at high slew rates while consuming small static current (e.g., 200nA). In addition, by adopting the interleaving operation in which one S/H samples the slope voltage while the other S/H provides the gamma voltage to the buffer simultaneously, the 1-H time can be fully used for both sampling and output driving. The interleaving S/H and buffer amplifier has three operation phases as shown in Fig. 3 (middle). In the sampling phase, 6-bit RDAC is connected to the SF for charging or discharging C_s. Then, the S/H samples the target voltage (V_{SH}) from the V_{DAC}, which sampling switch timing (Φ_{SPL}) is determined by the lower 4-bit data, and enters the holding phase. At this time, C_s is floating to hold V_{CS}, while the S/H including SF is deactivated to save power. During the holding phase, V_{CS} is same to V_{SH}-|$V_{GS,SF}$|-V_{CM} where |$V_{GS,SF}$| is the gate-source voltage of the SF input transistor and V_{CM} is the common-mode voltage. While one S/H is in the sampling or holding phase, the other S/H operates at the driving phase in which the SF, C_s, and buffer amplifier form a negative feedback loop to drive the panel. The driving timing is shown as Φ_{DRV}. At this time, the buffer output voltage can be expressed as $V_{out} = V_{CM} + V_{CS} + |V_{GS,SF}| = V_{SH}$, which indicates that the sample nonlinear gamma voltage can be used to drive the panel.

The proposed 10-bit column driver was fabricated in a 180nm 1.8V/5V process. Fig. 4 shows the display-data-dependent adaptive biasing buffer that uses the reduced bias current depending on the difference between previous and current output voltages by using MSB 3-bit data. In other words, the reduced bias current can be still sufficient to drive smaller output voltage transition. Measured buffer output waveforms in Fig. 4 (bottom right) verified the sufficient slew rates to drive the load of 30kΩ and 30pF even with adaptive biasing. The 1-H time was set to 7.2µs for UHD display panels. Average static current of the adaptive biasing buffer was 1.085µA, resulting in 12.5% power saving. Fig. 5 shows the measured DNL/INL of the proposed 10-bit column driver and inter-channel deviation of output voltages (DVO). The maximum DNL and INL were measured as 0.46 LSB and 0.89 LSB, respectively, where 1 LSB is 4.49 mV. The DVOs were measured from four different chips, and the maximum DVO from each chip was 7.47mV, 5.75mV, 4.28mV, and 4.64mV.

The silicon area of the proposed fully nonlinear 10-bit column driver was compared with conventional 8-bit and 10-bit RDAC-based column drivers in Fig. 6 (top). The proposed column driver occupied the area of 358×18 µm² per channel, which leads to the area shrinkage of 44% and 83.6% compared to conventional 8-bit and 10-bit column drivers, respectively. A comparison table with state-of-the-arts in Fig. 6 (bottom) shows that the proposed column driver can provide the fully nonlinear 10-bit gamma voltages between 0.2V and 4.8V to the panel, while achieving competitive area shrinkage and low static current consumption.

Acknowledgment

This work was supported by Samsung Display Co. Ltd, Korea.

References

[1] Y.-J. Jeon et al., "A Piecewise Linear 10 Bit DAC Architecture With Drain Current Modulation for Compact LCD Driver ICs," *IEEE JSSC*, 2009.

[2] H.-S. Kim et al., "A 10-Bit Column-Driver IC with Parasitic-Insensitive Iterative Charge-Sharing Based Capacitor-String Interpolation for Mobile Active-Matrix LCDs," *IEEE JSSC*, 2014.

[3] H.-M. Lee et al., "An Area and Power Efficient Interpolation Scheme Using Variable Current Control for 10-Bit Data Drivers in Mobile Active-Matrix LCDs," *IEEE TCE*, 2019.

[4] J.-S. Kim et al., "A Low-Area and Fully Nonlinear 10-Bit Column Driver with Low-Voltage DAC and Switched-Capacitor Amplifier for Active-Matrix Displays," *IEEE JSSC*, 2021.

[5] G.-G. Kang et al., "A 12-Bit Mobile OLED/µLED Display Driver IC with Cascaded Loading-Free Capacitive Interpolation DAC and 6.24V/µs-Slew-Rate Buffer Amplifier," *IEEE Symp. on VLSI*, 2021.

979-8-3503-3004-5/23 $31.00 © 2023 IEEE

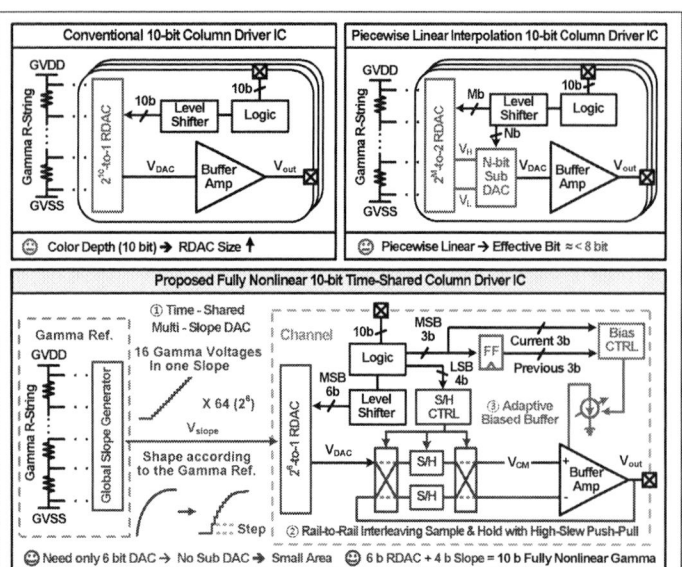

Fig. 1. Proposed fully nonlinear 10-bit time-shared column driver.

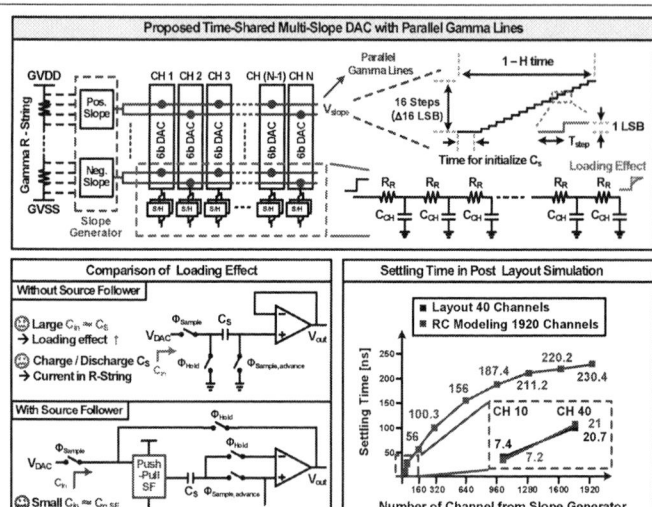

Fig. 2. Time-shared multi-slope DAC with parallel gamma lines and its loading effect analysis through the channels.

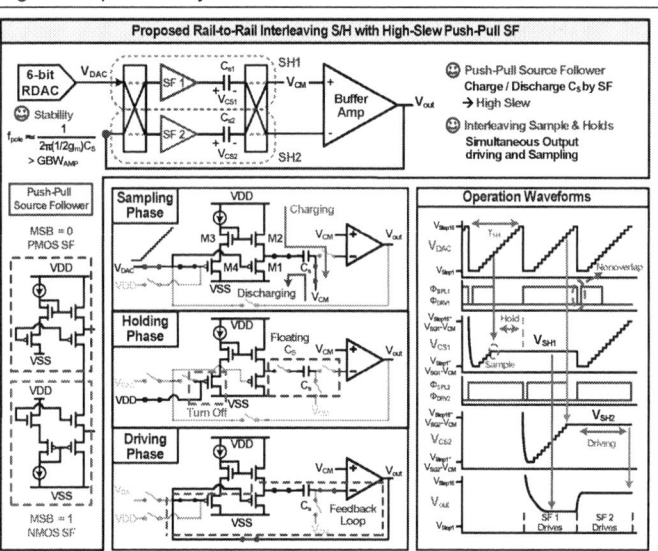

Fig. 3. Schematic diagrams and operation phases of the rail-to-rail high-slew interleaving S/H with high-slew push-pull SF.

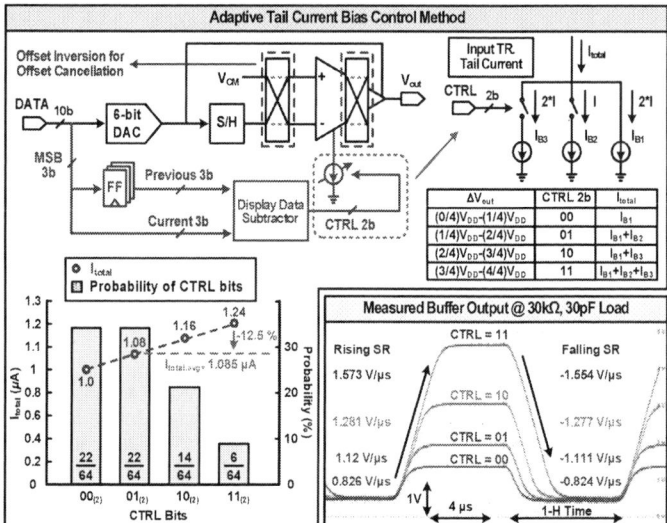

Fig. 4. Display-data-dependent adaptive biasing buffer amplifier and measured output voltage waveforms.

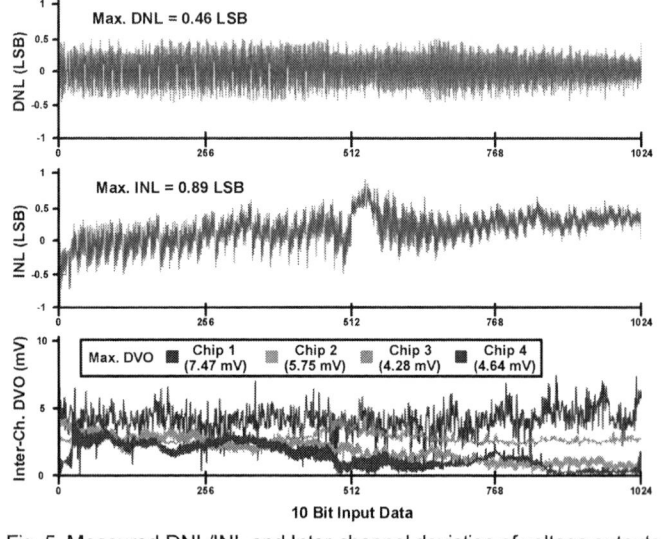

Fig. 5. Measured DNL/INL and Inter-channel deviation of voltage outputs (DVO) of the proposed 10-bit column driver.

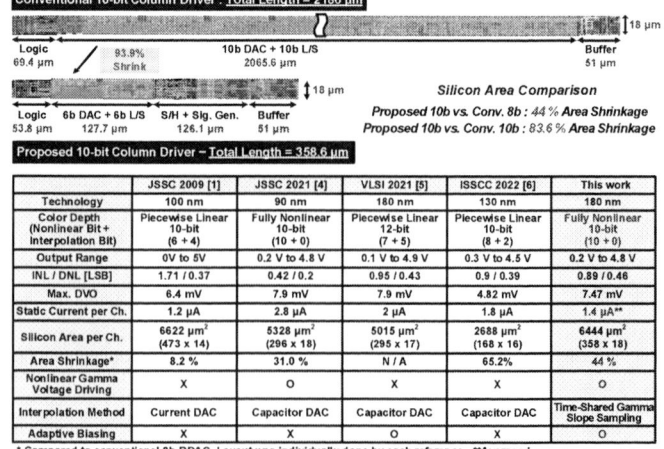

Fig. 6. Silicon area comparison and performance summary

	JSSC 2009 [1]	JSSC 2021 [4]	VLSI 2021 [5]	ISSCC 2022 [6]	This work
Technology	100 nm	90 nm	180 nm	130 nm	180 nm
Color Depth (Nonlinear Bit + Interpolation Bit)	Piecewise Linear 10-bit (6 + 4)	Fully Nonlinear 10-bit (10 + 0)	Piecewise Linear 12-bit (7 + 5)	Piecewise Linear 10-bit (8 + 2)	Fully Nonlinear 10-bit (10 + 0)
Output Range	0V to 5V	0.2 V to 4.8 V	0.1 V to 4.9 V	0.3 V to 4.5 V	0.2 V to 4.8 V
INL / DNL [LSB]	1.71 / 0.37	0.42 / 0.2	0.95 / 0.43	0.9 / 0.39	0.89 / 0.46
Max. DVO	6.4 mV	7.9 mV	7.9 mV	4.82 mV	7.47 mV
Static Current per Ch.	1.2 µA	2.8 µA	2 µA	1.8 µA	1.4 µA**
Silicon Area per Ch.	6622 µm² (473 × 14)	5328 µm² (296 × 18)	5015 µm² (295 × 17)	2688 µm² (168 × 16)	6444 µm² (358 × 18)
Area Shrinkage*	8.2 %	31.0 %	N / A	65.2%	44 %
Nonlinear Gamma Voltage Driving	X	O	X	X	O
Interpolation Method	Current DAC	Capacitor DAC	Capacitor DAC	Capacitor DAC	Time-Shared Gamma Slope Sampling
Adaptive Biasing	X	X	O	X	O

* Compared to conventional 8b RDAC. Layout was individually done by each reference **Averaged

Additional References:

[6] G.-W. Lim, et al., "A 10b Source-Driver IC with LSB-Stacked LV-to-HV-Amplify DAC Achieving 2688µm^2/channel and 4.8mV DVO for Mobile OLED Displays," *IEEE ISSCC*, 2022.

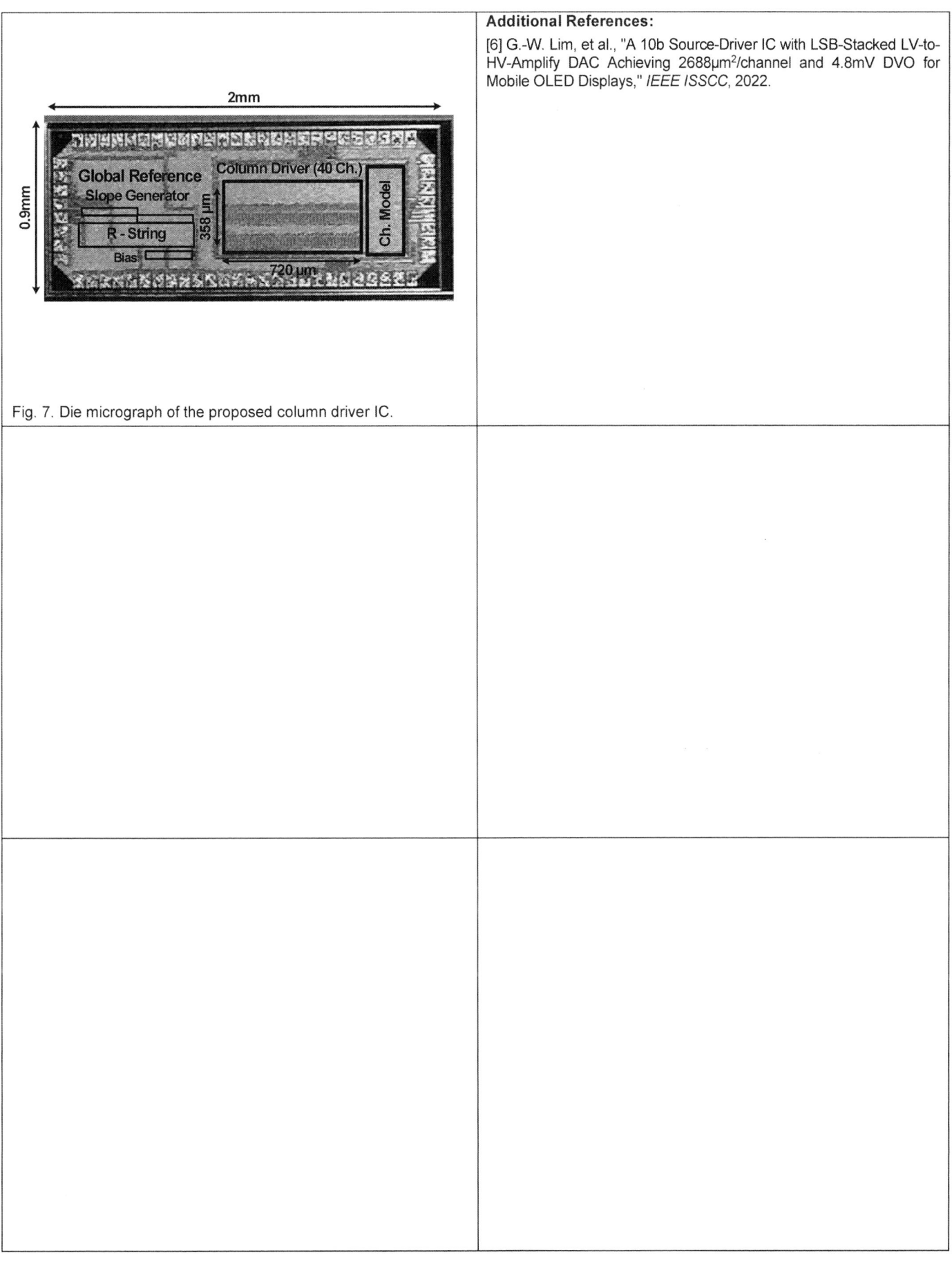

Fig. 7. Die micrograph of the proposed column driver IC.

A 5.37-TSOPS/W Reconfigurable Neuron Array with Dual-mode Neurons and Asynchronous Synapses for Energy-Efficient Inference and Biological Neural Network Simulation

Xiangao Qi[1], Yuqing Lou[1], Yongfu Li[1], Guoxing Wang[1], Kea-Tiong Tang[2], Jian Zhao[1]

[1]Shanghai Jiao Tong University

[2]National Tsing Hua University

Spiking neural networks (SNNs) have recently garnered increasing attention for both their potential to "understand" and "utilize" the human brain. Consequently, there is considerable promise in developing an SNN-based neuron array that targets both of these applications. However, such neuron arrays pose challenges that must be overcome [1]. Firstly, the primary applications of spiking neural networks - simulation and inference - require neuronal models with different levels of computational complexity [2]. On the one hand, complex models (e.g., Izhikevich (IZ) model) are needed for simulating large-scale neural networks to better understand brain function and create real-world applications. On the other hand, simple and well-studied models (e.g., Leaky Integrate and Fire (LIF) model) are used to improve the capabilities and energy efficiency of machine learning. However, due to the diverse circuit implementations of different neuron models, existing neural arrays are mainly built on simple LIF model and target only inference applications. Secondly, densely connected networks with non-uniform spike event distributions can cause Network-on-Chip (NoC) congestion, which in turn leads to inevitable delays in worst-case scenarios [3]. In particular, if spike events are processed in a serial manner, the potential delays makes it a challenging problem to optimize the throughput/area trade-off of synapses in large-scale neuron arrays.

This paper proposes a reconfigurable neuron array, which operates efficiently in LIF or IZ mode and is suitable for both inference and neural network simulation applications. The array addresses the above challenges with following solutions. Firstly, a dual-mode neuron core circuit is proposed on the basis of mathematical decomposition. The neuron core is initially an IZ neuron and can split into two LIF neurons when required, as shown in Fig. 2. Secondly, asynchronous fully-parallel synapses that realize real-time one-to-one connection between arbitrary neurons are proposed by conducting parallel synaptic connections with hardware cost linearly proportional to the neuron numbers. This synapses effectively mitigates the throughput/area trade-off in conventional SNN arrays. This CMOS neuron array is fabricated in a 55-nm process and incorporates 128 reconfigurable neuron cores. It achieves a throughput of 25.6 MSpike/s under a 0.6-V supply voltage. The power efficiency is 5.37 TSOPS/W. These key features allow the realization of on-chip inference with sub-10pJ/pixel energy efficiency for tasks including temporal coding and image sparse coding.

According to Fig. 1, the IZ model can be divided into two differential equations with a similar formation as the LIF model. Therefore, it is possible to implement an IZ neuron with two LIF neuron circuits. Fig.2 shows the circuit implementation of the proposed dual-mode neuron core consisting of two Differential Pair Integrator (DPI) based LIF neurons, a multiplier, and switches to configure the feedback loop between two operating modes. In LIF mode, the two LIF neurons work independently by breaking the feedback loop between the two neurons through the negative signal ϕ_{IZ}. Each neuron supports the integration of synaptic current onto C_{mem} using a DPI circuit (M_{1-4}), together with an on-chip programmable bias to enable bio-plausible (~ms) to accelerated (~µs) time constant. And the inactive circuit area in LIF mode is less than 10% of the whole core. In IZ mode, LIF_a acts as the membrane variable, and LIF_b acts as the slow variable, together with a translinear multiplier (M_{8-13}) for the implementation of a second-order feedback through the positive signal ϕ_{IZ}. Furthermore, We implement the event generator by sharing a dynamic comparator among a neuron group composed of 8 neuron cores. The comparator connects to neurons in the core successively and generates spikes by comparing V_{th} with each membrane voltage in the group. In this manner, no event generator is wasted in IZ mode. Each neuron core occupies an area of 600µm^2.

To mitigate the throughput/area trade-off in conventional synapses, a neuron array with 16 neuron groups connected through asynchronous fully-parallel synapse is proposed and shown in Fig. 3. Compared to fully-discrete and axon-sharing synapses, the proposed synapse realizes the highest throughput/area trade-off while allowing arbitrary intra-array connections within 10µs. This is leveraged by processing all connections in parallel. The proposed synapse consists of an asynchronous axon and 256 time-modulated pulse generators. The operation of the synapse consists of three steps: 1) Incoming spikes including 8-bit address and 3-bit weight information are transmitted and buffered at matching pulse generators in a time-multiplied manner through an AER interface within one period of less than 2µs; 2) Pulse generators produce pulse width weighted signals in parallel to activate corresponding synapses; 3) Weighted synaptic current is injected into neurons. This synapse comes at the cost of pulse generators with the same number of neurons in the array. Each pulse generator is implemented with a 3-bit buffer and control logic with an area of 10µm^2.

The prototype IC is fabricated using 55-nm CMOS technology, with a total silicon area of 1.47×0.32mm^2 as shown in Fig. 7. This work demonstrates a large-scale integration of 128 IZ neurons (or 256 LIF neurons) connected through asynchronous fully-parallel synapses. For neuron in IZ model, all typical behaviors are fulfilled with an average power efficiency of 0.6 pJ/SOP. For testing the neuron in LIF mode, the circuits were operated with a 100kHz input and 30MHz comparator clock. Each neuron in LIF mode produces a wide range of 100-1.8MHz spikes/neuron under the supply voltage of 0.6-1.2V. The neuron achieve the best energy efficiency of 0.18pJ per operation at the supply voltage of 0.6V. Fig. 6 compares the proposed array with current state-of-the-art neuromorphic processors. The proposed array obtains the best-of-class power and area efficiency, with a power efficiency of 5.37 TSOPS/W and a throughput of 25.6-MSpikes/s [4-7].

We verify the proposed neuron array with two benchmarks: 1) for IZ neuron model, networks simulating Pavlov's dog and temporal coding with two polarities are implemented as shown in Fig.4 [8]. Compared to LIF neuron, the utilization of phasic bursting and inhibition-induced behaviors in temporal coding reduces the complexity of network by avoiding the self-coupled connections and anti-color processing. For efficiency performance, the IZ neuron is over 100× faster and 10^4× more efficient than a computation simulation. 2) for LIF neuron model, biologically realistic sparse and independent local network (SAILnet) is implemented and shown in Fig. 5 [9]. The SAILnet mimics the feature extraction of the primary visual cortex simple cells to realize image sparse coding. However, the fully-connected network architecture impedes its efficient hardware implementation due to quadratically growing synapse overhead [10]. The proposed asynchronous fully-parallel synapse addresses this issue. At room temperature and 0.6V voltage supply, the array realizes the throughput of 46 Mpixel/s and energy efficiency of 9.25 pJ/pixel. The measured power consumption, compared to [10], is plotted in Fig. 5.

Acknowledgement:

This work was supported by the National Key Research and Development Program of China under Grant 2019YFB2204500. The Corresponding author is Jian Zhao.

References:

[1] K. Roy et al., "Towards spike-based machine intelligence with neuromorphic computing," Nature 2019:607-617.

[2] C. Frenkel et al., "A 0.086-mm^2 12.7-pJ/SOP 64k-synapse 256-neuron online-learning digital spiking neuromorphic processor in 28-nm CMOS," IEEE TBioCAS 2018:145-158.

[3] BV. Benjamin et al., " Neurogrid: A mixed-analog-digital multichip system for large-scale neural simulations," Proc. IEEE 2014:699-716.

[4] C. Frenkel et al., "ReckOn: A 28nm sub-mm^2 task-agnostic spiking recurrent neural network processor enabling on-chip learning over second-long timescales," ISSCC 2022:1-3.

[5] W. Wan et al., "33.1 A 74 TMACS/W CMOS-RRAM neurosynaptic core with dynamically reconfigurable dataflow and in-situ transposable weights for probabilistic graphical models," ISSCC 2020:498-500.

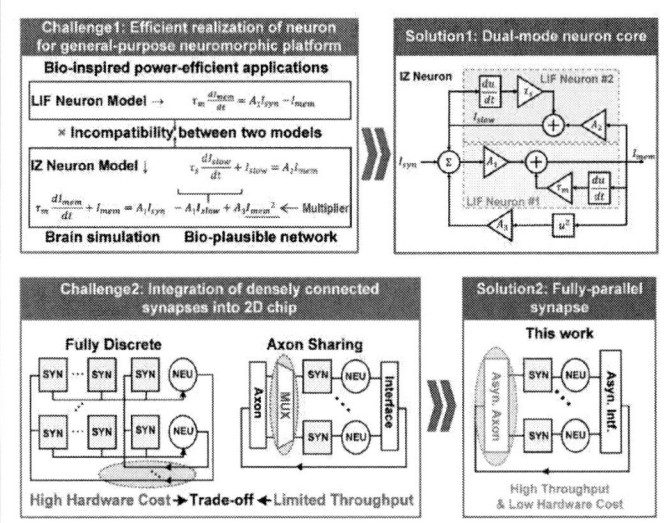

Fig. 1. Challenges (top) and the proposed solutions (middle and bottom).

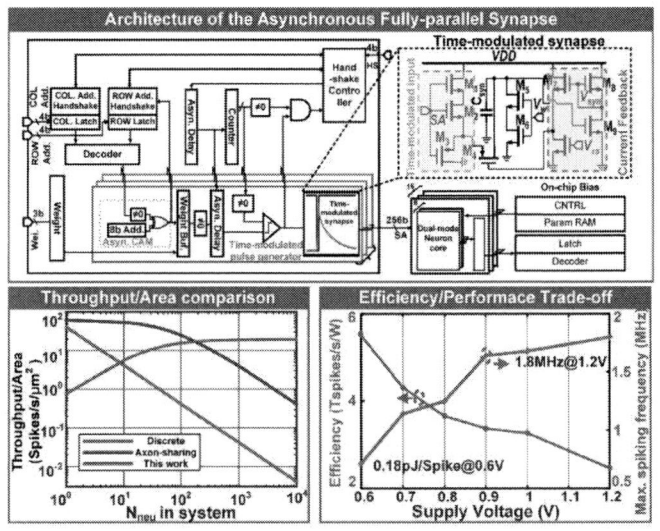

Fig. 3. Schematic of the asynchronous fully-parallel synapse (top) and measured efficiency/performance trade-off (bottom).

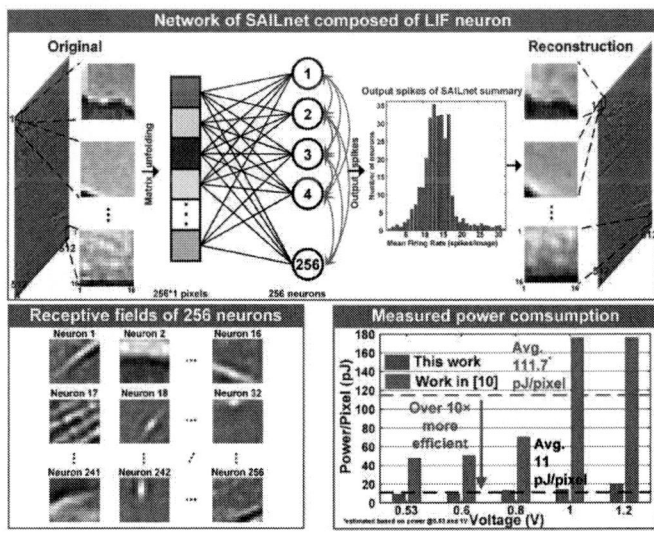

Fig. 5. Architecture of SAILnet (top) and measured results (bottom).

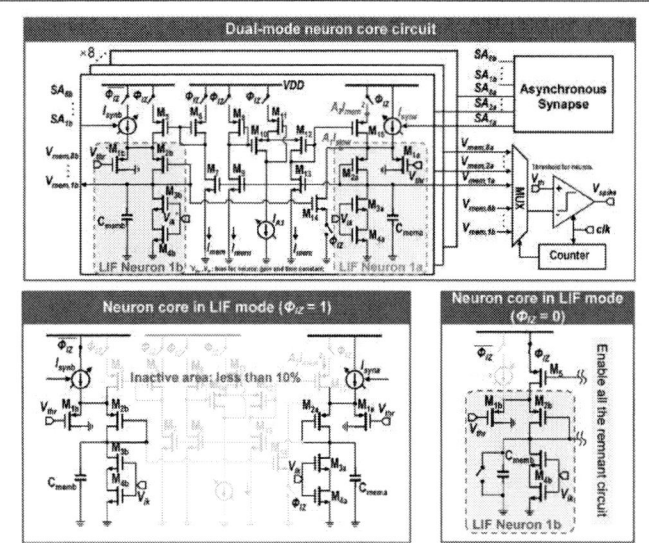

Fig. 2. Structure of the proposed dual-mode neuron core (top) and principle of dual-mode operation (bottom).

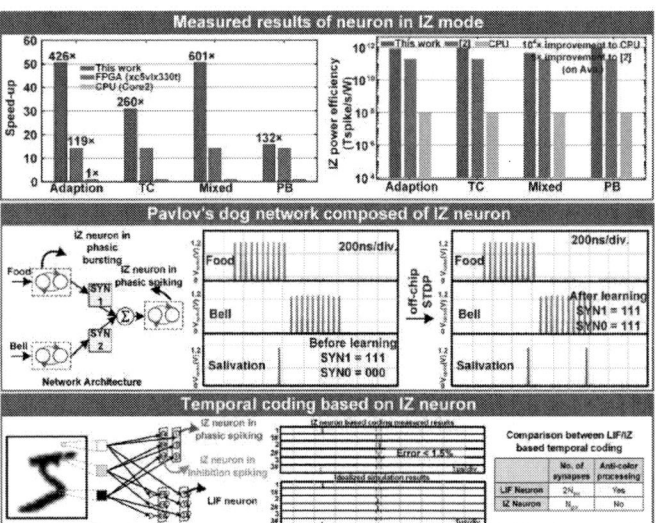

Fig. 4. Speed and power comparison, architecture and measured results of Pavlov's dog network, temporal coding network.

| | [1]
ISSCC'22 | [2]
ISSCC'20 | [3]
JSSC'19 | [4]
VLSI'17 | This work | |
					LIF Mode	IZ Mode
CMOS Tech. (nm)	28	130	10	40	55	
Circuit Type	Digital	MS	Digital	MS	MS	
Neuron Type	LIF	LIF	LIF	LIF	Dual-mode (LIF/IZ)	
Supply Voltage (V)	0.5	N.A.	0.525	0.9	0.6	
Throughput (Spikes/s)	32k	N/A	3G	N/A	25.6M	
No. of Neurons	272	256	4096	512	128	256
Neuron Area (μm²)	58	1200	50	1377	270	600
Energy Efficiency (pJ/SOP)	5.3	1.53	3.8	0.29	0.186	0.6
Power Efficiency Normalized to 55-nm* (TSOP/W)	0.187 0.09	0.65 1.53	0.26 0.045	3.43 2.5	5.37 (1.56x)	1.67
FoM** (TSpikes/W/mm²)	3.2k	0.54k	5.2k	2.5k	19.8k (3.8×)	4k

$* \ Nor. power \ efficiency = \frac{Power \ efficiency \times Tech.}{55 \ nm}$ $** \ FoM = \frac{Throughput \times peak - efficiency}{Area \ of \ synapse \ systme}$

Fig. 6. Comparison with state-of-art on-chip neuromorphic processors.

979-8-3503-3004-5/23 $31.00 © 2023 IEEE

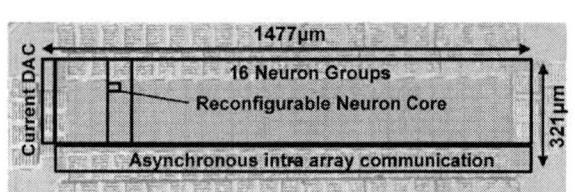

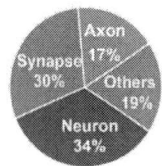

Chip Summary	
Technology	55-nm CMOS
Supply Voltage	0.6-1.2V
Frequency	100-1.8MHz
Power Efficiency	5.37TSOPS/W
Core Area	1.2×0.6mm
Throughput	25.6MSpikes/s

Power Breakdown

Axon 17%
Synapse 30%
Others 19%
Neuron 34%

Fig. 7. Die photo, summary table and power breakdown.

Additional References:

[6] GK. Chen *et al.*, "A 4096-neuron 1M-synapse 3.8-pJ/SOP spiking neural network with on-chip STDP learning and sparse weights in 10-nm FinFET CMOS," IEEE JSSC 2018:992-1002.

[7] FN. Buhler *et al.*, "A 3.43 TOPS/W 48.9 pJ/pixel 50.1 nJ/classification 512 analog neuron sparse coding neural network with on-chip learning and classification in 40nm CMOS," VLSI circuits 2017:C30-C31.

[8] Q. Yu *et al.*, "Rapid feedforward computation by temporal encoding and learning with spiking neurons," IEEE TNNLS 2013:1539-1552.

[9] J. Zylberberg *et al.*, " A sparse coding model with synaptically local plasticity and spiking neurons can account for the diverse shapes of V1 simple cell receptive fields," PLoS Comput. Biol. 2011: e1002250.

[10] P. Knag *et al.*, "A Sparse Coding Neural Network ASIC With On-Chip Learning for Feature Extraction and Encoding," IEEE JSSC 2015: 1070-1079.

A 28nm 386.5GOPS/W Coarse-grained DSP using Configurable Processing Elements for Always-on Computation with FPGA Implementation

Hedi Wang[1], Zengwei Wang[1], Yaolei Li[1], Chen Tang[1], Jinxu Gao[1], Huazhong Yang[1], Yongpan Liu[1]

[1]Tsinghua University, Beijing, China

In recent years, always-on chips have become prevalent in keyword spotting and face recognition [1]. These chips require minimal power consumption, enabling them to operate over long periods in battery-limited conditions. To function effectively, always-on chips must perform various signal processing tasks, such as MFCC feature extraction and image filtering [2]. Traditional designs commonly employ specialized ASIC circuits and DSPs to process signals, as illustrated in Fig. 1. ASICs prioritize low power consumption over versatility, which limits their practicality in real-world scenarios [3]. DSP-based solutions can be further categorized into general-purpose scalar DSPs and parallel SIMD DSPs. Scalar DSPs, characterized by substantial energy consumption on instruction control, tend to be relatively inefficient [4,5]. For the parallel SIMD DSPs, their hundred-milliwatt power consumption fundamentally violates the requirement of an always-on system, though they can provide superior computing performance [6]. Consequently, no existing hardware solution has achieved the optimal combination of versatility, low power consumption, and high energy efficiency.

This paper proposes a coarse-grained DSP that achieves a trade-off between versatility, power consumption, and energy efficiency for signal processing tasks in always-on chips. Coarse-grained refers to treating a vector as the smallest unit of computation. The work presents three key features: a time-serial vector computation architecture that automatically completes the calculation of each element in the vector one by one, a co-designed coarse-grained instruction set that reduces the number of instructions and provides programmability, and configurable processing elements (CPEs) that enable various signal processing algorithms. Figure 2 illustrates the overall architecture of the coarse-grained DSP, including a top controller, a 2KB instruction SRAM, 16 2KB data SRAM banks, and several CPEs. Each CPE comprises a thread controller and a set of configurable arithmetic units such as ALU, MAC, IIR, and others.

Figure 3 depicts the time-serial vector computation of the above DSP. In the first stage, instructions and data are written to the relevant memory space through the bus. An instruction is triggered by an external CPU and assigned by the top controller to a specified CPE. The CPE requests access to the corresponding SRAM banks and activates a computation thread if the required banks are free. In the second stage, the CPE starts to execute vector computation. The thread controller in CPE calculates an address every cycle and initiates storage access to the data memory. Input data is read from the SRAM and then fed to the arithmetic unit for pipelined calculations. For instance, in Fig. 3, the ALU performs a calculation of adding a vector to the constant 4, while the other arithmetic units are clock gated to avoid extra power consumption. The result of the arithmetic calculation is then written back to the corresponding address in the Data SRAM. This process is repeated in the CPE until a vector calculation is completed. During this process, the top controller can assign new instructions to other CPEs, enabling multiple CPEs to perform calculations in parallel, resulting in higher peak computing performance. In the third stage, the DSP returns an interrupt to the external CPU, indicating the finish of the computation thread, and results can be read out from the bus. Through the above three stages, the DSP can complete a vector calculation with only one coarse-grained instruction, thereby reducing the control overhead such as instruction fetching and decoding.

Figure 4 shows the architecture of the co-designed coarse-grained instruction set, with each instruction required to control one vector computation. The length of the coarse-grained instruction is specified as 8 32-bit words to make the instruction format as regular as possible to reduce the complexity of decoding. The instruction contains three types of elements: hardware mapping, access address control, and pipeline control. The hardware mapping element determines which CPE executes the instruction and specifies which arithmetic unit the CPE uses for the computation.

The access address control element gives the first address and the inner and outer loop steps of two source operation vectors and one destination vector, providing better addressing flexibility. The pipeline control element controls the read and write interval based on the different calculation delays of different arithmetic units, giving full play to the hardware performance of arithmetic units. For a vector addition of length N, coarse-grained DSP needs only O(1) instruction word length to accomplish the task, whereas a scalar DSP needs O(N) instructions. The reduction of instruction word length helps to improve the energy efficiency of coarse-grained DSP.

Figure 5 illustrates the architecture of CPE, the key module of DSP, which contains a thread controller and a set of configurable arithmetic units. The thread controller calculates a memory address at each cycle according to the parameters of the inner and outer loops and tells the arithmetic unit when to read and write data through the finite state machine. Although the thread controllers play an important role in vector control, they account for only 3.9% of the DSP power consumption and 3.4% of the chip area, indicating that the DSP control overhead is very low, and more resources can be used for signal processing. The arithmetic units in CPEs complete various signal processing algorithms according to the data from memory and the control signal from the thread controller. In this architecture, the thread controller acts as an intermediary to decouple the arithmetic units from the memory access control, allowing users to customize the DSP hardware structure according to the application scenario by configuring different arithmetic units. For example, in the audio filtering scenario, IIR and FIR units are configured for each CPE; in the neural network inference scenario, more CPEs are implemented and MAC units are configured for each CPE. The architecture enables coarse-grained DSP to have better hardware tunability. So far, the arithmetic units that have been implemented include ALU, MAC, IIR, FIR, format converter, data statistics, and FFT. These arithmetic units are optimized for low power consumption and high energy efficiency, and are sufficient to cover most signal processing needs.

This work is synthesized using Synopsys Design Compiler, and then is placed and routed under IC Compiler. The power analysis is performed on PrimeTime PX. Figure 6 illustrates the DSP layout and presents a comparison to previously used signal processing hardware. Compared with dedicated ASICs, coarse-grained DSP can be used for various vector calculations and are more versatile. Compared with the scalar DSPs, this work achieves 9.43 times higher energy efficiency through the coarse-grained mechanism. Additionally, the power consumption of this DSP is only 4.87mW, which is 3.5% of the parallel SIMD DSP, and is suitable for always-on scenarios. Furthermore, coarse-grained DSPs are also 2.72 times more energy efficient than parallel SIMD DSPs.

The implementation of this coarse-grained DSP on Xilinx Zynq UltraScale+ XCZU9EG FPGA is also explored. Resource utilization and power consumption are estimated using Vivado 2019.1 and are presented in Fig. 7, in comparison to existing FPGA-based processors. The dynamic power consumption of this work demonstrates 1.22 times more energy efficiency when compared to the scalar processors. And the total power efficiency of the DSP is comparable to that of SIMT processors. Operating at 100MHz clocks, the system successfully executes the MFCC feature extraction algorithm. Figure 7 also depicts the original audio waveform, the correct extraction results and the DSP extraction results. The coarse-grained DSP achieves an acceleration ratio of 1.63 times compared to calculations using only ARM cores. This acceleration is achieved through the parallel computation of multiple CPEs.

References:

[1] S. Kim *et al.*, "SNPU: Always-on 63.2 µW Face Recognition Spike Domain Convolutional Neural Network Processor with Spike Train Decomposition and Shift-and-Accumulation Unit," A-SSCC 2022.

[2] R. Guo *et al.*, "A 5.1pJ/Neuron 127.3us/Inference RNN-based Speech Recognition Processor using 16 Computing-in-Memory SRAM Macros in 65nm CMOS," VLSI Symp. 2019.

[3] J. Giraldo, *et al.*, "Vocell: A 65-nm speech-triggered wake-up SoC for 10µW keyword spotting and speaker verification," JSSC 2020.

[4] https://www.cadence.com/en_US/home/tools/ip/tensilica-ip/hifi-dsps/hifi-4.html.

979-8-3503-3004-5/23 $31.00 © 2023 IEEE

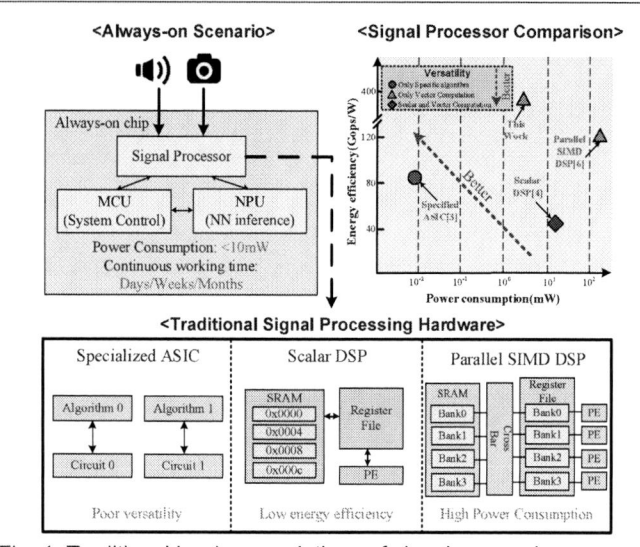

Fig. 1. Traditional hardware solutions of signal processing requirements in always-on chips

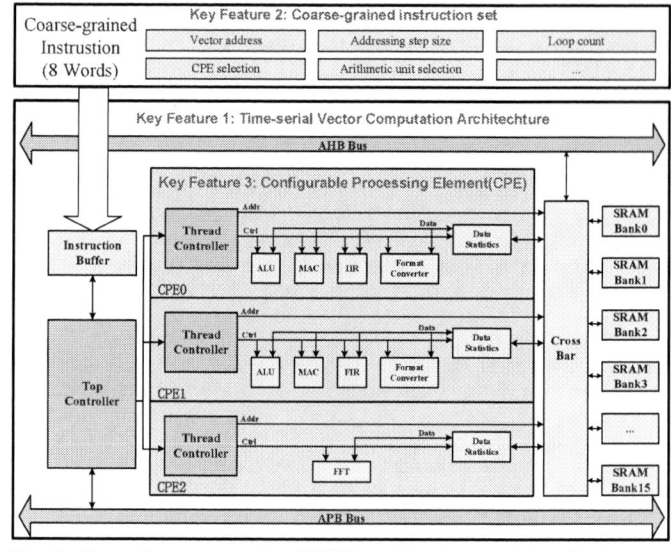

Fig. 2. Overall coarse-grained DSP architecture and key features

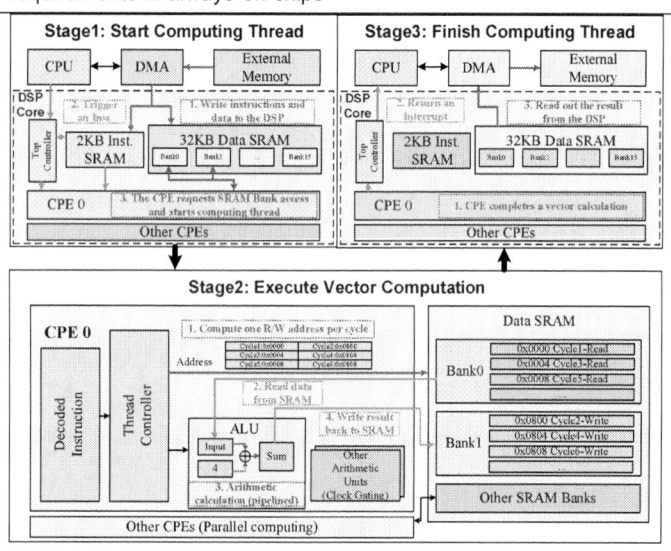

Fig. 3. Time-serial vector computation architecture and workflow

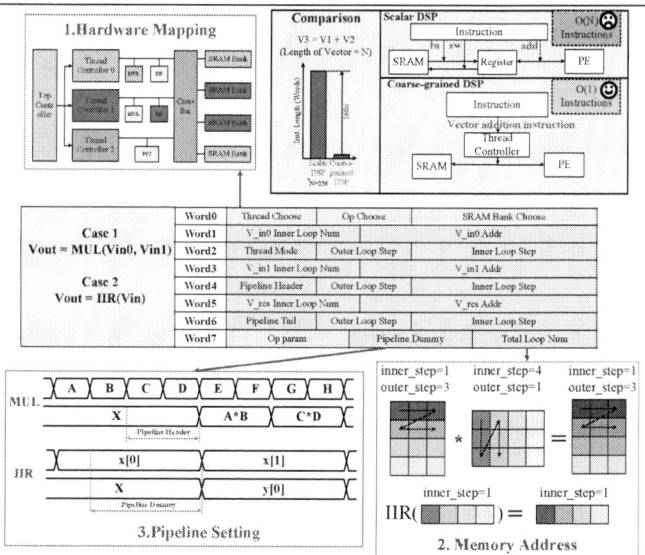

Fig. 4. Co-designed coarse-grained instruction set

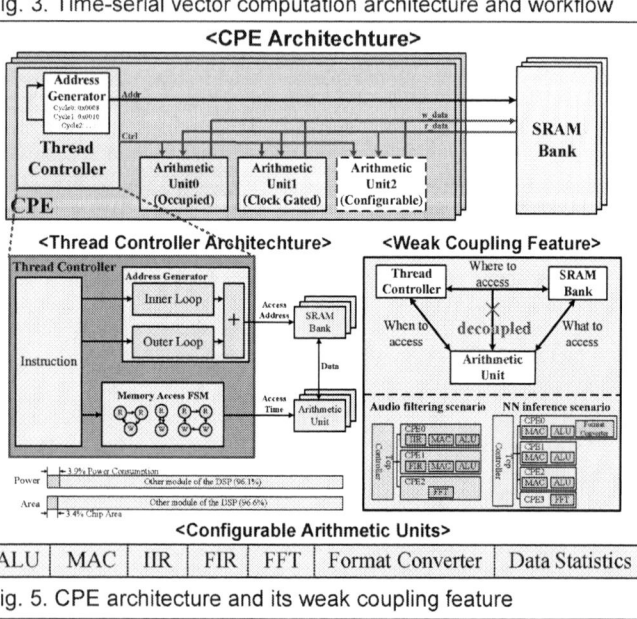

Fig. 5. CPE architecture and its weak coupling feature

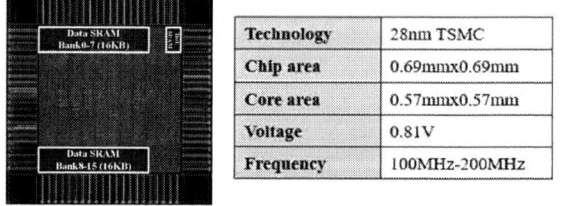

Technology	28nm TSMC
Chip area	0.69mmx0.69mm
Core area	0.57mmx0.57mm
Voltage	0.81V
Frequency	100MHz-200MHz

Specifications	[3] Vocell	[4] Hifi4	[6] Ara	This Work
Hardware type	ASIC	Scalar DSP	Parallel SIMD DSP	Coarse-grained DSP
Technology(nm)	65	28	22	28
Voltage (V)	0.6-1.2	1.13	0.8	0.81
Frequency (MHz)	0.25	200	1250	200
Area (mm²)	0.375	-	1.125	0.321
Power (mW)	0.009	78	138	4.87
Peak Perf.[1](MOPS)	0.75	3200	19640	1890
Peak Eff.[1] (GOPS/W)	83.3	41.0	142.4	387
Versatility	Only MFCC extraction	Scalar and vector computation	Only vector computation	Only vector computation

[1] Normalized to 16-bit operations

Fig. 6. DSP layout and comparison with other signal processing hardware

<DSP Demo Photo>

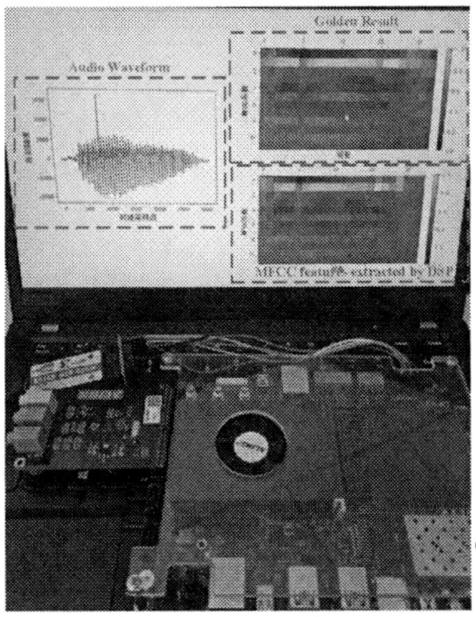

<FPGA Implementation Result>

Specifications	[7] Dual core IPPro	[8] FGPU	This Work
Hardware type	Scalar Processor	SIMT Processor	Coarse-grained DSP
FPGA Fabric	Zynq XC7Z020	Zynq ZC706	Zynq XCZU9EG
LUT	4987	21000	17633
FF	5519	35000	9222
BRAM	4.5	47	9.5
DSP	2	48	10
Frequency (MHz)	100	200	100
Peak Perf. (MOPS)	200	3200	950
Dynamic Power† (W)	0.015	-	0.070
Total Power† (W)	-	2.36	0.69
Versatility	Scalar and vector computation	Only vector computation	Only vector computation

† Estimated by Vivado Software.

<MFCC acceleration ratio>

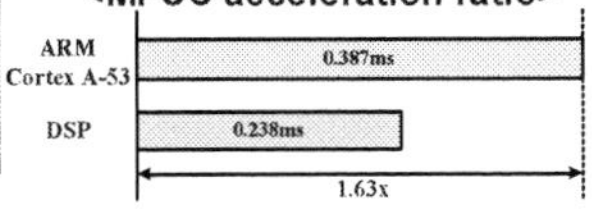

ARM Cortex A-53 — 0.387ms
DSP — 0.238ms
1.63x

Fig. 7. The summary of FPGA resource utilization and demo photo of the coarse-grained DSP on MFCC extraction

Additional References:

[5] M. Horowitz, *et al.*, "Computing's Energy Problem," ISSCC 2014.

[6] M. Cavalcante, *et al.*, "Ara: A 1 GHz+ Scalable and Energy-Efficient RISC-V Vector Processor with Multi-Precision Floating Point Support in 22 nm FD-SOI," T-VLSI 2019.

[7] F. Siddiqui *et al.*, "FPGA-based processor acceleration for image processing applications," Journal of Imaging 2019.

[8] M. Al Kadi *et al.*, "FGPU: An SIMT-architecture for FPGAs," FPGA 2016.

A 92.7%-Efficiency 6.78-MHz Dual-Output Energy-Resuscitating Resonant Regulating Rectifier for Wirelessly Powered Systems

Hyun-Su Lee and Hyung-Min Lee

School of Electrical Engineering, Korea University, Seoul, Korea

Wirelessly powered systems, such as implantable medical devices (IMD), widely rely on a two-stage regulating structure comprising an ac-dc rectifier and a dc-dc regulator to convert the wireless power. Recently, there have been advancements aimed at enhancing power efficiency and reducing the number of output capacitors, leading to the introduction of a one-stage resonant regulating rectifier (R^3) that efficiently converts ac input into stable dc output. Fig. 1 (top) depicts this R^3 configuration, while four types of conventional R^3 structures are also illustrated in Fig. 1 (middle). The topology reconfiguration modulation (TRM) R^3 enables the receiver to switch between two modes, resulting in output options of x1 or xN [1]. Also, the pulse density modulation (PDM) R^3 allows for up to five tuning modes by determining the number of pulses required for power transfer from the L_2C_2 tank to the load [2]. However, these discrete tuning methods deteriorate the output ripple, and the unselected pulses circulates in the L_2C_2 tank dissipating power through the effective series resistor (ESR) of the L_2C_2 tank. To mitigate this issue, alternative methods such as pulse width modulation (PWM) [3] and pulse width/density modulation (PWDM) [4] were investigated to control the pulse width continuously until the output voltage reaches the target voltage. However, these methods still suffer from the limitations related to the energy wastage, leading to heat generation and low power transfer to the load, which restrict their usage in IMD applications. To get the utmost out of this wasted period, the proposed energy-resuscitating resonant regulating rectifier (ER^4) invigorates this discarded period to provide additional regulated energy, resulting in increased power delivered to the load (PDL) and continuously regulated dual output as shown in Fig. 1 (bottom).

The overall structure of the proposed dual-output ER^4 is outlined in Fig. 2 (top). The resonant L_2C_2 tank receives ac power from the transmitter, and the cross-coupled NMOS pair ($N_{1,2}$) transforms the full-wave ac voltage into the half-wave ac voltages (V_{INN} and V_{INP}). The main power is then routed through power pass PMOS transistors ($P_{1,2}$) and subsequently through $R_{L1}C_{L1}$ load, generating the primary regulated output voltage (V_{REG}). Additionally, the auxiliary power is directed through the power pass PMOS transistors ($P_{9,10}$) and subsequently passes through $R_{L2}C_{L2}$ load, producing the additional regulated output voltage (V_{ER}). $V_{G1,G3}$ are generated by dual $CMP_{1,2}$, which controls $P_{1,2}$ as well as the main path's replica transistors ($P_{3,4}$), while $P_{3,4}$ have a thousandth of the width size of $P_{1,2}$. Here, the body voltages of $P_{1,3}$ share the same dynamic body biasing circuits comprising of $P_{5,6}$. The replica current is then directed through the pass transistor P_{15} and subsequently flows through the replica load (R_RC_R). Here, the replica output voltage (V_{REPL}) is equal to V_{REG} due to the feedback 3 (FB$_3$) comprising of AMP$_1$ and P_{15}. When the power supplied to the V_{REG} node exceeds a pre-defined threshold, which is V_{MAXPST} (>$I_{VREG} \cdot R_R/1000$), CMP_3 outputs Φ_{MEMEN}. Moreover, the dual output comparator is not only responsible for generating the gate control signals ($V_{G1,G3}$) but also $V_{G2,G4}$, which control $P_{9,10}$ to generate V_{ER}. Here, the body voltage of P_9 is supported by the dynamic body biasing with $P_{11,12}$. Without requiring LDO regulators, the two regulated outputs (V_{REG} and V_{ER}) are sensed and returned through two analog feedback loops, feedback 1 (FB$_1$) and feedback 2 (FB$_2$).

Fig. 2 (bottom-left) illustrates the timing diagram of the ER^4 system featuring the two gate control signals, $V_{G1,G2}$. The offset-controlled common-gate (CG) comparator compares V_{INN} and V_{REG} to set the falling timing of V_{G1} (t_S) and the ER-end-protection timing of V_{G2} (t_E). Additionally, the PWM-1 and PWM-2 controllers, each incorporating an analog feedback loop, determines the rising time of V_{G1} (t_1) and the rising time of V_{G2} (t_2), respectively, thus enabling the regulation of dual outputs, V_{REG} and V_{ER}. To limit the pulse width of V_{G1}, a pulse width memorizer (PWMem) detects a specific t_1, at which point the output power to the load $R_{L1}C_{L1}$ reaches a predetermined value, and subsequently designates this t_1 as t_{MEM}. To restrict the maximum end timing of V_{G2} to t_E, the ER-end-protector uses t_2 and t_E to prevent t_2 from becoming later than t_E, and subsequently outputs the ER protection voltage (V_{PROT}).

Fig. 2 (bottom-right) shows a PWM controller featuring the proposed current bleeding technique, implemented with the use of a resistor

R_{CB}. The purpose of R_{CB} is to prevent the delay from reaching infinity, thereby ensuring a secure startup condition for the entire regulation system. The internal structure of the dual output comparator is presented in Fig. 3 (top-left), accompanied by the diagram of three feedback loops in Fig. 3 (top-right). The FB$_1$, which includes PWM-1, a V_{REG}-Driver, and P_1, monitors V_{REG} and tracks its target voltage (V_{TG1}), while also generating t_1. The FB$_2$, consisting of PWM-2, a V_{ER}-Driver, an ER-end-protector, and P_9, monitors V_{ER} and tracks the target voltage of V_{ER} (V_{TG2}) and then yields t_2. Despite generating two outputs, the dual-output comparator maintains high power conversion efficiency (PCE) by employing only a single pair of CG comparators. The dual-output comparator also incorporates two protection techniques, namely PWMem and ER-end-protector, to regulate the pulse widths of $V_{G1,G2}$. These techniques prevent high surge currents to the load, thereby preventing both outputs from dropping to zero. The proposed PWMem consists of a delay line and two arrays of D flip-flops for reading and locking operation, while generating Φ_{PWMEM} as depicted in Fig. 3 (bottom).

Fig. 4 (left) illustrates that in theory, the ER^4 exhibits reduced ripples by $1/N_{PWDM}$ and increased PDL by $N_{PWDM} \times (T_{W1}+T_{W2})/T_{W1}$ when compared to the conventional PWDM R^3. Fig. 4 (right) displays the measured waveforms of ER^4 in one-regulation mode (Φ_2) and two-regulation mode (Φ_3) with two output capacitors of 1μF each. As the ER^4 transitions from Φ_2 to Φ_3, V_{REG} remains constantly regulated at 4.5V, while V_{ER} is activated and regulated to 3.3V, 1.8V, and 1.0V. V_{ER} can be generated as high as 4.5V, providing a wide voltage range of ΔV_{ER}=3.5V. In Fig. 5 (top-left), measured PCEs are shown under various V_{ER} load conditions for V_{REG} with 200Ω and 500Ω. As the V_{REG} output power increases, PCE also increases due to the constant logic power and the system optimized for the V_{REG} load condition of 200Ω. On the other hand, as the V_{ER} output power decreases, PCE increases because the conduction losses of $P_{9,10}$ become dominant compared to the switching losses. The ER^4 system can handle PDL up to 123mW with PCE of 87.24%. Fig. 5 (bottom-left) presents the measured PCEs at different V_{ER} voltage conditions. Since the output power of V_{REG} is much higher than that of V_{ER}, the PCE remains relatively similar for different V_{ER} load conditions. The maximum PCE was measured as 92.74% for the load condition of V_{REG} (200Ω) and V_{ER} (1kΩ), which corresponds to the output load power of 107.5mW. With V_{REG} load of 200Ω and varying V_{ER} conditions, PCE consistently exceeds 87.24%.

Fig. 5 (top-right) depicts the on-chip back/forward telemetry within the ER^4 chip, and the corresponding waveforms are displayed in Fig. 5 (bottom-right). During back telemetry where the bit error rate (BER) is less than 10^{-6} at 200kbps data rate and output power of 31mW, or during forward telemetry with BER of less than 10^{-4} at data rate of 167kbps, both V_{REG} and V_{ER} can still be reliably regulated at 4.5V and 3.3V, respectively. It is noteworthy that the ER^4 is the only dual-output R^3 system equipped with back/forward telemetry, as depicted in Fig. 6. Furthermore, the difference between $V_{OUT,MAX}$ and $V_{OUT,MIN}$ as well as the difference between V_{OUT1} and V_{OUT2} was higher than other works up to 3.5V. Fig. 7 shows the die micrograph, which was fabricated with 180nm process and occupies 0.77mm^2 active area.

Acknowledgment:

This work was supported by the Technology Innovation Program (RS-2022-00154983) funded by Ministry of Trade, Industry & Energy (MOTIE), Korea. Chip fabrication was supported by IC Design Education Center (IDEC), Korea.

References:

[1] H. S. Gougheri et al., "Adaptive Reconfigurable Voltage/Current-Mode Power Management with Self-Regulation for Extended Range Inductive Power Transmission," IEEE ISSCC, 2017.
[2] J. Lin et al., "A 6.78-MHz Single-Stage Regulating Rectifier with Hysteretic Control and Current-Wave Modulation," IEEE ASSCC, 2020.
[3] H.-M. Lee et al., "A Power-Efficient Wireless System with Adaptive Supply Control for Deep Brain Stimulation," IEEE JSSC, 2013.
[4] C. Kim et al., "A 144-MHz Fully Integrated Resonant Regulating Rectifier with Hybrid Pulse Modulation for mm-Sized Implants," IEEE JSSC, 2017.
[5] F. Marefat et al., "A 280μW 108dB DR Readout IC with Wireless Capacitive Powering Using a Dual-Output Regulating Rectifier for Implantable PPG Recording," IEEE ISSCC, 2020.
[6] Z. Luo et al., "A 90%-Efficiency 40.68MHz Single-Stage Dual-Output Regulating Rectifier with ZVS and Synchronous PFM Control for Wireless Powering," IEEE ISSCC, 2023.

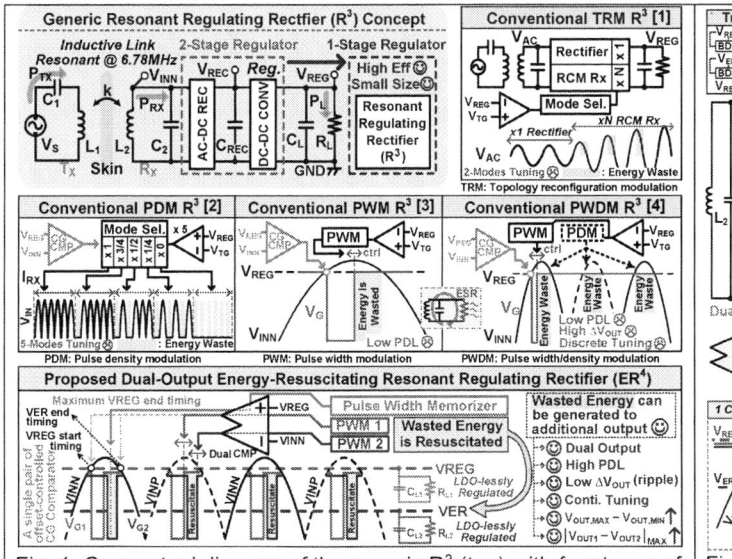

Fig. 1. Conceptual diagram of the generic R³ (top) with four types of conventional R³ (middle) and the proposed ER⁴ (bottom).

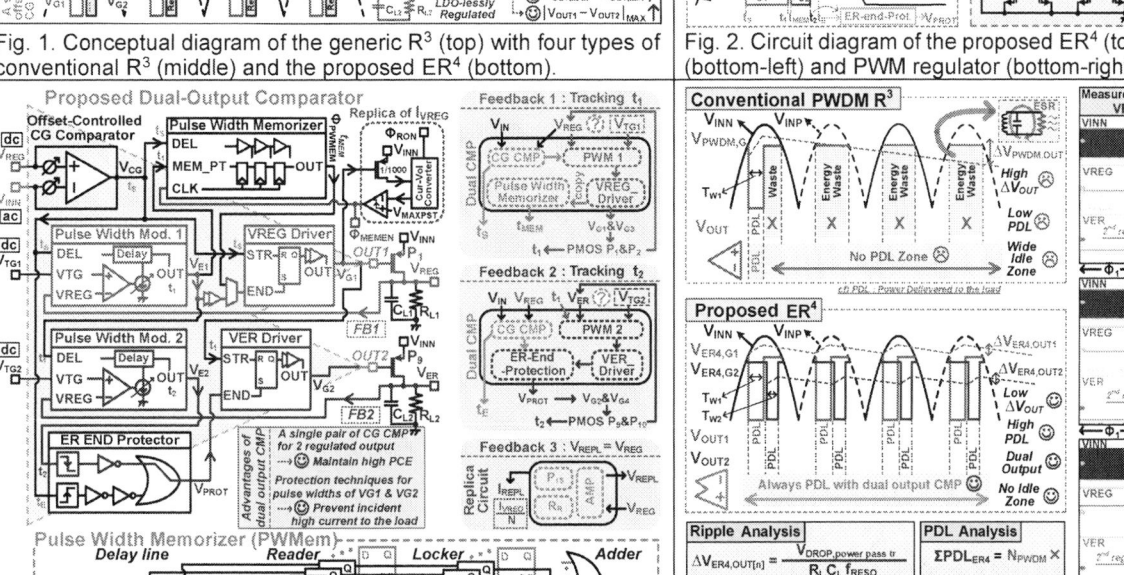

Fig. 2. Circuit diagram of the proposed ER⁴ (top) with timing diagram (bottom-left) and PWM regulator (bottom-right).

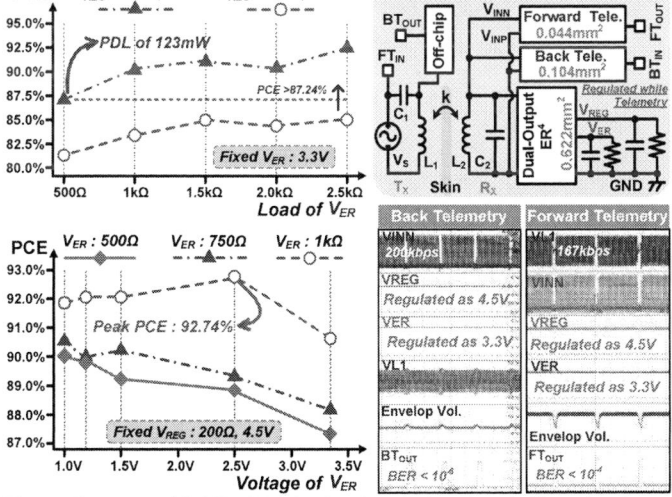

Fig. 3. Circuit diagrams of the dual output comparator (top-left) along with PWMem (bottom) and three feedback loops (top-right).

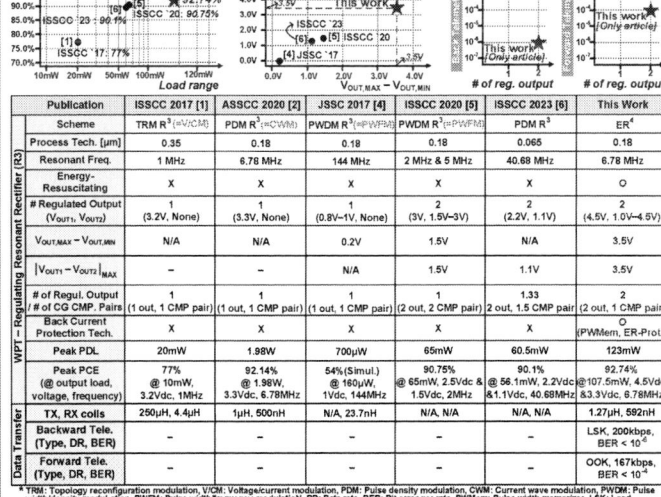

Fig. 4. Analysis and comparison of PDL and ripple (left) and measured waveforms of start-up mode operation (right).

Fig. 5. Measured PCE with V_ER load variation (top-left) and voltage variation (bottom-left); Back/forward telemetry conceptual diagram (top-right) and measured waveforms (bottom-right).

Fig. 6. PCE, V_OUT range, and BER plots for ER⁴ system (top) and performance comparison with state-of-the-art R³ works (bottom).

979-8-3503-3004-5/23 $31.00 © 2023 IEEE

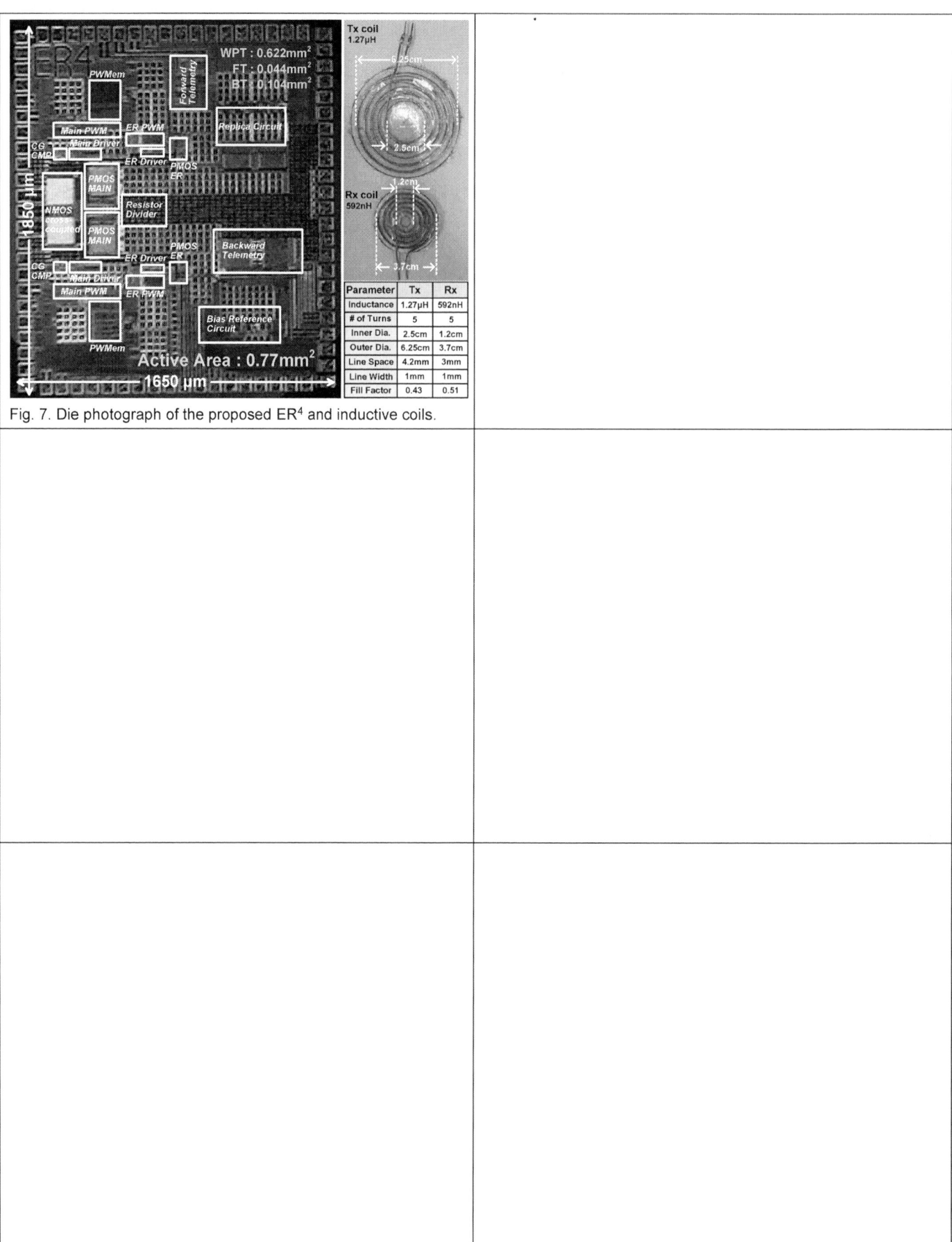

Fig. 7. Die photograph of the proposed ER⁴ and inductive coils.

2-20-V Wide Input Range Active Rectifier with Single-Multiplexing Adaptive Delay Compensation and Low-voltage Enhancement for Wireless Power Transmission

Chao Xie[1], Guangshu Zhao[2], Junliang Wei[3], Man-Kay Law[2], Milin Zhang[1]

[1]Tsinghua University, Beijing, China
[2]University of Macau, Macau, China
[3]Beijing Ningju Technology, Beijing, China

Wireless power transfer (WPT) has emerged as a vital technology for biomedical implants and wearable devices, which eliminates the need for physical connectors and the associated risks of infections, tissue damage, and device failure. A highly efficient power receiver is essential to suppress the device temperature rising during charging as well as to improve the charging rate [1]. However, the voltage of the receiving coil varies dramatically based on the power requirements, loadings, transmission distance, and axes misalignment [2]. Furthermore, the maximum power point tracking (MPPT) implementation for the power receiver system may double the system voltage tolerance if employing the fractional open-circuit voltage approach. Therefore, an active rectifier with a wide input range and high power conversion efficiency (PCE) is required to improve the system safety and robustness.

To protect chips from high voltage, previous works [3] have employed methods, such as controlling the power delivery through a wireless feedback loop or adjusting loadings, which unfortunately limited the output power. Although the input range of the active rectifier in [4] has been extended to 12V by utilizing high-voltage MOSFET, it sacrifices the PCE at low-voltage input due to a higher MOSFET threshold voltage. This problem becomes even worse when further extending the input range. However, a larger input range is crucial to increase power delivery and accommodate various loading conditions, particularly when considering MPPT implementation. In order to address this challenge, this paper proposes a 2-20-V wide input range active rectifier with a PCE over 90% across the entire input range by utilizing the proposed single-multiplexing adaptive delay compensation and low-voltage enhancement (LVE) techniques.

Fig. 1 illustrates the system block diagram of the proposed active rectifier, consisting of a core power path, an adaptive delay compensation circuit, a low-voltage enhancement circuit, and a power management unit. The core power path employs two high-side 20V PMOS switches and two low-side 20V NMOS switches. M_{n1}/M_{p2} and M_{n2}/M_{p1} are switched on in turn. In order to decrease the driving loss, the PMOS switches are driven directly by the rectifier input instead of extra drivers due to their higher threshold voltage and larger size than NMOS switches. The LVE technique is proposed to compensate for the gate-to-source voltage drop consumed by the voltage-limiting switches, improving the PCE at low input voltage. In addition, traditional active rectifiers without adaptive on/off control suffer from reduced PCE due to circuit delays in the gate voltage control path of the low-side NMOS switches. The delays lead to less conduction time and increased reverse current, especially for high-resonant-frequency WPT systems. Moreover, they vary according to the process, voltage, and temperature (PVT) conditions. In order to address this issue while decreasing the control loss, a single-multiplexing adaptive delay compensation technique is proposed to adjust the offset of the comparators automatically. Since the gate voltage control paths of M_{n1} and M_{n2} are symmetric, the on-transition offset and off-transition offset reference are generated by a single adaptive on/off loop and reused by two comparators.

Fig. 2 shows the proposed adaptive on/off loop and the circuit details of the unbalanced comparator. In the adaptive on/off loop, in order to decrease the power consumption, speed up the transition process, and accommodate low rectifier input voltage, the unbalanced comparator works at 1.8V, which contradicts the up to 20V wide input range of the rectifier. Therefore, the comparator input $V_{ac,lim}$ is clamped by a cascode voltage limiter below 1.8V. The comparator compares $V_{ac,lim}$ with the ground to toggle the power switch M_{n1} on or off. The sample and hold (S/H) circuits in the on-compensation and off-compensation paths sample $V_{ac,lim}$ at the rising and falling edges of V_{GN1}, respectively. The difference between the sampled voltage and the ground is amplified by the Gm-C integrator to adjust the offset of the unbalanced comparator until approximating to zero. The unbalanced comparator is based on a common-gate structure with switched offset. To eliminate the risk of multiple pulsing problems when M_{n1} is turned on prematurely and $V_{ac,1}$

goes lower than the ground again, a glitch offset is proposed to be generated by a diode-connected NMOS. The on/off offsets are also generated by the voltage-based method instead of the current-injection method to reduce power consumption. Nevertheless, the equivalent resistance of a gate-controlled NMOS exhibits uneven variations with gate voltage and becomes excessively large below 1V, sacrificing the effective output range of the Gm-C integrator. The loop stability is even worse due to the wide input range of the rectifier and the temperature rising during high-power delivery. In order to address these challenges, the offset voltage generator employs an NMOS in parallel with salicide-resistor pre-calibration to improve the adjustment linearity of the equivalent resistance. Fig. 2 also shows the switch timing sequences and offset variation of the comparator.

Fig. 3 shows the proposed low-voltage enhancement (LVE) circuit, its key low-power dynamic level shifter, and the self-start-up power management unit (PMU). The PMOS power switches are driven by the rectifier AC input with M_{p10} and M_{p11} clamping the voltage to 5V below $V_{OUT,AC-DC}$. However, the driving voltage of the PMOS power switches obviously decreases due to the gate-to-source voltage drop of M_{p10} and M_{p11}, leading to the decrease in voltage conversion ratio (VCR) and PCE, especially at low voltage scenarios. In order to compensate for this voltage drop, the gate voltages of M_{p1} and M_{p2} are pulled down to $V_{DD,N5V}$ by the LVE at the rising edges of V_{GN4} and V_{GN3}, respectively. A dynamic level shifter without static power dissipation is proposed to address the tradeoff between low power, fast speed, and high-voltage tolerance from 2 to 20V. The level shifter uses a pulse to trigger output flipping and a simple inverter-based latch to store the data. R_2 and R_3 pull up the PMOS gate voltage to minimize the static current to zero. R_4 and R_5 facilitate the transition at low voltage by increasing the current during the pulse. According to the simulation, the level shifter achieves an energy consumption of 5.38pJ/bit at 3V with a delay of 1.26ns. A self-start-up PMU is employed to power these circuits. In order to avoid the requirement of a 2-20-V bandgap reference, a self-start-up circuit for LDO is proposed, and the bandgap reference is powered by the LDO with an output of 5V.

The rectifier is fabricated in a 0.18 um CMOS process, occupying an area of 0.85 mm^2. Fig. 4 shows the simulated start-up waveforms of the rectifier, including AC input, DC output, the gate voltage (V_{GN1}) of the NMOS power switch, and the internal 1.8V voltage source. During cold start-up, energy is transferred to the output by the body diode of the power switches. After VDD$_{1V8}$ rises enough to ensure the unbalanced comparators and other digital circuits work properly, the rectifier begins active operation. The default configuration turns on the LVE and adaptive on/off delay compensation to maximize PCE at low voltage. When manually turning them off, $V_{OUT,AC-DC}$ decreases for lower PCE and output power. Fig. 4 also shows the measured transient AC input voltage of the rectifier with or without LVE and adaptive on/off loop. Voltage spikes occur without LVE and adaptive on/off loop due to reversed current caused by delayed switching time.

Fig. 5 shows the measured PCE and VCR of the active rectifier with different loads. Maximum PCE improvement of 18.14% is observed at 2V with a 500Ω load, while the VCR is increased by 16.15% with the same conditions. The peak PCE of 94.73% is achieved at 6V with a 500Ω load. The peak VCR of 98.29% is achieved at 14V with a 1KΩ load. The LVE should be turned off due to its reduced effectiveness when the amplitude of AC input voltage exceeds 5V. Fig. 6 summarizes the performance of the proposed active rectifier and compares it to the state-of-the-art designs. The chip achieves the largest input voltage range of 2-20V while maintaining the PCE over 90% and the VCR over 96% across the entire range with different loads.

Acknowledgements: This work is supported in part by the National Key R&D Program of China (2022YFB4400800), in part by the Natural Science Foundation of China through grant 92164202, in part by the Beijing Innovation Center for Future Chip, in part by the Beijing National Research Center for Information Science and Technology.

References:

[1] Y. Lu, *et al.*, "A 13.56 MHz CMOS Active Rectifier With Switched-Offset and Compensated Biasing for Biomedical Wireless Power Transfer Systems," *IEEE TBioCAS*, **8**(3), pp. 334-344, Jun. 2014.

[2] L. Cheng, *et al.*, "Adaptive On/Off Delay-Compensated Active Rectifiers for Wireless Power Transfer Systems," *IEEE JSSC*, **51**(3), pp. 712-723, Mar. 2016.

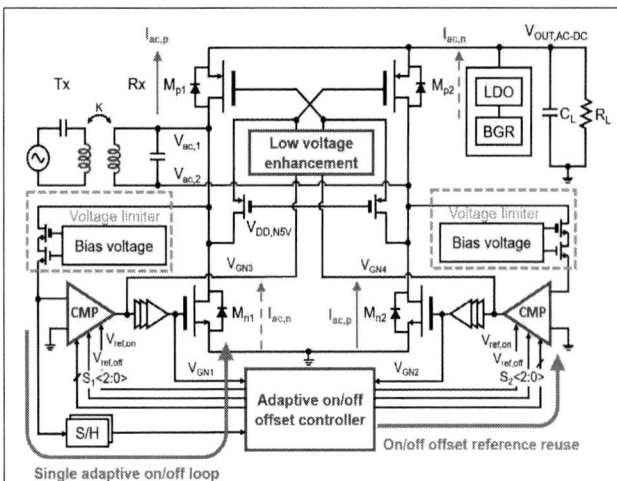

Fig. 1. System block diagram of the proposed rectifier with low-voltage enhancement (LVE) and single-multiplexing adaptive on/off.

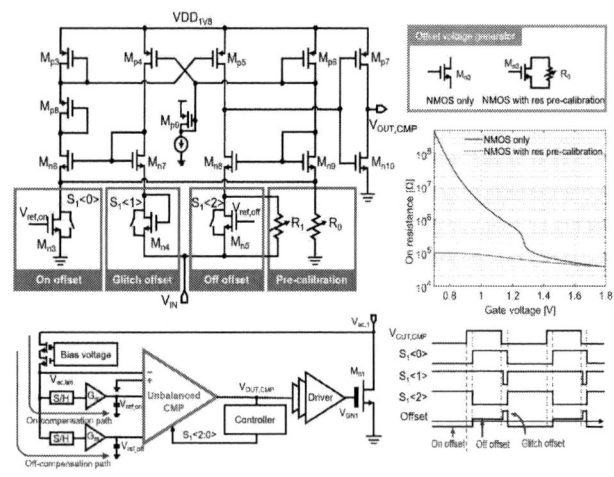

Fig. 2. Adaptive on/off loop, the unbalanced comparator implementation and offset adjustment linearity improvement.

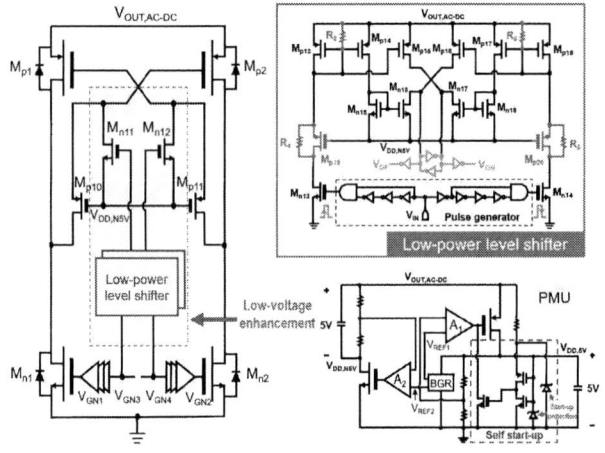

Fig. 3. Proposed low-voltage enhancement (LVE) technique, the low-power level shifter and the self-start-up PMU.

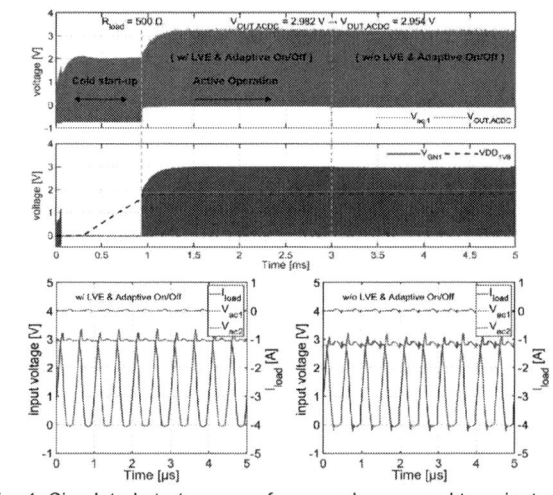

Fig. 4. Simulated start-up waveforms and measured transient waveforms of the rectifier AC inputs.

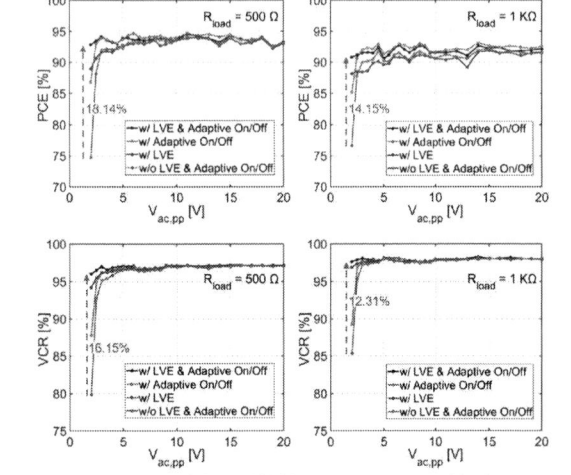

Fig. 5. Measured PCE and VCR improvement with the proposed LVE and adaptive on/off techniques with 500 Ω and 1 KΩ loads.

	TVLSI [5]	JSSC [6]	JSSC [7]	TCASI [8]	TCASI [4]	**This work**
Year	2018	2019	2020	2020	2021	**2023**
Technology	0.35 μm	0.18 μm	0.18 μm	0.18 μm	0.18 μm	**0.18 μm**
Area (mm²)	0.38	0.117	0.853	0.208	0.611	**0.85**
Frequency (MHz)	13.56	13.56	1-10	40.68	6.78	**1**
Offset compensation	Adaptive On/Off	Adaptive On/Off	Adaptive On/Off	Adaptive On/Off	Adaptive On/Off	**Adaptive On/Off**
Compensation scheme	Voltage	Delay	Current	Voltage + Time	Voltage + Time	**Voltage + pre-calibration**
Multiple pulsing problem	Risk	No	Risk	No	No	**No**
Input range (V)	2.9-5.4	1.0-2.5	1.8-5.0	2.5-4	3-12	**2-20**
$P_{out,max}$ (mW)	126.7	34.1	231.6	56.6	288	**766**
VCR (%)	90.1-92.7 (R_L=500Ω)	96.2-97.5 (R_L=2KΩ) 92.8-94.9 (R_L=510Ω)	88.6-95.1 (R_L=300Ω) 83.4-93.6 (R_L=100Ω)	73.2-84.1 (R_L=500Ω)	94.9-96.8 (R_L=2KΩ) 91.3-94.0 (R_L=1KΩ)	**97.57-98.29 (R_L=1KΩ) 96.01-97.21 (R_L=500Ω)**
PCE (%)	84.6-86.1 (R_L=500Ω) 85.8-89.0 (R_L=200Ω)	82.6-92.9 (R_L=1KΩ) 85.0-94.1 (R_L=510Ω)	84.4-91.5 (R_L=300Ω) 85.2-91.8 (R_L=100Ω)	70.7-80.9 (R_L=500Ω)	82.0-93.6 (R_L=1KΩ) 86.0-93.7 (R_L=500Ω)	**90.53-93.13 (R_L=1KΩ) 92.63-94.73 (R_L=500Ω)**

Fig. 6. Measured performance and comparison table with prior art.

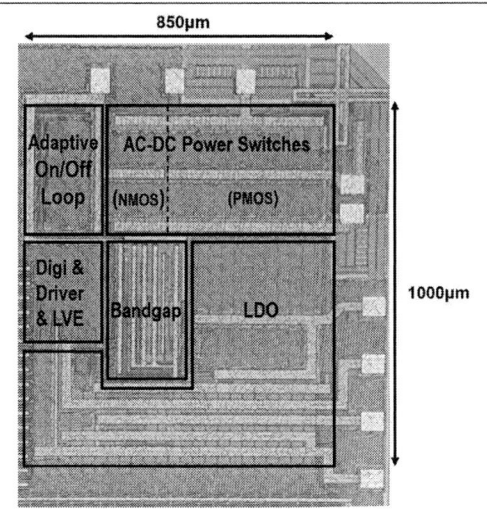

Fig. 7. Chip micrograph.

Additional References:

[3] X. Li, *et al.*, "A 13.56 MHz Wireless Power Transfer System With Reconfigurable Resonant Regulating Rectifier and Wireless Power Control for Implantable Medical Devices," *IEEE JSSC*, **50**(4), pp. 978-989, Apr. 2015.

[4] G. Namgoong, *et al.*, "3–12-V Wide Input Range Adaptive Delay Compensated Active Rectifier for 6.78-MHz Loosely Coupled Wireless Power Transfer System," *IEEE TCASI*, **68**(6), pp. 2702-2713, Jun. 2021.

[5] K. Noh, *et al.*, "A 13.56-MHz CMOS Active Rectifier With a Voltage Mode Switched-Offset Comparator for Implantable Medical Devices," *IEEE TVLSI*, **26**(10), pp. 2050-2060, Oct. 2018.

[6] Z. Xue, *et al.*, "A 13.56 MHz, 94.1% Peak Efficiency CMOS Active Rectifier With Adaptive Delay Time Control for Wireless Power Transmission Systems," *IEEE JSSC*, **54**(6), pp. 1744-1754, Jun. 2019.

[7] R. Erfani, *et al.*, "A 1–10-MHz Frequency-Aware CMOS Active Rectifier With Dual-Loop Adaptive Delay Compensation and >230-mW Output Power for Capacitively Powered Biomedical Implants," *IEEE JSSC*, **55**(3), pp. 756-766, Mar. 2020.

[8] L. Cheng, *et al.*, "A 40.68-MHz Active Rectifier With Hybrid Adaptive On/Off Delay-Compensation Scheme for Biomedical Implantable Devices," *IEEE TCASI*, **67**(2), pp. 516-525, Feb. 2020.

A 1.08ms Ultrafast Scanning Capacitive Touch-Screen Sensor Interface with Charge-Interpolated Common-Mode Compensation and Host-Based Adaptive Median Filtering

Jonghang Choi[1], Subin Kim[2], Yongjun Lee[1], Sanghyun Heo[3], Keum-Dong Jung[3], Young-Ha Hwang[4], Jun-Eun Park[1]

[1]Sungkyunkwan University, Korea, [2]LX Semicon, Korea, [3]Samsung Display, Korea, [4]Soongsil University, Korea

Recent advancements in display technologies and the increasing demand for enhanced user experiences in human-machine interfaces pose challenges for touch-screen sensors. While display-embedded touch-screen panels (TSPs) offer high transparency and a slim form factor, they suffer from increased parasitic capacitance, resulting in a poor signal-to-noise ratio (SNR) [1-2]. To meet the requirements of high scan rates in displays up to 240Hz, a system frame rate of touch-screen sensors needs to exceed 500Hz. Consequently, a mutual sensing time (MST) required to scan the entire mutual capacitance is significantly reduced with the increased frame rate. Furthermore, there is a need to support active stylus sensing without the need for additional sensor layers. Active stylus sensing involves periodic beacon communication through the TSP, further reducing the MST, as illustrated in Fig. 1. As a result, the available MST decreases to just milliseconds, leading to a significant degradation of the SNR [1-4]. Prior works employed analog noise filtering schemes to enhance the SNR. However, these methods are unable to provide sufficient noise filtering within the short MST condition. While an FFT-based sensing scheme was proposed in [5], it is challenging to implement with heavy-load narrow bandwidth TSP.

This work presents an ultrafast scanning analog front-end (AFE) and a host-based touch processing architecture, as depicted in Fig. 1. The proposed AFE achieves rapid sensing of the capacitance variation within an MST of 1ms. Instead of relying on the analog-intensive noise filtering schemes, noise reduction is carried out in the host processor using a soft-defined adaptive median filter (AMF). The host touch processing has gained attention recently due to its ability to leverage the powerful computational capabilities of the host processor, thereby reducing computational burden on the touch controller. By combining the ultrafast scanning AFE with host-processed filtering, the proposed system can operate within a very short MST, accommodating both the active stylus detection and a system frame rate higher than 500Hz, as shown in Fig. 1.

Figure 2 shows the overall block diagram of the ultrafast touch sensing system. The proposed AFE supports heavy-load TSPs with up to 33 transmitter (TX) channels and 16 receiver (RX) channels. To achieve a reduced MST, high-frequency excitation is employed in the TX, enabling an excitation signal driving with a frequency higher than 500kHz into the TSP channels, which typically have capacitances in the hundreds-of-pF range. However, high-frequency excitation in narrow-band TSPs can introduce phase errors during RX demodulation. To mitigate this issue, the AFE incorporates a sequential driving method and an I-Q demodulation structure. The initial stage of each RX sensing channel consists of a charge-interpolated common-mode compensator (CI-CMC) that not only cancels out common mode noise but also addresses channel-dependent display-coupled noise [6]. By utilizing the noise interpolation process, the CI-CMC offers superior noise rejection compared to conventional differential sensing [1-2]. A charge amplifier with an inverter-based pre-amplifier and a single-to-differential amplifier (SDA) enhance the touch sensitivity of the capacitance signal. Following I-Q demodulation, the demodulated signal is integrated and sampled by the following integrator. The I/Q capacitance signals are interleaved by the 16:2 multiplexer and converted to digital signals by two 12-bit successive-approximation-register (SAR) ADCs. The serializer transmits 6-bit MSB and 6-bit LSB data to the host processor in sequence. For an MST of 1ms, the required data rate of the host processing is approximately 6.2Mbps. The number of data lanes to the host processor can be further reduced by utilizing standard protocols such as I3C or SPI. The host processor handles the noise filtering and the touch coordinate

extraction based on the capacitance data received from the AFE. In this work, the host-based noise filtering was demonstrated using an ARM Cortex-A53 processor embedded in ZCU104. The proposed AMF was implemented in the host processor and the processing time for the AMF was measured to be 0.97ms in the worst-case scenario.

Figure 3 shows the detailed schematic of the CI-CMC comprising charge pumps and an auto-zeroed bootstrapping amplifier. The column-parallel CI-CMCs operates in two modes: 1) common-mode interpolation and 2) mutual capacitance C_M sensing modes. The even and odd sensing channels alternate between these two modes. For instance, during common-mode interpolation of odd channels, each bootstrapping amplifier forces the charge pump to ensure that its sensing channel is equal to V_{REF}. The required charge signals to reach V_{REF} are then interpolated between two adjacent odd channels, and these interpolated charge signals are subtracted from the even sensing channels. Consequently, the CI-CMC offers improved noise rejection against the channel-dependent display noise [6] and common-mode mismatches. The charge signals from the even sensing channels are subsequently converted to voltage signals using capacitively-coupled charge amplifiers. The inverter-based preamplifier enhances the dynamic range of weak touch signals. In the measurement results, the inverter-based preamplifier demonstrated an SNR improvement by 3.9dB.

Figure 4 presents the conceptual diagram of the host-processed AMF. Unlike conventional median or mean filters, which exhibit delayed detection of capacitance variation caused by touch input, the proposed AMF ensures immediate touch detection. When the filter detects a capacitance variation exceeding a threshold, the capacitance signal is filled in the filter window with one less than half of its size. In the subsequent frame, the AMF generates an updated baseline output if the input capacitance signal falls within the threshold range from the previous input signal. By utilizing the AMF, the measured RMS and peak-to-peak noises can be reduced by 78.2% and 80%, respectively. Fig. 4 shows the measured capacitance profile. Compared to a raw data and conventional filters, the AMF achieves significantly improved noise filtering performance and shorter touch detection delay.

The proposed touch sensing system prototype was fabricated in a 130nm CMOS process, operating within a supply range of 2.7 to 3.6V. The total power consumption was 39.6mW. The measurement results of the prototype with an 8-inch TSP are presented in Fig. 5. When the MST was set to 1.08ms, equivalent to an AFE frame rate of 930Hz, the measured SNRs were 35.1dB and 35.8dB for 8Φ metal pillar and multi-finger inputs, respectively. Figure 5 also shows the relationship between the number of AMF windows and the measured SNR. Moreover, even with a high frame rate of 930Hz, the prototype achieved enhanced immunity against touch-injection noise. When subjected to a 5V_{PP} noise injection at 100kHz [1,4,7], the SNR degradation was only 0.5dB with the assistance of the AMF.

Figure 6 shows the performance comparison with prior works. The prototype achieved ultrafast sensing performance with an SNR of 35.8dB and an MST of 1.08ms, equivalent to AFE frame rate of 930Hz. To compare SNRs in the short MST condition of 1ms, the estimated SNRs were evaluated using figure-of-merit (FoM) commonly employed in touch sensor interfaces [1-3]. Among the compared works, the prototype achieved the highest estimated SNR of 35.1dB. Fig. 7 shows the die microphotograph of the AFE. The total active area of the AFE is 2.84mm^2.

Acknowledgement

This work was partly supported by Samsung Display and Korea Evaluation Institute of Industrial Technology(KEIT) grant funded by the Korea government(MOTIE) (RS-2022-00144290). The EDA tool was supported by the IC Design Education Center(IDEC), Korea.

References:

[1] J.-E. Park *et al.*, "A Noise-Immunity-Enhanced Analog Front-End for 36×64 Touch-Screen Controllers With 20-V$_{PP}$ Noise Tolerance at 100 kHz," IEEE Journal of Solid-State Circuits 2019: 1497-1510.

[2] J. Park *et al.*, "A Mutual Capacitance Touch Readout IC With 64% Reduced-Power Adiabatic Driving Over Heavily Coupled Touch Screen," IEEE Journal of Solid-State Circuits 2019: 1694-1704.

979-8-3503-3004-5/23 $31.00 © 2023 IEEE

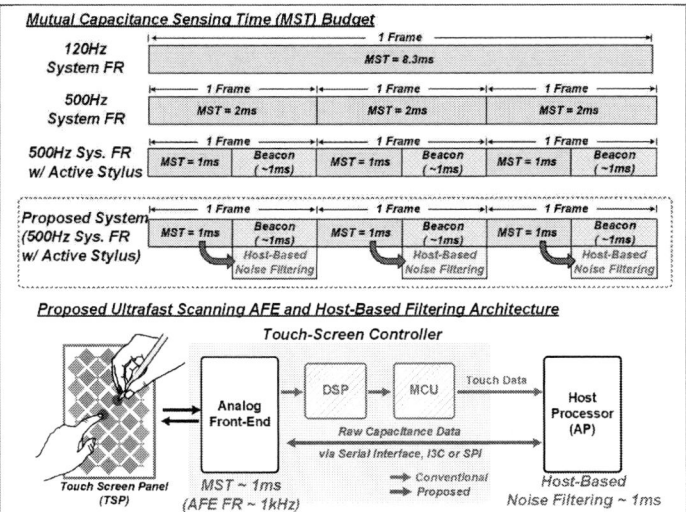

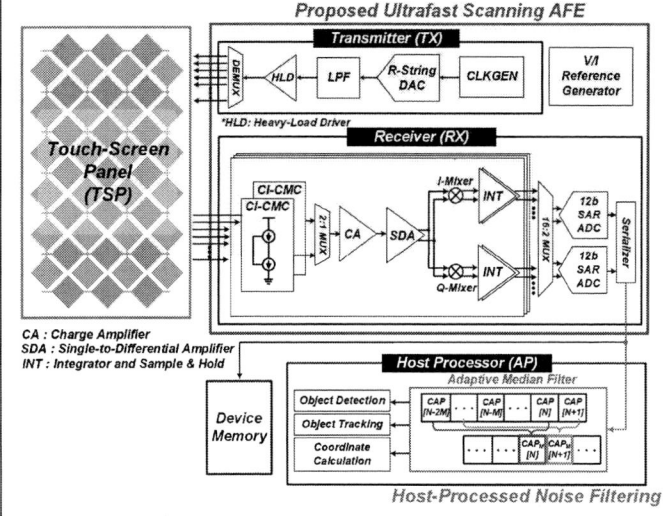

Fig. 1. Conceptual diagrams of mutual capacitance sensing time budget (upper) and proposed touch sensing system (lower).

Fig. 2. Block diagram of proposed touch sensing system.

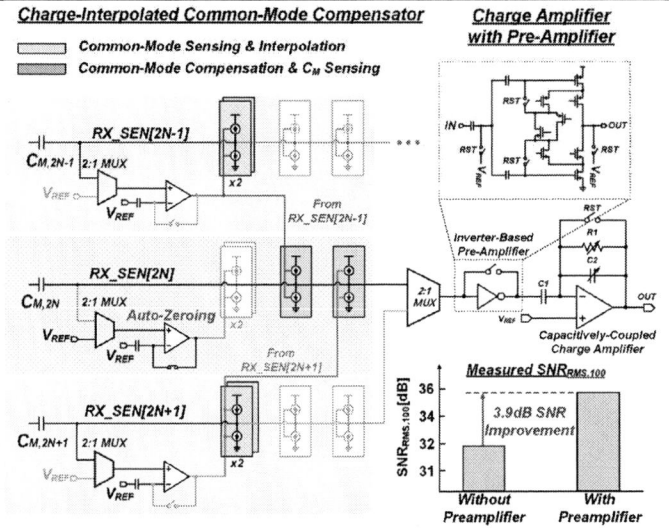

Fig. 3. Circuit implementation of CI-CMC and charge amplifier.

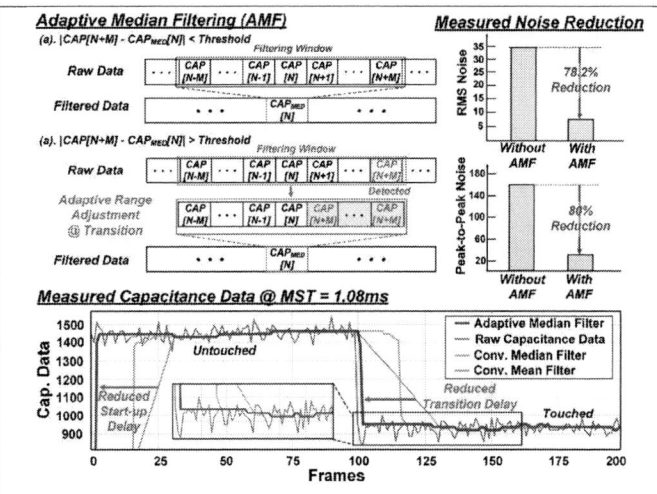

Fig. 4. Conceptual diagram of AMF (upper left) and measured RMS and peak-to-peak noise reduction after the noise filtering (upper right). Measured capacitance data profile depending on touch input and filters types.

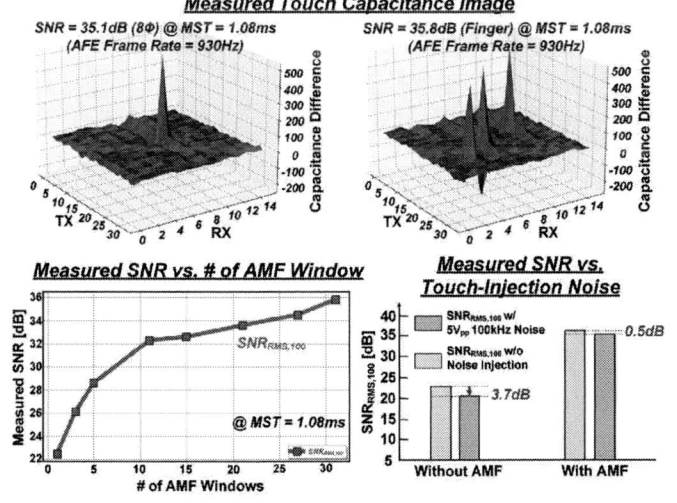

Fig. 5. Measured touch images (upper). Measured SNR versus the number of AMF windows (lower left). Measured SNR comparison depending on AMF and noise injection (lower right).

	JSSC 2019 [1]	JSSC 2019 [2]	JSSC 2018 [3]	ESSCIRC 2019 [4]	JSSC 2021 [7]	**This Work**
Process [nm]	180	180	180	80	180	**130**
TSP	12.2-inch TX: 36 RX: 64	5.8-inch TX: 16 RX: 33	10.1-inch TX: 28 RX: 50	5.8-inch TX: 16 RX: 33	6.7-inch TX: 8 RX: 17	**8-inch TX: 33 RX: 16**
Supply Voltage [V]	2.7 ~ 3.3	1.8/3.3	1.8/3.3	1.2/8	1.8/3.3	**2.7 ~ 3.6**
Power Consumption [mW]	67.7	17.8	6.9	13.46	7.6	**39.6**
Analog Front-End Frame Rate (FR) [Hz]	85 ~ 385	120	120	120	120	**930**
Min. Mutual Sensing Time (MST) [ms]	2.6	3.07	4.16	8.3	8.3	**1.08**
SNR [dB] @ FR	[1]36 @ 385Hz 54 @ 120Hz	[1]40.6 @ 326Hz 57 @ 120Hz	[1]33.7 @ 240Hz 53.3 @ 120Hz	40 @ 120Hz	37.5 @ 120Hz	**35.8 @ 930Hz**
[2]FoM [nJ/Step] @ FR	2.07 @ 385Hz	1.18 @ 326Hz	0.52 @ 240Hz	2.60 @ 120Hz	7.6 @ 120Hz	**1.60 @ 930Hz**
[3]Estimated SNR [dB] @ MST=1ms	27.7	30.9	21.3	21.6	19.1	**35.1**
Active Area [mm²]	36	7.1	1.96	4.87	0.36	**2.84**

[1]Based on the measurement results. [2]FoM [nJ/Step] = $\frac{Power}{2^{\frac{SNR-1.76}{6.02}} \times \# \ of \ TSP \ nodes \times FR}$

[3]Estimated SNR (@ MST = 1ms) = $6.02 \times log_2 \left[\frac{Power}{\frac{2}{FoM} \times \# \ of \ TSP \ nodes \ / \ 1ms} \right] + 1.76$ [dB]

Fig. 6. Performance summary and comparison with prior works.

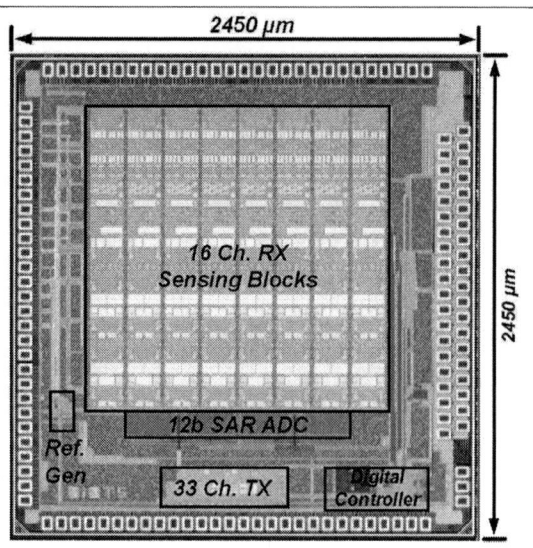

Fig. 7. Die microphotograph.

Additional References:

[3] H. Hwang *et al.*, "A 1.8-V 6.9-mW 120-fps 50-Channel Capacitive Touch Readout With Current Conveyor AFE and Current-Driven $\Delta\Sigma$ ADC," IEEE Journal of Solid-State Circuits, 2018: 204-218.

[4] Y. -H. Hwang *et al.*, "An Always-On 0.53-to-13.4 mW Power-Scalable Touchscreen Controller for Ultrathin Touchscreen Displays With Current-Mode Filter and Incremental Hybrid $\Delta\Sigma$ ADC," IEEE ESSCIRC 2019: 313-316.

[5] J.-S. An *et al.*, "A 3.9kHz-frame-rate capacitive touch system with pressure/tilt angle expressions of active stylus using multiple-frequency driving method for 65" 104x64 touch screen panel," IEEE ISSCC 2017: 168-169.

[6] J. Lee *et al.*, " "Robust Touch Screen Readout System to Display Noise Using Multireference Differential Sensing Scheme for Flexible AMOLED Display," MDPI Micromachines 2022.

[7] T. -G. Song *et al.*, "A 50.7-dB-DR Finger-Resistance Extracting Multi-Touch Sensor IC for Soft Classification of Fingers Contacted on 6.7-in Capacitive Touch Screen Panel," IEEE JSSC 2021: 3470-3485.

A Primary Driver with Real-Time Resonance Tracking for Wireless-Powered Implantable Medical Devices

Xiaodong Meng[1], Xing Li*[2], Yuan Yao[1], Chi-Ying Tsui[1], Wing-Hung Ki[1]

[1]Hong Kong University of Science and Technology, Hong Kong, China;
[2]Nanjing University of Aeronautics and Astronautics, Nanjing, China.

Inductively-coupled wireless powered transfer (WPT) is widely used in implantable medical devices (IMD) [1] to avoid batteries. Flexible coils [2] are desired in this application as they can be attached onto or be implanted into human body. However, flexible coil bending or incoming magnetic shielding material can cause mismatch between LC tank's resonant frequency (f_{LC}) and primary driver's switching frequency (f_{SW}). As the inductive link is a narrow-band system, the link gain and efficiency is very sensitive to the matching quality. To tune f_{LC} back to f_{SW}, auto-tuning techniques [3-5] were proposed. [3] uses a capacitor matrix scheme that need several capacitors and relays, and they are not integrated. [4-5] proposed the fractional-capacitance (FC) scheme, which only need two capacitors and two LDNMOS. By changing the on-duty of the capacitors, the effective primary capacitance can be tuned. But its system complexity is high and it operates at 2MHz. Besides, zero-voltage switching (ZVS) is not achieved at transient periods, which degrades the efficiency as the primary inductor is perturbed frequently in the IMD scenario.

Inspired by prior arts, we propose a 13.56MHz FC scheme that is controlled by a concise analog auto-tuning loop (ATL). It only needs one extra capacitor and one LDNMOS. As shown in Fig. 1 left, a H-bridge (S_{1-4}) class-D PA is used to drive the primary LC tank at phases ϕ_{1-2}. The primary LC tank consists of L_1, C_1, C_{FC} and S_{FC}. S_{FC} is implemented with high voltage LDNMOS and is controlled by the proposed ATL at phase ϕ_{FC}. Bootstrap technique is used to generate V_{BOOT1}-V_{SW1} and V_{BOOT2}-V_{SW2} domains, which drive S_{3-4} and S_{FC}. The conceptual waveforms are shown in Fig. 1 right. ϕ_{FC} is in quadrature with ϕ_{1-2}. When ϕ_{FC} = 0, S_{FC}'s gate voltage VG_{FC} = V_{SW2}. As S_{FC} is cut off, V_{FC}-V_{SW2} will follow V_{LC}. When ϕ_{FC} = 1, S_{FC}'s gate voltage VG_{FC} = V_{BOOT2}. As S_{FC} is ON, V_{FC}-V_{SW2} is 0V during this period. As the primary capacitance is C_1 when ϕ_{FC} = 0 and it is C_1+C_{FC} when ϕ_{FC} = 1, the positive trajectory of V_{LC} will have larger slope while the negative trajectory of it will have smaller slope. S_{FC} only switches when V_{FC}-V_{SW2} = 0V.

As shown in Fig. 2. The input signal CLK_{IN} is connected to the non-overlapping circuit to generate ϕ_{1-2}. Besides, CLK_{IN} is also wired to the ramp circuit to generate a synchronized triangular signal V_{Ramp}. The proposed ATL consists of three parts: target phase generator, current phase detector and phase comparator. The first generates a target phase ϕ_{FCt}, at which the system will be in resonance. The middle one detects the current phase ϕ_{FC} of the LC tank. The last compares ϕ_{FCt} and ϕ_{FC}, and minimize the difference between them.

In the target phase generator, V_{Ramp} is compared to a DC feedback signal V_{FB} with comparator CMP1 to output V_{cmp1}. The rising edge of V_{cmp} triggers a 5ns pulse FC_{End} while the falling edge of it triggers another 5ns pulse FCt_{Start}. FC_{End} will be applied to the reset pins of D-flip-flops DN1-2, which decide the falling edges of ϕ_{FCt} and ϕ_{FC}. FCt_{Start} will be applied to the set pin of DN1, which decides the rising edge of ϕ_{FCt}. ϕ_{FCt} is the targeted S_{FC} driving signal, which is symmetrical with the switching edges of ϕ_{1-2}. ϕ_{FC} is the real driving signal of S_{FC} and may be changed during auto-tuning. In the current phase detector, comparator CMP2 is used to detect the zero-crossing (ZC) point of S_{FC}. As V_{FC} is a high voltage node, it cannot be directly applied to CMP2's input. In result, a snubber (R_{FC}, D_{FC1-2}) is used to generate V_{FC_SNS}, which has the same polarity as V_{FC} but has less than 5V amplitude. CMP2 compares V_{SW2} and V_{FC_SNS} to generate V_{cmp2}. The rising edge of V_{cmp2} triggers a 5ns pulse FC_{Start}, which is applied to the set node of DN2 and decides the rising edge of ϕ_{FC}. In the phase comparator, ϕ_{FCt} and ϕ_{FC} are compared to generate ϕ_{Early} and ϕ_{Late}. If the on-time of ϕ_{FCt} is larger than that of ϕ_{FC}, ϕ_{Late} will be 1, which will enable a pull-down current to decrease V_{FB}. If the on-time of ϕ_{FCt} is smaller than that of ϕ_{FC}, ϕ_{Early} will be 1, which will enable a pull-up current to increase V_{FB}.

The target of the ATL is to move I_{LC}'s ZC point to V_{SW1}'s switching edge. The operation waveform is shown in Fig. 3. If $f_{SW} < f_{LC}$, the LC tank is in phase early condition and we should increase the primary capacitance. At time t_1, V_{Ramp} intersects with V_{FB} to trigger FC_{End},

*Corresponding author: Xing Li, email: xingli@nuaa.edu.cn

which resets both ϕ_{FC} and ϕ_{FCt} to 0. S_{FC} is cut off and V_{FC}-V_{SW2} starts its positive sinusoidal trajectory. When V_{FC}-V_{SW2} drops to 0V (t_2 moment), FC_{Start} is triggered, which sets ϕ_{FC} to 1 and turns on S_{FC}. A while later, V_{Ramp} drops to V_{FB} (t_3 moment), which triggers FCt_{Start} and sets ϕ_{FCt} to 1. The extension of ϕ_{FC}'s on-time over that of ϕ_{FCt} leads to a pulse of ϕ_{Early}, which enables a pull-up current to increase V_{FB}. A increased V_{FB} will set ϕ_{FC} to 1 earlier next cycle and leads to a larger S_{FC} on-time, meaning larger effective capacitance. If $f_{SW} > f_{LC}$, the LC tank is in phase late condition and we should decrease the primary capacitance. The behavior of ATL will be similar but with a opposite direction. Only when $f_{SW} = f_{LC}$, the LC tank is in resonant condition. Now I_{LC}'s ZC point coincides with V_{SW1}'s switching edge. Both ϕ_{Early} and ϕ_{Late} are 0 and the ATL will be "locked" in this condition. Unlike [4-5], the proposed ATL can maintain the ZVS of S_{FC} during both steady and transient periods.

At a switching frequency of 13.56MHz, the comparator delay is not negligible. As shown in Fig. 4 (left top), due to the rising edge delay t_r and falling edge delay t_f of CMP1, ϕ_{FCt} cannot be symmetric with ϕ_1's switching edges. To compensate t_r and t_f, we propose the adaptive offset controller. As shown in Fig. 4 (right), the rising edge of ϕ_{FCt} is used to trigger a 5ns pulse, which samples V_{Ramp} and store it as V_{Sp1}. The falling edge of ϕ_{FCt} is used to trigger a 5ns pulse, which samples V_{Ramp} and store it as V_{Sp2}. Due to t_r, V_{Sp1} is larger than V_{FB}, which decreases V_{OS+} and increases I_{OS+} to generate a positive offset to CMP1. Due to t_f, V_{Sp2} will be smaller than V_{FB}, which will decrease V_{OS-} and increase I_{OS-} to generate a negative offset. I_{OS+} and I_{OS-} will be applied to CMP1 alternatively when ϕ_1 = 1 and ϕ_1 = 0. At the steady state, V_{Sp1} and V_{Sp2} will be moved to close to V_{FB}, which indicates that t_r and t_f are compensated.

The testbench was set up according to Fig. 1. A flexible coil (720nH) was used as L_1 and a FR-4 coil (398nH) was used as L_2. The spacing between them was 10mm. C_1 was 79.2pF and C_{FC} was chosen to be twice as large (161pF). C_2 was 346pF and R_{Load} was 1kΩ. The measured key waveform at steady state is shown in Fig. 5 (top). I_{LC}'s ZC points coincide with V_{SW1}'s switching edge, indicating the resonance of the LC tank. Next we bent L_1 to observe the real-time behavior of the ATL (Fig. 5 bottom). Before L_1 bending, V_{FB} was 656mV, which gave a 31.2ns ϕ_{FC} on-time. After L_1 bending, V_{FB} moved up to 779mV and ϕ_{FC}'s on-time increased to 43.9ns. It means the effective capacitor was increased to compensate the smaller L_1. The amplitudes of V_{Load} before and after bending were 21.2V and 20.8V, meaning the resonance of primary LC tank was maintained.

As shown in Fig. 6 top, we measured the tuning range by varying L_1 from 500nH to 1.2µH. A primary driver without ATL was also tested as a reference set. It achieved a link gain of 22 when L_1=700nH. However, when L_1=1.2µH, the link gain of the reference set was dropped to 5 while that of the ATL set was slightly below 20. In sum, the proposed ATL can maintain link gain over a 85.7% L_1 variance range, which is equivalent to a frequency tuning range of 31.5%.

The comparison with state-of-the-arts is shown in Fig.6 bottom. This work was fabricated with 180nm BCD process and had an active area of 0.66mm^2 (Fig.7). It can operate at 13.56MHz and is suitable for IMD application. The measured tuning error between V_{SW1}'s falling edge and I_{LC}'s ZC point is 1.89ns. By selecting C_{FC} = $2C_1$, the theoretical tuning range is 1.73, which can be further extended by increasing the C_{FC}:C_1 ratio. Due to the large loop bandwidth of analog ATL, this work can settle within 4 - 6 cycles.

References:

[1] X. Li et al., "Wireless Power Transfer System Using Primary Equalizer for Coupling- and Load-Range Extension in Bio-Implant Applications," ISSCC, Feb. 2015.

[2] Y. Yao et al., "Polyimide-based flexible 3-coil inductive link design and optimization," APCCAS, Nov. 2018.

[3] Y. Lim et al., "An Adaptive Impedance-Matching Network Based on a Novel Capacitor Matrix for Wireless Power Transfer," TPE, Aug. 2014.

[4] H. Kennedy et al., "A Self-Tuning Resonant Inductive Link Transmit Driver Using Quadrature-Symmetric Phase-Switched Fractional Capacitance," ISSCC, Feb. 2017.

[5] H. Kennedy et al., "A High-Q Resonant Inductive Link Transmit Modulator/Driver for Enhanced Power and FSK/PSK Data Transfer Using Adaptive-Predictive Phase-Continuous Switching Fractional-Capacitance Tuning," ISSCC, Feb. 2019.

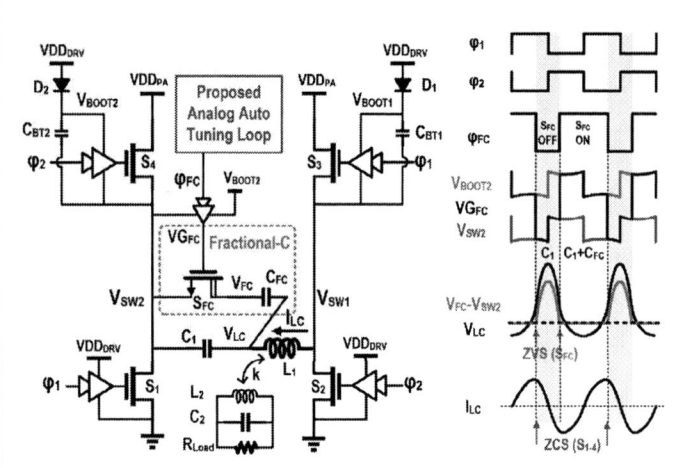

Fig. 1. Proposed primary driver with fractional capacitance scheme and conceptual waveform.

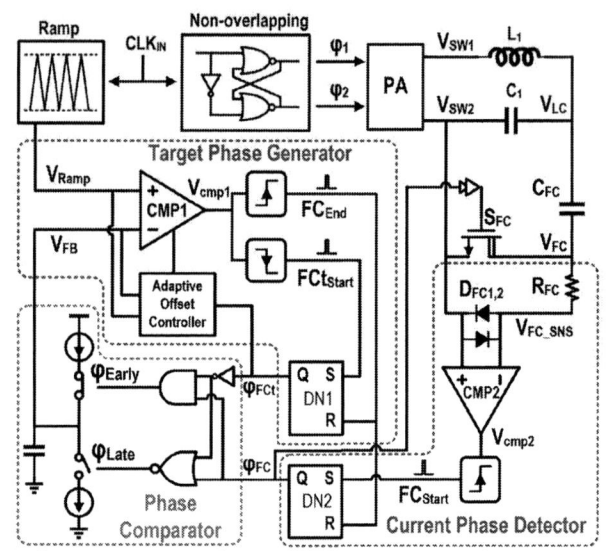

Fig. 2. System diagram of the proposed auto-tuning loop.

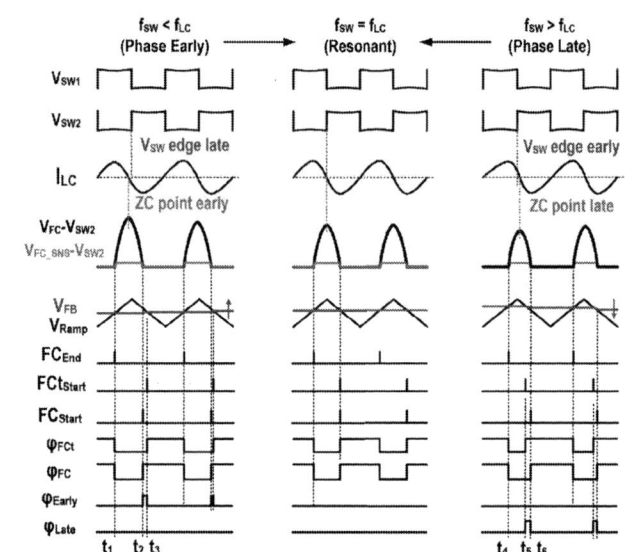

Fig. 3. Operation waveform of the proposed auto-tuning loop.

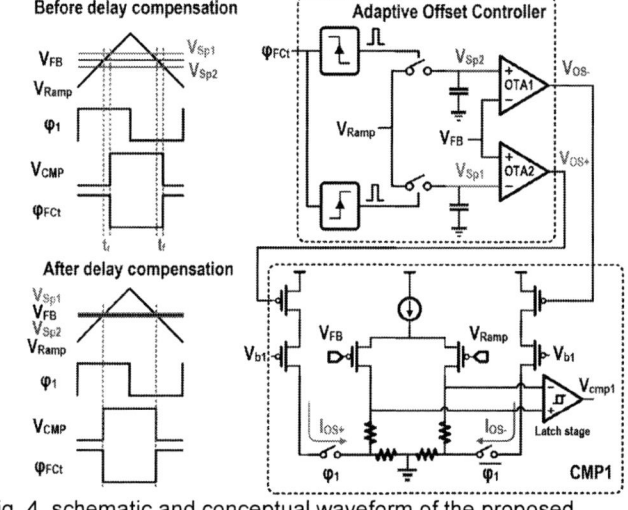

Fig. 4. schematic and conceptual waveform of the proposed adaptive offset controller for comparator delay compensation.

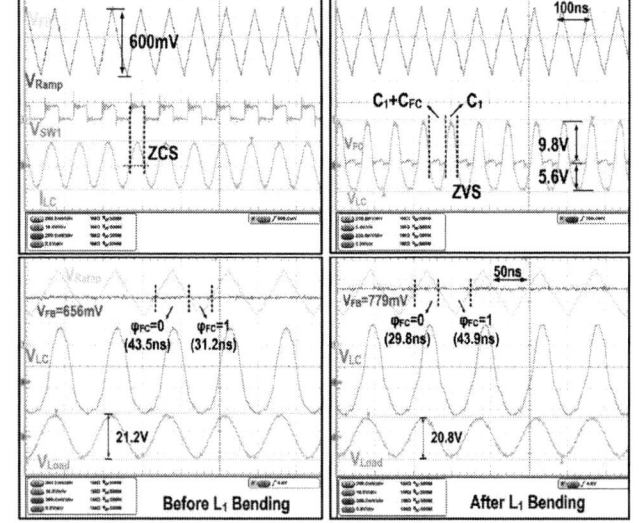

Fig. 5. Measured V_{Ramp}, V_{FB}, V_{SW1} and I_{LC} at steady state (top); measured V_{Ramp}, V_{FB}, V_{LC}, V_{Load} before and after coil bending.

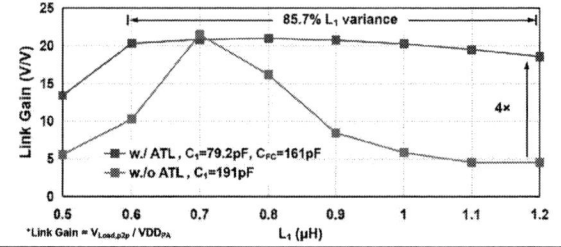

	ISSCC2017[4]	ISSCC2019[5]	This work
Active area	1.606 mm²	¹3.84 mm²	0.66 mm²
Max. f$_{SW}$	2 MHz	2 MHz	13.56 MHz
Tuning error	²4.83 ns	n/a	1.89 ns
Theoretical tuning range	1:1.41	1:1.41	1:1.73 (C$_{FC}$ = 2C$_1$)
Measured tuning range	n/a	13%	31.5%
Transient cycles	> 30 cycles	³4 cycles	4 - 6 cycles
PA output power	79mW	n/a	200mW
Support H-bridge?	N	N	Y

¹Estimated from the die photo
²Equivalently converted from phase error
³With pre-designed offset voltage step at fixed frequency range

Fig. 6. Measured link gain of the proposed ATL (top); comparison with prior-art fractional-capacitance works.

Additional References:

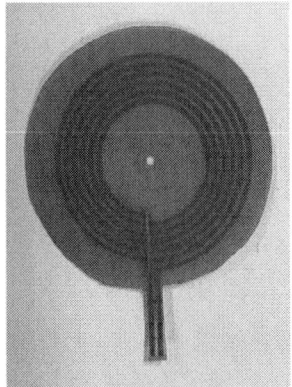

Fig. 7. Chip photo.

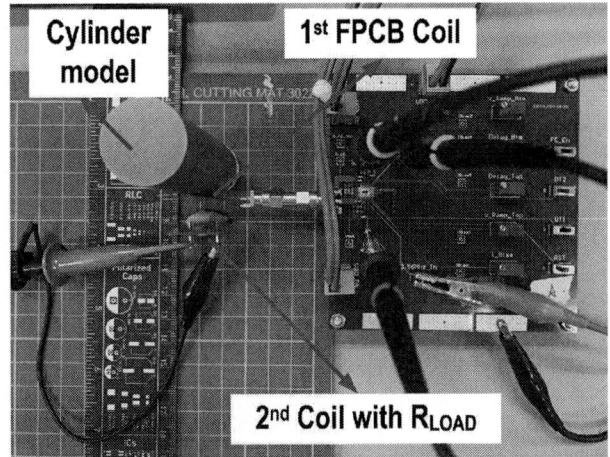

Fig. S1 Flexible coil as the primary inductor.

Fig. S2 Testbench setup scenario

Fully Integrated Reconfigurable Solar Energy Harvester for 100µA Burst Output Current Delivery With 78.6% Peak Energy Extraction Efficiency and Minimum Startup Incident Light Power of 0.27mW/cm²

Jiangchao Wu[1], Guangshu Zhao[1], Litao Zhang[1], Yu Jia[1], Yang Jiang[1], Pui In Mak[1], Rui P. Martins[1,2], Man Kay Law[1]

[1]University of Macau; [2]Universidade de Lisboa

The advancement in CMOS technology has facilitated the development of small form factor wireless sensing nodes, which can be coupled with energy harvesting (EH) for extending the operation lifetime. Even though the system power in energy-constrained EH wireless sensing nodes can be in the 10µA and even down to the sub-µA level [1-2], the peak current consumption during data transmission can increase up to 100µA and more [3]. To resolve this problem, prior works typically require a bulky off-chip storage capacitor for providing the peak output current (up to 50µF for 2mA in [4]), as shown in Fig. 1. Despite the inevitable increase in system volume and cost, such an approach also suffers from a poor energy utilization (only ~2% of the energy stored is extracted). Although the counter-flow switched-capacitor (SC) reconfiguration approach in [5] can achieve a peak energy extraction efficiency (EEE) of up to 70% using on-chip capacitors, the requirement of a stable voltage source can limit its application in EH systems, as the power generated from EH sources are highly dependent on the ambient condition. This work presents a fully-integrated single-chip solar EH system, which employs on-chip solar cells and a capacitive EH interface for generating a stable V_{out} of 0.85V with a V_{in} of 0.325~0.5V, featuring a high power conversion peak efficiency (PCE) of 79.3%. It can also provide an I_{out} of ~100µA using only 8×400pF on-chip capacitors, with a peak EEE of 78.6%.

Figure 2 presents the proposed solar energy harvester using on-chip solar cells. It consists of an on-chip solar cell array for generating the harvested voltage (V_{in}), 8 on-chip MIM/MOS capacitors for dual-branch implementation with 3× voltage boosting (i.e., SC branch A and B, $SC_{A,B}$), a mode controller for system reconfiguration, as well as level shifters for ensuring proper switch on/off operation. The system can be divided into three operating modes: 1) normal mode; 2) energy storage mode; and 3) burst output mode. In normal mode, both SC branches are employed for power delivery, and the output voltage is regulated at V_{min} using the pulse skipping loop. During energy storage mode, only the SC_A output (V_{o1}) is regulated for sustaining V_{out}, while that of SC_B is charged to $3V_{in}$ without regulation to increase the stored energy. This can enhance the deliverable energy for burst operation. In burst output mode, both SC_A and SC_B are reconfigured to deliver energy to the output, while maintaining V_{out} to be above V_{min} at the peak I_{out}. To ensure proper cold-start operation, switch SW_0 is employed to initiate the startup sequence, allowing $C_{6,8}$ to be charged up by V_{in} for control signal generations.

Figure 3 presents the implementation of the on-chip solar cells, together with the reconfigurable capacitors (C_{1-8}) contributing to $SC_{A,B}$. Similar to [6], this work utilizes vertically-stacked N+/Pwell/DNwell/Psub solar cells with parallel connections to improve energy harvesting efficiency. However, the work in [6] can only achieve a PCE of 67% due to the requirement of an extra SC inversion stage. In this work, we employed the Psub for generating a positive voltage for eliminating the first SC inversion stage to improve the PCE. Both $SC_{A,B}$ can be configured to work as a dual-branch Dickson SC converter in normal mode, or two independent SC Dickson branches in energy storage mode.

The capacitor voltages among SC_A and SC_B are $V_{C1,3,5,7} = [V_d, 2V_d, 3V_d, 3V_d]$ and $V_{C2,4,6,8} = [V_{in}, 2V_{in}, 3V_{in}, 3V_{in}]$ at the beginning of the burst output mode, respectively. As the highest capacitor voltage of $3V_{in}$ (i.e., $V_{C6,8}$) is larger than $V_{min}=3V_d$, its direct utilization for powering the output through series capacitor stacking would result in a substantial V_{out} change. To achieve the counter-flow methodology during burst output mode, it is necessary to establish two groups of four distinct voltages with a gradient prior to the counter-flow operation. This work enforces two primary rules during the reconfiguration process. The first rule entails the utilization of a maximum of two series stacked capacitors for delivering I_{out}. As the static comparator CMP_2 only consumes ~450nW, it exhibits a

relaxed speed with ~250ns delay. A rapid drop in V_{out} can therefore introduce misdetection and hence reconfiguration failure. Another rule is to ensure that all the generated voltages through stacking should fall into the range from $V_{min}+0.2V$ to $V_{min}+0.3V$. Specifically, the lower bound ensures successful detection by CMP_2, while the upper one prevents excessive V_{out} change. Notice that as the voltage drop across the switches can be significant (~100mV in simulation), it is necessary to take this into account in consolidating the reconfiguration steps to guarantee the system's robustness.

Figure 4 shows the detailed implementation for generating the required voltage gradient across capacitors from step 1 to step 5. As observed in step 1, $V_{C6,8}$ are discharged from $3V_{in}$ (i.e., the maximum voltage among all the capacitors) to $V_{min}=3V_d$. To abide by the second rule, step 1 also determines whether to connect $C_{3,4}$ to V_{out} or not depending on the instantaneous V_{in} (through comparing V_{in} with 0.425V). In step 2, $C_{3,4}$ is stacked together to deliver charge to V_{out}, and the resulting settled voltage after discharging becomes $V_{C3}=V_{C4}=3V_d/2$. $C_{1,5}$ and $C_{3,5}$ are similarly stacked and discharged in step 3 and 4, respectively. During step 5, we stack $C_{1,7}$ for providing energy to the output. Notice that we still maintain two sets of voltage gradients of $V_{C1,3,5,7} = [3(V_d+V_{in})/16, (19V_d+V_{in})/16, (41V_d-V_{in})/16, (45V_d-3V_{in})/16]$ and $V_{C2,4,6,8} = [3(V_d+V_{in})/8, (19V_d+V_{in})/16, (41V_d-V_{in})/16, 3V_d]$ at the end of step 5. In this work, there is a total of 12 steps for the entire burst output mode operation, and the subsequent step 6 to step 12 still follows the abovementioned rules. Theoretically, the achievable EEE can be up to ~80.6% with $V_{in}=0.4V$, and $V_{min}=0.85V$ after all the reconfiguration steps.

The proposed fully integrated solar EH system was fabricated in 0.18µm CMOS, occupying an active area of 4mm² (excluding pads), as shown in Fig. 7. All the switches are implemented with low-threshold transistors to reduce the conduction loss. We employed a total of 3.2nF on-chip capacitors, with each capacitor implemented using 100pF MIM in parallel with 300pF MOS capacitors. Figure 5 plots the measured output power of the on-chip solar cells under various illumination levels, exhibiting an external quantum efficiency (EQE) of ~16.9%. With an input light power of 6.79mW/cm², the measured peak PCE without regulation is 79.3% when $I_{out}=24µA$. The minimum required input voltage for a successful system cold-start is 0.325V (with a probe loading of 10MΩ//9.5pF), corresponding to an input light power of 0.27mW/cm². Figure 5 also shows the measured transient waveform to validate the capability of delivering a peak I_{out} of ~100µA at $V_{in}=0.4V$. The measured I_{out} during normal mode, energy storage mode, and burst output mode are 9.2, 8.98, and 99.1µA, respectively. As shown in Fig. 6, the energy extracted from the 3.2nF on-chip capacitors maintains above ~0.86nJ across 10 to 100µA, corresponding to a maximum EEE of 78.6%. When compared with the state of the art, this work achieves both a high PCE and EEE, while featuring a fully integrated solution for supplying a burst I_{out} of ~100µA without bulky off-chip storage capacitors.

Acknowledgment: This work was supported in part by the Macao FDCT under Grant 0148/2020/A3, 0040/2021/APD, and SKL-AMSV(UM)-2023-2025, and in part by the Research Committee of University of Macau under Grant MYRG2022-00024-IME. Corresponding author: Man-Kay Law.

References:

[1] M. K. Law et al., "Miniaturized energy harvesting systems using switched-capacitor DC-DC converters," *TCAS-II*, 2022.

[2] J. Wu et al., "A multimode CMOS vision sensor with on-chip motion direction detection and simultaneous energy harvesting capabilities," *IEEE Sensors Journal*, 2022.

[3] H. Huang et al., "A 2Mbps sub-100µW crystal-less RF transmitter with energy harvesting for multi-channel neural signal acquisition," in *A-SSCC*, 2019.

[4] H. C. Cheng et al, "A reconfigurable capacitive power converter with capacitance redistribution for indoor light-powered batteryless Internet-of-Things devices," *JSSC*, 2021.

[5] X. Wu et al, "A fully integrated counter-flow energy reservoir for 70%-efficient peak-power delivery in ultra-low-power systems," in *ISSCC*, 2017.

[6] Z. Chen et al., "A single-chip solar energy harvesting IC using integrated photodiodes for biomedical implant applications," *TBCAS*, 2017.

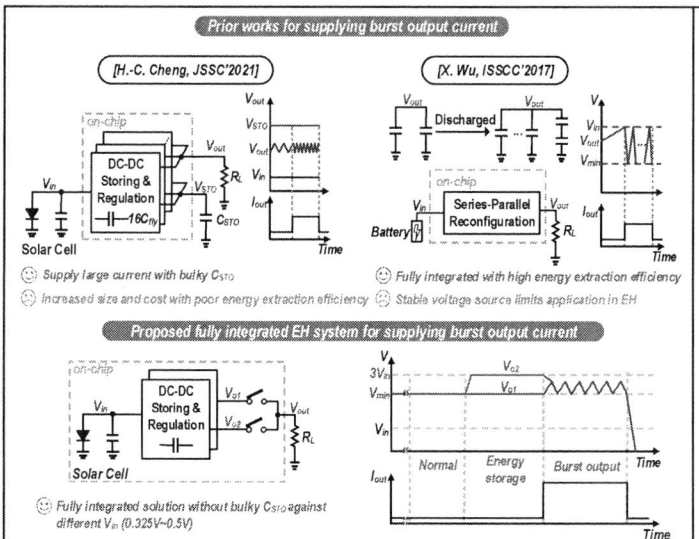

Fig. 1. Motivation and the prior works for supplying burst peak current during data transmission.

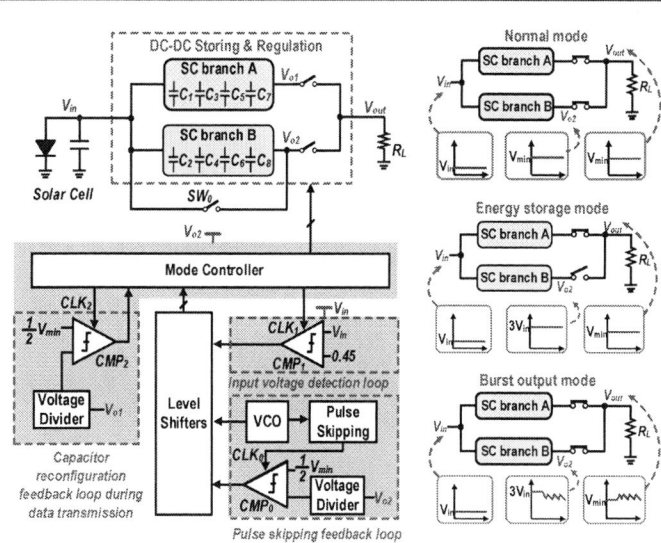

Fig. 2. Proposed reconfigurable harvester and its illustrative principle.

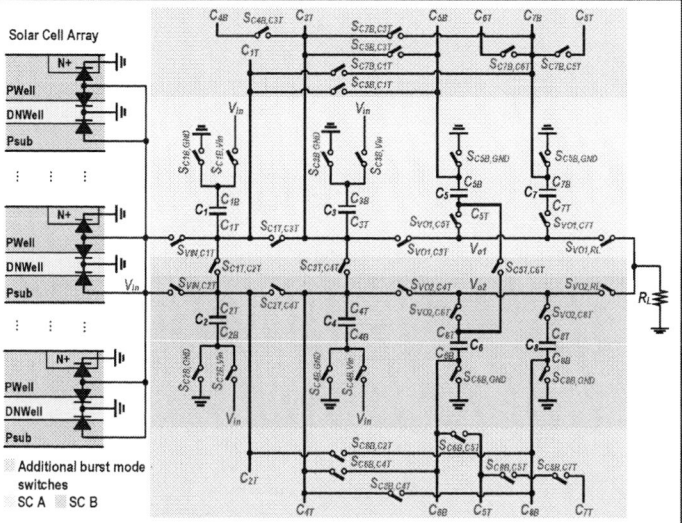

Fig. 3. Implementation of the on-chip solar cells and SC branches $SC_{A,B}$.

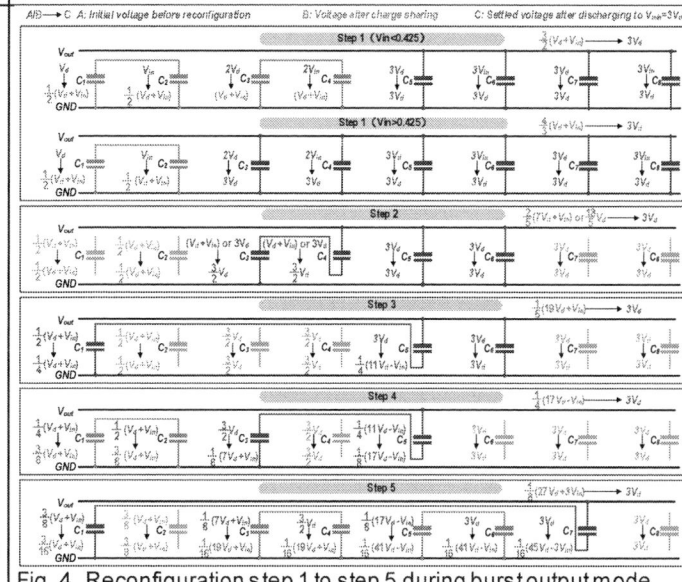

Fig. 4. Reconfiguration step 1 to step 5 during burst output mode.

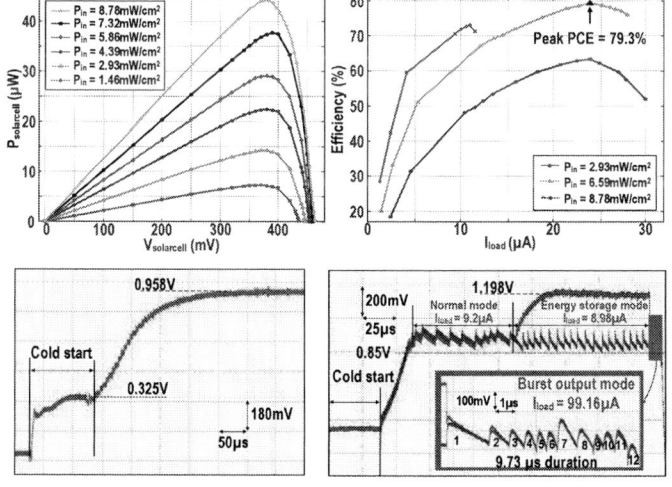

Fig. 5. Measured solar cell performance (top left), power conversion efficiency (top right), transient waveform during start up (bottom left) and the 3 operation modes (bottom right).

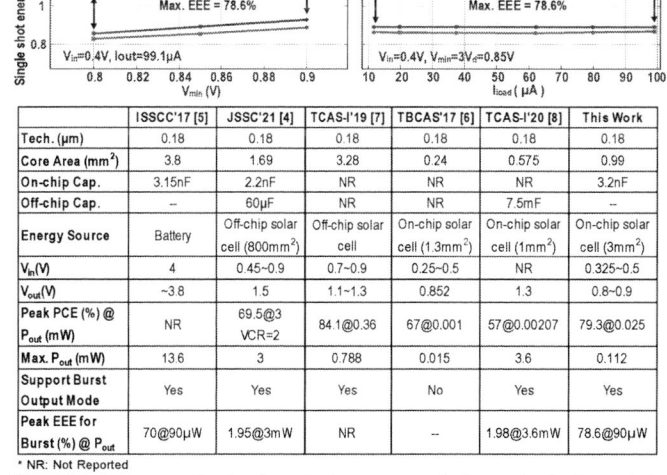

Fig. 6. Measured single shot energy delivered during data transmission (top) and comparison table (bottom).

	ISSCC'17 [5]	JSSC-21 [4]	TCAS-I'19 [7]	TBCAS'17 [6]	TCAS-I'20 [8]	This Work
Tech. (µm)	0.18	0.18	0.18	0.18	0.18	0.18
Core Area (mm²)	3.8	1.69	3.28	0.24	0.575	0.99
On-chip Cap.	3.15nF	2.2nF	NR	NR	NR	3.2nF
Off-chip Cap.	--	60µF	NR	NR	7.5mF	--
Energy Source	Battery	Off-chip solar cell (800mm²)	Off-chip solar cell	On-chip solar cell (1.3mm²)	On-chip solar cell (1mm²)	On-chip solar cell (3mm²)
V_{in}(V)	4	0.45-0.9	0.7-0.9	0.25-0.5	NR	0.325-0.5
V_{out}(V)	~3.8	1.5	1.1~1.3	0.852	1.3	0.8~0.9
Peak PCE (%) @ P_{out} (mW)	NR	69.5@3 VCR=2	84.1@0.36	67@0.001	57@0.00207	79.3@0.025
Max. P_{out} (mW)	13.6	3	0.788	0.015	3.6	0.112
Support Burst Output Mode	Yes	Yes	Yes	No	Yes	Yes
Peak EEE for Burst (%) @ P_{out}	70@90µW	1.95@3mW	NR	--	1.98@3.6mW	78.6@90µW

* NR: Not Reported

979-8-3503-3004-5/23 $31.00 © 2023 IEEE

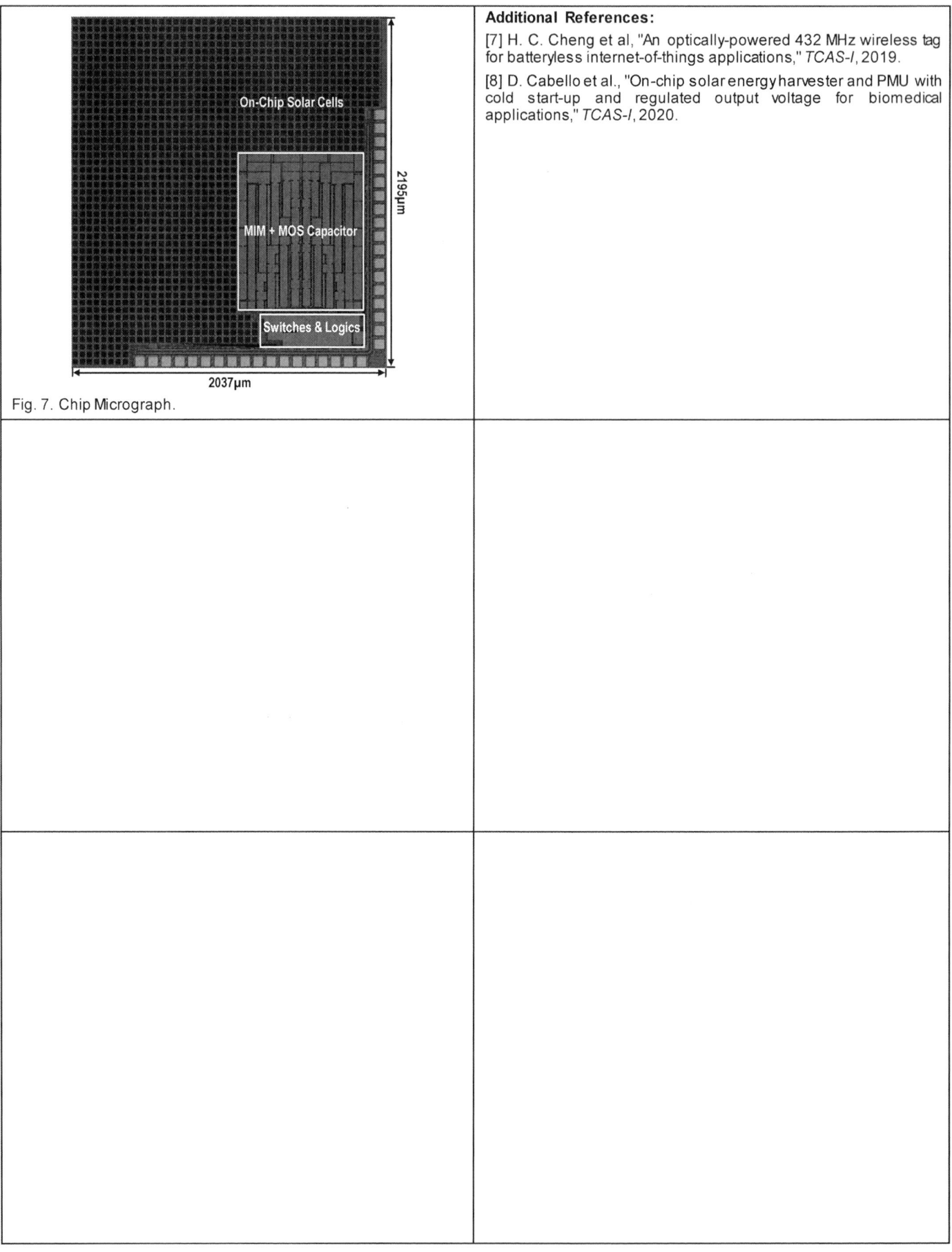

Fig. 7. Chip Micrograph.

Additional References:

[7] H. C. Cheng et al, "An optically-powered 432 MHz wireless tag for batteryless internet-of-things applications," *TCAS-I*, 2019.

[8] D. Cabello et al., "On-chip solar energy harvester and PMU with cold start-up and regulated output voltage for biomedical applications," *TCAS-I*, 2020.

An 80-dB Dynamic Range Hybrid Pulse-Phase Analog Front-End Circuit Cooperating With a Low-Resolution TDC for LiDAR

Zeli Li, Tianrui Lyu, Kaiyou Li, Haoxin Zheng, Zhuohang Ye, Shengzhao Su, Jianping Guo*, and Yang Liu

Sun Yat-sen University, Guangzhou, China

Light detection and ranging (LiDAR) systems sense the distance by illuminating a target and detecting the reflected optical signal. Depending on the modulation method, LiDAR systems can be categorized into three types: pulse time-of-flight (ToF) [1, 2], phase-shift (PS) [3], and frequency-modulated continuous-wave (FMCW), where the pulse and PS are more widely used than FMCW. Fig. 1 illustrates the principle of ToF and PS methods. The accuracy of ToF measurements, as shown in equation (1) of Fig. 1, depends on the precision of the flight time measurement, which typically requires a picosecond-level time-to-digital converter (TDC) for high precision ranging of millimeter level. As for PS ranging, the carrier frequency limits the maximum detectable ambiguity, which is known as 2π limitation as indicated by equation (2). Although multi-frequency ranging can be used to achieve high precision and extend the measurement scope [4], it increases the design complexity of the ranging system. A hybrid pulse-phase ranging system based on an under-sampling technique is proposed in [5]. However, the maximum detectable range is limited by the carrier frequency, resulting in slow phase discrimination compared to prior work.

In this paper, we present a hybrid pulse-phase LiDAR analog front-end (AFE) cooperating with a low-resolution TDC. An under-sampling technique expands the phase distinction in the time domain by aliasing high-frequency carriers to the Nyquist range, loosening the resolution constraint for the post-TDC. Meanwhile, two transistors for transient current attenuating (TCA) are utilized in a transimpedance amplifier (TIA) to achieve an 80-dB dynamic range (DR) without any gain control circuit [3]. Furthermore, by combining pulse ranging and PS ranging, the detecting range is extended beyond the 2π-limited range, avoiding the conflicts between precision and maximum detectable range.

The system architecture of the proposed AFE is shown in Fig. 1 top, and the working principle of the under-sampling technique in the hybrid pulse-phase AFE is depicted in Fig. 2 top. This calculation involves two main steps: pulse ranging and indirect ranging. In pulse ranging, a received 15-MHz repetitive pulse signal is processed through a pulse-receiving channel, and latched by an SR latch. This pulse-ranging step provides a rough distance measurement. In PS ranging, a 15.05-MHz carrier, along with the transmitted modulation signal, is under-sampled and low-pass filtered to obtain a 50-kHz aliased signal for both transmitted and received signals. Consequently, the phase difference between these signals is obtained and converted into an extended time interval, which precisely represents the residue time of flight. The accurate distance can be obtained by combining the coarse and fine quantities.

The proposed AFE block diagram is shown at the bottom of Fig. 2. It consists of a pulse-receiving channel, which includes a TIA, a single-to-differential amplifier, and a post-amplifier. The TIA circuit, illustrated at the top of Fig. 3, utilizes a three-stage push-pull inverter-based amplifier followed by a common-source buffer. The TCA prevents the output signal from saturation and distortion by sourcing or sinking the current to/from VDD and ground which can extend the dynamic range of the AFE and accelerate recovery time under large current (top right of Fig. 3). The single-to-differential amplifier converts the single-ended signal to a differential signal, simplifying timing discrimination and suppressing common-mode noise. Capacitors C_1 and C_2 in Fig. 2 are employed to isolate DC operating points between stages. The post amplifier amplifies the signal,

enabling detection even with weak input signals. Additionally, a level shifter and a comparator are used for time discrimination, preventing false triggering. The sample-filter circuit is shown in the bottom left of Fig. 3, and the sampling process is depicted in the bottom right of Fig. 3. The bootstrap switch circuit ensures a stable on-resistance, allowing V_{out} to closely follow V_{in} during the sampling period. Besides, the aliased low-frequency sinusoidal signals are input to comparators to obtain low-frequency square signals. Finally, the low-frequency square signals and the output of the SR latch are fed to the low-resolution TDCs.

The proposed TDC architecture in this paper offers a resource-efficient alternative to conventional high-precision TDCs used in ToF LiDAR (top of Fig. 4). By combining a pulse-ranging TDC and a phase-ranging TDC, this design significantly reduces complexity with only 28 LUTs and 31 FFs. With 5ns-resolution TDCs, the proposed AFE achieves an equivalent distance precision of 2.5 mm, compared to the 750-mm precision of conventional ToF methods (bottom right of Fig. 4). Although an FPGA is used for implementation and validation purposes, the TDC architecture can be implemented using an MCU for low-cost LiDAR systems.

This design was prototyped in a 180-nm CMOS process consuming an area of 0.6 mm^2. The noise voltage measured at the output buffer is displayed below (top left of Fig. 5), which is 1.646 mV$_{RMS}$. Therefore, the input reference noise current is calculated to be 224.9 nA$_{RMS}$. V_{start} and V_{drive} consist of a series of pulses simulating the transmitted signal. The walk error of pulse ranging accounts for the entire period and does not exceed 5% (bottom left of Fig. 5). As pulse ranging is primarily used to determine the number of periods, the lumped error of the proposed AFE is determined by the PS ranging, which is approximately ± 16.78 cm (bottom center of Fig. 5). When simulating a delay exceeding 2π, the resulting distance error remains below ± 0.3 cm (bottom right of Fig. 5).

At the bottom of Fig. 6, a performance comparison between the proposed AFE and the state-of-the-art is presented. Compared to [1] and [2], the proposed AFE does not require additional gain control, walk error compensation, or high-speed ADC/TDC. In comparison to [3], the proposed design overcomes the 2π ranging limitation, enabling the measurement of longer distances. Compared to [4], the proposed AFE eliminates the need for transmitting dual-frequency signals, reducing the complexity of the optical system design. Additionally, the proposed AFE offers a more integrated receiver chip, which suits a low-cost highly integrated LiDAR system. Fig. 7 shows the chip micrograph and the chip size is 0.6 mm^2.

Acknowledgement:

This work was supported by the Key Area R&D Program of Guangdong Province under Grant 2022B0701180001.

Reference:

[1] H. Zheng et al., "A linear dynamic range receiver with timing discrimination for pulsed TOF imaging LADAR application," *IEEE TIM*, vol. 67, no. 11, pp. 2684-2691, Nov. 2018.

[2] X. Wang et al., "A low walk error analog front-end circuit with intensity compensation for direct ToF LiDAR," *IEEE TCAS-I*, vol. 67, no.12, pp. 4309-4321, Dec. 2020.

[3] H. Zheng et al, "Analog front-end circuits with input current attenuation and heterodyne techniques for low-cost phase-shift LADAR receiver," *IEEE JSEN*, vol. 21, no. 6, pp. 7813-7824, Mar. 2021.

[4] D. Hanto et al., "Time of flight Lidar employing dual-modulation frequencies switching for optimizing unambiguous range extension and high resolution," *IEEE TIM*, vol. 72, pp. 1-8, 2023.

[5] Ohishi, Masahiro et al. "High-resolution rangefinder with a pulsed laser developed by an undersampling method." *Optical Engineering* 49 (2010): 064302.

[6] D. Portaluppi, et al., "Multi-channel FPGA time-to-digital converter with 10 ps Bin and 40 ps FWHM," *IEEE TIM*, vol. 71, pp. 1-9, 2022.

979-8-3503-3004-5/23 $31.00 © 2023 IEEE

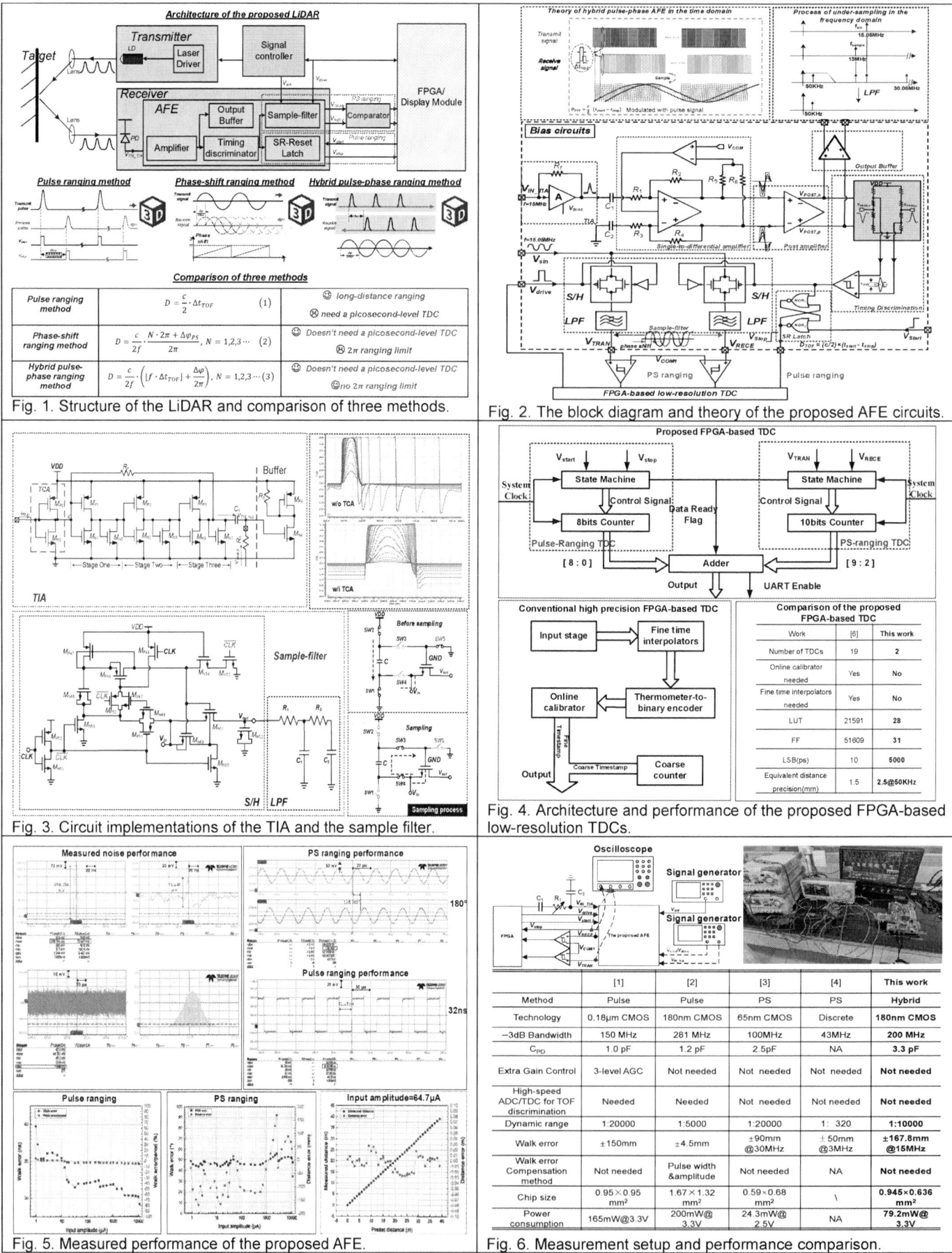

Fig. 1. Structure of the LiDAR and comparison of three methods.

Fig. 2. The block diagram and theory of the proposed AFE circuits.

Fig. 3. Circuit implementations of the TIA and the sample filter.

Fig. 4. Architecture and performance of the proposed FPGA-based low-resolution TDCs.

Fig. 5. Measured performance of the proposed AFE.

Fig. 6. Measurement setup and performance comparison.

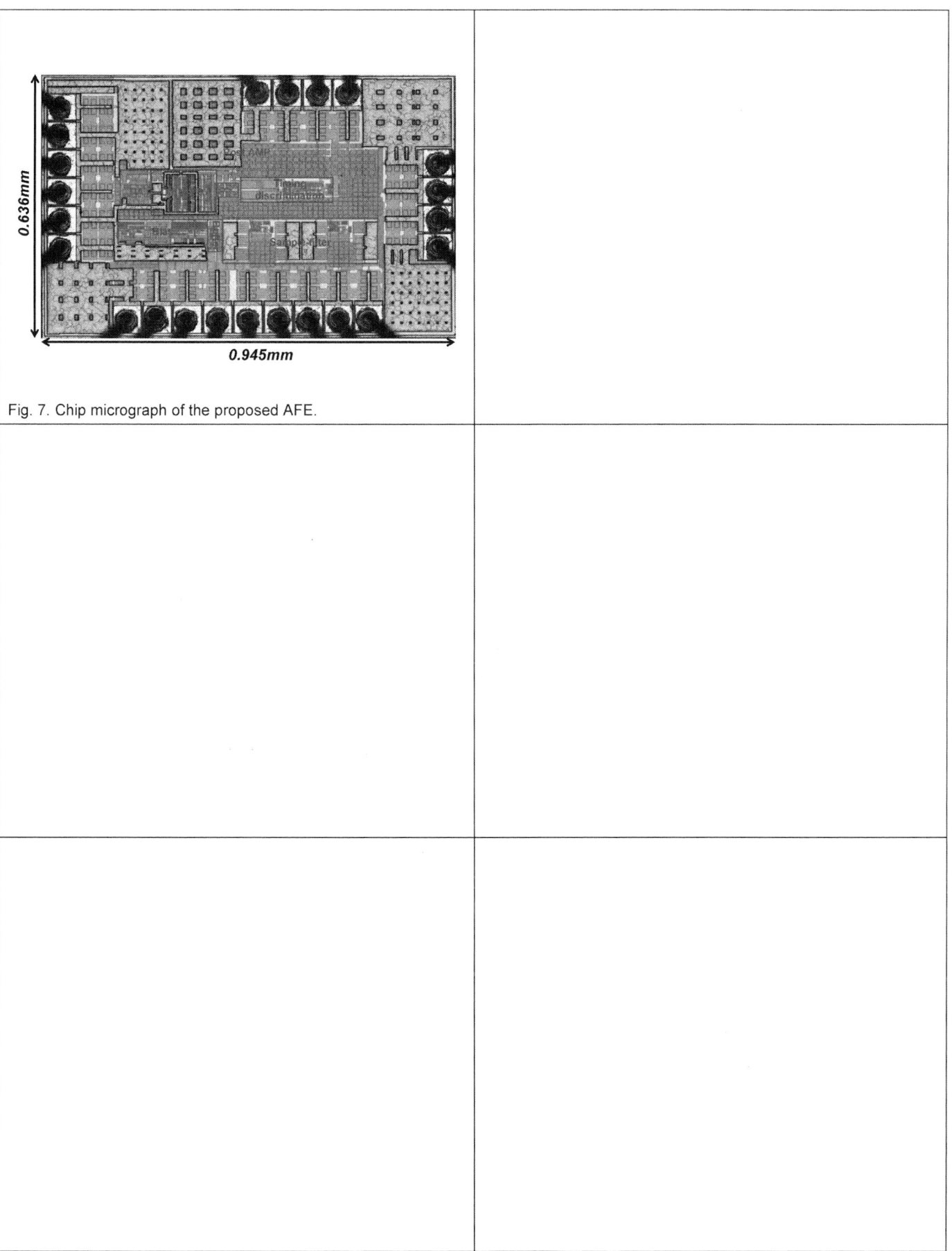

Fig. 7. Chip micrograph of the proposed AFE.

A 4-Element 4-Beam Ka-band Phased-Array Receiver Using Mesh Topology in 65 nm CMOS

Xiangrong Huang[1], Haikun Jia[1], Wei Deng[1], Chuanming Zhu[1], Zhihua Wang[1], Xuzhi Liu[2], Zhiming Chen[2], and Baoyong Chi[1]

[1]School of Integrated Circuits, BNRist, Tsinghua University

[2]Beijing Institute of Technology, Beijing 100081, China

Email: jiahaikun@tsinghua.edu.cn

Phased array technique has been widely applied in wireless and satellite communications due to its high spectrum efficiency and system reconfigurability. To maximize the communication capacity, multi-beam phased-array transceivers have drawn more attention in recent years, which enable concurrent communication with multiple independent users at different locations. To achieve the highest beamforming gain and the largest antenna aperture, the fully-connected phased-array is preferred. The conventional fully-connected network based multi-beam topology is shown in Fig. 1 top. As the increase number of elements (N) and beams (M), a total number of M x N amplitude-phase-control channels are needed. In [1], an 8-element 2-beam phased array is reported. However, this architecture is difficult to extend to more beams due to the complexity of local-oscillator (LO) distribution network. A 2-element 4-beam phased-array architecture with the differential passive combining network is proposed in [2]. However, the power-combining network occupies large chip area and the path layout will be more complex when extending to a larger scale.

To address the layout issue of all channels, a novel mesh-like layout is proposed in this paper. Figure 1 illustrates the architecture of the proposed 25.4-to-32.3 GHz 4-element 4-beam phased-array receiver. The mesh-like Wilkinson power splitters/combiners are placed in horizontal and vertical directions, respectively. A vector modulator based phase shifter (PS) and a couple-based variable-gain amplifier (VGA) constitute a PS-VGA channel. 16 channels form a 4x4 mesh-like matrix. Two LNAs are located at the top and bottom side of chip, respectively. Each 3-stage LNA drivers 4 PS-VGA channels. The output of 4 horizon PS-VGA channels are combined and then transmitted to an 1-stage driver amplifier (DA). The $\lambda/4$ transmission line (TL) which originally occupied a large area can be merged to the PS-VGA channel thanks to the multi-tap inductors. Therefore, the power splitters/combiners almost consume little extra chip area. Besides, it can be easier to extend to a larger scaled arrays such as 8-element and 8-beams due to the simple layout of TLs. At the intersection of horizontal and vertical transmission lines, layer hopping technique is applied to enhance the isolation between each element and beam.

Figure 2 describes the schematic of the compact LNA using folded three-coil transformer and dual-feedforward techniques [3]. The folded three-coil transformer brought the LNA good gain and low-noise results with compact chip size. To extend the bandwidth (BW) and provide better isolation between input and output, a differential third-stage is inserted and normal transformer-based network is utilized in the interstage and output match. According to the simulation results, ~10 GHz 3 dB BW and 2.4 dB minimum noise figure (NF) is achieved for a single LNA. The bottom part of Fig. 2 shows the optimization of $\lambda/4$ TL in Wilkinson combiner/splitter. The conventional $\lambda/4$ TL occupies too large chip area at Ka-band. In the proposed work, a multi-tap inductor replaces the original $\lambda/4$ TL. To reduce the magnetic coupling with surrounding passive devices, the multi-tap inductor is winded like an 8-shape. Then the 8-shaped multi-tap inductor can be placed into the PS-VGA channel which is beneficial for compact chip size. The 2-to-1 combiner utilizing optimized multi-tap based $\lambda/4$ TL has an ~0.4 dB improvement in insertion loss (IL) compared to the conventional topology, although there is a certain deterioration for in-band flatness.

Figure 3 shows the detailed implementation of a PS-VGA channel. The 2-stage VGA is composed of two 2-bit current-bleeding VGA-cells for coarse tuning and three 8-bit coupled-based transformers for fine tuning. The VGA-cell provides 3 attenuation states 3.6 dB, 6 dB, 8 dB, because the size of current bleeding transistors M2 and M3 are half and equal to the size of cascode transistor M1 (32 um/60

nm), respectively. The coupled-based transformer provides a 8 dB attenuation range for each stage while consume almost no extra chip area. The PS is mainly composed of a 2-stage hybrid coupler and vector-modulator (VM)-based gm-cells to ensure high phase resolution and low gain variation. Compared with a single-stage hybrid coupler followed by a transformer-based match network to the gm-cell, the 2-stage architecture achieves wider bandwidth. The simulated I/Q amplitude/phase imbalance is <±1 dB / ±2° over 20 to 40 GHz. Besides, additional matching networks is not required any more. Therefore, there is almost no difference in chip area consumption. The gm-cell is based on a Gilbert-unit. 6-bit gm-cell arrays are added in I/Q path respectively. By controlling the signal polarity, different I/Q magnitude is achieved and then combined in current domain.

The proposed 4-element 4-beam phased-array RX FE has been fabricated in 65 nm CMOS technology and occupies 3.5 mm×2.4 mm area including pads as shown in Fig.7. The entire chip consumes 656 mW from a 1.2V supply.

The transceiver is measured through an on-chip probing system. Figure 4 shows the measured small- and large-signal performance of the RX FE. The RX achieves 25.4 to 32.3 GHz 3 dB BW with 25.0 dB maximum gain and 2.9 dB minimum NF. The measured NF is less than 4 dB from 24.5 to 32.5 GHz. The IP1dB at 28GHz is above −31.6 dBm. The receiver shows a good small-signal gain consistency among the measured 4 element-beam channels. The small gain variation is less than 0.5 dB. The measurement results of amplitude control is also shown in Fig. 4. The proposed gain controller covers a 24 dB gain control range with a 0.1 dB resolution. Each VGA-cell provides 3 coarse levels of gain, including 3.7 dB, 6.0 dB and 7.6 dB. The first and third coupled-based transformers provide 1 dB tuning range with 0.1 dB step due to the relative low impedances (50Ω). The second stage provides 6 dB tuning range with 0.6 dB step. From 25 GHz to 33 GHz, the RMS gain error is less than 0.08 dB. The RMS phase variation is less than 1.2°. The gain control range can be further extended when it is necessary thanks to some redundant gain calibration bits.

Figure 5 shows the measured performance of the phase shifter. The proposed phase shifter covers a 360° phase shifting range with a 6-bit resolution. From 25 GHz to 33 GHz, the RMS phase error is less than 1.7°, and the gain variation under different phase shift states is less than ±0.5 dB. The RMS gain variation is less than 0.04 dB.

Compared with state-of-the-art silicon-based multi-beam phased-array RX FEs at similar frequencies in Fig. 6, our RX FE demonstrates the lowest NF, the best phase shift and gain control performance in aspects of resolution and RMS error. Also, our TRX FE achieves the relatively small chip area per element per beam and has the potential for easier for extension of phased-array scale thanks to the proposed mesh-like topology.

Acknowledgment:

This work was supported by the National Natural Science Foundation of China (No. 62074090) and the Beijing Innovation Center for Future Chips (ICFC).

References:

[1] S. Mondal et al., "A 25–30 GHz fully-connected hybrid beamforming receiver for MIMO communication," JSSC, vol. 52, no. 12, pp. 1275–1287, May 2018.

[2] N. Peng et al., "A Ka-Band CMOS 4-Beam Phased-Array Receiver with Symmetrical Beam-Distribution Network," ISSCL, vol. 3, pp. 410-413, 2020.

[3] X. Huang et al., "28 GHz Compact LNAs with 1.9 dB NF Using Folded Three-Coil Transformer and Dual-Feedforward Techniques in 65nm CMOS," RFIC, Denver, CO, USA, 2022, pp. 223-226.

[4] S. Mondal et al., "A Reconfigurable 28-/37-GHz MMSE-Adaptive Hybrid-Beamforming Receiver for Carrier Aggregation and Multi-Standard MIMO Communication," JSSC, vol. 54, no. 5, pp. 1391-1406, May 2019.

[5] N. Li et al., "A 4-Element 7.5-9 GHz Phased Array Receiver with 8 Simultaneously Reconfigurable Beams in 65 nm CMOS Technology," RFIC, Los Angeles, CA, USA, 2020.

979-8-3503-3004-5/23 $31.00 © 2023 IEEE

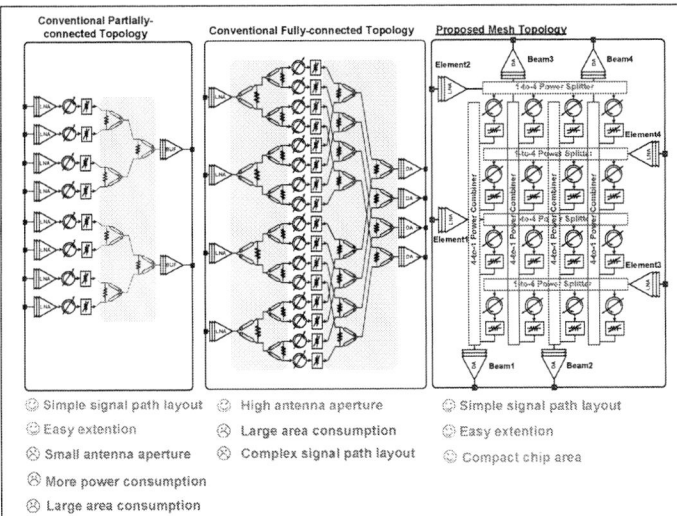

Fig. 1. System architecture diagram of the proposed 4-element 4-beam Phased array RX FE and the comparison with conventional partially/fully-connected topology.

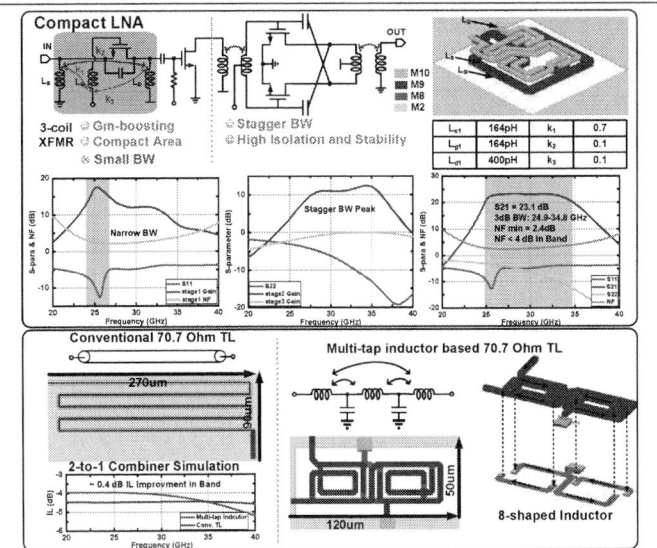

Fig. 2. Detailed schematics of compact LNA and Wilkinson power combiner/splitter.

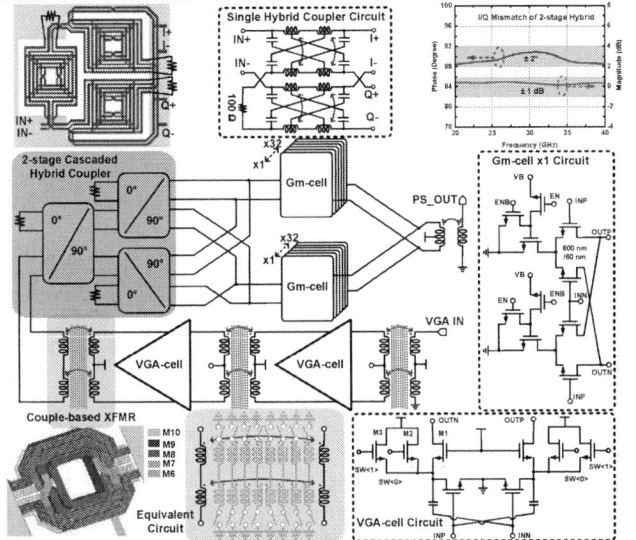

Fig. 3. Detailed schematics of vector-modulation based PS and couple-based VGA.

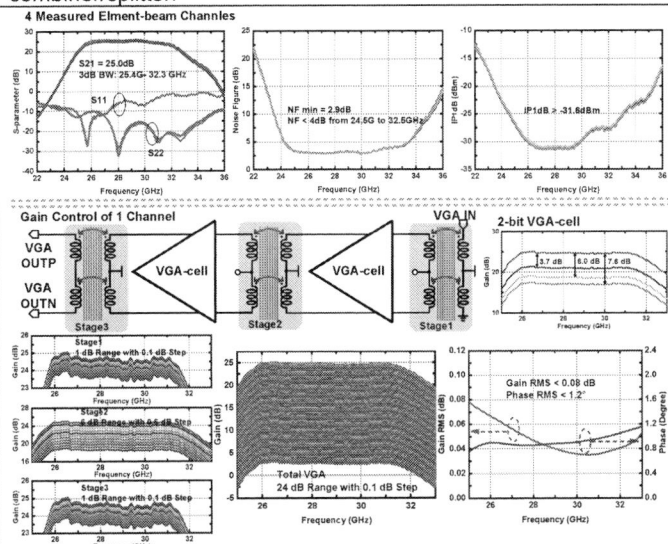

Fig. 4. Measured results of S-parameters, NF and IP1dB of a single channel and the amplitude control of VGA.

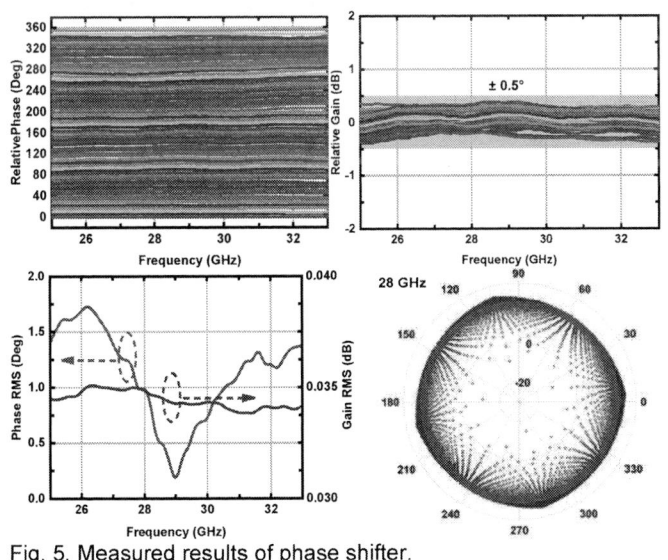

Fig. 5. Measured results of phase shifter.

	JSSC 2018 [1]	ISSCL 2020 [3]	JSSC 2019 [4]	RFIC 2020 [5]	This Work
Technology	65 nm CMOS	65 nm CMOS	65 nm CMOS	65 nm CMOS	65 nm CMOS
Frequency (GHz)	25-30	27-31	27-29.75/35-38.75	7.5-9	25.4-32.3
Elements	8	2	4	8	4
Beams	2	4	2	4	4
Phase Control (°)	360	360	360	360	360
Phase RMS Err. (°)	N/A	<4	N/A	<2	<1.7
Gain Control (dB)	30	17	N/A	33.75	24
Gain RMS Err. (dB)	N/A	<0.4	N/A	<0.3	<0.08
Max. Gain (dB)	34	3.2	33/26.5	20	25.0
NF (dB)	7.3	N/A	5.7/8.5	3.6	2.9
IP1dB (dBm)	-29	N/A	-30/-23	-23	-31.6
Power (mW)	340	40	310	860	656
Power/Elements/Beams (mW)	21.25	5	38.75	27	41
Chip Area (mm²)	6.16	10.4	4.8	19.62	8.4
Area/Elements/Beams (mm²)	0.385	1.3	0.563	0.613	0.525
					0.341*

*Core area per element per beam excluding LNA and DA

Fig. 6. Comparison with the state-of-the-art phased array RX FEs.

979-8-3503-3004-5/23 $31.00 © 2023 IEEE

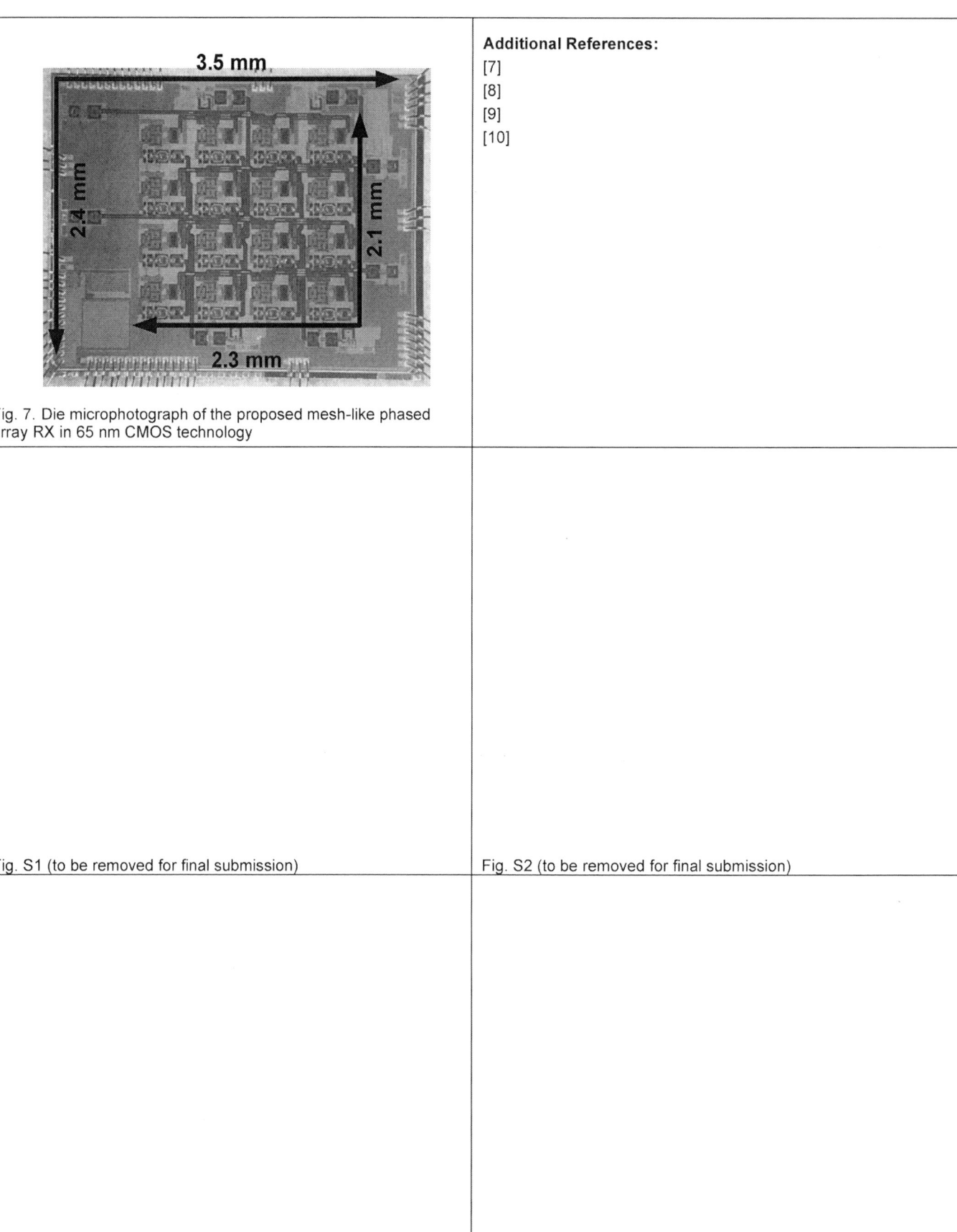

Fig. 7. Die microphotograph of the proposed mesh-like phased array RX in 65 nm CMOS technology

Additional References:

[7]
[8]
[9]
[10]

Fig. S1 (to be removed for final submission)

Fig. S2 (to be removed for final submission)

Fig. S3 (to be removed for final submission)

Fig. S4 (to be removed for final submission)

A Compact E-band Load-Modulation Balanced Power Amplifier Using Coupled Transmission-line Output Network Achieving 22.1-dBm Psat and 34.9%/12.2% Efficiency at Psat/6-dB PBO

Xiangrong Huang, Haikun Jia, Wei Deng, Zhihua Wang, and Baoyong Chi

School of Integrated Circuits, BNRist, Tsinghua University

Email: jiahaikun@tsinghua.edu.cn

Among the mm-wave frequency bands, E-band is considered to have great potential for next-generation wireless communication due to its wide available bandwidth. One of the well-known challenges for mm-wave circuit in CMOS is to design a power amplifier (PA) with relatively high output power and enhanced power-added efficiency (PAE) at both saturated output power (Psat) and power back-off (PBO) levels. Especially for high-frequency band. Several innovative PA design approaches have been presented to address the issue in recent years. Most of them are based on a Doherty concept [1]-[2]. However, most reported Doherty PAs are either intrinsically narrow-band or only tunable over several bands with considerable nonlinearity and amplitude/phasing tuning. Recently, a new concept has been introduced, namely the load-modulated balanced amplifier (LMBA) [3], to address the narrow bandwidth of traditional Doherty PAs.

Figure 1 illustrates the block diagram and main working principle of a typical LMBA. Different from the conventional balanced amplifier (BA) architecture where the isolation port of the output quadrature coupler is terminated by a 50-Ω resistor load, the isolation port of LMBA is loaded by a Class-C amplifier, known as an auxiliary power amplifier (AUX PA) or control-signal path. When the amplitude and phase of the AUX PA satisfy certain conditions, the load impedance of BA will be effectively modulated. However, as the BA and AUX PA are biased for Class-AB and Class-C operations, respectively, using an unequal power splitter is desirable. Therefore, extra phase shifters (PS) are applied in [3]-[4] to tune the phase of AUX PA. Both previous designs consume relatively large chip areas. Moreover, most of the presented works are designed in low-frequency mm-wave such as Ka-band. Due to the more serious parasitic effect and high passive loss in E-band, the extra PS introduces significant loss. In addition, it is much more difficult to design an output match-network with large-size transistors for high Psat in E-band. To address these challenges, an E-band LMBA using a transmission-line (TL) for phase compensation of AUX PA and Coupled transmission-line (CTL) for output-power matching is proposed here. To the best of the author's acknowledgment, this is the first reported LMBA in CMOS that works at E-band.

Figure 2 illustrates the detailed schematic of this work. The AUX PA path is placed opposite the Main PA path. This floorplan has two advantages over the common floorplan where the three paths are stacked in a column as shown Fig. 1. First, the signal path from AUX PA to the isolation port of the output 90° hybrid is significantly shorter, which reduces the passive loss, especially in the high-frequency mm-Wave band. Second, the height of the whole of a single PA is smaller which is more suitable for a single channel of large-scale phased arrays. What's more, the vacant part below the AUX PA can be replaced by an LNA and T/R switch, constituting a basic RF front-end (FE). A 90° hybrid splits the input power into two paths. The Main PA path uses a parallel hybrid splitter/combiner to achieve a 2-way power-combining structure. A single transmission line is inserted between the input power splitter and the input port of the Main PA to tune the phase difference between the Main PA and AUX PA path while no extra PS is needed. The PA core stage employs a common-source structure with a size of 64 um/60 nm. Capacitive neutralization (~20 fF) is used to enhance stability and power gain.

Figure 3 presents the EM structure of the input network. To obtain the optimum phase relationship between the Main PA and AUX PA, a parametric sweep simulation is conducted. An ideal TL is inserted between the THR port of the input hybrid coupler and the input port of the Main PA. According to the simulation results, the proposed LMBA will achieve an optimized performance including the highest Psat and PAE, if a 90°-120° phase delay can be applied. Therefore, the PS covering the whole 360° in [3] is not necessary. A simple TL

is sufficient to compensate for the phased difference. The TL introduces a 110° phase delay with only 0.45 dB insertion loss (IL) around 65 GHz. Another benefit is that without an extra PS, the input power network is easy to design and area-efficiency which is also beneficial for phased arrays.

Figure 4 shows the output power combining network, which consists of three coupled transmission-line-based baluns and an output hybrid coupler. Utilizing the CTL instead of regular transformer (XFMR) based baluns is due to the following reasons. The size of output differential transistors is large to obtain enough Psat, which means severe parasitic capacitance (Cpar). While the output balun usually needs high passive-efficient, which requires high magnetic coupling (K) and quality factor (Q), to ensure output power and PAE. However, by reducing the winding-to-winding spacing and increasing the winding width, the inter-winding capacitance between two inductors implemented by two stacked metal layers, is also very large, which leads to a low self-resonance frequency (SRF) and serious mismatch of the Q. The Cpar at both locations lead to a small inductor value especially in the high-frequency band. Even worse, the Cpar between two inductors will couple the common-mode signal to the output which will deteriorate the output power and PAE. Since the CTL is implemented using the same layer of metal, the parasitic sidewall capacitance is greatly smaller than that of stacked transformers. Besides, the CTL has fewer design parameters, including even/odd-mode impedance Z_{0e}/Z_{0o} and electrical length θ, which makes it easier for iterative design. According to the simulation results, compared with regular XFMR, the CTL can provide better performance of in-band flatness, about 1 dB, and 1% improvement in the power gain and PAE, respectively.

The proposed PA is implemented in a 65nm bulk CMOS technology with a core area of 0.7 mm×0.4 mm. The die micrograph is shown in Fig. 7. Figure 5 shows the small-signal S-parameters and large-signal continuous-wave (CW) measurements. The peak S21 is 17.8 dB with a 3 dB small-signal bandwidth from 57.7 to 74.5 GHz. The output is well matched over a wide frequency range, thanks to the balanced structure. S22 is well below -10dB over the entire bandwidth. The CW measurements from 55 to 75 GHz are also shown in Fig. 5. Over a frequency range of 58 to 74GHz, the PA achieves a saturated output power (Psat) > 21.2dBm and peak PAE (PAEmax) > 31.7%. At 65 GHz the PA achieves a measured PAE max of 34.9 % with 22.1dBm Psat and 30.9% PAE at 21.2dBm OP1dB. At 6dB PBO from Psat, the PA achieves a PAE of 12.1%.

Figure 6 shows a comparison of E-band PAs. The proposed LMBA advances both output power and efficiency. In summary, the presented architecture uses a single TL for phase compensation and CTL output network, making it suitable for phased-array applications.

Acknowledgment:

This work was supported by the National Natural Science Foundation of China (No. 62074090) and the Beijing Innovation Center for Future Chips (ICFC).

References:

[1] T. Huang et al., "A 26-to-60GHz continuous Coupler-Doherty linear power amplifier for over-anoctave back-off efficiency enhancement," ISSCC, pp. 354-356, Feb. 2021.

[2] N. Mannem et al., "A reconfigurable series/parallel quadrature-coupler-based Doherty PA in CMOS SOI with VSWR resilient linearity and back-off PAE for 5G MIMO arrays," ISSCC, pp. 364-366, Feb. 2020.

[3] V. Qunaj et al., "A Doherty-like load-modulated balanced power amplifier achieving 15.5dBm average pout and 20% average PAE at a data rate of 18Gb/s in 28nm CMOS," ISSCC, pp. 356-358, Feb. 2021

[4] C. Chappidi et al., "Load-modulated balanced mm-Wave CMOS PA with integrated linearity enhancement for 5G applications," IEEE IMS, pp. 1101-1104, Aug. 2020.

[5] S. Callender et al., "A compact 75 GHz PA with 26.3% PAE and 24 GHz bandwidth in 22 nm FinFET CMOS," in Proc. IEEE RFIC, Philadelphia, PA, USA, Jun. 2018, pp. 224–227.

979-8-3503-3004-5/23 $31.00 © 2023 IEEE

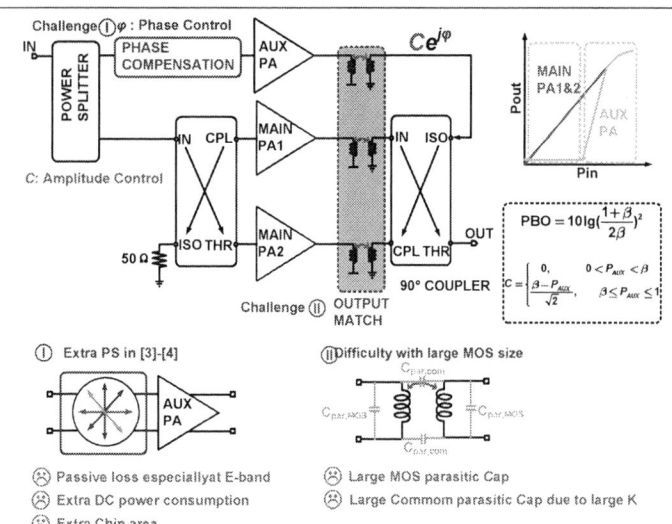

Fig. 1. Block diagram, working principle and main design challenge of the LMBA at E-band.

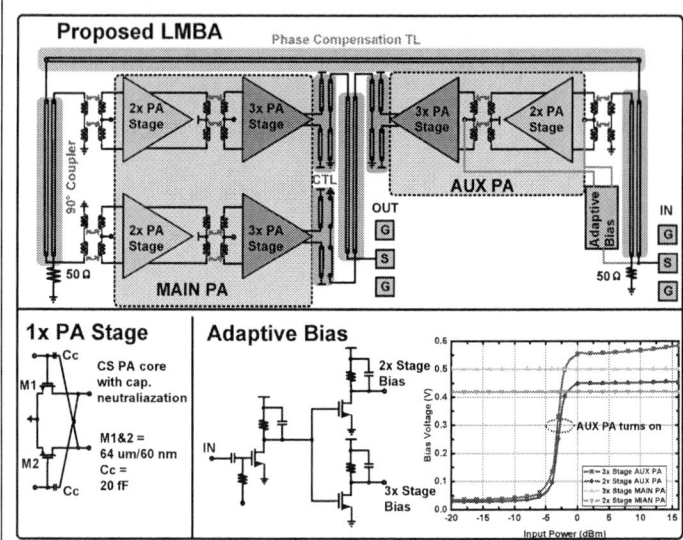

Fig. 2. Schematic representation of the proposed LMBA and the adaptive bias circuits

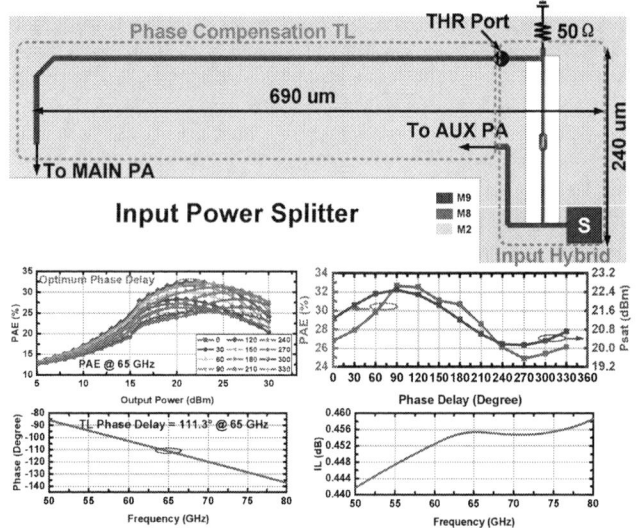

Fig. 3. Top-view of the designed input network used for phase control and their EM simulation results.

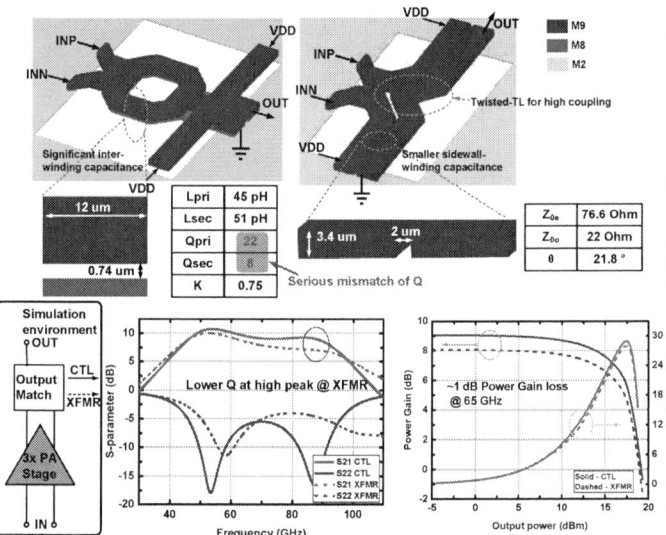

Fig. 4. Top-view of the designed CTL output network and EM simulation results.

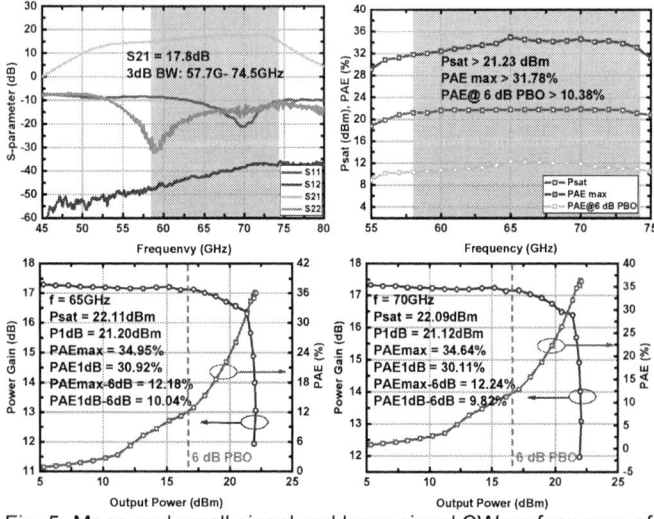

Fig. 5. Measured small-signal and large-signal CW performance of the proposed LMBA.

	RFIC 2018 [5]	RFIC 2018 [6]	TMTT 2019 [7]	RFIC 2019 [8]	TMTT 2020 [9]	This Work
Technology	22 nm Fin-FET	28 nm CMOS	0.12 um SiGe	22 nm FD-SOI	65 nm CMOS	65nm CMOS
Architecture	Single Path CS	Two Paths Series Combined	Single Path S.E.	Single Path CS + CG	Single Path CS	Two Paths CS LMBA
Frequency (GHz)	62-86	67-80	60-77	64-82	70-75	57.7-74.5
BW (GHz)	24	13	17	18	5	16.8
VDD (V)	1	1	2	2	1.2	1.2
Gain (dB)	16.6	13.8	25.3	17.8	26	17.8
Psat (dBm)	12.8	15.7	15.8	17.8	14.29	22.1
P1dB (dBm)	5.7	11.8	13.6	13.3	12.03	21.1
PAE max	26.3	8.9%	18.1%	17.3%	22.3%	34.9%
PAE 1dB	11.6	N/A	13.0%	8.1%	13.3%	30.9%
PAE @ Psat-6dB	N/A	N/A	4.7%	4.5%	5.6%	12.1%
PAE @ P1dB-6dB	N/A	N/A	4.1%	2.5%	5.1%	10.0%
Core Area (mm²)	0.054	0.42	0.25	0.02	0.025	0.28

Fig. 6. Comparison table of E-band PAs.

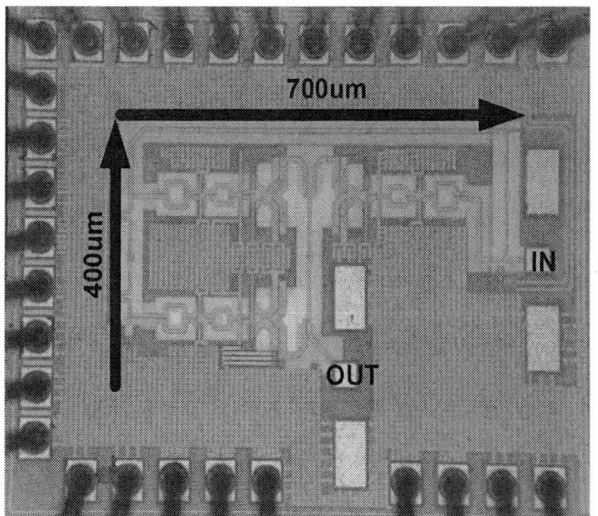

Fig. 7. Die microphotograph of the proposed LMBA in 65 nm CMOS technology

Additional References:

[6] N.-C. Kuo et al., "An E-band QPSK transmitter element in 28-nm CMOS with multistate power amplifier for digitally-modulated phased arrays," in Proc. IEEE RFIC, Jun. 2018, pp. 184–187.

[7] E. Wagner et al., "Single and power-combined linear E-band power amplifiers in 0.12-µm SiGe with 19-dBm average power 1-Gbaud 64-QAM modulated waveforms," IEEE Trans. Microw. Theory Techn., vol. 67, no. 4, pp. 1531–1543, Apr. 2019.

[8] U. Celik et al., "An E-band compact power amplifier for future array-based backhaul networks in 22 nm FD-SOI," in Proc. IEEE RFIC, Boston, MA, USA, Jun. 2019, pp. 187–190.

[9] L. Chen et al., "A Compact E-Band Power Amplifier With Gain-Boosting and Efficiency Enhancement," in IEEE Transactions on Microwave Theory and Techniques, vol. 68, no. 11, pp. 4620-4630, Nov. 2020, doi: 10.1109/TMTT.2020.3017728.

Fig. S1 Measurement environment for on-wafer tests.

Fig. S2 (to be removed for final submission)

Fig. S3 (to be removed for final submission)

Fig. S4 (to be removed for final submission)

A Crystal-less Clock Generator for Low-power and Low-cost Sensor Transceivers with 12.9MHz-to-3.3GHz Range, 16.67ppm/°C Inaccuracy from -25°C to 85°C, and 0.25us Settle-time

Sangdon Jung[1,2], Jehyeok Yu[1], Minsu Park[1], and Jung-Hoon Chun[1]

[1]Department of Semiconductor and Display Engineering, Sungkyunkwan University, Suwon, South Korea

[2]Samsung Electronics, Hwaseong, South Korea

In the field of sensor system design, the need for low-power and cost-effective solutions is growing. Existing clock generation technologies for sensor transceivers, however, rely on external references or high-cost components, introducing economic constraints. To address these issues, there is on-going research on utilizing on-chip RC oscillators (RCOs) to generate clocks without external references [1, 2]. The objective is to deliver precise frequency outputs that are immune to variations in process, voltage, and temperature. Particularly, resistors with Proportional to Absolute Temperature (PTAT) and Complementary to Absolute Temperature (CTAT) characteristics are often used to balance temperature changes [3]. However, due to high fabrication and measurement costs, these techniques may not suitable for low-cost sensor systems. Therefore, there is a compelling need for the design of accurate RCOs that employs a single type of resistor.

In the domain of low-power sensor system and transceiver clock technologies, Duty Cycle Operation (DCO) and Dynamic Frequency Scaling (DFS) are predominantly employed. DCO enhances power efficiency by optimizing the on/off ratio, although prolonged settling times might lead to unnecessary power usage. DFS, on the other hand, adjusts the clock frequency to effectively manage power consumption. To maximize the potential of DCO and DFS technologies, it is crucial to develop crystal-less clock generators with fast settling times and a wide frequency range [4].

In this paper, we introduce a crystal-less clock generator, comprised of a relaxation RCO with a single p-poly resistor as a reference, a Fractional-N PLL providing a wide frequency range and swift settling time, along with an on-chip temperature sensor. The relaxation RCO, while offering rapid startup and settling time in contrast to closed-loop switched-cap RCO, suffers from a logic delay (T_D) drawback that considerably impacts frequency variation with temperature as depicted in Fig. 1. To diminish the T_D error rate, we strategically set the RCO's target frequency as low as possible and fully cover the entire target frequency range through multiplication in the PLL. The proposed PLL structure (W-FPLL) achieves a wide frequency range from 12.9 MHz to 3.3 GHz and guarantees settling within one and a half cycles of F_{REF} during output frequency transitions. The on-chip Temperature Sensor Unit (TSU) measures the temperature and converts it into a 14-bit digital value that is subsequently incorporated into the Delta-Sigma Modulator (DSM) of the W-FPLL to counteract temperature-induced frequency variations. Additionally, Piecewise Linear Calibration is applied to enhance calibration accuracy.

Fig. 2 presents the comprehensive block diagram of the proposed crystal-less clock generator, comprising three main components. First, an on-chip RCO generates the reference frequency. This RCO is calibrated to a relatively low 6-MHz frequency to ensure that the output frequency is influenced solely by the resistor's temperature dependency. Second, a Fractional-N PLL serves to multiply the frequency. Utilizing a ring-VCO for broad frequency operation, it incorporates components such as a resistive V-to-I converter, a 6-bit DAC, a Current Controlled Oscillator (CCO), a Gain Calibration Unit (GCU), and a Timing Control Unit (TCU). A Pulse-swallow programable divider is utilized to facilitate operation across an expansive frequency range while achieving fast phase locking. Lastly, an on-chip temperature sensor is integrated. This sensor generates two frequencies, F_{CONST} and F_{CTAT}, based on the CTAT and constant voltages produced by the BGR. By comparing these two frequencies, the temperature value is extracted as a digital code. The output value is averaged to remove the noise. The temperature-induced frequency variation of the RCO is counteracted by the synthesized digital block through the DSM within the W-FPLL. The DSM control code, DSM [9:0], is configured in a piecewise linear fashion to nullify the frequency variation of the RCO.

Fig. 3 illustrates the calibration procedures of the proposed clock generator. Process variations are mitigated through a 1-point calibration at 35°C. In the first step, the on-chip RCO resistor, implemented using a 7-bit ladder, is configured to yield a 6-MHz output at 35°C with an accuracy represented by a 1 LSB error per 0.1% precision. In the second step, the CCO of the on-chip TSU is calibrated. Again, through a 1-point calibration, the F_{CTAT} and F_{CONST} are set to 3 MHz and 1 MHz, respectively, at 35°C, with 0.1% precision. The third step calibrates the VCO gain of the W-FPLL for fast settling. The fourth step entails calculating the gain needed to compensate for temperature-induced frequency variations by feeding the output value of the TSU into the DSM of the W-FPLL. While linear calibration is straightforward, it does not fully account for the nonlinearity of the output frequency variation with temperature. To address this, the DSM gain is adjusted based on the temperature coefficient of the on-chip RCO's p-poly resistance in the final step.

Fig. 4 illustrates the proposed fast-settling technique. The initial step involves a calibration process, aiming to attain a constant gain as dictated by the frequency control bit, F_{CB}. The 6-bit F_{CB} code is anchored at its median value (=32), and the resistor ladder values of the resistive V-to-I converter are adjusted; the VCO output frequency is tuned to 1.5GHz. To expedite the calibration process, the GCU employs a binary search logic. After calibration, the output frequency gain, based on the F_{CB}, stabilizes at 46.875 MHz/LSB. For subsequent DCO and DFS operations, the F_{CB} and the Main Divider Factor (M) values are altered to align with the target frequency. However, frequency changes induce ripples in the VCO control voltage, prompting undesired phase shifts. To counteract this, the TCU monitors changes in the M and F_{CB} values, establishes a time window for one and a half cycles of F_{REF}, and adjusts the M and F_{CB} values midway through this window to reach the target frequency. During this process, a phase shift is observed, necessitating a temporary suspension of the phase lock operation by resetting the PFD, CP, and Main Divider. The reset status of the paused blocks is then released, synchronized with the falling edge of F_{REF}, thereby resuming the phase-lock operation. Simultaneously, the pulse-swallow divider, synchronized with the rising edge of the VCO, produces F_{DIV}. As a result, following the reset release, the first VCO edge emerges as the rising edge of F_{DIV}. The maximal phase error between F_{DIV} and F_{REF} is deemed negligible as it is no more than one cycle of the VCO. After initial calibration, the clock generator can achieve the settling state within one and a half cycles of F_{REF}.

Fig. 5 shows the on-chip temperature sensor and its measurement results. A chopping technique is applied to the comparator to curtail frequency variations induced by offset. A time window is created with the F_{CONST} clock, and during this window, the F_{CTAT} clock undergoes counting to transcribe temperature information into a digital code. The resultant digital codes inherently include noise from the BGR and CCO, which is mitigated through 8 to 32 rounds of averaging. By measuring three samples, using a window spanning 4096 cycles of F_{CONST} and 32 rounds of averaging, a maximum temperature error of 1.5°C over the range of -25°C to 85°C was achieved.

Fig. 6 presents the output clock jitter histogram and the calibrated output frequency. The target output frequency was set to 3.3 GHz by applying a Main Divider Factor (M) of 550. The on-chip RCO exhibited a frequency precision of 409 ppm/°C from -25°C to 85°C. By applying three segments of piecewise linear gain, a temperature coefficient of 16.67 ppm/°C was achieved.

As shown in Fig. 7, the W-FPLL occupies just 0.07mm², contributing to a total area 0.102 mm². Operating at 3.3 GHz, the W-FPLL consumed 5.2 mW, leading to a total power consumption of 6.35 mW. A comparison with previous studies confirms the successful implementation of both crystal-less and fast-settling characteristics.

Acknowledgement:

This work was supported in part by the Korea government (MSIT, No. 2022-0-01171), and in part by Samsung Electronics Co., Ltd (MEM210728_0001).

References:

[1] Çağrı Gürleyük, et al., "A 16 MHz CMOS *RC* Frequency Reference With ±90 ppm Inaccuracy From -45°C to 85°C," in IEEE JSSC, vol. 57, no. 8, pp. 2429-2437, August 2022.

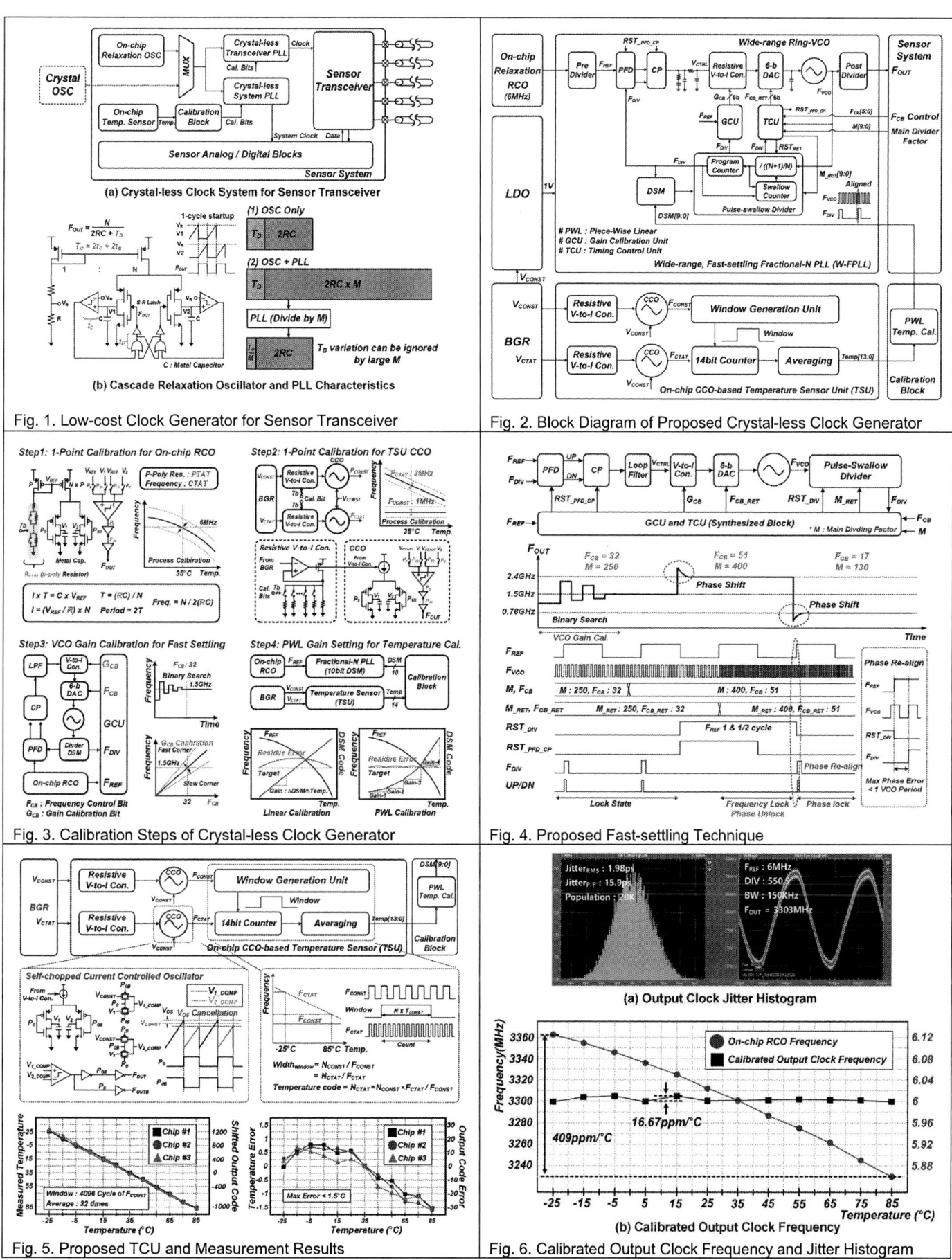

Fig. 1. Low-cost Clock Generator for Sensor Transceiver

Fig. 2. Block Diagram of Proposed Crystal-less Clock Generator

Fig. 3. Calibration Steps of Crystal-less Clock Generator

Fig. 4. Proposed Fast-settling Technique

Fig. 5. Proposed TCU and Measurement Results

Fig. 6. Calibrated Output Clock Frequency and Jitter Histogram

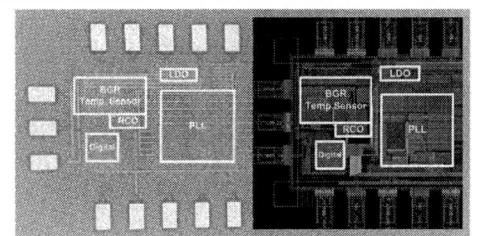

Block	Size (mm²)	Power (mW)
RCO	0.005	0.4
Temp. Sensor	0.012	0.43
BGR	0.004	0.12
LDO	0.003	0.1
Digital	0.008	0.1
PLL	0.07	5.2
Total	0.102	6.35

		This Work	[4] '19 ASSCC	[5] '15 ISSCC	[6] '14 ISSCC	[7] '13 ISSCC
Technology		28nm	7nm	14nm	20nm	28nm
Architecture		RCO + Frac.-N PLL + Temp. Sensor	Frac.-N PLL	Frac.-N PLL	Frac.-N PLL	Frac.-N PLL
Crystal Oscillator		Don't need	Need	Need	Need	Need
Freq. Range	(GHz)	0.012 – 3.3	0.009 – 2.4	0.032 – 2.0	0.025 – 1.6	0.032 – 2.0
Reference	(MHz)	6	26	32	-	10 – 30
Fast-settling Technique		Yes	No	No	No	No
Power	(mW)	5.2@3.3G (PLL) 6.35@3.3G (Total)	2.0@2.4G	2.1@2.0G	3.1@1.6G	5.3@2.0G
Efficiency	(mW/GHz)	1.57	0.84	1.05	1.94	2.65
Area	(mm²)	0.07 (PLL) 0.102 (Total)	0.032	0.009	0.012	0.026

Fig. 7. Die Photo and Comparison Table

Additional References:

[2] W. Choi *et al.*, "A 0.9V 28MHz dual-*RC* frequency reference with 5pJ/Cycle and ±200 ppm Inaccuracy from −40°C to 85°C," ISSCC Dig. Tech. Papers, pp. 434-436, Feb. 2021.

[3] J. Jung et al., "A 1.08-nW/kHz 13.2-ppm/°C Self-Biased Timer Using Temperature-Insensitive Resistive Current," in IEEE JSSC, vol. 53, no. 8, pp. 2311-2318, August 2018.

[4] S. Jung et al., "A 9.4MHz-to-2.4GHz jitter-power reconfigurable fractional-N ring PLL for multi-standard applications in 7nm FinFET CMOS technology," IEEE A-SSCC., pp. 87-90, Nov. 2019.

[5] M. Song *et al.*, "A 0.009mm² 2.06mW 32-to-2000MHz 2nd-Order ΔΣ Analogous Bang-Bang Digital PLL with Feed-Forward Delay-Locked and Phase-Locked Operations in 14nm FinFET Technology," ISSCC Dig. Tech. Papers, pp. 226-228, Feb. 2015.

[6] J. Liu *et al.*, "A 0.012mm² 3.1mW Bang-Bang Digital Fractional-N PLL with a Power-Supply-Noise Cancellation Technique and a Walking-One-Phase-Selection Fractional Frequency Divider," ISSCC Dig. Tech. Papers, pp. 268-270, Feb. 2014.

[7] T. -K. Jang *et al.*, "A 0.026mm² 5.3mW 32-to-2000MHz Digital Fractional-N Phase-Locked Loop Using a Phase-Interpolating Phase-to-Digital Converter," ISSCC Dig. Tech. Papers, pp. 270-271, Feb. 2013.

A 6.78-MHz Wireless Power Transfer System with Voltage-/Current-Mode 0X/1X Regulating Rectifier and Global-Loop Power Control

Bing-Jen Wu, Dao-Han Yao, and Po-Hung Chen

National Yang Ming Chiao Tung University, Hsinchu, Taiwan

Wireless power transfer (WPT) has great potential to be adopted in various applications, such as implantable medical devices (IMDs) and radio frequency identification (RFID) tags. However, the transmission distance and received power level vary due to coil misalignment and environmental conditions. Figure 1 compares the conventional WPT system with the proposed design. [1] employs a full-bridge rectifier with local-loop regulation to eliminate the regulation stage in the receiver (R_X). The global-loop power control (GLPC) is also combined to improve the power conversion efficiency (PCE), but the R_X voltage conversion efficiency (VCE) is less than one, limiting the transmission distance and input power range. Resonant-type receivers [2]–[6] resonate the LC tank for several cycles to accumulate the energy, which improves the VCE and minimum input power level of the R_X. [2] presents the resonant voltage mode operation to isolate the R_X coil from the output for resonating, but an extra off-chip inductor is needed. [3] demonstrates a resonant current mode (RCM) R_X at 13.56 MHz. However, VCE improvement is limited, and the output voltage is not regulated. [4, 5] present the voltage-current mode (VCM) R_X that combines the RCM and voltage mode (VM) operations to extend the input-voltage range while regulating the output, but the operating frequency is only 1 MHz. Without GLPC, transmitters deliver too much power than the required output under a light load condition, which limits efficiency.

This work proposes a 6.78-MHz VCM regulating rectifier with GLPC. The output voltage is regulated through local-loop control by employing 0X/1X mode regulation in the VCM rectifier, and the 0X/1X mode information in the R_X is transferred back to the T_X for GLPC. The demodulated backward data information controls the DC-DC converter output through digital pulse-width modulation (DPWM). Thus, the output power of T_X is adjusted automatically according to the loading conditions to minimize unnecessary power, resulting in an improved PCE of the system.

Figure 2 illustrates the block diagram of the proposed WPT system and the VCM operation principle in the R_X. The transmitter is composed of a duty ratio detector (DRD) for calculating the R_X 0X/1X duty ratio, a dc-dc converter with DPWM for adaptively changing the V_{PA}, and a Class-D power amplifier for driving the T_X coil. Resonant frequencies of receiver and transmitter coils are both set to 6.78 MHz to realize resonant inductive coupling for a higher coupling efficiency. The receiver is composed of a VCM 0X/1X regulating rectifier for voltage regulation, a clock recovery for generating a 6.78-MHz clock signal, 0X/1X local-loop control for regulating the output, and a pulse generator with timing calibration for switching the M_N power transistor at right timing to improve efficiency.

Under a light load condition, the rectifier first operates in the resonating mode (RM) to accumulate energy through the L_2C_2-tank for several cycles (T_{RES}) by turning on the M_N. It then switches to CM by turning off the M_N to charge the output through i_{L2}. On the other hand, under a heavy load condition, the received voltage (V_R) amplitude is relatively higher and the R_X operates in RM for a very short period. When the VR amplitude is accumulated to higher than $V_{OUT} + V_D$, the rectifier automatically operates in VM to deliver the output power. After few cycles, the rectifier becomes CM to transfer all the energy stored in the inductor to the output. The turn-on and turn-off timings of M_N have to be considered carefully to avoid M_N body diode conduction during CM operation. M_N starts turning off and turning on when V_R and V_{AC} become zero, respectively. The pulse generator with timing calibration senses the V_R and V_{AC} to adjust the turn-on and turn-off timing every cycle. Therefore, even if the output loading and transmission distance change, the proposed VCM rectifier can operate at a correct timing in few cycles. Figure 2 also illustrates the operation principle of the 0X/1X mode operation. The regulating rectifier enters the 1X mode to charge the output and 0X mode to separate the output from the L_2C_2-tank when the V_{OUT} drops below the preset lower bound and exceeds the preset upper bound,

respectively. Thus, the proposed rectifier realizes rectification and regulation within a hysteresis window in a single stage.

Figure 3 illustrates the operation principle of the proposed T_X. The reflected impedance (Z_{REF}) is highly related to the RX 0X/1X operation mode. During the 0X mode, T_X draws more current due to a low Z_{REF}, but the 0X mode R_X does not deliver power to the load. Therefore, the system PCE degrades significantly if the 0X mode duration (T_{0X}) becomes longer. It is worth mentioning that T_{0X} only depends on the load, whereas the 1X mode duration (T_{1X}) depends on the load and the power received by R_X. Therefore, the D_{1X} needs to be well controlled to obtain a high duty ratio by adjusting the T_X power transmission. In this work, D_{1X} is controlled within 80%–95% adaptively by GLPC for a high PCE. With the help of duty ratio detector, the D_{1X} is successfully detected. And the transmission power is controlled by V_{PA}, which is determined by digital pulse width modulation circuit.

The proposed T_X and R_X chips are fabricated in a 0.18-μm CMOS process with a high voltage option. Figure 4 shows the measured waveforms of the VCM 0X/1X regulating rectifier under different loading conditions. Under light load conditions, the received power and voltage amplitude ($V_{R,peak} = 0.6V$) is low in the RM. After few cycles, R_X becomes CM operation and regulates the V_{OUT} at 1.8 V, which demonstrates 3V/V VCE. Under heavy load conditions, the rectifier operates in the V_M from the second cycle and becomes CM in the fourth cycle. All these operating modes are selected automatically according to the input and output conditions of R_X. During light-to-heavy load transient, the R_X operates in the 1X mode for a long period due to local-loop control, and GLPC adjusts the V_{PA} to increase the T_X output power. The output is regulated back to 1.8 V in 5.5 ms and GLPC decreases V_{PA} to reduce T_X power since D_{1X} becomes lower than 80%. During heavy-to-light load transient, the overshoot is negligible due to the 0X mode operation in R_X. Thanks to GLPC, the proposed WPT system keeps D_{1X} within 80%–95% for a high PCE.

Figure 5 shows the measured dependences of the PCE and input power (P_{IN}) on the output current (I_{OUT}). The peak efficiencies of the transmitter, receiver, and system are 75%, 70.6%, and 25.7%, respectively. Compared to the conventional VM operation, the input power is reduced by 78%, and the I_{OUT} range is extended by 500%. Figure 6 compares the performance of the proposed resonant VCM rectifier with state-of-the-art designs. Thanks to the VCM operation, the proposed VCM rectifier achieves the 3V/V VCE. Figure 7 shows the die micrograph of T_X and R_X chips.

References:

[1] F.-B. Yang, et al., "Structure-Reconfigurable Power Amplifier (SR-PA) and 0X/1X Regulating Rectifier for Adaptive Power Control in Wireless Power Transfer System," IEEE JSSC, vol. 56, no. 7, pp. 2054-2064, Jul. 2021.

[2] S.-U. Shin, et al., "A 13.56MHz Time-Interleaved Resonant-Voltage-Mode Wireless-Power Receiver with Isolated Resonator and Quasi-Resonant Boost Converter for Implantable Systems," ISSCC, pp. 154-155, Feb. 2018.

[3] S.-W. Hung, "A 13.56MHz Current-Mode Wireless Power and Data Receiver with Efficient Power Extracting Controller and Energy-Shift Keying Technique for Loosely Coupled Implantable Devices," ISSCC, pp. 486-488, Feb. 2020.

[4] H. Gougheri, et al., "Adaptive Reconfigurable Voltage/Current-Mode Power Management with Self-Regulation for Extended-Range Inductive Power Transmission," ISSCC, pp. 374-375, Feb. 2017.

[5] H. Gougheri, et al., "An Inductive Voltage-/Current-Mode Integrated Power Management with Seamless Mode Transition and Energy Recycling," IEEE JSSC, vol. 54, no. 3, pp. 874-884, Mar. 2019.

[6] H. Gougheri, et al., "A Dual-Output Reconfigurable Shared-Inductor Boost-Converter/Current-Mode Inductive Power Management ASIC with 750% Extended Output-Power Range, Adaptive Switching Control, and Voltage-Power Regulation," IEEE TBioCAS, vol. 13, no. 5, pp. 1075-1086, Oct. 2019.

979-8-3503-3004-5/23 $31.00 © 2023 IEEE

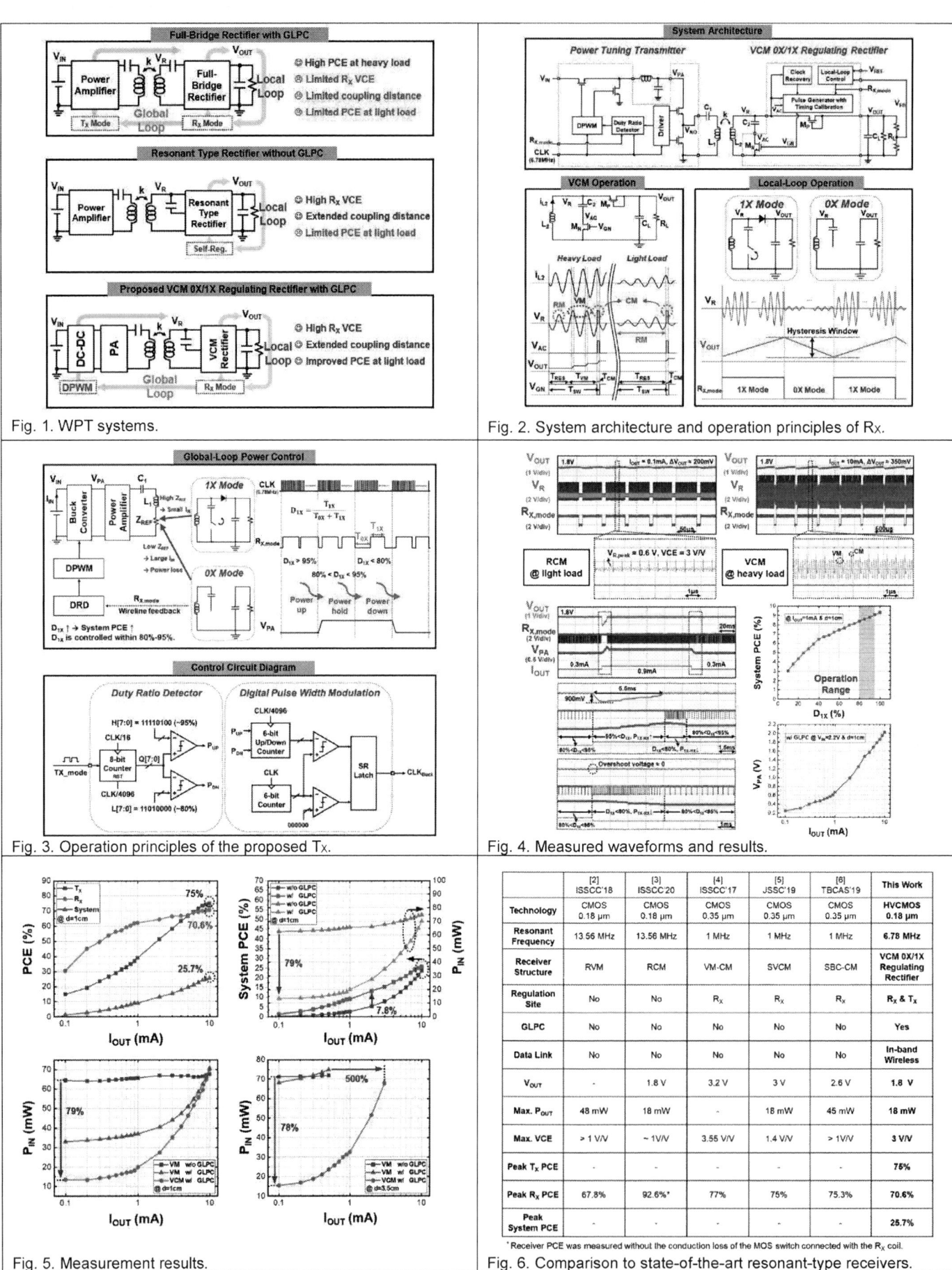

Fig. 1. WPT systems.

Fig. 2. System architecture and operation principles of Rx.

Fig. 3. Operation principles of the proposed Tx.

Fig. 4. Measured waveforms and results.

Fig. 5. Measurement results.

Fig. 6. Comparison to state-of-the-art resonant-type receivers.

	[2] ISSCC'18	[3] ISSCC'20	[4] ISSCC'17	[5] JSSC'19	[6] TBCAS'19	This Work
Technology	CMOS 0.18 μm	CMOS 0.18 μm	CMOS 0.35 μm	CMOS 0.35 μm	CMOS 0.35 μm	HVCMOS 0.18 μm
Resonant Frequency	13.56 MHz	13.56 MHz	1 MHz	1 MHz	1 MHz	6.78 MHz
Receiver Structure	RVM	RCM	VM-CM	SVCM	SBC-CM	VCM 0X/1X Regulating Rectifier
Regulation Site	No	No	Rx	Rx	Rx	Rx & Tx
GLPC	No	No	No	No	No	Yes
Data Link	No	No	No	No	No	In-band Wireless
V_{OUT}	-	1.8 V	3.2 V	3 V	2.6 V	1.8 V
Max. P_{OUT}	48 mW	18 mW	-	18 mW	45 mW	18 mW
Max. VCE	> 1 V/V	~ 1V/V	3.55 V/V	1.4 V/V	> 1V/V	3 V/V
Peak T_X PCE	-	-	-	-	-	75%
Peak R_X PCE	67.8%	92.6%*	77%	75%	75.3%	70.6%
Peak System PCE	-	-	-	-	-	25.7%

*Receiver PCE was measured without the conduction loss of the MOS switch connected with the R_X coil.

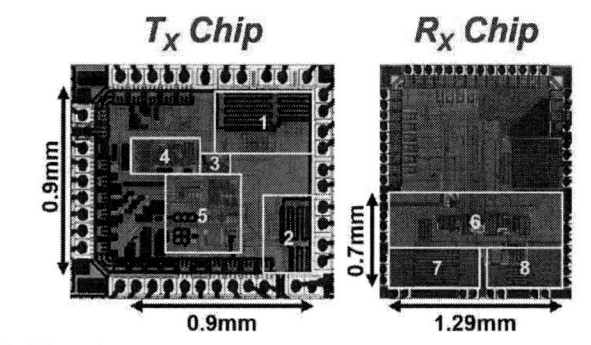

1: Buck Converter
2: Class-D Power Amplifier
3: Duty Ratio Detector
4: Digital Pulse Width Modulation
5: TX Control Circuits
6: RX Control Circuits
7: Power PMOS
8: Power NMOS

Fig. 7. Chip micrograph.

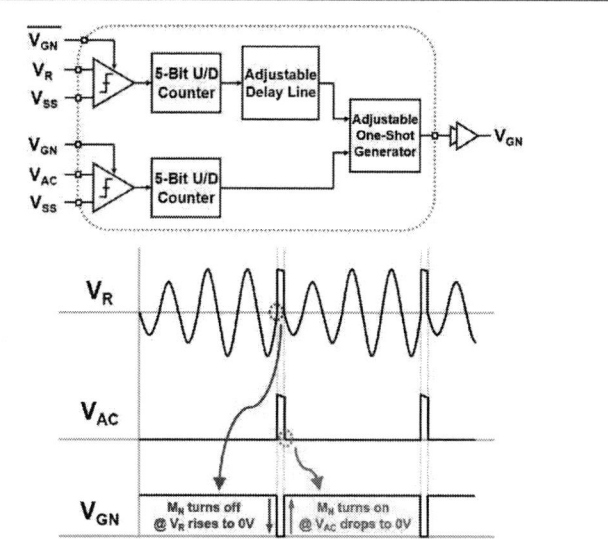

Fig. S1 Operation principles of the pulse generator with timing calibration.

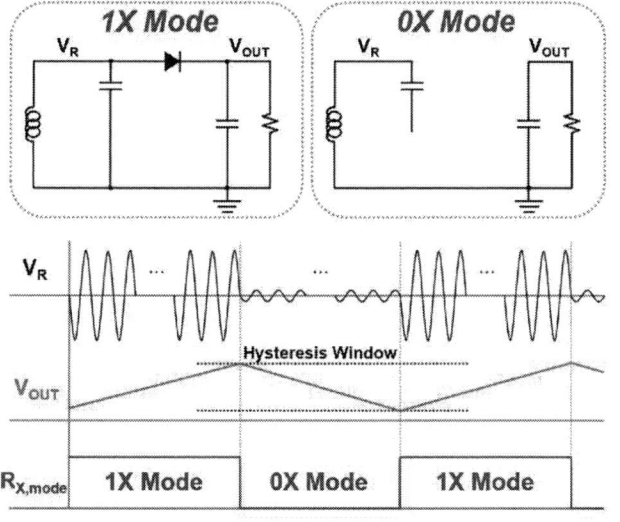

Fig. S2 Operation principles of the conventional VM.

A 40nm 2.76µJ/Op Energy-Efficient Secure Post-Quantum Crypto-Processor for CRYSTALS-KYBER on Module-LWE

Aobo Li, Jiahao Lu, Dongsheng Liu*, Xiang Li, Shuo Yang, Tianze Huang, Jiaming Zhang, Siqi Xiong, Chenjun Yang

Huazhong University of Science and Technology

Most information is transmitted through untrusted channels, and people use traditional asymmetric encryption methods/algorithms to prevent information leakage. Post-quantum cryptography (PQC) alternatives are refining to counter the possibility of brute-forcing encryption in finite time with progressively feasible quantum computers. As the winner of the NIST PQC project global competition, CRYSTALS-KYBER (Kyber) has a balance of security and efficiency. This paper proposes an efficient ASIC for chip-level applications of Kyber. Configurable operators and data generators are designed. Separation of secure memory and public memory from bus arbiter ensures leak-proof of critical information. A unified command and control system schedule execution process and data interaction. The proposed Kyber processor is fabricated and properly verified at a 40nm process, achieving high energy efficiency for the execution of the Kyber key encapsulation mechanism (KEM).

Figure 1 shows the chip architecture of the Kyber processor. The data generator uses security hash algorithm 3 (SHA-3) as a pseudo-random number generator (PRNG). SHA3-256/512 in SHA-3 is used as the hash function. Extension output function SHAKE-128 and SHAKE-256 are adopted to generate bitstream for sampling and coefficients that conform to a certain probability distribution. After the data is generated, it is stored in the SRAM. The degree of privacy of different data determines the storage location. Secure RAM is used to store private information that is prohibited from being accessed outside the chip, including private keys that cannot be leaked, noise vectors, and random seeds. The public RAM stores data and unimportant content for the key exchange process. Due to the existence of the bus arbiter, the data paths of the secure RAM and the public RAM are isolated. Secure RAM access can only be performed by special instructions. Cryptographic calculations are performed through 4-channel butterfly units with two RAMs. Directly connected and symmetrical computing pathways ensure efficient computing. The whole system is controlled by instructions, and a unified interactive interface is responsible for data interaction. A small SRAM is planned as the space of instruction and design for test (DFT). The ROM is used to store the NTT rotation factor, as well as some solidified configuration instructions.

The key operation steps of the Kyber key encapsulation mechanism in the key exchange process are illustrated in Figure 2. The server and client complete the trusted exchange process of message m by exchanging public key pk and ciphertext ct. Each perform a large number of polynomial multiplications and forward/inverse number-theoretic transformations (NTT/INTT). A butterfly arithmetic unit (BAU) is used to calculate NTT. By adjusting the position of the modular multiplier, modular adder, and modular subtractor in the data path, BAU can be adjusted to two structures, Cooley-Tukey (CT) and Gentleman-Sande (GS). Adding the step of modulo division by 1/2 in the GS structure can save n^{-1} multiplication in INTT. The NTT performed by the CT structure will cause the bit-order polynomial sorting becoming bit-reverse. The INTT of the GS structure restores the bit-reverse order to the bit-order. This process requires no additional rearrangement and address translation. Since the polynomial dimension in Kyber is 256 formed by the odd-even combination of two 128-th polynomials, the odd sequence and the even sequence are stored in blocks. Each address can load 4 coefficients, and every 32 addresses are defined as a block. The memory of 1024 addresses is planned as 32 blocks. Block operations facilitate the scheduling of calculations and the simplification of address control.

Figure 3 presents the circuit of the modular multiplier for polynomial multiplication. A single-channel butterfly arithmetic unit is formed by a fast Barrett modular multiplier. Flexible data selection supports configuration of the previous CT and GS structures. The calculation of the modulo division method 1/2 can be simply implemented by the process of shifting and adding after selection. Another circuit for a Montgomery modular multiplier is also implemented and evaluated. In the Design compiler (DC), we synthesize the two modular multipliers in the same constraint conditions and under the 40nm digital standard cell library. The area reports demonstrates that the Barrett modular multiplier has a smaller superiority. The histogram analyzes the period and energy of the realized NTT circuit respectively. The 4.9x cycle savings is achieved on a 256-point NTT implementation compared to [1]. Compared with [1] and [2], the energy consumption is reduced by 46% and 15%.

Figure 4 shows the circuit structure of SHA-3 and sampler. SHA-3 completes the iteration of the internal state through the keccak core. Each function takes 24 cycles to complete an iteration. Cascading two keccak cores can cut the time in half. But this will also cause increased area and additional delay. We evaluated the area and delay of one keccak and two keccaks using the DC. Two keccak cores yield acceptable area and latency increases. The iterated data within keccak generates different coefficients through rejection sampling and binomial sampling. This design produces a polynomial that consumes 26.5nJ of energy, and 40% reduction compared to [1].

Figure 5. (a) shows the change of load characteristics of power consumption as frequency increasing under different chip core supply voltages. Figure 5. (b) is the highest frequency change process of the chip under different supply voltages. The frequency performance of the chip is stable between 0.9V-1.2V, but drops sharply when it is lower than 0.9V. Figure 5.(c) and Figure 5.(d) list the area and power breakdown of the chip. SRAM, SHA-3 and NTT occupy the largest resource overhead. The left table shows the key generation (KeyGen), encapsulation (Enc) and decapsulation (Dec) time at 100MHz. The right shows the energy consumption for completing the entire KEM (KeyGen+Enc+Dec) under several typical voltages at 100MHz.

Figure 6 shows the comparison of the ASIC results between previous related work and this work. The design is manufactured at 40nm, and the total chip core circuit area is only 0.6mm², using 532 kGE (NAND2 equiv.). Compared to the similar [3], we consume only 22% circuit area. And compared with [1], 14KB SRAM is occupied, and the memory size is reduced by 65%. Compared with [2], NTT performance is 5x faster and energy consumption is reduced by 15%. The execution energy of the Kyber scheme is 2.76µJ@1.1V, saving 62% compared with [3]. Compared with [1] and [2], the speed is 4.7x and 28.4x faster, respectively. The results show that our chip is energy-efficient and achieves a good balance between area and performance. Figure 7 shows the micrograph and specification of the chip.

The chip implements an energy-efficient Kyber post-quantum cryptographic processor architecture. The optimization of storage strategy and computing circuit brings positive efficiency improvement to the chip. The securely isolated storage method and instruction control ensure the anti-theft and security of the chip in the application situation, which will empower the suitability of post-quantum cryptography on the Internet of Things and terminals.

Acknowledgements:

This work is supported by the National Key Research and Development Program of China (No. 2021YFA0715502), the National Natural Science Foundation of China (No. 61874163, 62104076, 62134002), the National Key Analog Integrated Circuit Laboratory Project of China (No. JCKY2021210C004), the Introduced Innovative R&D Team of Dongguan (No.201760712600139), and the Laboratory Open Fund of Beijing Smart-chip Microelectronics Technology Co. Ltd,.

References:

[1] U. Banerjee et al.,"2.3 An Energy-Efficient Configurable Lattice Cryptography Processor for the Quantum-Secure Internet of Things" ISSCC, pp. 46-48, Feb. 2019.

[2] Ghosh, A et al., "A 334uW 0.158 mm2 Saber Learning with Rounding based Post-Quantum Crypto Accelerator. " CICC, pp. 1-2, Apr.,2022.

[3] T. Shimada and M. Ikeda, "High-speed and energy-efficient crypto-processor for post-quantum cryptography CRYSTALS-Kyber," ASSCC, pp. 12-14, Nov., 2022.

979-8-3503-3004-5/23 $31.00 © 2023 IEEE

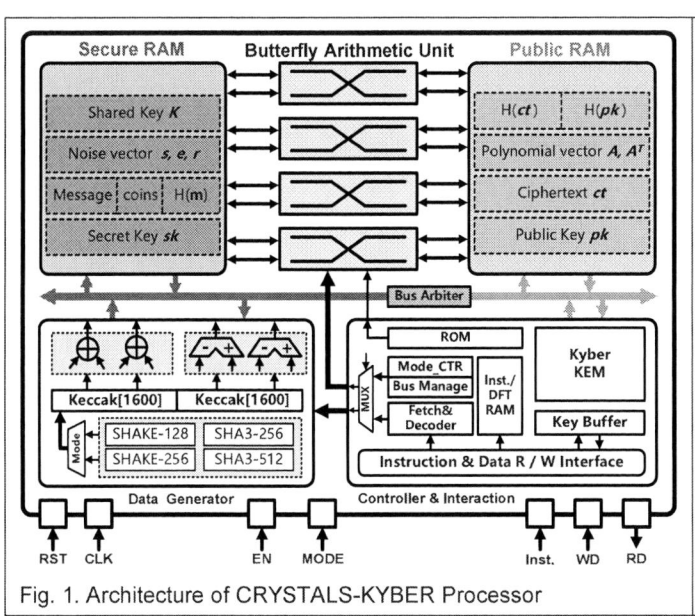

Fig. 1. Architecture of CRYSTALS-KYBER Processor

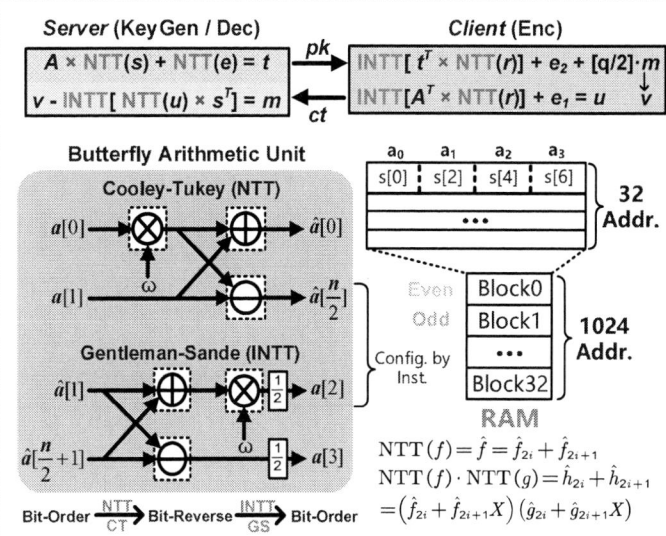

Fig. 2. Main Arithmetic Operation of Kyber and NTT Configuration

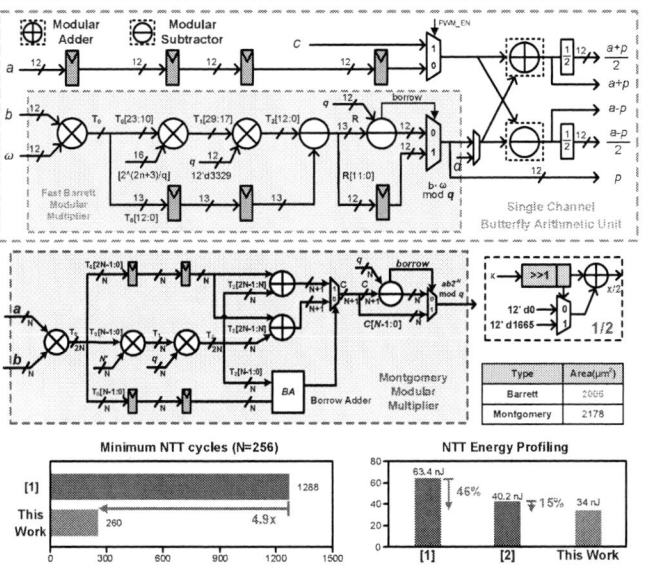

Fig. 3. Modular Arithmetic Circuits and NTT Profiling

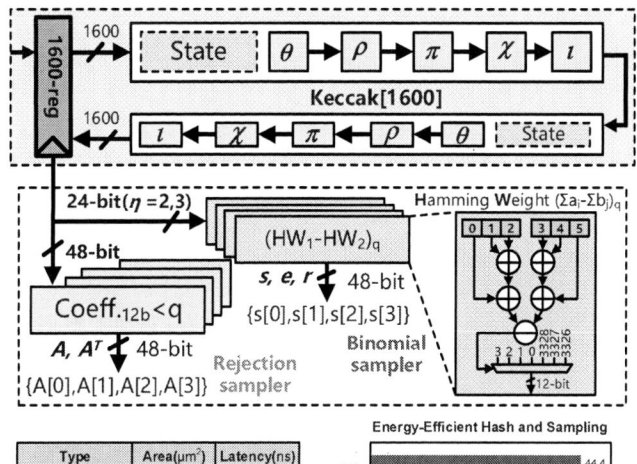

Fig. 4. Hash of Two Keccak Core and Sampling with Evaluation

Fig. 5. Chip Characteristics and Measured Results

KEM Scheme	KeyGen	Enc.	Dec.
Time(μs)	24	47.42	51.03

Supply Voltage	1.0V	1.1V	1.2V
Energy(μJ)	2.29	2.76	3.29

Table I Results Comparison with Related ASIC Works

	ISSCC'19 [1]	CICC'22 [2]	ASSCC'22 [3]	This work
Technology	40nm	65nm	65nm	40nm
Supply Voltage(V)	0.68-1.1	0.7-1.1	0.46-1.2	0.8-1.2
Frequency (MHz)	12-72	40-160	25-336	80-115
Crypto primitives	Ring-LWE	Module-LWR	Module-LWE	Module-LWE
Total Area (mm²)	0.28*	0.158*	2.74	0.6**
Logic Gates (kGE)	106*	--	1370	532**
Memory Size (KB)	40.25	10.1875	--	14
Power(mW)	0.516*	0.3339*	21-542	1.14-27.57**
NTT performance (N=256)				
Cycles	1288	1298	150	260
Energy(nJ)	63.4	40.21	--	34
Kyber scheme (keygen + encaps. + decaps.)				
Cycles	347826	57014	3602	12245
Energy(μJ)	26.6	1.748***	7.18	2.76

* Not contain memory area and power, only logic circuit in coprocessor
** Contains all, including memory, logic, interface and DFT
*** Measured at 0.7V without memory and SHA-3 power consumption

Fig. 6. Results Comparison with Other Designs

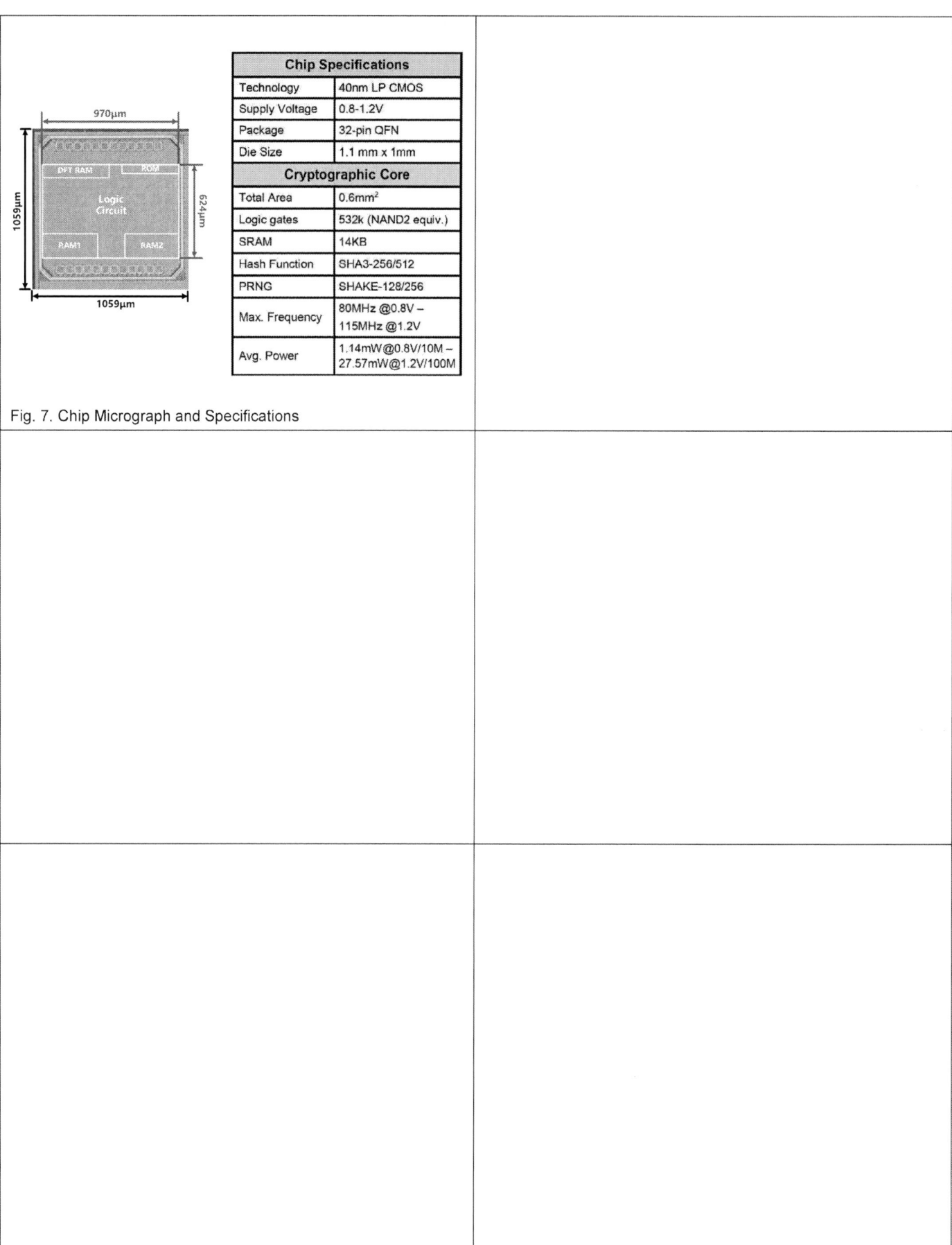

Fig. 7. Chip Micrograph and Specifications

A Resistor/Trimming-Less Self-Biased Current Reference Class with Area down to 3,500 μm^2, 42.8 pW Power and 10.4% Accuracy across Corner Wafers in 180 nm

Luigi Fassio, Hoang Hong Hanh, Massimo Alioto

National University of Singapore, Singapore

Current references are fundamental building blocks in sensor interfaces and other analog circuitries [1]-[9]. In applications with low-cost targets, their area and testing effort need to be minimized, and trimming must be eliminated. In power-constrained systems, their net power (i.e., not going into their output bias current I_{REF}) and minimum supply voltage V_{min} need to be kept low, so as to minimize the power burden and remove supply scaling limitations at the system level. Unfortunately, lower net power generally leads to poorer accuracy due to PVT variations, requiring innovation to make the power-accuracy tradeoff more favorable while avoiding trimming.

In prior ultra-low power current references, OTA-based references operating in sub-threshold and sub-V supply were demonstrated, although at the cost of large area (>50,000 μm^2) due to the need for a resistor [5], [6], or of high process sensitivity (~25%) being based on gate leakage [8], which is notoriously process-dependent and not really usable in technologies older than 65 nm. CMOS current references typically have relatively high minimum supply V_{min} (e.g., 1.1-1.35 V) since they mitigate their temperature dependency by operating around the zero-temperature coefficient point, which usually lies somewhere above the threshold voltage [3], [4]. This limitation was alleviated in [9] in FDSOI technology, although at large silicon area. Large area is a general issue in several current references, which can easily reach and exceed the 50,000-100,000 μm^2 range [2], [7] (Fig. 5). In this work, a new class of trimming-less and resistor-less current references is introduced for low-cost/low-power targets, compact area, and favorable power-accuracy tradeoff. This class of references is exemplified by two instances with the lowest area reported to date (3,500-8,100 μm^2 in 180 nm) by 2.6-34X over prior art. At the same time, they respectively achieve the second lowest net power and the best accuracy including corner wafers (not considered in prior art, with the exception of [1]).

Fig. 1 shows the schematic of the proposed self-biased current reference and the adopted transistor sizes. Transistor $M7_1$ operating in the triode region acts as a resistor with its gate terminal being biased by transistors $M1_1$, $M2_1$ and $M3_1$. In particular, $M1_1$ performs implicit voltage regulation at its source terminal (i.e., it decouples all other transistors from V_{DD}), and hence for all current reference bias transistors, enhancing the line sensitivity (LS). $M2_1$ is a zero V_{GS}-connected transistor generating a pA-range bias current to bias all other bias transistors. Based on the current of $M2_1$, the diode-connected transistor $M3_1$ generates the bias voltage for transistor $M7_1$ in the triode region. The bias voltage $V_{BIAS-TRIODE}$ effectively tracks the process corner, increasing (decreasing) at the SS (FF) corner, effectively compensating the weaker (stronger) transistor $M7_1$ in triode region at the same process corner. The latter approximates a resistor and serves as current sensing element in the loop through $M4_1$ and $M6_1$. The latter feedback loop is the key mechanism that improves the robustness of the output reference current I_{REF1} against voltage/temperature variations, as enabled by operation of $M7_1$ in triode (in place of a physical resistor). Indeed, an increase in I_{REF1} would lead to an increase in the voltage drop across $M7_1$ and hence the gate-source voltage of $M4_1$. In turn, this increases its current and hence reduces the internal voltage V_1. As a consequence, the current of $M6_1$ is reduced, contrasting the original increase of I_{REF1} due to voltage/temperature variations. The feedback loop is inherently robust against process corners since it uses only NMOS transistors in its core (native devices only regulate their source voltage, without interfering with I_{REF1}). The feedback loop is made nearly independent of temperature via proper sizing of $M2_1$-$M4_1$ for minimum temperature coefficient, whereas $M2_1$ sizing directly adjusts the output reference current I_{REF1}. Such simple and clear dependencies define a general and iterative transistor sizing procedure for systematic design.

As a second instance of the proposed class of current references, the OTA-based reference in Fig. 2 again uses $M7_2$ in triode region as a resistor and sensing element. $M1_2$-$M4_2$ form a classic 2T voltage reference with $M1_2$ as LS enhancement. From the analysis in Fig. 2, the internal voltage V_{REF} has a temperature dependence that can be adjusted with sizing. The voltage V_{REF} is subject to process variations influenced by the corner variations of RVT (regular threshold voltage ≈ 600 mV) and MVT (lower doping channel, medium threshold voltage ≈ 450 mV) devices, being $M4_2$ of the first type while $M2_2$ and $M3_2$ of the second type. Again, the sensing element $M7_2$ interacts with the feedback loop that now contains the OTA (NMOS input, two-stage, Miller-compensated), which sets the gate voltage of $M5_2$ to keep V_{SENSE}=V_{REF} and hence maintains a nearly-constant current against voltage/temperature variations (mismatch is not compensated, and its slight impact is mitigated through transistor up-sizing, compared to Fig. 1). The bias voltage required to keep $M7_2$ (RVT device) operating in the triode region is provided by the diode- connected $M6_2$ (MVT device). The voltage V_{SENSE} generated across the transistor $M7_2$ in triode region linearly depends on the reference current I_{REF2}, and is hence influenced by the threshold voltages of both MVT and RVT devices, effectively compensating for process variations in the reference voltage V_{REF}. The topology in Fig. 2 has a lower minimum supply V_{min} of 0.6 V compared to the topology in Fig. 1, due to the much lower V_{REF} compared to $V_{BIAS-TRIODE}$ in Fig. 1.

The 180nm testchip (Fig. 7) occupies an area of 3,500 μm^2 for the self-biased (8,100 μm^2 for the OTA-based) reference. The self-biased reference operates down to V_{min}=0.9 V and consumes 42.8 pW at room temperature, while the OTA-based has lower V_{min}=0.6 V and higher net power (1.95 nW, Figs. 3 and Figs. 5), averaging across 15 dice (5 TT, 5 SS and 5 FF, with mixed SF and FS corner wafers providing essentially the same results since the proposed current reference class is realized with NMOS transistors only). Power expectedly increases exponentially with temperature similar to the leakage temperature dependency (Fig. 3). At V_{min}, the TC is 440 ppm/°C for the self-biased and 550 ppm/°C for the OTA-based reference in the -10 to 100°C temperature range (Fig. 4). Fig. 4 shows that the overall process sensitivity (PS) including measured corners is 8% for the self-biased and 11.5% for the other proposed reference. The reference current expectedly increases (decreases) slightly at the FF (SS) corner in both topologies. In the self-biased structure, the temperature coefficient (TC) is process-dependent and shows the best TC performance with the FF corner, while the SS corner represents the worst-case scenario. On the other hand, the TC variation in the OTA-based structure is primarily due to mismatch (Fig. 4), and is expectedly slightly worse than the self-biased reference due to the lower drain-source transistor voltage. From Table I, the self-biased reference has the lowest area by 2.6-16X over prior art [1]-[8]. In spite of having the second-lowest net power after [3] (at 10.9X lower area), it has a competitive accuracy thanks to a moderate TC (1.6X higher than [1] at 23.8X lower power and 4.9X lower area) and LS (7X lower than [1]). The overall accuracy of 10.4% is the second best compared to [1]-[8], even if [2]-[8] do not include corner wafer measurements. This makes the proposed self-biased reference a compelling solution for low-cost and low-power design targets. Similar considerations hold for the OTA-based reference with lowest area by 1.2-6.8X over prior art [1]-[8] and only 2.3X larger than the self-biased solution). Overall, the OTA-based reference can be immersed in systems with lower supply voltage, at the expense of a slight accuracy degradation. Fig. 5 shows the statistical distribution of the power consumption and a more extensive comparison with the state-of-art. From this figure, the two current references belonging to the proposed class are well-suited for tightly-constrained low-cost systems, as they offer a highly-favorable tradeoff between temperature coefficient and area. Between the two, the self-biased reference exhibits superior accuracy.

Acknowledgements: Singapore Hybrid-Integrated Next-Generation μ-Electronics (SHINE) Center, TSMC for fabrication.

References:

[1] Q. Dong, et al., "A 1.02nW PMOS-Only, Trim-Free Current Reference with 282ppm/°C from -40°C to 120°C and 1.6% within-Wafer Inaccuracy," in 2017 ESSCIRC, pp. 19–22.

[2] T. Hirose, et al., "A Nano-Ampere Current Reference Circuit and its Temperature Dependence Control by using Temperature Characteristics of Carrier Mobilities," 2010 ESSCIRC, pp. 114–117.

[3] M. Choi, I. Lee, T.-K. Jang, D. Blaauw, and D. Sylvester, "A 23pW, 780ppm/°C Resistor-less Current Reference Using Subthreshold MOSFETs," 2014 ESSCIRC pp. 119–122.

979-8-3503-3004-5/23 $31.00 © 2023 IEEE

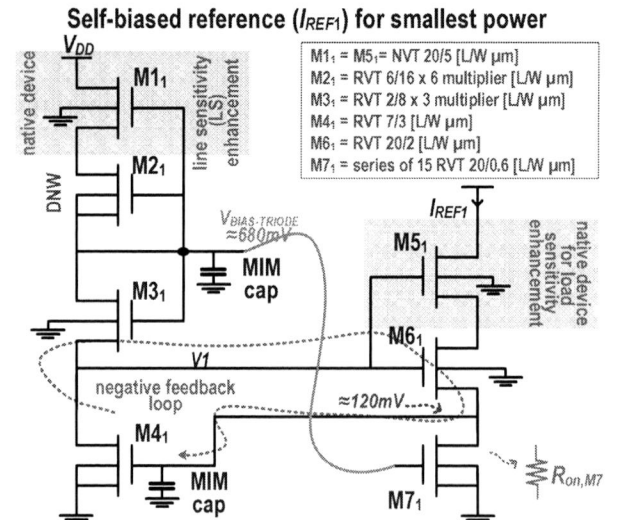

Fig. 1. Proposed class of current references exemplified by the self-biased topology generating current I_{REF1}.

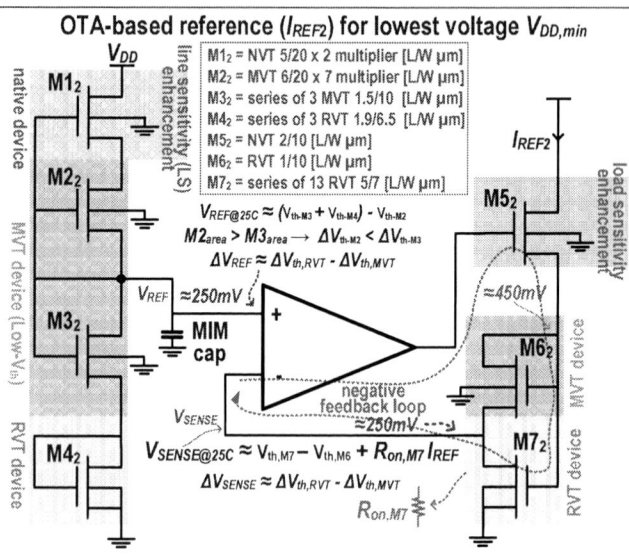

Fig. 2. Proposed class of current references exemplified by the OTA-based reference generating current I_{REF2}.

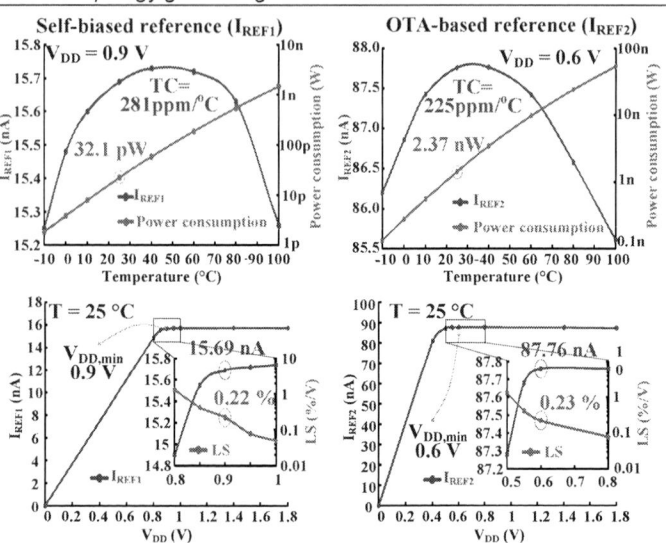

Fig. 3. Temperature coefficient and power consumption (top), line sensitivity (bottom) for the 2 proposed solutions on typical sample.

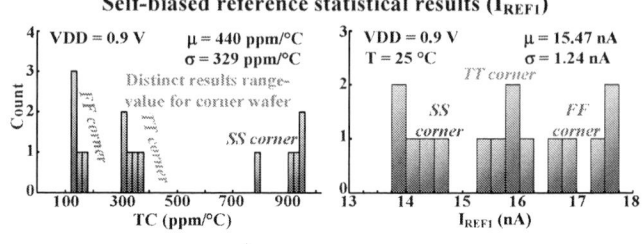

Fig. 4. Temperature coefficient and reference current statistical results for self-biased (top) and OTA-based (bottom) solutions across 15 dice (5 TT, 5 SS and 5 FF).

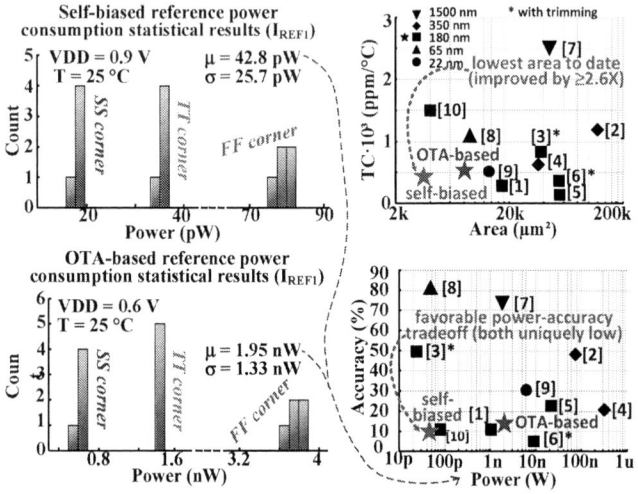

Fig. 5. Power consumption statistical results for self-biased (top-left) and OTA-based (bottom-left) solutions across 15 dice (5 TT, 5 SS and 5 FF). Relative comparison with the state of the art (right).

Table I. Comparison with state-of-the-art voltage references.

	self-biased	OTA-based	ESSCIRC 2017 [1]	ESSCIRC 2014 [3]	TCAS-I 2019 [5]	ISSCC 2017 [6]	ESSCIRC 2017 [8]
technology [nm]	180	180	180	180	180	180	65
type	CMOS	OTA	CMOS	CMOS	OTA	BGR-OTA	OTA
resistor	NO	NO	NO	NO	YES	YES	gate leakage
active area [μm²]	3,500	8,100	17,000	38,200	55,000	55,000	9,100
V_{min} = min V_{DD} [V]	0.9	0.6	1.5	1.2	0.7	1.3	0.6
I_{REF} [nA]	15.47	82.95	35	0.02	9.97	6.64	0.01
net power (w/o I_{REF}) [pW]	42.8	1950	1020	23	28000	9300	49.7
power utilization** $I_{REF}I/I_{TOT}$	99.7%	96.2%	98.1%	51%	24.9%	48.1%	10.9%
TC [ppm/°C]	440	550	282	780	149.8	283	1077
temperature range tested [°C]	-10 100	-10 100	-40 120	0 80	-40 125	0 110	-40 120
LS [%/V]	0.43	0.41	3.00	0.58	0.60	1.16	1.79
PS [%]	8	11.5	8.6	15	7.2	1.19	24.9
corner wafer testing	YES	YES	YES	NO	NO	NO	NO
trimming	NO	NO	NO	YES	YES²	YES	NO
accuracy[1] [%]	10.4	14.4	11.2	>40%	>40%	22.6	5.5	>40%

**Power utilization = $\frac{V_{DD}I_{REF}}{V_{DD}I_{TOT}} \times 100 = \frac{I_{REF}}{I_{TOT}} \times 100$ [1] accuracy [%] = PS(%)·(1 if corner wafer, 3 otherwise)+LS (%)·0.4V+
+TC(%/°C)·50°C [i.e., 3-σ within-lot variations for fair comparison, as most references do not report corner wafer, 0.4-V harvested voltage or micro-battery change 50°C temperature deviation] ² batch trimming

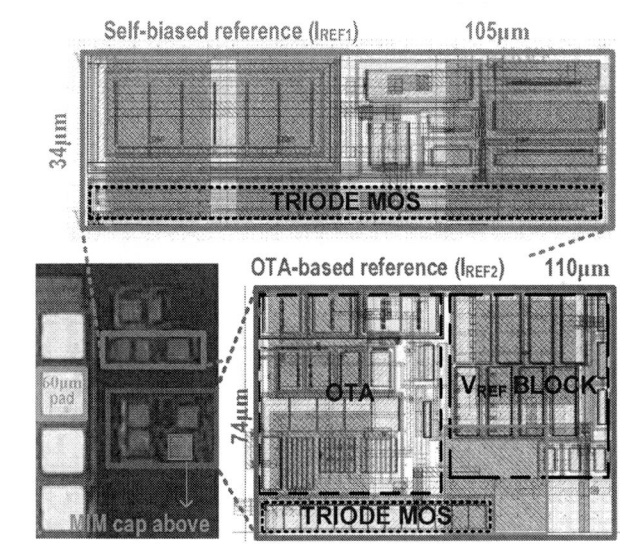

Fig. 7. Chip micrograph and layout of both topologies.

Additional References:

[4] D. Osipov et al., "Compact Extended Industrial Range CMOS Current References," IEEE Trans. on Cir. and Sys. I: Reg. Papers, vol. 66, no. 6, pp. 1998–2006, Jun. 2019.

[5] L. Wang et al., "A 0.7-V 28-nW CMOS Subthreshold Voltage and Current Reference in One Simple Circuit," IEEE Trans. on Cir. and Sys. I: Reg. Papers, vol. 66, no. 9, pp. 3457–3466, Sept. 2019.

[6] Y. Ji et al. "A 9.3nW All-in-One Bandgap Voltage and Current Reference Circuit," 2017 IEEE ISSCC, pp. 100–101.

[7] E. M. Camacho-Galeano et al., "A 2-nW 1.1-V self-biased current reference in CMOS technology," in *IEEE Transactions on Circuits and Systems II: Express Briefs*, vol. 52, no. 2, pp. 61-65, Feb. 2005.

[8] H. Wang et. Al., "A 1.6%/V 124.2 pW 9.3 Hz relaxation oscillator featuring a 49.7 pW voltage and current reference generator," 2017 ESSCIRC, pp. 99-102.

[9] M. Lefebvre et al., "A 0.9-nA Temperature-Independent 565-ppm/°C Self-Biased Current Reference in 22-nm FDSOI," ESSCIRC 2022 pp. 469-472.

[10] L. Fassio et al., "A 0.6-to-1.8V CMOS Current Reference With Near-100% Power Utilization," in *IEEE Transactions on Circuits and Systems II: Express Briefs*, vol. 68, no. 9, pp. 3038-3042, Sept. 2021

A 11.6-to-13.6GHz, 47-fsrms Jitter Integer-N PLL with Simultaneous Differential 2ND-Harmonic Output using a -128dBc/Hz Phase Noise at 1MHz offset Quad-Core VCO in 65nm CMOS

Ruichang Ma[1], Haikun Jia[1], Hongzhuo Liu[1], Wei Deng[1,2], Zhihua Wang[1], and Baoyong Chi[1]

[1]School of Integrated Circuits, BNRist, Tsinghua University, China

[2]Research Institute of Tsinghua University in Shenzhen, China

Email: jiahaikun@tsinghua.edu.cn

PLL frequency synthesizers with low jitter are essential for mm-Wave high-speed wireless communication systems and data converters. By doing away with the frequency doubler, low jitter synthesizers with the 2ND harmonic output can simplify their architecture. Additionally, to produce high-quality local oscillator signals for 2ND harmonic output, it is beneficial to suppress the fundamental frequency component at 2ND harmonic output. In general, lowering the voltage-controlled oscillator's phase noise (PN) and reducing the PLL's bandwidth to suppress in-band noise are the two most popular ways to minimize PN of PLL. The phase noise of N-core VCO reduce PN by 10logN, which is an effective method for obtaining low PN. Additionally, increasing the reference clock frequency hence decreasing the loop frequency division ratio can decrease PN. Literature [1] used a reference clock frequency of 500MHz with 175mW to implement the PLL of 52-fsrms Jitter. Ref. [2] adopted a folded transformer-based series resonance CMOS VCO, the PN of which achieved -130.8dBc/Hz@1MHz, consumed 242mW while reducing the core area. Ref. [3] uses a clock frequency of 250MHz to reduce the overall contribution of VCO through wideband PLL, while suppressing the noise introduced by the reference clock through the design of double-sampling phase detector (SPD). Ref. [4] uses the high gain of phase-frequency detector (PFD) with time amplification to suppress the gain of CP, thus achieving the performance of 60-fsrms Jitter. This study presents the construction of a PLL based on a 4-core resistance-synchronized VCO that achieves 47-fsrms Jitter and produces 2ND harmonic with fundamental frequency suppression.

Figure 1 depicts the proposed PLL architecture based on a quad-core VCO. The single-core VCO achieves about PN of -116dBc/Hz using common mode resonance at twice oscillation frequency by adding a 8-shaped coil inside the Class-F transformer, which couples to the drain and source in the common-mode while keeps transparent in differential-mode, extracted 2ND harmonic output while obtaining low PN. In order to create a good synchronization effect and lessen effect on PN of the VCO, the quad-core VCO connected the gate, drain of each core together with small resistors. The gain of SPD is increased by a reference buffer working in the high voltage range of 2.5V based on the conventional SPD-analog PLL structure, which suppressed OTA noise. High operating voltage for reference buffer may make larger SPD gain, although it will increase the power of reference buffer, there is a trade-off between them. As shown in Figure 2, the configurable 5bit-OTA helps to adjust the loop bandwidth more easily while ensuring the stability of the loop, which is determined by the low pass filter (LPF) at this condition. In addition, the input clock frequency of the multi-mode divider (MMD) can be reduced using an inverter based divide-by-2 [4], which stores charges with the parasitic capacitance of inverter. Like the structure of TSPC, leakage make the lowest operating frequency higher than static DFF, but that is enough for working at tens of GHz. The simulation results show that the high-speed divide-by-2 can work from 5 to 15GHz while consuming 2mW. It should be noted that reducing the width of

the finger of the MOSFET in the inverter is helpful to improve the operating frequency of the divider.

Figure 3 shows the layout realization of 2ND harmonic extraction and notch filter at fundamental frequency for quad-core VCO. The second-order harmonic component of the fundamental frequency signal is contained in each core's 8-shaped inductor, which uses a cross-transmission line structure to lessen the impact of the transmission line on adjacent area. The fundamental signal component is decreased by notch after 2ND harmonic has been combined. The simulation results show that the center frequency of notch is around 12.5GHz and that its maximum trap magnitude is 22dB.

The proposed PLL is fabricated in 65nm CMOS process with an core area of 1.08 mm^2. The measured frequency range of oscillator is 15.9% from 11.6 to 13.6GHz. Figure 4 shows the phase noise performance of free-running VCO and locked PLL at 11.6GHz from the fundamental output. The PN of free-running VCO at 1MHz offset is -123.25dBc/Hz, and the integrated jitter of locked PLL from 10KHz to 40MHz is 47-fsrms, while the PN of 2ND harmonic output at 1MHz offset is -111.57dBc/Hz, which is 6dB lower than that of VCO, and the integrated jitter from 10KHz to 40MHz is 49.14-fsrms, and the reference spur at 11.6GHz is -67.55dB. The low-jitter 2ND harmonic extraction demonstrating the proposed 2ND harmonic extraction and the notch filtering method.

As shown in figure 5, the reference spur is less than -62dB over the entire frequency range. And the phase noise fluctuation range of the VCO at 1MHz and 10MHz frequency offset is around 5dB, and the integrated jitter at both fundamental output and the 2ND harmonic output over the entire range is smaller than 60-fsrms. The PN deterioration of the VCO at high frequency causes the jitter degradation.

Figure 6 compares the reported PLL to ultra-low jitter in 60-fsrms. This design uses a 100MHz reference clock to achieve relatively low integrated jitter and FoM, and simultaneously extracted low integrated jitter at 2ND harmonic output with fundamental frequency suppression.

Acknowledgement

This work was supported by the Beijing Municipal Science and Technology Project (No. Z221100007722024) and the Beijing Innovation Center for Future Chips (ICFC).

References:

[1] M. Mercandelli et al., "12.5GHz Fractional-N Type-I Sampling PLL Achieving 58fs Integrated Jitter," ISSCC, pp. 274-276, Feb 2020.

[2] S. Zhang et al., "A 100 MHz-Reference, 10.3-to-11.1 GHz Quadrature PLL with 33.7-fsrms Jitter and -83.9 dBc Reference Spur Level using a -130.8 dBc/Hz Phase Noise at 1MHz offset Folded Series-Resonance VCO in 65nm CMOS," CICC, pp.1-2, May 2023.

[3] Y. Zhao et al., "A 19-GHz PLL with 20.3-fs Jitter," 2021 Symposium on VLSI Circuits, Kyoto, Japan, 2021, pp. 1-2, June 2023.

[4] Y. Zhao, et al., "A 56GHz 23mW Fractional-N PLL with 110fs Jitter," ISSCC, pp. 1-3. Feb 2022.

[5] R. Ma, et al., "A 12.5-to-15.4GHz, -118.9dBc/Hz PN at 1MHz offset, and 191.0dBc/Hz FoM VCO with Common-Mode Resonance Expansion and Simultaneous Differential 2ND-Harmonic Output using a Single Three-Coil Transformer in 65nm CMOS," CICC, pp. 1-2, May 2022.

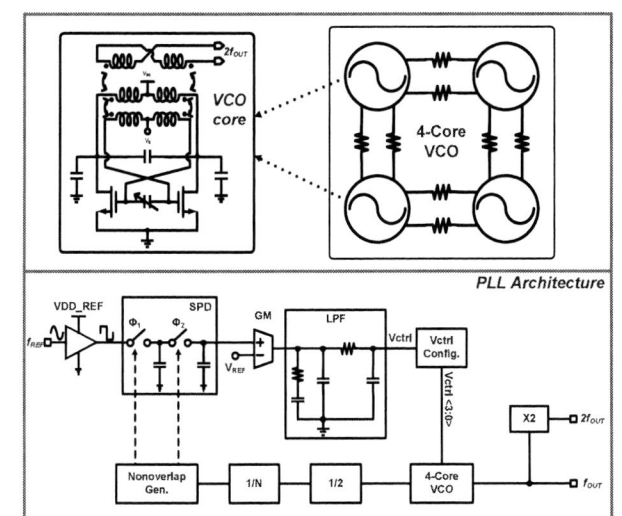

Fig. 1. Architecture of proposed PLL and res synchronic 4-core VCO.

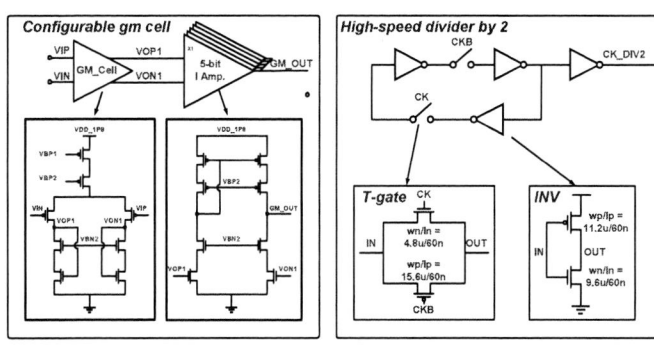

Fig. 2. Schematic of proposed configurable gm cell and High-speed divider by 2.

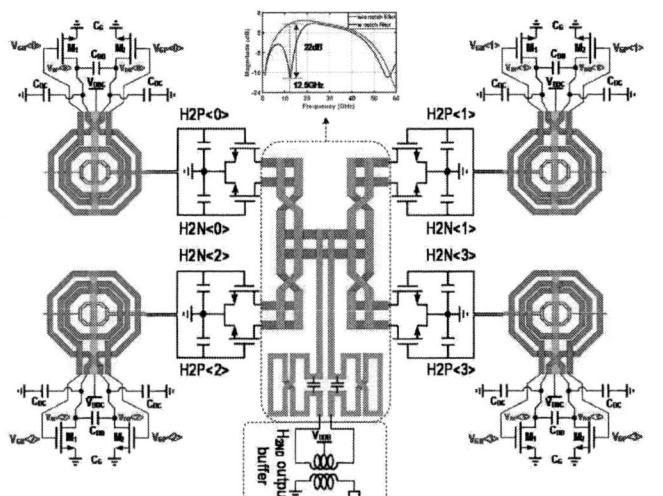

Fig. 3. Detail schematic of the proposed 2ND-harmonic extraction of 4-core VCO and simulation result of notch filter.

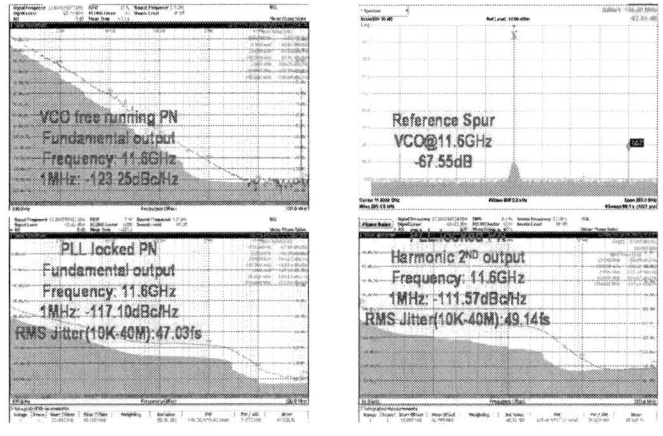

Fig. 4. Left: Measured phase noise of VCO and PLL at 11.6GHz from fundamental output. Right: Measured reference spur and phase noise of 2ND harmonic output.

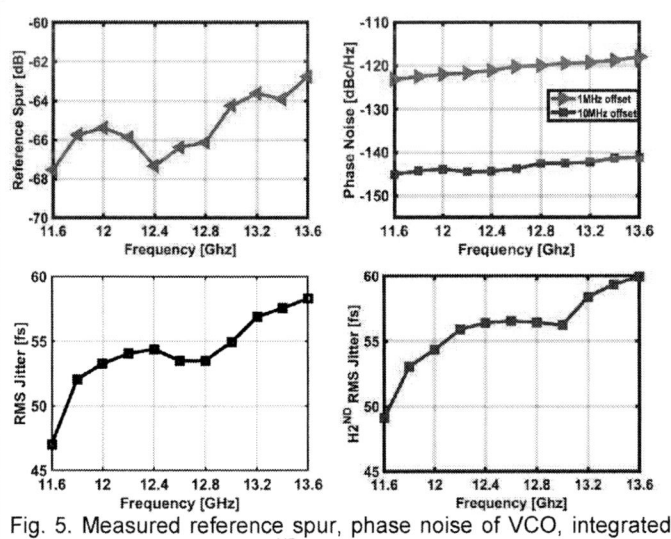

Fig. 5. Measured reference spur, phase noise of VCO, integrated jitter of VCO output and H2ND over the PLL output frequency range.

	This work	ISSCC20	ISSCC22	VLSI21	CICC22
Technology	65nm	28nm	65nm	28nm	65nm
Architecture	SPD	SSPD	CPPLL	Double SPD	SPD
Type	Integer N	Integer N	Integer N	Integer N	Integer N
Ref.Freq [MHz]	100	500	200	250	100
Output Freq. [GHz]	11.6-13.6 23.2-27.2	11.9-14.1	24-24.8	19	10.3-11.1
RMS Jitter [fs]	47	51.7	60	20.3	33.7
Integ. Range [Hz]	10K-40M	1K-100M	10K-30M	10K-100M	10K-40M
Spur [dB]	-67.55	-64.6	-47	-66	-83.9
2ND Harmonic Output	yes	No	No	No	No
2ND Harmonic RMS Jitter [fs]	49	-	-	-	-
Power [mW]	94	18	14.48	12	242
FoM [dB]	-246.8	-253.2	-252.8	-263	-245.6
Core Area [mm²]	1.08	0.23	0.35	0.06	0.7

FoM = 20log10(jitter$_{RMS}$/1s)+10log10(Power/1mW)

Fig. 6. Comparision summary and comparison with state of the art.

979-8-3503-3004-5/23 $31.00 © 2023 IEEE

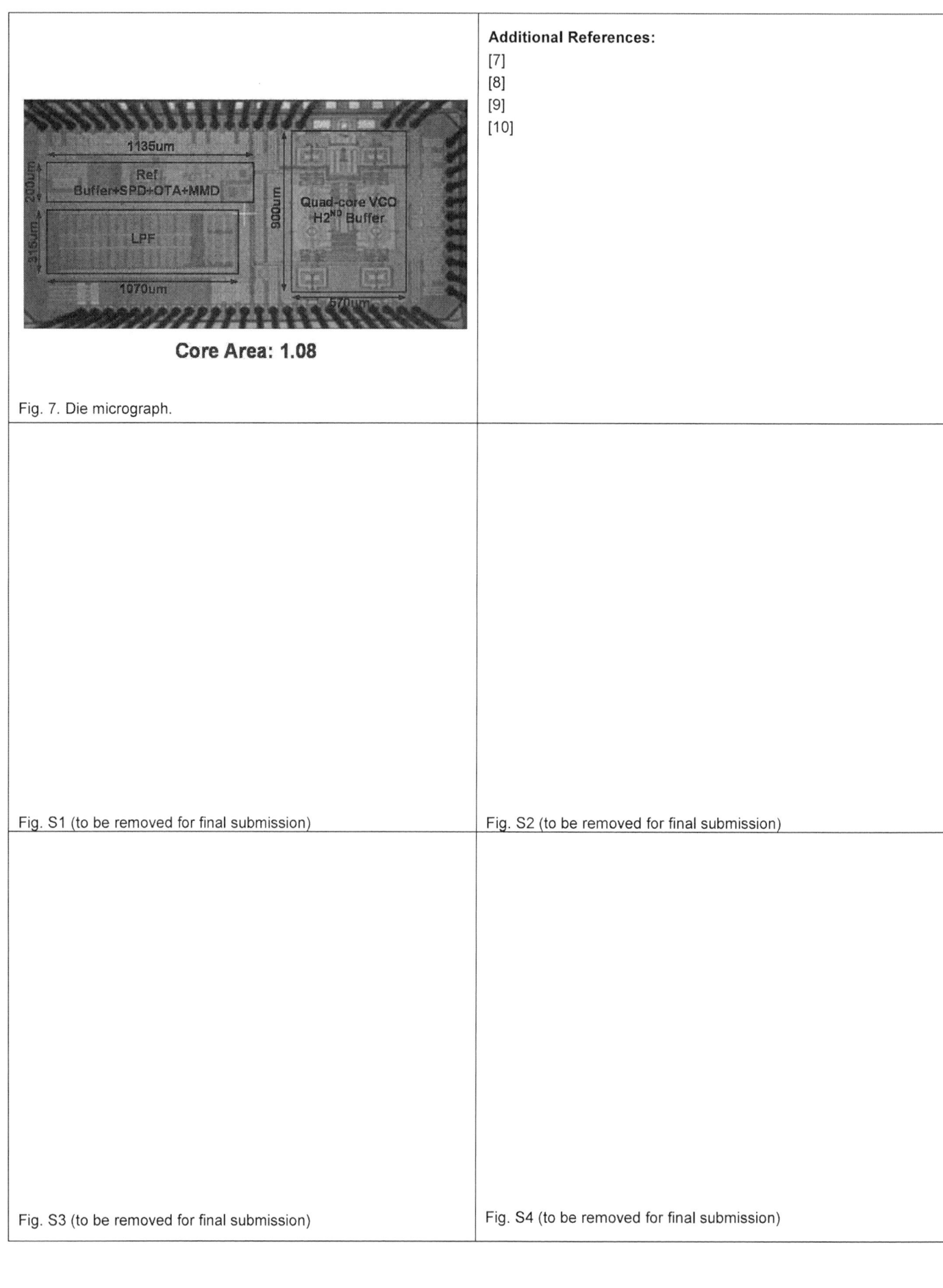

Fig. 7. Die micrograph.

Additional References:
[7]
[8]
[9]
[10]

Fig. S1 (to be removed for final submission)

Fig. S2 (to be removed for final submission)

Fig. S3 (to be removed for final submission)

Fig. S4 (to be removed for final submission)

A 27-to-31.6 GHz 8-Element Phased-Array Transmitter Front-End with Inter-Element-Interference Cancellation Scheme in 65 nm CMOS

Dongze Li, Wei Deng, Xintao Li, Ruiheng Qiu, Haikun Jia, Xiangrong Huang, Ziyuan Guo, and Baoyong Chi

School of Integrated Circuits, Tsinghua University, China

Email: wdeng@tsinghua.edu.cn

Millimeter-wave (mm-wave) phased-arrays have demonstrated great potential in 5G, satellite communication, radar applications. However, their compact element arrangement makes them highly susceptible to the influence of other elements, especially in the case of large array elements with a narrow beamwidth. The strength of signal in transmitters (TXs) is commonly stronger than that in the receivers (RXs), resulting in greater influence. In this work, an inter-element-interference cancellation (IEIC) scheme is introduced to cancel coupling caused by phase-shifting between elements. A 27-to-31.6 GHz 8-element phased-array transmitter front-end (FE) is proposed to validate this scheme. By implementing this scheme, the coupling can be reduced by 20 dB, proving its effectiveness.

Figure 1 illustrates the interference effect and block diagram of the proposed phased-array TX FE with the IEIC scheme. The phase shift in element 3 leads to interference in other elements due to coupling. Simulation results show the calculated radiation pattern of a 16-element linear array with random error efforts. The phase and gain error may cause 10-dB degradation in sidelobe, thereby compromising the beamforming quality. In the TX, each element consists of a power amplifier (PA), a vector-modulated phase shifter (VMPS), a driver amplifier (DA), and a switch-type attenuator (ST-ATT). In the IEIC scheme, output signals including useful signals along with unwanted coupling from other elements are introduced into a vector network analyzer (VNA) to extract the phases and amplitudes of signals. After then, the coupling among elements is calculated. After remapping the address of the phase shifter control bits in the codebook, the reconfigured control bits are renewed and sent to SPI in TX FE chip.

The detail interpretation of the proposed IEIC scheme and the procedure are presented in Figure 2. In the case of an 8-element TX, the phase control bits are expanded to form a 1*8 matrix S_n to represent the interference among elements. Here 'n' corresponds to the element number. For example, in the case of element 1 (ele.1), A_{11} in S_1 represents the influence of ele.1's control bits on its own phase while A_{12} represents the influence of ele.2's control bits on the ele.1's phase. The matrix S_n captures the impact of each element's control bits on its own phase and the phases of other elements in the array.

$$S_n = [A_{n1} \quad A_{n2} \quad A_{n3} \quad A_{n4} \quad A_{n5} \quad A_{n6} \quad A_{n7} \quad A_{n8}]^T \quad (1)$$

The detailed procedure is as follows. Firstly, choose element 1 and measure all phase shift data with keeping phases of all other elements steady. For the selected phase, choose selected vector $\vec{V_A}$ and write its control bits address into the codebook as the initial codebook. Secondly, measure the phase data $\vec{V_E}$ with inter-element-interference by keeping the phase control bits of the element-under-test unchanged while iterating through all phase states of the other elements. The error Vector is defined as: $\vec{a} = \vec{V_E} - \vec{V_A}$. Thirdly, obtain the error compensation vector $\vec{b}$ by keeping the amplitudes the same and the phase differing by 180 degrees from $\vec{a}$. The phase data with interference-compensation can be expressed as: $\vec{V_C} = \vec{V_A} + \vec{a}$. Search for the nearest control bits addresses of $\vec{V_A}$ to find $\vec{V_C}$. Write the control bits into the correction codebook. Then S_1 is finished. Finally, repeat above procedures for each element to get correction codebook of all 8 elements.

Figure 3 shows schematics of the proposed high-accuracy VMPS and amplifiers. The VMPS incorporates a transformer-based quadrature hybrid coupler (QHC) to generate 4-way IQ signals. To broaden the working bandwidth, a cascade-QHC is introduced. Simulation results indicate phase and amplitude mismatches of less than 3° and 0.6 dB, respectively. The quadrature weighting is implemented by Gm-array, which consists of 6 binary-multiplied-sized Gm-cells to achieve discrete normalized ac current within range of −63 to +63. The basic gain-cell of the PA and driver amplifier DA is formed by a cross-coupled differential amplifier. Cap-arrays are introduced in magnetically coupled resonators to help tune the designed bandwidth and ripple, as shown in the simulation results.

The proposed 8-element phased-array TX has been fabricated in 65-nm CMOS technology and occupies 3.4 mm×4.5 mm area including pads as shown in Fig.7. The performance of the TX is measured through an on-chip probing system with 120 mW per element power consumption.

Figure 4 depicts performance of the IEIC scheme. Results indicate that with IEIC, the sidelobe can be decreased. Based on [1], the coupling can be expressed as:

$$\frac{C}{Y} = \frac{\sqrt{Y^2 + (Y + A_{err})^2 - 2Y * (Y + A_{err})cos\theta_a}}{Y} \quad (2)$$

Y is equal to $|\vec{V_A}|$ and $A_{err} = |\vec{V_E}| - |\vec{V_A}|$ in (2). Due to the symmetrical placement of the TX FE, other elements share the same electromagnetic environment as element 1 and 2. The results in Figure 2 indicate that only adjacent elements exhibit significant interference. By leveraging the IEIC scheme, the coupling can be reduced by 16 dB and 20 dB for element 1 and element 2, respectively.

The measured S-parameters and large-signal performance are presented in Fig. 5. The peak gain is 20.4 dB and BW is 4.6 GHz from 27-to-31.6 GHz. The gain variation among 8 elements is less than 1.4 dB. The S22 shows good return loss below −10 dB throughout the band of interest. The output 1dB gain-compression-point (OP1dB) and saturation power are around 10.7 dBm and 13.6 dBm, respectively. The performance of the phase and gain control is also depicted in Figure 5. With the proposed high-performance phase shifter, a 360° phase-shifting range with 6-bit resolution is achieved. From 27-to-31.5 GHz, the RMS phase error is less than 2.2° and the minimum value is only 0.2°. The ST-ATT has a dynamic range of 31.5 dB. The measured RMS gain error is less than 0.31 dB and the minimum value is only 0.17 dB.

The TX is also measured using modulation data. For sat-com application, the EVM is better than −30.46 dBc with 600Msym/s 16APSK signal and −30.03 dBc with 600Msym/s 32APSK signal. For 5G NR application, the EVM is better than −33.00 dBc with 200Msym/s 64QAM signal and −33.19 dBc with 200Msym/s 128QAM signal.

Compared with state-of-the-art silicon-based phased-array TX FEs at similar frequencies in Figure 6, the proposed TX FE exhibits superior performance in terms of phase shift and gain control, as well as modulation capabilities. The proposed IEIC scheme helps to reduce the coupling by 20 dB without consuming additional power. The result demonstrates the effectiveness of the scheme in mitigating inter-element interference.

Acknowledgement:

This work was partly supported by the National Key R&D Program of China under Grant No. 2020YFB1807300, the Shenzhen Science and Technology Program (No. JCYJ20200109113601723 and No. JCYJ20210324115610028), the Tsinghua-Samsung Joint Research Project, and the Beijing Advanced Innovation Center for Integrated Circuits.

References:

[1] T. Yu et al., "A 22–24 GHz 4-element CMOS phased array with on-chip coupling characterization," JSSC, vol. 43, no. 9, pp. 2134-2143, Sep. 2008.

[2] A. C. -W. Wong, et al., "A 4Rx, 4Tx Ka-band transceiver in 40nm bulk CMOS technology for satellite terminal applications," RFIC, 2021: 211-214.

[3] A. Li, et al., "A Fully Integrated 27.5-30.5 GHz 8-Element Phased-Array Transmit Front-end Module in 65 nm CMOS," A-SSCC, 2019: 153-156.

[4] J. Pang, et al., "A 28-GHz CMOS Phased-Array Beamformer Utilizing Neutralized Bi-Directional Technique Supporting Dual-Polarized MIMO for 5G NR," JSSC, vol. 55, no. 9, pp. 2371-2386, Sep. 2020.

979-8-3503-3004-5/23 $31.00 © 2023 IEEE

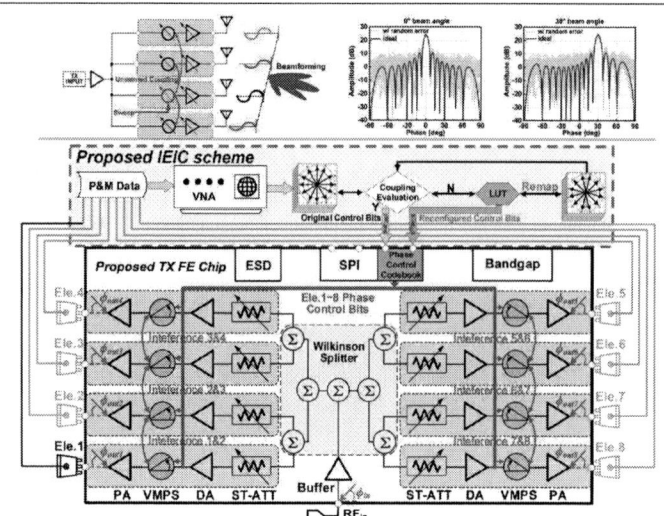

Fig. 1. Interference effect and block diagram of the proposed phased-array TX FE with the IEIC scheme.

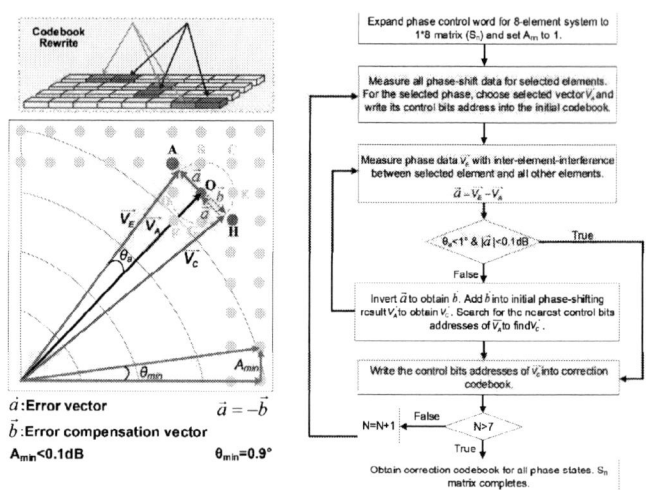

Fig. 2. Procedure of proposed IEIC scheme.

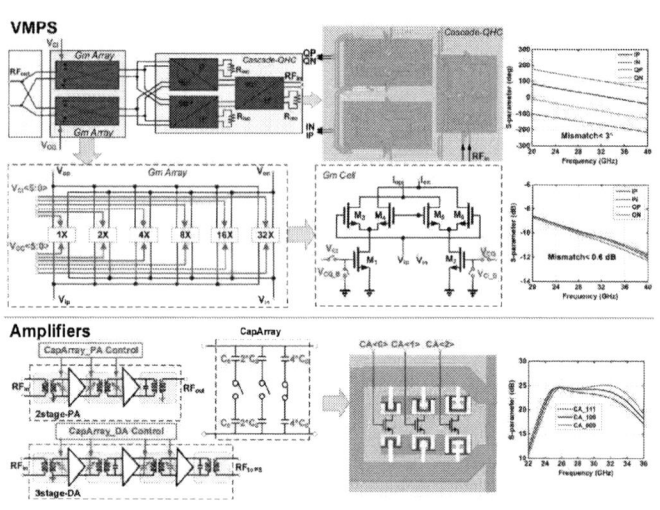

Fig. 3. Schematics of the proposed high-accuracy VMPS and amplifiers.

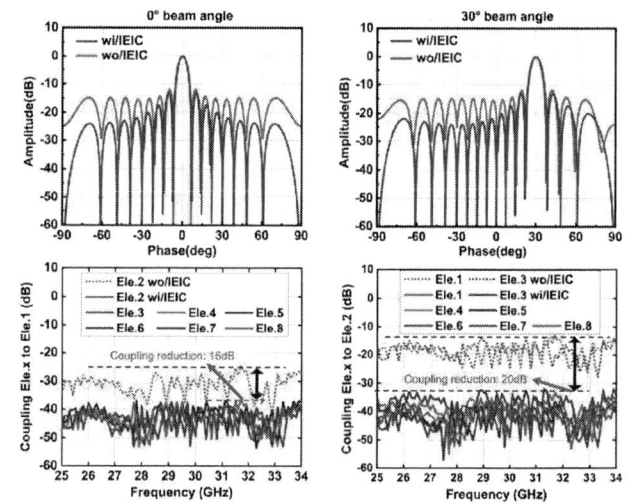

Fig. 4. Performance based on measured results of IEIC scheme.

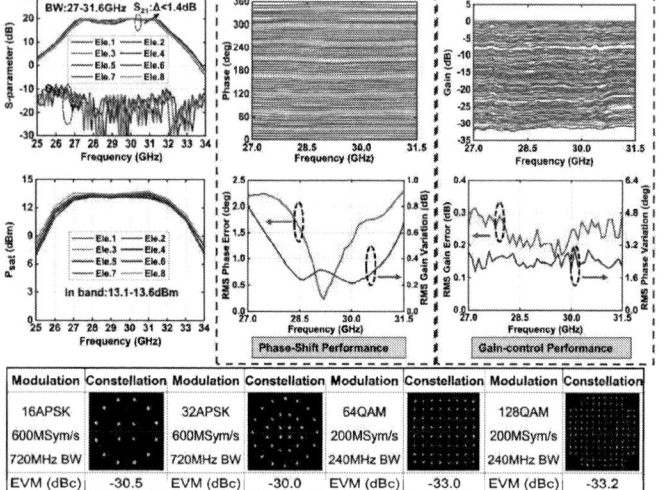

Fig. 5. Measured results of S-parameters, P1dB, Psat, phase-shift, gain-control and modulation.

Fig. 6. Performance comparison.

	[2] ASSCC 19	[3] RFIC 21	[4] JSSC 20	This Work
Technology	65 nm CMOS	40 nm CMOS	65 nm CMOS	65 nm CMOS
Elements	8	4	8	8
Frequency (GHz)	27.5-30.5	27-30	26.5-29.5	27-31.6
Bandwidth (GHz)	3	3	3	4.6
Gain (dB)	36	N/A	19.5	20.4
Eff. Phase bit	6	5	5	6
RMS Phase min. err. (deg)	1.25	3	0.4	**0.2**
Eff. Gain bit	5	N/A	N/A	6
Gain Range (dB)	15.5	N/A	N/A	**31.5**
RMS Gain min. err. (dB)	0.26	N/A	N/A	**0.17**
OP1dB (dBm)	12.7	N/A	11.5	10.7
Psat (dBm)	16.0	10	15.1	13.6
Modulation	N/A	QPSK 45Msym/s −20.2 dBc	64QAM 2.5Gsym/s −24.2 dBc	128QAM 200Msym/s **−33.2 dBc**
Coupling Cancellation (dB)	Not Supported	Not Supported	Not Supported	20
PDC for cancellation	Not Supported	Not Supported	Not Supported	0
PDC/ele. (mW)	249	105	252	120
Total Area (mm²)	7.0×3.2	10.7	4×3	3.4×4.5

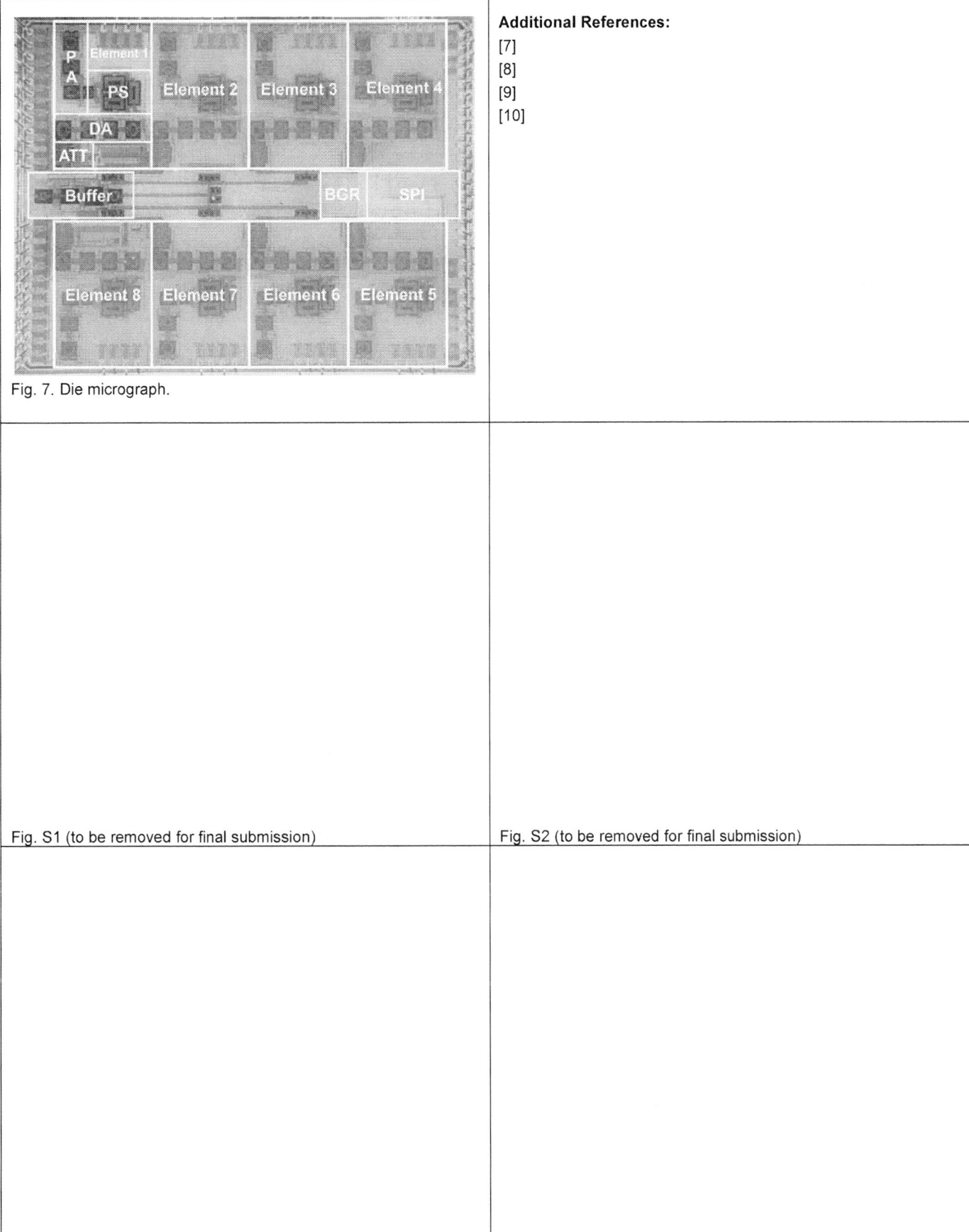

Fig. 7. Die micrograph.

Additional References:
[7]
[8]
[9]
[10]

Fig. S1 (to be removed for final submission)

Fig. S2 (to be removed for final submission)

Fig. S3 (to be removed for final submission)

Fig. S4 (to be removed for final submission)

A 69µW Dual-Mode Electrochemical miR-21 Detection SoC with 3rd-Harmonic Direct-Solution Technique

Xiangdong Feng[1*], Xin Liu[1*], Yangfan Xuan[1], Fengheng Li[1], Yuxuan Luo[1], Ping Wang[1], and Bo Zhao[1]

[1]Zhejiang University, Hangzhou, China

*Equally Credited Authors (ECAs)

MicroRNA-21 (miR-21) serves as an early biomarker for multiple diseases, including multiple tumors [1] and central nervous system disorders. The electrochemical miR-21 detector acts as a critical regulator of diverse cellular processes, such as cell differentiation, proliferation, invasion, apoptosis, and metabolism. However, the massive and frequent detections during epidemics have challenged miR-21 detection in device cost and testing time. There are two design challenges in conventional biomarker detection circuits (Fig.1, top left & middle): 1) The sinusoidal stimulation source in electrochemical impedance spectroscopy (EIS) mode dominates the system power consumption, limiting the lifetime in long-term massive detections. 2) The frequency sweeping procedure in EIS takes a long time in the detection of a large amount of samples.

To address the issues, we demonstrate a dual-mode electrochemical system-on-chip (SoC) for miR-21 detection (Fig.1, bottom), which merges the EIS and chronoamperometry (CA) circuits into a single stimulation-and-detection chain: 1) The 3rd harmonic of a square wave is explored to stimulate the sample (Fig.1, top right), resulting in a much lower total power (69µW) than conventional chips based on sinusoid stimulators. Then, the chip and battery can be reused in a long term with massive disposable probes in frequent detections. 2) The R-C parameters in the equivalent Randles circuit model can be obtained with a readout speed of 5kHz, which eliminates the need for the time-consuming frequency sweeping procedure and then shortens the testing time in massive detections.

Fig.1 illustrates the proposed technique for miR-21 detection. The fundamental frequency and the 3rd harmonic of the stimulation square wave are explored to directly solve the charge-transfer resistance R_{CT}, ionic resistance R_S, and double-layer capacitance C_{DL} in the Randles circuit model, resulting in significant reduction in testing time and energy consumption compared to conventional frequency sweeping methods. In prior EIS-and-CA chips such as [2], an off-chip waveform generator was used for the CA mode, and a power-hungry filter was required to provide a sine wave for EIS stimulation. In addition, the two modes utilized separated detection circuits, leading to 2.25 mW/pixel power consumption. In this work, a low-power square-wave generator and the current detection circuit are shared by both the EIS and CA modes, where the impedance changes in EIS mode is converted into current signal in CA mode, which is based on the previous study in [3]. In this way, the 69µW low power consumption enables the proposed chip to be powered by a miniature battery, which can also present a long lifetime. Moreover, the 5kHz readout speed avoids the long delay in massive detections.

Fig.2(a) detailed the system architecture of the proposed miR-21 detection SoC imbedded with the proposed 3rd-harmonic direct-solution technique, as well as an miR-21 probe. The sensor probe is stimulated by a circuit chain composed by a frequency-locked loop (FLL), a frequency divider, and a bio-sample driver. Then, a voltage signal is induced by the sensor probe, which is processed by a low noise amplifier (LNA), a synchronous SAR ADC, and a digital processing unit. The chip also integrates a power-management unit (PMU) consisting by a bandgap, an LDO for analog submodules, and an LDO for digital submodules. In the digital processing unit, the values of R_{CT}, R_S, and C_{DL} of the Randles circuit model shown in Fig.2(b) are solved by the equations in Fig.2(c) as well as the frequency-domain impedance analysis. The EIS results are output through an SPI in real time. In addition, the CA measurement can be conducted through time-domain transient-current analysis by setting the stimulation frequency to several Hz. As a result, the FLL should present a wide locking range to adopt to both the EIS and CA modes.

Fig.3 provides the circuit details of the wideband FLL for stimulation in both EIS and CA modes. The EIS mode should be stimulated by a signal with a frequency of several kHz, while the CA mode requires Hz-level stimulation. As a result, the FLL should cover the wide frequency range, which is designed with the combination of coarse tuning and fine tuning. The coarse tuning is realized by programable current source, and the fine tuning is controlled by varactors. An automatic current selection logic (Fig.3, right bottom) is designed to realize the self-adaptive frequency tuning. In addition, a supply-insensitive period detector in [7] and a switched-capacitor integrator are utilized to lock the oscillator frequency corresponding to R_{REF} and C_1 (Fig.3, left bottom), forming a negative feedback. As the control voltage of the varactor (V_C) reaches the up or down limit, the current selection logic automatically generates an event-driven clock to increase or decrease the current, which keeps the control voltage within the expected range. The FLL includes 256 current branches to cover the frequency ranges of both EIS and CA modes.

Fig.4 demonstrates the measurement results and relative errors in both EIS and CA modes. As shown in Fig. 4(a), our chip realizes <1% magnitude error and <2.5° phase error at the fundamental frequency, as well as <3% magnitude error and <4° phase error at the 3rd harmonic compared to Keysight 4294A. Fig. 4(b) displays the measurement result of harmonic-based impedance solution, where a relative error of <4% for both R_{CT} and R_S is achieved. To validate the current measurement function of the proposed chip, we simultaneously measure the current response versus different concentrations of potassium ferricyanide solution, using the commercial electrochemical workstation CHI660E and our chip, respectively. The results in Fig.4(c) indicates that the chip measurement results match favorably with the CHI660E measurement results.

Fig.5 displays the in-vitro biological measurement results of miR-21 in both EIS and CA modes. The charge-transfer resistance between the electrode and the potassium ferricyanide solution rises as the miR-21 probe presents to be complementary with the targeting miR-21. Fig.5(a) illustrates the Nyquist plot of the results tested by CHI660E for a golden reference. The diameter of the semicircle (R_{CT}) increases with the targeting miR-21 concentration. Fig. 5(b) shows the R_{CT} solved by the proposed chip, which is in consistent with the CHI660E testing result. As the concentration of the target miR-21 varies from 0pM to 100pM, R_{CT} increases from 29.8 kΩ to 58.3 kΩ. In addition, Fig.5(c) and Fig.5(d) demonstrate the CA measurement results of CHI660E and our chip, which also indicates a good match. Finally, the case with one base pair (bp) mismatch is also tested in Fig. 6, which can be distinguished by the proposed chip.

The chip performance is summarized and compared with the state-of-the-art electrochemistry chips and DNA analyzers in Fig.6. The proposed 3rd-harmonic direct-solution technique achieves: 1) the lowest power consumption (69µW) and 2) the highest readout speed (5kHz) among all the reported works. Therefore, the chip will present a long lifetime and rapid testing in massive frequent miR-21 detections.

Acknowledgement: This work was supported by NSFC 61974130. Corresponding author: Bo Zhao (zhaobo@zju.edu.cn).

References:

[1] Y. Feng *et al.*, "Emerging role of microRNA-21 in cancer," Biomedical report, Aug. 2016.

[2] S. Kumashi *et al.*, " A CMOS Multi-Modal Electrochemical and Impedance Cellular Sensing Array for Massively Paralleled Exoelectrogen Screening," TBioCAS, 2021.

[3] X. Feng *et al.*, " A Square-Wave Stimulated DNA Analyzer Chip Featuring 120µW Power Consumption and Simultaneous Dual-Frequency Detection," TCAS-II, 2022.

[4] J. Chien *et al.*, "Design and Analysis of a Sample-and-Hold CMOS Electrochemical Sensor for Aptamer-Based Therapeutic Drug Monitoring," JSSC, 2020.

979-8-3503-3004-5/23 $31.00 © 2023 IEEE

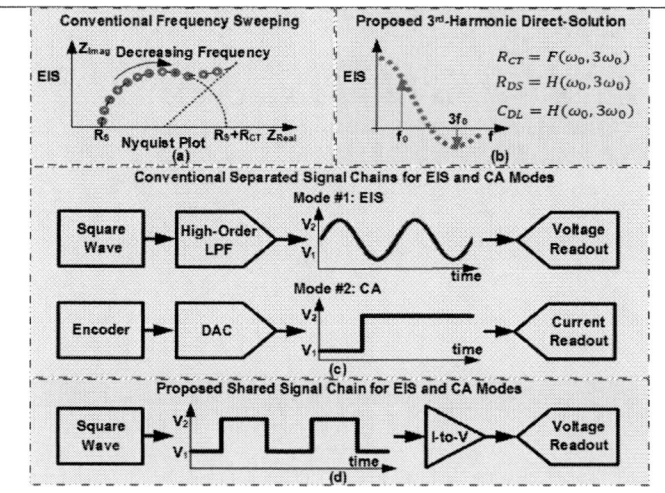

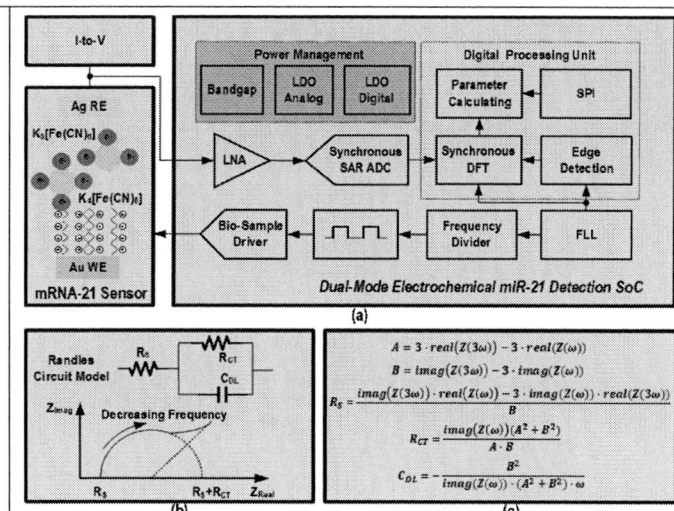

Fig. 1. Conventional and proposed techniques: (a) Conventional frequency-sweeping technique for EIS. (b) Proposed 3rd-harmonic direct-solution technique for EIS. (c) Conventional two-path structures (EIS and CA). (d) Proposed dual-mode structure.

Fig. 2. (a) Chip architecture and miR-21 sensor probe. (b) Randles circuit model. (c) Calculation formulas for 3rd-harmonic direct solution technique.

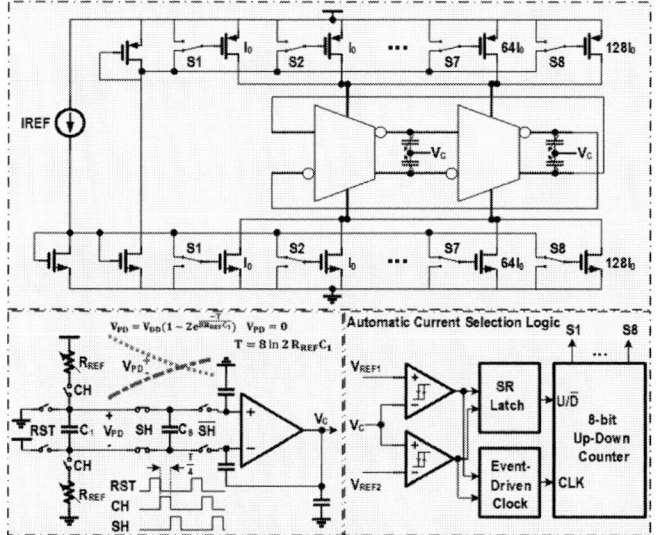

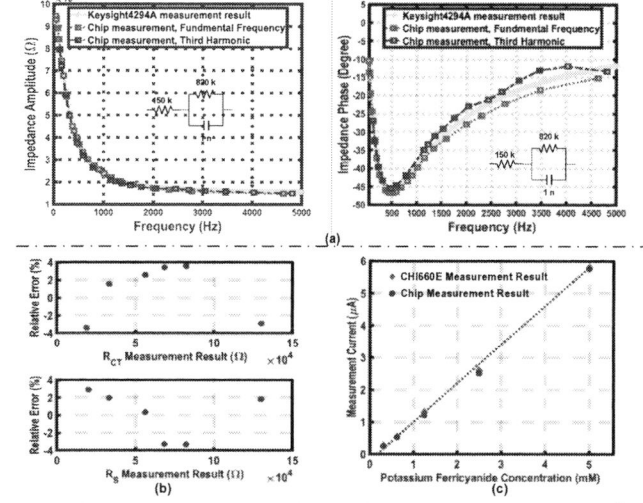

Fig. 3. Wideband frequency-locked loop (FLL) for both EIS and CA modes.

Fig. 4. Measurement results of (a) EIS with fundamental and 3rd harmonic tones, (b) relative error of parameters, and (c) CA current linearity.

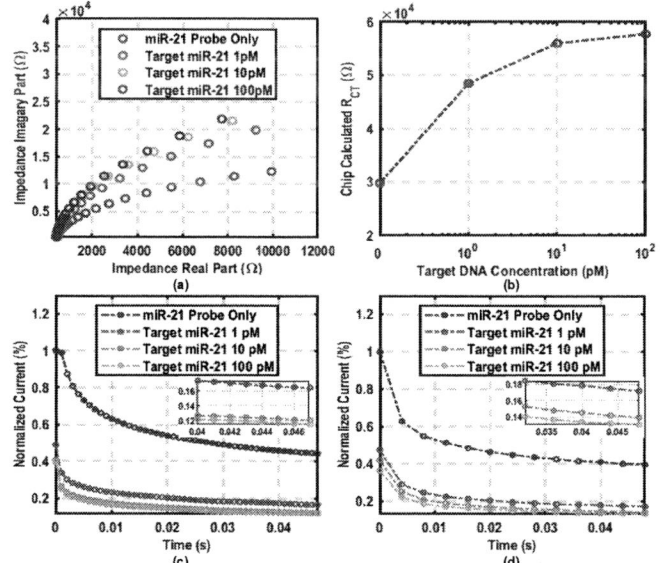

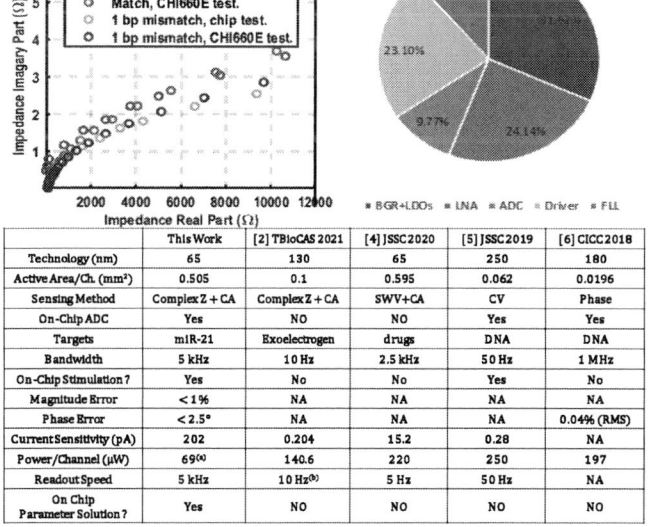

Fig. 5. Measurement results of (a) miR-21 hybridization using CHI660E, (b) miR-21 hybridization by proposed chip, (c) CA mode by CHI660E, and (d) CA mode by proposed chip.

	This Work	[2] TBioCAS 2021	[4] SSC 2020	[5] JSSC 2019	[6] CICC 2018
Technology (nm)	65	130	65	250	180
Active Area/Ch. (mm²)	0.505	0.1	0.595	0.062	0.0196
Sensing Method	Complex Z + CA	Complex Z + CA	SWV+CA	CV	Phase
On-Chip ADC	Yes	NO	NO	Yes	Yes
Targets	miR-21	Exoelectrogen	drugs	DNA	DNA
Bandwidth	5 kHz	10 Hz	2.5 kHz	50 kHz	1 MHz
On-Chip Stimulation?	Yes	No	No	Yes	No
Magnitude Error	< 1%	NA	NA	NA	NA
Phase Error	< 2.5°	NA	NA	NA	0.04% (RMS)
Current Sensitivity (pA)	202	0.204	15.2	0.28	NA
Power/Channel (μW)	69(a)	140.6	220	250	197
Readout Speed	5 kHz	10 Hz(b)	5 Hz	50 Hz	NA
On Chip Parameter Solution?	Yes	NO	NO	NO	NO

(a) Excluding digital processing power for fair comparison.   (b) Limited by bandwidth.

Fig. 6. Measurement result of miR-21 mismatch, power breakdown, and comparison table.

979-8-3503-3004-5/23 $31.00 © 2023 IEEE

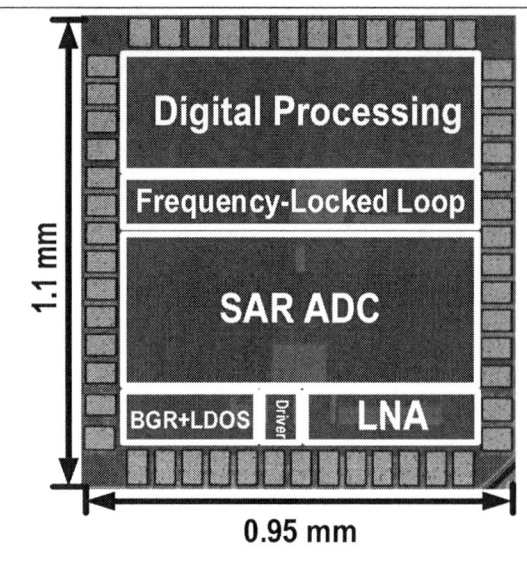

1.1 mm

0.95 mm

Fig. 7. Die micrograph.

Additional References:

[5] A. Manickam *et al.*, " A CMOS Electrochemical Biochip With 32×32 Three-Electrode Voltammetry Pixels," JSSC, 2019.

[6] C. Hsu *et al.*, " A 16×20 Electrochemical CMOS Biosensor Array with In-Pixel Averaging Using Polar Modulation," CICC, 2018.

[7] J. Lee *et al.*, "A 4.7MHz 53μW fully differential CMOS reference clock oscillator with −22dB worst-case PSNR for miniaturized SoCs," ISSCC 2015.

A Transient Enhancement Digital LDO with Adaptive Ripple Cancelation Based on Optimal Compensation Period Approximation

Angxiao Yan, Wei Deng, Haikun Jia, Shiwei Zhang, and Baoyong Chi

School of Integrated Circuits, Tsinghua University, Beijing, China

Email: wdeng@tsinghua.edu.cn

The ripple suppression performance of low-dropout regulators (LDOs) is crucial when cascading them after a DC/DC converter to achieve high power efficiency [1,2]. Analog LDOs (A-LDOs) and Hybrid LDOs (H-LDOs) achieve decent ripple suppression but suffer from heavy analog components, leading to design complexity and scalability issues [3-5]. Recently, a digital LDO (D-LDO) employing a time-domain digital-pre-distortion (DPD) technique was introduced [6], which provides a digital scheme for ripple suppression. However, the ripple suppression performance severely degrades when the ripple period is a non-integer multiple of the clock period. To address this issue, this paper introduces a period multiplication technique based on an optimal integer approximation scheme to minimize the period residual when conducting ripple cancelation. The measurement results indicate that the proposed D-LDO achieves equally decent ripple suppression performance in various decimal situations of T_{RIPPLE}/T_{CLK}. Moreover, this work proposes a transient-response-enhancement architecture based on a threshold-adjustable inverter to compensate for the output-voltage undershoot/overshoot.

Fig. 1 shows the conceptual diagram of the proposed period multiplication method for adaptive ripple cancelation. When the ripple period is an integer multiple (4 in this example) of the clock period, the input voltage sampling points within the compensation period remain at fixed voltage levels over time. However, when the ripple period is a non-integer multiple (3.5 in this example) of the clock period, the compensation period is different from the ripple period (half a clock period difference in this example). Therefore, the transient input voltage corresponding to each sampling point goes through rapid variation over time. The DPD mechanism is unable to completely converge in the non-integer-multiple case. To ensure there is no relative slipping between the compensation period and ripple period so that the target values of the DPD are constants, instead of assigning the compensation period as the nearest integer of T_{RIPPLE}/T_{CLK}, this work proposes a period multiplication method to choose an optimal integer as the compensation period. In the example shown in the upper of Fig. 1, when T_{RIPPLE}/T_{CLK} is 3.5, the optimal-integer compensation period is 3.5*2 = 7. Since 2 ripple cycles have the same time length as 7 clock cycles, there is no relative slipping between the compensation period and the ripple period. This guarantees the complete convergence of the DPD.

Fig. 2 depicts the detailed architecture of the proposed D-LDO. The ripple pattern is extracted using an AC-coupling trans-impedance amplifier (TIA). Subsequently, the time duration of 1024 ripple cycles is digitized based on the clock period. The fractional part of the resulting T_{RIPPLE}/T_{CLK} is sent to a look-up table. The look-up table outputs an optimal multiplier ranging from 1 to 10, which is used to multiply the obtained T_{RIPPLE}/T_{CLK}. This multiplication produces an optimal compensation period (an integer multiple of T_{CLK}) that minimizes the difference from an integer multiple of T_{RIPPLE}. Consequently, the relative slipping between the ripple period and compensation period is minimized, ensuring good convergence under various decimal multiples of T_{RIPPLE}/T_{CLK}. The transient enhancement is implemented using an AC-coupling undershoot/overshoot detection mechanism based on a threshold-adjustable inverter. The threshold-adjustable inverter comprises a master inverter biased by an adjustable resistance and a slave inverter that shares the same V_{TH} as the master inverter. The DC input of the compensation paths is set by another two adjustable resistances. The input voltage of the undershoot compensation path, which is pulled down during undershoot events, has a DC value higher than the path threshold. On the other hand, the input voltage of the overshoot compensation path, which is pulled up during overshoot events, has a DC value lower than the path threshold. The charging/discharging switch of the power NMOS gate is activated when undershoot/overshoot compensation path is triggered by the rapid change of the output voltage.

Fig. 3 illustrates the working principle of the proposed optimal integer approximation. For each fractional situation of T_{RIPPLE}/T_{CLK} obtained during the ripple period digitization, there exists a corresponding multiplier ranging from 1 to 10. This multiplier, when multiplied with T_{RIPPLE}/T_{CLK}, results in the minimum difference from a nearby integer. The mapping between the fractional part and the corresponding multiplier is pre-determined and stored in the look-up table mentioned earlier, as depicted in the upper part of Fig. 3. As shown in the lower part of Fig. 3, The maximum period residual across all decimal multiples of T_{RIPPLE}/T_{CLK} is effectively reduced from $0.5*T_{CLK}$ to less than $0.1*T_{CLK}$, which is small enough for the adaptive fitting mechanism to track the variation of the target compensation values.

The proposed D-LDO is fabricated in a 65nm technology. Its die photo is shown in Fig. 7. The core area of the proposed D-LDO is 0.066mm². The measured peak current efficiency is 98.1%. Fig. 4 shows the measured ripple suppression performance of the proposed D-LDO. It is seen that the ripple suppression performance is further improved by more than 10dB under a 2.105263MHz ($f_{CLK}/9.5$) supply ripple across a swing range of 150mV to 300mV. Fig. 4 (lower and upper right) plots the ripple suppression performance under different ripple frequencies and decimal situations of f_{CLK}/f_{RIPPLE}. It is seen that the proposed optimal compensation period approximation method achieves equally decent ripple suppression performance in both integer and non-integer situations of f_{CLK}/f_{RIPPLE}. Fig. 5 (upper) shows the measured transient output waveform under supply ripple frequencies of $f_{CLK}/9.5$ and $f_{CLK}/9.7$ with a 200mV input swing. Fig. 5 (lower left) shows the measured transient response with a 100-ns edge current step of 0.74mA to 34mA. Fig. 5 (lower middle) shows the measured transient response with a 100-ns edge current step of 55.8mA to 0.74mA. Thanks to the proposed transient response enhancement mechanism, the undershoot is improved from 343mV to 180mV, and the overshoot is improved from 414mV to 270mV. Fig. 5 (lower right) shows the measured load regulation. A comparison with prior art designs is provided in Fig. 6. The proposed D-LDO is designed with zero output capacitance. However, a 4-pF capacitance is introduced by a probe for time-domain measurement.

Acknowledgements

This work was supported by the National Natural Science Foundation of China under Grant 62131013, the Tsinghua-Samsung Joint Research Project, and the Beijing Advanced Innovation Center for Integrated Circuits.

References:

[1] J. Lin et al., "A High-efficiency and Fast-transient Digital-low-dropout Regulator with the Burst Mode Corresponding to the Power-saving Modes of DC-DC Switching Converters," ISSCC, Feb. 2018.

[2] I. Bukreyev et al., "Four Monolithically Integrated Switched-Capacitor DC–DC Converters With Dynamic Capacitance Sharing in 65-nm CMOS," TCAS-I, Jun. 2018.

[3] M. Huang et al., "An Analog-Proportional Digital-Integral Multiloop Digital LDO With PSR Improvement and LCO Reduction," JSSC, Jun. 2020.

[4] S. B. Nasir et al., "Switched-Mode-Control Based Hybrid LDO for Fine-Grain Power Management of Digital Load Circuits," JSSC, Feb. 2018.

[5] X. Liu et al., "A Modular Hybrid LDO with Fast Load-Transient Response and Programmable PSRR in 14nm CMOS Featuring Dynamic Clamp Tuning and Time-Constant Compensation," ISSCC, Feb. 2019.

[6] A. Yan et al., "A Capacitor-less Digital LDO using Ripple-Frequency-Adaptive Time-domain Digital Pre-distortion Technique," A-SSCC, Nov. 2022.

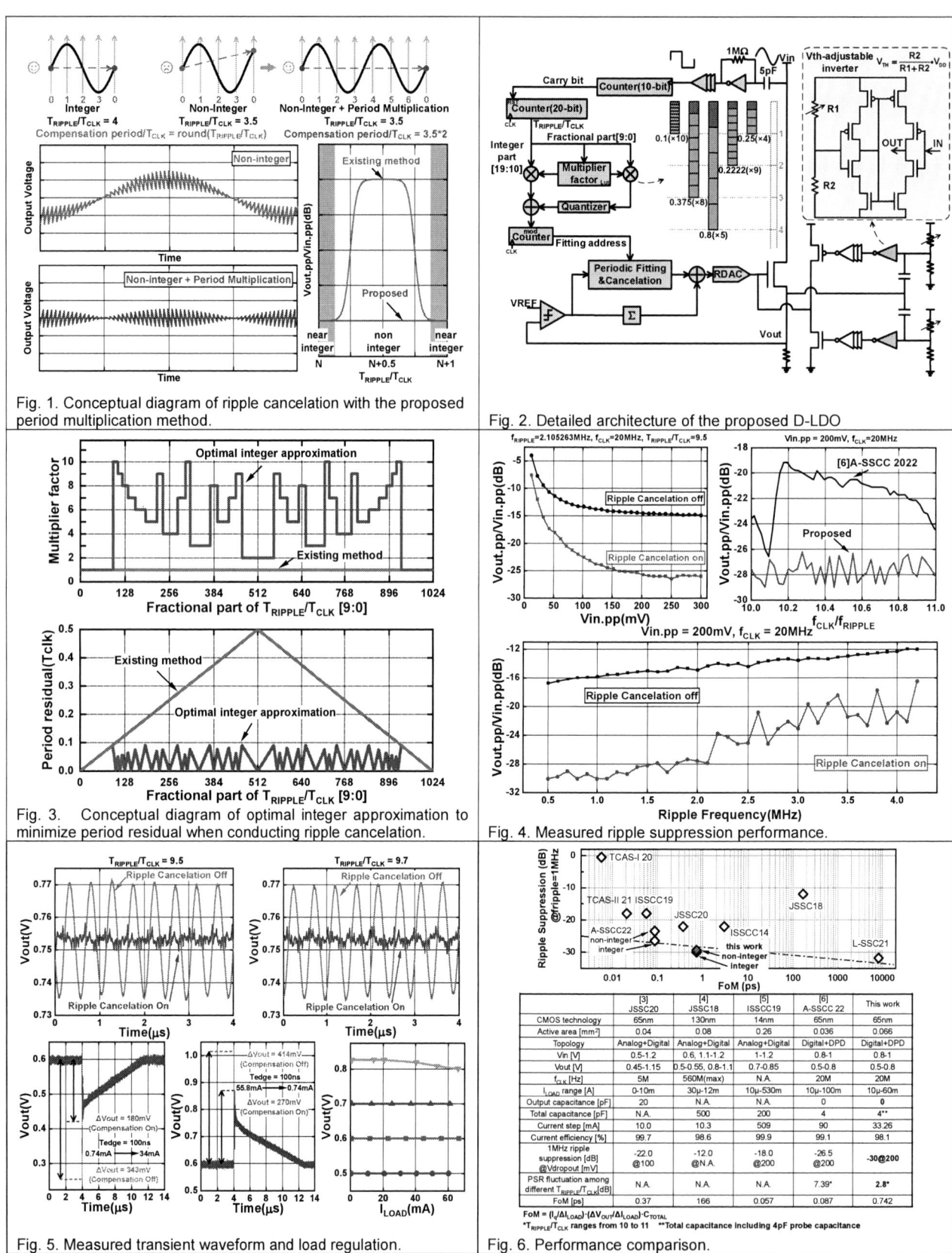

Fig. 1. Conceptual diagram of ripple cancelation with the proposed period multiplication method.

Fig. 2. Detailed architecture of the proposed D-LDO

Fig. 3. Conceptual diagram of optimal integer approximation to minimize period residual when conducting ripple cancelation.

Fig. 4. Measured ripple suppression performance.

Fig. 5. Measured transient waveform and load regulation.

Fig. 6. Performance comparison.

	[3] JSSC20	[4] JSSC18	[5] ISSCC19	[6] A-SSCC 22	This work
CMOS technology	65nm	130nm	14nm	65nm	65nm
Active area [mm²]	0.04	0.08	0.26	0.036	0.066
Topology	Analog+Digital	Analog+Digital	Analog+Digital	Digital+DPD	Digital+DPD
Vin [V]	0.5-1.2	0.6, 1.1-1.2	1-1.2	0.8-1	0.8-1
Vout [V]	0.45-1.15	0.5-0.55, 0.8-1.1	0.7-0.85	0.5-0.8	0.5-0.8
f_CLK [Hz]	5M	560M(max)	N.A.	20M	20M
I_LOAD range [A]	0-10m	30μ-12m	10μ-530m	10μ-100m	10μ-60m
Output capacitance [pF]	20	N.A.	N.A.	0	0
Total capacitance [pF]	20	500	200	4	4**
Current step [mA]	10.0	10.3	509	90	33.26
Current efficiency [%]	99.7	98.6	99.9	99.1	98.1
1MHz ripple suppression [dB] @Vdropout [mV]	-22.0 @100	-12.0 @N.A.	-18.0 @200	-26.5 @200	**-30@200**
PSR fluctuation among different T_RIPPLE/T_CLK [dB]	N.A.	N.A.	N.A.	7.39*	2.8*
FoM [ps]	0.37	166	0.057	0.087	0.742

FoM = (I_q/ΔI_LOAD)·(ΔV_OUT/ΔI_LOAD)·C_TOTAL

*T_RIPPLE/T_CLK ranges from 10 to 11   **Total capacitance including 4pF probe capacitance

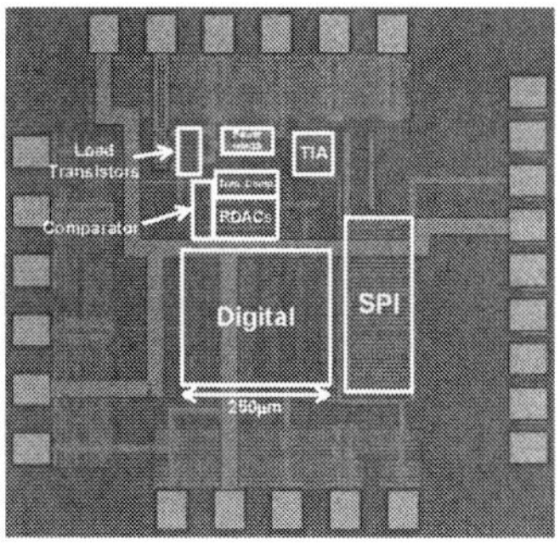

Fig. 7. Chip die micrograph.

Task-Specific Low-Power Beamforming MIMO Receiver Using 2-Bit Analog-to-Digital Converters

Timur Zirtiloglu[1], Peter Crary[1], Eyyup Tasci[1], Arslan Riaz[1], Yonina C. Eldar[2], Nir Shlezinger[3], and Rabia Tugce Yazicigil[1]

[1]ECE Department, Boston University, USA, [2]Faculty of Math and CS, Weizmann Institute of Science, Israel. [3]School of ECE, Ben-Gurion University of the Negev, Israel.

MIMO communication systems deliver higher data capacity and wide-area coverage. However, they traditionally apply brute-force data acquisition with spatial signal processing using high-resolution ADCs and costly analog pre-processing circuits, leading to high power consumption and hardware complexity. This work presents the first integrated task-specific MIMO receiver (TSMRx) with beamforming using low-bit ADCs and achieving spatial blocker rejection implemented in 65nm CMOS technology, leveraging the recently proposed task-specific digital signal processing techniques [1-2]. We co-design analog pre-processing circuits and task-specific digital signal processing to achieve RF front-end power savings while maintaining an accurate task (signal of interest) recovery. The analog pre-processing front end utilizes highly-reconfigurable Noise-Cancelling Constant-Gm Vector Modulators (VMs) that perform current-domain processing with embedded beamforming while exploiting sparsity and variable (coarse or fine) VM resolution. The TSMRx system achieves successful task recovery using low-resolution (2-bit) ADCs with coarse-quantized (4-point) VMs and a 25%-sparse analog combiner while consuming only 75-115 mW from a 1.3 V supply for 0.5-3 GHz, the lowest reported among the works [3-6] in a similar technology node and frequency range.

The system diagram in Fig. 1 shows that the received input signals are processed by an analog combiner with N antenna elements and P baseband processing chains ($P<N$), by performing a matrix-vector multiplication with an analog pre-processing matrix A. Each element of matrix A corresponds to a Constant-Gm VM with a downconversion adapted from [3]. Baseband outputs are sampled and quantized by low-resolution ADCs and digital processing matrix B is applied in the back end. The task-specific algorithm aims to find an optimal analog combiner matrix A and digital matrix B such that task recovery error is minimized, solving the optimization problem to minimize the loss measure given by (1) [2]:

$$L(A) = MSE(A) + \gamma_I IntRej(A) + \gamma_S ||A||_{1,1} \tag{1}$$

The first term represents the accuracy loss due to finite ADC resolution, the second term is the spatial interference penalty, and the third term accounts for the analog combiner sparsity. The digital processing matrix B is calculated as a linear MMSE estimate once A is chosen. Iterative optimization algorithm that leverages proximal gradient descent also takes into account finite VM resolution, compensating for both low-resolution ADCs and VMs, resulting in overall system power consumption reduction.

The TSMRx chip block diagram shown in Fig. 2 has 8 antenna inputs and 2 baseband outputs. Highly-reconfigurable 16 Constant-Gm VMs embedding noise-cancellation (NC) consist of 15 parallel downconversion slices shown in Fig. 3, composed of Common-Gate (CG) and Common-Source (CS) LNTAs, current-domain passive mixers, and directivity switches. Each slice performs downconversion and summation in the current domain followed by baseband transimpedance amplifiers (TIAs) and recombination amplifiers with a 6-bit control for noise cancellation. Noise analysis in Fig. 3 illustrates that the proper recombination ratio (δ) selection minimizes the CG device noise contribution. Compared to the earlier Constant-Gm VM topology in [3], the embedded NC technique effectively decouples the input impedance from the noise figure and the gain since the CG devices are used for impedance matching [7]. The TSMRx also has an on-chip 25% duty-cycle clock generator with external clock input and a digital control block. The baseband outputs are digitized off-chip for offline task-specific recovery using two bits.

To reduce the RF front-end power consumption, our design features the following hardware system techniques leveraging the task-specific approach: 1) VM quantization can be adjusted from 15 slices (fine quantization with high resolution at the expense of higher power, 256 points) to 8 slices (coarse quantization with low resolution at low power, 4 points), resulting in the constellations shown in Fig. 4, enabling up to 47% power reduction in each VM at the cost of task recovery accuracy and NF, 2) analog combiner matrix A can be sparsified, which is equivalent to maintaining a sleep mode for the selected VMs. Algorithmic simulations showed that 25% sparsity, i.e., turning off 4 out of 16 VMs, gives a practical design point in power and performance trade-off. To exploit the sparsity of VMs, we control the bias voltage of the CS NMOS device as shown in Fig. 3. The CG LNTA is always on for input impedance matching, and since it consumes 7× lower power than the CS LNTA, it does not result in a significant power consumption overhead. Passive mixer drivers are also disabled for the slices in sleep mode for further power savings. The TSMRx reconfigurability enables four operational modes as shown in Fig. 4, combining the number of slices (SL) with different sparsity levels (SP). We achieve up to 60% VM power reduction for 8 SL and 25% SP. However, baseband amplifiers and LO distribution consume additional power, resulting in up to 30% system power reduction.

The TSMRx was implemented in 65 nm CMOS with a 9 mm² total die area (1.49 mm² active area) and operated from a 1.3 V supply. The test setup was built using commercial power splitters/combiners and phase shifters to emulate desired signals and interferers coming from various angles. The baseband outputs are captured using an oscilloscope and processed offline in MATLAB, where signal quantization and digital back-end processing for task-specific recovery are performed. Fig. 5 shows the measured gain (overall and 1-VM), NF, and S11 performance for the 0.5-3 GHz range. The NF results show a 5 dB noise reduction compared to a CG-only mode. The measured Array Factor plots show successful blocker cancellation in both task-agnostic (conventional) and task-specific modes (proposed) with the capability to form beams and nulls in various directions. EVM measurements were performed to demonstrate the successful task recovery using low-bit ADCs in the presence of an interferer. Fig. 5 shows the task-specific recovery of a QPSK signal, with corresponding input signal levels and angles of arrival, with a 26% EVM for the 8 SL 25% SP case using 2-bit ADCs compared to an unsuccessful recovery (>100% EVM) for a conventional task-agnostic approach under the same hardware conditions.

Fig. 6 compares the performance of the TSMRx chip to the state-of-the-art [3-6]. The 15 SL 0% SP mode shows good linearity and blocker tolerance, while the 8 SL 25% SP achieves the lowest RF front-end power consumption relative to the works in similar technology nodes while maintaining a comparable NF. In addition, using Walden FoM [8] and the reported performance of ADCs [9], we estimate more than 10-15x times ADC power reduction by using only two bits, compared to conventional systems with high resolution (10 or more). Fig. 7 shows the chip micrograph and the measurement setup.

References:

[1] N. Shlezinger et al., "Hardware-limited task-based quantization," IEEE TSP, vol. 67, no. 20, 2019

[2] T. Zirtiloglu et al., "Power-Efficient Hybrid MIMO Receiver with Task-Specific Beamforming using Low-Resolution ADCs," 2022 ICASSP.

[3] S. Golabighezelahmad et al., "A 0.7–5.7 GHz Reconfigurable MIMO Receiver Architecture for Analog Spatial Notch Filtering Using Orthogonal Beamforming," IEEE JSSC, vol. 56, no. 5, 2021.

[4] M. Bajor et al., "A Flexible Phased-Array Architecture for Reception and Rapid Direction-of-Arrival Finding Utilizing Pseudo-Random Antenna Weight Modulation and Compressive Sampling," IEEE JSSC, vol. 54, no. 5, 2019.

[5] L. Zhang et al., "Arbitrary Analog/RF Spatial Filtering for Digital MIMO Receiver Arrays," IEEE JSSC, vol. 52, no. 12, 2017.

[6] J. Poojary et al., "6.4 A 1-to-3GHz Co-Channel Blocker Resistant, Spatially and Spectrally Passive MIMO Receiver in 65nm CMOS with +6dBm In-Band/In-Notch B1dB," 2021 IEEE ISSCC.

[7] J. Zhou et al., "Low-Noise Active Cancellation of Transmitter Leakage and Transmitter Noise in Broadband Wireless Receivers for FDD/Co-Existence," IEEE JSSC, vol. 49, no. 12, 2014.

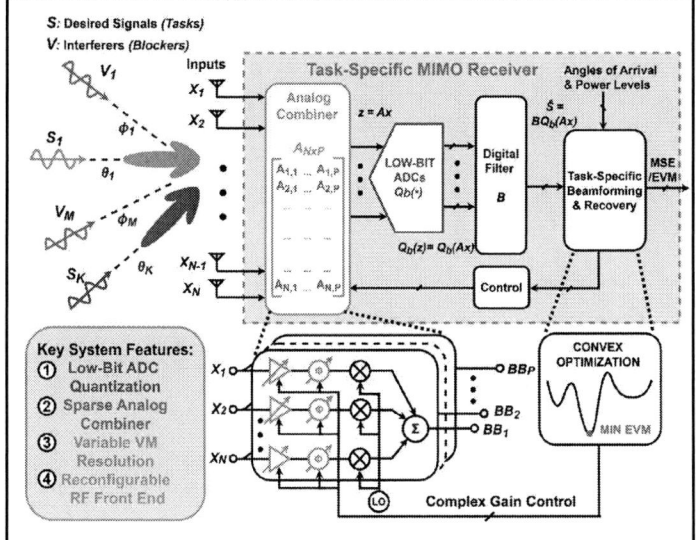

Fig. 1. Proposed Task-Specific MIMO receiver system diagram.

Fig. 2. The TSMRx system architecture with RF front-end details.

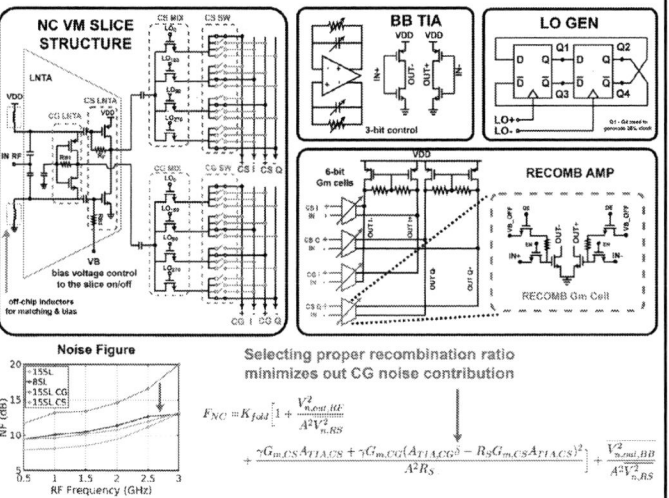

Fig. 3. Circuit implementation of the TSMRx blocks (top) and **measured** noise figure and noise analysis (bottom).

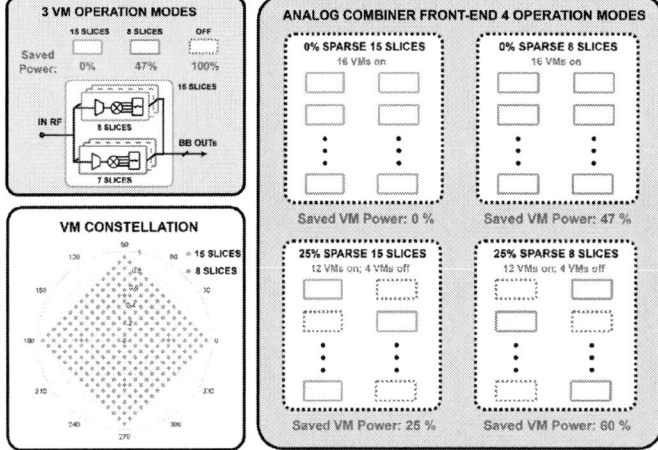

Fig. 4. Constant-Gm NC VM configurations with 15 and 8 slices and its corresponding complex constellation (left), analog combiner configuration for various VM and sparsity settings, and corresponding **measured** VM power savings (right).

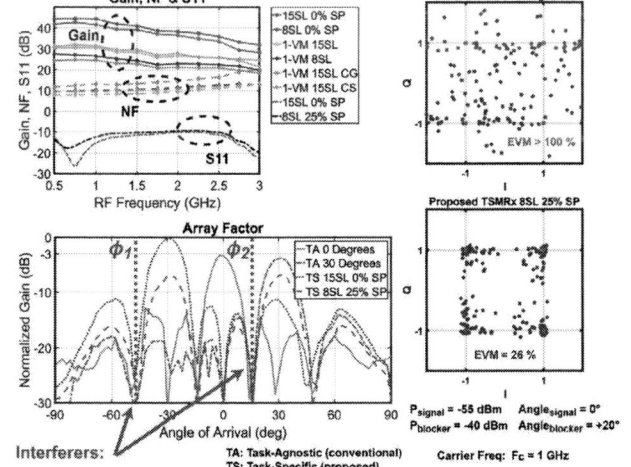

Fig. 5. Measured gain, NF, and S11 plots (top left); Measured array factor plots (bottom left) and recovery performance with 2-bit ADCs (right) compared to conventional approach under the same conditions.

	This Work		ISSCC'21 [6]	JSSC'21 [3]	JSSC'19 [4]	JSSC'17 [5]
	15 SL 0% SP	8 SL 25% SP				
Technology (nm)	65		65	22	65	65
Array Elements	8 x 2		4 x 4	4 x 4	8 x 2	4 x 4
Freq Range (GHz)	0.5 - 3		1 - 3	0.7 - 5.7	1 - 3	0.1 - 3.1
Supply Voltage (V)	1.3		1.4	0.8	1.2	1.2
RF Power (mW)	104 - 165	75 - 115	130 - 242	77 - 139	158[1]	116 - 147
Gain (dB)	44.4[2]	39.2[2]	10[3]	41	32[3]	43
Noise Figure[2,3] (dB)	8	10	9.8	7	6.4[1]	4.6
OIP3[4]/IIP3[5]/B1dB[5] (dBm)	5/8.9/-5.5	5.5/6/-18.5	17.4/22.8/6	22/17/-11.5	-/-/-11.3	11/-5/-25
Spatial Suppression (dB)	>28	>23	>27	>29	>25[6]	51 - 56
ADC Resolution	Low (2 bits)		High			

[1]at 1.5 GHz, [2]at 1 GHz, [3]single element, [4]in-band/in-beam, [5]in-band/in-notch, [6]estimated from plot

Fig. 6. Performance comparison table with the state-of-the-art. TSMRx performance is summarized for 15 Slices VMs with 0% sparsity (15 SL 0% SP) and low-power 8 Slices VMs with 25% sparsity (8 SL 25% SP).

979-8-3503-3004-5/23 $31.00 © 2023 IEEE

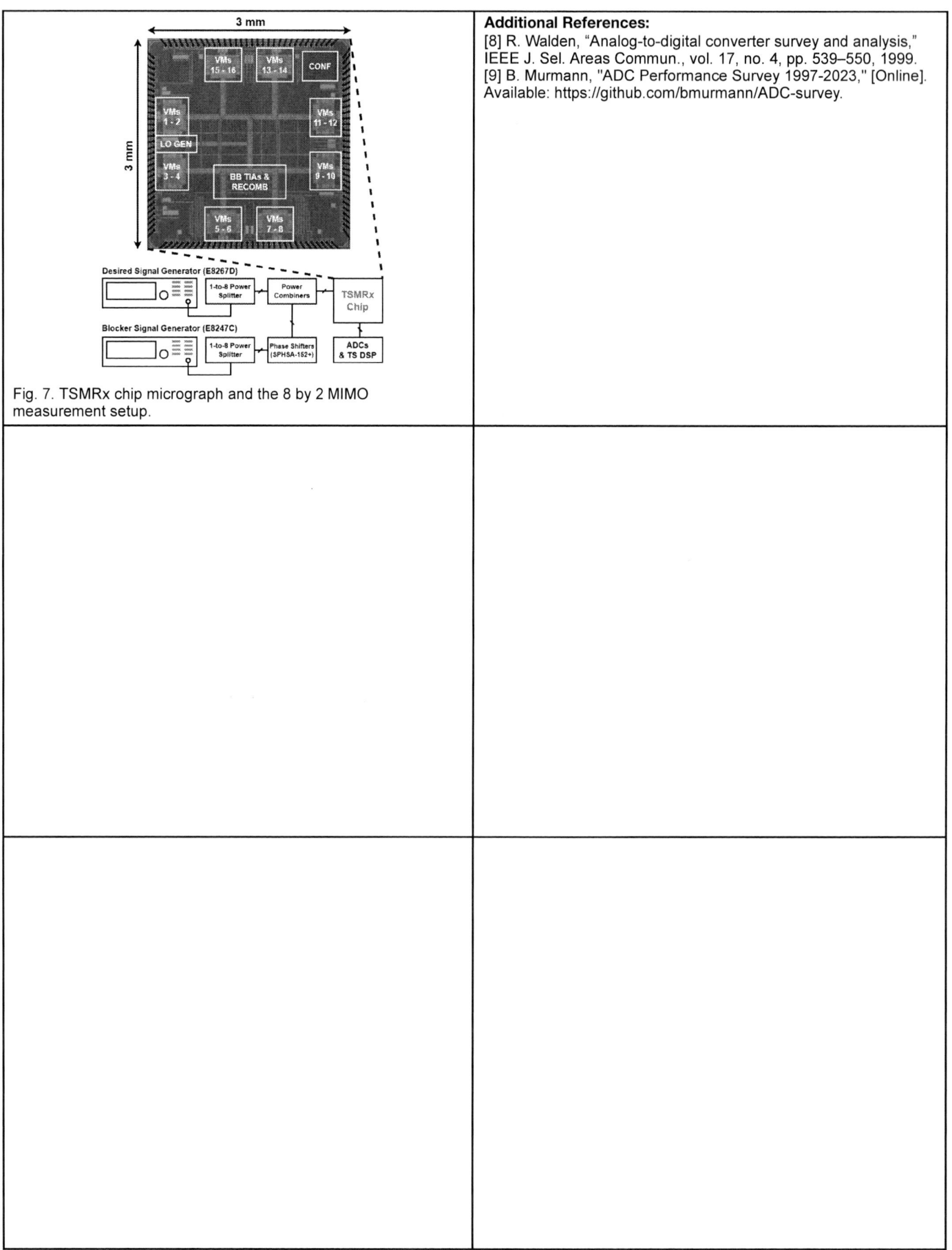

Fig. 7. TSMRx chip micrograph and the 8 by 2 MIMO measurement setup.

Additional References:
[8] R. Walden, "Analog-to-digital converter survey and analysis," IEEE J. Sel. Areas Commun., vol. 17, no. 4, pp. 539–550, 1999.
[9] B. Murmann, "ADC Performance Survey 1997-2023," [Online]. Available: https://github.com/bmurmann/ADC-survey.

A 2D Beam-Steerable 252–285-GHz 25.8-Gbit/s CMOS Receiver Module

Takeshi Yoshida[1], Shinsuke Hara[1,2], Tatsuo Hagino[2], Mohamed H. Mubarak[2], Akifumi Kasamatsu[2], Kyoya Takano[1,3], Yoshiki Sugimoto[4], Kunio Sakakibara[4], Shuhei Amakawa[1], Minoru Fujishima[1]

[1]Hiroshima University, [2]National Institute of Information and Communications Technology, [3]Tokyo University of Science, [4]Nagoya Institute of Technology

There have recently been numerous reports on sub-THz transmitters (TX) and receivers (RX) for high-speed wireless communications, covering chip-level TX and/or RX designs and their modularization including antenna integration. Compared to these, there still are much fewer reports on beam steering [1]–[7]. At sub-THz frequencies (here we focus on 200 GHz or higher), beam steering based on a phased array of TXs/RXs (as opposed to phased array of oscillators) is extremely challenging because of the very short wavelengths (λ = 1.5 mm or shorter in air). Ideally, antennas should be arrayed with a pitch of λ/2 or shorter. This necessitates that the short side of the TX/RX chip be shorter than λ/2, if the chips are to be arrayed horizontally to form a 1D array. Note that because of high losses of transmission lines and waveguides at these frequencies, arraying of small phase shifters only is not usually an option, and TX/RX chips need to be arrayed. As such, antenna pitches reported in [3]–[6] were all greater than λ/2. Only [7] demonstrated an antenna array with a λ/2 pitch. This was realized by vertically stacking antennas built into printed circuit boards (PCBs). However, all these were 1D arrays, and no 2D beam steering seems to have been reported above 200 GHz. Herein we present a 2D beam-steerable CMOS RX module operating at 252–285 GHz. It combines a 1D phased array for horizontal beam steering and a motorized planocylindrical lens for mechanical vertical beam steering.

Figure 1 shows the block diagram of the four-element 1D phased-array RX consisting of four down-converter chips and an IF amplifier chip. Each down-converter chip includes a double-balanced mixer, a differential buffer amplifier, and an LO phase shifter. The mixer receives a 300-GHz-band RF signal via an on-PCB waveguide transition. The LO signal is 216 GHz, and it comes from an external signal source at 12 GHz multiplied by 18. The down-converted signal goes off chip, and the four phase-shifted signals are combined via on-PCB passive components. The combined IF signal goes into the IF amplifier chip.

Four waveguide transitions are built into a PCB, arranged in a line as shown in Fig. 2 (bottom). The down-converter chips are flip-chip mounted on the PCB with staggered layout. The IF amplifier chip is also mounted on the same PCB. Attached on the other side of the PCB is a metallic waveguide block with four parallel WR-3.4 rectangular waveguides. The waveguides are connected to a 1D array of horn antennas, serving as primary radiators. The antenna pitch is about a wavelength (λ). The horizontally polarized horn antennas radiate into the movable planocylindrical lens, motor-driven for vertical beam steering.

Figure 3 shows the schematic diagrams of the double-balanced mixer and the following differential buffer amplifier. The down-conversion mixer adopts the FET resistive mixer with a zero drain-to-source bias. The backgates of the MOSFETs are reverse biased at −1 V, which makes the mixer somewhat low-noise. The differential buffer amplifier consists of two stages of common-source amplifiers with neutralizing cross-coupled capacitors.

The IF amplifier consists of five common-source stages as shown in Fig. 4. The latter four stages have PMOS variable resistors, making the frequency response of the IF amplifier tunable [8]. The PMOS resistors lower the Q factors of the interstage matching networks and helps make each stage more wideband.

Figure 5 (top left) shows the conversion gain and the single sideband (SSB) noise figure (NF) of one of the four parallel receiver paths. The SSB NF was simulated at the chip level and does not include PCB-and waveguide-related losses. On the other hand, the SSB NF and the conversion gain (CG) were measured at the module level, so the reference plane was at a WR-3.4 waveguide flange. However, the discrepancy between the simulated and measured SSB NF is larger than expected even when considering the difference in the setups. The reason is yet to be found. Figure 5 (top right) shows the simulated and measured vertical (H-plane) beam patterns when the lens is moved vertically by ±9 mm. Figure 5 (middle right) shows the simulated and measured horizontal (E-plane) beam patterns by scanning the 1D phased array. The "measured" pattern was obtained by first measuring the beam pattern of each of the four horns in the array fed using eccentric waveguide with an antenna near-field measurement system, and then "synthesizing" the phased-array beam patterns from the individual beam patterns with phase shifts. The scan range is 28°. Figure 5 (bottom) compares phased-array TXs/RXs operating above 200 GHz. Only ours realizes 2D beam steering. The total power consumption of our CMOS chips is 2.8 W.

Figure 6 (top) shows the wireless constellation measurement setup. The transmitter consists of an arbitrary waveform generator (AWG), a signal source (SG), an up-converter, a bandpass filter (BPF), and an off-the-shelf amplifier. The highest data rates achieved with 16QAM and QPSK are 52 and 44 Gbit/s, respectively, over a distance of 6 cm without beam steering. 2D beam steering experiment was also conducted over a distance of 60 cm as shown in Fig. 6 (bottom right). The transmitting antenna is a 40 dBi lens horn. The TX beam was mechanically steered horizontally, and the RX beam was steered two-dimensionally. The modulation format was QPSK, and the constellation was recovered only when the transmitting and receiving antennas were aligned. A data rate of 25.8 Gbit/s was achieved.

Micrographs of the down-converter chip and the IF amplifier chip are shown in Fig. 7. The chips were fabricated using a 40-nm CMOS technology, and they measure 3.52×1.37 mm² and 2.15×0.96 mm², respectively. The front and the back sides of the PCB are also shown in Fig. 7. The waveguide block is attached to the front side, where the waveguide transitions (WGT) can be seen. The chips are mounted on the back side.

We thank Y. Asano, Y. Sako, and S. Yabuki for contributing to the design of the CMOS chips. This work was financially supported by the Ministry of Internal Affairs and Communications of Japan (JPJ000254).

References:

[1] M. Elkhouly et al., "220–250-GHz Phased-Array Circuits in 0.13-µm SiGe BiCMOS Technology," *IEEE TMTT*, vol. 61, no. 8, pp. 3115–3127, Aug. 2013.

[2] S. Rey et al., "Performance evaluation of a first phased array operating at 300 GHz with horn elements," *EUCAP*, pp. 1629–1633, March 2017.

[3] T. Merkle *et al.*, "Testbed for phased array communications from 275 to 325 GHz," *CSICS*, pp. 1–4, Oct. 2017.

[4] G. D. Ntouni et al., "Real-time experimental wireless testbed with digital beamforming at 300 GHz," *EuCNC*, pp. 271–275, June 2020.

[5] X. D. Deng *et al.*, "A 320-GHz 1×4 Fully Integrated Phased Array Transmitter Using 0.13-µm SiGe BiCMOS Technology," *IEEE TTST*, vol. 5, no. 6, pp. 930–940, Nov. 2015.

[6] I. Abdo et al., "300-GHz-band four-element CMOS-InP hybrid phased-array transmitter with 36° steering range," *IEEE MWTL*, vol. 33, no. 6, pp. 887–890, June 2023.

[7] I. Abdo et al., "A Bi-Directional 300-GHz-Band Phased-Array Transceiver in 65-nm CMOS With Outphasing Transmitting Mode and LO Emission Cancellation," *IEEE JSSC*, vol. 57, no. 8, pp. 2292–2308, Aug. 2022.

[8] S. Yabuki, S. Fujimoto, S. Amakawa, T. Yoshida, and M. Fujishima, "29-to-65-GHz CMOS amplifier with tunable frequency response," *RFIT*, pp. 63–65, Aug. 2022.

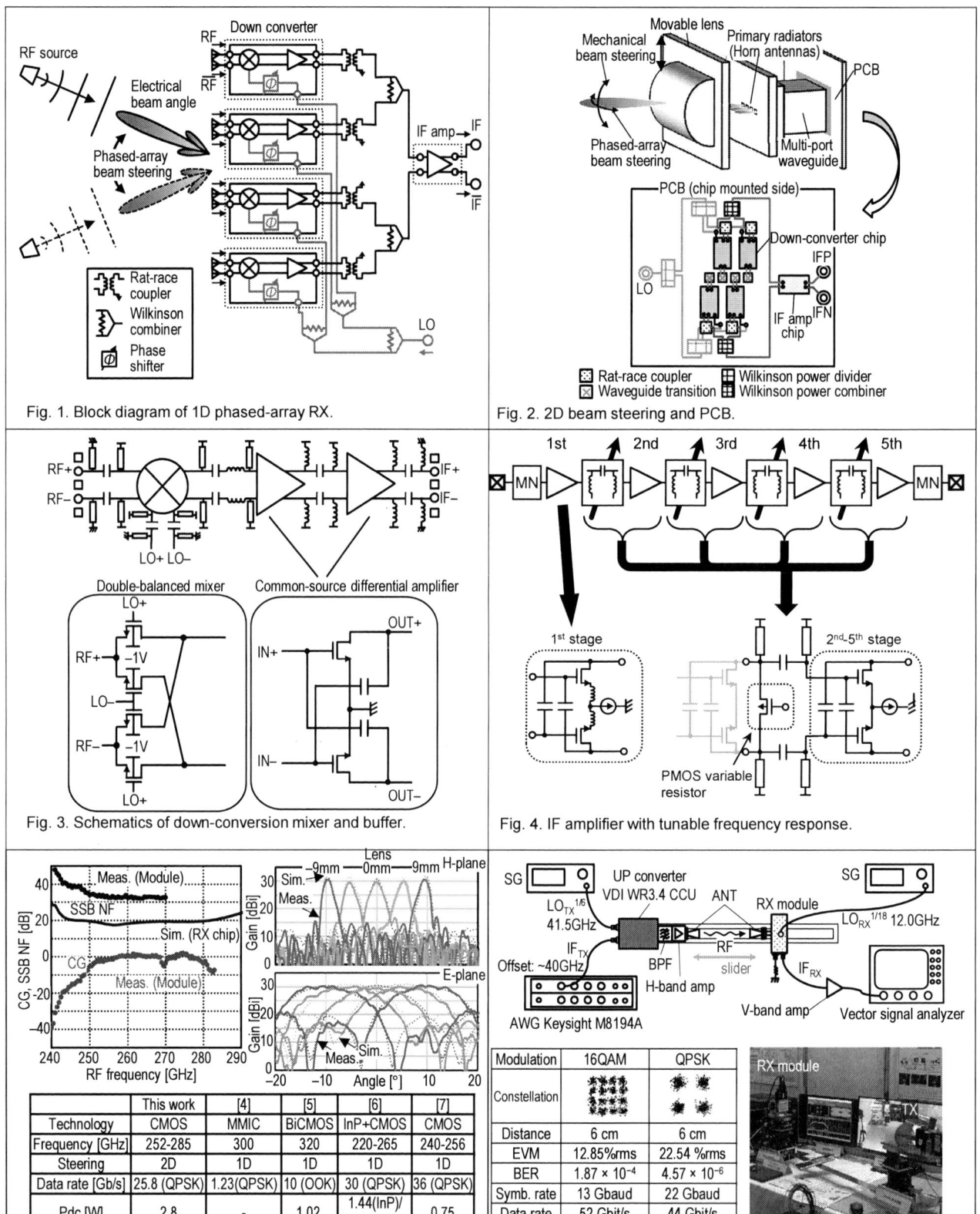

Fig. 1. Block diagram of 1D phased-array RX.

Fig. 2. 2D beam steering and PCB.

Fig. 3. Schematics of down-conversion mixer and buffer.

Fig. 4. IF amplifier with tunable frequency response.

	This work	[4]	[5]	[6]	[7]
Technology	CMOS	MMIC	BiCMOS	InP+CMOS	CMOS
Frequency [GHz]	252-285	300	320	220-265	240-256
Steering	2D	1D	1D	1D	1D
Data rate [Gb/s]	25.8 (QPSK)	1.23(QPSK)	10 (OOK)	30 (QPSK)	36 (QPSK)
Pdc [W]	2.8	-	1.02	1.44(InP)/1.2(CMOS)	0.75

Fig. 5. Frequency response, beam pattern, and comparison.

Modulation	16QAM	QPSK
Constellation		
Distance	6 cm	6 cm
EVM	12.85%rms	22.54 %rms
BER	1.87×10^{-4}	4.57×10^{-6}
Symb. rate	13 Gbaud	22 Gbaud
Data rate	52 Gbit/s	44 Gbit/s

Fig. 6. Measurement setup, results, and 2D steering photo.

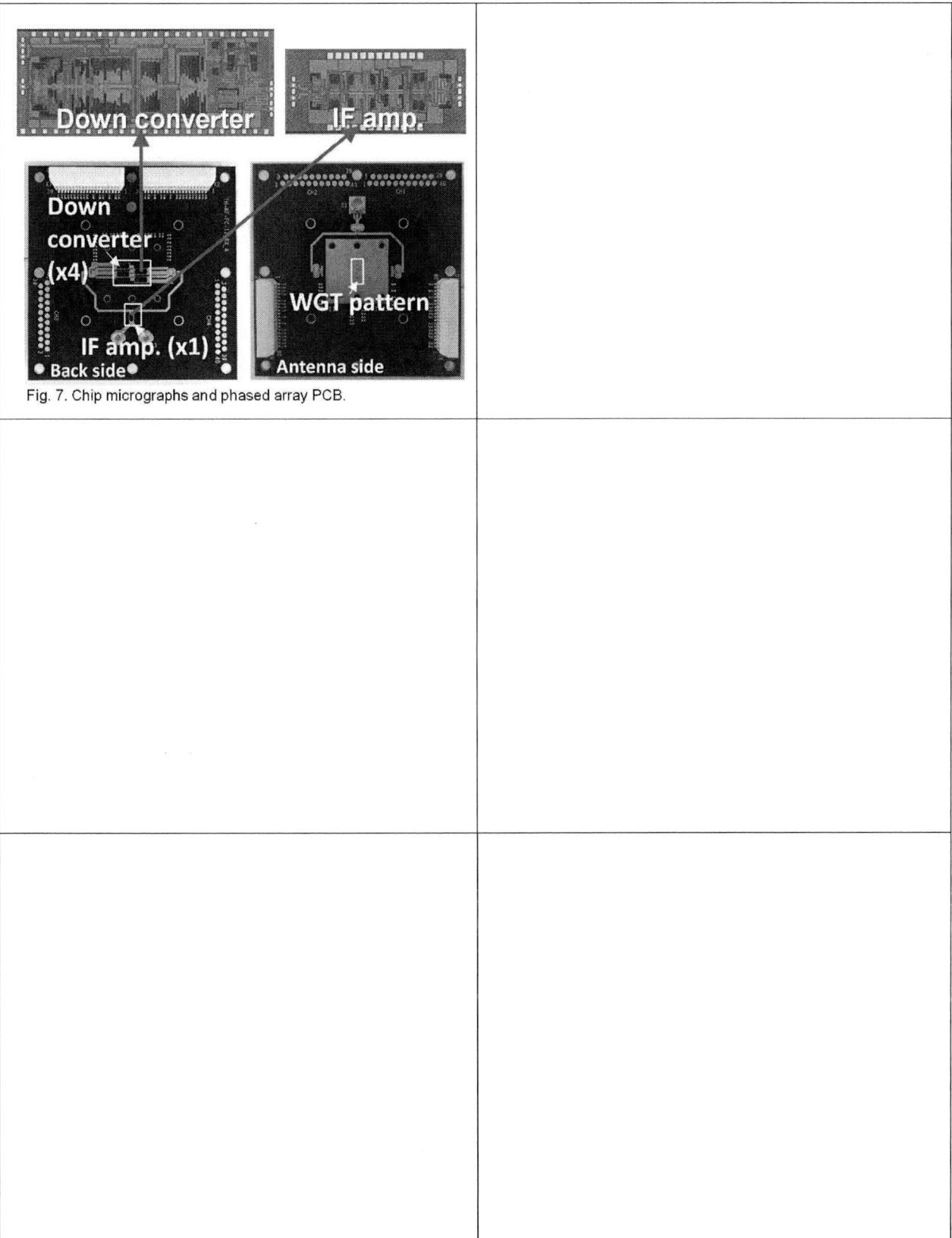

Fig. 7. Chip micrographs and phased array PCB.

A 2pA/√Hz Current-Conveyor-Assisted Ultrasound Receiver with 25pF CMUT Parasitic Capacitance

Gichan Yun[1], Kyeongwon Jeong[1], Haidam Choi[1], Seunghyeon Nam[1], Chaerin Oh[1], Hyunjoo Jenny Lee[1], Sohmyung Ha[2], and Minkyu Je[1]

[1]KAIST, Daejeon, Korea, [2]New York University Abu Dhabi, Abu Dhabi, United Arab Emirates

Ultrasound (US) imaging has emerged as a promising solution for medical diagnosis with its low cost and biocompatibility. Recent demands for multi-dimensional imaging in medical diagnosis brought a challenge of operating multiple US transducers, especially within miniaturized probes. There have been some studies of US systems for the application of intracardiac echocardiography (ICE) [1] and transesophageal echocardiography (TEE) [2] that require highly integrated systems within miniaturized probes. The miniaturization led the capacitive micromachined ultrasonic transducer (CMUT) to be a desirable choice for its compatibility with CMOS and wide bandwidth (BW) [3]. One main characteristic of the CMUT is that it generates current signals, making a transimpedance amplifier (TIA) preferable to the low noise amplifier (LNA). However, the parasitic capacitance (C_P) at the input of the TIA highly affects the performance due to the nature of converting the input-referred voltage noise at the virtual ground to the current noise by its impedance. Therefore, power consumption should be increased with a large C_P to maintain low noise. Moreover, the input multiplexer (MUX) is often used for the US imaging system to reduce overall power consumption [4]. The input switches are implemented using bulky high-voltage MOS transistors to protect the low-voltage devices of the receiver (RX) circuit from the high-voltage pulse of the transmitter (TX) circuit. It also attributes additional C_P at the input. Note that using individual transducers for TX and RX requires a larger area. Hence, the TIA should be carefully optimized for CMUT-based US imaging systems. In this work, we present an analog front-end (AFE) assisted by a current conveyor (CC) to improve noise performance while maintaining the power consumption. The CC buffers the input current and isolates the C_P of the input from the high transimpedance gain stage, relieving the overall AFE burden.

Fig. 1 shows the overall diagram of our system. We adopt a 64-channel one-dimensional (1-D) CMUT array with a C_P of 25pF. A sub-group comprises an element-level TX (8 channels) and a single AFE with an analog-to-digital converter (ADC), including an 8-to-1 MUX. The main challenge of the 1-D CMUT array is the considerable C_P due to the sizeable component size [5]. Prior works employed a resistive-feedback (RF) TIA and capacitive-feedback (CF) TIA to sense the current signal from the CMUT. The RF TIA is the most commonly used structure for its low input impedance, moderate noise figure (NF), and compact area [5–7]. However, the feedback resistor (R_F) directly adds thermal noise to the input and forms a pole with the input parasitic, which deteriorates stability and might affect the BW. It leads to a trade-off between noise and BW, which makes low-noise design challenging for large C_P. On the other hand, the CF TIA can avoid noise from the resistor by using the noiseless capacitive gain. The capacitive divider amplifies the input current while adding extra C_P to the input. Since the capacitor should be large enough to achieve the desired NF, the area can be excessive for multichannel implementation [6]. Moreover, both the input and output stages should operate with a sufficiently high bias current to satisfy the noise and BW requirements.

In the proposed RX, the CC is placed prior to the RF TIA. The CC senses the input current and transfers the amplified signal to the TIA. Although CC adds additional noise, it can significantly mitigate the requirements of the RF TIA by isolating the C_P. The simulated effective C_P of TIA in our proposed design is about 70fF, which has minimal effects on the TIA. However, the BW and noise limitations of the CC can cause problems (Fig. 2). The input impedance of the conventional CC is inversely proportional to the transconductance (g_m) of the input MOS transistors. The input impedance should be low enough since it generates a pole when combined with the input parasitic. The BW of the complete RX chain should be wide enough to process the US signal with a center frequency of 7.5MHz, which means that the g_m of the input MOS transistors should be over 1mS. One way to increase the g_m is by increasing the bias current. Since current noise increases proportionally to the bias current, unlike

voltage noise, the overall input-referred noise floor would reach around hundreds of pA/√Hz to achieve the desired BW, which is too high for measuring the signal from the CMUT. To address this issue, we employ a feedback amplifier, which boosts the g_m of the input MOS transistors to use a smaller bias current, leading to low input-referred current noise. Fig. 2 plots the noise and BW performances with and without the feedback amplifier. Using the feedback amplifier, the BW of the LNA, which is the CC-assisted RF TIA, becomes about 40 times wider, and its noise is suppressed by a factor of 10 at the center frequency of the CMUT while consuming only 1.21mW.

The overall AFE design is described in Fig. 3. It consists of a CC, TIA, programmable-gain amplifier (PGA), and ADC buffer. The open-loop amplifiers used in the AFE are also shown, which are based on the current-reuse structure for low noise. The PGA and ADC buffer are implemented with the capacitively coupled instrumental amplifier (CCIA) structure. The gain of the CC is set to 6dB by manipulating the current-mirroring ratio between the input and output stages to suppress the current noise generated by the TIA. The gain of the overall RX chain can vary from 82dBΩ to 118dBΩ for time-gain compensation (TGC). It compensates for the signal attenuation caused by penetrating the tissue to relieve the dynamic range requirement of the ADC. The overall AFE achieves at least 18MHz BW for all possible gain settings, sufficient for US signal processing. The input-referred current noise of the AFE with the highest gain is 2.0pA/√Hz at 7.5MHz and 2.3pA/√Hz over BW.

The 10b SAR ADC operating at 30MS/s is adopted to digitize the analog US signal to avoid interferences or crosstalks occurring when connected to the back-end system through long wires (Fig. 4). The ADC achieves 57.6dB SNR with 1.6mW power consumption. The area and power breakdowns of the RX are depicted in Fig. 4. The total active area is 0.8452mm^2, mainly occupied by the high-voltage MUX, and the power consumption is 3.62mW, dominated by the ADC and LNA. The element-level TX with a class-D structure is also included in the chip to excite the transducer with 30V. Both the US TX and RX are fabricated in a 180nm BCD process to utilize high-voltage MOS transistors for the pulser and T/R switch. The measured TX output with a pulse width of 67ns is also shown. It can efficiently excite the transducer with a 7.5MHz center frequency.

The prototype US TX and RX ICs are verified with a single-element CMUT pulse-echo test (Fig. 5). The CMUT is placed in the oil tank, and the transmitted US signal is reflected by the quartz placed at a distance of 9mm. The CMUT is excited with a 30V pulse through the bias-tee, and the RX senses the reflected signal. The time-of-flight is expected to be 12µs considering the speed of the US in oil (c = 1460m/s). The reverberation occurs due to the reflection at the transducer, which results in signal repetitions every 12µs. Fig. 6 summarizes the proposed RX performances in comparison with other state-of-the-art CMUT-based US RX. Thanks to the proposed CC-assisted US RX, the power consumption and input-referred noise are highly reduced even for a large C_P of the CMUT. Our work achieves the best modified noise efficiency factor (NEF'), which is normalized with the C_P of the CMUT. The high-voltage TX, T/R switch, and high-speed ADC are also fabricated on chip with maintaining overall low power consumption.

Acknowledgements: This research was supported by the Korean government under grants 2021M3F3A2A01037365 and 20014214. The chip fabrication was supported by the IDEC, Korea.

References:

[1] E. Kang *et al.*, "A 2pA/√Hz transimpedance amplifier for miniature ultrasound probes with 36dB continuous-time gain compensation," ISSCC 2020:354–356.

[2] C. Chen *et al.*, "A front-end ASIC with receive sub-array beamforming integrated with a 32×32 PZT matrix transducer for 3-D transesophageal echocardiography," IEEE JSSC vol. 52, no. 4, 2017:994–1006.

[3] K. Chen *et al.*, "A column-row-parallel ASIC architecture for 3-D portable medical ultrasonic imaging," IEEE JSSC vol. 51, no. 3, 2016:738–751.

[4] M. Tan *et al.*, "A front-end ASIC with high-voltage transmit switching and receive digitization for 3-D forward-looking intravascular ultrasound imaging," IEEE JSSC vol. 53, no. 8, 2018:2284–2297.

979-8-3503-3004-5/23 $31.00 © 2023 IEEE

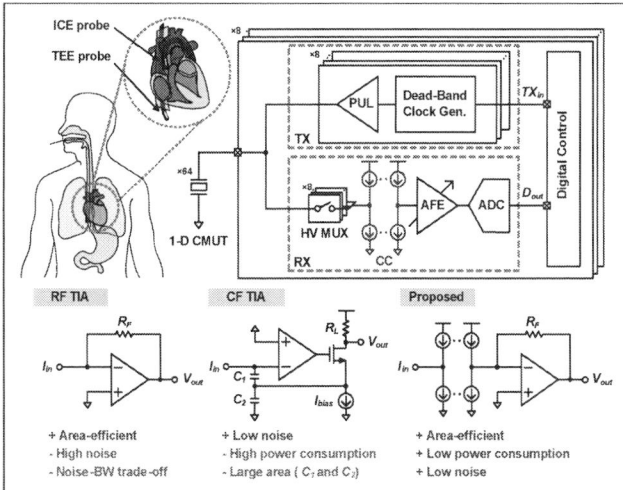

Fig. 1. Overall block diagram of the proposed US imaging system and comparison among different transimpedance stages.

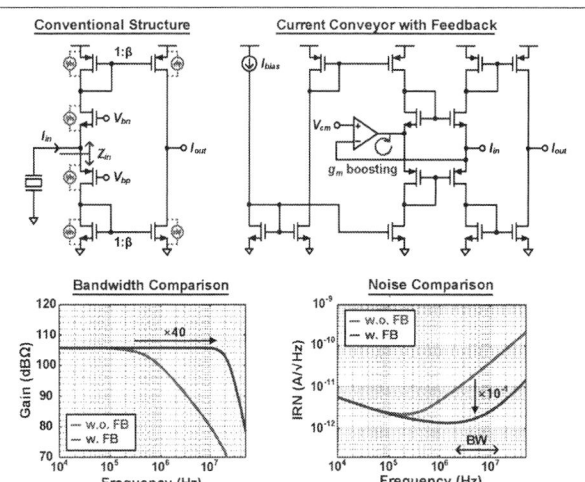

Fig. 2. Circuit diagram of the current conveyor and the comparison between the designs with and without feedback.

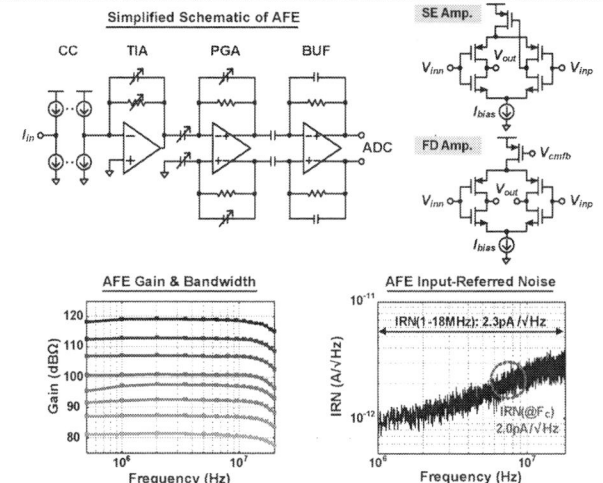

Fig. 3. Circuit diagrams and measurement results of the analog front-end.

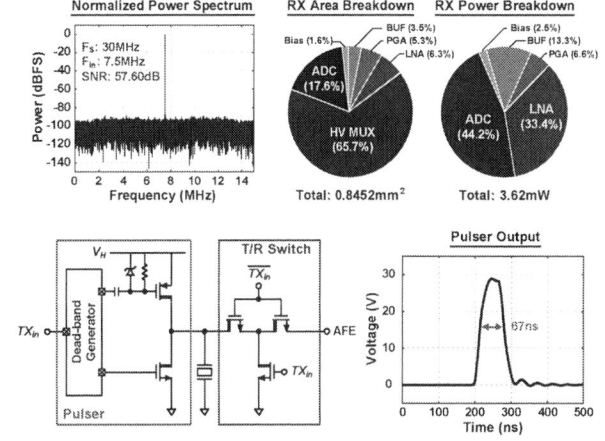

Fig. 4. Measured power spectrum and area/power breakdown of the US RX IC (top); Circuit diagram of the US TX IC and T/R switch and the output measurement result for the US TX IC (bottom).

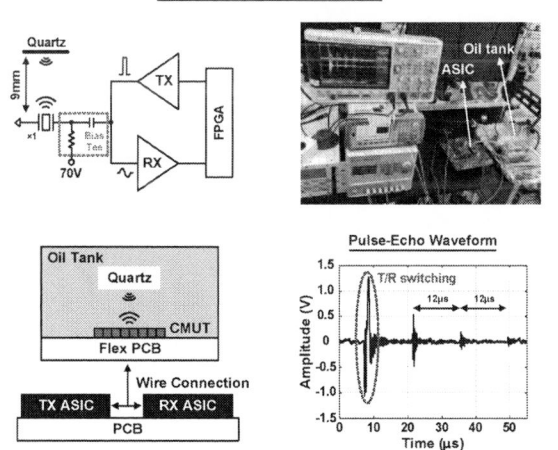

Fig. 5. Pulse-echo experiment: test setup and measured transient output waveform of the received pulse-echo signal.

	JSSC 2016 [3]	TBCAS 2017 [7]	ISSCC 2020 [1]	TBCAS 2021 [5]	This Work
Process	180nm HV CMOS	65nm CMOS	180nm HV BCDMOS	40nm CMOS	180nm HV BCDMOS
Transducer	CMUT	CMUT	CMUT	CMUT	CMUT
CMUT Cap. [pF]	2	5.5	15	25	25
Center Freq. [MHz]	5	5	5	Wideband	7.5
LNA Architecture	RF TIA	RF TIA	CF TIA	RF TIA	CC + RF TIA
LNA BW [MHz]	11	7.5	7	18	18
IRN density @F_C [pA/√Hz]	0.41 @5MHz	4.8 @2.5-7.5MHz	2.0 @5MHz	*3.1 @3-18MHz	*2.0 @7.5MHz
LNA Gain [dBΩ]	104-116	79-97	70-106	78-90	88-106
LNA Power [mW]	1.4	0.18	4.9	1.45	1.21
NEF' [pA/pF·√(mW/Hz)]	0.24	0.37	0.30	0.15	0.088
LNA Area [mm²]	<0.016	0.0035	0.12	<0.025	0.045
ADC/HV Tx on Chip	X/O	O/X	X/O	O/X	O/O
Sampling Rate [MS/s]	-	30	-	80	30
SNR [dB]	-	50.5	-	36.3	**57.6
Rx Power/Ch. [mW]	1.4	0.24	4.9	5.0	3.62

NEF': $i_{n,in}$·√Power$_{LNA}$/C$_{CMUT}$ where $i_{n,in}$ is the input-referred current noise
*Measured with entire AFE **Measured from ADC

Fig. 6. Performance summary and comparison.

979-8-3503-3004-5/23 $31.00 © 2023 IEEE

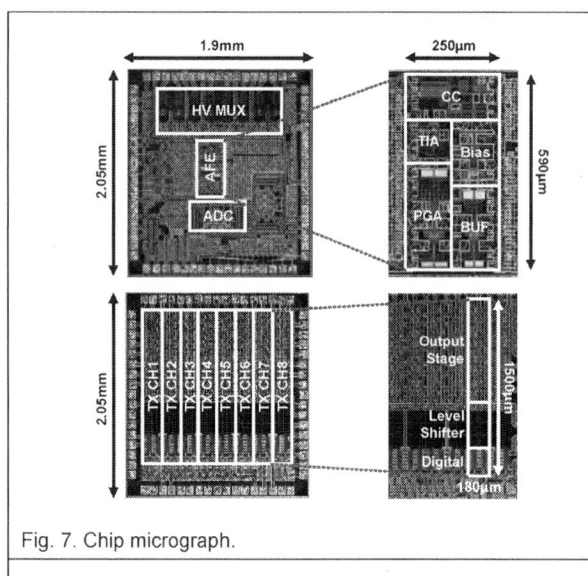

Fig. 7. Chip micrograph.

Additional References:

[5] M. Zhou *et al.*, "An RX AFE with programmable BP filter and digitization for ultrasound harmonic imaging," IEEE TBCAS vol. 15, no. 6, 2021:1430–1440.

[6] M–C. Chen *et al.*, "A pixel pitch-matched ultrasound receiver for 3-D photoacoustic imaging with integrated delta-sigma beamformer in 28-nm UTBB FD-SOI," IEEE JSSC vol. 52, no. 11, 2017:2843–2856.

[7] H. Attarzadeh *et al.*, "A low-power high-dynamic-range receiver system for in-probe 3-D ultrasonic imaging," IEEE TBCAS vol. 11, no. 5, 2017:1053–1064.

A 226 GHz Coupled Harmonic VCO with 9.34% Tuning Range Utilizing Three-coil Transformer with Switched Inductor in 65nm CMOS

Yue Liang, Qin Chen, Xu Wu, Xiangning Fan, Lianming Li

Southeast University, Nanjing, China

Purple Mountain Laboratories

Low-power CMOS process offers a low-cost commercial solution for millimeter-wave and terahertz (THz) system applications. However, their limited maximum oscillation frequency (fmax) remains a bottleneck for the fundamental signal sources with large output power and low phase noise. To tackle these problems, harmonic coupled oscillators are popularly adopted to increase the output power by power combination while reducing the phase noise by $10\log_{10}N$ dB (N is the number of coupling cores). Several topologies [1-3] of the THz oscillator core and the corresponding coupling schemes are listed and compared in Fig. 1 along with their different features. The oscillator in [1] adopts a source degeneration varactor for tuning, however, due to its self-feeding topology, the transistor gate and drain are biased together, restricting its performance. Also, the λ/4 transmission lines required for both power supply and grounding increase its area. The oscillator in [2] optimizes the output harmonic power by cascading doublers with the fundamental oscillators, resulting in a larger area and limited tuning range. Reference [3] introduces a concise structure, and achieves a wide tuning range through the biasing voltages, intending to avoid the varactors to preserve the tank quality factor (Q).

Herein, we present a quad-core coupled harmonic voltage-controlled oscillator (VCO) with a tuning range of 9.34% and a phase noise of -113.3 dBc/Hz at 10 MHz offset. As depicted in Fig. 1 (bottom right), the proposed coupled VCO topology features separated biases for the gate and drain of the transistor and achieves a compact floorplan without extra components for power supply or core coupling. Besides, thanks to the three-coil transformer with a switched inductor, this oscillator achieves a broader frequency tuning range without deteriorating the phase noise and output power.

Figure 2 (left) shows the scheme of our proposed 226GHz wideband coupled harmonic VCO. Four differential oscillator cores operate at the fundamental frequency (f_0 = ~113 GHz) and are coupled out of phase by transmission lines (T-line) TL_D and transformers TR_G. The oscillation tank of one oscillator core consists of inductor L_G (equivalent inductor seen toward TR_G), capacitor C_1, and T-line TL_D. The second-harmonic signal ($2f_0$) is extracted, combined in phase at the center of the TL_D, and then output to the pad. Grounded co-planar waveguides (GCPW) are utilized as the power combiner and to match from the output port to the oscillator core at $2f_0$. Figure 2 (right) depicts the equivalent circuits of one oscillator core at f_0 and $2f_0$. At f_0, the differential oscillator core resembles a 2-stage ring oscillator, and its oscillation tank is divided into two identical Π-type networks inserted between two transistors. The value of these tank components can be synthesized from the optimal operating conditions of the transistors [4]. Two adjacent cores share a T-line TL_D (Z_0, θ) with a virtual ground in the middle, and each core still has the original TL_D. The inductor L_G is provided by our proposed three-coil transformer TR_G (to be interpreted later). The $2f_0$ signal generated at the drain of transistors travels in the common mode path as shown in Fig. 2 (right down) of the differential oscillator core, part of which is fed to the gate through the parasitic gate-drain capacitor $2C_{gd}$ of the transistor and the capacitors $2C_1$ of the oscillator tank, and part of which is output through the T-line. To mitigate potential common-mode oscillation, it is necessary to introduce a resistor, denoted as R_{GATE}, into the gate bias circuitry. The R_{GATE} value is set to 50 ohms to balance its effect on the phase noise and the impedance $Z_{F,2fo}$ seen at $2f_0$, and ultimately the power of the output $2f_0$ signal.

Figure 3 (left) presents the top and 3D views of the proposed three-coil transformer. The primary coil L_P and the secondary coil L_S form a 1:1 transformer with a coupling coefficient of k_{PS} to couple the two adjacent cores. The tertiary coil L_T is placed in the center of the transformer to improve the tuning range. All three coils utilize the top

two thick metal layers to minimize losses and enhance Q. Thanks to the symmetric layout, the equivalent inductances L_G seen from the two ports are equal. The area of the transformer for obtaining L_G is reduced compared to that of the single-coil inductor since the signal coupling polarity of mode "A" used in this oscillator (as marked in Figure 3) makes the transformer equivalent inductance $1+k_{PS}$ times as large as $L_{P/S}$. In contrast, in mode "B" a smaller inductance of $(1-k_{PS})L_{P/S}$ is generated, resulting in a higher tank resonance frequency and lower active gain, and the oscillation condition is difficult to satisfy. Therefore the two cores are unambiguously coupled in mode "A". Without any increase in area and excessive mismatch between two cores, a switching transistor M_{SW} is connected to the tertiary coil L_T and controls the influence of L_T on L_P, L_S, and k_{PS}, resulting in a change of L_G and subsequently altering the tank resonance frequency. The simulation results (Fig. 3 right) show that L_G varies by 13.1% with the switching transistor in the ON and OFF states, while the equivalent Q remains above 12.

The proposed 226GHz quad-core harmonic VCO was fabricated in 65 nm CMOS process. The chip micrograph is shown in Fig. 7. The total chip area is 0.314 mm², whereas the core circuit occupies only 0.0559 mm² excluding the power combiner and pads. The power supply is applied by the integrated DC bias-tee of the WR-3 waveguide probe. From a 1.2-V power supply, the VCO draws a DC current of 113 mA. The spectrum and phase noise are measured using R&S Phase Noise Analyzer FSWP with a harmonic mixer. Fig. 4 shows the screenshots of the measured spectrum and phase noise. The best measured phase noise is -93.5 dBc/Hz at 1 MHz offset and -113.3 dBc/Hz at 10 MHz offset. The measured oscillation frequencies under 1.2-V supply voltage are shown in Fig. 5. This oscillator achieves a frequency tuning range of 9.34% from 215.4 to 236.5 GHz by the inductor switching voltage and biasing voltage.

Figure 6 presents the measured results in comparison with the state-of-the-art relevant oscillators. To the best of the author's knowledge, this coupled harmonic VCO achieves the highest figure-of-merit-tuning (FoM$_T$) among quad-core coupled VCOs. Thanks to the proposed three-coil transformer, this oscillator simultaneously achieves an enlarged tuning range and a lower phase noise due to the stronger coupling between two cores. Moreover, benefiting from the topology of the oscillating cores and their coupling, the circuit is the most compact except for [3], which is combined in a star shape and limits the connections to other circuit blocks.

Acknowledgment:

The authors would like to thank Ms. Haiyan Lu for her measurement support. This work was supported by the National Key Research and Development Program of China under Grant 2018YFE0205900 and Major Key Project of PCL (PCL2021A01-2).

References:

[1] M. Adnan *et al.*, "A 247-to-263.5GHz VCO with 2.6mW Peak Output Power and 1.14% DC-to-RF Efficiency in 65nm Bulk CMOS," IEEE ISSCC 2014: 262–264.

[2] A. Nikpaik *et al.*, "A 219-to-231 GHz Frequency-Multiplier-Based VCO With ~3% Peak DC-to-RF Efficiency in 65-nm CMOS," IEEE JSSC, vol. 53, no. 2, pp. 389–403, 2018.

[3] H. Jalili *et al.*, "A 230-GHz High-Power and Wideband Coupled Standing Wave VCO in 65-nm CMOS," IEEE JSSC, vol. 55, no. 3, pp. 547–556, 2020.

[4] Y. Liang *et al.*, "A Compact 240 GHz Differential Fundamental Oscillator with -94.2dBc/Hz Phase Noise and 5.4% DC-to-RF Efficiency in 22nm FDSOI," ESSCIRC 2023, in press.

[5] S. K. Nath *et al.*, "A 253-280 GHz Wide Tuning Range VCO with -3.5 dBm Peak Output Power in 40-nm CMOS," ESSCIRC 2021: 299-302.

[6] A. Ghorbani-Nejad *et al.*, "Optimum Conditions for Efficient Second-Harmonic Power Generation in mm-Wave Harmonic Oscillators," IEEE JSSC, vol. 57, no. 7, pp. 2130-2142, 2022.

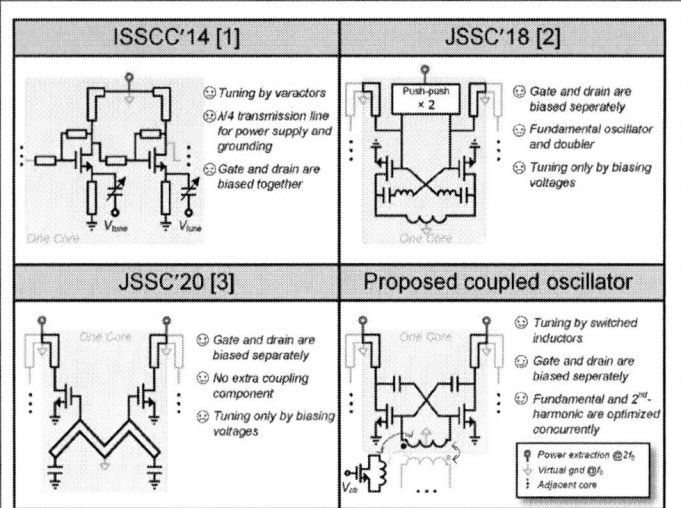

Fig. 1. Comparison of existing THz coupled 4-core VCO topologies with our proposed oscillator.

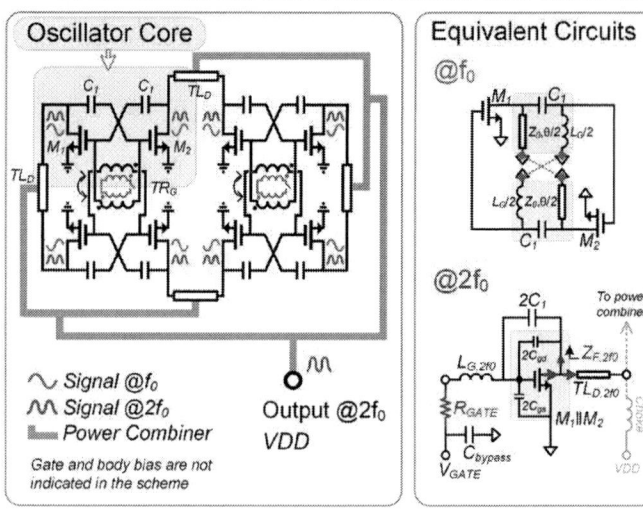

Fig. 2. The scheme of the proposed coupled VCO (left) and the equivalent circuits of the oscillator core at the fundamental and 2nd-harmonic frequency (right).

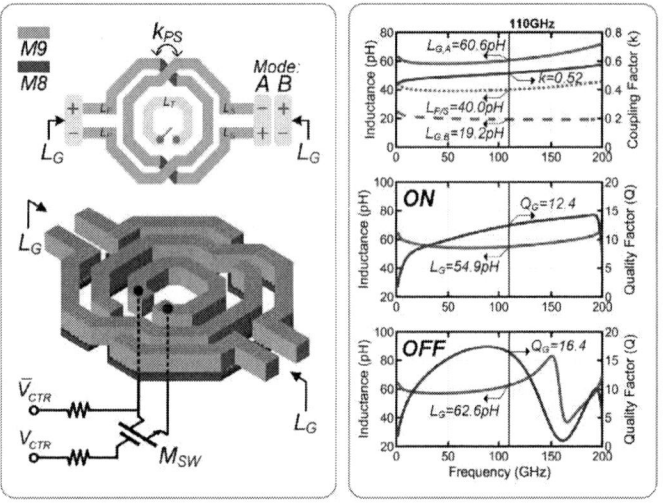

Fig. 3. The top view and two coupling modes (top left), 3D view (bottom left), and simulated results (right) of the proposed three-coil transformer.

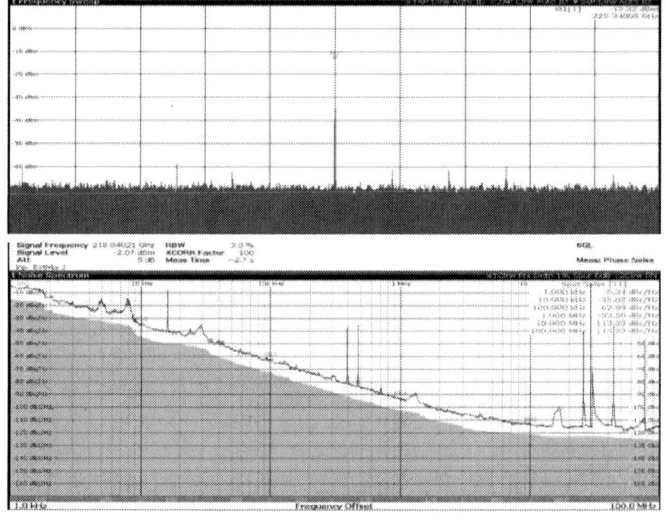

Fig. 4. Measured spectrum (top) and phase noise (bottom).

Fig. 5. Measured frequency versus the gate bias voltage and phase noise at 10 MHz offset across the tuning range.

Reference	ISSCC '14[1]	JSSC '18[2]	JSSC '20[3]	ESSCIRC '21[5]	JSSC '22[6]	This Work
Oscillator Type (No. of cores)	Harm. (2nd) 4	Fund. +Doub. 4	Harm. (2nd) 4	Harm. (2nd) 1	Harm. (2nd) 1	Harm. (2nd) 4
Freq. (GHz)	247-263.5	219-231	219-238	253-280	201-209*	**215.4-236.5**
Tuning Range (%)	6.5	5.33	8.35	10.16	3.8	**9.34**
Output Power (dBm)	-4.9* @1V 0* @1.2V 4.1 @1.6V	3 @1.2V	0 @1.2V 3.4 @1.5V	-3.5 @1.1V	2.5 @1.2V	**0.57 @1.2V**
DC Power (mW)	227	68	195	33	39.8	**135.6**
Efficiency (%)	0.95* @1V 0.99* @1.2V 1.14 @1.6V	2.95 @1.2V	0.85 @1.2V 1.16 @1.5V	1.35 @1.1V	5.6 @1.2V	**0.84 @1.2V**
PN (dBc/Hz) @1MHz	-94	-93.8	-84.1	-74*	-82	**-93.5**
@10MHz	NA	-99*	-105.8	-91.8	NA	**-113.3**
FoM# @1MHz	-174.8	-177.3	-170.2	-164.0	-166.3	**-178.7**
@10MHz	NA	-164.3	-171.9	-161.8	NA	**-178.5**
Area (mm²)	0.436	0.526	0.21	0.09	0.096	**0.314**
Technology (fmax)	65nm CMOS (NA)	65nm CMOS (233GHz)	65nm CMOS (NA)	40nm CMOS (NA)	65nm CMOS (240GHz)	**65nm CMOS (240GHz)**

* Estimated from measurement results plot.

\# $FoM_T = PN - 20\log_{10}(f_0/\Delta f) + 10\log_{10}(P_{DC}/1mW) - 20\log_{10}(TR/10\%)$

Fig. 6. Performance summary and comparison with the state-of-the-art THz VCOs.

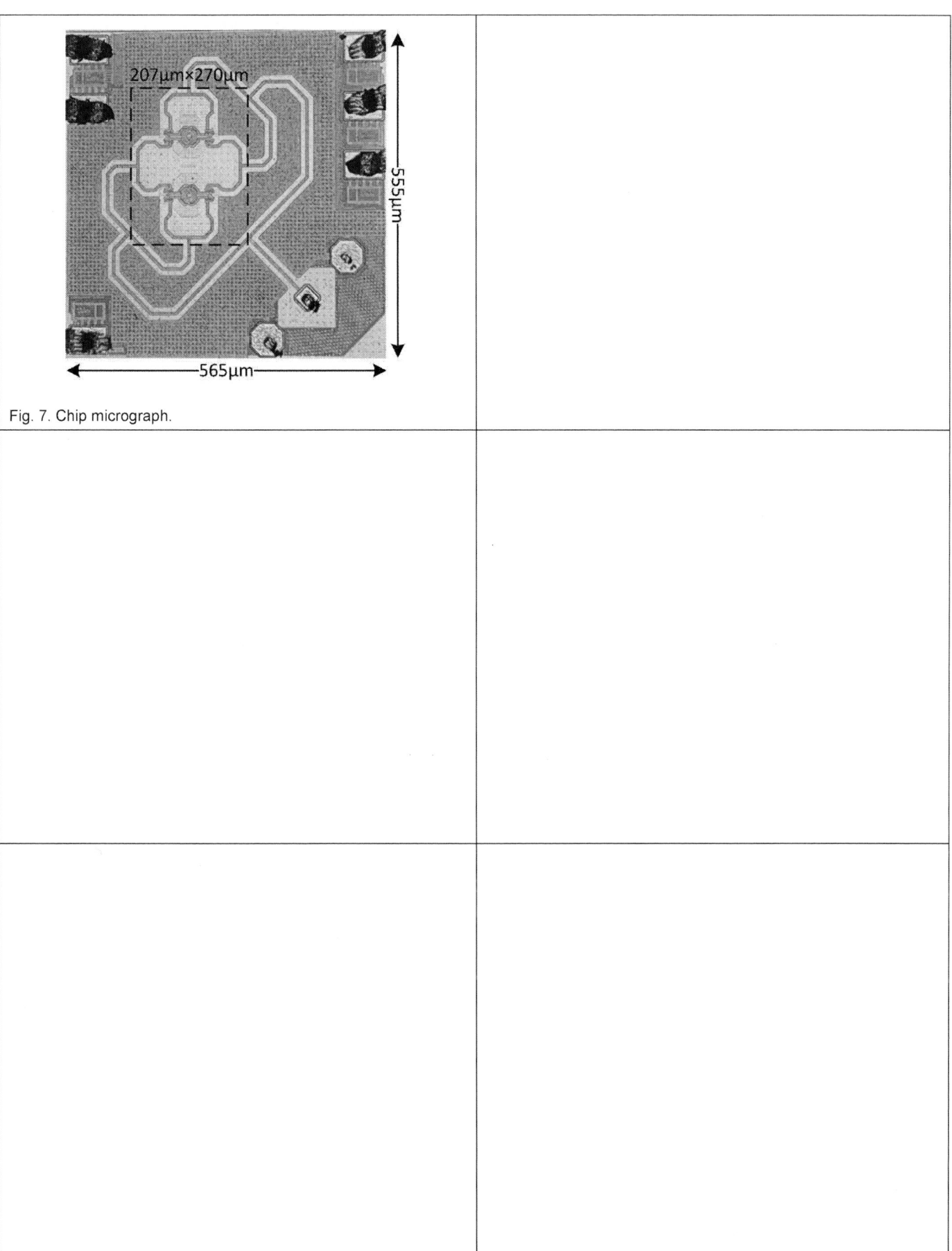

Fig. 7. Chip micrograph.

A 74.0 dB-SNDR 175.4 dB-FoM Pipelined-SAR ADC using a Cyclically Charged Floating Inverter Amplifier

Changjoo Park, Jeongmyeong Kim, Kyounghun Kang,
Minkyu Yang, Byeongmin Moon, Siheon Lee, and Wanyeong Jung

KAIST, Daejeon, South Korea

Energy and area efficiency are crucial performance indicators for small mobile IoT devices in terms of operating time and unit cost. Typically, mobile devices equipped with high-resolution sensors that operate in a range of a few megahertz of bandwidth require a pipelined-SAR ADC for digitizing analog signals [1]. However, a residue amplifier in the ADC needs to perform at a low noise level to achieve high resolution. This poses a deteriorated energy efficiency of the mobile devices since a substantial amount of energy is consumed in the residue amplifier to reduce thermal noise [1]-[4]. Recent floating inverter amplifiers (FIAs) [1], [4] in Fig. 1 (left) improve energy efficiency by utilizing the current-reuse technique.

However, such FIAs have a limit to improving energy and area efficiency because of their large reservoir capacitor (C_{RES}). During amplification phase (Φ_{AMP}), a vast amount of charge stored in the large C_{RES} generates an exponentially decaying large bias current (I_{AMP}) to the amplifier. This substantial current increases the amplifier's overdrive voltage (V_{OV}), and thus deteriorates the amplifier's energy efficiency [5] by decreasing the g_m/I_D.

Moreover, its dynamic behavior renders the conventional FIA's gain difficult to control precisely. Closed-loop multi-stage FIAs can be used for stabilizing gain [1], but at the cost of stability issues and deteriorated energy efficiency compared to an open-loop single-stage FIA.

In this paper, we present a cyclically charged FIA (Fig. 1 right) that features stably reconfigurable bias current by using a floating current source. The large single C_{RES} in a conventional FIA is divided into several small capacitors (C_{RES-P}). In the reset phase, all C_{RES-P} are charged to the V_{DD}. During amplification phase (Φ_{AMP}), some of the divided capacitors are connected to the amplifier to supplies charge, while other capacitors are recharged for next cycles. Each capacitor sequentially changes connection one by one and supplies charge alternately. This switched-capacitor operation functions as a floating current source for the amplifier, supplying a reconfigurable and stable bias current (I_{AMP}). The ripples of the supply voltage (ΔV_S) and the bias current can be significantly decrease with the increasing number of discharging C_{RES-P} capacitors (N_C).

The cyclically charged FIA improves energy efficiency by maintaining optimal levels of stable bias current and overdrive voltage (V_{OV}) during amplification. As shown in Fig. 2, the amplifier is powered by a stably controlled bias condition, which allows amplifier to have an improved g_m/I_D and enhances the FIA's energy efficiency. Simulation results with the same amplifier transistors and load capacitance (C_O=250 fF) show that while the conventional FIA consumes 253 fJ and achieves input-referred noise (IRN) of 134 μV_{rms}, the presented FIA consumes only 151 fJ and has a lower IRN of 120 μV_{rms} for reaching same amplifier gain of 8 V/V. In terms of amplifier's energy efficiency (=noise power × energy), the cyclically charged FIA achieves 1.90 nJ × μV_{rms}^2, which is ~2.08x improved from the conventional FIA's 3.96 nJ × μV_{rms}^2. Although its amplification time is longer than that of the prior FIA, this is not an issue for designing a high-resolution ADC operating at a few MS/s. Moreover, compared to the large reservoir capacitance of 5 pF used in the conventional FIA, this work's FIA uses a significantly smaller total reservoir capacitance of 0.2 pF, resulting in enhanced area efficiency.

The cyclically charged FIA also enables precise open-loop gain control by using Vernier floating current source. Fig. 3 (left) shows the FIA using two floating current sources for precise gain control. Φ_{RES1} and Φ_{RES2} are signals controlling charge/discharge of C_{RES1} and C_{RES2}, respectively, where C_{RES1} is designed slightly larger than C_{RES2}. When amplification starts, the array of C_{RES1} repeats charging/discharging alternately and provides power to the array. After a certain number of C_{RES1} switching (N_{SW1}), C_{RES1} stops and C_{RES2} begins charge/discharge alternately for N_{SW2} times. By properly adjusting the N_{SW1} and N_{SW2}, the amplifier's differential gain can be controlled with an accuracy equivalent to the charge stored

in ΔC_{RES} (= C_{RES1} − C_{RES2}), which is much smaller than unit capacitors of C_{RES1} and C_{RES2}. Fig. 3 (top-right) shows the FIA's simulated gain controllability with various combinations of N_{SW1} and N_{SW2}. The FIA's gain gradually increases with the increase in $N_{SW1}+N_{SW2}$ or N_{SW1}. As a result, the FIA can be configured to a desired open-loop gain with the maximum gain error of 0.28 % over output swing range under 150 mV (simulation), which is within the required gain error of 0.39% for the ADC in this work. (Fig. 3, bottom-right). When used for a pipelined ADC, this gain control method resolves the ADC's stability issues and improves its energy efficiency as the ADC no longer requires a high-loop-gain closed-loop.

Fig. 4 represents the overall architecture and timing diagram of a 13-bit pipelined-SAR ADC using the cyclically charged FIA. The ADC consists of a 6-bit coarse-stage and a 7-bit fine-stage ADC with 2-bit fine-stage redundancy and 1-bit inter-stage redundancy. An open-loop residue amplifier is implemented with the FIA powered by dual floating current sources. The charge/discharge operation of each reservoir capacitor (C_{RES1}, C_{RES2}) is controlled by a C_{RES} switching controller based on a leakage-based VCO [6]. The inter-stage gain is controlled by counting the numbers of oscillation cycles of the C_{RES1}, C_{RES2} groups (N_{SW1}, N_{SW2}) using an asynchronous counter. The cyclically charged FIA operates as a residue amplifier (Mode$_{RA}$ = low) and reused as a pre-amplifier (Mode$_{RA}$ = high) of the coarse-stage comparator, eliminating the need for calibrating mismatch between the two different amplifiers [7]. When operating as a residue amplifier, the FIA drives the relatively large fine-stage split C-DAC array [8] sized for ADC's target noise level using many operation cycles. On the other hand, when operating as a pre-amplifier, the FIA drives a small load capacitance and operates with a fewer cycles for low-power and high-speed operation. This enables optimized energy efficiency for each operating mode. A global SAR logic controls the ADC's operation with an external clock.

The 13-bit pipelined-SAR ADC is fabricated in a 65-nm CMOS process, and it occupies 0.18 mm^2 and 0.0083 mm^2 for the residue amplifier as shown in Fig. 7. The Fig. 5 (top) shows the ADC's DNL and INL. The ADC achieves 74.0 dB SNDR and 85.5 dB SFDR at a Nyquist sampling rate (F_S) of 2 MS/s. The ADC with the open-loop FIA features stable performances over different chips and sampling rate through setting proper gain control values (N_{SW1}, N_{SW2}). This shows that the ADC is suitable for applications that require a scalable sampling rate, such as event-driven IoT devices [1]. Fig. 6 shows the ADC performance summary and comparison with prior arts. The ADC consumes 72.3 μW at 2 MS/s and achieves a good Schreier FoM of 175.4 dB. Considering the ADC's analog blocks, especially the residue amplifier based on the cyclically charged FIA, consume small portions of power (Fig. 5 bottom), this confirms high energy efficiency of the presented FIA. Moreover, the FIA uses total reservoir capacitance of 0.76 pF, occupying a smaller area than the prior FIA-based ADCs using large capacitances of 32.5 pF [1] and 3.2 pF [4].

Acknowledgement:

This work was supported by Samsung Electronics and the National Research Foundation of Korea (NRF) grant funded by the Korea government (MSIT) (No. RS-2023-00213676). The chip fabrication and EDA tool were supported by the IC Design Education Center (IDEC), Korea.

References:

[1] X. Tang, et al., "27.4 A 0.4-to-40MS/s 75.7dB-SNDR Fully Dynamic Event-Driven Pipelined ADC with 3-Stage Cascoded Floating Inverter Amplifier," ISSCC 2021: 376-378.

[2] H. Yoon, et al., "A 65-dB-SNDR Pipelined SAR ADC Using PVT-Robust Capacitively Degenerated Dynamic Amplifier," JSSC 2023: 961-971.

[3] Y. Lim et al., "A 1 mW 71.5 dB SNDR 50 MS/s 13 bit Fully Differential Ring Amplifier Based SAR-Assisted Pipeline ADC," JSSC 2015: pp. 2901-2911.

[4] L. Shen et al., "A Two-Step ADC With a Continuous-Time SAR-Based First Stage," JSSC 2019: 3375-3385.

[5] X. Tang et al., "An Energy-Efficient Comparator With Dynamic Floating Inverter Amplifier," JSSC 2020: 1011-1022.

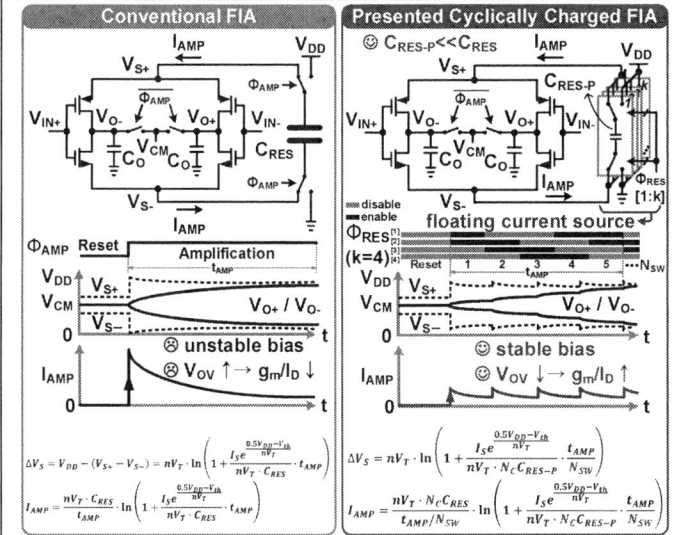

Fig. 1. Operation principle comparison between the conventional FIA (left) and the presented cyclically charged FIA (right).

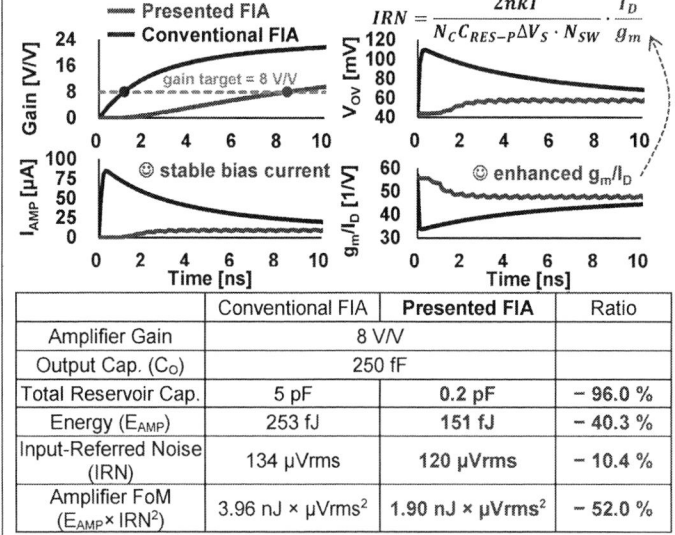

$$IRN = \frac{2nkT}{N_C C_{RES-P} \Delta V_S \cdot N_{SW}} \cdot \frac{I_D}{g_m}$$

	Conventional FIA	**Presented FIA**	Ratio
Amplifier Gain	8 V/V		
Output Cap. (C_O)	250 fF		
Total Reservoir Cap.	5 pF	0.2 pF	− 96.0 %
Energy (E_{AMP})	253 fJ	151 fJ	− 40.3 %
Input-Referred Noise (IRN)	134 μVrms	120 μVrms	− 10.4 %
Amplifier FoM ($E_{AMP} \times IRN^2$)	3.96 nJ × μVrms²	1.90 nJ × μVrms²	− 52.0 %

Fig. 2. Simulated performance comparison of the conventional and presented cyclically charged FIAs.

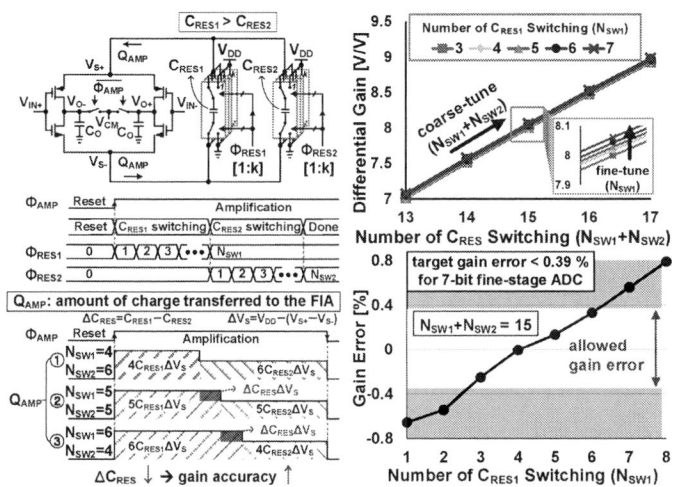

Fig. 3. Schematic and timing diagram of the cyclically charged FIA with a Vernier floating current source (left), FIA differential gain controllability (top-right) and gain error (bottom-right).

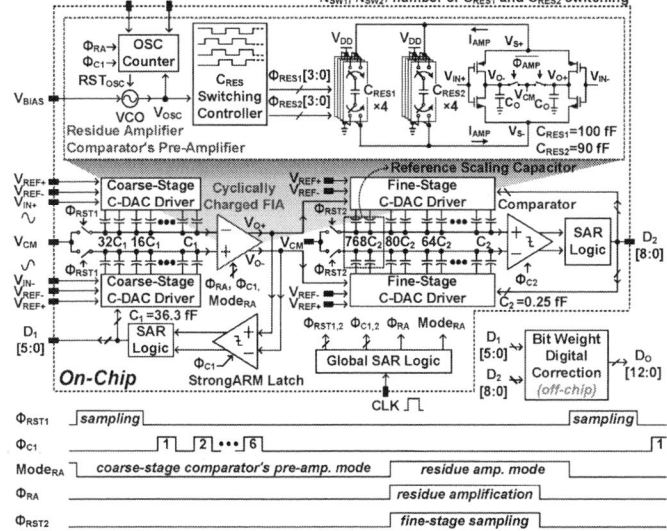

Fig. 4. Block and timing diagram of the pipelined-SAR ADC.

Fig. 5. Measured static/dynamic performance & power breakdown.

	This Work	ISSCC 2021 [1]	JSSC 2019 [4]	JSSC 2015 [3]	CICC 2017 [7]
Architecture	Pipelined-SAR ADC				
Process [nm]	**65**	40	40	65	130
Residue Amp (RA) Type	**Cyclically Charged FIA**	FIA	FIA	Ring Amp.	Dynamic Amp.
RA Topology	**Open-Loop**	Closed-Loop	Open-Loop	Closed-Loop	Open-Loop
V_{DD} [V]	**1.2**	1.2	1.1	1.2	1.2
F_S [MS/s]	**2**	40	2	50	10
Power [mW]	**0.072**	0.82	0.025	1.0	0.28
SFDR [dB]	**85.5**	81.4	79.4	84.6	75.4
SNDR [dB]	**74.0**	75.7	71.7	70.9	63.2
FoM* [dB]	**175.4**	179.6	177.8	174.9	166.4
Area [mm²] (active)	**0.18 (0.023)**	0.056	0.01	0.054	0.15
Reservoir Capacitance [pF]	**0.76**	32.5	3.2	n/a	n/a

$* FoM = SNDR[dB] + 10 \cdot \log_{10}(F_S/(2 \cdot Power))$

Fig. 6. Pipelined-SAR ADC performance summary & comparison.

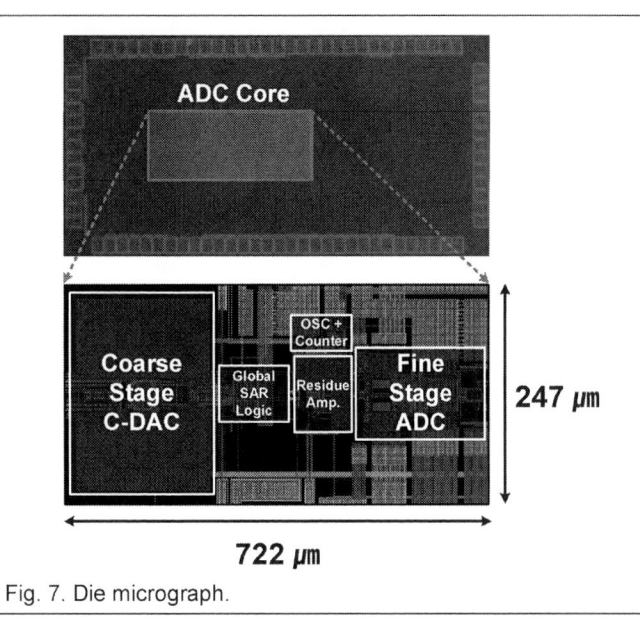

Fig. 7. Die micrograph.

Additional References:

[6] I. Lee *et al.*, "A Constant Energy-Per-Cycle Ring Oscillator Over a Wide Frequency Range for Wireless Sensor Nodes," JSSC 2016: 697-711.

[7] M. Gandara *et al.*, "A pipelined SAR ADC reusing the comparator as residue amplifier," CICC 2017: 1-4.

[8] B. P. Ginsburg *et al.*, "500-MS/s 5-bit ADC in 65-nm CMOS With Split Capacitor Array DAC," JSSC 2007: 739-747.

A Hybrid Integrated W-Band 4-Element Phased-Array Transceiver Front-end Achieving 21.6% Full TX Peak PAE at 14.8dBm Output Power and <1°/dB Phase/Gain Resolution in 65-nm CMOS Technology

Wei Zhu[1], Jian Zhang[2], Jiazhi Ying[1], Xiangjie Yi[2], Yan Wang[2], and Houjun Sun[1]

[1]School of Integrated Circuits and Electronics, Beijing Institute of Technology, Beijing, China

[2]Institute of Microelectronics of Tsinghua University, Beijing, China
Email: zhuwei@bit.edu.cn

W-band phased arrays for radar and B5G/6G communication systems have obtained an ever-increasing attention [1-6]. Due to the challenges of lower active gain and higher passive loss, most W-band transceiver (TRX) front-end (FE) do not integrate T/R switches, attenuators (ATTs) and high resolution phase shifters (PSs) [1-6], which will significantly reduce the system performance and increase the die area. To address this challenge, we have reported a coupled-line (CL) -based implementation mechanism of a W-band T/R switch, PS and ATT in [7]. In order to further advance the evolution of the W-band TRX FE and address the issues of TX/RX isolation, power efficiency as well as die area. This work presents a hybrid integrated W-band 4-element phased-array TRX FE in 65-nm CMOS technology. A hybrid integrated architecture is proposed to integrate the functions of phase shifting and attenuation into active amplifiers to reduce the IL and die area of the PSs and the ATTs, and greatly improve the power and area efficiency of the TRX FE. A CL-based isolation enhanced T/R switch is proposed to increase the TX/RX isolation while remaining low IL. Benefiting from the proposed techniques, the proposed W-band TRX FE prototype achieves (1) 10.2%/21.6% OP1dB/peak PAE at 10.8/14.8dBm output power (Pout), (2) <1°/dB phase/gain resolution, (3) 53% area reduction compared to the previous version in [7].

Figure 1 (top) shows the concept of the proposed nearly PS- and ATT-free hybrid integrated phased array TRX FE architecture. The PS and ATT are all hybrid integrated in the matching networks of active amplifiers, leading to reduction in IL and die area. We refer the hybrid integrated PS and ATT as PSATT. The T/R switches are also hybrid integrated in the TX/RX matching networks to further improve the FE performance. Fig.1 (bottom) shows the block diagram of the proposed W-band 4-element phased array TRX FE. The high-resolution phase shifting bits, the ATT as well as the T/R switches are all hybrid integrated in the matching networks of TX and RX, leading to significant improvements in power and area efficiency.

Figure 2 shows the detailed schematic of the proposed low noise amplifier (LNA) and power amplifier (PA) as well as proposed hybrid integrated PSATT and isolation enhanced T/R switch. The proposed T/R switch is improved based on the CL-based T/R switch in [7]. It is mainly consisted of four main CLs L1- L4. The L1 and L2 perform the load matching of PA, while the L1 and L3 perform the source matching of LNA. The coupling of L4 and L3 provides the passive voltage gain to realize gm-boosting of the LNA. A transistor switch M3 is introduced to perform TX/RX isolation. The improvement compared to the T/R switch in [7] is the introduction of two auxiliary CLs L5- L6 with transistor switches M1- M2 to switch the magnetic coupling strength between L1 and L2 as well as L1 and L3 for isolation enhancement. In TX mode where CTR_A1/2 = 0/1, off capacitance of M1 serves as part of the PA output matching network, providing high-efficiency power transfer from TX to ANT port, while the on resistance of M2 performs TX/RX isolation enhancement. The RX mode follows the same operation mechanism. Simulation results reveals that with the introduction of L5 and L6, the switch isolation is improved by ~3dB at 95GHz, with the IL penalty of only ~0.1dB.

Figure 3 shows the schematic of the proposed hybrid integrated PSATT. The ATT part is improved based on the ATT in [7]. It is consisted of 2 main CLs and 8 auxiliary CLs with transistor switches M1-16A/B for gain control by adjusting the coupling strength between the adjacent main CLs. When CTR_A<n> is switched to 0 or 1, the coupling strength difference can be generated. Compared to the previous version in [7], the improvement lies in that each CL of this ATT part is controlled by multiple pairs of transistor switches instead of just one pair, resulting in higher resolution attenuation and higher

precision array beam control and calibration. The PS part of the PSATT is consisted of four switchable artificial resonators Z0- Z3, which are connected to the main CLs to achieve high-resolution phase shifting while maintaining constant amplitude. Each artificial resonator consists of cascaded inductor and 4-bit capacitor bank. High resolution phase shifting can be achieved by tuning the impedance of the resonator. The Z0 and Z1 are designed to resonate at around 130GHz and each bit provides a coarse phase shifting of ~2°, while Z2 and Z3 are designed to resonate at around 160GHz and each bit provides a fine phase shifting of ~0.9°. Similar to ATT in [7], two meandered single-ended PSATT can be combined to form a differential PSATT, so it can be employed as an inter-stage matching network without occupying additional die area, greatly improving TRX FE area efficiency. Simulation results show that the IL of the PSATT is only 1.6dB at 94GHz. Considering the inherent ~1dB IL of the W-band inter stage matching network, the proposed PSATT almost eliminates the IL of W-band ATTs and PSs, greatly improving TRX FE power efficiency. Each stage of the PSATT can provide phase resolution of <1° in the range of 32° and gain resolution of <0.4dB in the range of 8dB. Similar to phase interpolator (PI) in [7], the PI in this work is implemented with a single-bit vector-modulator. It has a 90° phase-shifting step in the range of 360° to map the high phase resolution of PSATTs to full 360° range.

Figure 7 shows the die micrograph of the proposed 4-element phased-array TRX FE. The TRX FE is fabricated in a 65-nm CMOS technology. The chip size is 2.5×2.5 mm^2 including pads. Benefiting from the proposed hybrid integration architecture, the core area of each TRX FE element unit is only 0.47mm^2, which is 53% smaller than the previous version in [7]. The performance of the TRX FE is measured using an on-chip probing system, as shown in Fig. 4(a).

Figure 4 shows measured phase and amplitude responses of the RX. The gain and phase response of the PI are first characterized, the PI achieves a precise 90° phase shift step over 360° range with an RMS phase error of <1° over 85-110GHz, showing precise phase mapping capability. The gain variation of the PI is <±0.5 dB. Then the PSATT and the RX is characterized, as shown in Fig. 4(c), the three stages of PSATT achieves a phase resolution of ~0.9° and a phase shift range of ~96° to cover the 90° phase step of PI. The RX achieves a phase resolution of ~0.9° and a phase shifting range of 366° with an RMS phase error of <0.42° over 92-96GHz. The gain variation of the RX during the ~366°phase shift is less than ±1.2dB. The RX also achieves 28.5dB gain control range with average 0.4dB gain step at 95GHz, the RMS gain error and the phase variation are <0.22dB and <±6° over 85-110GHz, as shown in Fig. 4(d).

Figure 5(a) shows the measured small and large signal performance of the RX. The RX achieves 91.5-to-98.4GHz BW$_{-3dB}$ with 20.5dB maximum gain and 6.3dB minimum noise figure (NF). The IP1dB at 95GHz is −14.2dBm. Fig. 5(b) shows the measured small and large signal performance of the TX. The TX achieves 92.2-to-97.3GHz BW$_{-3dB}$ with 25.5dB maximum gain. At 95GHz, The TX achieves 14.8dBm Psat, 10.5dBm OP1dB, 21.6%/10.2% peak/OP1dB PAE at highest gain setting as well as >9.5dBm OP1dB and >6.2% OP1dB PAE at lowest gain setting, the proposed PSATT exhibits excellent large signal linearity characteristics. The Psat BW$_{-1dB}$ of the TRX FE covers 93.1-to-103.1/92.5-to-107.9GHz with 14.8/16.4dBm maximum Psat with 1/1.2V supply voltage, respectively. Fig. 5(c) shows the measured reliability and lifetime of the proposed TRX FE under constant RF stresses (at OP1dB) with different excessive supply voltages. By extrapolation, the TRX FE show predicted lifetime of 61K hours when they continuously work at 10.8dBm Pout and 1V supply voltage.

Table I shows the performance comparison of this work with the state-of-the-art W-band TRX FEs. Our proposed TRX FEs exhibit the highest FE integration with the lowest layout overhead. Benefiting from the proposed hybrid integrated architecture, our TRX FE achieves the highest phase/gain resolution, the highest OP1dB/Psat, and up to 10.2%/21.6% TX OP1dB/peak PAE.

References:

[1] A. Townley et al., "A 94-GHz 4TX–4RX Phased-Array FMCW Radar Transceiver with Antenna-in-Package," JSSC, vol. 52, no. 5, pp. 1245–1259, May 2017.

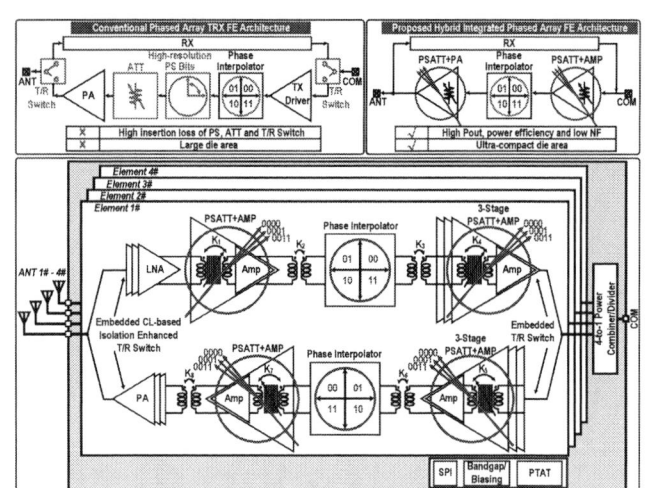

Fig. 1. Concept of the proposed hybrid integrated phased array TRX FE architecture (top) and the block diagram of the proposed hybrid integrated W-band 4-element phased array TRX FE (bottom).

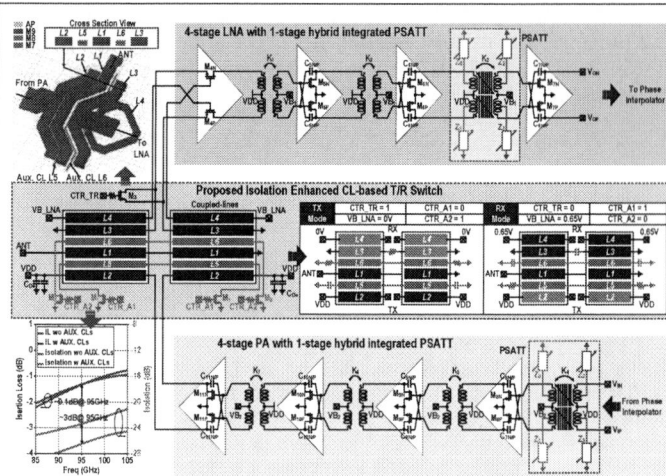

Fig. 2. Detailed schematic of the LNA and PA with proposed hybrid integrated CL-based PSATT as well as proposed isolation enhanced CL-based T/R switch.

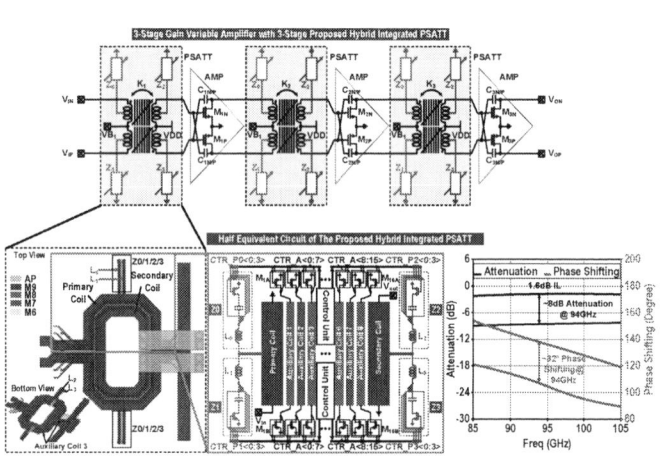

Fig. 3. Detailed schematic and physical implementation of the proposed hybrid integrated PSATT.

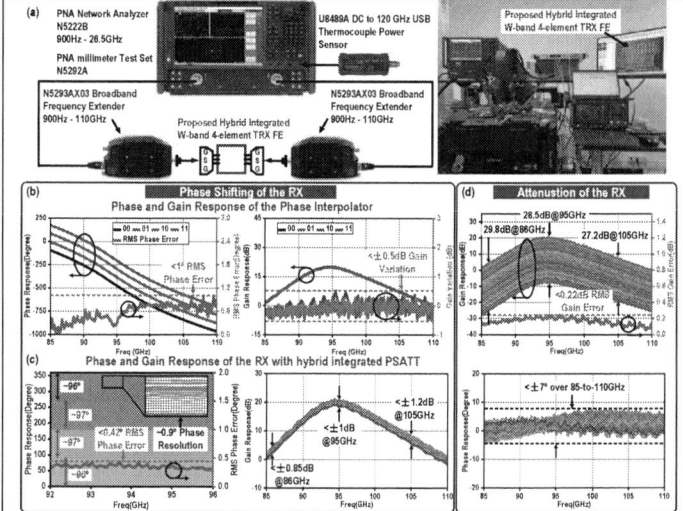

Fig. 4. (a) Small and large signal measurement setup and photograph. Measured (b)-(c) phase and (d) amplitude responses of the RX with hybrid integrated PSATT.

Fig. 5. (a) The measured small and large signal performance of the RX. (b) The measured small and large signal performance of the TX. (c) The measured reliability and lifetime of the proposed TRX FE.

	This Work	[7] ISSCC 2021	[1] JSSC 2017	[2] JSSC 2019	[3] JSSC 2017	[4] JSSC 2018	[5] TMTT 2015	[6] JSSC 2022
Technology	65nm CMOS	65nm CMOS	130nm SiGe BiCMOS	180nm SiGe BiCMOS	130 nm InP HBT	130nm SiGe BiCMOS	65nm CMOS	130nm SiGe BiCMOS
Integration	4T4R TRX FE	4T4R TRX FE	TRX	TRX	TRX FE	TRX	4T4R TRX	4T4R TRX
Frequency (GHz)	92.2-97.2	92-97	94	80-100	94	94	87.8-98.9	93
Eff. Phase Bit	8.6	8.9	5.9	5	3.6	7.8	7	5.6
RMS Phase Err.(°)	<0.42	<0.7	N/A	N/A	N/A	0.3	1.3	0.39
Eff. Gain Bit	6.1	6	--	--	--	1	--	--
Gain Range (dB)	28	14.5	--	--	--	12	--	--
RMS Gain Err. (dB)	<0.22	<0.31	--	--	--	--	--	--
RX NF (dB)	6.3	<9	9.5	6.5-8	<9.3	<7	10	7.7
RX IP1dB (dBm)	-14.2	-20.1	N/A	N/A	-23	-44.5	N/A	-28
TX Gain (dB)	25.5	34.9	N/A	14	22	35.5	17	26
TX OP1dB (dBm)	10.8	9.5	N/A	N/A	N/A	4.6	N/A	11
TX Psat (dBm)	14.8	15.1	6.4	6-8	5	7.8	9	14
TX PAEpeak (%)	21.6	12.3	<12	N/A	N/A	N/A	10.2	N/A
CoreArea /Element (mm²)	0.47	0.99	2.04	3.03	2.7	1.25	1.89	0.9

Fig. 6. Performance comparison with recently reported state-of-the-art phased-array TRX FEs at similar frequencies.

979-8-3503-3004-5/23 $31.00 © 2023 IEEE

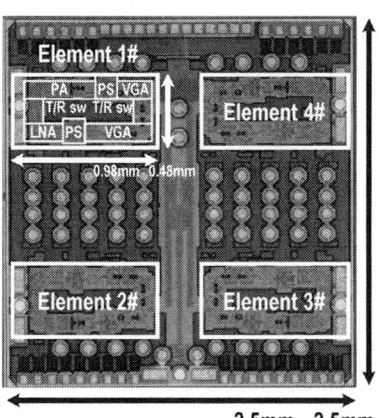

Fig. 7.. Die photograph of the W-band 4-element TRX FE in a 65-nm CMOS technology. The chip occupies 2.5mm×2.5mm die area excluding pads. The core area of each TRX FE element is only 0.98mm×0.48mm (0.47mm^2).

Additional References:

[2] S. Shahramian et al., "A Fully Integrated 384-Element, 16-Tile, W -Band Phased Array with Self-Alignment and Self-Test," JSSC, no. 9, pp. 2419–2434, Sep. 2019.

[3] S. Kim et al., "An Ultra-Low-Power Dual-Polarization Transceiver Front-End for 94-GHz Phased Arrays in 130-nm InP HBT," JSSC, vol. 52, no. 9, pp. 2267–2276, Sep. 2017.

[4] W. Lee et al., "Fully Integrated 94-GHz Dual-Polarized TX and RX Phased Array Chipset in SiGe BiCMOS Operating up to 105 °C," JSSC, vol. 53, no. 9, pp. 2512–2531, Sep. 2018.

[5] A. Natarajan et al., "W-Band Dual-Polarization Phased-Array Transceiver Front-End in SiGe BiCMOS," IEEE TMTT, vol. 63, no. 6, pp. 1989–2002, Jun. 2015.

[6] H. Li et al., "W-band Scalable 2×2 Phased-Array Transmitter and Receiver Chipsets in SiGe BiCMOS for High Data-Rate Communication," JSSC, vol. 57, no. 9, pp. 2685–2701, Sep. 2022.

[7] W. Zhu, et al. , "A 1V W-Band Bidirectional Transceiver Front-End with <1dB T/R Switch Loss, <1°/dB Phase/Gain Resolution and 12.3% TX PAE at 15.1dBm Output Power in 65nm CMOS Technology," ISSCC 2021:226-228.

248 GHz Compact Mixer-Last Direct-Conversion Transmitter with I/Q Imbalance and LO Feedthrough Calibration Capability

Seunghoon Lee, Junhyeong Kim, Ho-Jin Song

POSTECH, Pohang, Korea

The Terahertz (THz) band, encompassing a range from 0.1 to 10 THz, has attracted research interest due to its expansive frequency bandwidth and minimal hardware footprint requirements. Unfortunately, the output power of transmitter (TX) in the THz band is low, while the free-space path loss is high, resulting in an short operating range for wireless communication systems [1]. The THz communication via a low-loss plastic waveguide emerges as a formidable alternative to existing solutions such as copper-based wired transmission or optical fibers [2-3]. Extensive research has been conducted on developing plastic waveguide links integrating frequency shift keying modulators in bulk CMOS technology [2-3]. However, the utilization of low-order modulation schemes inherently limits the attainable data rate. The digital I/Q RF-DAC can be a feasible candidate due to its capability to generate high-order data modulation schemes while retaining a cost-effectiveness and compact form factor. In this work, as a prototype, a 248 GHz direct conversion transmitter (DCT) with a push–push sub-harmonic double-balanced mixers is presented, which is capable of IQ and LO feed-through (LOFT) calibration for low error vector magnitude (EVM) degradation. It achieves a maximum data rate of 20-Gb/s with low power consumption and a compact size.

The conceptual schematic representation of THz plastic waveguide links is shown in Fig. 1, where TX and receiver (RX) integrated with an on-chip antenna, are packaged on a low-cost printed circuit board (PCB). Through the medium of high permittivity waveguide materials, signals can be conveyed from the TX to the RX. The proposed DCT employs a sub-harmonic double-balanced mixer topology, reducing the LO chain's complexity. However, this type of TX faces significant challenges due to I/Q imbalance and LOFT, which can degrade EVM. Especially, the layout asymmetry and process variations in mixers can degrade image rejection ratio (IRR) and LOFT as low as 13.7 dB and 19.2 dB, respectively. For the purpose of IQ calibration, the TX incorporates a low-loss phase-tunable 45° generator and variable gain amplifiers. Additionally, in order to minimize LOFT, a current tuning circuit is integrated into the differential sub-harmonic mixers.

Fig. 2 shows a schematic diagram of the 248 GHz TX architecture. The TX comprises two main parts: the up-conversion sub-harmonic mixers and LO chain. A push-push sub-harmonic double-balanced mixer topology is used to ensure a compact design. Contrary to a conventional fundamental mixer, frequency doublers can be eliminated from the LO chain in this case. The LO chain contains frequency multipliers (×4 and ×2), a Wilkinson power divider, a transmission line (TL) with varactors, and 2-stage buffer amplifiers. An $LO^{1/16}$ input signal is supplied from an off-chip signal generator to a frequency multiplier chain, which upconverts this signal to an $LO^{1/2}$ signal. A Wilkinson power divider and a TL with variable capacitors are used to split the $LO^{1/2}$ signal with a phase difference of 45° along with phase controllability. Two-stage neutralized differential amplifiers are used to ensure that sufficient $LO^{1/2}$ power supplied for the sub-harmonic mixer and IQ amplitude calibration. The IQ sub-harmonic mixers are driven by differential $LO^{1/2}$ signals, and the baseband IQ signals are converted into RF frequency signals. Push-push sub-harmonic double-balanced mixers are employed for the IQ modulators with a quad-mixer stage structure resembling a frequency doubler. This configuration doubles not only the frequency of $LO^{1/2}$ signal but also the phase of $LO^{1/2}$ signal in the mixer stage. The proposed mixers exhibit an enhanced conversion gain compared to conventional architecture, with an approximate 1 dB improvement attributed to the decreased junction capacitance in quad-transistors, achieved by lowering the body voltage. An LC series circuit, inserted between the source of quad-transistors and the ground, ensures the flow of LO signals to the ground, effectively preventing a 2 dB conversion gain reduction caused by imperfect drain-source isolation of quad-transistors. Two current sources are added at the baseband side of each IQ modulator to minimize the LOFT by introducing an artificial offset.

A phase-controlled push-push topology, followed by a frequency doubler, is implemented in the LO multiplier stage to efficiently produce the eighth harmonic, as shown in Fig. 3. This topology has been previously shown to be more effective than the three-stage cascaded frequency doublers in terms of size and DC-RF efficiency. The simulated peak output power is -2.46 dBm at 136 GHz with a harmonic suppression greater than 24 dB and a power consumption of 31 mW. A Wilkinson power divider and a TL terminated with capacitive loads, which are used in the 0/45° signal generator for low loss and phase controllability. By increasing the TL impedance and adding shunt capacitors, the Q path line length can be reduced, substituting a conventional 64° TL with a 40° TL and 9.1 fF shunt PMOS variable capacitors on each side at 124 GHz. Within the 110–150 GHz range, any IQ phase mismatch can be effectively calibrated by adjusting the capacitance of the PMOS varactors between 6.1 and 13 fF. The IQ amplitude mismatch is less than 2 dB at the frequency where the IQ phase mismatch is 0°, and this mismatch can be compensated by adjusting the gate bias in the Q-channel two-stage neutralized differential amplifiers following the TL.

Fig. 4 shows the measurement results of TX. A WR 3.4-band extension module and a network analyzer were used to measure the RF signals. The reference plane was placed at the RF probe tips. The measured IRR and LOFT suppression after calibration were improved by approximately 20 dB at 246 GHz and maintained above 25 dB and 28 dB, respectively, in the range of 242 GHz to 252 GHz. The figure also shows the input/output characteristics of the DCT including PCB line loss and bonding wire loss with the BB IQ signals at 1 GHz. The power conversion gain and output P1dB were measured under single side band mode. The conversion gain and output P1dB of the DCT were -14.78 dB and -19.81 dBm, respectively. The 3-dB bandwidth was 10 GHz from 243 GHz to 253 GHz.

Fig. 5 presents the setup and measurement results of the digital modulation signal measurement. The baseband IQ signals, comprising digital communication signals, were generated by using an AWG. The $LO^{1/16}$ signal that was produced by a signal generator was supplied through an on-wafer probe. The RF output signal was subsequently down-converted to 8 GHz through an external sub-harmonic mixer, measured with a Keysight DSOV254A, and finally demodulated utilizing the VSA 89600 software. The constellations and EVM were measured for the QPSK, 16-QAM, and 32-QAM. A peak data rate of 20 Gb/s was measured at 248 GHz using the 16-QAM. Fig. 6 compares this work with the state-of-the-art H-band TX. Our work achieves 20 Gb/s data rate with a direct conversion structure that enables both IQ and LOFT calibration. This has been accomplished with a power consumption of 96.3 mW and a compact chip size of 0.86 mm².

Fig. 7 shows the die micrograph. The chip size and core area are 0.86 and 0.40 mm², which is fabricated in a 40nm CMOS process.

Acknowledgement:

This research was supported by Institute for Information & Communications Technology Planning & Evaluation (IITP) grant funded by the Korea government (MSIT) (Grant No. 2018-0-00823, 2019-0-00060`). The chip fabrication was supported by the ID Design Education Center (IDEC), Korea.

References:

[1] H.-J. Song, "Terahertz wireless communications: Recent developments including a prototype system for short-range data downloading," IEEE Microw. Mag., vol. 22, no. 5, 2021:88–99.

[2] R. Han et al., "Filling the Gap: Silicon Terahertz Integrated Circuits Offer Our Best Bet," IEEE Microw. Mag., vol. 20, no. 4, 2019:80-93.

[3] N. Van Thienen, W. Volkaerts and P. Reynaert, "A Multi-Gigabit CPFSK Polymer Microwave Fiber Communication Link in 40 nm CMOS," IEEE J. Solid-State Circuits, vol. 51, no. 8, 2016:1952-1958.

[4] M. H. Eissa et al., "Wideband 240-GHz transmitter and receiver in BiCMOS technology with 25-Gbit/s data rate," IEEE J. Solid-State Circuits, vol. 53, no. 9, 2018:2532–2542.

[5] S. V. Thyagarajan, S. Kang, and A. M. Niknejad, "A 240 GHz fully integrated wideband QPSK receiver in 65 nm CMOS," IEEE J. SolidState Circuits, vol. 50, no. 10, 2015:2268–2280.

979-8-3503-3004-5/23 $31.00 © 2023 IEEE

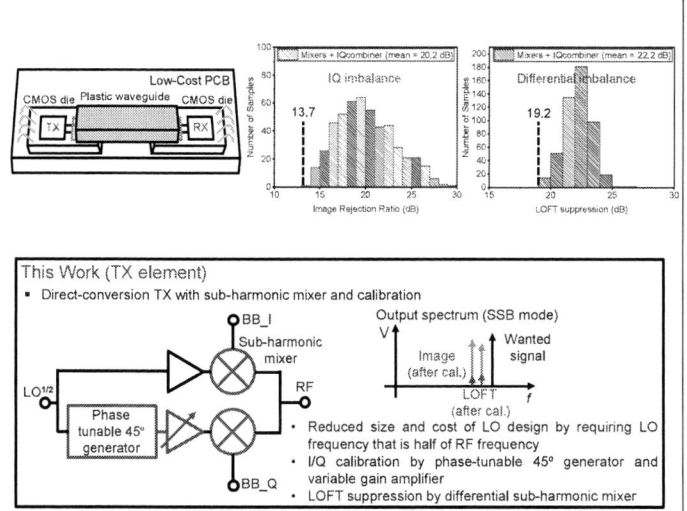

Fig. 1. Conceptual THz plastic waveguide links and proposed sub-harmonic TX with I/Q imbalance and LOFT calibration capability.

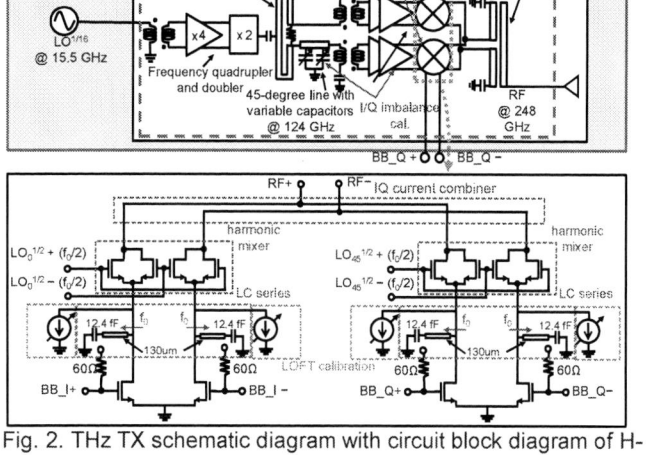

Fig. 2. THz TX schematic diagram with circuit block diagram of H-band sub-harmonic mixers.

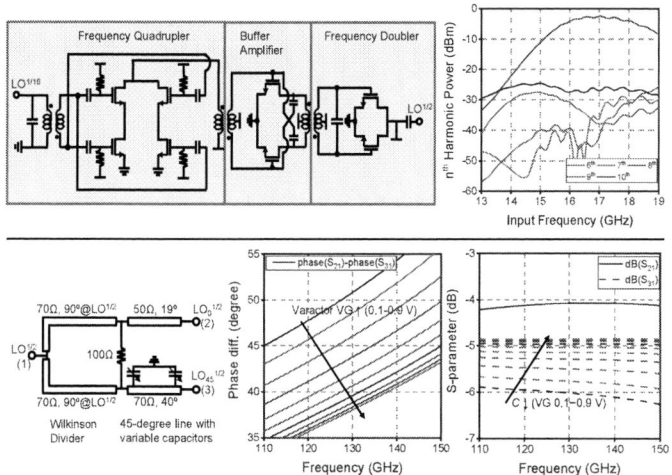

Fig. 3. Schematic diagram and simulated nth harmonic output power of frequency multipliers; Schematic of low-loss 0/45° generator along with simulated phase difference and gain.

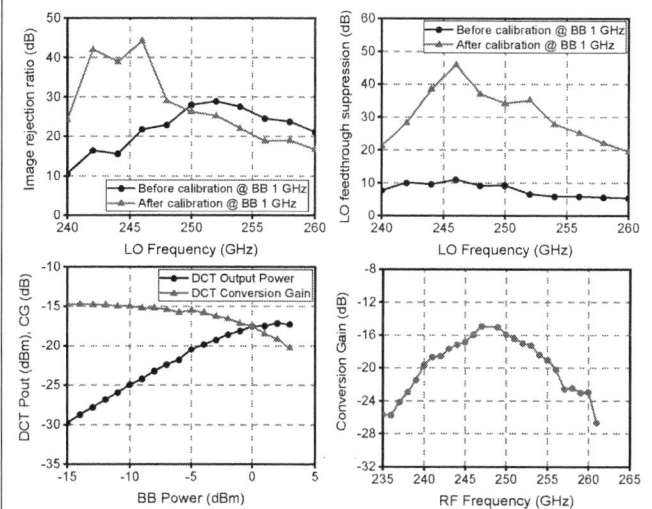

Fig. 4. Measured IRR and LOFT before and after calibration; The RF output power, conversion gain (f_{LO}, f_{BB}=248, 1 GHz), and conversion gain vs RF frequency at a fixed LO (f_{LO} = 248 GHz).

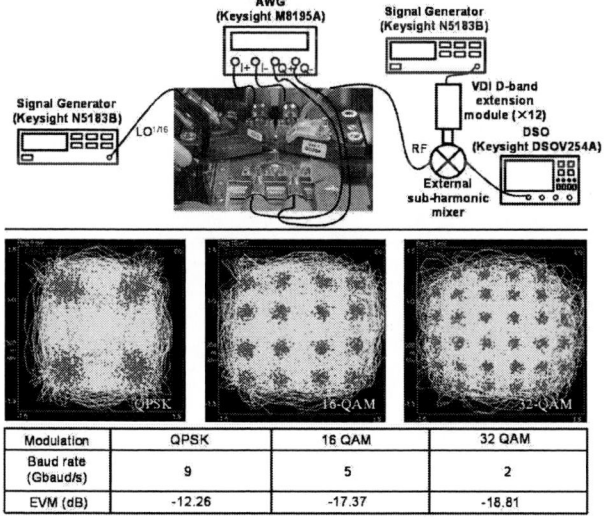

Modulation	QPSK	16 QAM	32 QAM
Baud rate (Gbaud/s)	9	5	2
EVM (dB)	-12.26	-17.37	-18.81

Fig. 5. Measurement setup and measured constellations for various modulation formats.

	JSSC 2018' [4]	JSSC 2015' [5]	ISSCC 2017' [6]	IMS 2020' [7]	JSSC 2022' [8]	This work
Technology	130nm SiGe	65nm CMOS	40nm CMOS	65nm CMOS	65nm CMOS	40nm CMOS
Frequency (GHz)	240	240	300	288	256	248
TX architecture	Direct-conversion TX	Multiplier last	Cubic mixer last	Mixer last	Mixer last	Mixer last
Integrated TX circuit blocks	LO mult. (×8), LO amp., IQ gen., BB amp., IQ mixer, PA.	LO mult. (×2), ILFM, IQ gen., IQ mixer, PA, Frequency multiplier (×3), On-chip antenna.	LO mult. (×3), SSB mixer, Amplifiers, Square mixers.	LO mult. (×3), LO amp., IF amp., sub-harmonic mixer.	LO mult. (x4), LO amp., PS, IF amp., sub-harmonic mixer.	LO mult. (×8), LO amp., IQ gen., IQ sub-harmonic mixer.
I/Q calibration	-	-	-	-	Yes	Yes
LOFT calibration	-	-	-	-	Yes	Yes
Pout (dBm)	-0.8	0	-5.5	-13"	-13"	-17.2
Conversion gain (dB)	10.5	-	-8"	8"	-	-14.78
Data Rate (Gb/s)	25	16	105	34	52	20
P_{DC} (mW)	375	220	1400	270	750	96.3
Area (mm²)	1.25	2	5.2	1.9	4.17'''	0.86

'BER=10⁻⁶, "Estimated from the figures. '''TRX area.

Fig. 6. Performance summary and comparison.

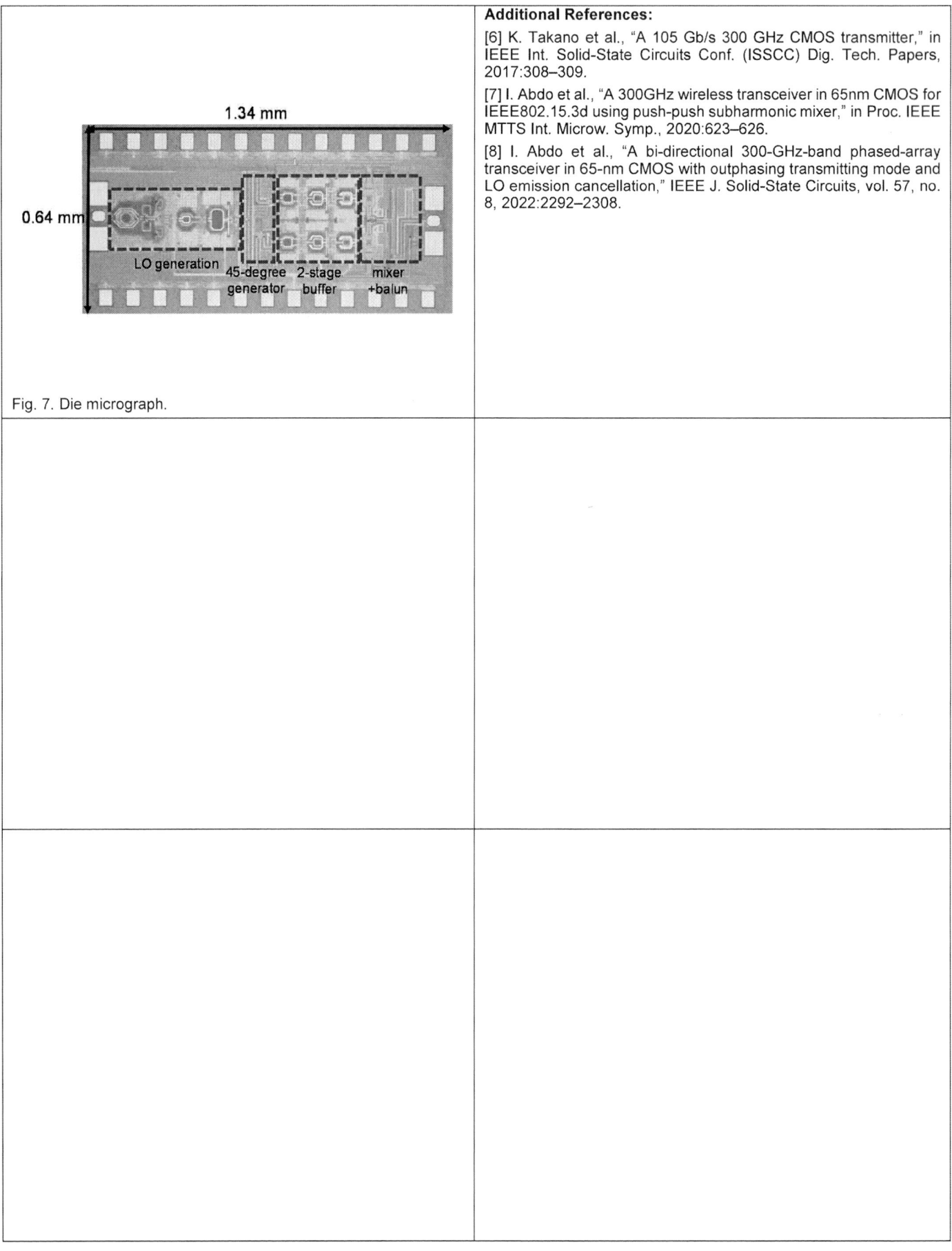

Fig. 7. Die micrograph.

Additional References:

[6] K. Takano et al., "A 105 Gb/s 300 GHz CMOS transmitter," in IEEE Int. Solid-State Circuits Conf. (ISSCC) Dig. Tech. Papers, 2017:308–309.

[7] I. Abdo et al., "A 300GHz wireless transceiver in 65nm CMOS for IEEE802.15.3d using push-push subharmonic mixer," in Proc. IEEE MTTS Int. Microw. Symp., 2020:623–626.

[8] I. Abdo et al., "A bi-directional 300-GHz-band phased-array transceiver in 65-nm CMOS with outphasing transmitting mode and LO emission cancellation," IEEE J. Solid-State Circuits, vol. 57, no. 8, 2022:2292–2308.

A Cross-Coupled Hybrid SC Converter with Extended VCR Range and Intrinsic Loss Balance Achieving 90% Average Efficiency with 1.5% Variation over Full Li-ion Battery Input Range and 0.95A/mm^2 Peak Current Density

Qiaobo Ma[1], Huihua Li[1], Xiongjie Zhang[1], Anyang Zhao[1], Yang Jiang[1], Man-Kay Law[1], Rui P. Martins[1,2], and Pui-In Mak[1]

[1]University of Macau, Macau, China,
[2]Instituto Superior Técnico, Universidade de Lisboa, Portugal

The demand for portable electronics powered by Li-ion batteries has experienced a significant surge. However, these devices face a tradeoff between high performance and long-lasting power. Designing power converters that achieve high efficiency, high output current (I_{OUT}), and high power density while covering the full Li-ion battery voltage range presents significant challenges. To address the tradeoff between efficiency and power density, various hybrid switched-capacitor (SC) converters have been proposed [1-5]. In particular, the use of dual-path (DP) structures can reduce the inductor DC current ($I_{L,DC}$), leading to lower DCR losses and enabling smaller-sized inductors. However, compared to conventional buck converters, DP-based topologies generally have a limited voltage conversion ratio (VCR) range, making it difficult to cover the full-range conversion from Li-ion battery input (2.7 to 4.2V) to a typical point-of-load level (e.g., 1 to 1.8V), which corresponds to a VCR range of 0.25 to 0.7. Although a design presented in [1] addresses this issue by adding an extra power switch and an additional switching phase, it compromises the effectiveness of $I_{L,DC}$ reduction as the VCR decreases, resulting in a noticeable drop in conversion efficiency. The topology proposed in [3] extends the VCR range from 0.5-1 to 0.33-1 by incorporating an additional SC cell. However, it still falls short of meeting the targeted conversion scenario.

In this work, we propose a cross-coupled (XC) SC hybrid topology to further extend the VCR range while maintaining effective $I_{L,DC}$ reduction, enabling the use of small-sized inductors for a smaller form factor. The proposed converter also achieves an intrinsic balance between inductor DCR loss and switches/line conduction loss, ensuring a consistently high efficiency with minimal variation across the desired VCR range. In addition, we introduce a two-quadrant level shifter circuit to meet the driving requirements for the power switches that operate in positive-negative floating domains over the wide conversion regions. Under a full-range input of 2.7 to 4.2V, the proposed converter achieves a peak efficiency of 91% and a maximum I_{OUT} of 4.2A using two 4.7-μH inductors with 175mΩ DCR in 16.7mm^3 each. Further, it maintains an average efficiency of 90% with a variation of only 1.5% under a 1.5V output and 600mA I_{OUT}.

Figure 1 presents a comparative analysis of two previous dual-path (DP) based topologies. Both converters exhibit the same VCR expression of 1/(2-D), ranging between 0.5 and 1, where D is the duty ratio. To extend the VCR range, we introduce a method that utilizes additional SC cells to leverage the charging voltage (V_{CF}) of the flying capacitors (C_F) from V_{IN}-B_1V_{OUT} to V_{IN}-(B_1+1)V_{OUT} during the de-energizing phase of the inductor, as depicted in Fig. 1. This approach enhances the de-energizing slope of the inductor, resulting in a wider VCR range while reducing the $I_{L,DC}$. As shown, employing single-stage SC cells enables the theoretical VCR range extension to 0.25-1, which meets the application requirements. However, the use of additional SC cells leads to increased hardware cost and switching loss. To further improve cost-effectiveness and efficiency, we propose a XC hybrid SC topology that facilitates the sharing of the two interleaved SC cells to attain the same function.

Figure 2 provides a detailed depiction of the operational phases of the proposed XC hybrid topology. The transition between the two operation modes on either side of VCR = 0.5 can be automatically achieved through normal pulse-width modulated D control. Noted that during phase IV, both inductors are energized simultaneously, while the bottom plate of $C_{1a/b}$ remains floating to ensure stable voltage across $C_{1a/b}$ and suspend charge sharing. Furthermore, the proposed topology inherently maintains a balance between inductor DCR and switch/line conduction losses, exhibiting a complementary relation throughout the VCR range. As demonstrated in Fig. 2, the analysis reveals that as VCR increases, the reduction in $I_{L,DC}$ weakens, but the current through the switches also decreases, and

vice versa. The loss analysis, validated by both modeling and circuit-level simulations in Fig. 2, confirms this property. Furthermore, Fig. 3 presents modeled efficiency comparisons based on the same inductor volume and switch area, demonstrating the efficiency advantage of the proposed topology under wide-range VCRs and I_{OUT} conditions with minimal variation.

Figure 3 provides an overview of the power stage and gate driver implementation. The power stage consists of two branches, each containing all 5V switches ($S_{1a/b~6a/b}$). The switches $S_{1a/b-3a/b}$ are subjected to a voltage stress of 2V_{OUT}, while switches $S_{4a/b~6a/b}$ operate at V_{OUT}. Specifically, during phase 1A*, the source/drain voltage of $S_{3a/b}$ alternates between high and low levels, necessitating the use of dynamic body bias. We introduce three categories of gate drivers (GDs) to accommodate different switch operation conditions. The floating-domain (FD) GD is utilized for $S_{1a/b}$, $S_{4a/b}$, and $S_{6a/b}$. It includes a FD level shifter (LS), a gate buffer, and a bootstrap capacitor (C_{BST}). Notably, the LS in the GD of $S_{1a/b}$ is implemented using the proposed 2QLS circuit, enabling two-quadrant (2Q) operation, allowing the floating domain to shift between positive and negative levels. During phases 1A* and 2B, $S_{3a/b}$ is turned off, and a stack GD is employed for gate pull-down. Its turn-on is enabled using an FDGD. Buffer-chain-based GD drives $S_{2a/b}$ and $S_{5a/b}$.

Figure 4 depicts the proposed 2QLS circuit. Considering the wide VCR range, the source terminal voltages of $S_{1a/b}$ ($V_{SSH1} = V_{IN} - 2V_{OUT}$) can become negative. The 2QLS circuit features autonomous switching between negative and positive modes, controlled by $SIG_{A/B}$ signals. In the positive mode, when V_{DDH1} surpasses V_{DD5}, $S_{P1/P2}$ are alternately turned off. In the negative mode, when V_{SSH1} is below GND, $S_{N1/N2}$ are alternately turned off. This configuration prevents leakage current from flowing between V_{DD5} and GND, reducing the overall quiescent current.

The proposed XC hybrid SC converter was fabricated in a 180-nm BCD process using 5V NMOS switches, occupying an area of 4.42mm^2. To meet the requirements of a wide VCR range and high load current, we selected 10μF and 22μF capacitors for $C_{1a/b}$ and $C_{2a/b}$, respectively, minimizing charge sharing losses. Fig. 5 presents the measured steady-state efficiency across the full input and output voltage ranges, as well as efficiency under various I_{OUT} conditions. The efficiency exhibits a low variation of 1.5% and 4% over the entire conversion range, maintaining constant V_{OUT} and V_{IN}. Fig. 5 also displays the measured waveforms of inductor voltage and current. The measured $I_{La/b,DC}$ approximates 0.15I_{OUT}, in line with the theory.

Figure 6 presents the measured performance of our converter compared to state of the art. Benefiting from the intrinsic loss balance property, our design achieves a wide VCR range and attains a high average efficiency of 90% with only 1.5% variation at 1.5V output. At 3.3V V_{IN}, the average efficiency remains high at 89.3% with a 4% variation, outperforming other designs. Besides, the 4 conduction paths with the converter enable a high current density of 950mA/mm^2 and a power density of 1.71W/mm^2. Fig. 7 shows the die micrograph.

Acknowledgment:
This work was supported in part by the Macao FDCT under Grant 0034/2021/A1, 0148/2020/A3, and SKL-AMSV(UM)-2023-2025, and by the NSFC under Grant 62104269, and by University of Macau under Grant MYRG2022-00249-IME and MYRG2022-00024-IME.

References:

[1] Y. Huh, *et al.*, "A Hybrid Structure Dual-Path Step-Down Converter With 96.2% Peak Efficiency Using 250-mΩ Large-DCR Inductor," *IEEE JSSC*, 2019.
[2] N. Tang *et al.*, "Fully Integrated Switched-Inductor-Capacitor Voltage Regulator With 0.82-A/mm^2 Peak Current Density and 78% Peak Power Efficiency," *IEEE JSSC*, 2021.
[3] K. Hata, *et al.*, "Always-Dual-Path Hybrid DC-DC Converter Achieving High Efficiency at Around 2:1 Step-Down Ratio," *IEEE APEC*, 2021.
[4] J.-Y. Ko, *et al.*, "A 4.5V-Input 0.3-to-1.7V-Output Step-Down Always-Dual-Path DC-DC Converter Achieving 91.5%-Efficiency with 250mΩ-DCR Inductor for Low-Voltage SoCs," *IEEE Sym. on VLSI Circuits*, 2021.
[5] X. Yang et al., "A 5V Input 98.4% Peak Efficiency Reconfigurable Capacitive-Sigma Converter with Greater than 90% Peak Efficiency for the Entire 0.4~1.2V Output Range," 2022 *IEEE ISSCC*, 2022.

979-8-3503-3004-5/23 $31.00 © 2023 IEEE

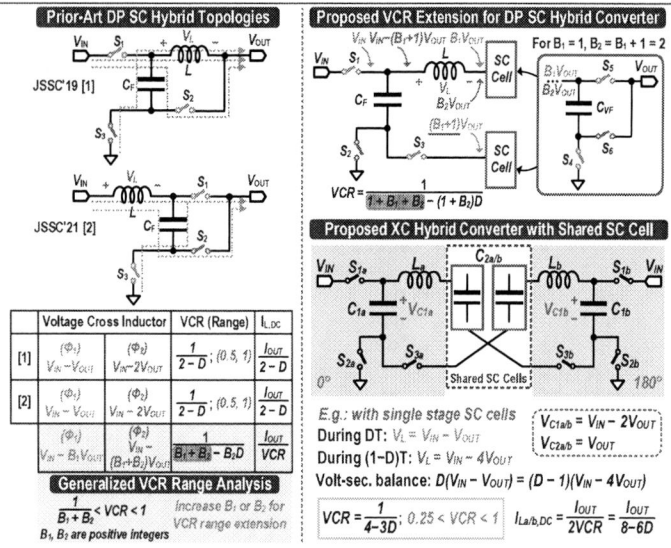

Fig. 1. Proposed XC hybrid topology with extended VCR range.

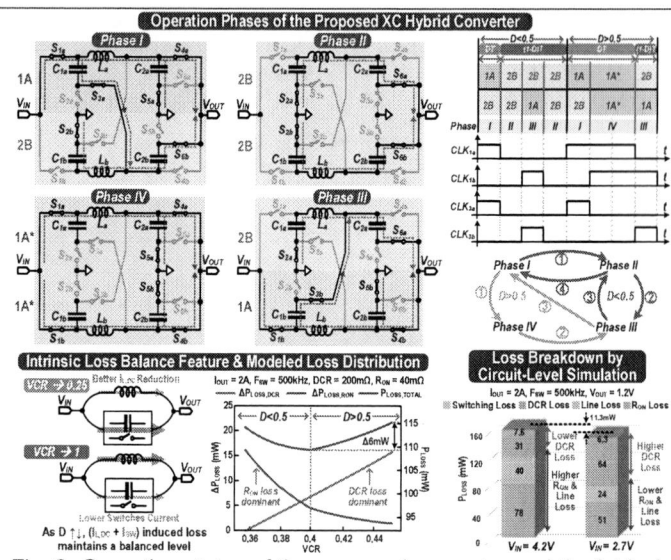

Fig. 2. Operation states of the proposed converter and the intrinsic loss balance feature analysis with simulation validation.

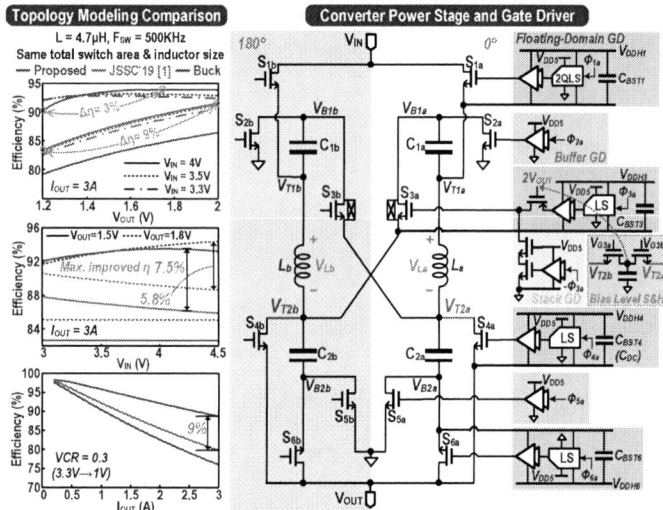

Fig. 3. Modeled topology feature comparison, and converter power stage overview with the proposed gate driving scheme.

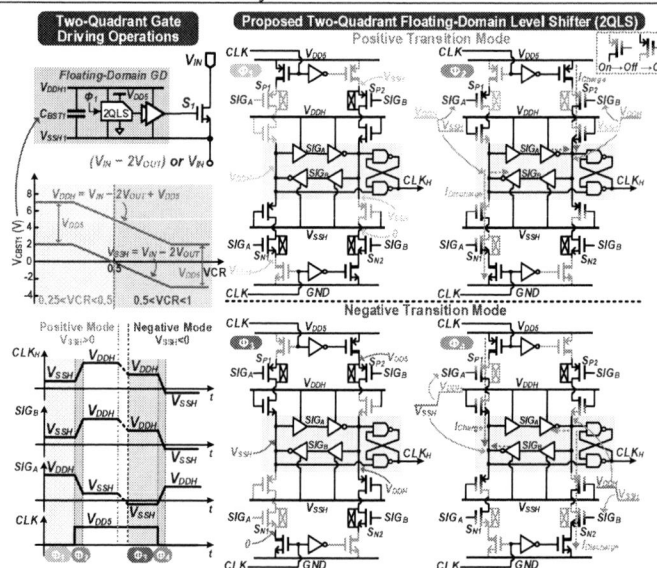

Fig. 4. Proposed two-quadrant floating-domain level shifter.

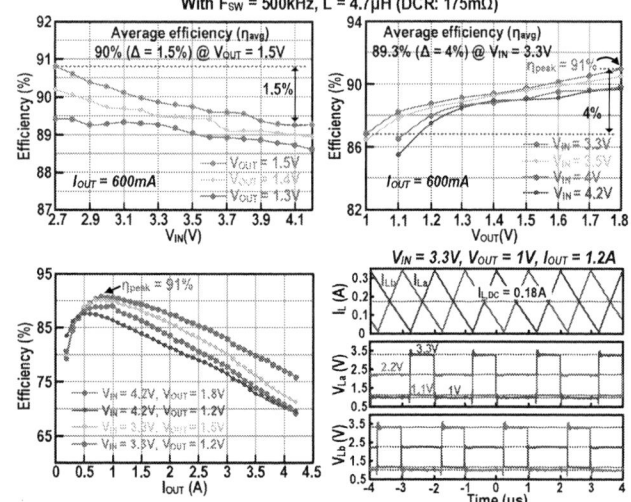

Fig. 5. Measured steady-state efficiency under full-range input and output voltages and different load currents, and measured inductor voltage and current waveforms.

	This Work	VLSI'21[4]	JSSC'19[1]	JSSC'23 Cai	CICC'22 Jung	JSSC'21[2]
Process	180nm BCD	180nm CMOS	180nm CMOS	65nm CMOS	130nm BCD	65nm CMOS
Topology	XC-SC Hybrid	ADP-Buck	DPDC	CPL-Buck	ADPR	SCI
V_{IN} [V]	2.7 ~ 4.2	3.4 ~ 4.5	4.5	3 ~ 4.2	2.8 ~ 4.2	1.2
V_{OUT} [V]	1 ~ 1.8	0.3 ~ 1.7	0.8 ~ 4	0.6 ~ 1	0.9 ~ 1.1	0.6 ~ 0.9
F_{SW} [MHz]	0.5	1	1	2	1	450
Max. I_{OUT} [A]	4.2	1.6	1.6	1.2	0.35	0.533
Max. P_{OUT} [W]	7.56	2.72	5.12	1.2	0.385	0.475
VCR Range	0.25 ~ 1	0 ~ 0.5	0 ~ 1	0 ~ 0.5	0 ~ 0.5	0 ~ 0.5
Max. I_{LDC} reduction	$0.125I_{OUT}$	$0.5I_{OUT}$	$0.5I_{OUT}$	$0.33I_{OUT}$	$0.33I_{OUT}$	$0.5I_{OUT}$
Inductor DCR [mΩ]	175	250	250	64	288	NA.
Inductor [H]	2L: 4.7μ	1L: 4.7μ	1L: 4.7μ	1L: 0.47μ	1L: 4.7μ	1L: 0.85n
Capacitor [F]	2C$_{FLY}$: 10μ 2C$_{FLY}$: 22μ C$_{OUT}$: 2×22μ	2C$_{FLY}$: 10μ C$_{OUT}$: 10μ	2C$_{FLY}$: 10μ C$_{OUT}$: 10μ	2C$_{FLY}$: 10μ C$_{OUT}$: 10μ	2C$_{FLY}$: 4.7μ C$_{OUT}$: 10μ	On-chip C$_{total}$ 4.82n
Chip Area	4.42mm²	4.08mm²	3.24mm²	2.72mm²	3.55mm²	0.65mm²
I_{OUT} Density [mA/mm²]	950	392	493	441	98.6	820
P_{OUT} Density [W/mm²]	1.71	0.67	1.58	0.44	0.11	0.73
Average Efficiency (Full Conv. Range)@I_{OUT}	90% @600mA	*86% @400mA	*87.8% @1A	*87% @400mA	*93.5% @100mA	*72% @300mA
Efficiency Variation (Full Conv. Range)	1.5%	*13.5%	*21%	*8%	1.6%	*13%
Peak Efficiency @ VCR (I_{OUT})	91% @0.54(600mA)	91.5% @0.38(400mA)	96.2% @0.72(400mA)	92.9% @0.27(600mA)	95.4% @0.284(75mA)	78% @0.75(400mA)

*Estimated from the reported data

Fig. 6. Performance summary and comparison with state of the art.

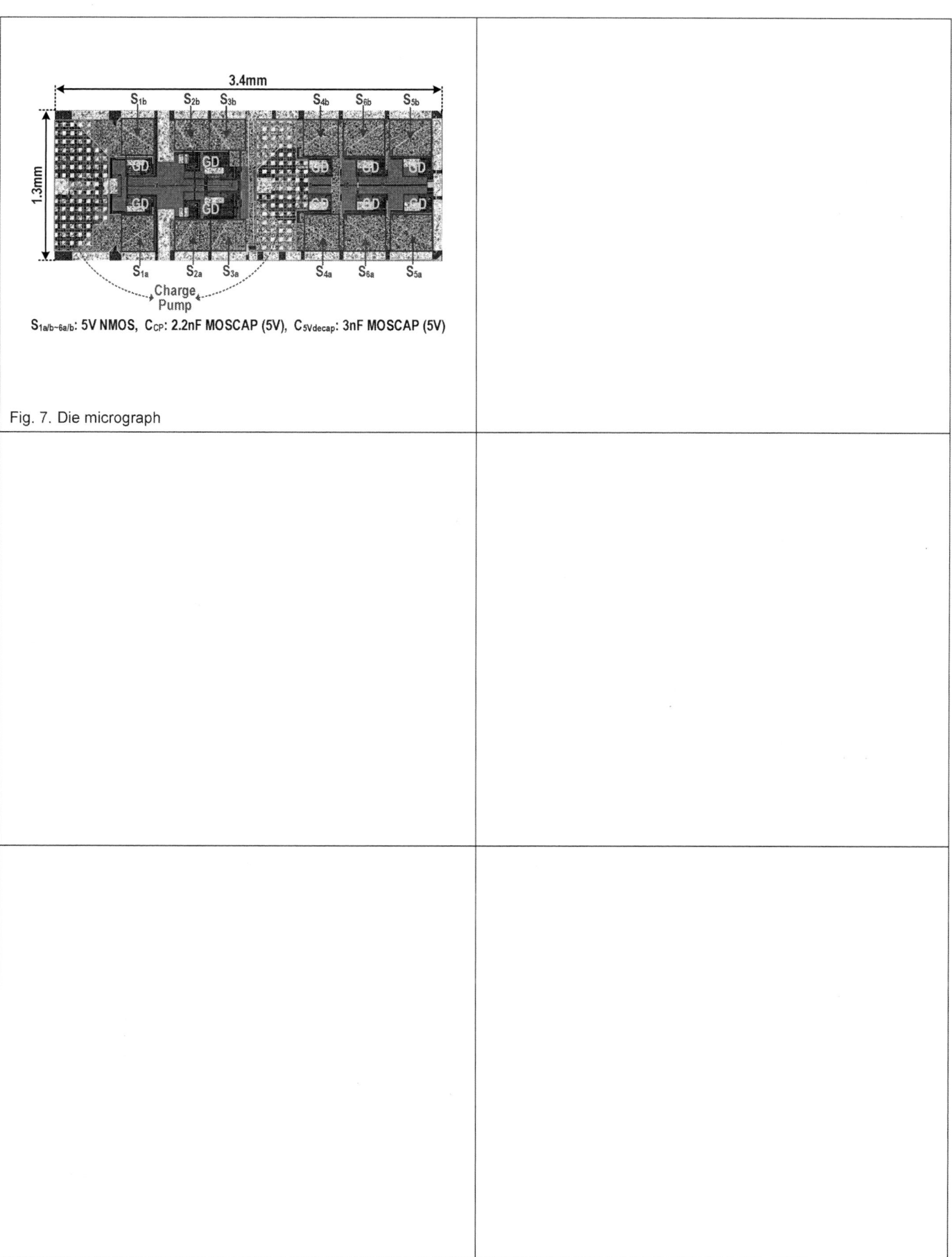

$S_{1a/b \sim 6a/b}$: 5V NMOS, C_{CP}: 2.2nF MOSCAP (5V), $C_{5Vdecap}$: 3nF MOSCAP (5V)

Fig. 7. Die micrograph

A DRAM Bandwidth-Scalable Sparse Matrix-Vector Multiplication Accelerator with 89% Bandwidth Utilization Efficiency for Large Sparse Matrix

Hyunji Kim, Eunkyung Ham, Sunyoung Park, Hana Kim, and Ji-Hoon Kim

Ewha Womans University, Seoul, Korea

Sparse matrix-vector multiplication (SpMV) plays a crucial role in diverse engineering applications, including scientific/engineering modeling, machine learning, and information retrieval, as depicted in Fig. 1 [5]. To efficiently store sparse matrices and minimize memory waste, the widely employed COO (Coordinate) compression format stores only the coordinates (row index, column index) of the non-zero elements in the matrix along with their corresponding values. However, the memory-intensive nature of SpMV operations, combined with irregular memory access patterns and limited data reuse resulting from the COO format, pose significant challenges for achieving high-performance implementations [6]. To assess the performance and efficiency of FPGA-based SpMV accelerators [1-4], which are typically optimized for specific hardware platforms, *Bandwidth Utilization* (BU) serves as a key metric for fair comparisons across different hardware specifications [6].

Fig. 2 illustrates the expected continuous increase in off-chip DRAM bandwidth. Given the well-known memory-bound nature of SpMV operations, the importance of a scalable SpMV accelerator becomes evident, capable of efficiently utilizing the maximum off-chip memory bandwidth to achieve peak performance, even with the anticipated DRAM bandwidth advancements. However, previous SpMV accelerators developed for specific platforms exhibit limitations in their ability to scale with off-chip memory bandwidth [1, 3-4]. These accelerators also suffer a considerable performance decline, up to 50% in worst-case scenarios, primarily due to underutilization of available bandwidth, falling short of the theoretical limit. This issue is particularly pronounced when handling large sparse matrices [4]. To quantify the improvement in this regard, *BU efficiency* is introduced as a quantitative measure and determined by the ratio of actual BU to theoretical BU. This paper presents an optimized SpMV accelerator that ensures scalability to off-chip DRAM bandwidth while avoiding performance degradation when processing large matrices.

Fig. 3 depicts the architecture and operation of the proposed SpMV accelerator, which includes a data distributor, multiple Processing Element (PE) lines, and an output updater. The number of PE lines is determined based on the ratio of the maximum available off-chip memory bandwidth to the data width. PE lines consist of a matrix buffer, a multi-bank vector buffer, and two MAC units. Vector elements are assigned to two banks based on their indices. The proposed SpMV accelerator utilizes offline pre-processing, involving matrix partitioning and rearrangement in blocks. Matrix partitioning divides the input matrix into blocks limited by the vector and PSUM buffers. Subsequently, in the matrix rearrangement step, the non-zero elements in the sparse matrix are rearranged to mitigate performance degradation and convert the matrix into COO compressed format stored in off-chip memory. The blocks are stored in the order of completed rows in the matrix, as shown in Fig. 3 (e.g., Block 0, Block 1, Block 2, etc.). SpMV operation follows the storage sequence. In each block operation, the data distributor loads data into the vector buffer and matrix data. Two independent PEs within the PE line perform multiplication by reading matrix elements and corresponding vectors. The results undergo accumulation within the PE line. When the row index changes, they are stored in the PE line's output FIFO. The output updater saves the data from the PE line output FIFO into the PSUM buffer, adding the previously stored partial sum before storing it.

Fig. 4 provides a detailed explanation of the purpose and specific steps involved in the offline pre-processing stage. Offline pre-processing is crucial to address potential issues such as bank conflicts resulting from the random distribution of non-zero elements in the input matrix, as well as Read-After-Write (RAW) dependencies when accessing the same accumulator in the MAC unit concurrently.

The rearrangement process in offline pre-processing aims to eliminate bank conflicts by pairing matrix elements with column indices corresponding to different banks. Additionally, to avoid RAW dependencies, rows are assigned to different MAC units during the pairing process. An example of offline pre-processing applying these rules is shown at the bottom of Fig. 4. The input matrix is divided into multiple blocks, and each block undergoes the rearrangement process. In this process, a different part of the block is assigned to each PE line. After that, the actual rearrangement process is started. The first step is to assign rows to each MAC unit. Subsequently, the process involves selecting the first remaining element in a row assigned to MAC0 and finding another element in a row assigned to MAC1 that can share the vector element with the previously chosen element. If no corresponding element is found, the algorithm seeks a bank-conflict-free element. This process is repeated until all elements in a row are processed. Once all elements in one row are finished, the MAC unit is assigned another row that has not been allocated yet. Reusing the vector element between MAC units increases PE utilization and helps achieve load balancing between banks, improving overall performance.

Fig. 5 shows the data size comparison result and *BU* result of the proposed SpMV accelerator. The sparse matrix used in the evaluation is at the top of Fig. 5 which is from the University of Florida sparse matrix collection [8]. For the fair comparison, the evaluation parameter is set as same as in [4]. The total available on-chip SRAM size is set to 4MB. In 4MB, the vector buffer is assigned 0.25MB, PSUM buffer is assigned the remaining 3.75MB. The data width of the off-chip memory data interface is set in three cases 64B/cycle, 128B/cycle, and 256B/cycle. Each case has 2 PE lines, 4 PE lines, and 8 PE lines. For non-zero elements, the COO Compressed format includes two INT32 indexes (row, col) and one FP64 value. As illustrated in Fig. 5, the offline pre-processing step incurs a negligible impact on the data size (less than 2%) as well as the latency of the SpMV operation. It has been observed that the proposed SpMV accelerator comes close to achieving the theoretical *BU* of 0.125 in most matrices and has achieved a maximum performance improvement of 28.3% in comparison to previous work [4].

Fig. 6 shows the comparison table. Most previous works lack scalability to DRAM bandwidth, while the proposed accelerator exhibits scalability and achieves higher performance. The proposed SpMV accelerator achieves an average BU efficiency of 89% and can achieve a performance improvement of up to 43.8% compared to previous works. Due to the lack of memory bandwidth adaptivity in the internal architectures of previous works, the scalability of the proposed SpMV accelerator allows for high BU efficiency even on HBM with high memory bandwidth. In the case of [2], it exhibits scalability but achieves performance 32% lower than the theoretical performance. Additionally, although using lower precision may allow for reading more data at once and potentially achieving higher performance, this is not the case for [1, 2]. Hence, this paper outperforms [1, 2] when using the same data precision. Fig. 7 shows the demonstration setup and problem formulation equation for the proposed SpMV accelerator evaluation.

Acknowledgement:

This research was supported by Institute of Information & communications Technology Planning & Evaluation (IITP) grant funded by the Korea government (MSIT) (No. 2020‐0‐01308, Intelligent Mobile Processor based on Deep-Learning Micro Core Array) and the EDA tool was supported by the IC Design Education Center (IDEC), Korea.

References:

[1] Fowers et al. "A high memory bandwidth fpga accelerator for sparse matrix-vector multiplication." IEEE FCCM 2014.

[2] Jain et al. "A domain-specific architecture for accelerating sparse matrix vector multiplication on fpgas." IEEE FPL 2020.

[3] Li et al. "Optimized Data Reuse via Reordering for Sparse Matrix-Vector Multiplication on FPGAs." IEEE ICCAD 2021.

[4] Liu et al. "Towards High-Bandwidth-Utilization SpMV on FPGAs via Partial Vector Duplication." IEEE ASP-DAC 2023.

979-8-3503-3004-5/23 $31.00 © 2023 IEEE

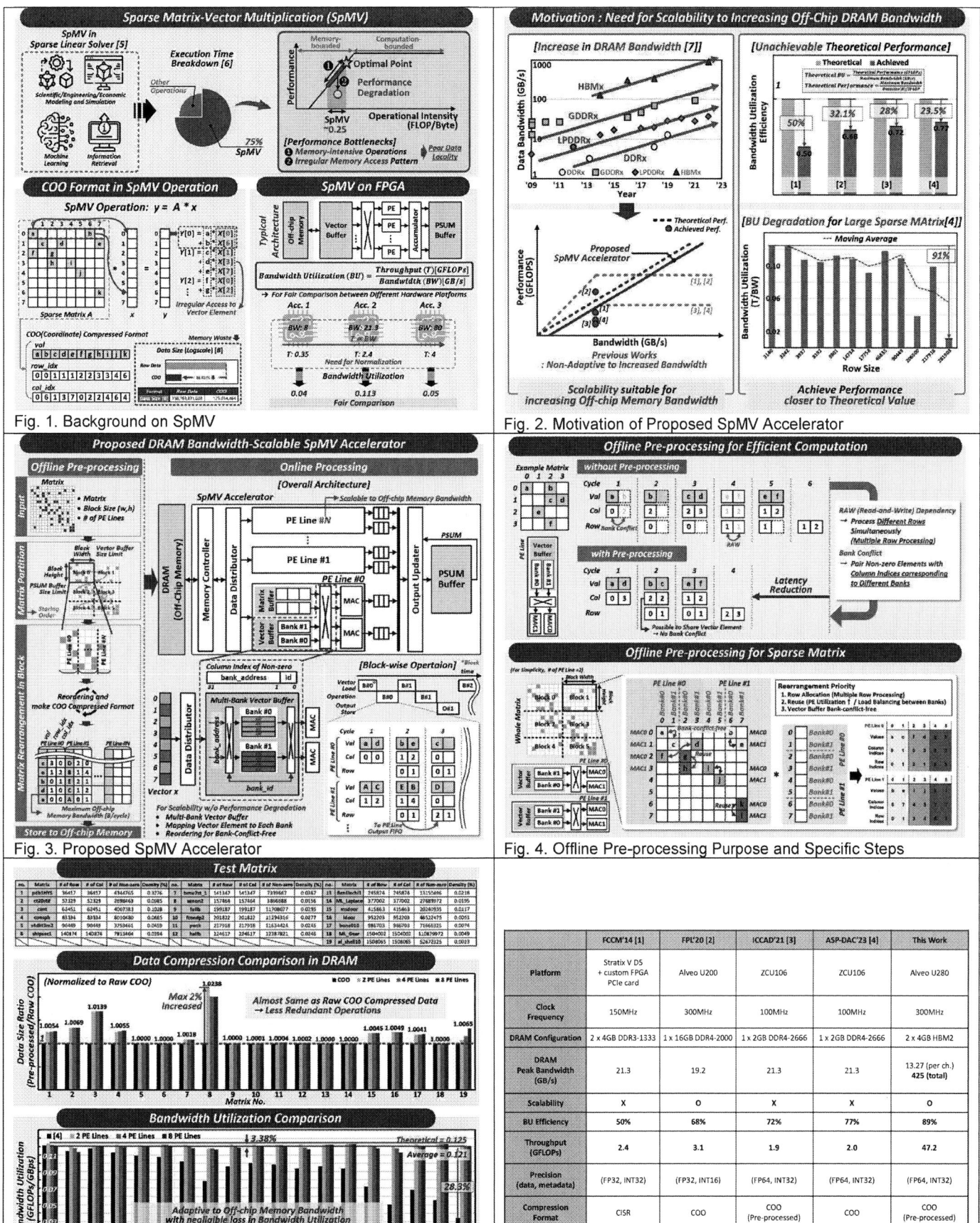

Fig. 1. Background on SpMV

Fig. 2. Motivation of Proposed SpMV Accelerator

Fig. 3. Proposed SpMV Accelerator

Fig. 4. Offline Pre-processing Purpose and Specific Steps

Fig. 5. Evaluation Result

Fig. 6. Comparison Table

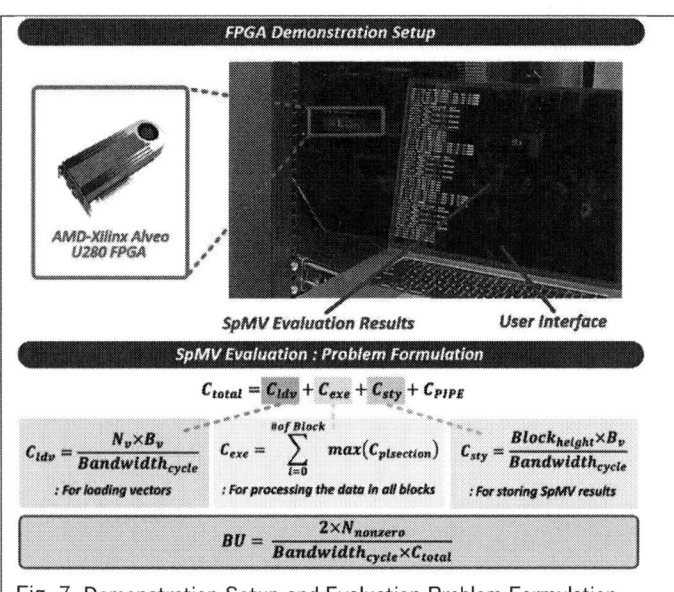

SpMV Evaluation : Problem Formulation

$$C_{total} = C_{ldv} + C_{exe} + C_{sty} + C_{PIPE}$$

$$C_{ldv} = \frac{N_v \times B_v}{Bandwidth_{cycle}}$$

: For loading vectors

$$C_{exe} = \sum_{i=0}^{\#of\ Block} max(C_{plsection})$$

: For processing the data in all blocks

$$C_{sty} = \frac{Block_{height} \times B_v}{Bandwidth_{cycle}}$$

: For storing SpMV results

$$BU = \frac{2 \times N_{nonzero}}{Bandwidth_{cycle} \times C_{total}}$$

Fig. 7. Demonstration Setup and Evaluation Problem Formulation

Additional References:

[5] Manguoglu et al. "A domain-decomposing parallel sparse linear system solver." Journal of computational and applied mathematics 236.3 (2011): 319-325.

[6] Lu et al. "Redesk: A reconfigurable dataflow engine for sparse kernels on heterogeneous platforms." IEEE ICCAD 2019.

[7] S. Mirabbasi, L. C. Fujino and K. C. Smith, "Through the Looking Glass—The 2023 Edition: Trends in solid-state circuits from ISSCC," in IEEE Solid-State Circuits Magazine, vol. 15, no. 1, pp. 45-62, winter 2023, doi: 10.1109/MSSC.2022.3222727.

[8] Timothy et al. The University of Florida Sparse Matrix Collection. ACM Transactions on Mathematical Software 38, 1, Article 1. 2011.

ESD Protection is about Circuit Design: Practices and Perspectives

Weiquan Hao, Xunyu Li, Zijin Pan, Runyu Miao and Albert Wang*

Dept. of ECE, University of California, Riverside, CA 92521, USA

It is worth-noting that electrostatic discharge (ESD) protection design is an IC-specific circuit-level design task, not merely on devices. This keynote reviews proven CAD-based ESD protection circuit design methods and offers future ESD design perspectives.

To prevent thermal failure and voltage breakdown induced by ESD current and voltage pulses, the two basic ESD protection principles are (1) to discharge ESD pulse without overheating and (2) to clamp pad voltage to below the concerned breakdown voltage (BV) [2]. Fig. 1 depicts key concepts for on-chip ESD protection designs. An ideal ESD device is a well-controlled switch with its ESD discharging I-V curve characterized by the ESD-Critical Parameters, including triggering voltage (V_{t1}), current (I_{t1}) and time (t_1); holding voltage (V_h) and current (I_h); discharge resistance (R_{ON}); and thermal breakdown voltage (V_{t2}) and current (I_{t2}); which must be accurately designed to comply with the ESD Design Window [1]. Full-chip ESD protection circuit design must ensure a low-R_{ON} path existing between any two pads on a chip during any ESD events. Advanced IC technologies lead to ESD Design Window Shrinking. ESD Design Overhead exists, including ESD-induced parasitic capacitance (C_{ESD}), leakage (I_{leak}), noise, noise-coupling, large ESD device size and full-chip layout floor planning difficulty, must be carefully addressed at chip level [2]. CAD-based full-chip ESD protection design methodologies are hence critical, which are reviewed in this keynote [3].

2D/3D TCAD-based mixed-mode ESD simulation-design methodology can address the complex transient-electro-thermal-process-device-circuit-layout multiple-coupling effects, a hall-mark ESD design challenge [4]. TCAD supports numerical ESD simulation at both device and circuit levels without limited by ESD compact models, often unavailable or inaccurate. Fig. 2 shows the conceptual TCAD ESD simulation schematic where the ESD stimuli adopted from any ESD test standards. In the example shown, an inverter with a gate-coupled NMOS (gcNMOS) ESD power clamp was simulated [5]. TCAD confirms that the HBM (human body model) ESD pulse discharges through the ESD clamp as designed, causing temperature increase mainly in gcNMOS. Circuit node voltage analysis confirms no gate breakdown. A minor gate voltage overshoot observed suggests further design modification is needed to cover the lower BV_G corner in practices as needed.

For ordinary IC designers, ECAD ESD circuit simulation is more practical since TCAD requires technology recipes and in-depth process and device knowledge. SPICE simulation is common for ECAD ESD simulation that requires accurate ESD device modeling. ESD behavior modeling is an effective alternative to ESD compact modeling, which is not always available and accurate. Recent advanced in ESD modeling has benefited from machine learning algorithms [6]. Schematic-level ESD circuit simulation utilizes two ESD failure criteria: (1) transient ESD discharging currents in all devices and paths, and (2) instant voltage at all circuit nodes. Fig. 3 describes an example of an I/O buffer in 28nm CMOS (BV_{GS}=8.52V, BV_{DS}=7.01V), which features pull-up (PU) ESD ($I_{t2-F} \approx$ 2.3A, V_{t1-R} ~7.98V) and pull-down (PD) ESD ($I_{t2-F} \approx$ 1.8A) diodes at I/Os and a power clamp [7]. ESD-critical parameters in both directions are essential. The series-R (R_{bus1}≈1.5Ω, R_{bus2}≈0.3Ω, R_1≈1Ω, R_2≈1Ω, R_3≈1.15Ω, R_4≈0.6Ω) in all ESD paths should be included due to large ESD currents. In an exemplar negative input 2kV HBM ESD zapping case (B⇐A), simulation confirms the designed ESD discharging path (ADEFGB) passes both ESD current and voltage criteria. A parasitic ESD discharge path (ACB) was discovered, resulting in a noticeable current (~40mA), as confirmed in the run-time thermal image, which suggests further design optimization will be beneficial.

Since ESD-induced parasitic effects cannot not be zeroed out, which will affect circuit performance, particularly for high-frequency and wideband RF ICs and high-data-rate ICs, ESD-RFIC co-design becomes essential to advanced ICs, which requires careful full-chip ESD circuit simulation involving 3 steps: Step-1 to optimize ESD devices by TCAD, Step-2 to characterize ESD devices for their parasitic effects (e.g., C_{ESD}, I_{leak}, noise), and Step-3 to include the ESD parasitics in core circuit simulation to evaluate its negative impacts (mostly I/O impedance matching corruption) and conduct I/O re-matching co-design to balance the ESD and RFIC performance [8-10]. Fig. 4 presents co-design of an 5GHz LNA with 5kV ESD protection in 180nm RFCMOS (die photo in Fig. 7a) including 4 splits [10]. Measurement shows that using a parasitic-rich grounded-gate NMOS (ggNMOS) ESD causes significant degradation in LNA specs (e.g., gain, S_{11}, NF), which can be reduced by using an optimized diode ESD. It is clearly observed that co-design can almost eliminate the ESD impact on the LNA by careful I/O rematching. Fig. 5a-c depict exemplar co-design of SP10T switch for 4G global phones [11]. Fabricated in 180 SOI CMOS (die photo in Fig. 7b), the SP10T uses series-shunt topology and features full-chip 8kV ESD protection. MOSFET stacks are used to support high voltage of 30.5V (33dBm for GSM), which desires even voltage drop (V_{ds}) across each MOSFET, typically realized using feed-forward capacitors (FFC). Testing shows that ESD protection seriously affects the V_{ds} distribution, which was re-evened by careful ESD co-design. Millimeter-wave-ESD co-design is studied using a 28GHz distributed travelling-wave SPDT TRx switch (Fig. 5d/e), designed in 22nm FDSOI (die photo in Fig. 7c) with 9kV ESD protection [12]. Measurement reveals that insertion loss was seriously affected by ESD protection, which was substantially recovered by co-design. It is also obvious that the co-design method needs further improvement in millimeter-wave bands.

Beyond schematic-level ESD simulation, full-chip ESD protection physical design verification is the ultimate goal in advanced IC designs [3], making ESD design similar to the common IC design flow, which has drawn significant research on new ESD CAD technologies [13, 14]. Fig. 6 illustrates an ideal full-chip ESD design verification flow where the unique and key functions include not only ESD-flavored DRC/LVC/ERC/xRC checking features, but also and more importantly extracting any ESD devices (intentional and parasitic) of arbitrary shapes and any possible ESD discharge paths, estimating their ESD-critical parameters, and generating full-chip ESD netlists, all from layout data (GDSII). The CAD flow then allows comprehensive full-chip ESD circuit simulation and complex ESD zapping test simulation, achieving ESD-function-based full-chip ESD physical design verification at both schematic and layout levels enabled by a smart ESD parametric check mechanism [15]. Fig. 6b shows that multiple ESD discharge paths were extracted from the GDSII data for an exemplar IC in 28nm CMOS with 2kV HBM ESD protection. ESD functions must be analyzed to determine which ESD discharge path takes the role and ESD failure criteria (e.g., branch current and node voltage) will be used to smart-check if an ESD design works for the chip in design phase, before IC tapeout.

Though CAD-based ESD design simulation methods have been widely used, challenges constantly emerge for ESD protection for emerging technologies and advanced ICs. For example, CDM ESD failures become notoriously random in large, complicated and advanced chips because CDM ESD is internal-oriented, hence, the classic pad-based HBM ESD protection method may not work in CDM ESD event, because the charges, stored randomly inside an IC, will have to route internally before being discharged into a grounding pad, causing unexpected internal CDM ESD failures even though a robust pad-based ESD protection network exists [16]. A new internal-distributed CDM ESD protection concept was proposed [17]. For emerging 3D packaging and heterogeneous integration (HI) technologies, ESD phenomena are more complicated (e.g., CDM randomness), which require holistic cross-layer ESD protection design techniques including understanding new ESD phenomena, developing new CAD techniques, and exploring non-traditional ESD protection mechanisms, drawing active R&D efforts [18].

References:

[1] A. Wang, *Practical ESD Protection Design*, Wiley-IEEE, 2022.

[2] Z. Pan, *et al*, "ESD Protection Designs: Topical Overview and Perspective", TDMR 2022: 356-370.

[3] Z. Pan, *et al*, "On-Chip ESD Protection Design Methodologies by CAD Simulation", TODAES 2023 (in press).

[4] H. Feng, *et al*, "A Mixed-Mode ESD Protection Circuit Simulation-Design Methodology", JSSCC 2003: 995-1006.

[5] H. Feng, *MS Thesis*, IllinoisTech, 2001.

979-8-3503-3004-5/23 $31.00 © 2023 IEEE

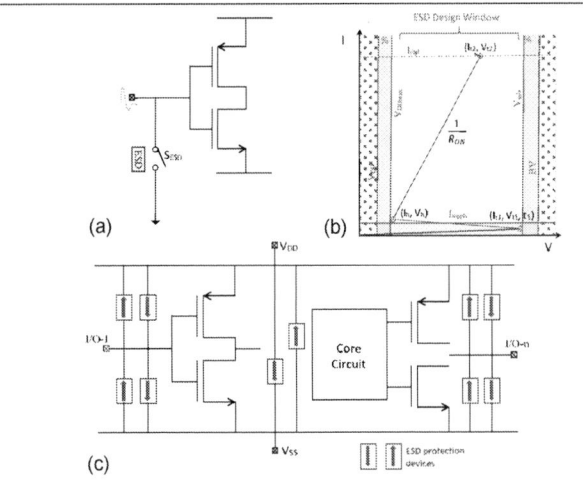

Fig. 1. Basic ESD protection concepts: (a) an ESD device is a controlled switch, (b) a snapback ESD discharge I–V curve with ESD-Critical Parameters has to comply with ESD Design Window, including safety margins, and (c) full-chip ESD proteciton circuit network.

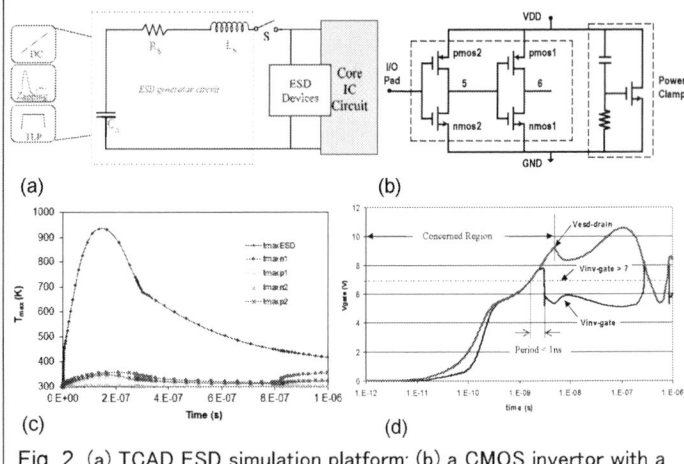

Fig. 2. (a) TCAD ESD simulation platform; (b) a CMOS invertor with a gcNMOS ESD power clamp, (c) significant temperate variation in ESD clamp only during HBM ESD event, and (d) invertor gate voltage during HBM ESD event remains lower than typical BV_G of 8V with a minor overshoot over BV_{G-min} of 7V.

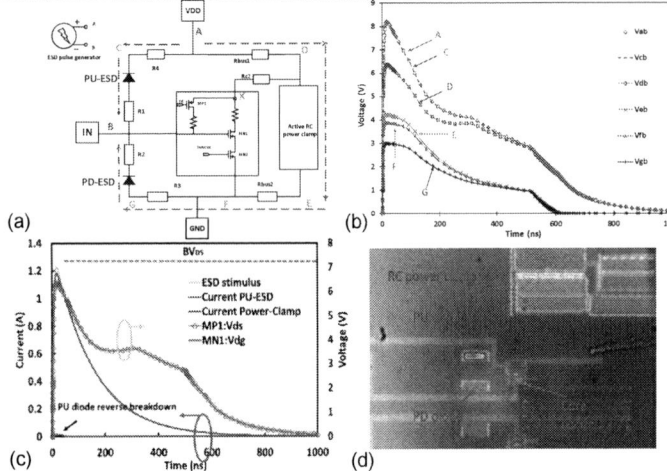

Fig. 3. (a) A buffer circuit with ESD proteciton is zapped by a −2kV HBM pulse as IN pad (B⇐A), (b) transient node voltage w.r.t. B, (c) ESD discharges mainly through power clamp via ADEFGB path, with 40mA in PU-ESD (ACB) as confirmed in thermal image (d).

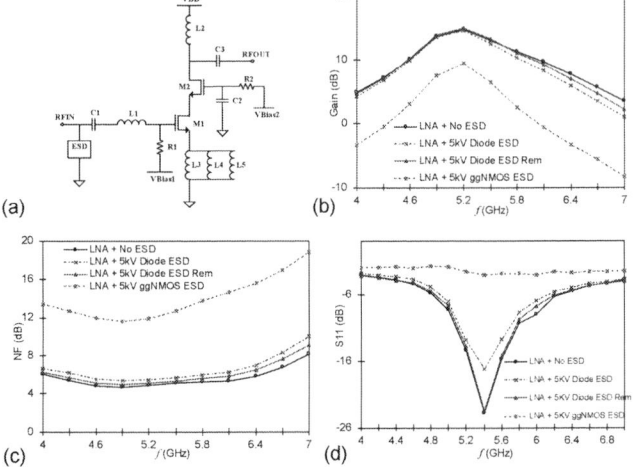

Fig. 4. (a) An example of 5kV–ESD–protected 5GHz LNA in 180nm RFCMOS using RLC I/O matching. Testing shows LNA Specs degradation due to ESD parasitic effects, (b) gain, (c) noise figure, and (d) S_{11}. ESD-RFIC co-design (Rem) can recover LNA Specs.

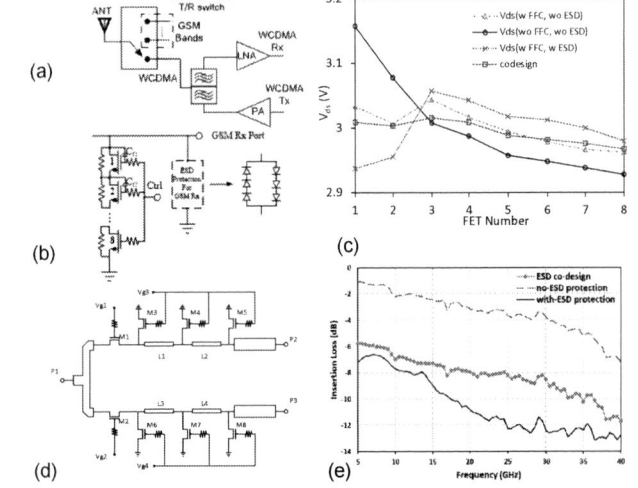

Fig. 5. SP10T example: (a) diagram, (b) ESD proteciton, and (c) ESD impact on V_{ds}; and 28GHz SPDT example: (d) schematic, and (e) ESD impact on intersion loss. All from measurements.

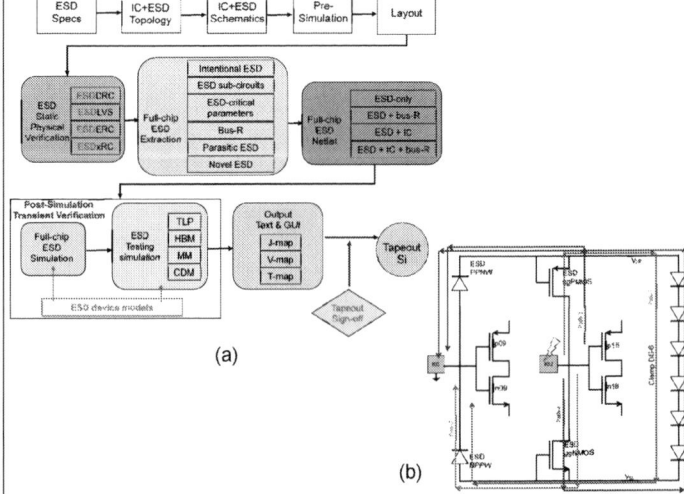

Fig. 6. (a) Desired ESD CAD flow for full-chip ESD physical design verification. (b) ESD CAD analysis example under IO2-to-IO1 HBM ESD zapping identifies four possible ESD discharge paths.

979-8-3503-3004-5/23 $31.00 © 2023 IEEE   280

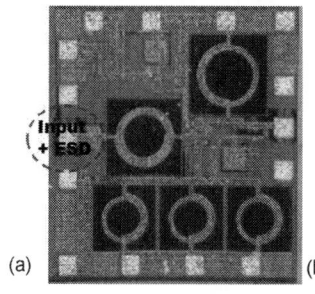

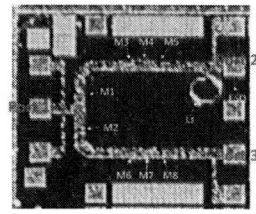

Fig. 7. Die photos for (a) 5GHz LNA, (b) SP10T, and (c) 28GHz SPDT.

Additional References:

[6] W. Liang, *et al*, "Novel ESD Compact Modeling Methodology Using Machine Learning Techniques for Snapback and Non-Snapback ESD Devices", EOSESD 2021: 455-464.

[7] F. Zhang, *et al*, "A Full-Chip ESD Protection Circuit Simulation and Fast Dynamic Checking Method Using SPICE and ESD Behavior Models", TCAD 2019: 489-498.

[8] G. Chen, *et al*, "RF Characterization of ESD Protection Structures", RFIC 2004: 379-382.

[9] A. Wang, *et al*, "A Review on RF ESD Protection Design", TED 2005: 1304-1311.

[10] X. Guan, *et al*, "ESD-RFIC Co-Design Methodology", RFIC 2008: 467-470.

[11] X. S. Wang, *et al*, "Concurrent Design Analysis of High-Linearity SP10T Switch with 8.5kV ESD Protection", JSSCC 2014: 1927 – 1941.

[12] M. Di, et al, "A Study of ESD-mmWave-Switch Co-Design of 28GHz Distributed Travelling Wave Switch in 22nm FDSOI for 5G Systems", J-EDS 2021: 1290-1296.

Additional References:

[13] R. Zhan, *et al*, "ESDExtractor: A New Technology-Independent CAD Tool for Arbitrary ESD Protection Device Extraction", TCAD 2003: 1362-1370.

[14] R. Zhan, *et al*, "ESDInspector: A New Layout-level ESD Protection Circuitry Design Verification Tool Using a Smart-Parametric Checking Mechanism", TCAD 2004: 1421-1428.

[15] A. Wang, "A Parameter Checking Method for on-Chip ESD Protection Circuit Physical Design Layout Verification", U. S. Patent #7,243,317B2, 2007.

[16] M. Di, *et al*, "Pad-Based CDM ESD Protection Methods Are Faulty", J-EDS 2020: 1297-1304.

[17] M. Di, *et al*, "Non-Pad-Based in Situ in-Operando CDM ESD Protection Using Internally Distributed Network", J-EDS 2021: 1248-1256.

[18] X. Li, *et al*, "Holistic ESD Protection Co-Design: Challenges", ICEPT 2023 (in press).

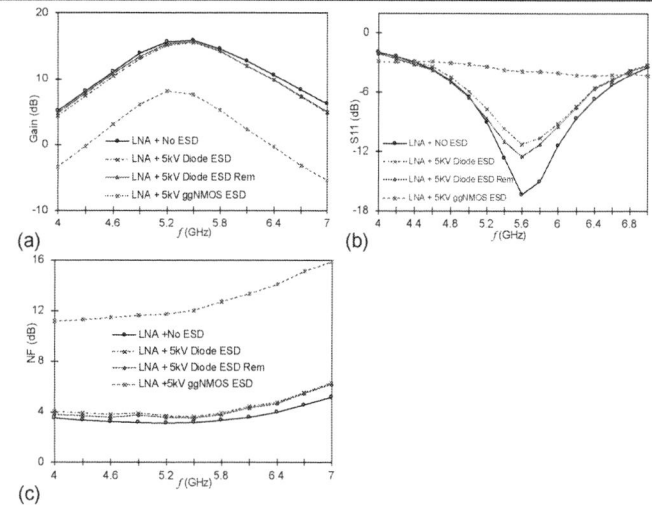

Fig. S2 Post-simulation for the circuit example in Fig. 4 matches measurement very well, (a) gain, (b) S_{11} and (c) NF, suggesting that ESD-IC co-design is very practical and useful.

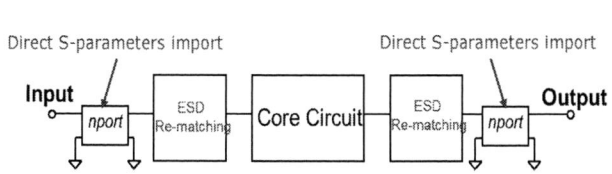

Fig. An ESD-RFIC co-design platform integrates measured ESD parasitic parameters into core IC simulation to evaluate its impacts on core circuit performance. Careful co-design, typically I/O impedance re-matching, can achieve simultaneous performance optimization of both core IC and ESD protection. Some CAD tools allow direct inserting S-parameters into the simulation deck.

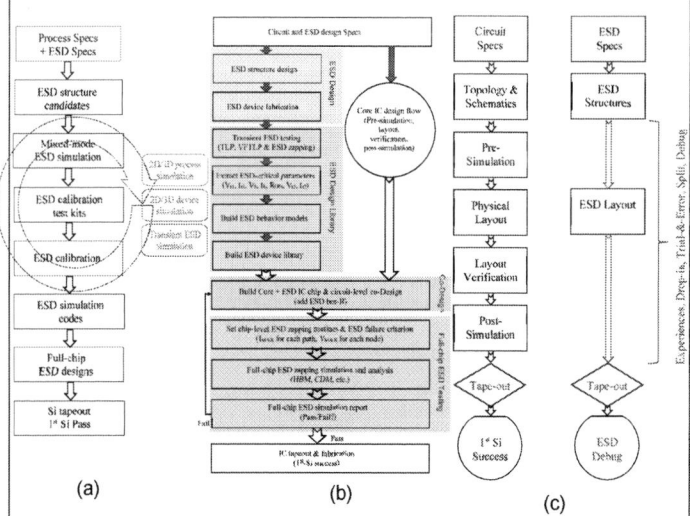

Fig. S4 (a) CAD-based ESD design flows: (a) TCAD, (b) ECAD, and (c) layout-based ESD physical design verification.

5G Lessons Learned and An Outlook on 6G

Bongtae Kim, bkim@etri.re.kr

ETRI, 218 Gajeong-ro, Yuseong-gu, Daejeon, 34129, Korea

Keywords: Beamforming, CNF, 4G, 5G, LTE, MEC, MIMO, Network Slicing, NFV, NR, ORAN, 6G, Terahertz, VNF, VR, XR

Abstract

This paper presents the challenges facing 5G in progress in Korea. It discusses matters to consider in technical research and business development in the future. It outlines 6G service vision, technical issues, and candidate technologies.

Introduction

Korea's mobile telecommunications industry has been the driving force behind the growth of the national economy. The mobile communication service, equipment, and terminal manufacturing industries have grown together as new technology development, network construction, new service launch, service activation, network expansion, and terminal distribution spread are linked in a virtuous cycle. In recent years, however, as the digital economy accelerates, telecommunication networks are increasingly moving away from revenue sources due to increased complexity and changes in the industrial ecosystem.

After 5G commercial roll-out, activation has been delayed, which could adversely affect the virtuous cycle structure of the mobile communication industry. In particular, since 5G mobile communication is highly utilized in other industries as a whole, the impact of the delay in activation on all industries can be significant..

5G lessons learned

The official name of 5G, called IMT-2020, was designated by the ITU [1], and the 3GPP named 5G new radio (NR) as the name of wireless access technology [2]. ITU-R has the authority to evaluate and approve the mobile communication standards. 3GPP, a private standard organization, prepares the actual standards. The global standard for 5G was confirmed in 3GPP's Release 15 in June 2018.

5G standards specifies three main usage scenarios, they are the enhanced mobile broadband (eMBB), the massive machine type communication (mMTC), and the ultra-reliable low latency communication (uRLLC).

eMBB is an extension of 4G mobile broadband services to responding to high traffic demand for user-dense hotspots at busy times, the data rate per cell is increased up to 20 Gbps by enhancing frequency efficiency of 30 bit/s/Hz. mMTC is an improved version of narrow band long term evolution (NB-LTE), inheriting its virtues of wide coverage, low cost, and battery longevity. An industrial IoT cellular application can connect sensors up to one million per square kilometer area. uRLLC aims to provide wireless communication with 1 millisecond of air latency and one error per hundred-thousand bits, which enables mission critical applications such as remote control and autonomous driving.

These three distinct usage scenarios requires that three different systems be supported by the time division duplexing system for eMBB, the improved NB-LTE system for mMTC, and the frequency division duplexing system for uRLLC. The three integrated into one system solution for 5G could be a financial burden for mobile network operators [3].

The mmWave (24 ~ 300 GHz) band is used in 5G for high-speed data transmission. Millimeter waves have narrow coverage due to their physical characteristics and relatively weak penetration through obstacles [3], which are resolved with beamforming and massive multiple input multiple output (MIMO) technologies.

The beamforming technology can precisely control signals on numerous antennas to concentrate energy in a specific direction, it concentrates radio energy to increase distance and minimize interference between beams. The massive MIMO technology obtains high energy efficiency by installing multiple antennas in the base station with a large number of multi-input multi-outputs [4].

New telecom equipment price tends to increase in the early stage, but 5G equipment cost per cell increases to a large due to the use of MIMO. In addition, the electricity maintaining the target signal-to-noise level for wider bandwidth also increases. The 5G transmit power is three times or more than the 4G because of wider bandwidth [3].

Despite controversy about the introduction of 5G, the first 5G commercial services were launched in April 2019. It has been four years, while some challenges faces by 5G are being resolved, 5G deployment is in progress. By June 2023, GMSA identified 535 mobile operators investing in 5G [5], 25 percent of global subscribers are expected to use 5G in 2025 [8].

Among Korea domestic subscribers, 5G subscribers were 25.7 million by August 2022, accounting for 33.9 percent of all subscribers, and the number of subscribers has increased by an annual average of 53.2 percent, while 5G smartphone traffic accounted for 73.7 percent of total mobile phone traffic [10]. As listed in a related study [9,10], the slowdown in the spread of 5G in Korea and the causes are as follows.

First, at or after 5G commercialization, 5G flagship services were not created. Instead of the various services originally presented as 5G vision, only mobile Internet multimedia services, which are representative services of 4G LTE, are available at a faster rate. The initial average download speed of 5G in Korea was 800 Mbps, more than 10 times that of 4G, compared to the initial 4G download speed of 75 Mbps [2].

As the subscriber numbers have been maximized, major business models of mobile operators that focused on the consumer subscription sales based on monthly traffic usage per subscriber, has shifted to unlimited data packages. Data fees paid per gigabyte by 5G subscribers in Korea are more than 70 percent cheaper and 5G subscribers use more than 50 percent more data than LTE subscribers [8].

Second, facility investment is being sluggish. Base stations should be established nationwide to expand coverage of 5G services. Mobile operators are avoiding 5G investment and adjusting the pace of deployment due to uncertainties such as consumers' willingness to pay and lack of initial demand. By April 2022, the number of 28 GHz base station deployments appears to have fulfilled about 11 percent of the network deployment obligation [10].

As base stations are not properly established, terminals supporting 28 GHz band have not been released. Augmented reality (AR) and virtual reality (VR) services have been released, but limitations of the device form factor itself, such as discomfort and ease of use when worn for a long time, are hindering the spread of the service [8].

Services differentiated from 4G, such as low-latency communication and large-scale internet of things, has mostly not been activated. Factors can be found in various aspects, including device and service technology immaturity, low market demand, and regulatory challenges, but it has not created 5G market driver other than increasing the data transmission rate [11].

While the execution has not been the case, standards are on the right track. 5G-Advanced standards (Release 18 and Release 19) are in progress to be applied after the mid-2020s. For 6G, ITU-R has completed the 6G framework recommendation in June 2023. When the standards are completed, the earliest commercialization may take place in 2028, while full-scale 6G commercial service could begin around 2030 [8].

An outlook on 6G

5G's three representative modes, eMBB, uRLLC, and mMTC, will be further evolved in 6G. Services not possible with 5G technology and innovative new services not being introduced due to technology limitations will be provided in earnest. Those includes autonomous driving, urban air mobility (UAM), extended reality (XR), hologram, and digital twin [11]. Most of these services requires ultra-high speed data transmission, which needs the vast amount of frequency bandwidth.

Frequency spectrum

Mobile communications has evolved over multiple generations around every 10 years, in which the data rate for each generation has increased 10 times more of the previous generation. This has been achieved by increasing the system bandwidth with its highest possible spectral efficiency. Since the lower frequency bands has almost been allocated, the wider bandwidth can be obtained at higher frequencies. Typically, the low band is assigned for the primary mobile applications, and the high band is assigned for the wireless broadband applications. Triple the spectral efficiency for 6G can be accomplished mainly by further enhancing algorithms for the massive MIMO large-scale antenna system [3].

In order to secure the frequency required for 6G, various approaches are being discussed. Those include new band exploration, spectrum clearing and reallocation, and spectrum sharing [7]. The 6G candidate bands are grouped into a low-band (< 1 GHz), a medium band (1-24 GHz), and a high band (24-300 GHz). The medium band can be divided into the lower mid-band (1-7 GHz) and the upper mid-band (7-24 GHz). The high band can be divided into the mmWave band (24-92 GHz) and the sub-THz band (92-300 GHz).

The low-band is used in large outdoor area and used for deep indoor coverage. The mid-band, the band from 1 GHz to 7 GHz is generally called mid-band today, is in various rulemaking processes including 3GPP, GSA, and ITU-R [7]. The mmWave band is used in both 5G and 6G, and the sub-THz band is a new frontier band for 6G.

Meanwhile, the terahertz band (100 GHz – 10 THz) may provide further vast amount of bandwidth, which will provide tens of GHz-wide bandwidth. This could make it possible to meet the 1 Tbps transmission rate requirements of 6G. With the advance of related technologies, 6G can utilize up to 3 THz [7].

Coverage expansion

In wireless system design, there are two distinct design issues. One is to increase the data rate, and the other is to increase coverage. In the short distance wireless system such as WiFi and Bluetooth, the main design efforts focus on achieving the highest possible spectral efficiency, in other words, the highest data rate. In contrast, the cellular system operates through long distance with many obstacles, where the design focus is to maximize the handset power efficiency.

According to the Shannon formula, the power and the spectrum cannot be optimized simultaneously. That is, the power efficient operating region and the spectral efficient operating region stretch on the opposite directions [3]. Depending on design objectives, the system operating region needs to be expanded toward the necessary direction. For the ultra-high speed data rate applications such as hologram, efforts should put more focus on expanding to the spectral efficiency direction, but for mMTC applications, completely new approaches are necessary to improve power efficiency and battery longevity [3].

In 6G, the satellite system integration including direct-to-cell services is being promoted [11], which could expand and improve the coverage greatly. In particular, low-earth-orbit satellite systems have recently made progress in resolving both link budget and cost problems, which have been pointed out as limitations of other type satellite systems, brightening the prospects for this integration.

There are, however, many difficulties ahead of actual integration. Among them are a number of technical issues to be resolved, such as ultra large cell size over a wide area, moving cells, large propagation delays, and large path loss, as well as multilateral consultations with interested countries [3,6,11].

Transformation into a computing network

Telecommunications network can be divided into the access network to which subscribers first connect, and the core network in which the operator's service systems are concentrated. In the 5G core network, the virtual network function (VNF) based services are provided through isolation of physical hardware resources. In 6G, the core network function is to be redesigned with more granular services based on cloud-native network function (CNF) by eliminating abstraction technology [11].

Network slicing that cuts networks according to individual services and uses them as multiple virtual networks. 5G allows each data service to allocate network resources independently, ensuring quality without being affected by other services. In 5G, network slicing can only be applied to the core networks, so its usefulness was limited, but in 6G, end-to-end personalized services are to be provided as network slicing extends to wireless access network.

The wireless access network is a place where the most changes take place to compromise with the costs related to maintain service quality from subscriber's service initiation to termination. The mobile edge computing (MEC), that the part of core computational resource is relocated to the network edge, offloads the core computation to improve performance especially for immersive applications such as XR and hologram. Along with MEC, subscribers can choose an appropriate network slice to meet their application specific data processing needs.

In the 6G pervasive computing environment, various forms of AI embedding entities from network access to service processing will be utilized. In this process, it collects and utilizes the vast amount of data generated by the connection of billions of objects, machines, and humans.

The open radio access network (ORAN) separates the hardware and the control software of base stations through virtualization technology. It allows equipment from different venders to be linked through interface standardization, which lower barriers to equipment market entry. This can expedite for many small and medium enterprise (SMEs) to develop new equipment development and deployment.

Trustworthiness

As the telecom systems become increasingly open and controllable externally, the attack surface may also increase. This could make a system vulnerable to security threats. Subscriber's confidential information could be leaked with open source software code that are not properly prepared for security attacks, which poses a serious threat to subscriber privacy.

Compromised subscriber devices may damage telecom system's security. Trustworthiness requirements has to include the secure system environment ensuring the secure operation of software code and the secure-by-design approach that any system constituents can be trusted [6].

Conclusions

As in 5G lessons learned, it is needed more efforts finding the 6G market driver in advance. In 6G, the data transmission rates will continue to increase with more bandwidth at higher frequencies and enhancing the massive MIMO technology. To expand coverage, more communication modes including satellites are to be integrated into cellular operations. Telecommunication networks will advance to be more intelligent so that a number of innovative new services can be realized.

References

[1] ITU-R M.2083, "IMT Vision — Framework and Overall Objectives of the Future Development of IMT for 2020 and Beyond," Sep. 2015.

[2] 3GPP TR 38.807, "Study on Requirements for NR beyond 52.6 GHz," Mar. 2019.

[3] Qi Bi, "Ten Trends in the Cellular Industry and an Outlook on 6G," IEEE Comm. Mag., pp 31-36, Dec. 2019.

[4] E. G. Larsson et al., "Massive MIMO for Next Generation Wireless Systems," IEEE Comm. Mag., vol. 52, no. 2, Feb. 2014.

[5] GSMA, 5G Market Snapshot, July 2023.

[6] 6G - The Next Hyper-Connected Experienced for All, Samsung Research, July 2020.

[7] 6G – Spectrum Expanding the Frontier, Samsung Research, overview-c22-740248, July 2020.

[8] GSMA, 2019 Mobile Industry Impact Report: Sustainable Development Goals, Sep. 2019.

[9] J.E. Yu et al., "An Empirical Study on the Subscriber's Usage and Attitude in the Korean Mobile Service Market," ETRI ET Trends, vol. 37, no. 3, June 2022.

[10] J. Yeo et al., "A Study on Prediction of Future Market Environment Changes and Industry Revitalization Factors in Mobile Telecommunications," KISDI Institutional Project Report, Dec. 2022.

[11] SKT White Paper, 5G Lessons Learned, 6G Core Requirements, 6G Network Technology Trend, 6G Spectrum, July 2023.

Semiconductor Chip Design in a Legoland

Chih-Ming Hung

MediaTek

Moore's Law and Dennard scaling have been ticking and propelling the semiconductor industry for decades. They have also pushed the digitization and pixelization of our everyday life. The industry has attained a great achievement in terms of Power consumption, Performance, silicon Area, chip and system Cost as well as Time to market (PPACT). Unfortunately, the pace and benefits of CMOS scaling have now slowed down, yet the explosive demand for chip PPACT is on a completely opposite trajectory due to the rise of Generative AI (GAI), next-generation wireline and wireless (5G-Advanced and 6G) communication, automotive electronics and so forth. On the energy supply side, the worldwide electricity generation increasing at ~6%/year is far less than the growth of the power consumption for just computing alone [1] making the supply vs demand unsustainable. Emerging technologies, such as new material, compound semiconductor devices, and novel circuits and systems both in the Cloud and at the edge are coming to the rescue. But splitting functions then stitching diverse chiplets together is still at the dawn of the new homogeneous and heterogeneous integration (HI) era. It should also be noted that HI should not be limited to semiconductor chips in a package. Some promising progress will be illustrated ranging from electrical, optical, and passive component integration to support co-packaged optics (CPO), Antenna-in-Package/Module (AiP/AiM), socket waveguides, etc. While the advance in devices and manufacturing is essential, there are lots of innovations needed for architectures, systems, algorithms, and circuits to develop the huge variety of applications. Tools like Computer-Aided Design (CAD), Electronic Design Automation (EDA) and automatic software code generation are supposed to help human engineers to find optimal solutions in a timely manner. However, the pace of enhancement for those tools is around an order of magnitude slower than the Moore's Law. The emergence of machine learning and inference, commonly called ML or AI, is showing a great potential [2, 3] especially when the engineers need to deal with high-dimensional complex options. In fact, many designs assisted by ML have surpassed the PPACT achieved by human engineers using conventional software tools. Following the new large AI foundation model framework, "ChipGPT" could become an essential tool in the future. So far, we have only cracked the tip of the iceberg for all the aforementioned fronts. A lot more challenges and opportunities still lie ahead to be explored.

MOSFET scaling has brought us all the way from µm-scale transistors down to atomic-level device components with more than 1000x shrinkage. It's still alive but significantly slowed. During the golden era with near-perfect exponential scaling (i.e. around 2^{-1} for dimension and power, 2^{-2} for area, and 2^{+1} for speed performance when migrating from one node to the next), System-on-Chip (SoC) dominated every product development not only digital but also RF and analog. A fully integrated GSM SoC including RF transceivers, local power management, a modem, an application processor, and numerous hardware accelerators shipping near a billion chips is a classic example [4, 5]. Fast forward by 20 years, SoC's including GPUs, MCUs, high-speed interface, tons of accelerators, and so on still dominates the world [6]. However, the number of new products integrating chiplets either homogeneously or heterogeneously is increasing [7, 8]. The "wafer-level chip" [9] may also be classified as homogeneous chiplets on a silicon carrier. There seems to be no clear verdict between SoC and HI. It's necessary to first examine some of the challenges for building SoC's and microsystems with HI. Although not exhaustive, the following list attempts to provide some common considerations.

1. Yield: The features integrated on modern electronic chips increase at warp speed. Fig. 1 shows an application processor of a smartphone. Many new modules are integrated onto new chips (e.g. multi-modal GAI inference), and every module advances its features and performance over time (e.g. 8MP camera becomes 100+MP one with rich of AI-enabled functions). As a result, the number of transistors is increased significantly faster than what the Moore's Law can offer. Considering manufacturing defect is approximately inversely proportional to the die size [10], it would be better to split a large chip into a few smaller chiplets.

2. Transistor cost and PPA tradeoff: Fig. 2 shows the transistor cost and power consumption trend versus process nodes. The former is generally not gaining further benefits out of the CMOS scaling leaving the decision of having SoC or HI to base on e.g. power, speed, and yield. Typically, it would be more attractive for high-end logic chips to consider the SoC strategy because the most advanced node can still offer some PPA advantages. The overhead, such as die-2-die (D2D) interface power consumption, can also be eliminated. For AIoT products, HI may be the choice to have a low cost and the flexibilities to integrate functions from heterogeneous chiplets.

3. Chip footprint and volume: A SoC with rich features on a large die is against the trend of miniaturization for modern electronics. The physical distance among timing-critical IPs also becomes long creating new challenges for on-chip network fabrics. One way to mitigate is to segregate the SoC into chiplets and build 3D interconnect to shorten the routing length while simultaneously reduce the footprint and volume. The technology for die thinning or complete removal of silicon carrier is being developed to support the objective.

4. Interconnect loading: While people pay most attention to transistor performance which has generally plateaued, the impact from back-end-of-line (BEOL) metal system which is crucial for RF and high-speed signaling is often underestimated. A ratio between effective and physical distance on chip may be formulated as the following:

$$(eff.\, distance)/(phy.\, distance) \propto (RLC\, /\, unit\, length)$$

The approximate ratio over several process generations is drawn in Fig. 3. For old nodes where metal thickness << minimum metal width and the maximum operating speed is limited, the ratio is relatively flat because the intrinsic RC product of each metal layer is generally constant and the effect from L is negligible. The ratio increases when the metal width is scaled down faster than the thickness while the effect of L is also increasing. For recent process nodes, the ratio is flattening again not because of technology advancement but the fact that the metal system received limited scaling. For processor designers, a better approach to reach higher performance and energy efficiency is to pursue parallel computing architectures instead of pushing a higher clock rate. The strategy is further supported by chiplet architectures including the use of interconnect bridges which have a better metal system.

5. Heterogeneous requirements: For digital logic, the observed trends for power consumption at a constant speed and for speed at a constant power consumption are shown in Fig. 4. Although the two performance indexes cannot be obtained simultaneously nowadays, they can often be traded with each other. However, such a tradeoff is not applicable to e.g. embedded memory where the capacity and access speed have both been increased dramatically for the new AI/ML workloads. Consequently, high-density memory chiplets fabricated with a dedicated process technology become mandatory. Similar situations can be observed for RF and analog functions because the BEOL schemes for the advanced nodes are typically optimized for digital designs. Alternatively, the passive elements can be built on a different die [11] but the operating frequency might be limited due to the D2D interconnect.

6. Development cost: Designing a SoC involves specification, architecture, circuit and system design, physical design, software development, and productization. Fig. 5 shows the overall cost hike over process nodes. The hardware and the complete development cost using a 3nm process technology is expected to exceed $400M and $800M, respectively leaving only large chip companies with high-volume products to adopt advanced nodes. Note that the backend design (verification and physical implementation) takes around two thirds of the hardware development cost. That is, even a new

SoC was based on a previous design using the same process, and the design has been highly modular, only a limited reduction in the development cost can be obtained because the backend design is highly unique to each chip. The high cost also reduces the feasibility of having dark silicon meaning several Stock-Keeping Units (SKUs) can be created out of the same chip by enabling and disabling features. An intuitive solution may be to extend the modular approach and encapsulate each module as a chiplet. Innovations to maintain sufficient flexibility and reusability are needed

Fig. 6 shows a family of high-speed networking chips which enable the backbone of the data centers and cloud computing. It can be observed that these chips are built with only a few foundation chiplets. The chip family has highly scalable features spanning from fully connected high-bandwidth network routing to a stand-alone retimer function. Further integrating 800Gbps optical modules each in a small socket assembled on a large package substrate, a microsystem simultaneously supporting co-packaged optics, optical linear-drive pluggable and high-speed LR SERDES for 800G pluggable has been demonstrated (Fig. 7). The obvious benefits are higher yield and better PPACT by using the best suited process technology for each chiplet. It also provides serviceability so that one can replace the optical modules in the field. A large socket and delicate surrounding structures were designed to house the large substrate and to support the mechanical stress, high-speed electrical interconnect, thermal dissipation and repairability of the optical modules in the field.

HI comes with tremendous challenges. When starting a new design, the complexity is exponentially higher than making a single chip because not only the considerations for each chiplet are multiplied but also the interactions among chiplets need to be sorted out and optimized. To name a few, one needs to partition the functions into separate chiplets, select process technology for each of them, optimize routing congestion and signal/power integrity (SI/PI), determine and design in 2.xD or 3D integration, plan for thermal solutions and production, etc. all with a high degree of freedom which could overwhelm human designers' capacity. New ML-based EDA tools are likely a must to enable HI.

The scope of HI should be broadened beyond semiconductor chiplets. Fig. 8 shows a chip containing a silicon IC and 16 antennas for a MIMO radar which can detect a small vehicle at 200m away. Conceptually, the passive components including impedance matching, vertical transitions for RF signals, and the antenna elements should be treated as a chiplet. With a different specification, a Frequency Selective Surface (FSS) can be deigned and integrated to increase the effective antenna gain thus a greater radar detection distance (Fig. 8). The HI concept can be extended to the module and socket level as shown in Fig. 9. The mmWave signal is emitted from an in-package antenna array, guided with two 90-degree turns to reach the receiver outside the socket. Fabricated with metal and the typical alignment tolerance, the waveguide exhibits negligible loss.

The demand of better PPACT in the post Moore's Law era has spurred the trend of designing more specialized IPs and hardware accelerators. To be sustainable, those IPs need to be designed with maximal reusability like elementary Lego blocks so that the integration engineers can focus on system design and easily pick modules from a library of chiplets. While the process seems promising, there are many obstacles. On the flip side, there are equally many innovation opportunities. To illustrate a few:

1. New owner: Most of the system development today has a high level of independence or isolation among architecture, hardware, and software. When integrating a microsystem using chiplets which closely interact with e.g. a million interconnect wires (compared to a few hundred or thousand today), a new top-down approach and ownership is needed. Traditionally, system developers may not need to be proficient in the upper stream technologies, such as transistor physics. Now, a new integration engineer or team need to comprehend to make all key decisions.

2. Interface standardization: Observing that the unified interface for Lego blocks has enabled so many possibilities limited by

one's imagination, it would make sense to have interface standards for key types of chiplets. The end-to-end interface energy efficiency is also far from ideal. Although standardization might sacrifice a little bit of performance, the gain in the microsystem level could be substantially positive.

3. Cooling: 3D heterogeneous integration has been limited to around 2-3 layers of chips today when homogeneous integration for e.g. vertical memory cells are excluded. One of the key obstacles is heat removal concerning not only the total heat generation but also the high local heat density. Those call for research on thermal interfacial material, micro fluid channels, micro/nano vapor chambers, etc.

4. Material and structure: All chiplets lie flat today, either facing up or down. Vertical chiplet placement facing east/west, north/south, or free angles might create new avenues for integration. Without the constraint of deposition and etching on a plane, material for interconnect may go beyond copper and signals may not be limited to electrical.

5. Power delivery and IO access: Although 3D chip stacking may eliminate some of the external IOs, the reduced footprint of the chiplets would again increase the competition between power delivery and RF / high-speed pins. They cannot be prioritized because both are also collaborative to achieve low noise and good SI. Communication among stacked chiplets can be another challenge. Methods without a sea or a long string of through silicon/molding/etc. vias are an area of interest. Innovations in material, power regulation networks, and packaging techniques are needed.

6. RF and optics: The signal loss due to metal, dielectric material and fabrication imperfection is exacerbated when the operating frequency is increased. The flexibility of dielectric thickness at low cost is also required, which plays a critical role for passive elements including antennas. Metal and via pitches should generally be in the order of 1/10 to 1/20 of the wavelength to confine and guide the electromagnetic energy. In the case of mmWave and THz circuits, the design rules for standard low-cost flipchip assembly need to be significantly tightened to compete with wafer-level assembly. Novel material, manufacturing, and structures with low loss and low cost demands innovations. Similarly, carrying optical signals in a package or inside a socket requires active research to enable miniaturization of optical systems.

7. Validation and verification: Historically, test cost does not scale. That means when chips get bigger, more complex, and more flexible, the test cost would surpass the transistor cost. Even worse, it's more difficult to achieve high test coverage resulting in yield loss and high cost. ML is seen as a must-have tool and methodology to mitigate the concerns not only to increase the test coverage but also to reduce the number of test vectors. We have only scratched the surface so far. Another obvious concern with chiplets is that the conventional known-good-die (KGD) approach no longer works because KGDs may not operate completely as expected after assembly and the tests done after assembly cannot recover the loss of materials. Novel FA techniques need to be developed for 3D HI.

8. AI/ML-assisted EDA: Design productivity may be improved by both simulation speed and expert (ML) systems. The former often involves large meshing data, large matrices to compute, etc. The progress in the past decades is amazing, but the pace is significantly slower than the growth of design complexity. Physics-aware ML has shown great promises, such as 10-100x speed up for IR drop estimation. There are thousands of EDA utilities that could adopt ML to help. Further with the introduction of GAI, ML can assist designers to assess the best floorplan, write algorithms and codes, synthesize their netlist, and so on. We are just at the beginning of making a "ChipGPT".

9. Security: When the number of chiplets is increased, the attack surface is generally enlarged resulting in a higher vulnerability. Furthermore, when the chiplets are produced by multiple vendors, security could become a major concern. The

security problem is now moving from system level down to the integrated microsystem level. Research is needed to enable zero trust throughout the supply chain.

To summarize, the scaling of semiconductor process technology has been decelerating. The challenges prompted engineers to focus on new system architectures and new circuit designs together with homogeneous and heterogeneous integration. In fact, it has been debated for decades between SoC and HI except that now Moore's Law and Dennard scaling cannot shield SoC camp as well as before. That in turn spurs innovations for the HI camp to take off. Several exciting examples have been illustrated, yet many more challenges have been raised. Overall, between SoC and HI, there is no one size fits all. In the near future, HI does offer lots of promises although it doesn't broadly bring back the Moore's Law and MOSFET scaling. Novel research to advance HI including adopting machine learning and GAI are accelerating so that designers can leap from playing semiconductor Tetris to truly enjoying the semiconductor Legoland.

Acknowledgement:

MediaTek teams and all partners contributing to the mentioned works.

References:

[1] https://www.src.org/about/decadal-plan/decadal-plan-full-report.pdf: 123.

[2] J. Lee *et al.*, "A Learning-Based Algorithm for Early Floorplan With Flexible Blocks," A-SSCC 2022:1-2

[3] W. Chang *et al.*, "Muzero-Guided Simulated Annealing for Nanometer Circuit Placement,", DAC 2022: July 12th

[4] C. Hung *et al.*, "A Low-Cost SAW-less GSM/GPRS SoC with Integrated Connectivity and 32-kHz Crystal Removal in 55nm," VLSI Circuits Symposium 2013:74-75

[5] K. Muhammad *et al.*, "A Low-Cost Quad-Band Single-Chip GSM/GPRS Radio in 90nm Digital CMOS," RFIC 2009:197-200

[6] J. Choquette *et al.*, "The A100 Datacenter GPU and Ampere Architecture," ISSCC 2021:48-49

[7] M. Park *et al.*, "A 192-Gb 12-High 896-GB/s HBM3 DRAM with a TSV Auto-Calibration Scheme and Machine-Learning-Based Layout Optimization," ISSCC 2022:444-445

[8] L. Su, *et al.*, "Innovation For the Next Decade of Compute Efficiency," ISSCC 2023:8-12

[9] S. Lie *et al.*, "Cerebras Architecture Deep Dive: First Look Inside the HW/SW Co-Design for Deep Learning : Cerebras Systems," Hot Chips 2022

[10] L. Loh, *et al.*, "Fertilizing AIoT from Roots to Leaves,", ISSCC 2020:15-21 and visual

[11] M. Tsai *et al.*, "A Fully Integrated Multimode Front-End Module for GSM/EDGE/TD-SCDMA/TD-LTE Applications Using a Class-F CMOS Power Amplifier," ISSCC 2017:216-217

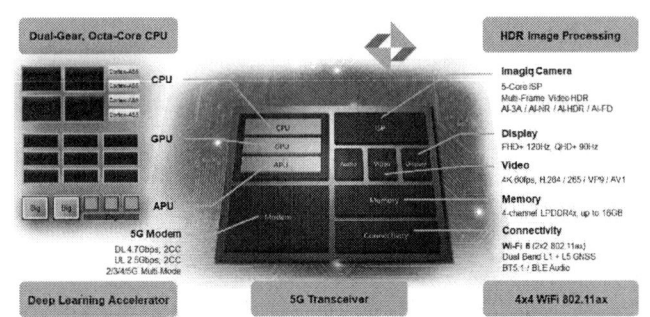

Fig. 1. An application processor for smartphones built rich functions, such as muti-core multi-cluster CPU, GPU, APU, multimedia processors, image signal processors, wireless modems, high-speed wireline interface.

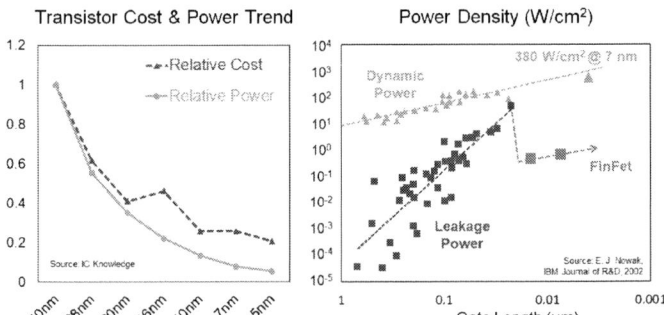

Fig. 2. Left: normalized transistor cost and digital cell power trend versus process nodes. Right: normalized dynamic and leakage power consumption trend versus process nodes.

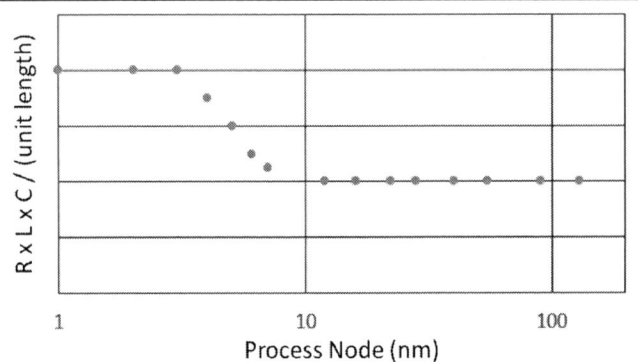

Fig. 3. Normalizing to the physical routing distance, the effective routing length is increased when metal pitch is scaled over process generations.

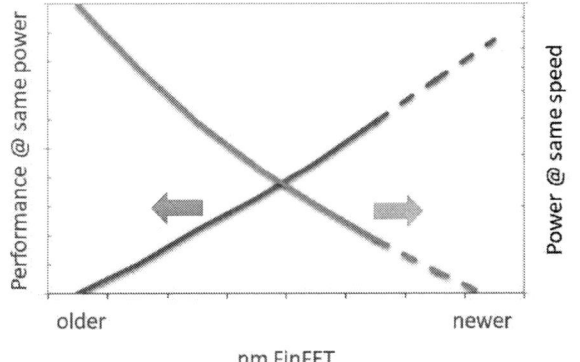

Fig. 4. Trend lines for performance @ same power and power @ same speed.

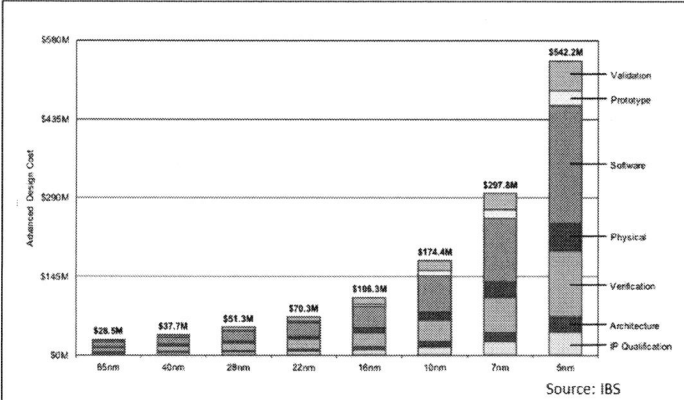

Fig. 5. A breakdown of design cost from 65nm and 5nm nodes, which accelerates exponentially.

Fig. 6. A family of high-speed networking chips built with very few types of chiplets.

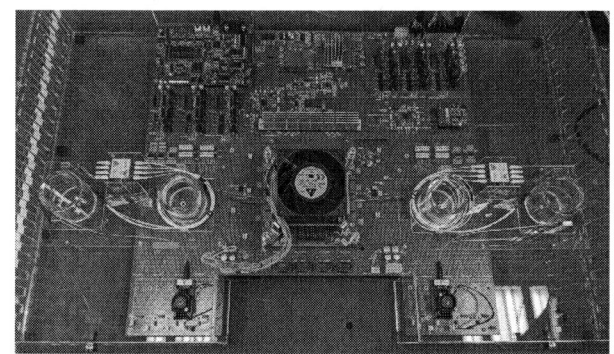

Fig. 7. A microsystem simultaneously supporting co-packaged optics, optical linear-drive pluggable and high-speed LR SERDES for 800G pluggable.

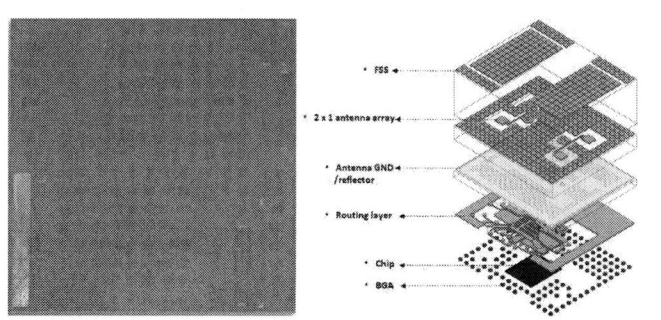

Fig. 8. Front and back sides of a radar chip with in-package antenna arrays.

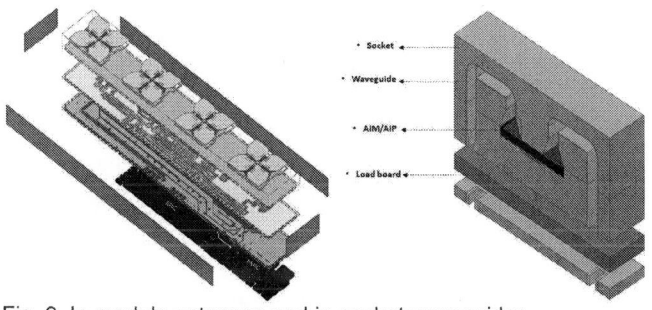

Fig. 9. In-module antennas and in-socket waveguides.

979-8-3503-3004-5/23 $31.00 © 2023 IEEE

Architecture Challenges for Heterogeneous Processors in Embedded SoCs

Masayuki Ito

NSITEXE Inc, Japan

AI-based edge devices such as robots, self-driving vehicles, and unmanned construction equipment are increasingly being used to develop advanced social activities while saving labor. Research continues to improve AI algorithms used for object recognition, determining actions and behaviors, and new concepts to support the sophistication of these edge devices. In addition to conventional CPUs and GPUs, there is a need for processors that can flexibly adapt to changing AI algorithms while efficiently executing the legacy processing required for edge devices. Currently, the main method for edge processors is to combine high-performance CPU and GPU and achieve a large amount of repetitive operations, but the significant increase in power consumption on the edge side is raising the power consumption of devices, affecting heat exhaust design and reliability, and is also one of the causes of increased CO2 emissions. On the other hand, in the method of using dedicated hardware for fixed-function, systems that require a high level of safety, such as automated vehicles, generally have a long product development life cycle time and a strict approach to quality assurance, so mechanisms that have already been proven in the market are often used without modification. It is often the case that multiple AI processes, small to medium scale but with a mixture of old and new AI processes, will be used in the system. In such cases, processor technology with high flexibility in terms of functionality is more suitable than fixed-function hardware optimized for specific processing.

As a challenge for these next-generation processors, it is necessary to develop processor architectures with low power consumption, high performance, and flexibility, and to develop low-power implementation technologies. In addition to parallelization by threading each graph structure to be processed to increase parallel performance, the architecture must be equipped with features such as improved scalar processing capability, and memory bandwidth to support high parallelism. Furthermore, to support full-scale AI edge computing, it is necessary to not only repeat a single AI process, but also perform a combination of conventional control, signal processing, and other processing. In particular, low power consumption requires consideration of not only architectural methods but also memory methods that match these methods. The range of applications of AI edge computing is very wide, including not only AI-related processing such as recognition and decision making, but also conventional control and signal processing applications, and many applications require real-time response from the viewpoint of edge device control, along with increased dynamic execution and graph processing.

Figure 1 shows the major processing flow of an automated driving system and its processing characteristics. Processing is divided into four major blocks, from sensor input on the left to driving control on the right: The first is image processing of sensor input data and pre-recognition processing, which requires a large amount of highly regular data processing; the second is object recognition and sensor fusion processing, which is slightly less regular but still requires a large amount of data processing; and the third is decision processing, from self-position estimation to path planning, which requires relatively smaller amount of data but is more complex. The third is decision processing from self-position estimation to path planning, which requires smaller amount of data than the previous processes, but still requires a high level of complexity. Thus, in automatic operation processing, it is necessary to perform processing with significantly different processing characteristics depending on the processing step.

Such processing characteristics are not limited to automated operation processing for automobiles. In machine control as well, control has conventionally been performed from open-loop control to feedback control, but now Full Model Control is being considered, and the processing characteristics are diverse and the amount of processing is extremely large. The relationship between these processing characteristics (Apps) and the hardware that processes them is shown in the upper part of Figure 2. When there was little variation in processing characteristics, a hardware configuration combining CPUs and GPUs with dedicated HW-IP was sufficient for processing. However, as processing characteristics become more complex and the amount of processing increases exponentially, the same hardware configuration is no longer sufficient to handle the processing, and power consumption increases due to mismatched characteristics. Developing dedicated HW-IP for each process to improve processing efficiency and power efficiency is not realistic in terms of development cost and time. To overcome this situation, a next-generation processor with high processing efficiency and power efficiency that is flexible and programmable, as shown in the lower right of Figure 2, is desirable.

Figure 3 illustrates the processing characteristics of these next-generation processors compared to conventional server processing characteristics. The left side schematically shows the Server workload, where processing indicated by circles is allocated to the CPU, processing indicated by triangles is allocated to the dedicated HW-IP (accelerator), and processing indicated by long circles is allocated to the GPU. Since the overall amount of processing in server processing is large, the communication time between each piece of hardware can be tolerated to a certain extent, so strict real-time performance is not required, and the hardware can be realized with a hetero-processor consisting of a multi-core CPU, GPU, and dedicated HW-IP group as a component in the SoC. On the other hand, the Embedded system shown on the right side of Figure 3 is a heterogeneous system. On the other hand, in the Embedded workload shown on the right side of Figure 3, the processing volume of each task is relatively small, and the hardware on the left side of Figure 3 has a large communication overhead between processing IPs, which increases the total processing time and reduces the real-time performance. As a processor that satisfies these embedded characteristics, we propose a micro-heterogeneous processor that consists of an SCU for CPU processing, a vector processor (VPU) with data parallelism, and a convolution unit (CVU) for AI-specific convolution processing, all tightly coupled within a single processor IP. The micro-heterogeneos processor, which is a tightly coupled combination of SCUs, VPUs, and CVUs for AI processing, is suitable for this purpose.

NSITEXE is developing such a next-generation processor as the Akaria Processor Platform. As shown in Figure 4, the Akaria Platform includes a 32-bit and 64-bit version of RISC-V developed in-house as the Standard Processor, and multi-core configurations, vector extensions, and AI accelerators as Extension Units that can be tightly coupled to the Standard Processor. The Extension Units are available in multi-core configurations, vector extensions and AI accelerators. These Extension Units are programmable, domain-specific micro-hetero processors that achieve high parallelism in Task Parallelism, Data Parallelism, and Neural Network processing. The Akaria Processor Platform enables flexible, high real-time processing characteristics, high power efficiency, and area efficiency at the same time.

Figure 5 shows a hardware block diagram of the DR4100MD, the largest configuration of the Akaria Platform, which consists of four clusters, each cluster consisting of two 64-bit RISC-V processors sharing a single Vector Processing Unit. In addition, the cluster is equipped with a Convolution Unit and Intelligent DMA. Functional and implementation evaluations of the DR4100MD have been conducted. In terms of functionality, we have implemented part of the DR4100MD in FPGAs to run applications such as basic benchmarking and model predictive control. We have also built an evaluation system to control a robot arm and implemented a self-position estimation process and path planning algorithm for the robot arm. In addition, as PPA evaluation, power efficiency (TOPS/W) and area efficiency (TOPS/mm^2) were evaluated by performing layout in a 7nm process library environment. The standard neural network VGG16 was used as the evaluation program for the power efficiency evaluation. Figure 6 shows the evaluation results. As shown in the table on the left, 15.1 TOPS/W was achieved in the power efficiency evaluation, which is the world's highest level for a 7nm generation commercial processor. In the area efficiency evaluation, 1.36 TOPS/mm^2 was achieved, which is the world's highest level for this processor architecture.

979-8-3503-3004-5/23 $31.00 © 2023 IEEE

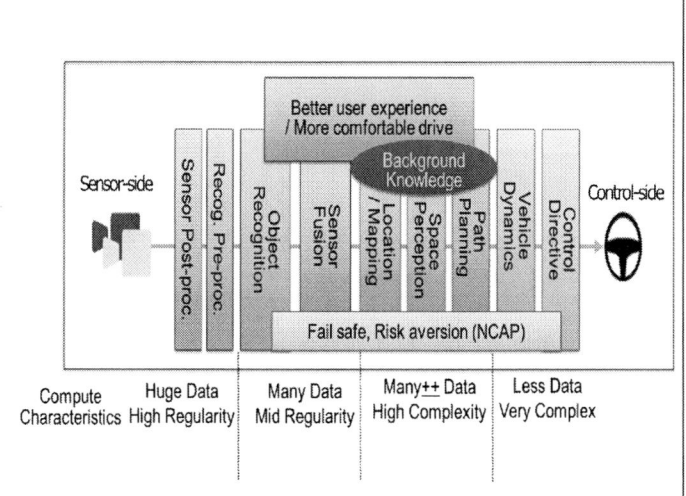

Fig. 1. AD System requires multiple semiconductor IPS

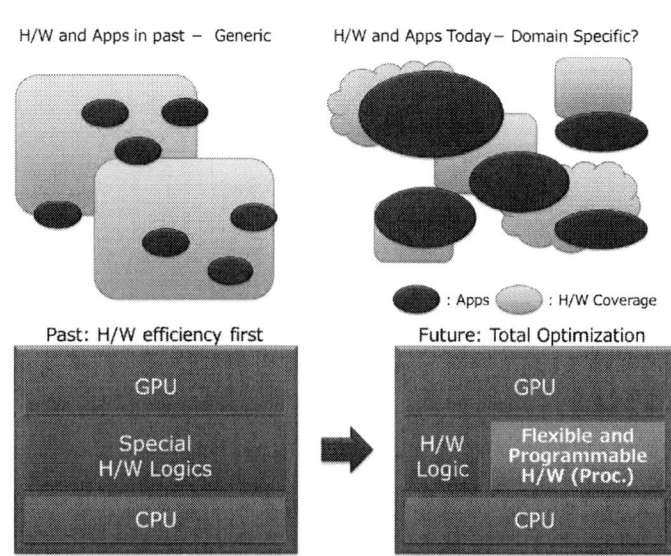

Fig. 2. Hardware and Applications Relationship: Past and Today

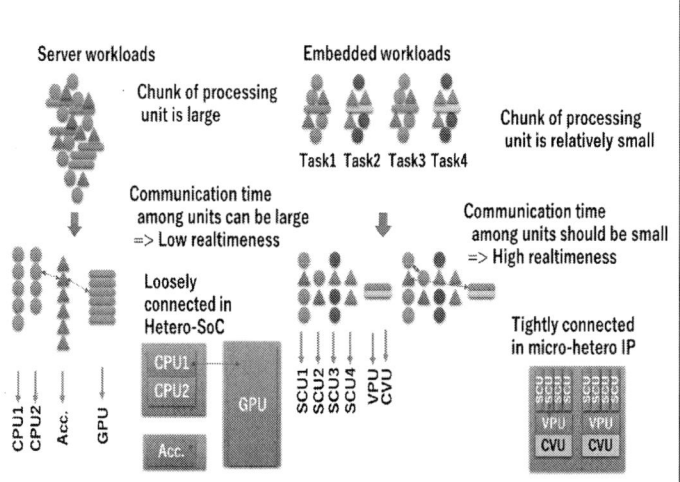

Fig. 3. Characteristics of Micro-hetero Processor

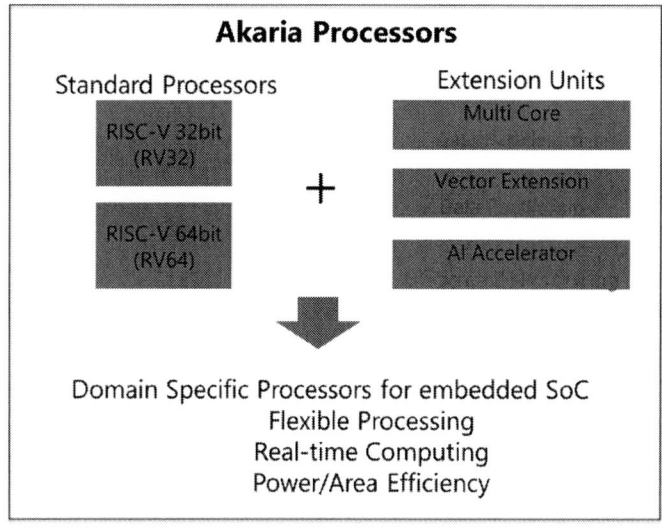

Fig. 4. Akaria Platform: Micro-hetero processor system

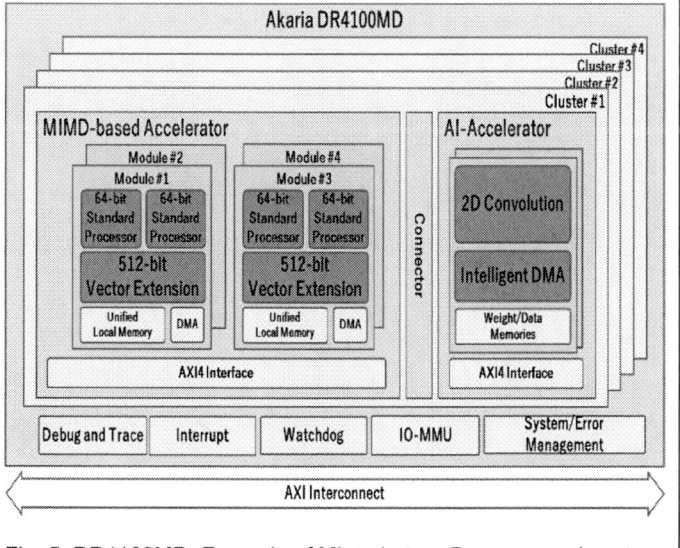

Fig. 5. DR4100MD: Example of Micro-hetero Processor subsystem

Power Efficency	Evaluation Result
Execution time	14.5(msec)
Power	0.141(W)
# of Operations	0.031(TOP)
TOPS	2.14 (TOPS)
TOPS/W	15.1(TOPS/W)

Area Efficency	Evaluation Result
Logic Area	0.95(mm²)
SRAM Area	3.42(mm²)
Total Area	4.37(mm²)
(Logical)TOPS @500MHz	5.96(TOPS)
TOPS/mm²	1.36(TOPS/mm²)

Fig. 6. Power and Area Evaluation Results

AUTHOR INDEX

Agrawal, M 192
Agrawal, P 192
Ahn, Hyun-A 79
Ahn, Jaewoong 195
Alioto, Massimo 237
Amakawa, Shuhei 255
An, Junyeol 195
Anand, R 192
Anupreetham, Anupreetham 103
Aoyama, Takuma 22
Bae, Seung-Jun 79
Baek, Jinhyeok 79
Baek, Kwang-Hyun 115
Baek, Kyungmin 67
Bakalzuk, G 192
Barton, Taylor 28
Cao, Hanzhang 37
Cao, Peng 94
Cao, Yu 103
Chae, Youngcheol 100, 106, 121
Chang, Jun 31
Chang, Mau-Chung Frank 28
Chang, Shao-Ting 55
Chang, Sheng-Kai 55
Chatterjee, R 192
Chen, Daigao 183
Chen, Dihu 168, 180
Chen, Jixin 58, 64
Chen, Po-Hung 231
Chen, Qin 261
Chen, Tai-Jung 177
Chen, Tianbao 13
Chen, Weiliang 52
Chen, Wen 154
Chen, Wendi 160
Chen, Xiaofeng 34
Chen, Yong 139
Chen, Yutang 168, 180
Chen, Zhe 58, 64
Chen, Zhiming 222
Cheng, Kuang-Wei 55
Cheng, Xuxu 49
Chi, Baoyong 25, 189, 222, 225, 240, 243, 249
Chiang, Shiuh-Hua Wood 28
Cho, Seonghwan 124
Cho, Youngjae 4
Choi, Ga-Ram 79
Choi, Haidam 258

Choi, Jiwon 148
Choi, Jonghang 210
Choi, Jung-Hwan 16, 115
Choi, Jun-Han 16
Choi, Michael 4
Choi, Seung Hun 195
Choi, Woo-Seok 67
Choi, Woo-Young 10
Choi, Youngdon 16, 115
Choo, Kyojin 76
Chun, Jung-Hoon 228
Crary, Peter 252
Dang, Ziyue 183
Deng, Dazheng 13
Deng, Lishuo 127
Deng, Mingxing 25
Deng, Wei 25, 189, 222, 225, 240, 243, 249
Diao, Pengfei 160
Ding, Ruixue 31, 151
Dong, Zirui 85
Dong, Zizheng 183
Duan, Zongming 43
Eimitsu, Masatomo 22
Eldar, Yonina C. 252
Fan, Xiangning 261
Fan, Yuanqi 13
Fassio, Luigi 237
Feng, Liqun 163
Feng, Xiangdong 246
Fujishima, Minoru 255
Gao, Hao 43
Gao, Jinxu 201
Gemmeke, Tobias 91
Geng, Yujie 166
Go, Young-Gil 79
Gong, Hao 160
Gu, Junjie 43
Gu, Peng 157
Guan, Pingda 25
Guo, Jianping 168, 180, 219
Guo, Mohan 157
Guo, Ruiqi 34
Guo, Yuhao 73
Guo, Ziyuan 243
Ha, Sohmyung 258
Hagino, Tatsuo 255
Ham, Eunkyung 276
Han, Changxuan 154

Han, Donghyeon	186
Han, Kefeng	43
Hanh, Hoang Hong	237
Hao, Lujie	189
Hao, Weiquan	279
Hara, Shinsuke	255
Hassan, Ahmed	103
He, Jian	183
He, Junxian	109
Heo, Sanghyun	210
Hiraki, Mitsuru	19
Hisakado, Takashi	145
Hong, Feifan	160
Hong, Seongyon	46
Hong, Wei	58, 64
Hong, Zhiliang	94
Hsieh, Yi-Yen	177
Hu, Ang	130
Hu, Tuo	1
Hu, Xianwu	73
Hu, Yang	13, 34
Huang, Gongxing	133
Huang, Leyang	58
Huang, Lijie	133
Huang, Tianze	234
Huang, Tongde	37
Huang, Xiangrong	189, 222, 225, 243
Hung, Chih-Ming	284
Hwang, Jung-Hye	118
Hwang, Young-Ha	210
Igarashi, Yasuto	19
Ihm, Jeong-Don	16
Iijima, Hiroaki	22
Iizuka, Yoichi	82
Ikeda, Shinichi	22
Im, Dongseok	186
Ishikura, Hiromichi	19
Islam, Mahfuzul	145
Ito, Masao	19
Ito, Masayuki	288
Iwata, Akira	22
Jang, Taekwang	124
Je, Minkyu	258
Jeong, Deog-Kyoon	67
Jeong, Han-Ki	16
Jeong, Kyeongwon	258
Jia, Haikun	25, 189, 222, 225, 240, 243, 249
Jia, Yu	216
Jiang, Yang	216, 273
Jin, Jin	37
Jo, Wooyoung	97, 148
Jo, Yongjun	174

Ju, Peijuan	160
Juluri, R.	192
Jung, Jaehong	40
Jung, Jae-Woo	79
Jung, Jinook	16
Jung, Keum-Dong	210
Jung, Sangdon	228
Jung, Wanyeong	264
Junsen, He	115
Kang, Jubin	118
Kang, Kyounghun	264
Kasamatsu, Akifumi	255
Kasichainula, Kishore	103
Kawakami, Shinya	22
Ki, Wing-Hung	213
Kim, Bongtae	282
Kim, Eunkyoung	136
Kim, Gyusik	40
Kim, Hana	276
Kim, Ho-Jin	4
Kim, Hyunji	276
Kim, Janghoo	79
Kim, Jeongmin	195
Kim, Jeongmyeong	264
Kim, Ji-Hoon	276
Kim, Jiwoong	195
Kim, Joomyoung	40
Kim, Junhyeong	270
Kim, Kahyun	67
Kim, Ki-Duk	195
Kim, Kihan	79
Kim, Sangjin	46, 97
Kim, Sangyeob	46, 148
Kim, Seong-Jin	118
Kim, Seungjin	40
Kim, Sol	136
Kim, Subin	210
Kim, Tae-Sung	16
Kim, Tony Tae-Hyoung	112, 115, 174
Kim, Ye-Dam	4
Kim, Yong-Hun	79
Kim, Yong-Hyun	79
Kim, Yong-Jo	124
Kiryu, Toshiki	19
Kitamura, Masahiro	19
Ko, Hyungjong	16
Kono, Takashi	19
Krishnan, Gokul	103
Kuan, Yen-Cheng	28
Kwak, Myoungbo	16
Kwon, Ohjo	195
Kwon, Yigi	121

Kye, Chan-Ho 76
Lai, Jinze 133
Law, Man-Kay 207, 216, 273
Lee, Bo-Hyeon 79
Lee, Changuk 121
Lee, Dong-Hun 4
Lee, Donghun 79
Lee, Eonhui 67
Lee, Hyeyeon 106
Lee, Hyung-Min 195, 204
Lee, Hyunjoo Jenny 258
Lee, Hyun-Su 204
Lee, Inhee 106
Lee, Jae-Woo 79
Lee, Jingu 97
Lee, Jongwoo 40
Lee, Kyunghea 136
Lee, Kyungmin 40
Lee, Myung-Jae 100
Lee, Sangwoo 121
Lee, Seunghoon 270
Lee, Siheon 264
Lee, Yongjun 210
Lee, Yoonmyung 106
Li, Aobo 130, 234
Li, Dongze 243
Li, Fengheng 246
Li, Guike 183
Li, Huihua 273
Li, Kaiyou 219
Li, Keran 127
Li, Leliang 183
Li, Lianming 261
Li, Xiang 130, 234
Li, Xing 213
Li, Xintao 243
Li, Xueqing 52
Li, Xunyu 279
Li, Yanlei 73
Li, Yaolei 201
Li, Yixi 139
Li, Yongfu 198
Li, Zehao 112
Li, Zekun 58
Li, Zeli 219
Li, Zhenghao 49
Li, Ziyu 127
Liang, Hongzhi 31
Liang, Xingdong 73
Liang, Yue 261
Liao, Huabing 189
Liao, Qianxian 163

Liao, Xiaofei 160
Liao, Yu-Te 171
Lim, Daihyun 79
Lim, Won-Mook 16
Lin, Haichuan 166
Lin, Heng-Li 171
Lin, I-Fan 171
Lin, Yang 70
Lin, Yukai 73
Lin, Zhi-Wei 55
Lin, Ziyi 189
Liu, Dongsheng 130, 234
Liu, Han 183
Liu, Hongzhuo 240
Liu, Huaiyu 70
Liu, Jian 139, 142, 183
Liu, Jinlin 1
Liu, Leibo 13, 34, 52
Liu, Liangliang 157
Liu, Liyuan 109, 139, 142, 183
Liu, Shubin 31, 88, 151
Liu, Weitian 43
Liu, Xiang 160
Liu, Xiaolong 37
Liu, Xin 246
Liu, Xuzhi 222
Liu, Yan 70
Liu, Yang 219
Liu, Ye 183
Liu, Yongpan 201
Lou, Yuqing 198
Lozada, Kent Edrian 4
Lu, Donglai 183
Lu, Jiahao 130, 234
Lu, Liqun 58, 64
Lu, Yu 1
Lu, Yuncheng 112, 174
Luo, Xiongshi 49
Luo, Xun 154
Luo, Yuxuan 168, 180, 246
Lyu, Liangjian 85
Lyu, Tianrui 219
Ma, Qiaobo 273
Ma, Ruichang 25, 240
Ma, Zizhao 73
Maezono, Akihide 82
Mak, Pui-In 61, 216, 273
Martins, Rui P. 216, 273
Matsuoka, Yoshiki 22
Meng, Xiaodong 213
Miao, Runyu 279
Min, Byounghan 121

Min, Hao	1
Mitsugi, Jin	1
Moon, Byeongmin	264
Moon, Daesik	79
Morishita, Fukashi	82
Mubarak, Mohamed H.	255
Mukherjee, M.	192
Na, Yoondeok	100
Nair, Gopikrishnan Raveendran	103
Nalla, Pragnya Sudershan	103
Nam, Bu-Il	136
Nam, Seunghyeon	258
Nonaka, Makoto	19
Ogata, Toma	19
Oh, Chaerin	258
Oh, Jonghyun	103
Oh, Seunghyun	40
Oh, Tae-Young	79
Otomo, Goichi	22
Pan, Quan	49
Pan, Zijin	279
Park, Changjoo	264
Park, Dahwan	118
Park, Gwangtae	186
Park, Jae-Koo	10
Park, Jaewoo	16
Park, Jongjun	186
Park, Jun-Eun	210
Park, Junhong	136
Park, Jun-Seo	79
Park, Minsu	67, 228
Park, Sung Min	40
Park, Sunyoung	276
Park, Wonhoon	148
Park, Yongjae	118
Qi, Gengzhen	61
Qi, Nan	139, 142, 183
Qi, Xiangao	198
Qian, Junyi	127
Qin, Haoqi	43
Qin, Yubin	13
Qiu, Ruiheng	243
Rahmanony, O	192
Rajmohan, K	192
Ren, Heyu	85
Rhee, Woogeun	163
Rho, Dae-Won	10
Riaz, Arslan	252
Roh, Kyoung-Jun	16
Ryu, Junha	186
Ryu, Seung-Tak	4
Sahu, D	192

Saito, Wataru	82
Sakakibara, Kunio	255
Sancheti, A	192
Sato, Kiyohito	22
Seo, Donguk	106
Seo, Jae-Sun	103
Seok, Mingoo	103
Shan, Weiwei	127
Sharma, Vishal	174
Shen, Xinyu	139, 142
Shen, Yi	31, 151
Shen, Yuke	151
Shen, Zhengguo	127
Shi, Bao	189
Shi, C.-J. Richard	85
Shi, Cong	109
Shigeta, Yoshinori	22
Shimogawa, Toyohiro	19
Shiraishi, Mikio	22
Shlezinger, Nir	252
Smith, Shea	28
Sohn, Young-Soo	79
Son, Insang	118
Song, Ho-Jin	270
Srinivas, V	192
Su, Shengzhao	219
Sugimoto, Yoshiki	255
Sun, Hao	34
Sun, Houjun	267
Sun, Shiyan	25
Sung, Yoo-Chang	79
Suzuki, Tomoaki	22
Tajalli, Armin	28
Takami, Kazuhiko	82
Takano, Kyoya	255
Tan, Zhichao	133
Tang, Chen	201
Tang, Dawei	58, 64
Tang, Fang	109
Tang, Kea-Tiong	198
Tang, Xian	168, 180
Tasci, Eyyup	252
Tian, Li	88
Tian, Min	109
Tong, Tzu-Wei	177
Tsai, Yu-Chu	171
Tsuchiya, Shigehiro	22
Tsui, Chi-Ying	213
Um, Soyeon	46
Vimal, E	192
Wabnitz, Malte	91
Wada, Osami	145

Wang, Albert	279	Yan, Angxiao	25, 249
Wang, Bo	112, 166	Yan, Na	43
Wang, Cheng	166	Yan, Xiangxi	157
Wang, Dawei	7	Yan, Zheng	58
Wang, Guoxing	70, 198	Yang, Bingzheng	154
Wang, Haibing	109	Yang, Chenjun	234
Wang, Hanyang	1	Yang, Chia-Hsiang	177
Wang, Hao	37	Yang, Chun	58
Wang, Hedi	201	Yang, Huazhong	52, 201
Wang, Lei	34	Yang, Jingsen	85
Wang, Ping	246	Yang, Mengru	157, 160
Wang, Ruitao	7	Yang, Minkyu	264
Wang, Shuang	52	Yang, Seung-Jae	10
Wang, Tao	94	Yang, Shuo	130, 234
Wang, Tengxiao	109	Yang, Xiaolong	13
Wang, Yan	7, 267	Yao, Dao-Han	231
Wang, Yang	13, 34	Yao, Yuan	213
Wang, Yihao	73	Yayama, Kosuke	19
Wang, Yu	73	Yazicigil, Rabia Tugce	252
Wang, Zeming	73	Ye, Zhuohang	219
Wang, Zengwei	201	Yeo, Injune	103
Wang, Zhihua	25, 163, 222, 225, 240	Yerramsetty, A.	192
Wei, Cong	133	Yi, Xiangjie	7, 267
Wei, Jingchuan	13, 34	Yi, Yongran	157
Wei, Junliang	207	Yin, Rui	43
Wei, Rongshan	133	Yin, Shouyi	13, 34
Wei, Shaojun	13, 34	Ying, Jiazhi	267
Wen, Gan	73	Ying, Zhongyuan	1
Wen, Kui	151	Yoo, Changsik	16, 79
Woo, Young-Jin	106	Yoo, Hoi-Jun	46, 97, 148, 186
Wu, Bing-Jen	231	Yoon, Chi-Weon	136
Wu, Hongzhi	49	Yoon, Dong-Hyun	115
Wu, Jiangchao	216	Yoon, Hyunchul	121
Wu, Nanjian	109, 139, 142, 183	Yoon, Jayang	136
Wu, Weitao	49	Yoshida, Takeshi	255
Wu, Wen	37	You, Xiaohu	157, 160
Wu, Xu	261	Youn, Jae-Youn	16
Xiao, Xi	183	Yu, Jehyeok	228
Xie, Chao	207	Yu, Shuangming	109
Xie, Yufeng	73	Yun, Daeho	67
Xiong, Siqi	234	Yun, Gichan	258
Xu, Chenyu	157, 160	Yun, Seokho	100
Xu, Dongfan	49	Zeng, Xiaoyang	73
Xu, Hao	43	Zhang, Jiaming	130, 234
Xu, Jiawei	94	Zhang, Jian	7, 267
Xu, Peng	94	Zhang, Litao	216
Xu, Yiqing	142	Zhang, Milin	207
Xu, Yongming	1	Zhang, Rui	58, 64
Xuan, Yangfan	246	Zhang, Shiwei	249
Xue, Jiamin	25	Zhang, Shutao	91
Yamaguchi, Shun	145	Zhang, Xianghui	88
Yamane, Atsushi	82	Zhang, Xin	112, 174

Zhang, Xiongjie 273
Zhang, Xueli 94
Zhang, Yanbo 88, 151
Zhang, Yidan 142
Zhang, Zhao 139, 142, 183
Zhang, Zhaoyu 139
Zhang, Zhihan 183
Zhao, Anyang 273
Zhao, Bo 246
Zhao, Dixian 157, 160
Zhao, Guangshu 207, 216
Zhao, Jian 198
Zhao, Yuxiao 1
Zhao, Zhiren 13
Zheng, Haoxin 219
Zhong, Liping 49
Zhong, Zhengqing 109
Zhou, Jie 154
Zhou, Peigen 58, 64
Zhou, Xichuan 109
Zhou, Yang 13
Zhu, Chuanming 189, 222
Zhu, Wei 7, 267
Zhu, Zhangming 31, 88, 151
Zirtiloglu, Timur 252
Zou, Xuecheng 130

IEEE
445 Hoes Lane
Piscataway, NJ 08854-4141

ISBN 979-8-3503-3004-5